T0201489

FINE-TUNING IN THE PHYSICAL UNIVERSE

Is the Universe fine-tuned for complexity, life, or something else? This comprehensive overview of fine-tuning arguments in physics, with contributions from leading researchers in their fields, sheds light on this often-used but seldom-understood topic. Each chapter reviews a specific subject in modern physics – such as dark energy, inflation, or solar-system formation – and discusses whether any parameters in our current theories appear to be fine-tuned and, if so, to what degree. Connections and differences between these fine-tuning arguments are made clear, and detailed mathematical derivations of various fine-tuned parameters are given. This accessible yet precise introduction to fine-tuning in physics will aid students and researchers across astrophysics, atomic and particle physics, and cosmology, as well as all those working at the intersections of physics and philosophy.

DAVID SLOAN is a lecturer in physics at Lancaster University. Following his PhD at Penn State, he has held positions at Utrecht, Oxford and Cambridge Universities.

RAFAEL ALVES BATISTA is a research fellow at Radboud University. He obtained his PhD from the University of Hamburg. He was previously a postdoctoral researcher at the Universities of São Paulo and Oxford.

MICHAEL TOWNSEN HICKS is a researcher in the philosophy of science and philosophy of physics at the University of Birmingham. He completed his PhD at Rutgers University and has held postdoctoral positions at the Universities of Oxford and Cologne.

ROGER DAVIES is the Philip Wetton Professor of Astrophysics at the University of Oxford and Director of the Hintze Centre for Astrophysical Surveys. He is former Head of the Physics Department at Oxford, a past President of the Royal Astronomical Society and current President of the European Astronomical Society.

FINE-TUNING IN THE PHYSICAL UNIVERSE

Edited by

DAVID SLOAN
Lancaster University

RAFAEL ALVES BATISTA
Radboud University, Netherlands

MICHAEL TOWNSEN HICKS
University of Birmingham

ROGER DAVIES
University of Oxford

CAMBRIDGE
UNIVERSITY PRESS

CAMBRIDGE
UNIVERSITY PRESS

University Printing House, Cambridge CB2 8BS, United Kingdom

One Liberty Plaza, 20th Floor, New York, NY 10006, USA

477 Williamstown Road, Port Melbourne, VIC 3207, Australia

314–321, 3rd Floor, Plot 3, Splendor Forum, Jasola District Centre, New Delhi – 110025, India

79 Anson Road, #06–04/06, Singapore 079906

Cambridge University Press is part of the University of Cambridge.

It furthers the University's mission by disseminating knowledge in the pursuit of
education, learning, and research at the highest international levels of excellence.

www.cambridge.org
Information on this title: www.cambridge.org/9781108484541
10.1017/9781108614023

© Cambridge University Press 2020

First published 2020

Printed in the United Kingdom by TJ Books Limited, Padstow Cornwall

A catalogue record for this publication is available from the British Library.

Library of Congress Cataloging-in-Publication Data
Names: Sloan, David, 1982– editor. | Batista, Rafael Alves, 1987– editor. |
Hicks, Michael Townsen, 1985– editor. | Davies, Roger L., editor.
Title: Fine-tuning in the physical universe / edited by David Sloan,
Rafael Alves Batista, Michael Townsen Hicks, Roger Davies.
Description: Cambridge ; New York, NY : Cambridge University Press, 2020. |
Includes bibliographical references and index.
Identifiers: LCCN 2020007781 (print) | LCCN 2020007782 (ebook) |
ISBN 9781108484541 (hardback) | ISBN 9781108614023 (epub)
Subjects: LCSH: Multiverse. | Cosmology–Mathematical models. | Standard
model (Nuclear physics) | Physical constants. | Physical laws. | Nuclear models.
Classification: LCC QB981 .F485 2020 (print) | LCC QB981 (ebook) | DDC 523.1/2–dc23
LC record available at https://lccn.loc.gov/2020007781
LC ebook record available at https://lccn.loc.gov/2020007782

ISBN 978-1-108-48454-1 Hardback

Contents

List of Contributors   *page* ix
Acknowledgements   xi

Part I   Introduction   1

1   Fine-Tuning, Complexity, and Life in the Multiverse   3
    MARIO LIVIO AND MARTIN REES
    1.1   Introduction   3
    1.2   Prerequisites for Complexity   7
    1.3   The Multiverse   12
    1.4   Conceptual Shifts   15
    References   17

2   Hierarchy of Fine-Structure Constants   20
    BERNARD CARR
    2.1   Preface   20
    2.2   Cosmic Uroborus   22
    2.3   Physical Constants   26
    2.4   Types of Fine-Tuning   28
    2.5   Scales of Structure and Natural Tunings   29
    2.6   Non-anthropic Tunings   34
    2.7   Weak Anthropic Principle   35
    2.8   Strong Anthropic Principle: Astrophysical Coincidences   38
    2.9   Strong Anthropic Principle: Cosmological Coincidences   45
    2.10   Cosmology and the Complexity Principle   54
    2.11   Interpretations of Anthropic Principle   56
    2.12   Concluding Remarks   58
    References   61

Part II   Cosmological Fine-Tunings   65

3   Naturalness, Fine-tuning, and Observer Selection in Cosmology   67
    JOHN A. PEACOCK
    3.1   Overview   67
    3.2   Defining Fine-Tuning   68
    3.3   Probabilistic Framework   71
    3.4   Anthropic Principles   77
    3.5   The Puzzle of Dark Energy   82
    3.6   The Anthropic Vacuum   97
    3.7   Semi-anthropic Galaxy Formation   102
    3.8   Outlook   106
    References   108

4   Cosmic Inflation: Trick or Treat?   111
    JEROME MARTIN
    4.1   Introduction   111
    4.2   The Standard Model of Cosmology   113
    4.3   Fine-Tuning Puzzles of the Standard Model   120
    4.4   Inflation   125
    4.5   Is Inflation Fine-Tuned? Choosing the Free Parameters
           of the Inflationary Potential   139
    4.6   Inflationary Initial Conditions   142
    4.7   The Multiverse   152
    4.8   Conclusion   165
    References   167

5   Is the Universal Matter-Antimatter Asymmetry Fine-Tuned?   174
    GARY STEIGMAN AND ROBERT J. SCHERRER
    5.1   Introduction and Overview   175
    5.2   Definition of Fine-Tuning of the Baryon Asymmetry Parameter 178
    5.3   The Case against a Symmetric Universe   179
    5.4   Particle Physics Models for Generating the Universal Matter-
           Antimatter Asymmetry   187
    5.5   The Baryon Asymmetry Parameter and Primordial Nucleosynthesis 189
    5.6   Relation between the Baryon Asymmetry Parameter and the
           Observable Cosmological Parameters   192
    5.7   Alternate Universes with Different Baryon Asymmetry Parameters 195
    5.8   Summary and Conclusions   199
    References   200

6   Structure Formation   203
     ADRIANNE SLYZ
     6.1   The Emergence of Structure in the Universe   203
     6.2   The Early Stages of Evolution: Linear Regime   209
     6.3   The Final Stretch: Non-linear Growth   220
     6.4   The Fine-Tuning of Structure Formation   231
     References   233

     Part III   Fine-Tuning in Particle and Nuclear Physics   235

7   Nuclear Physics and Its Impact on Primordial and Stellar Nucleosynthesis   237
     JEAN-PHILIPPE UZAN
     7.1   Introduction   237
     7.2   Strategy   241
     7.3   From the Standard Model of Particle Physics to Nuclear Physics   243
     7.4   Primordial Nucleosynthesis   260
     7.5   Stellar Nucleosynthesis   278
     7.6   Discussion   291
     References   293

8   Fine-Tunings at Particle Scales   307
     GIULIA ZANDERIGHI
     8.1   Introduction   307
     8.2   Historical Examples   309
     8.3   Quantifying Fine-Tuning   317
     8.4   Fine-Tuning in the Electroweak Sector   323
     8.5   Fine-Tuning in the Strong Sector   333
     8.6   Anthropic Arguments   339
     8.7   Final Remarks   341
     References   341

9   Dark Matter   345
     EDWARD W. KOLB
     9.1   Overview of the Current Composition of the Universe   345
     9.2   A Brief History of the Discovery of Dark Matter   348
     9.3   Dark Matter Bestiary   350
     9.4   Role of Dark Matter in the Formation of Structure   359
     9.5   Testing the WIMP Hypothesis   366
     9.6   What If Dark Matter Is Not a WIMP?   376
     9.7   Final Remarks on Dark Matter and Fine-Tuning   377
     References   378

Part IV Fine-Tuning for Life 381

10 Fine-Tuning: From Star to Galaxies Formation 383
    JOSEPH SILK
    10.1 Introduction 383
    10.2 Stellar Basics 384
    10.3 Stellar Mass-Scales 386
    10.4 Galactic Scales 396
    10.5 Supermassive Black Holes 400
    10.6 Conclusions 407
    References 408

11 How Special Is the Solar System? 412
    MARIO LIVIO
    11.1 Introduction 412
    11.2 Lack of Super-Earths 416
    11.3 Lack of Close-In Planets or Debris 416
    11.4 On the Formation of Super-Earths 417
    11.5 The Potential Significance of Asteroid Belts for Life 421
    11.6 How Rare Are Extrasolar Intelligent Civilisations? 444
    References 448

12 On the Temporal Habitability of Our Universe 458
    ABRAHAM LOEB
    12.1 Introduction 458
    12.2 The Habitable Epoch of the Early Universe 461
    12.3 CEMP Stars: Possible Hosts to Carbon Planets in the Early Universe 463
    12.4 Water Formation during the Epoch of First Metal Enrichment 476
    12.5 An Observational Test for the Anthropic Origin
          of the Cosmological Constant 484
    12.6 The Relative Likelihood of Life as a Function of Cosmic Time 489
    References 497

13 Climbing Up the Theories of Nature: Fine-Tuning and Biological Molecules 511
    GEORGE ELLIS, JEAN-PHILIPPE UZAN, AND DAVID SLOAN
    13.1 Introduction 511
    13.2 Quantum Physics and Observables 516
    13.3 Influence on the Structure of Molecules 519
    13.4 Applications 521
    13.5 The Issue of Scaling 529
    13.6 Extrapolations and Conclusions 532
    13.7 Conclusion 536
    References 536

Index 539

Contributors

Bernard Carr
School of Physics and Astronomy, Queen Mary University of London

George Ellis
Applied Mathematics Department, University of Cape Town

Edward W. Kolb
Kavli Institute for Cosmological Physics and the Enrico Fermi Institute, The University of Chicago

Mario Livio
Department of Physics and Astronomy, University of Nevada, Department of Particle Physics and Astrophysics, Faculty of Physics, The Weizmann Institute of Science, Rehovot, Israel

Abraham Loeb
Astronomy department, Harvard University

Jerome Martin
Paris Institute of Astrophysics

John A. Peacock
Institute for Astronomy, University of Edinburgh

Martin Rees
Institute of Astronomy, University of Cambridge

Robert J. Scherrer
Department of Physics and Astronomy, Vanderbilt University, Nashville, TN

Joseph Silk
Paris Institute of Astrophysics, Pierre and Marie Curie University, Paris,
Beecroft Institute of Particle Astrophysics and Cosmology, University of Oxford,
Department of Physics and Astronomy, The Johns Hopkins University, Baltimore, MD

David Sloan
Department of Physics, Lancaster University

Adrianne Slyz
Department of Physics – Astrophysics, University of Oxford

Gary Steigman
Center for Cosmology and AstroParticle Physics & Department of Physics,
Ohio State University

Jean-Philippe Uzan
Paris Institute of Astrophysics, Pierre and Marie Curie University, Paris,
Sorbonne University, The Lagrange Institute of Paris

Giulia Zanderighi
Theoretical Physics Department, CERN, Geneva

Acknowledgements

We would like to thank the John Templeton Foundation for its generous support over the course of this project. Without its commitment to enable research into big questions of science, projects such as this would not be possible. We are particularly grateful to Daniel Darg, Ashley Zauderer. and Melissa Elgendy for championing our cause, and to the anonymous referees and review boards whose feedback overwhelmingly improved our efforts.

We are particularly grateful to all our chapter authors for their participation and active backing of the project. The success of this project is a direct consequence of the manner in which they gave both their time and their expertise. It is both humbling and heartening that as junior researchers we found leading researchers with enthusiasm for our goals and a willingness to share their knowledge freely.

This work would also not have been possible without significant effort from those at Oxford University who provided invaluable administrative and technical support: Melissa Lee, Khalil Chamcham, Chris Doogue, Jane Kent and Jayne Whittern and, above all, Leanne O'Donnell, who worked above and beyond any reasonable level to perform miracles on a weekly basis.

We thank those working academically around the project who donated time and professional guidance. Joe Silk was instrumental in getting the project together in the first place. Pedro Ferreira provided endless advice on academic and managerial aspects, without which we would have been lost. On technical implementations, the input of Chris Lintott was invaluable.

Finally, our friends and families have supported us tirelessly. Jasmine, Helena, Celeste, Nikki, and Ioana, we cannot be thankful enough.

Part I

Introduction

1

Fine-Tuning, Complexity, and Life in the Multiverse

MARIO LIVIO AND MARTIN REES

Abstract

The physical processes that determine the properties of our everyday world, and of the wider cosmos, are determined by some key numbers: the constants of microphysics and the parameters that describe the expanding Universe in which we have emerged. We identify various steps in the emergence of stars, planets, and life that are dependent on these fundamental numbers and explore how these steps might have been changed – or completely prevented – if the numbers were different. We then outline some cosmological models where physical reality is vastly more extensive than the Universe that astronomers observe (perhaps even involving many big bangs) – which could perhaps encompass domains governed by different physics. Although the concept of a multiverse is still speculative, we argue that attempts to determine whether it exists constitute a genuinely scientific endeavour. If we indeed inhabit a multiverse, then we may have to accept that that there can be no explanation other than anthropic reasoning for some features of our world.

1.1 Introduction

At their fundamental level, phenomena in our Universe can be described by certain laws – the so-called laws of nature – and by the values of some three dozen parameters (e.g., [38]). Those parameters specify such physical quantities as the coupling constants of the weak and strong interactions in the Standard Model of particle physics and the dark-energy density, the baryon mass per photon, and the spatial curvature in cosmology.

What actually determines the values of those parameters, however, is an open question. Many physicists believe that some comprehensive 'theory of everything' yields mathematical formulae that determine all these parameters uniquely. But growing numbers of researchers are beginning to suspect that at least some parameters are, in fact, random variables, possibly taking different values in different members of a huge ensemble of universes – a multiverse (see, e.g., [23] for a review). Those in the latter camp take the view that the question 'Do other universes exist?' is a genuine scientific one. Moreover, it is one that may be answered within a few decades. We address such arguments later in this chapter, but first we address the evidence for *fine-tuning* of key parameters.

A careful inspection of the values of the different parameters has led to the suggestion that at least a few of those constants of nature must be fine-tuned if life is to emerge. That is, relatively small changes in their values would have resulted in a universe in which there would be a blockage in one of the stages in emergent complexity that lead from a 'big bang' to atoms, stars, planets, biospheres, and eventually intelligent life (e.g., [2, 3, 6, 25]).

We can easily imagine laws that were not all that different from the ones that actually prevail but would have led to a rather boring universe – laws which would have led to a universe containing dark matter and no atoms; laws where there were hydrogen atoms but nothing more complicated and, therefore, no chemistry (and no nuclear energy to keep the stars shining); laws where there was no gravity; laws where there was a universe where gravity was so strong that it crushed everything; laws where the cosmic lifetime was so short that there was no time for evolution; or laws where the expansion was too fast to allow gravity to pull stars and galaxies together.

Some physicists regard such apparent fine-tunings as nothing more than statistical flukes. They would claim that we should not be surprised that nature seems 'tuned' to allow intelligent life to evolve – we would not exist otherwise. This attitude has been countered by John Leslie, who gives a nice metaphor. Suppose you were up before a firing squad. A dozen bullets are fired at you, but they all miss. Had that not happened, you would not be alive to ponder the matter. But your survival is still a surprise – one that it's natural to feel perplexed about.

Other physicists are motivated by this perplexity to explore whether 'fine-tuning' can be better understood in the context of parallel universe models. In this connection, it's important to stress that such models are consequences of several much-studied physical theories – for instance, cosmological inflation and string theory. The models were not developed simply to remove perplexity about fine-tuning.

Before we explore some prerequisites for complexity, it is instructive to examine a pedagogical diagram that demonstrates in a simple way the properties of a vast range of objects in our Universe. This diagram (Figure 1.1), adapted from Carr and Rees [5], shows the mass vs size (on a logarithmic scale) of structures from the subatomic scale to the cosmic scale. Black holes, for example, lie on a line of slope 1 in this $\log M - \log R$ plot. A black hole the size of a proton has a mass of some 10^{38} protons, which simply reflects how weak the force of gravity is. Solid objects such as rocks or asteroids, which have roughly the atomic density, lie along a line of slope 3, as do animals and people. Self-gravity is so weak that its effects are unnoticeable up to objects even the size of most asteroids. From there on, however, gravity becomes crucial – causing, for instance, planets to be spherical – and by the time objects reach a mass of about $0.08 M_\odot$, they are sufficiently squeezed by gravity to ignite nuclear reactions at their centres and become stars. The bottom-left corner of Figure 1.1 is occupied by the subatomic quantum regime. On the extreme left is the 'Planck length' – the size of a black hole whose Compton wavelength is equal to its Schwarzschild radius. Classical general relativity cannot be applied on scales smaller than this (and indeed may break down under less extreme conditions). We then need a quantum theory of gravity. In the absence of such a theory, we cannot understand

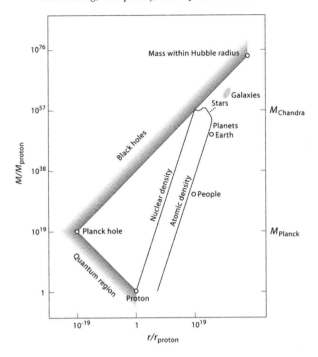

Figure 1.1 This diagram summarises the scales of stars, planets, black holes, and other bodies in a log-log plot of mass against radius. Ordinary lumps lie on the line of slope 3. The mass, in units of the proton mass, scales roughly as the cube of the radius. That line would eventually cross the black hole line (of slope one) at a mass of about 100 million suns. However, it is curtailed before it can do so. The reason is that any mass above about that of Jupiter (containing more than 10^{54} atoms) would be crushed by gravity to a higher density than an ordinary solid. If G were different, the shape of the diagram would not change much, but the number of powers of 10 between the scale of stars and of atoms would scale as the inverse 3/2 power.

the Universe's very beginnings (i.e., what happened at eras comparable to the Planck time of 10^{-43} seconds).

Despite this unmet challenge, it is impressive how much progress has been made in cosmology. In the early 1960s, there was no consensus that our Universe had expanded from a dense beginning. But we have now mapped out, at least in outline, the evolution of our Universe, and the emergence of complexity within it, from about a nanosecond after the Big Bang. At that time, our observable Universe was roughly the size of the solar system, and characterised by energies of the order of those currently realised at the Large Hadron Collider (LHC) near Geneva. Nucleosynthesis of the light elements gives us compelling corroboration of the hot and dense conditions in the first few seconds of the Universe's existence (see Chapter 7; see also, e.g., [8] for a recent review).

The cosmic microwave background (CMB) provides us with not only an astonishingly accurate proof for a black-body radiation state that existed when the Universe was

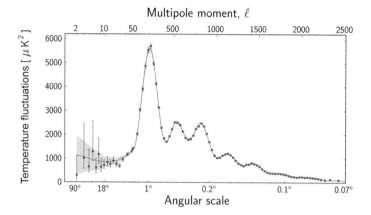

Figure 1.2 The fluctuations in the microwave background on different angular scales. The data come from the Planck spacecraft. The angular scale of the strongest peak is consistent with a flat universe, and the relative heights of the other peaks determine the baryon and dark matter densities.

400,000 years old but also a detailed map of the fluctuations in temperature (and density), $\Delta T/T \sim 10^{-5}$, from which eventually structure emerged. Peaks in the power spectrum of the CMB fluctuations, mapped with great accuracy by the WMAP and Planck satellites, can, even without any other information, offer precise determinations of a few cosmological parameters (e.g. [13, 30]), such as the fractions of baryonic matter, dark matter, and so-called dark energy in the cosmic energy budget (Figure 1.2).

The latter is a mysterious form of energy latent in empty space which has a negative pressure and causes the cosmic expansion to accelerate. It was discovered through observations of Type Ia supernovae [29, 31]. Since then, however, its existence has been confirmed through other lines of evidence, including the CMB, baryon acoustic oscillations, and the integrated Sachs-Wolfe effect (see [28] for a brief review). The simplest hypothesis is that the dark energy has the same properties as the cosmological constant 'lambda' which Einstein introduced in his original equations, but it is possible that it has more complicated properties. In particular, it could change with time and could correspond to just one of many possible vacua. In addition, many lines of evidence have led to the realisation that some form of gravitating dark matter outweighs ordinary baryonic matter by about a factor of five in galaxies and clusters of galaxies. Here are four: (1) flat rotation curves in galaxies extending out beyond the stellar disk; (2) the motions of galaxies in clusters of galaxies; (3) the temperature of the hot gas in clusters of galaxies; (4) gravitational lensing. All of these measure the depth of the gravitational potential well in galaxies or clusters and reveal the presence of mass that does not emit or absorb light. While all the attempts to detect the constituent particles of dark matter have so far been unsuccessful (see Chapter 9; see also, e.g., [11] for a review), this may not be so surprising when we realise that there are some 10 orders of magnitude between the currently observed mass-energies and the GUT unification energy where these particles could hide. Moreover, there are other options such as axions and ultra-low-mass bosons.

Dark matter provided the scaffolding on which the large-scale structure formed. In fact, while some uncertainties about the details remain (see, e.g., [6]), computer simulations can generally reproduce the types of structures we observe on the galactic and cluster scale while starting from the fluctuations observed by Planck and WMAP (see, e.g., [1]).

Similarly, a combination of hydrodynamics, thermodynamics, and nuclear physics has led to a fairly satisfactory understanding of the main processes involved in stellar structure, star formation, evolution, and stellar deaths (e.g., [17, 18]), as well as the formation of planetary systems. Thanks to observations in the past two decades (especially by the Kepler Space Observatory), we now know that the Milky Way contains about one Earth-size habitable-zone planet for every six M-dwarfs [9], which makes the prospects of finding extrasolar life (at least in simple form) with planned or proposed telescopes more promising [26, 35, 36].

Given our current understanding of the evolution of our Universe and of galaxies, stars and planets within it, we may attempt to identify the prerequisites for life. However, since our knowledge of the processes involved in the emergence of life lags far behind our comprehension of fundamental physical processes, we shall only list those very basic requirements that we think should apply to any generic form of complexity.

1.2 Prerequisites for Complexity

There are (at least) five prerequisites for the emergence of complexity in a universe; these prerequisites would not be fulfilled in a counterfactual universe where the fundamental constants are too different from their actual values.

'Counterfactual' exercises of this type are useful for developing an intuition about the role of physical constants in the evolution of the Universe and in the emergence of complexity. Similar studies are used by historians to explore various 'what if?' scenarios, such as speculating what might have happened had Archduke Franz Ferdinand of Austria not been shot by a Serb nationalist in Sarajevo in 1914. Biologists similarly wonder about how the history of life on Earth might have changed had the dinosaurs not been wiped out by an asteroid impact.

If the acceptable range of values for some parameter is small, we would define it as 'fine-tuned'. We shall briefly discuss the extent to which this is the case for some key parameters.

1.2.1 Constraints on Gravity

As numerical simulations of structure formation in the Universe have demonstrated, gravity enhances density fluctuations (see Chapter 6). In our Universe, gravity caused the denser regions to lag behind the cosmic expansion and form the sponge-like structure that characterises the Universe on its largest scales. Eventually, gravity led to the formation of galaxies at the density peaks, of stars, and of planets. Stellar evolution also represents one continuous battle with gravity, the latter pushing the stellar central densities and temperatures to higher and higher values. On the surface of planets, gravity played crucial roles in keeping an atmosphere bound and bringing different elements into contact to initiate the chemical reactions that eventually led to life. But gravity in our Universe is a very weak force – the ratio of the repulsive electric force between two protons to their gravitational mutual attraction is $e^2/Gm_p^2 \sim 10^{36}$. The reason gravity becomes important on the scale of large asteroids and higher is that large objects have a net electric charge that is close to zero, so gravity wins once sufficiently many atoms are packed together.

Figure 1.1 allows us to make a first attempt to examine what would happen in a universe in which the values of some 'constants of nature' are different. How would Figure 1.1 be different if gravity were not so weak? The general structure of the diagram would remain the same, but there would be fewer powers of 10 between the subatomic and cosmic scales. Stars, which effectively are gravitationally bound nuclear fusion reactors, would be smaller in such a universe and would have shorter lives. If gravity were much stronger, then even small solid bodies (such as rocks) might be gravitationally crushed. If gravity's strength were such that it would still have allowed tiny planets to exist, life forms the size of humans would be crushed on the planetary surface. Overall, the universe would be much smaller, and there would be less time for complexity to emerge. In other words, to have what we may call an 'interesting' universe (in the sense of complexity), we must have many powers of 10 between the microscale and the cosmic scale, and this requires gravity to be very weak. It is important to note, however, that gravity does not need to be fine-tuned for complexity to emerge. In fact, a universe in which gravity is ten times weaker than in our Universe, may be even more 'interesting' in that it would allow bigger stars and planets and more time for life to emerge and evolve.

1.2.2 CP Violation – More Matter Than Antimatter

The Big Bang in our Universe created a slight excess (by about one part in three billion) of matter over antimatter (see Chapter 5). It has been shown that for such an imbalance to be created, baryon number and CP symmetry (charge conjugation and parity) had to be violated in the Big Bang and interactions had to be out of thermal equilibrium (the so-called Sakharov conditions [32]). Had the matter-antimatter imbalance not existed, particles and antiparticles would have all been annihilated to form only radiation (what we observe today as the CMB) – leaving no atoms and therefore no galaxies, no stars, no planets, and no life. Within the Standard Model of particle physics the most promising source of CP violation appears to be in the lepton sector, where it generates matter-antimatter asymmetry via a process known as leptogenesis. If, however, CP violation in the lepton sector will be experimentally determined to be too small to explain the matter–anti-matter imbalance (as was the case with the Cabibbo-Kobayashi-Maskawa matrix in the quark sector [22]), physics beyond the Standard Model would be required.

1.2.3 Fluctuations

'Curvature fluctuations' were imprinted into the Universe at a very early era. Their amplitude is almost independent of scale. Many theorists suspect that they originated as quantum fluctuations during an inflationary phase, when the presently observable universe was of microscopic size. The physics of this ultra-early era is, of course, still speculative and uncertain. However, we know from observations that the fluctuations gave rise to temperature fluctuations that grew to $\Delta T/T \sim 10^{-5}$ at the time of recombination.

These fluctuations were crucial for the emergence of complexity. If the early Universe had been entirely smooth, then, even with the same microphysics, the Universe today would have been filled only with cold hydrogen and helium. Stars, galaxies, and, indeed, people would never have formed. The parameter that measures the 'roughness' of the Universe is called Q. At recombination, the temperature fluctuations across the sky $\Delta T/T$ are of order Q. There is no firm theoretical argument that explains why it has the observed value of about 10^{-5} (see, e.g., [37, 38] for a discussion). Computer simulations have offered a huge boost to the credibility of our current ΛCDM model by showing that under the action of gravity and gas dynamics, the fluctuations observed in the CMB would evolve into galaxies with the morphological properties and luminosity functions observed, grouped into clusters whose statistical properties also match the observations.

But what would happen in a counterfactual universe where Q were different from its actual value but all other cosmic parameters stayed the same? If the amplitude of the fluctuations were larger, say $Q \sim 10^{-4}$, masses of about $10^{14} M_\odot$ would condense at a cosmic age of about 300 million years. At that time, Compton cooling on the (then warmer) microwave background would allow the gas to collapse into huge disc galaxies. The virial velocity in large-scale systems scales as $Q^{1/2}c$, and these giant galaxies would find themselves (after some 10^{10} years) in clusters with masses of $\gtrsim 10^{16} M_\odot$. A universe with $Q \sim 10^{-4}$ would have an even larger range of non-linear scales than ours. It would offer more spectacular cosmic vistas; and the only reason why it might be somewhat less propitious for life is that stars in the galaxies would be more close-packed, rendering it less likely that a planetary system could remain undisrupted by a passing star for long enough to permit biological evolution. However, if Q were even larger ($Q \gtrsim 10^{-3}$), conditions would be very unfavourable for life. Enormous masses of gas (far larger than a cluster of galaxies in our present Universe) would condense out early on, probably collapsing to massive black holes – an environment too violent for life.

(Incidentally, any observers who could exist in a high-Q universe would find it far more challenging to interpret and quantify their surroundings. Because Q is small in our actual universes, even the largest non-linear structures are very small compared to the cosmic horizon [they are smaller by a factor of order $Q^{1/2}$]. We can therefore observe a large number of independent patches and define average smoothed-out properties of the Universe – the mean density, etc – and use the standard homogeneous cosmological models as a good approximation. By analogy, a sailor watching ocean waves can meaningfully describe their statistical properties because even the longest wavelength is small compared

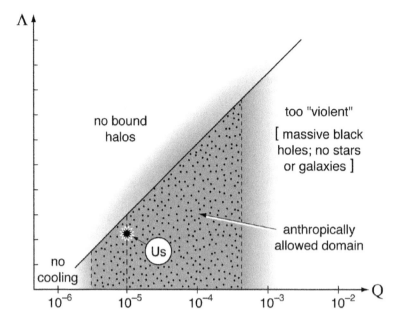

Figure 1.3 Plot of the cosmological constant Λ versus amplitude of fluctuations in cosmic microwave background Q. The shaded-dotted region shows conditions that allow for the existence of complexity.

to the distance of the horizon. In contrast, an astronomer in a high-Q universe would resemble a climber in a mountain landscape, where one peak could dominate the view, and averages are not well defined.)

What about the other extreme, a 'smoother' universe with $Q \lesssim 10^{-6}$? In this case, the disruptive dark energy would push protogalaxies apart before they had a chance to collapse. Even if the dark energy were not there, any galaxies that formed in a lower-Q universe would be small and rather loosely bound (and forming later than in our actual Universe). At $Q \lesssim 10^{-6}$, stars would still form, but material enriched in heavy elements and ejected via stellar winds or supernovae may escape from the shallow gravitational potential wells, not allowing for second-generation stars and planetary systems to form. For values of Q that are significantly smaller than 10^{-6}, there would be inefficient radiative cooling, and therefore, stars would not form within a Hubble time. The conclusion from this discussion (summarised also in [25]; see Figure 1.3) is that for a universe to be conducive for complexity and life, the amplitude of the fluctuations should best be between 10^{-6} and 10^{-4} and, therefore, not particularly finely tuned.

1.2.4 Non-Trivial Chemistry

For life to emerge, the Universe requires nuclear fusion. Fusion not only powers the stars; nucleosynthesis at the hot stellar centres also forges elements such as carbon, oxygen, iron, and phosphorus, all of which are essential for life as we know it. In general, many of the elements in the periodic table participate in the complex chemistry required for the formation of planets and the evolution of their biospheres.

To obtain the nuclear fusion reactions that lead to the creation of the periodic table requires a certain balance between the strength of the electromagnetic force (that repels two protons from each other) and the strong nuclear force (that attracts them). This balance, in our Universe, where the strong nuclear force is about a hundred times stronger than the electromagnetic force, is responsible for the fact that we do not have atomic numbers higher than 118. Had the ratio of the two interactions been much smaller, carbon and heavier elements could not have formed, but the necessary tuning is not excessive.

Similarly, much has been written about Fred Hoyle's prediction of the existence of a 7.65 MeV resonant level of ^{12}C [14, 16]. However, while the prediction itself was indeed remarkable, the degree of fine-tuning required for the energy of that level is not fantastic (e.g., [27, 33]; see [10] for a recent study of this energy level).

The topic of chemistry actually allows us to examine a much more extreme counterfactual universe – a 'nuclear-free universe' – in which hydrogen is the only element that exists. Surprisingly, on the large scale, such a universe would not look much different from ours. Gravity would ensure that galaxies would still form, and even stars would shine (albeit generally for shorter times) by releasing their gravitational energy as they contract to form white dwarfs and black holes. Even Jupiter-like planets composed of solid hydrogen could exist. Of course, no complexity or life of the types we are familiar with will emerge in such a universe (only perhaps something similar to Fred Hoyle's science fiction concept of *The Black Cloud* [15]).

1.2.5 'Tuned' Cosmic Expansion Rate

The results from the Planck satellite depicted in Figure 1.2 (in combination with observations of baryon acoustic oscillations, lensing reconstruction and a prior on the Hubble constant) give for the cosmic energy budget $\Omega_m \sim 0.3$, $\Omega_\Lambda \sim 0.7$, with baryons making less than 5% of this budget [30]. If the cosmic acceleration is indeed driven by a cosmological constant (energy of the physical vacuum, with an equation of state parameter $w = P/\rho = -1$), then the acceleration will continue forever (see Chapter 3). It is clear, however, that if the dark-energy density would have dominated over the matter density (dark matter + baryons) much earlier in the life of our Universe, galaxies would never have formed (this is also dependent on the value of Q; see discussion in the next section). This means that for complexity to arise, some constraints are needed on the ratios of Ω_m/Ω_Λ and Ω_b/Ω_{DM} (where Ω_b denotes the baryon fraction and Ω_{DM} the dark matter fraction). The second ratio is crucial because even though dark matter dominates over baryonic matter in our Universe, without the latter, there would be no stars, no planets, and no life.

As an aside we should note that the nature of the dark energy that propels the cosmic acceleration is one of the most fascinating puzzles in modern cosmology (and one that may not be solved until we have a better understanding of the granular structure of space-time on the Planck scale). Despite its importance for fundamental physics, the dark energy hardly affected any astrophysical phenomena in our Universe; in contrast, the evolution of our Universe so far – the emergence of and morphology of galaxies, clusters, and so forth – has been dominated by the effects of dark matter.

1.3 The Multiverse

As far as we can tell, the laws of physics and the values of the cosmological parameters are the same throughout our entire observable Universe. But the observable Universe is limited by the horizon, which is determined by the finite age of our Universe. What lies beyond this Hubble volume? The homogeneity and isotropy of our observable Universe, with the absence of any perceptible gradient across it (to the 10^{-5} level) suggest (though, of course, do not prove) that the same laws continue to apply thousands of times further. Indeed, many arguments suggest that galaxies beyond the horizon outnumber those we see by a vast factor – perhaps so vast that all combinatorial options would occur repeatedly, and we'd all, far beyond the horizon, have avatars.

Furthermore, some models for the inflationary phase lead to what has been dubbed 'eternal inflation' [24, 39]. According to these models, our Big Bang could be just one 'pocket universe' in a huge ensemble – one island of space-time in a vast archipelago. This scenario also fits well with the 'landscape' concept of string theory, in which there are some 10^{500} metastable vacua solutions, of which our Universe is but one [4, 19]. So the question arises: how large is physical reality?[1]

The first thing to realise is that because we live in an accelerating universe, galaxies are disappearing over an 'event horizon', so we will not observe their far future (rather, as we cannot observe the fate of an object that falls into a black hole after it has crossed the horizon). If the acceleration continued, then after about a trillion years, observers in the remnant of the Milky Way (or its merged product with the Local Group) would not be able to see (again, even in principle) any galaxy other than their own. This does not mean that those galaxies whose light would have been stretched beyond the cosmic scale would not exist.

Moreover, galaxies that are already beyond our current horizon will never become observable, even in principle. Yet most researchers would be relaxed about claims that these galaxies exist in the same way that, in the middle of the ocean, you expect that an ocean extends beyond the terrestrial horizon. These never-observable galaxies would have emerged from the same Big Bang as we did. But suppose that we imagine separate Big Bangs. Are space-times completely disjoint from ours any less real than regions forever

---

[1] It's perhaps necessary, especially in addressing philosophical readers, to inject a clarification at the start. Many would define 'the Universe' as 'everything there is' – and if that's the definition, then there plainly cannot be more than one. If there are other domains (perhaps originating in other Big Bangs, and perhaps differing from our observable domain in size, content, or dimensionality), then we should really define the whole enlarged ensemble as 'the Universe'. We then need a new word – 'metagalaxy', for instance – to denote the domain to which cosmologists and astronomers have observational access. However, so long as this whole idea remains speculative, it is probably best to leave the term 'universe' undisturbed, with its traditional connotations, even though this then demands a new term, 'multiverse', for the whole (still hypothetical) ensemble.

unobservable which are the aftermath of 'our' Big Bang? Surely not – so these other universes, too, should count as parts of physical reality.

Similarly, while we cannot observe any free quarks, we believe that quarks exist, because the Standard Model of particle physics has successfully passed many experimental tests. Likewise, we are disposed to believe in what Einstein's theory tells us about the metric within black holes (inside the event horizon), because general relativity has gained high credibility by being tested in numerous observations and experiments.

If we can develop a theory that makes numerous predictions that are testable (and confirmed) in the observable part of the Universe, then we should be prepared to accept its predictions in unobservable parts.

We currently have no theories of microphysics that are 'battle tested' above the energies reachable in the biggest particle accelerators. These energies are exceeded throughout the first nanosecond after the Big Bang. Theorists who model the inflation era therefore make assumptions about the physics. Some such assumptions predict eternal inflation; others do not (see Chapter 4). Some predict the landscape scenario; others do not. The details of this physics are already somewhat constrained (by, for instance, the observed properties of the fluctuations in the CMB), but we are still far from being able to prove or disprove a model like Linde's 'eternal inflation' [24]. We should therefore be open-minded about how far the aftermath of 'our' Big Bang extends beyond our horizon and also about whether other Big Bangs exist as part of physical reality. Once we are willing to entertain the notion that a multiverse may exist, an even more intriguing question arises: are the laws of physics and the values of the physical constants the same in other members of this ensemble of universes, or are they different?

If they are different, then what we call 'laws of nature' may be no more than local by-laws governing just our cosmic patch. Moreover, many of these pocket universes could be stillborn or sterile. That is, the physical laws prevailing in them (or the values of the parameters) may preclude the emergence of any kind of complexity, and life, in particular. They simply may not satisfy one or more of the prerequisites we discussed in Section 1.2.

The mere possibility that physical reality can encompass such a multiverse provides a strong motivation to develop the lines of thought outlined in Section 1.2 to explore a variety of counterfactual universes, with different values of physical constants, to ascertain which ranges of parameters would allow complexity. The identification and selection of such 'biophilic' universes constitutes what has been dubbed anthropic reasoning (e.g., [2, 12, 25, 34, 41]). Obviously, we humans find ourselves not in a typical member of the multiverse but in a typical domain in the subset of universes that allows complexity and life to emerge and evolve [12]. Copernican humility can only be taken so far. We live on an ordinary terrestrial planet orbiting an ordinary star in its habitable zone. The observable Universe may contain as many as two trillion galaxies [7], and our universe may be only one member of an ensemble of some 10^{500} universes. But our Universe is not 'typical'.

To give a simple analogy (which we believe was first suggested by physicist Leonard Susskind), suppose you wake up in the morning and think, 'What am I?' It seems that a natural answer may be 'I am an insect', since insects have the largest biomass of terrestrial animals. It is estimated that at any time, there are some 10^{19} insects alive. The reason that this argument is false is that by being able to wonder 'what am I?' we have already selected a small subset of the animal kingdom. On the other hand, we can argue that the probability that the answer to 'what am I?' is 'I am Leonardo da Vinci' is still very small.

The ability to actually determine the ranges of all the parameters that allow complexity to develop and the probability for its emergence is currently beyond what physics can achieve, since it requires a knowledge of all the probability distributions and the correlations among them. What we can currently do is a 'poor man's' simplified version of this daunting task – going beyond the discussion of Section 1.2, where parameters were varied one at a time, and analysing the 'anthropic' domain in a two-parameter diagram. For instance, we can depict different values of the dark energy (assumed to be due to a cosmological constant Λ) and the amplitude of the fluctuations Q (Figure 1.3; Ref. [25]). Structures can only form so long as gravity overwhelms the repulsive effect of Λ. A higher value of Q implies earlier formation of structure, and therefore, higher values of Λ would still be anthropically allowed in such a universe.

Another two-parameter example (see [38]) takes Q and the density of dark matter as two parameters that could vary. If the dark matter density were higher than in our actual Universe, the cosmic expansion would become matter dominated at an earlier stage, allowing more time for the growth of structures from initial fluctuations. So the anthropically allowed values of Q would then extend downward. (This contrasts with the effect of higher values of Λ, which extend the allowable Q upwards).

We are currently far from having any theory that determines the values of Λ or Q or the dark matter density (and we know even less about the relative likelihood of various combinations of these constants or how they might be correlated). We are even further from having a cosmological model that can put a 'measure' on the probability density of various combinations. But if we did, we would then have another way of testing – and, in principle, refuting – whether the 'fine-tuning' was due to anthropic selection. We could do this by examining whether we existed in a 'typical' part of the anthropically allowed multiverse or whether the tuning was even more 'special' than anthropic constraints required. This line of reasoning can be illustrated by a simple analogy:

Even if we knew nothing about how stars and planets formed, we would not be surprised to find that our Earth's orbit was fairly close to circular: had it been highly eccentric, water would boil when the Earth was at perihelion and freeze at aphelion – a harsh environment unconducive to our emergence. However, a modest orbital eccentricity, up to 0.1 or so, is plainly not incompatible with life. But suppose it had turned out that the Earth moved in a near-perfect circle with an eccentricity of 0.000001. Some quite different explanation would then be needed: anthropic selection from orbits whose eccentricities had a 'Bayesian prior' that was uniform in the range 0–1 could plausibly account for an eccentricity of 0.1, but not for one as tiny as this.

The methodology requires us to decide what range of values is compatible with our emergence. It also requires a specific theory that gives the relative Bayesian priors for any particular value within that range. With this information, one can then ask if our actual

Universe is 'typical' of the subset in which we could have emerged. If it is a grossly atypical member even of this subset (not merely of the entire multiverse), then we would need to abandon our hypothesis that the numbers were anthropically selected.

Most physicists would consider the 'natural' value of 'dark energy' in the 'landscape' to be large, because it is a consequence of a very complicated microstructure of space. Current evidence suggests that the 'dark energy' has an actual value 5–10 times below the anthropically allowed maximum (other parameters being constrained to their actual values). That would put our Universe between the 10th or 20th percentile of universes in which galaxies could form. In other words, our Universe is not significantly more special, with respect to Λ, than our emergence demanded. But suppose that (contrary to current indications), observations showed that Λ made no discernible contribution to the expansion rate and was thousands of times below the threshold, not just 5–10 times. This 'overkill precision' would (like the precisely circular orbit in the analogy given earlier) raise doubts about the hypothesis that Λ was equally likely to have any value and suggest that it was zero for some fundamental reason (or that it had a discrete set of possible values, and all the others were well above the threshold).

In principle, we could, when theoretical models were more advanced, analyse other important parameters of physics in the same way, to test whether our Universe is typical of the habitable subset that could harbour complex life. The methodology requires us to decide what values are compatible with our emergence. It also requires a specific theory that gives the probability of any particular value. For instance, in the case of Λ, is there a set of discrete vacua or a continuum of values? In the latter case, we need to know whether all values are equally probable or whether the probability density is clustered at a low value.

1.4 Conceptual Shifts

The introduction of the multiverse and of anthropic reasoning has generated considerable controversy, sometimes even accompanied by passionately negative reactions from a number of physicists. We have already discussed the first main objection – the sentiment that envisaging causally disconnected, unobservable universes is in conflict with the traditional 'scientific method'. We have emphasised that modern physics already contains many unobservable domains (e.g., free quarks, interiors of black holes, galaxies beyond the cosmological event horizon). If we had a theory that applied to the ultra-early Universe but gained credibility because it explained, for instance, some features of the microphysical world (the strength of the fundamental forces, the masses of neutrinos, and so forth) we should take seriously its predictions about 'our' Big Bang and the possibility of others.

We are far from having such a theory, but the huge advances already made should give us optimism about new insights in the next few decades. Indeed, even at this early stage, eternal inflation and the landscape scenario already make some predictions that are, in principle, testable. For example, in eternal inflation, our Universe is expected to have a very small (10^{-4}) negative curvature (a 'bubble'). Therefore, future measurements of spatial curvature (including measurements of the 21 cm transition) could falsify eternal inflation (e.g. [21]). Similarly, accelerator experiments can (in principle) generate conditions in which a number of metastable vacuum solutions are possible, thereby testing the premises of the landscape scenario. It is also possible (although the probability for such an event is very low), for another inflating bubble to pop close and collide with our bubble Universe, leaving an imprint in our CMB (e.g., [40]). These simple examples demonstrate that even though the multiverse idea is still in its infancy, this scenario constitutes a bona fide topic (albeit speculative) of scientific discourse, rather than metaphysics.

We have also pointed out that an anthropic explanation can be refuted, if the actual parameter values are far more 'special' than anthropic constraints require.

Many physicists still hope that a unique, self-consistent theory of the Universe will unambiguously determine the values of all the physical parameters. This is a lofty goal, but the history of science has already demonstrated that a quest for first-principle explanations for everything can fail. The great astronomer Johannes Kepler tried to find answers to two questions: (1) Why were there precisely six planets (only six were known at his time) and (2) What was it that determined the spacings among planetary orbits? Eventually, he thought he found the answer, and he published it in his book *Mysterium Cosmographicum* (originally published in 1597; [20]). Kepler's answer was impressive and at the same time absolutely wrong. He constructed a model of the solar system in which the five Platonic solids were embedded one inside the other and together in a surrounding sphere. This created exactly six spaces (like the number of planets), and by choosing the order of the solids in a particular way, the spacing agreed with the relative sizes of the orbits to within 10%. The model was impressive because Kepler used mathematics to explain observed phenomena. It was completely wrong because Kepler did not understand at the time that neither the number of the planets nor their orbits were fundamental phenomena that required first-principles explanations. Rather, we understand today that both the number of planets and their orbits are accidental variables whose values are determined by the environmental conditions in which the planetary system formed. Earth's orbit is special only insofar as it is in the habitable zone around the Sun.

The same may be true for at least some of the parameters of our Universe, such as the values of Λ and Q. These may be random variables in the multiverse, whose only 'explanations' are offered by anthropic arguments. In view of our current ignorance as to what is truly fundamental and what is not, we should keep an open mind to all options.

More specifically, some 'constants' may be truly universal and others not. As an analogy (which we owe to Paul Davies), consider the form of snowflakes. Their ubiquitous sixfold symmetry is a direct consequence of the shape of water molecules. But snowflakes display an immense variety of patterns because each is moulded by its micro-environments: how each flake grows is sensitive to the fortuitous temperature and humidity changes during its growth as it falls towards the ground. If physicists achieved a fundamental theory, it would tell us which aspects of nature were direct consequences of the bedrock theory (just as the symmetrical template of snowflakes is due to the basic structure of a water molecule) and which (like the distinctive pattern of a particular snowflake) were contingencies, taking many values across the multiverse.

If we indeed live in a multiverse, this would be a fifth (and, in some sense, the grandest) Copernican revolution. First, Copernicus showed that we are not at the centre of the solar system; Harlow Shapley showed that the solar system is not at the centre of our galaxy; the Kepler Space Observatory showed that there are billions of planetary systems in the Milky Way; Edwin Hubble and his namesake telescope have shown that there are trillions of galaxies in the observable Universe; and now we realise that our observable domain may be only a tiny part of an unimaginably large and diverse ensemble. The next few decades will hopefully shed some light on the reality of the multiverse.

One thing, however, is clear. Our cosmic horizons have expanded precisely as fast as human knowledge. Every one of the five Copernican revolutions marked an incredible human achievement. In that sense, we remain of central significance to our Universe.

References

[1] Bahé, Y. M., *et al.* 2017. 'The Hydrangea Simulations: Galaxy Formation in and around Massive Clusters'. *Monthly Notices of the Royal Astronomical Society,* **470**(Oct.), 4186–4208.

[2] Barrow, J. D., and Tipler, F. J. 1986. *The Anthropic Cosmological Principle.* Oxford University Press.

[3] Bostrom, N. 2002. *Anthropic Bias: Observation Selection Effects in Science and Philosophy.* Routledge.

[4] Bousso, R., and Polchinski, J. 2000. 'Quantization of Four-Form Fluxes and Dynamical Neutralization of the Cosmological Constant'. *Journal of High Energy Physics,* **6**(June), 006.

[5] Carr, B. J., and Rees, M. J. 1979. 'The Anthropic Principle and the Structure of the Physical World'. *Nature,* **278**(Apr.), 605–612.

[6] Carter, B. 1974. 'Large Number Coincidences and the Anthropic Principle in Cosmology'. Pages 291–298 of Longair, M. S. (ed), *Confrontation of Cosmological Theories with Observational' Data.* IAU Symposium, vol. 63.

[7] Conselice, C. J., *et al.* 2016. 'The Evolution of Galaxy Number Density at $z < 8$ and Its Implications'. *Astrophysical Journal,* **830**(Oct.), 83.

[8] Cyburt, R. H., *et al.* 2016. 'Big Bang Nucleosynthesis: Present Status'. *Reviews of Modern Physics,* **88**(1), 015004.

[9] Dressing, C. D., and Charbonneau, D. 2015. 'The Occurrence of Potentially Habitable Planets Orbiting M Dwarfs Estimated from the Full Kepler Dataset and an Empirical Measurement of the Detection Sensitivity'. *Astrophysical Journal,* **807**(July), 45.

[10] Epelbaum, E., *et al.* 2013. 'Viability of Carbon-Based Life as a Function of the Light Quark Mass'. *Physical Review Letters,* **110**(11), 112502.

[11] Freese, K. 2017. 'Status of Dark Matter in the Universe'. *International Journal of Modern Physics D,* **26**(Jan.), 1730012.

[12] Garriga, J., Livio, M., and Vilenkin, A. 2000. 'Cosmological Constant and the Time of Its Dominance'. *Physical Review D,* **61**(2), 023503.

[13] Hinshaw, G., *et al.* 2013. 'Nine-Year Wilkinson Microwave Anisotropy Probe (WMAP) Observations: Cosmological Parameter Results'. *Astrophysical Journal Supplements*, **208**(Oct.), 19.

[14] Hoyle, F. 1954. 'On Nuclear Reactions Occuring in Very Hot STARS. I. the Synthesis of Elements from Carbon to Nickel'. *Astrophysical Journal Supplements*, **1**(Sept.), 121.

[15] Hoyle, F. 2010. *The Black Cloud*. Penguin Modern Classics. Penguin Books Limited.

[16] Hoyle, F., *et al.* 1953. 'A State in C^{12} Predicted from Astrophysical Evidence'. *Physical Review*, **92**, 1095.

[17] Iben, Jr., I. 2013a. *Stellar Evolution Physics, Volume 1: Physical Processes in Stellar Interiors*. Cambridge University Press.

[18] Iben, Jr., I. 2013b. *Stellar Evolution Physics, Volume 2: Advanced Evolution of Single Stars*. Cambridge University Press.

[19] Kachru, S., *et al.* 2003. 'De Sitter Vacua in String Theory'. *Physical Review D*, **68**(4), 046005.

[20] Kepler, J., and Aiton, E. J. 1981. *Mysterium Cosmographicum*. Abaris Books.

[21] Kleban, M., and Schillo, M. 2012. 'Spatial Curvature Falsifies Eternal Inflation'. *Journal of Cosmology and Astroparticle Physics*, **6**(June), 029.

[22] Kobayashi, M., and Maskawa, T. 1973. 'CP-Violation in the Renormalizable Theory of Weak Interaction'. *Progress of Theoretical Physics*, **49**(Feb.), 652–657.

[23] Linde, A. 2017. 'A Brief History of the Multiverse'. *Reports on Progress in Physics*, **80**(2), 022001.

[24] Linde, A. D. 1986. 'Eternally Existing Self-Reproducing Chaotic Inflanationary Universe'. *Physics Letters B*, **175**(Aug.), 395–400.

[25] Livio, M., and Rees, M. J. 2005. 'Anthropic Reasoning'. *Science*, **309**(Aug.), 1022–1023.

[26] Livio, M., and Silk, J. 2017. 'Where Are They?' *Physics Today*, **70**(3), 50–57.

[27] Livio, M., *et al.* 1989. 'The Anthropic Significance of the Existence of an Excited State of C-12'. *Nature*, **340**(July), 281–284.

[28] Mortonson, M. J., Weinberg, D. H., and White, M. 2013. 'Dark Energy: A Short Review'. *arXiv e-prints*, Dec., arXiv:1401.0046.

[29] Perlmutter, S., *et al.* 1999. 'Measurements of Ω and Λ from 42 High-Redshift Supernovae'. *Astrophysical Journal*, **517**(June), 565–586.

[30] Planck Collaboration *et al.* 2016. 'Planck 2015 Results. XIII. Cosmological Parameters'. *Astronomy and Astrophysics*, **594**(Sept.), A13.

[31] Riess, A. G., *et al.* 1998. 'Observational Evidence from Supernovae for an Accelerating Universe and a Cosmological Constant'. *Astronomical Journal*, **116**(Sept.), 1009–1038.

[32] Sakharov, A. D. 1967. 'Violation of CP Invariance, C Asymmetry, and Baryon Asymmetry of the Universe'. *Soviet Journal of Experimental and Theoretical Physics Letters*, **5**(Jan.), 24.

[33] Schlattl, H., *et al.* 2004. 'Sensitivity of the C and O Production on the 3α Rate'. *Astrophysics and Space Science*, **291**(Apr.), 27–56.

[34] Smolin, L. and Susskind, L. Smolin vs. Susskind Debate. Edge. www.edge.org/conversation/lee_smolin-leonard_susskind-smolin-vs-susskind-the-anthropic-principle.

[35] Snellen, I. A. G., *et al.* 2013. 'Finding Extraterrestrial Life Using Ground-Based High-Dispersion Spectroscopy'. *Astrophysical Journal*, **764**(Feb.), 182.

[36] Stark, C. C., *et al.* 2015. 'Lower Limits on Aperture Size for an ExoEarth Detecting Coronagraphic Mission'. *Astrophysical Journal*, **808**(Aug.), 149.

[37] Tegmark, M. and Rees, M. J., 1998. 'Why Is the Cosmic Microwave Background Fluctuation Level 10^{-5}?' *Astrophysical Journal*, **499**(May), 526–532.

[38] Tegmark, M., *et al.* 2006. 'Dimensionless constants, Cosmology, and Other Dark Matters'. *Physical Review D*, **73**(2), 023505.

[39] Vilenkin, A. 1983. 'Birth of Inflationary Universes'. *Physical Review D*, **27**(June), 2848–2855.

[40] Wainwright, C. L., *et al.* 2014. 'Simulating the Universe(s) II: Phenomenology of Cosmic Bubble Collisions in Full General Relativity'. *Journal of Cosmology and Astroparticle Physics*, **10**(Oct.), 024.

[41] Weinberg, S. 1987. 'Anthropic Bound on the Cosmological Constant'. *Physical Review Letters*, **59**(Nov.), 2607–2610.

2

Hierarchy of Fine-Structure Constants

BERNARD CARR

Abstract

Modern physics describes the vast range of scales of structure in the Universe, with a hierarchy of forces providing connections between the microscopic and macroscopic domains. It also predicts various natural relationships between these scales which might otherwise be regarded as coincidental. However, there are numerous other relationships or 'fine-tunings' between the constants of physics – including the dimensionless coupling constants – which are unexplained by conventional physics and seem necessary for the emergence of observers. This chapter distinguishes between (1) natural tunings which arise between various scales of structure as a result of standard physics; (2) weak anthropic tunings, such as Dicke's argument for the age of the Universe, which regard the constants as given but require that we observe at a special time and place; and (3) strong anthropic tunings, which postulate relationships between the physical constants themselves. The last arise from both astrophysical and cosmological considerations and seem to be necessary for the development of complexity in the Universe. We consider possible interpretations of these fine-tunings, including the multiverse proposal, and some associated philosophical issues.

2.1 Preface

Nearly 40 years ago, I wrote an article in the journal *Nature* with Martin Rees [22], bringing together all of the known constraints on the physical characteristics of the Universe – including the fine-tunings of the physical constants – which seemed to be necessary for the emergence of observers. Such constraints had been dubbed 'anthropic' by Brandon Carter [24] – after the Greek word for 'human' – although it is now appreciated that this is a misnomer, since there is no reason to associate the fine-tunings with humankind in particular. We considered both the 'Weak' Anthropic Principle (WAP) – which accepts the laws of nature and physical constants as given and claims that the existence of observers then imposes a selection effect on where and when we observe the Universe – and the 'Strong' Anthropic Principle (SAP) – which (in the sense we used the term) suggests that the existence of observers imposes constraints on the physical constants themselves.

Anthropic claims – at least in their strong form – were regarded with a certain amount of disdain by physicists at the time and in some quarters still are. Although we took the view that any sort of explanation for the observed fine-tunings was better than none, many regarded anthropic arguments as going beyond legitimate science. The fact that some people of a theological disposition interpreted them as evidence for a Creator – attributing teleological significance to the Strong Anthropic Principle – doubtless enhanced that reaction. However, attitudes have changed considerably since then. This is not so much because the status of the anthropic arguments themselves have changed – as we will see, some of them have become firmer and others weaker. Rather, it is because there has been a fundamental shift in the epistemological status of the Anthropic Principle. This arises because cosmologists have come to realise that there are many contexts in which our Universe could be just one of a large ensemble of 'parallel' universes in which the physical constants vary. This ensemble is sometimes described as a 'multiverse'.

The multiverse proposals have not generally been motivated by an attempt to explain the anthropic fine-tunings; most of them have arisen independently out of developments in cosmology and particle physics. Nevertheless, it now seems clear that the two concepts are interlinked. For if there *are* many universes, this begs the question of why we inhabit this particular one, and – at the very least – one would have to concede that our own existence is a relevant selection effect. Indeed, since we necessarily reside in one of the life-conducive universes, the multiverse picture reduces the Strong Anthropic Principle to an aspect of the weak one. For this reason, many physicists would regard the multiverse as providing the most natural explanation of the anthropic fine-tunings

Many of the arguments were summarised in the book *Universe or Multiverse?* [21], which I edited in 2007. This grew out of a series of conferences which were sponsored by the Templeton Foundation. The first was entitled 'Anthropic Arguments in Fundamental Physics and Cosmology' and held at the home of Martin Rees, in Cambridge in 2001. It was funded by a grant awarded to myself, Robert Crittenden, Martin Rees, and Neil Turok for a project entitled 'Fundamental Physics and the Problem of Our Existence' as part of the Templeton 'Cosmology & Fine-Tuning' programme. The second meeting – with the same title as the book – was held at Stanford University in 2003 and came at a critical point in the development of the subject. It included contributions from some of the key players in the field and these provided the main part of the book. The third meeting was in 2005 and entitled 'Expectations of a Final Theory'. It was again held at Cambridge but this time hosted by Trinity College. Most of the focus was on the exciting developments in particle physics – in particular, M-theory and the string landscape scenario, which perhaps provide a plausible theoretical basis for the multiverse paradigm.

It should be stressed that *Universe or Multiverse?* was not a proselytising work and this is signified by the question mark in the title. The proponents predominated numerically, but many of the contributors were sceptical. Perhaps the most remarkable aspect of this book is that it testified to the large number of eminent physicists at the time who found the subject interesting enough to write about. It is unlikely that such a volume could have been produced even a decade earlier. The shift in attitude in the five years between the first and the third meetings is reflected in a quote from Frank Wilczek's contribution to the book [92]:

The previous gathering had a defensive air. It prominently featured a number of physicists who subsisted on the fringes, voices in the wilderness who had for many years promoted strange arguments about conspiracies among fundamental constants and alternative universes. Their concerns and approaches seemed totally alien to the consensus vanguard of theoretical physics, which was busy successfully constructing a unique and mathematically perfect Universe. Now the vanguard has marched off to join the prophets in the wilderness.

The focus of the present volume is fine-tuning, so the main purpose of this chapter is to update the discussion of my 1979 paper with Martin Rees. There will be rather little discussion of the multiverse proposal, even though this has played such an important role in establishing the respectability of anthropic arguments, since this is covered in other chapters. The first sections will discuss the Cosmic Uroborus (to put the topic in historical context), the physical constants (including the hierarchy of fine-structure constants of the title), and the different types of tunings. The middle sections will describe these tunings in more detail – the natural tunings connecting the various scales of structure in the Universe, the weak anthropic tunings and strong anthropic tunings arising in both astrophysics, and cosmology. The final sections will discuss possible interpretations of Anthropic Principle and some philosophical issues (including whether anthropic arguments qualify as legitimate science). Despite the negative response, it may induce in some quarters, the A word ('anthropic') will be used throughout this chapter.

2.2 Cosmic Uroborus

The history of physics might be regarded as a process in which the development of new instruments, like the telescope and the microscope, has allowed us to extend observations outwards to progressively larger scales and inwards to successively smaller ones. The outward journey explores the macroscopic domain and is associated with astronomy, while the inward journey explores the microscopic domain and is associated with particle physics. This process has revealed ever larger and smaller levels of structure in the Universe. Of course, a lot of interesting physics – including the whole domain of biophysics – is associated with complex structures in the intermediate mesoscopic domain, and that will be particularly relevant to this chapter.

The journey has also led to the discovery of the forces which determine the nature of these structures – gravity and electromagnetism on large scales, the weak and strong forces on short scales – and the associated laws of nature. These forces link the macroscopic and microscopic domains so that the outward and inward journeys are not disconnected but constantly throw light on each other. Both journeys have also led to new conceptual ideas and changes in our world view. The outer one has led to the shifts from geocentric to heliocentric to galactocentric to cosmocentric world views and to the radical change of view of space and time entailed in relativity theory. The inner one has led to atomic theory, quantum theory, and a progressively unified view of the forces of nature and the fundamental constituents of matter. Any final paradigm of physics must amalgamate relativity and quantum theory in some way.

So physics has revealed a unity about the Universe which makes it clear that everything is connected in a way which would have seemed inconceivable a few decades ago. This unity is succinctly encapsulated in the image of the Cosmic Uroborus (the snake eating its own tail) shown in Figure 2.1. The pictures drawn around the snake represent the different

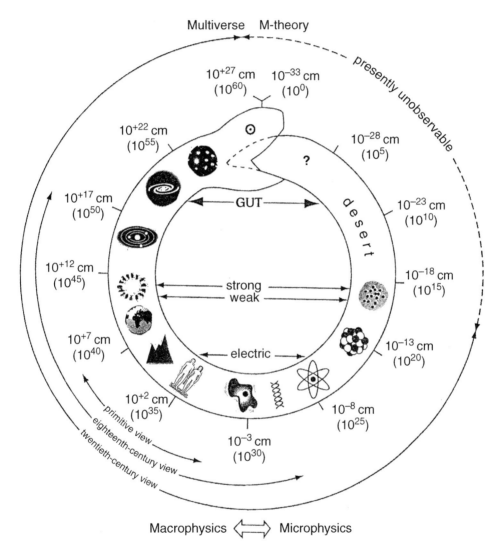

Figure 2.1 The image of the Cosmic Uroborus summarises the different levels of structure in the physical world, the intimate link between the microphysical and macroscopic domains, and the evolution of our understanding of this structure. From Reference [21].

types of structure in the Universe. Near the bottom are humans. As we move to the left, we encounter successively larger objects: a mountain, a planet, a star, a solar system, a galaxy, a cluster of galaxies, and, finally, the entire observable Universe. As we move to the right, we encounter successively smaller objects: a cell, a DNA molecule, an atom, a nucleus, a quark, the GUT scale, and, finally, the Planck length (the scale at which quantum gravity effects become important). The numbers at the edge indicate the scale of these structures in centimetres and also in units of the Planck length (in brackets). As we move clockwise from the tail to the head, the scale increases through 60 decades: from the smallest meaningful scale allowed by quantum gravity (10^{-33}cm) to the scale of the observable Universe (10^{27} cm). So one can regard the Uroborus as a clock in which each minute corresponds to a decade in scale.

The horizontal lines in Figure 2.1 correspond to the various interactions and illustrate the subtle connections between microphysics and macrophysics. For example, the electric line connects an atom to a planet because the electric force binds the electron to the nucleus in an atom and also determines the structure of solid objects. The strong and weak lines connect a nucleus to a star because the strong force which holds nuclei together also provides the energy released in the nuclear reactions which power a star, and the weak force which causes nuclei to decay also prevents stars from burning out too soon. The line associated with the grand unified theory (GUT) connects with large-scale structure because the density fluctuations which led to this originated when the temperature of the Universe was high enough for GUT interactions to be important.

The Big Bang might be regarded as the ultimate micro-macro link since it implies that the entire observable Universe was once compressed to a tiny volume. This is why the head of the snake meets the tail. Since light travels at a finite speed, we can never see further than the distance light has travelled since the Big Bang; this is about 40 billion light years, three times the age of the Universe times the speed of light because the cosmic expansion helps light travel further. More powerful telescopes probe to earlier times rather than larger distances. This is why early universe studies have led to an exciting collaboration between particle physicists and cosmologists.

As discussed by Kolb (Chapter 9), cosmologists now have a fairly complete picture of the history of the Universe. As one goes back in time, galaxy formation occurred at a billion years after the Big Bang, the background radiation last interacted with matter at 400,000 years, the Universe's energy was dominated by its radiation content before about 60,000 years, light elements were generated through cosmological nucleosynthesis at around 3 minutes, antimatter annihilated with matter at about 10^{-5} s (before which there was just a tiny fractional excess of matter), electroweak unification occurred at 10^{-9} s, the highest energy which can be probed experimentally was reached at 10^{-12} s, grand unification and inflation occurred around 10^{-35} s, and the quantum gravity era (the smallest meaningful time) was at 10^{-43} s.

The last few decades have seen two important developments on the outer front, and these are described in other chapters. First, the detection of temperature anisotropies in the cosmic background radiation and ever more precise measurements of their dependence on angular scale have confirmed the quantum origin of the density fluctuations. Second, although one expects the expansion of the Universe to slow down because of gravity, observations of distant supernovae suggest that it is accelerating. We do not know for sure what is causing this, but it must be some exotic form of dark energy, most probably related to the cosmological constant. These discoveries have led to the concordance ΛCDM model and the popularity of the inflationary scenario. Another idea that has become topical is that our entire Universe may be just one member of huge ensemble of universes called the multiverse, a notion which is particularly relevant to the theme of this book.

On the inner front, we have learnt that it may be possible to incorporate gravity into the unification of forces, leading some physicists to proclaim that we are on the verge of obtaining a Theory of Everything (TOE). However, in order to describe all the subatomic interactions, this requires extra wrapped-up dimensions of the kind proposed by Kaluza and Klein to explain electromagnetism. For example, superstring theory suggests there could be six additional spatial dimensions, and the way they are compactified is described by the Calabi-Yau group. There were originally five different superstring theories, but it was later realised that these are all parts of a single, more embracing model called M-theory, which has seven extra dimensions [93]. In one particular variant of M-theory, proposed by Randall and Sundrum [63], the 11th dimension is extended so that the physical world is viewed as a four-dimensional brane in a higher-dimensional bulk. We do not experience these extra dimensions directly – their effects only become important on the smallest and largest scales (i.e., at the top of the Uroborus).

Taken together, scientific progress on both the outer and inner fronts – culminating in the Big Bang model – can be regarded as a triumph. Indeed, as indicated by the arcs in Figure 2.1, the history of science might be regarded as an expansion of our awareness to ever larger and smaller scales. However, this achievement has come at a price. The anthropocentric view which prevailed at the start of the journey has been demolished, and the more we probe the Universe, the more irrelevant humans seem to become. According to the Newtonian paradigm, the cosmos operates likes a giant machine, oblivious to whether life or any form of consciousness is present, so the laws of physics and the characteristics of the Universe are independent of whether anybody actually observes them. Modern developments have reinforced this notion. We are completely insignificant not only as judged by scale but also in terms of duration. If the history of the Universe were compressed into a year, Homo sapiens would have existed for only the last few minutes.

Curiously, in recent decades, cosmology has brought about a reversal in this trend. This is because it seems that, in some respects, the Universe has to be the way that it is because, otherwise, it could not produce life, and we would not be here speculating about it. Indeed, since my paper with Rees, this idea has been explored in numerous works (e.g., [11, 15, 45, 54, 64]), and further references will be given in the chapter. Although this notion is referred to as the Anthropic Principle, it is not inevitable that it relates to the presence of observers, and it might be better described as the Complexity Principle. However, my personal hunch is that it does relate to observers and that the ultimate explanation will relate to the unification at the top of the Cosmic Uroborus, with its possible invocation of other universes and extra dimensions.

2.3 Physical Constants

The progress of physics has revealed a host of physical constants, all of which have been measured to varying degrees of precision – for example, the speed of light (c), Planck's constant ($\hbar$), the gravitational constant (G), the charge of the electron (e), and the masses of various elementary particles like the proton (m_p) and the electron (m_e). Our physical theories relate some of these constants, so it is important to identify the ones which are fundamental, in the sense that the others can be derived from them. As physics has become progressively unified, the number of fundamental constants has reduced, but there is still a large number of them. For example, the Standard Model of Particle Physics has 26, and the Standard Model of Cosmology has six. Some of these constants are listed in Table 2.1, which is taken from Barnes [8]. However, neither of these standard models can be complete, so one may expect further reductions. Indeed, one might hope that some Final Theory would determine them all uniquely, although there is no evidence for this. More likely, some of the constants will turn out to be contingent.

Certain combinations of these constants have a special physical significance. For example, $\hbar/m_p c$ is a length-scale of about 10^{-13} cm and specifies the size of the proton. If one divides this by c, one gets a timescale t_p of about 10^{-23} s, and this is the time light takes to traverse a proton; it is also the timescale associated with strong interactions. Another combination, $\hbar^2/m_e e^2$, gives a scale of about 10^{-8} cm, and this specifies the size of an atom. Dividing m_p by the cube of this, gives a density of about 1 g cm^{-3}, and this characterises the density of atomic material like ordinary solids and liquids. Dividing m_p by the cube of the size of the proton gives a density of 10^{14} g cm^{-3}, and this characterises nuclear density.

It is particularly interesting to take combinations of the constants which are dimensionless, in the sense that they are pure numbers. For example, the electric 'fine structure' constant,

$$\alpha \equiv e^2/\hbar c \approx 1/137, \tag{2.1}$$

determines the strength of the electric interaction and plays a crucial role in any situation where electromagnetism is important. Likewise, the gravitational fine structure constant,

$$\alpha_G \equiv Gm_p^2/\hbar c \approx 6 \times 10^{-39}, \tag{2.2}$$

determines the strength of the gravitational interaction and plays an important role in determining the structure of large objects (like stars). The fact that α_G is so much smaller than α reflects the fact that the gravitational force between two protons is so much smaller than the electric force between them. Gravity dominates the structure of large bodies only because these tend to be electrically neutral, so the electric forces cancel out.

Table 2.1 *Constants of Standard Models of particle physics and cosmology, taken from Reference [8]. Note that the electric, weak and strong coupling constants indicated are different from the low-energy definitions of Eqs. (2.1), (2.5), and (2.6).*

Quantity	Symbol	Value in our universe
Speed of light	c	299,792,458 m s^{-1}
Gravitational constant	G	6.673×10^{-11} m^3 kg^{-1} s^{-2}
(Reduced) Planck constant	$\hbar$	$1.05457148 \times 10^{-34}$ m^2 kg s^{-1}
Planck mass-energy	$m_{Pl} = \sqrt{\hbar c/G}$	1.2209×10^{22} MeV
Mass of electron; proton; neutron	$m_e; m_p; m_n$	0.511; 938.3; 939.6 MeV
Mass of up; down; strange quark	$m_u; m_d; m_s$	(Approx.) 2.4; 4.8; 104 MeV
Ratio of electron to proton mass	β	$(1836.15)^{-1}$
Gravitational coupling constant	$\alpha_G = m_p^2/m_{Pl}^2$	5.9×10^{-39}
Hypercharge coupling constant	α_1	$1/98.4$
Weak coupling constant	α_2	$1/29.6$
Strong force coupling constant	$\alpha_s = \alpha_3$	0.1187
Fine structure constant	$\alpha = \frac{\alpha_1 \alpha_2}{\alpha_1 + \alpha_2}$	1/127.9 (1/137 at low energy)
Higgs vacuum expectation value	v	246.2 GeV
QCD scale	Λ_{QCD}	$\approx$ 200 MeV
Yukawa couplings	$\Gamma_i = \sqrt{2}m_i/v$	Listed in [82]
Hubble constant	H	71 km/s/Mpc (today)
Cosmological constant (energy density)	Λ (ρ_Λ)	$\rho_\Lambda = (2.3 \times 10^{-3} eV)^4$
Amplitude of primordial fluctuations	Q	2×10^{-5}
Total matter mass per photon	ξ	$\approx$ 4 eV
Baryonic mass per photon	ξ_{baryon}	$\approx$ 0.61 eV

Many physical quantities can be expressed very simply in terms of α and α_G. For example, the radius of a hydrogen atom and the (Rydberg) energy required to ionise it are

$$a_o \sim \alpha^{-1} r_e \sim 10^{-8} \text{cm}, \quad E_o \sim \alpha^2 m_e c^2 \sim 10 \text{eV}, \qquad (2.3)$$

and the Planck length and Planck mass can be expressed as

$$R_P = \left(\frac{G}{\hbar c^3}\right)^{1/2} = \alpha_G^{1/2} r_p \sim 10^{-33} \text{cm}, \quad M_P = \left(\frac{\hbar c}{G}\right)^{1/2} = \alpha_G^{-1/2} m_p \sim 10^{-5} \text{g}. \quad (2.4)$$

These are the scales at which quantum gravity effects become important.

The value of α_G is of particular interest because simple physics (discussed later) shows that most of the scales appearing in Figure 2.1 can be expressed as powers of α_G. Also, we will see that the main-sequence lifetime of stars is roughly $\alpha_G^{-1} t_p$, so the size of the observable Universe R_U must be roughly $\alpha_G^{-3/2} \sim 10^{60}$ in Planck units. This explains the clocklike feature of the Uroborus (i.e., the factor of 60). If α_G were different, the form of the Uroborus would remain the same, but all the scales would change.

On scales smaller than atoms, two more fundamental interactions come into play: the strong and the weak force. Although these are many orders of magnitude stronger than the gravitational force, they are both short range, becoming negligible beyond distances of 10^{-13} cm and 10^{-15} cm, respectively. For this reason, they do not play an important role in determining the structure of objects larger than atoms. The associated coupling constants are also energy dependent, so Reference [22] only indicates their low-energy values. The weak force has a (low-energy) dimensionless coupling constant

$$\alpha_W \equiv (gm_e^2 c/\hbar^3) \sim 10^{-11}, \tag{2.5}$$

where $g \sim 10^{-49}$ erg cm^3 is the Fermi constant. Thus, its interaction strength is intermediate between that of gravity and electricity. Reference [22] describes the strong force by the (scalar) coupling constant

$$f^2 \approx 15. \tag{2.6}$$

However, it must be stressed that the strong coupling constant is strongly energy dependent:

$$\alpha_S(E) = \frac{12\pi}{(33 - 2n_f)\ln(E^2/\Lambda_{\text{QCD}}^2)}, \tag{2.7}$$

where n_f is the number of quarks and Λ_{QCD} is the QCD scale. This goes to zero for $E \gg \Lambda_{\text{QCD}}$, corresponding to asymptotic freedom. It diverges as $E \to \Lambda_{\text{QCD}}$, but QCD theory breaks down there, and lattice gauge theory gives $\alpha_S \approx 0.1$ at the appropriate energy.

2.4 Types of Fine-Tuning

The AP claims that there are various tunings between the physical constants – including the dimensionless coupling constants – which are necessary for the emergence of observers. Indeed, there could be enough of these to determine *all* the physical constants, at least to some precision. Current physics does not explain their values, but even if it did, it would be remarkable that the values predicted turned out to be the ones required anthropically. However, the term 'tuning' is used in different senses, so we first clarify the distinction between them.

- *Natural tunings.* Standard physics implies that the mass- and length-scales of many natural objects depend on α and α_G. This implies many apparent coincidences between these scales, which might be surprising if one did not understand the underlying physics. For example, the size of cell is 10^{-3} cm, which is roughly the geometric mean of the R_P and R_U.
- *Non-anthropic tunings.* Some tunings in particle physics have no obvious anthropic aspects. For example, the Higgs mass (250 GeV) is roughly the geometric mean of the dark-energy mass and the Planck mass, but this does not seem to have anthropic significance. Wilczek's classification of fundamental parameters distinguishes between four types of tunings; he terms these enlightenment, knowledge, ignorance, and temptation, with only the last being anthropic [91].

- *Weak anthropic tunings.* Given the constants of physics, there is an inevitable selection effect on when and where observers can exist, and this may involve the coupling constants. For example, an argument of Robert Dicke [40] suggests that observers can only exist when the age of the Universe is roughly the lifetime of a main-sequence star, $t_{MS} \sim \alpha_G^{-1} t_p \sim 10^{10}$ yr. The WAP is a logical necessity, although the nature of the tunings may still be a surprise.
- *Strong anthropic tunings.* There are also coincidences between the physical constants themselves which seem necessary for observers. However, most of these coincidences merely involve prerequisites for observers (such as planets, stars, galaxies, and chemistry), so the description 'anthropic' is misleading. They are better regarded as conditions for *complexity*. The SAP is much more controversial than the WAP because it refers to counterfactual universes in which the constants are different. However, if one accepts the existence of a multiverse, the SAP essentially becomes an example of the WAP.

2.5 Scales of Structure and Natural Tunings

Straightforward physics shows that, to an order of magnitude, α and α_G determine the mass and the size of nearly every macroscopic object in the Universe. This is illustrated in Figure 2.2, from which one can read off the scales associated with the Universe itself, galaxies, stars, planets, asteroids, exploding black holes, humans, atoms, protons, and the Planck length. The scales are given in grams and centimetres and also in terms of α, α_G, and the size and mass of the atom. A derivation of some of these results is given later.

Some of these dependencies are also summarised in Table 2.2. The proton and Planck scales trivially follow from Eq. (2.3) and (2.4), but the others are less obvious. For example, stars have a mass of roughly $\alpha_G^{-3/2} \sim 10^{60}$ and a radius of roughly $\alpha_G^{-1/2} \sim 10^{20}$ in atomic units (see also Chapter 10); the largest planets (like Jupiter) have a mass and radius smaller than this by factors of $\alpha^{3/2} \sim 10^{-3}$ and $\alpha^{1/2} \sim 10^{-1}$, respectively; the mass and size of a living creature – if we assume that it must live on a planet with a suitable temperature and a life-supporting atmosphere and does not shatter whenever it falls down – must be of order $(\alpha/\alpha_G)^{3/4} \sim 10^{27}$ and $(\alpha/\alpha_G)^{1/4} \sim 10^9$. Some of the scales also depend on the proton-to-electron mass ratio, but this can be written as

$$\beta \equiv m_e/m_p \approx 10\alpha^2, \tag{2.8}$$

due to a coincidence in nuclear physics (discussed later).

It should be stressed that the dependencies on α and α_G summarised in Figure 2.2 and Table 2.2 are all consequences of standard physics. No anthropic argument has been introduced, except in deriving the scale of the Universe (which we discuss later). At first sight, the dependencies might seem surprising, but the qualitative reason can be understood as follows: any stable structure in the Universe reflects a balance between either gravity or intermolecular forces, which are trying to hold it together, and various other effects – such as pressure or quantum interactions – which are trying to blow it apart. Since gravity depends on G, quantum effects on $\hbar$, and electrical effects on e, a balance between these

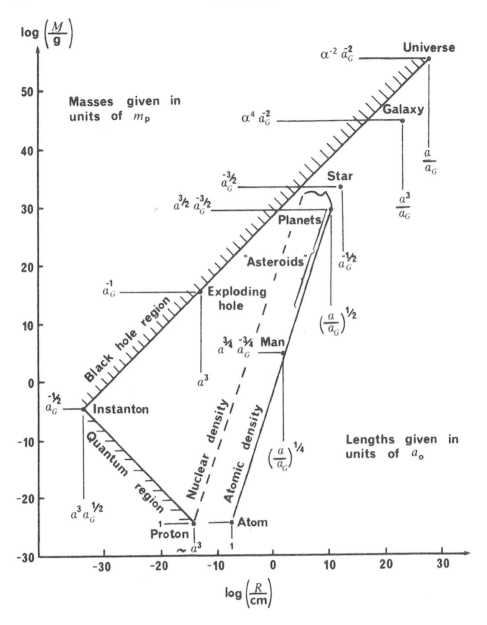

Figure 2.2 Dependence of mass and length scales of various objects on α and α_G, from Reference [22].

Table 2.2 *Scales of some objects in atomic units.*

	Mass/m_p	Size/a_0
Universe	$\alpha^{-2}\alpha_G^{-2}$	$\alpha\alpha_G^{-1}$
Galaxy	$\alpha^4\alpha_G^{-2}$	$\alpha^3\alpha_G^{-1}$
Star	$\alpha_G^{-3/2}$	$\alpha_G^{-1/2}$
Jupiter	$\alpha^{3/2}\alpha_G^{-3/2}$	$\alpha^{1/2}\alpha_G^{-1/2}$
Human	$\alpha^{3/4}\alpha_G^{-3/4}$	$\alpha^{1/4}\alpha_G^{-1/4}$
Proton	1	α^3
Planck	$\alpha_G^{-1/2}$	$\alpha^3\alpha_G^{1/2}$

factors is bound to involve the sort of combinations shown in Figure 2.2, However, one can only deduce the actual powers of α and α_G by detailed calculation.

We first consider the quantum and black hole regions in Figure 2.2. These are associated with the Compton wavelength and Schwarzschild radius for an object of mass M:

$$R_C = \frac{\hbar}{Mc} = r_p\left(\frac{M}{m_p}\right), \quad R_S = \frac{2GM}{c^2} = \alpha_G\left(\frac{M}{m_p}\right)r_p. \tag{2.9}$$

The Hawking temperature of a black hole of mass M is given by [43]

$$kT_H = \frac{\hbar c^3}{8\pi GM} \sim \left(\frac{M}{M_P}\right)^{-1}M_P \sim \alpha_G^{-1}\left(\frac{M}{m_p}\right)^{-1}m_p, \tag{2.10}$$

and it evaporates on a timescale

$$t_{\text{evap}} \sim \left(\frac{M}{M_P}\right)^3 N(M)^{-1}t_P \sim \alpha_G^2\left(\frac{M}{m_p}\right)^3 N(M)^{-1}t_p, \tag{2.11}$$

where $N(M)$ is the number of particle species with rest mass less than T_H. This means that a black hole evaporating at the present epoch has a mass

$$M_* \sim \alpha_G^{-2/3}\left(\frac{t_o}{t_p}\right)^{1/3}m_p \sim \alpha_G^{-1}m_p \sim 10^{15}\text{g}, \tag{2.12}$$

where we have used the relation $t_o \sim \alpha_G^{-1}t_p$ at the second step. This implies that its radius is just of order r_p.

We next show that stars have a mass in the range $(0.1–10)M_C$, where

$$M_C \equiv \alpha_G^{-3/2}m_p \sim 1M_\odot. \tag{2.13}$$

A gravitationally bound cloud represents a balance between gravity and thermal and electron degeneracy pressure. The virial theorem therefore implies that the temperature T is given by

$$kT + \frac{\hbar^2}{2m_e d^2} \sim \frac{GMm_p}{R} \sim \left(\frac{N}{N_o}\right)^{2/3} \frac{\hbar c}{d}, \tag{2.14}$$

where N is the number of protons, $N_o \equiv \alpha_G^{-3/2}$, and d is the mean particle separation. As the cloud collapses, T first increases as d^{-1}, being given by

$$kT \sim \frac{GMm_p}{R}, \tag{2.15}$$

but it then decreases due to the degeneracy term, reaching a maximum

$$kT_{\max} \sim \left(\frac{N}{N_o}\right)^{4/3} m_e c^2. \tag{2.16}$$

The cloud forms a star if $kT_{\max}$ exceeds the nuclear ignition threshold, $\chi m_e c^2$ with $\chi \sim 10^{-2}$, so this implies $N > 0.1 N_o$. On the other hand, the star will be unstable if it is radiation-pressure dominated. Using Eq. (2.14), the ratio of the radiation and matter pressures is

$$\frac{P_{\text{rad}}}{P_{\text{mat}}} \sim \frac{aT^4}{NkT/R^3} \sim 0.01 \left(\frac{N}{N_o}\right)^2, \tag{2.17}$$

so the upper limit is around $10\,M_C$, although a more precise calculation gives a limit of $50 M_C$. More details can be found in Chapter 10.

During its main-sequence phase, T adjusts itself so that the nuclear energy generation rate balances the luminosity, which is the radiative energy content divided by the photon leakage time. This corresponds to

$$L \sim \frac{acT^4 R^2}{\kappa M}, \tag{2.18}$$

where κ is the opacity. After its nuclear-burning (main-sequence) phase, the star resumes its collapse, and so Eq. (2.14) predicts zero temperature (corresponding to an electron-degeneracy-supported white dwarf) when the particle separation reaches

$$d_{\min} \sim \left(\frac{N}{N_o}\right)^{-2/3} r_e \quad \Rightarrow \quad R \propto N^{1/3} d_{\min} \propto M^{-1/3}. \tag{2.19}$$

There is no stable white dwarf configuration if the electrons become relativistic (i.e., for $kT_{\max} \sim m_e c^2$) since the degeneracy term in Eq. (2.14) then goes like d^{-1} rather than d^{-2}. Therefore, M_C also gives the upper limit on the mass of a white dwarf (the Chandrasekhar mass).

Note that the mass-scale at which a collapsing cloud stops fragmenting is

$$M_{\text{frag}} \sim q^{-1/2} \left(\frac{kT}{m_p c^2}\right)^{1/4} M_C \sim 10^{-2} q^{-1/2} M_C, \tag{2.20}$$

where T is the cloud's temperature and q is the ratio of its luminosity to that of a black body with this temperature. The important point is that M_{frag} has only a weak dependence on T and is at least an order of magnitude smaller than M_C.

Solid objects represent a balance between electron degeneracy energy and electrostatic binding energy. By comparing the degeneracy term in Eq. (2.14) with e^2/d, one obtains the atomic density line

$$\rho_o \sim m_p/a_o^3 \sim e^6 m_e^3 m_p \hbar^{-6} \sim 1\,g\,cm^{-3},\qquad(2.21)$$

which meets the white dwarf line at

$$M \sim \left(\frac{\alpha}{\alpha_G}\right)^{3/2} m_p \sim \alpha^{3/2} M_C \sim 10^{30} g,\quad R \sim \left(\frac{\alpha}{\alpha_G}\right)^{1/2} a_o \sim 10^{10}\,cm.\qquad(2.22)$$

This gives the maximum size of a planet and also lies on the Rydberg ($kT \sim \alpha^2 m_e c^2$) isotherm. The minimum mass of a planet, the mass range of a habitable planets and the maximum scale of a living creature (prescribed by the requirement that it does not fall apart when it falls through its own height) can be derived with similar arguments but is not given here.

We next discuss the galaxy scale [67, 71]. Let us assume that galaxies originate from over-dense regions in the gaseous primordial material, and that they have a mass M and radius R_B when they stop expanding and become bound. After binding, protogalaxies will virialise at a radius $\sim R_B/2$. Thereafter, they will deflate on a cooling timescale, with the virial temperature (2.15). Providing kT exceeds the Rydberg energy, the dominant cooling mechanism is bremsstrahlung with the associated cooling time

$$t_{cool} \sim \frac{1}{n\alpha\sigma_T c}\left(\frac{kT}{m_e c^2}\right)^{1/2} \sim \frac{G^{1/2} m_e^{3/2} m_p R}{\alpha^3 n^{1/2}},\qquad(2.23)$$

using $\sigma_T \sim \alpha^2 r_e^2$ and $n \sim M/m_p R^3$. The free-fall timescale, on the other hand, is

$$t_{ff} \sim (GM/R^3)^{-1/2} \sim (Gnm_p)^{-1/2},\qquad(2.24)$$

and this exceeds t_{cool} when R falls below a mass-independent value

$$R_g \sim \alpha^4 \alpha_G^{-1}\left(\frac{m_p}{m_e}\right)^{1/2} a_o \sim \alpha^3 \alpha_G^{-1} a_o \sim 100\,kpc,\qquad(2.25)$$

where we have used Eq. (2.8). Until a massive cloud gets within this radius, it will contract quasi-statically and cannot fragment into stars.

This argument applies only if $kT > E_o \sim \alpha^2 m_e c^2$ at the radius R_g, so one also needs the mass to exceed

$$M_g \sim \alpha_G^{-2}\alpha^5 (m_p/m_e)^{1/2} m_p \sim \alpha_G^{-2}\alpha^4 m_p \sim 10^{12} M_\odot.\qquad(2.26)$$

Gas clouds with mass below M_g cool more efficiently, owing to recombination and can never be pressure supported. On the other hand, clouds with mass above M_g are inhibited from fragmentation and may remain as hot pressure-supported clouds. This type of argument can be refined, but one always obtains a mass M_g, above which any fluctuations are likely to remain amorphous and gaseous. These considerations suggest that M_g and R_g characterise the mass and radius of a galaxy. However, of all the scale estimates in Figure 2.2, this is the least certain.

2.6 Non-anthropic Tunings

There are a number of tunings which seem surprising but do not have anthropic siginificance. Indeed, Wilczek's has classifed tunings into four types, according to whether or not they seem to involve a selection effect and whether or not we are able to calculate them [91]. He describes these tunings as 'enlightenment' (e.g., $m_p \ll M_P$), 'knowledge' (e.g., $\theta_{QCD} \ll 1$), 'ignorance' (e.g., most Standard Model parameters), and 'temptation' (e.g., anthropic relations).

A particular example of 'ignorance' is assocated with the Higgs boson. This gives elementary particles their masses, but its mass is also affected by those particles and is particularly sensitive to the mass of the top quark. The measured Higgs mass, $m_H \approx 125$ GeV, seems very finely tuned. This is because the Higgs particle is also related to the vacuum state. The Universe should come to rest in the lowest energy vacuum state, but it seems to be caught in a small trough. If the Higgs mass and top quark mass were slightly different, it would either be in a completely stable vacuum or an unstable vacuum that would have decayed a long time ago. So the Universe seems to be located on boundary, as illustrated in Figure 2.3, which is taken from Reference [30]. However, it is possible that supersymmetry

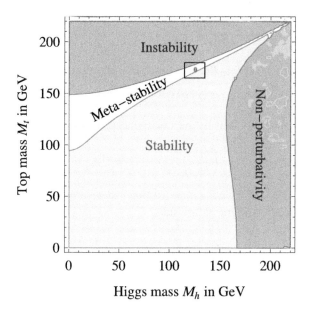

Figure 2.3 Fine-tuning of mass of Higgs particle and top quark; from Reference [30].

or some some symmetry associated with the axion may explain this. It may also be relevant that m_H is roughly the geometric mean of the Planck scale (10^{19} GeV) and the dark-energy scale (10^{-4} eV).

2.7 Weak Anthropic Principle

The 'Weak Anthropic Principle' (WAP) accepts the constants of nature as given and then shows that our existence imposes a selection effect on when we observe the Universe. In the ΛCDM picture, this may represent just a narrow window between the early radiation-dominated period and the late dark-energy-dominated period.

2.7.1 Dicke's Argument

As a simple example of a weak anthropic argument, consider the question: Why is the Universe as big as it is? The mechanistic answer is that, at any particular time, the size of the observable Universe is the distance travelled by light since the Big Bang, which is about 10^{10} light years. There is no compelling reason the Universe has the size it does; it just happens to be 10^{10} yr old.

There is, however, another answer to this question, which Robert Dicke [40] first gave in 1961. In order for life to exist, there must be carbon, and this is produced by cooking inside stars. The process takes about 10^{10} yr, so only after this time can stars explode as a supernovae, scattering the newly baked elements throughout space, where they may eventually become part of life-evolving planets. On the other hand, the Universe cannot be much older than 10^{10} yr, else all the material would have been processed into stellar remnants. Since all the forms of life we can envisage require stars, this suggests that it can only exist when the Universe is aged about 10^{10} yr. So the very hugeness of the Universe, which seems at first to point to our insignificance, is actually a prerequisite of our existence. This is not to say that the Universe itself could not exist with a different size, only that we could not be aware of it then.

We can express this result in terms of fundamental constants because standard physics predicts that the lifetime of a star is of order $\alpha_G^{-1} \approx 2 \times 10^{38}$ times the proton timescale $t_p = h/m_p c^2 \sim 10^{-23}$ s. (Since α_G^{-1} is so huge, we sometimes approximate it as 10^{40}, but we need a more precise value in this section.) We now present the argument for this. The luminosity of a star whose opacity is dominated by electron scattering (as applies for large stars) but not so large as to be radiation-pressure dominated is

$$L \sim (p_{\text{rad}}/p_{\text{mat}})L_E, \tag{2.27}$$

where the ratio of the pressures is given by Eq. (2.17) and

$$L_E = 4\pi GMm_p c/\sigma_T \quad \text{with} \quad \sigma_T = \alpha^2 r_e^2 \tag{2.28}$$

is the Eddington luminosity. A characteristic timescale, first discussed by Salpeter [69], is that over which an object of luminosity L_E would radiate away its entire rest mass:

$$t_E = \frac{c\sigma_T}{4\pi Gmp} \sim 0.1 \left(\frac{\alpha^2}{\alpha_G}\right) \left(\frac{m_p}{m_e}\right)^2 t_p. \tag{2.29}$$

If $\eta \sim 10^{-2}$ is the fraction of a star's rest mass that can be released through nuclear burning, the main-sequence lifetime is, thus,

$$t_{MS} \sim \eta \left(\frac{p_{rad}}{p_{mat}}\right)^{-1} t_E \sim 10 \left[\eta \alpha^2 \left(\frac{m_p}{m_e}\right)^2\right] \left(\frac{M}{M_C}\right)^{-2} \alpha_G^{-1} t_p, \tag{2.30}$$

the quantity in square brackets being of order unity. For a radiation-dominated star ($M > 50 M_\odot$), the M-dependence disappears, and t_{MS} levels off with the value of $10^{-2}\alpha_G^{-1} t_p \sim 2 \times 10^{13}$ s. Since the characteristic mass of a star is M_C, the Dicke argument requires

$$t_o > t_{MS} \sim 10\alpha_G^{-1} t_p \sim 2 \times 10^{16} \text{s}. \tag{2.31}$$

This is a factor of 20 shorter than the current age of the Universe $t_o \approx 4 \times 10^{17}$s. However, Eq. (2.30) is really only appropriate for an upper main-sequence star because the opacity is increased at lower masses. For a solar-mass star, t_{MS} exceeds $\alpha_G^{-1} t_p$ by a factor of around $m_p/m_e \sim 10^3$, so it is better approximated as $\alpha_G^{-1} t_e$.

Since t_{MS} is very sensitive to the value of M, Dicke's argument is not very precise. It would be more convincing if one could show that life requires stars of around a solar mass. For example, it might be difficult for life-bearing planets to evolve around much more massive stars, because they are too short-lived or because one needs some lower mass convective stars to have planets. Alternatively, even if the *first* stars were massive, perhaps some of the elements vital for life must be generated through the s-processes associated with later-forming, less massive stars. Whatever the appropriate value of M, t_o cannot be much bigger than observed, else most of the Universe would have been processed into stellar remnants. Therefore, observers are most likely to exist at an epoch $t_o \sim t_{MS}$.

2.7.2 Cosmological Consequences of Dicke's Argument

Dicke's argument for solar-mass stars implies the approximate relation

$$ct_o \sim \alpha_G^{-1}(\hbar/m_e c) \sim (\alpha/\alpha_G)a_o, \tag{2.32}$$

so the ratio of the size of the observable Universe to the size of an atom is comparable to the ratio of the electric and gravitational forces between protons. There is no other explanation for this well-known coincidence within conventional physics, but Dirac [32] suggested the unconventional explanation that α_G is *always* given by

$$\alpha_G \sim \hbar/(m_e c^2 t) \sim (t/t_e)^{-1}. \tag{2.33}$$

Assuming that $\hbar$, c, and m_e are constant in time, this requires that G decreases like t^{-1}, so Dirac invoked Eq. (2.33) as the basis for a new cosmology. However, such a variation of G is inconsistent with observation.

The relation (2.32) also implies that the number of protons in the observable Universe must be of order α_G^{-2}, thereby explaining another well-known coincidence. We now describe this argument in more detail. The total mass associated with the observable Universe (i.e., the mass within the horizon volume) is $\sim \rho_o c^3 t_o^3$, where ρ_o is the present matter density. This is given by the Friedmann equation:

$$\rho_o = \frac{3H_o^2}{8\pi G} + \frac{Kc^2}{16\pi G},$$ (2.34)

where K is the scalar curvature of the Universe. Providing the K term is smaller than the others, we deduce that the mass of the Universe is

$$M_U \sim c^3 t_o^3 G^{-1} H_o^2 \sim \alpha_G^{-2} \left(\frac{m_p}{m_e}\right) m_p \sim \alpha^{-2} \alpha_G^{-2} m_p,$$ (2.35)

where we have used $t_o \sim H_o^{-1}$ and Eq. (2.8). The fact that the number of protons in the Universe is of order α_G^{-2} is thus explained, providing one can neglect the K term in Eq. (2.34). Some people have argued that K must be zero by appealing to Mach's principle, but – as discussed later – there may also be anthropic reasons for expecting that the K term is small.

Given the expression for the size of the Universe, the dependencies derived in Section 2.6 allow one to predict amusing relationships between the different mass-scales discussed earlier:

- Human $\sim$ (planet $\times$ atom)$^{1/2}$
- Planck $\sim$ (exploding hole $\times$ proton)$^{1/2}$
- Exploding hole $\sim$ (Universe $\times$ proton)$^{1/2}$.
- The number of stars in the galaxy $\sim \alpha^4 \alpha_G^{-1/2}$.
- The number of galaxies in the Universe $\sim \alpha^{-6}$.

These relationships are all consequences of standard physics, providing one accepts the WAP prediction for the size of the Universe. Of course, they are only order-of-magnitude relationships, since the objects involved (humans, stars, galaxies, etc.) all span a range of sizes and masses.

2.7.3 Does WAP Suffice?

Dicke's argument helps us to understand why the preceding large number coincidences prevail. It accepts the constants of Nature as given and then shows that our existence imposes a selection effect on when we observe the Universe. As such, it is no more than a logical necessity: saying that we have to exist at a certain time is no more surprising than saying we have to exist in a certain place (e.g., close to a star). It might be surprising to find what the selection effects are, but their existence is not surprising in principle. In fact, most physicists would agree with the weak version of the Anthropic Principle.

The problem comes when we consider whether constants such as G are themselves determined by the requirement that life should arise, a notion we have referred to as the 'Strong Anthropic Principle'. That the weak principle may not be the whole story is also suggested by the fact that all the scales shown in Figure 2.2 are relative. If α and α_G differed from what we observe them to be, all the scales would change, but the basic relationships between them would be the same. For example, one could envisage a hypothetical universe in which all microphysical laws were unchanged, but G was (say) 10^6 times stronger. Planetary and stellar masses ($\propto \alpha_G^{-3/2}$) would then be lowered by 10^9, but hydrogen-burning main-sequence stars would still exist, albeit with lifetimes ($\propto \alpha_G^{-1}$) of 10^4 yr rather than 10^{10} yr. Moreover, Dicke's argument would still apply: a hypothetical observer looking at the Universe when $t_o \sim t_{MS}$ would find the number of particles in the Universe 10^{12} times lower than in ours, but he would still find the 'large number' coincidences described earlier. If one fixed α_G but allowed α to change, the effects would be less extreme but still noticeable.

What are the arguments against the 'cognisability' of this kind of small-scale speeded-up universe? One rather loose constraint on α_G comes from biological considerations. We have seen that the number of stars in the observable Universe is of order $\sim \alpha_G^{-1/2}$. If we regard stars – or at least their associated solar systems – as potential sites for life, this is also the number of places where life may have arisen. However, this is not a *sufficient* condition for life because there is a whole set of extra conditions, each of which may be very improbable (cf. the Drake equation). For example, we need the star to have a planet, we need the planet to be at a suitable distance from the star, it needs to have a suitable atmosphere and chemistry, and there must be the appropriate conditions for the first self-replicating cells to arise.

Clearly, therefore, the overall probability (P) of life arising at any particular site must be very small. So if we want there to be life somewhere in the observable Universe, we need the number of sites for life times the probability P to exceed 1. This implies that α_G must be less than P^2. For example, if P were 10^{-15}, one would need $\alpha_G < 10^{-30}$. This is not a very precise argument, but it gives a qualitative reason why α_G needs to be small. We now discuss more specific anthropic arguments that pin down α_G more narrowly.

2.8 Strong Anthropic Principle: Astrophysical Coincidences

The Strong Anthropic Principle (SAP) claims that there are tunings between the physical constants themselves. Some of them involve the four dimensionless coupling constants, while others involve various cosmological parameters, and the tuning is sometimes remarkably precise. Although one might hope that some final theory of physics or cosmological evolution would explain these relationships, they are not predicted by current physics.

2.8.1 Convective and Radiative Stars

One of the first and most striking SAP tunings was given by Brandon Carter and relates to the existence of stars with convective and radiative envelopes [24]. In the first case, the heat generated in its core by nuclear reactions is transported to the surface primarily by way of large-scale motions of the stellar material itself. This tends to be the case for sufficiently small stars (red dwarfs). By contrast, larger stars (blue giants) tend to be radiative in the sense that the heat gets out primarily via the flow of radiation. The dividing line between the two types is some critical mass which lies in the range around $\alpha_G^{-3/2} m_p$ in which stars actually exist only because of the remarkable coincidence

$$\alpha_G \sim \alpha^{20}. \tag{2.36}$$

Were G slightly larger, all stars would be blue giants; were it slightly smaller, all stars would be red dwarfs. This does not pin down the actual values of α and α_G, but it does specify a scaling law between them, and it explains why α_G is so much smaller than α. This is perhaps the most striking coincidence because of the high power of α involved. It is also the condition that the number of stars in a galaxy be comparable to the number of galaxies in the Universe.

Let us now derive relation (2.36). If radiation transport is unable to maintain a star's surface temperature T_S above the ionisation temperature $\sim 0.1\alpha^2 m_e c^2/k$, a convective outer layer develops, and this supplements the heat transport so that it does. The value of T_S that can be maintained by radiative transport, assuming a central temperature

$$T_C \sim 10^{-2}\alpha^2 m_p c^2/k \sim 10^7 \text{ K}, \tag{2.37}$$

is

$$T_S \sim (L/aR^2)^{1/4} \sim \tau^{-1/4} T_C, \tag{2.38}$$

where we have used Eq. (2.18). Here, τ is the optical depth through the star, which can be expressed as

$$\tau \sim nR\sigma_T \sim \frac{M\sigma_T}{m_p R^2} \sim \frac{k^2 \sigma_T T_C^2}{G^2 M m_p^3}, \tag{2.39}$$

where we have assumed electron-scattering opacity and used Eq. (2.15) at the last step. Thus, T_S exceeds the ionisation temperature, providing M exceeds

$$M_* \sim \alpha_G^{-2}\alpha^{10} m_p \sim \alpha_G^{-1/2}\alpha^{10} M_C. \tag{2.40}$$

The mass M_* divides the (convective) red dwarfs from the (radiative) blue giants and is comparable to the mass M_C in which main-sequence stars actually exist only because of relation (2.36). Page has discussed the argument in more detail [59], interpreting it as a constraint on the electron charge ($e \propto \sqrt{\alpha}$) and showing that it constrains e to 3%.

The relationship $\alpha_G \sim \alpha^{20}$ is clearly satisfied numerically, but current physics does not explain *why* this relationship should pertain. Its anthropic significance is that only radiative stars can end their lives as supernovae, which is required to disseminate heavy elements. Otherwise, all stars would be chemically homogeneous due to convective mixing and not develop the 'onion-skin' shell structure which characterises presupernova models. So α_G cannot be much smaller than α^{20}. On the other hand, Carter suggested that the formation of planetary systems may be associated with convective stars. This was based on the observational feature that red dwarfs have much less angular momentum than blue giants, and a loss of angular momentum could be a consequence of planet formation. This argument is no longer compelling because there are other ways of losing angular momentum. A better argument might be that only convective stars generate winds in their early phase intense enough to blow away the gaseous envelope of nearby planets, thereby facilitating the formation of solid planets with non-hydrogen atmospheres. In either case, one infers that no planets, and hence no life, would form if α_G were much larger than α^{20}. Even if the anthropic significance of the relation $\alpha_G \sim \alpha^{20}$ is disputed, the existence of both blue giants and red dwarfs certainly requires this.

If we had one more relationship between α and α_G, we could predict the value for each of them. Another relationship does, in fact, exist. It does not come from an anthropic argument but from an argument in quantum field theory. It has been suggested that all space intergals in quantum electrodynamics should be cut off at the Planck length, thereby reducing otherwise divergent integrals to finite functions of the parameter $\alpha \log \alpha_G$. Various arguments [31] suggest a self-consistent electrodynamics is possible only if this parameter has some specific value of order unity – i.e., one requires

$$\alpha^{-1} \sim \log \alpha_G^{-1}. \tag{2.41}$$

This relation, together with the convective star condition, implies that α must be about 10^{-2} and α_G must be about 10^{-40}, as observed. In view of the simple dependencies on α and α_G of the different scales of structure in the Universe, this suggests that the appearance of our Universe is determined, not merely in part, but to a very large degree by our existence.

In an important series of papers, Fred Adams and colleagues have explored these relationships in much more detail [1–5]. They argue that star constraints are not as strict as sometimes claimed and that the possible parameter space spans several orders of magnitude. A summary of their results is contained in Figure 2.4 and its caption. This does not include relation (2.36) because the connection between convective stars and planets and between radiative stars and supernovae is disputed.

2.8.2 Supernovae

The value of α_W is involved in an interesting anthropic constraint involving supernovae [24]. We have seen that supernovae are essential for life because they disseminate heavy elements throughout the Universe. The reason a star explodes after burning its nuclear fuel is that its core gets hot enough when it collapses to generate a surge of neutrinos, which then blow off the envelope as a result of weak interactions. For this model to work, one requires the timescale on which neutrinos interact with nuclei in the envelope to be comparable

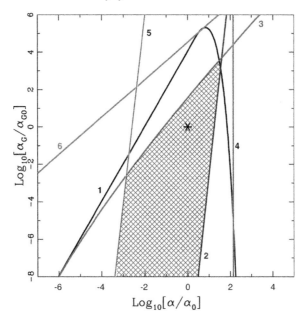

Figure 2.4 The shaded region shows the (α, α_G) plane allowed by the constraints of Reference [3]. Curve 1 indicates that the star is stable, curve 2 that its temperature is high enough to allow habitable planets, and curve 3 that it lives long enough for biological evolution to occur. For planets to be smaller than stars, α must be to the left of line 4. For stars to be smaller than galaxies, it must be to the right of line 5. Planets can support a biosphere and remain non-degenerate below line 6. From Reference [3].

to the dynamical timescale. If it were much longer, the envelope would be essentially transparent to the neutrinos; if it were much shorter, the neutrinos would be trapped in the core and could not escape to deposit their momentum in the less tightly bound surrounding layers.

Let us now examine this argument in more detail. The weak interaction cross section is

$$\sigma_W \sim c^{-4}\hbar^{-4}g^2(kT)^2 \sim \alpha_W^2 r_e^2 \left(\frac{kT}{m_e c^2}\right)^2, \tag{2.42}$$

so the weak interaction and dynamical timescales are comparable if

$$(nc\sigma_W)^{-1} \sim (Gnm_p)^{-1/2}, \tag{2.43}$$

where n is the nucleon number density. Most of the neutrinos are pair-produced by $e^+ + e^- \rightarrow \nu + \bar{\nu}$, so for this to be possible, one requires $kT \sim m_e c^2$. One also expects the density at the bounce to be of order the nucleon density which is $n \sim (\hbar/m_p c)^{-3}$. Putting these values for n and T into Eq. (2.43) gives

$$\alpha_G \sim \alpha_W^4 (m_p/m_e)^2. \tag{2.44}$$

The last factor is large but would be absent if we used the electron rather than proton mass in the definition of α_G. We know that this relationship holds numerically but the AP explains *why* it must hold. So if we accept that α_G is determined anthropically, we must also accept that α_W is so determined. As we will see later, the relation $\alpha_G \sim \alpha_W^4$ also plays a crucial role in Big Bang nucleosynthesis.

2.8.3 Triple-α

Perhaps the most famous anthropic tuning – and the most sensitive constraint on the value of α_S – concerns the generation of carbon (a prerequisite for our form of life) in the helium-burning phase of red giant stars via the 'triple-α' reaction. The way a star makes carbon is by first combining two α particles to make a beryllium nucleus and then adding another α particle to form a carbon nucleus:

$$\mathrm{He}^4 + \mathrm{He}^4 \rightarrow \mathrm{Be}^8, \quad \mathrm{Be}^8 + \mathrm{He}^4 \rightarrow \mathrm{C}^{12}.$$

The trouble is that beryllium is unstable (otherwise, the 'helium flash' in giants would lead to a catastrophic explosion), and it used to be thought that it would decay before the extra α particle could combine with it. For many years, therefore, it was difficult to understand why there is any carbon in the Universe. Then Fred Hoyle [48] realised that there must be a resonance (i.e., an enhanced interaction rate) in the second step, which allows the carbon to form before the beryllium disappears – i.e., C^{12} must have a state with energy just above the sum of the energies of Be^8 and He^4. There is, however, no similar favourably placed resonance in 10^{16}; otherwise, almost all the carbon would be transmuted into oxygen.

Once the suggestion was made, the resonance was looked for in the laboratory and rapidly found. So this might be regarded as the first confirmed anthropic prediction, although Kragh [53] takes a different view. Indeed, the fine-tuning required is so precise that Hoyle concluded that the Universe has to be a 'put-up job'. At the time, it was not possible to quantify this coincidence, but more recent work has studied this more carefully [6, 25, 37, 56]. In particular, studies by Oberhummer *et al.* – calculating the variations in oxygen and carbon production in red giant stars as one varies the strength and range of the nucleon interactions – indicates that the nuclear interaction strength must be tuned to at least 0.5% [58].

2.8.4 Constraints from Chemistry

As discussed in detail by Barrow and Tipler [11], many features of chemistry are sensitive to the value of α_S. For example, if α_S were increased by 2%, all the protons in the Universe would combine at cosmological nucleosynthesis to form diprotons (nuclei consisting of two protons). In this case, there would be no hydrogen and, hence, no hydrogen-burning stars. Since stars would then have a much reduced main-sequence time, there might not be time for life to arise. If α_S were increased by 10%, the situation would be even worse because everything would go into nuclei of unlimited size, and there would be no interesting chemistry. The lack of chemistry would also apply if α_S was decreased by 5% because all deuterons would then be unbound, and one could only have hydrogen.

There are several other coincidences involving f and the masses of various elementary particles which seem to be necessary for chemistry. These involve the mass ratios

$$m_e/m_N = 1/1837, \quad m_\pi/m_N = 1/7, \quad \Delta/m_N = 1/730, \tag{2.45}$$

where $\Delta = m_N - m_p$. The important features of nuclear physics depend upon the following four coincidences [24]:

$$
\begin{aligned}
&(a) \ f^2 \approx 2m_N/m_\pi, \\
&(b) \ \Delta/m_e \approx 2, \\
&(c) \ \alpha \approx \Delta/m_\pi, \\
&(d) \ f \approx 1/(3\alpha^{1/2}).
\end{aligned}
\tag{2.46}
$$

(a) implies that strong interactions are only marginally strong enough to bind nucleons into nuclei. If f were slightly weaker, only hydrogen would exist; if it were slightly stronger, nuclei of almost unlimited size would exist. (b) implies that neutrons are unstable to β-decay in isolation but not in the presence of relativistic degenerate electrons. (c) implies that the electrostatic energy in light nuclei $\sim \alpha m_e c^2$ is comparable to the neutron-proton mass difference. (d) implies that the electrostatic energy is small compared to nuclear binding energy in light nuclei but comparable to it for nuclei with $Z \sim (f\alpha)^{-3} \sim 30$, so such large nuclei are unstable to electromagnetic disruption. Note that the combination of (a), (b), (c), and (d) implies the relation (2.8).

If the relations indicated by Eq. (2.46) were not satisfied to at least 10% accuracy, elements vital to life would not exist, so one might ascribe anthropic significance to these relations. Kahn [49] has pointed out that $m_e \ll m_p$ may also be a prerequisite for complex chemistry, since this ensures that ions are localised to a precision $(m_e/m_p)^{1/4}$ times their mean spacing. More recent constraints on α and α_S from chemistry – also expressed in terms of the electron-proton mass ratio – are summarised in Figure 2.5, which is taken from Barnes [8].

From the modern perspective, α_S, m_p and m_e are no longer fundamental quantities; the QCD interaction strength and quark masses would be regarded as more significant. Nevertheless, fine-tuning is still required at some level. As indicated in Figure 2.6, which is taken from Hogan [46], the masses of the light fermions that make up the stable matter of which we comprise – the up and down quarks and the electron – have values in a narrow window that allows the existence of a variety of nuclei other than protons and also atoms with stable shells of electrons that are not devoured by their nuclei. For example, the stability of free protons requires [41]

$$\alpha < (m_d - m_u)/141 \text{ MeV} \approx 1/50. \tag{2.47}$$

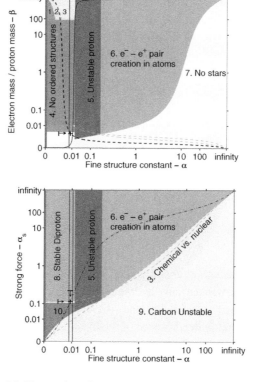

Figure 2.5 Fine-tuning of α, α_S, and m_e/m_p; from Reference [8].

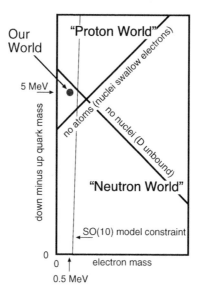

Figure 2.6 Constraints on electron mass and the down-up quark mass difference from the stability of hydrogen and deuterium. The thin vertical line shows the constraint for a particular SO(10) grand unified scenario. From Reference [46].

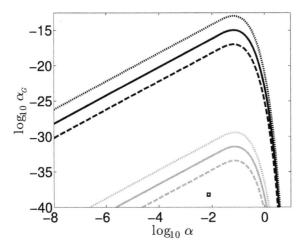

Figure 2.7 Limits on stable stars in (α_G, α) space for H-burning stars (black lines) and D-burning stars (grey lines) from Reference [9]. This shows that D-burning stars are stable in a much larger region of parameter space.

Since life requires stable nuclei other than protons and neutrons, these fundamental parameters of the Standard Model are good candidates for quantities whose values are determined anthropically.

Recent literature has focused on a fine-tuning associated with the diproton [33]. Although hydrogen burns slowly in the Sun (because it is moderated by weak inteactions), it would burn explosively if the Sun were made of heavy hydrogen (as in a nuclear bomb), and increasing f by 6% would bind the diproton [26, 62]. Barr and Khan [10] calculate the equivalent condition on the light quarks masses as

$$m_u + m_d < 0.75 \, (m_u + m_d)_{obs} = 5.3 \text{ eV}. \tag{2.48}$$

On the other hand, the deuteron would be strongly unbound for

$$m_u + m_d < 1.4 \, (m_u + m_d)_{obs} = 9.9 \text{ eV}, \tag{2.49}$$

in which case one would need much higher central temperatures in stars. Although Barrow and Tipler argue that a bound diproton would result in all the hydrogen being consumed at Big Bang nucleosynthesis, more detailed calculations show that this is not the case [57]. There could also be long-lived stars [14], although Barnes argues [9] that the strongest anthropic bound on stars in such a universe still comes from their lifetime, all stars burning out within 10^6 yr unless $\alpha_G < 10^{-30}$. Figure 2.7 is taken from his paper.

2.9 Strong Anthropic Principle: Cosmological Coincidences

The second set of fine-tunings is associated with Big Bang nucleosynthesis and the formation of galaxies and their subsequent fragmentation into stars. They involve various cosmological parameters, such as the current matter density in units of the critical density Ω_o, the amplitude of the density fluctuations Q on entering the cosmological horizon and the photon-to-baryon ratio S. Although most cosmologists might prefer to believe that these parameters were determined by processes in the early Universe rather than being prescribed freely as part of the initial conditions, even small deviations from the observed values would exclude the formation of cosmic structures.

2.9.1 Cosmological Nucleosynthesis

Let us first recall the process of cosmological nucleosynthesis, as discussed by Uzan in this volume (see Chapter 7). The prediction that one turns 24% of the mass of the Universe into helium is one of the great triumphs of the Big Bang picture. However, the only reason one gets an interesting amount of helium is because the neutron-to-proton ratio freezes out with an interesting value. The freeze-out occurs at 1 s when $T \sim 10^{10}$ K because the weak interactions become slower than the cosmological expansion rate then. However, it is only because the freeze-out temperature and neutron-proton mass difference are comparable that one gets the observed amount of helium production. It turns out that the condition for this is roughly $\alpha_G \sim \alpha_W^4$, precisely the condition required for supernovae.

Let us now describe this argument in more detail. The weak interactions proceed at a rate $n\sigma_W c$, where $n \propto T^3$ is the particle number density and $\sigma \propto T^2$ is given by Eq. (2.42). Therefore, the weak rate scales as T^5, whereas the cosmic expansion rate $\sim (GaT^4/c^2)^{1/2}$ scales as T^2. So freeze-out occurs at a temperature

$$kT_F \sim \alpha_G^{1/6} \alpha_W^{-2/3} c^2 m_e^{4/3} m_p^{-1/3}. \tag{2.50}$$

Since virtually all the frozen-out neutrons burn into helium, the resultant helium abundance is

$$Y \approx \frac{2n_N/n_P}{n_N/n_P + 1} \quad \text{where} \quad \frac{n_N}{n_P} \approx \exp\left(-\frac{\Delta c^2}{kT_F}\right). \tag{2.51}$$

Y is 24% rather than 0% or 100% only because $kT_F \approx \Delta c^2 \approx 2m_e c^2$, where we have used Eq. (2.46b). From Eq. (2.50) this derives from the 'coincidence'

$$\left(\frac{Gm_e^2}{\hbar c}\right)^{1/4} \sim \alpha_W. \tag{2.52}$$

The number on the left is the quarter power of the electron gravitational fine structure constant, so this is precisely equivalent to condition (2.44). In the Weinberg-Salam theory of weak and electromagnetic interactions, α_W to α are related by

$$\alpha_W \sim \alpha \left(\frac{m_e}{m_W}\right)^2, \tag{2.53}$$

so Eq. (2.52) may also be written in the form

$$\alpha_G \sim \left(\frac{m_e}{m_W}\right)^8 \left(\alpha^2 \frac{m_p}{m_e}\right)^2 \sim \left(\frac{m_e}{m_W}\right)^8, \tag{2.54}$$

where we have used Eq. (2.8).

It is not clear to what extent this coincidence can be interpreted anthropically. Life could probably not exist if Y were 100% (as would be the case if α_W were slightly smaller), since there would be no water. Also, the lifetime of a helium star is less than that of a hydrogen star and might not be long enough to permit evolution of life. However, it is not clear that a universe with no primordial helium would preclude life. On the other hand, the same constraint in the supernova context applies in both directions.

2.9.2 Density Parameter

Before the discovery of dark energy, there were well-known anthropic reasons for why the total matter density parameter Ω_o had to lie within an order of magnitude of 1 in order for the geometry of the universe to be nearly flat. If Ω_o were much larger than 1, the Universe would recollapse in a time $\Omega_o^{-1/2} H_o^{-1}$, and this would be less than the main-sequence time of a star. On the other hand, if Ω_o were much smaller than 1, density fluctuations would stop growing at the time $\Omega_o H_o^{-1}$, and this means that – for reasonable initial density fluctuations – galaxies would never bind at all. This argument requires that Ω_o be in the range 0.01–100. Similar arguments apply in a universe with dark energy, except that there are additional anthropic constraints on the value of the effective cosmological constant.

This is only a weak constraint on Ω_o, but in 1973, Collins and Hawking [27] gave a more precise argument, based on the observed isotropy of the microwave background. They showed that the set of spatially homogeneous cosmological models which approach isotropy at late times is of measure zero in the space of all spatially homogeneous models. So only a small set of initial conditions could give rise to a universe as isotropic as observed today, and these correspond to models which are spatially flat. Only such models would expand long enough for galaxies and intelligent life to form, suggesting that the observed isotropy might be a reflection of our own existence.

In the 1980s, early universe studies were revolutionised by the introduction of the inflationary scenario [39, 55, 85]. This *requires* that Ω_o be very close to 1 and – as discussed by Martin in this volume (Chapter 4) – observations seem to support this model. Therefore, anthropic considerations may no longer seem relevant. However, in the simplest model – with a single scalar field – inflation only works if the form of the vacuum potential $V(\phi)$ allows a sufficient number of expansion e-folds, which means that $V(\phi)$ must itself be fine-tuned [80]. Similar considerations apply in more complicated inflationary models, including quantum cosmological models [42], where universes are expected to collapse very quickly unless one imposes anthropic selection effects [44].

2.9.3 Photon-to-Baryon Ratio

As discussed by Steigman and Scherrer in this volume (Chapter 5), another important cosmological parameter is the entropy per baryon. Since the entropy is dominated by the microwave background, this is equivalent to the photon-to-baryon ratio S, which is of order $10^8 \Omega_B^{-1}$, where $\Omega_B \approx 0.05$ is the baryon density parameter. The value of S is associated with an interesting coincidence: if $S \sim 10^9$, the matter and radiation densities are comparable at the time they thermally decouple. We first examine this coincidence in more detail.

According to the hot Big Bang model, the background radiation dominated the density of the Universe until a redshift z_{eq}, which depends on S and the current CMB temperature T_o:

$$1 + z_{eq} \sim \frac{m_p c^2}{S f_B k T_o}, \quad T_o \sim \frac{\hbar c}{k} \left(\frac{S f_B}{G m_p t_o^2} \right)^{1/3}, \tag{2.55}$$

where $f_B \equiv \Omega_B / \Omega_M \approx 0.2$ and $\Omega_M \approx 0.25$ is the total matter density. In the matter-dominated era, $t \sim t_o(1 + z)^{-3/2}$, and so matter-radiation equality occurs at

$$t_{eq} \sim S^2 f_B^2 \alpha_G^{-1/2} t_p \sim 10^{12} \text{ s.} \tag{2.56}$$

On the other hand, the CMB thermally decoupled from the matter when T fell below $T_{dec} \sim 0.1 \alpha^2 m_e c^2 / k$, which corresponds to the somewhat later time

$$t_{dec} \sim t_o \left(\frac{\alpha^2 m_e}{k T_o} \right)^{-3/2} \sim S^{1/2} f_B^{1/2} \alpha_G^{-1/2} \alpha^{-6} t_p \sim 10^{13} \text{ s.} \tag{2.57}$$

(These equations use the fact that the radiation density and temperature fall like R^{-4} and R^{-1}, respectively.) The requirement that t_{eq} and t_{dec} be close to each other therefore corresponds to the coincidence

$$S \sim \alpha^{-2} f_B^{-1} \left(\frac{m_p}{m_e} \right) \sim \alpha^{-4}, \tag{2.58}$$

where we use Eq. (2.8) in the last step.

It is not clear that this coincidence has any anthropic significance, but there are certainly anthropic aspects to the values of t_{eq} and t_{dec}. For example, there is an upper limit on S if one requires that the Universe be radiation dominated at cosmological nucleosynthesis to avoid all the hydrogen going into helium. From Eq. (2.50), the weak freeze-out time is

$$t_F \sim \hbar c^2 \alpha_G^{-1/2} (k T_F)^{-2} m_p \sim \alpha_G^{-5/6} \alpha_W^{4/3} \left(\frac{m_p}{m_e} \right)^{8/3} t_p, \tag{2.59}$$

so the condition $t_{eq} > t_F$ corresponds to

$$S > f_B^{-1} \left(\frac{m_p}{m_e} \right)^{4/3} \left(\frac{\alpha_W^4}{\alpha_G} \right)^{1/6} \sim 10^4. \tag{2.60}$$

This does not pin down the value of S very precisely, but it does impose an interesting lower limit.

Other anthropic constraints on S are associated with galaxy formation (see also Chapter 6). In the standard Big Bang model, galaxies cannot form until the background radiation density falls below the matter density at t_{eq}, but this occurs before the Dicke timescale (2.30) (i.e., the main-sequence lifetime of a star) only if

$$S < \alpha^{-1}\alpha_G^{-1/4} f_B^{-1} \sim 10^{11}. \qquad (2.61)$$

Since the Jeans mass in the period t_{eq} to t_{dec} has a value $M_J \sim \alpha_G^{-3/2} S^2 f_B^2 m_p$, this exceeds the galaxy mass given by Eq. (2.26) only for

$$S > \alpha_G^{-1/4} \alpha^{5/2} \left(\frac{m_p}{m_e}\right)^{1/4} \sim \alpha_G^{-1/4} \alpha^2 \sim 10^6. \qquad (2.62)$$

Eqs. (2.61) and (2.62) constrain S rather tightly.

One can strengthen Dicke's weak anthropic argument for the age of the Universe by the general requirement – independent of considerations of stellar physics – that observers can exist only for $t_o > t_{dec}$, when thermodynamic disequilibrium is possible. This corresponds to the condition

$$S < 10^4 \alpha_G^{-1} \alpha^{12} f_B^{-1} \sim 10^{17} \qquad (2.63)$$

but this only gives only a very weak upper limit on S. An interesting twist on this argument has been provided by Aguirre [7], who describes anthropic constraints on 'cold' cosmological models (i.e., models with an initial photon-to-baryon ratio much smaller than currently observed). He points out that such models could provide life-supporting conditions with very different values of the cosmological parameters.

In the context of limits (2.61) and (2.62), we note that there are a number of scenarios which predict $S \sim \alpha_G^{-1/4}$. This could apply, for example, if the radiation was generated by a first generation of large pregalactic objects [18]. Such objects have a characteristic lifetime t_{MS} given by Eq. (2.30), so after t_{MS}, one would expect the radiation density to be ξF times the matter density, where F is the fraction of the Universe which goes into the stars and ξ is the fraction of each star's rest mass released through nuclear energy generation. The value of t_{eq} associated with the generated S is thus $(\xi F)^{3/2} t_{MS}$, and Eqs. (2.56) and (2.30) imply

$$S \sim \left[\alpha\left(\frac{m_p}{m_e}\right) F^{3/4} \xi^{3/4}\right] \alpha_G^{-1/4}. \qquad (2.64)$$

Since the term in square brackets is of order unity, one would expect $S \sim \alpha_G^{-1/4}$, as observed. The same relationship between S and α_G would apply if the radiation were generated by black holes accreting at the Eddington limit. Such scenarios are no longer mainstream, but they illustrate that the S coincidences may be explained naturally, without recourse to anthropic considerations.

Nowadays, most cosmologists believe the value of S results from of baryon-violating processes in the early Universe, most probably at the GUT epoch, around 10^{-34} s after the Big Bang. However, in most GUT models, S is predicted to be of order α^{-n}, where n is an integer, so the anthropic constraint $S < \alpha_G^{-1/4}$ merely translates into the constraint $\alpha_G < \alpha^{4n}$. If $n = 5$, this just gives the convective star condition [15].

2.9.4 Cosmological Constant

As discussed by Peacock in this volume (Chapter 3), another striking feature of the Universe is that its expansion appears to be accelerating under the influence of some form of 'dark energy'. The source of this energy is uncertain, but it may be associated with a cosmological constant, denoted by Λ in Table 2.1. One possibility is that Λ arises through quantum vacuum effects. We do not know how to calculate these, but the most natural value would be the Planck density (which is 120 orders of magnitude larger than observed value). For example, in the 'string lanscape' variant of M-theory there could be 10^{500} vacuum states [12], with the associated Λ covering the full range from minus to plus the Planck value [77, 78]. There is also the remarkable coincidence that the vacuum and matter densities are comparable at the present epoch, even though their ratio is time dependent.

As first emphasised by Weinberg [88, 89] and later studied by Efstathiou [35], Vilenkin [86], and Peacock [61], this may provide the strongest anthropic fine-tuning of all, since a priori Λ could be 120 orders of magnitude larger than observed. This is because the growth of density perturbations is quenched once Λ dominates the cosmological density, so if bound systems have not formed by then, they never will. This is not the only possible explanation for the smallness of Λ, but it may be the most plausible one. The crucial issue is whether the number of vacuum states is sufficiently large and their spacing sufficiently small to satisfy the anthropic constraints [51], but this is still unresolved. As discussed in the next section, the precise form of the Λ constraint depends on the amplitude of the primordial density fluctuations at the horizon epoch.

It should be stressed that a cosmological constant is not the only possible explanation for the cosmic acceleration. An alternative explanation is to invoke dark energy in the form of a scalar field – termed 'quintessence' [75] – and this may better explain the near equality of the vacuum and matter densities. However, some anthropic fine-tuning may be required even in this case. Kallosh [50] gives some examples from her studies of M/string theory where anthropic reasoning helps to shed light on the properties of dark energy. These issues are discussed in more detail by Peacock in Chapter 3.

2.9.5 Density Fluctuations

The precise form of the anthropic upper limit on Λ depends on the amplitude of the density fluctuations, Q, but this also has anthropic aspects. If Q is too low, no galaxies form because cooling is ineffective. If Q is too high, there is excesssive black hole formation and galaxies

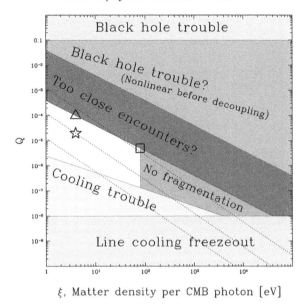

Figure 2.8 Constraints on Q as a function of ξ; from Reference [82].

are too dense for long-lived planetary systems because of disruption by neighbouring stars. For $\Lambda = 0$, Tegmark and Rees showed that this leads to a constraint of the form [81]

$$\alpha^{-1} \ln(\alpha^{-2})^{-16/9} \alpha_G (\beta/\xi)^{4/3} \Omega_B^{-2/3} < Q < \alpha^{16/7} \alpha_G^{4/7} \beta^{12/7} \xi^{-8/7}, \qquad (2.65)$$

where $\xi = f_B^{-1} \alpha_G^{-1/2} S^{-1}$ is the mass of non-relativistic matter per photon in Planck units. As illustrated in Figure 2.8, this corresponds roughly to the range $10^{-6} < Q < 10^{-4}$. Later, Tegmark *et al.* [82] expanded this analysis to consider variations of Q and Λ together, and the results are illustrated in Figure 2.9. Several other authors have also considered this problem [1, 12, 13, 38].

2.9.6 Dark Matter

As discussed by Kolb in this volume (Chapter 9), the existence of dark matter with 25% of the total cosmological density is now firmly established. The anthropic significance of this is unclear. One needs dark matter in order to amplify the fluctuations at decoupling enough to provide galaxies, but one could also achieve that by increasing the value of Q. There are still many possible dark matter candidates, but several of them are associated with anthropic constraints.

If WIMPs provide the dark matter, then their density will be comparable to the baryon density provided the coincidence $\alpha_G \sim \alpha_W^4$ is satisfied [84]. This is the same relationship which arises in the context of supernovae and cosmological nucelosynthesis, essentially

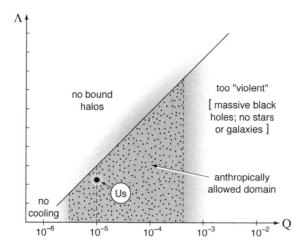

Figure 2.9 Constraints on Λ as a function of Q; from Reference [66].

because all three arguments involve the weak freeze-out condition. So there is at least an indirect sense in which this condition is anthropic.

If axions provide the dark matter, then anthropic arguments may also explain why their density is comparable with the baryon density [91]. The axion is a dark matter particle associated with the breaking of Peccei-Quinn (i.e., strong charge-parity – CP) symmetry at a time of order 10^{-30} s after the Big Bang. Large values of the symmetry-breaking energy scale, associated with large values of the Peccei-Quinn 'misalignment' angle, are forbidden in conventional axion cosmology. However, if inflation occurs after the breaking of Peccei-Quinn symmetry, then the CP-violating factor θ may vary across the different inflationary domains. So large values are permitted, providing we inhabit a domain where θ is small. Although such regions may occupy only a small volume of the multiverse, they contain a large fraction of potential observers.

In recent years, primordial black holes (PBHs) have become a popular dark matter candidate [17], partly because of the failure to find more conventional candidates but also because of the possibility that the coalescing black holes detected by LIGO could be primordial. PBHs are a natural dark matter candidate because they form in the radiation-dominated era and are therefore unrestricted by the nucleosynthesis bound on the baryonic density. However, it does require fine-tuning of the collapse fraction. In the standard scenario, the fraction of the Universe going into PBHs of mass M is only $\beta \sim 10^{-9}(M/M_{\odot})^{1/2}$ at formation. Reference [20] discusses a possible anthropic resolution of this problem by invoking PBH formation at the QCD epoch. Since the horizon mass then is around M_C, this would explain why dark matter and stars have comparable masses. The collapse fraction is $\beta \sim S^{-1}$ because a baryon asymmetry of $O(1)$ is generated in the hot expanding shell around each PBH and this is then diluted by a factor β when the baryons disperse, the factor being tuned so that the times of matter-radiation equality and decoupling are comparable.

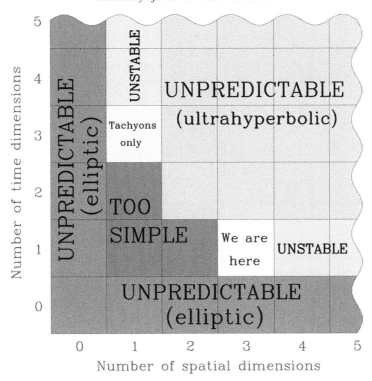

Figure 2.10 Illustrating why we live in 3 + 1 dimensions. When the partial differential equations of nature are elliptic or ultrahyperbolic, physics has no predictive power. In the remaining (hyperbolic) cases, $n > 3$ admits no stable atoms and $n < 3$ may lack sufficient complexity for observers (e.g., no gravity). From Reference [79].

2.9.7 Number of Dimensions

Since many models of particle physics invoke extra spatial dimensions, the issue of why we live in a world with three spatial dimensions naturally arises. While there may well be some physical explanation for this, it clearly has anthropic aspects. For example, there would be no gravity with two spatial dimensions, and planetary orbits would be unstable with four of them. There are also constraints on the number of time dimensions, associated with causality. Other arguments for the number of space and time dimensions have been given by Tegmark [79], and Figure 2.10 is taken from his paper.

In the context of higher-dimensional models, it is interesting that brane cosmology [63] may give a natural explanation for the sort of power-law relations between the coupling constants which arise in the anthropic arguments. This is because the variation in the gravitational coupling constant would be associated with the change in the volume of the bulk, whereas the variation in the other coupling constants would be depend on the change in the volume of the brane or some manifold of intermediate dimensionality. In this

Table 2.3 *Fine-tunings associated with history of Big Bang.*

$\log(t/s)$	Event	Condition	Anthropic significance
+17.5	Present epoch	$\Omega_o < 10$	Else premature recollapse
+17.0	Planet formation	$\alpha_G < \alpha^{20}$	Need convective stars
+16.5	Metals from stars	$\alpha_G \sim \alpha_W^4$	Need supernovae
		$\alpha_G > \alpha^{20}$	Need radiative stars
	Carbon from stars	$|\Delta\alpha_S| < 0.005\alpha_S$	Need triple-α resonance
+16	Galaxy formation	$\Omega_o > 0.1$	Over-dense regions must bind
+11	End of radiation era	$S < \alpha_G^{-1/4}$	Must precede galaxy formation
+2	Big Bang nucleosynthesis	$\alpha_G < \alpha_W^4$	Else all hydrogen $\rightarrow$ helium
		$\Delta\alpha_S < 0.02\alpha_S$	Else all hydrogen $\rightarrow$ diprotons
		$\Delta\alpha_S > 0.05\alpha_S$	Else deuterons unbound
−30	Axion production	$\theta \ll 1$	Need enough baryons
−34	Baryosynthesis	$\alpha_G > \alpha^{4n}$	Need enough photons
−35	Inflation	$V'' \ll V$	Need enough inflation

case, relationships like $\alpha_G \sim \alpha_W^4 \sim \alpha^{20}$ could just reflect the relative number of internal and external dimensions [15].

2.10 Cosmology and the Complexity Principle

The crucial role of the various fine-tunings in the evolution of the Universe is summarised in Table 2.3. This indicates the times of various key steps in the history of the Big Bang and indicates the fine-tunings associated with each of them. These might be regarded as a prerequisite for the large variety of structures appearing in Figure 2.2.

It is interesting to put these considerations into a broader historical context. In the nineteenth century, the second law of thermodynamics was taken to imply that the Universe must eventually undergo a 'heat death', with life and all other forms of order inevitably deteriorating. However, developments in cosmology have led to a reversal of this view. According to the Big Bang theory, the history of the Universe reveals an increasing rather than decreasing degree of organisation, and modern physics – without any violation of the second law of thermodynamics – is able to explain this.

Some of the types of organisation which exist in the Universe are summarised in the Pyramid of Complexity, introduced by Reeves [68] and reproduced in Figure 2.11. This shows the different levels of structure as one goes from quarks to nucleons to atoms to simple molecules to biomolecules to cells and finally to living organisms. This hierarchy of structure reflects the existence of the strong force at the lower levels and the electric force at the higher ones. As one ascends the pyramid, the structures become more complex – so that the number of different patterns becomes larger – but they also become more fragile.

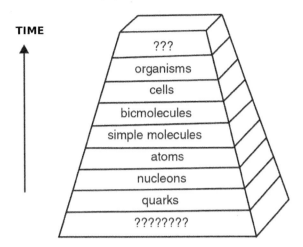

TIME

???

organisms

cells

bicmolecules

simple molecules

atoms

nucleons

quarks

????????

Figure 2.11 This summarises the different levels of structure which exist in the Universe and how this arose during the Big Bang; from Reference [21].

The pyramid becomes narrower as one rises because the fraction of matter incorporated into the objects decreases as the degree of organisation increases. No violation of the second law of thermodynamics is involved because local pockets of order can be purchased at the expense of a global increase in entropy, the fraction of matter going into these structures decreasing as one ascends the pyramid.

The Pyramid of Complexity emerges as the Universe expands and cools, and the Big Bang model explains *when* these structures arise. At early times, the Universe is mainly in the form of quarks. Neutrons and protons appear at a few microseconds, light nuclei at several minutes, atoms at a million years, and – following the formation of stars and planets – molecules and cells at 10 billion years. The Big Bang model also explains *why* the pyramid arises. The key point is that structures arise because processes cannot occur fast enough in an expanding universe to maintain equilibrium. If it had its way, each type of force would always form the objects which are most stable from its own perspective (e.g., the strong force would turn all nuclei into iron, the electric force would turn all atoms into noble gases, and gravity would turn all matter into black holes). However, all variety would be lost if this were the case, and it is only the disequilibrium entailed by the rapid expansion of the Universe which prevents this.

For example, the reason all nucleons do not go into iron as a result of cosmological nucleosynthesis is that the Universe is expanding too fast for most nuclei to interact with each other at this time. The reason gravity does not turn all stars into black holes is because the pressure associated with nuclear energy release and eventually quantum effects support them against gravity. The forces may eventually attain their goal but only after an enormous length of time and, even then, only for a limited period. Thus, even if the Universe does eventually end up in black holes, on a still longer timescale, these black holes will evaporate into radiation. As emphasised in Table 2.3, it is only the anthropic fine-tuning of the coupling constants that allows the ascension of the lower levels of the pyramid.

Figure 2.11 suggests that the anthropic fine-tunings are more related to the emergence of complexity than life; they could equally well be regarded as prerequisites for inanimate objects like motor cars or TV sets. However, here on Earth at least, the development of observers seems to have occurred relatively quickly once the first signs of life arose, so it is conceivable that this applies more generally. Provided there are no extra 'biological' fine-tunings required for the higher levels of the pyramid to arise, the evolution of complexity may inevitably (and fairly rapidly) lead to life. In this case, the distinction between life and complexity is not so clear-cut. The former is just a particular realisation of the latter and may naturally emerge from it, so the question of what constitutes an observer may not be so crucial.

2.11 Interpretations of Anthropic Principle

Anthropic arguments have always been controversial because they seem to exclude the more usual type of physical explanation for the values of the constants. Three very different views of the Anthropic Principle are illustrated by the following quotes. One is from the protagonist Freeman Dyson [34]:

I do not feel like an alien in this Universe. The more I examine the Universe and examine the details of its architecture, the more evidence I find that the Universe in some sense must have known we were coming.

This contrasts with the view of the antagonist Heinz Pagels [60]:

The influence of the anthropic principle on contemporary cosmological models has been sterile. It has explained nothing and it has even had a negative influence. I would opt for rejecting the anthropic principle as needless clutter in the conceptual repertoire of science.

An intermediate stance is taken by Brandon Carter [24]:

The anthropic principle is a middle ground between the primitive anthropocentrism of the pre-Copernican age and the equally unjustifiable antithesis that no place or time in the Universe can be privileged in any way.

The rising popularity of the multiverse picture has encouraged a drift towards Carter's view, but the A word is still taboo in some quarters.

As far as is known, the relationships discussed in this chapter are not predicted by any unified theory, and even if they were, it would be remarkable that the theory should yield exactly the coincidences required for life. One therefore needs *some* form of explanation, even if this veers into the border of science and philosophy. Three interpretations of the anthropic coincidences have been suggested, and these are illustrated in Figure 2.12.

The first possibility is that the coincidences reflect the existence of a 'beneficent being' who tailor-made the Universe for our benefit. One could envisage our Universe as occupying a point in some multidimensional space of coupling constants, with the tailor putting a

 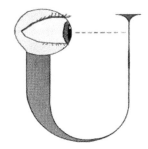 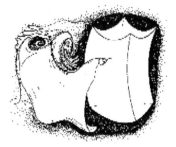

Figure 2.12 Three explanations of fine-tunings (from left): selection effect in a multiverse, consciousness collapsing the wave function of the Universe, fine-tuner choosing the coupling constants.

pin in the optimal spot. Such an interpretation is logically possible and appeals to theologians [47]. Indeed some people now use the term 'Strong Anthropic Principle' to imply that the Universe was created with the *purpose* of creating life. Not surprisingly, most physicists are uncomfortable with this interpretation.

The second possibility, proposed by Wheeler [90], is that the Universe does not properly *exist* until consciousness has arisen. This is based on the notion that the Universe is described by a quantum mechanical wave function and that consciousness is required to collapse this. Once the Universe has evolved consciousness, one might regard it as reflecting back on its Big Bang origin, thereby forming a closed circuit which brings the world into existence. Even if consciousness really does collapse the wave function (which is far from certain), this explanation is also somewhat metaphysical.

The third possibility is that there is not just one universe but a large ensemble of them, all with different (possibly random) coupling constants. As reviewed by Tegmark [83], there are many versions of the multiverse proposal, although not all of them necessarily entail a variation in the physical constants across the ensemble. As stressed by Rees [65], a key issue in assessing the multiverse proposal is whether some of the physical constants are contingent on accidental features of symmetry breaking and the initial conditions of our Universe or whether some future Theory of Everything will determine all of them uniquely. In the first case, there would be room for the Anthropic Principle, but in the second case, any fine-tunings would have to be regarded as coincidental. There might, in principle, be other universes in this case, but they would all have the same values for the constants, so there would be little point in invoking them.

If one grants the existence of a multiverse, the question of whether our Universe is typical or atypical within the ensemble then arises. Anthropic advocates usually assume that life forms similar to our own will be possible in only a tiny subset of universes. More general life forms may be possible in a somewhat larger subset (e.g., if one envisages cold cosmological models of the kind discussed by Aguirre [7]), but life will not be possible everywhere. One may not have the *same* anthropic relation in every universe, but one will have *some* relation.

On the other hand, by invoking a Copernican perspective, Smolin has argued that *most* of the universes should have properties like our own, so that we are typical [72, 73]. His approach invokes a form of cosmological natural selection: the formation of black holes is supposed to generate new baby universes in which the constants are slightly mutated. In this way, after many generations, the parameter distribution will be peaked around those values for which black hole formation is maximised. This proposal involves very speculative physics since we have no understanding of how the baby universes are born, but it has the virtue of being *testable* since one can calculate how many black holes would form if the parameters were different. Note that Smolin's proposal makes no reference to observers, so it would not be regarded as anthropic in the usual sense of the term, but it still invokes a multiverse.

But how legitimate is it to invoke the existence of other universes for which there may never be any direct evidence? Smolin stresses [74] that the multiverse proposal is legitimate only if one has a theory which independently predicts it, and such a theory, to be scientific, must be falsifiable. He argues that the notion of a multiverse is neither falsifiable nor testable. However, not everybody concedes this point. For example, Rees points out [66] that one way of testing the multiverse proposal is to calculate the probability distribution for various parameters across the different universes. One would then be surprised if the value of some constant was on the tail of the distribution. In particular, if Weinberg's explanation for the value of Λ is correct, one would be surprised if its value was *much* less than the anthropic limit.

Even if one accepts that a multiverse exists and gives rise to anthropic selection effects, there is still considerable ambiguity as to how one interprets the selection effect. If the Anthropic Principle can be interpreted as a Complexity Principle, what qualifies as an observer in anthropic considerations? It would be most natural to associate the anthropic constraints with life in general rather than humans in particular. In fact, Davies explicitly associates them with a 'life principle' [28]. But in this case, does the mere existence of consciousness suffice, or is some minimum threshold of intelligence required? We have seen that this may not be so crucial since – whatever threshold one selects – it may be attained relatively quickly once the first signs of life arise [52].

2.12 Concluding Remarks

In concluding, I will make a few philosophical points, most of which relate to the common criticisms of anthropic arguments.

- In so much as the WAP is a selection effect, it is uncontroversial. It is clear that the presence of observers implies a non-random sample of the underlying population, and one needs to allow for this to avoid spurious correlations. However, one should not infer from this that the WAP is trivial, since some of the selection effects are unexpected and depend upon subtle aspects of physics. But what is being selected for? Even if it is life, one must avoid being too anthropocentric since it is clearly not restricted to humans

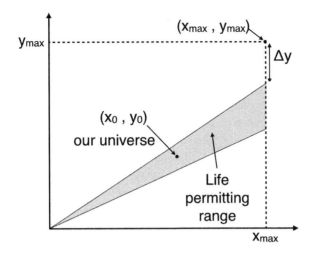

Figure 2.13 Illustrating that anthropic constraints may specify a wedge in parameter space if two constants (x, y) are allowed to vary. From Reference [8].

and may not even be carbon based. According to Smolin [72, 73], it just relates to the abundance of black holes and is irrelevant to life.

- A common objection to the SAP is that the apparent tunings involved could just be coincidental. In this context, one might compare the contrasting views of Victor Stenger [76] and Luke Barnes [8]. Clearly, this depends on the number of tunings and their precision, but it is not straightforward to assess the probabilities involved, especially since anthropic arguments do not explain exact values. In assessing the weight of the evidence, it is also relevant to know whether the tunings are post hoc or predicted. Most are post hoc, but it could be argued that the triple-α and Λ tunings were predictions.

- Many of the strong tunings discussed in this chapter are treated in isolation (i.e., with only one constant being varied at a time). But what happens if one allows some constants to covary? Stenger argues that, for two parameters, the constraints correspond to a 'wedge', as illustrated in Figure 2.13, which means that the probability is not just the product of the one-dimensional probabilities. However, Barnes points out that the wedge description does not always apply, as illustrated by Figure 2.8. In general, when considering multiple constraints, one needs to consider the intersection rather than the union of wedges. Indeed, if there were too many constraints, life would presumably be excluded altogether.

- Anthropic arguments usually only consider small variations of the constants in the 'island' of parameter space around the observed values. However, there could be other islands of life, so this may not suffice. This does not invalidate the Anthropic Principle, providing the total area of the islands is small. As Richard Dawkins remarks [29], 'However many ways there may be of being alive, there are vastly more ways of being dead'. On the other hand, there are no small islands in Smolin's proposal since our Universe should be representative of the most populated region of parameter space.

- One must distinguish between different levels of explanation. The Anthropic Principle – viewed as a selection effect – provides some insight into the Universe, but it does not provide an ultimate explanation. For comparison, consider the question of why distant quasars are so luminous [8]. Although a selection effect is involved, in the sense that it is easier to see very bright objects at large distances, this does not *explain* quasars. The luminosity explains the visibility, but one needs to invoke accreting supermassive black holes to explain the luminosity. In the present context, the analogue of the supermassive black hole is the multiverse or whichever model discussed in Section 2.11 one prefers. Perhaps the ultimate explanation will relate to what happens at the top of the Cosmic Uroborus in Figure 2.1, in which case the Final Theory will need to refer to observers in some way.

- The Anthropic Principle need not imply that the conditions in the Universe are optimal for life or that life is pervasive. Indeed, Carter has argued that it may be very rare [23], using the following argument. The time for life to arise on Earth seems to have been a few billion years and, therefore, comparable to the age of the Earth itself. However, one would not expect that a priori since the time for life to evolve t_L is disconnected from the cosmological timescale (t_o). Were t_L much smaller than t_o, then life would be very abundant in the Universe, but that contradicts the evidence. Were t_L much larger than t_o, life would be very rare, but presumably t_L has some distribution so that – in an infinite universe – there will always be some regions where life arises much sooner than expected. The WAP merely requires that we reside in one of the rare Hubble volumes where life has arisen after only 10^{10} yr. Carter therefore infers that we could be the only life form within the Hubble scale and that the Universe is infinite.

- A new twist arises if the (so-called) constants vary in time even in our Universe. This may be expected in theories of particle physics which invoke compactified extra spatial dimensions, since the constants may be related to the size of these dimensions, and this could change during the Universe's history. Some astronomers claim to have found evidence for a variation in α – of about seven parts in a million – by studying absorption lines in several hundred quasars [87]. Sandvik *et al.* attempt to model this effect and suggest that α should remain constant during both the early radiation-dominated phase of the Universe and the late curvature-dominated or Λ-dominated phases [70]. However, it could still vary over the intermediate matter-dominated phase, and this would make it difficult to satisfy the anthropic constraints on α for an extended period unless the curvature or cosmological constant were very small. Indeed, the anthropic constraints may only be satisfied at a particular epoch.

Finally, one should address the issue of whether anthropically inspired speculations – such as the multiverse – should be regarded as science or philosophy. So long as there is no independent evidence for other universes, the notion certainly fails to meet the usual criteria of science [36]. On the other hand, one might argue that it still qualifies if it is predicted by a legitimate physical theory, such as the string landscape scenario. The problem is that such theories may themelves be untestable and therefore dismissed as mathematics rather than physics. In my dialogue with George Ellis on this issue [16], I conceded that the multiverse does not currently meet the criteria for science but argued that it might eventually do so. For this reason, I suggested that the multiverse proposal might be described as 'metacosmology' rather than 'cosmology' [19], this representing a grey area between physics and philosophy. From a historical perspective, the advent of new data has constantly promoted cosmological speculations from philosophy to physics, so today's metacosmology may well become tomorrow's cosmology.

References

[1] Adams, F. C., Coppess, K. R., and Bloch, A. M. 2015. 'Planets in Other Universes: Habitability Constraints on Density Fluctuations and Galactic Structure'. *Journal of Cosmology and Astroparticle Physics*, **9**(Sept.), 030.

[2] Adams, F. C. 2008. 'Stars in Other Universes: Stellar Structure with Different fundamental constants'. *Journal of Cosmology and Astroparticle Physics*, **0808**, 010.

[3] Adams, F. C. 2016. 'Constraints on Alternate Universes: Stars and Habitable Planets with Different Fundamental Constants'. *Journal of Cosmology and Astroparticle Physics*, **1602**(02), 042.

[4] Adams, F. C. 2019. 'The Degree of Fine-Tuning in Our Universe and Others'. *Physics Reports*, **807**, 1–111.

[5] Adams, F. C., and Grohs, E. 2017a. 'On the Habitability of Universes without Stable Deuterium'. *Astroparticle Physics*, **91**, 90–104.

[6] Adams, F. C., and Grohs, E. 2017b. 'Stellar Helium Burning in Other Universes: A Solution to the Triple Alpha Fine-Tuning Problem'. *Astroparticle Physics*, **87**, 40–54.

[7] Aguirre, A. 2001. 'Cold Big-Bang Cosmology as a Counterexample to Several Anthropic Arguments'. *Physical Review D*, **64**(8), 083508.

[8] Barnes, L. A. 2012. 'The Fine-Tuning of the Universe for Intelligent Life'. *Publications of the Astronomical Society of Australia*, **29**(June), 529–564.

[9] Barnes, L. A. 2015. 'Binding the Diproton in Stars: Anthropic Limits on the Strength of Gravity'. *Journal of Cosmology and Astroparticle Physics*, **12**(Dec.), 050.

[10] Barr, S. M., and Khan, A. 2007. 'Anthropic Tuning of the Weak Scale and of m_u/m_d in Two-Higgs-Doublet Models'. *Physical Review D*, **76**(4), 045002.

[11] Barrow, J. D., and Tipler, F. J. 1988. *The Anthropic Cosmological Principle*. Oxford University Press.

[12] Bousso, R., and Polchinski, J. 2000. 'Quantization of Four-Form Fluxes and Dynamical Neutralization of the Cosmological Constant'. *Journal of High Energy Physics*, **6**(June), 006.

[13] Bousso, R., Hall, L. J., and Nomura, Y. 2009. 'Multiverse Understanding of Cosmological Coincidences'. *Physical Review D*, **80**(6), 063510.

[14] Bradford, R. A. W. 2009. 'The Effect of Hypothetical Diproton Stability on the Universe'. *Journal of Astrophysics and Astronomy*, **30**(June), 119–131.

[15] Carr, B. 1991. 'Anthropic Principles as Constraints on Cosmological Models'. *Journal of the British Interplanetary Society*, **44**(Feb.), 63–70.

[16] Carr, B., and Ellis, G. 2008. 'Universe or Multiverse?'. *Astronomy and Geophysics*, **49**(2), 2.29–2.33.

[17] Carr, B., Kühnel, F., and Sandstad, M. 2016. 'Primordial Black Holes as Dark Matter'. *Physical Review D*, **94**(8), 083504.

[18] Carr, B. J. 1977. 'Black Hole and Galaxy Formation in a Cold Early Universe'. *Monthly Notices of the Royal Astronomical Society*, **181**(Nov.), 293–309.

[19] Carr, B. J. 2014. *Metacosmology and the Limits of Science*. Copernicus Center Press. Pages 407–432.

[20] Carr, B., Clesse, S., and García-Bellido, J. 2019. 'Primordial Black Holes, Dark Matter and Hot-Spot Electroweak Baryogenesis at the Quark-Hadron Epoch'. *arXiv e-prints*, Apr, arXiv:1904.02129.

[21] Carr, B. J., (ed). 2007. *Universe or Multiverse?* Cambridge University Press.

[22] Carr, B. J. and Rees, M. J. 1979. 'The Anthropic Principle and the Structure of the Physical World'. *Nature*, **278**, 605–612.

[23] Carter, B. 1983. 'The Anthropic Principle and Its Implications for Biological Evolution'. *Philosophical Transactions of the Royal Society of London Series A*, **310**(Dec.), 347–363.

[24] Carter, B. 1974. 'Large Number Coincidences and the Anthropic Principle in Cosmology'. *IAU Symposium*, **63**, 291.

[25] Coc, A., *et al.* 2009. 'Constraints on the Variations of Fundamental Couplings by Stellar Models'. *Mem. Societa Astronomica Italiana*, **80**, 809.

[26] Cohen, B. L., 2008. 'Understanding the Fine Tuning in Our Universe'. *The Physics Teacher*, **46**(May), 285–289.

[27] Collins, C. B., and Hawking, S. W. 1973. 'Why Is the Universe Isotropic?'. *The Astrophysical Journal*, **180**(Mar.), 317–334.

[28] Davies, P. C. W. 2007. 'Universes Galore: Where Will It All end?'. in B. J. Carr, ed. *Universe or Multiverse?*. Cambridge University Press. Pages 487–505.

[29] Dawkins, R. 1988. *The Blind Watchmaker*. W. W. Norton and Co.

[30] Degrassi, G. *et al.* 2012. 'Higgs Mass and Vacuum Stability in the Standard Model at NNLO'. *Journal of High Energy Physics*, **8**(Aug.), 98.

[31] Rosen, G. 1971. 'Quantum Theory of Gravitation and the Mass of the Electron'. *Physical Review*, **D4**, 275–277.

[32] Dirac, P. A. M. 1937. 'The Cosmological Constants'. *Nature*, **139**, 323.

[33] Dyson, F. J. 1971. 'Energy in the Universe'. *Scientific American*, **225**(Sept.), 50–59.

[34] Dyson, F. J. 1979. 'Time without End: Physics and Biology in an Open Universe'. *Reviews of Modern Physics*, **51**(July), 447–460.

[35] Efstathiou, G. 1995. 'An Anthropic Argument for a Cosmological Constant'. *Monthly Notices of the Royal Astronomical Society*, **274**, L73.

[36] Ellis, G., and Silk, J. 2014. 'Scientific Method: Defend the Integrity of Physics'. *Nature*, **516**(Dec.), 321–323.

[37] Epelbaum, E., *et al.* 2013. 'Viability of Carbon-Based Life as a Function of the Light Quark Mass'. *Physical Review Letters*, **110**(11), 112502.

[38] Garriga, J., and Vilenkin, A. 2006. 'Anthropic Prediction for Λ and the Q Catastrophe'. *Progress of Theoretical Physics Supplement*, **163**, 245–257.

[39] Guth, A. H. 1981. 'Inflationary Universe: A Possible Solution to the Horizon and Flatness Problems'. *Physical Review D*, **23**(Jan.), 347–356.

[40] Dicke, R. H. 1961. 'Dirac's Cosmology and Mach's Principle'. *Nature*, **192**(11), 440–441.

[41] Hall, L. J., and Nomura, Y. 2008. 'Evidence for the Multiverse in the Standard Model and Beyond'. *Physical Review D*, **78**(3), 035001.

[42] Hartle, J. B., and Hawking, S. W. 1983. 'Wave Function of the Universe'. *Physical Review D*, **28**(Dec.), 2960–2975.

[43] Hawking, S. W. 1974. 'Black Hole Explosions?'. *Nature*, **248**(Mar.), 30–31.

[44] Hawking, S. W., and Hertog, T. 2006. 'Populating the Landscape: A Top-Down Approach'. *Physical Review D*, **73**(12), 123527.

[45] Hogan, C. J. 2000. 'Why the Universe Is Just so'. *Reviews of Modern Physics*, **72**(Oct.), 1149–1161.

[46] Hogan, C. J. 2007. 'Quarks, Electrons, and Atoms in Closely Related Universes'. in B. J. Carr, ed. *Universe or Multiverse?*. Cambridge University Press. Pages 221–230.

[47] Holder, R. 2004. *God, the Multiverse, and Everything: Modern Cosmology and the Argument from Design*. Ashgate.

[48] Hoyle, F. 1953. 'A State in C12 Predicted from Astrophysical Evidence'. *Phys. Rev.*, **92**, 1095.

[49] Kahn, F. D. 1972. *Life in the Universe*. University of Virginia Press. Page 71–89.

[50] Kallosh, R., and Linde, A. 2003. 'M Theory, Cosmological Constant, and Anthropic Principle'. *Physical Review D*, **67**(2), 023510.

[51] Kane, G. L., Perry, M. J., and Zytkow, A. N. 2002. 'The Beginning of the End of the Anthropic Principle'. *New Astronomy*, **7**(Jan.), 45–53.

[52] Kauffman, S. 1995. *At Home in the Universe: The Search for the Laws of Self-Organization and Complexity*. Oxford University Press.

[53] Kragh, H. 2010. 'An Anthropic Myth: Fred Hoyle's Carbon-12 Resonance Level'. *Archive for History of Exact Sciences*, **64(6)**, 721–751.

[54] Lewis, G. F., and Barnes, L. A. 2016. *A Fortunate Universe*. Cambridge University Press.

[55] Linde, A. D. 1986. 'Eternally Existing Self-Reproducing Chaotic Inflanationary Universe'. *Physics Letters B*, **175**(Aug.), 395–400.

[56] Livio, M., *et al.* 1989. 'The Anthropic Significance of the Existence of an Excited State of C-12'. *Nature*, **340**(July), 281–284.

[57] MacDonald, J., and Mullan, D. J. 2009. 'Big Bang Nucleosynthesis: The Strong Nuclear Force Meets the Weak Anthropic Principle'. *Physical Review D*, **80**(4), 043507.

[58] Oberhummer, H., Csoto, A., and Schlattl, H. 2000. 'Stellar Production Rates of Carbon and Its Abundance in the Universe'. *Science*, **289**, 88.

[59] Page, D. N. 2009. 'Anthropic Estimates of the Charge and Mass of the Proton'. *Physics Letters B*, **675**, 398–402.

[60] Pagels, H. R. 1985. *Perfect Symmetry: The Search for the Beginning of Time*.

[61] Peacock, J. A. 2007. 'Testing Anthropic Predictions for Λ and the Cosmic Microwave Background Temperature'. *Monthly Notices of the Royal Astronomical Society*, **379**(Aug.), 1067–1074.

[62] Pochet, T., *et al.* 1991. 'The Binding of Light Nuclei, and the Anthropic Principle'. *Astronomy and Astrophysics*, **243**(Mar.), 1–4.

[63] Randall, L., and Sundrum, R. 1999. 'An Alternative to Compactification'. *Physical Review Letters*, **83**, 4690–4693.

[64] Rees, M. 2000. *Just Six Numbers: The Deep Forces That Shape the Universe*. Weidenfeld & Nicolson.

[65] Rees, M. 2001. *Our Cosmic Habitat*. Princeton University Press.

[66] Rees, M. J. 2007. 'Cosmology and the Multiverse'. in B. J. Carr, ed. *Universe or Multiverse?*. Cambridge University Press. Pages 57–76.

[67] Rees, M. J., and Ostriker, J. P. 1977. 'Cooling, Dynamics and Fragmentation of Massive Gas Clouds: Clues to the Masses and Radii of Galaxies and Clusters'. *Monthly Notices of the Royal Astronomical Society*, **179**(June), 541–559.

[68] Reeves, H. 1991. *The Hour of Our Delight: Cosmic Evolution, Order, and Complexity*. W.H. Freeman.

[69] Salpeter, E. E. 1964. 'Accretion of Interstellar Matter by Massive Objects'. *The Astrophysical Journal*, **140**, 796–800.

[70] Sandvik, B., Barrow, J. D., and Magueijo, J. 2002. 'A Simple Cosmology with a Varying Fine Structure Constant'. *Physical Review Letters*, **88**(3), 031302.

[71] Silk, J. 1977. 'On the Fragmentation of Cosmic Gas Clouds. I. The Formation of Galaxies and the First Generation of Stars'. *The Astrophysical Journal*, **211**(Feb.), 638–648.

[72] Smolin, L. 1992. 'Did the Universe Evolve?'. *Classical and Quantum Gravity*, **9**(Jan.), 173–191.

[73] Smolin, L. 1997. *The Life of the Cosmos*. Oxford University Press.

[74] Smolin, L. 2007. *Scientific Alternatives to the Anthropic Principle*, in B. J. Carr, ed. *Universe or Multiverse?* Cambridge University Press. Pages 323–366.

[75] Steinhardt, P. J., and Turok, N. 2006. 'Why the Cosmological Constant Is Small and Positive'. *Science*, **312**(May), 1180–1183.

[76] Stenger, V. J. 2011. *The Fallacy of Fine-Tuning: Why Universe Is Not Designed for Life*. Prometheus Books.

[77] Susskind, L. 2005. *The Cosmic Landscape: String Theory and the Illusion of Intelligent Design*. Little, Brown and Company.

[78] Susskind, L. 2007. 'The Anthropic Landscape of String Theory', in B. J. Carr, ed. *Universe or Multiverse?*. Cambridge University Press. Pages 247–266.

[79] Tegmark, M. 1997. 'On the Dimensionality of Spacetime'. *Classical and Quantum Gravity*, **14**(Apr.), L69–L75.

[80] Tegmark, M. 2005. 'What Does Inflation Really Predict?' *Journal of Cosmology and Astroparticle Physics*, **4**(Apr.), 001.

[81] Tegmark, M., and Rees, M. J. 1998. 'Why Is the Cosmic Microwave Background Fluctuation Level 10^{-5}?'. *The Astrophysical Journal*, **499**(May), 526–532.

[82] Tegmark, M., *et al.* 2006. 'Dimensionless Constants, Cosmology, and Other Dark Matters'. *Physical Review D*, **73**(2), 023505.

[83] Tegmark, M. 2007. 'Parallel Universes', in B. J. Carr, ed. *Universe or Multiverse?*. Cambridge University Press. Pages 459–491.

[84] Turner, M. S., and Carr, B. J. 1987. 'Why Should Baryons and Exotic Relic Particles Have Comparable Densities?'. *Modern Physics Letters A*, **2**, 1–7.

[85] Vilenkin, A. 1983. 'Birth of Inflationary Universes'. *Physical Review D*, **27**(June), 2848–2855.

[86] Vilenkin, A. 1995. 'Predictions from Quantum Cosmology'. *Physical Review Letters*, **74**(Feb.), 846–849.

[87] Webb, J. K., *et al.* 2001. 'Further Evidence for Cosmological Evolution of the Fine Structure Constant'. *Physical Review Letters*, **87**(9), 091301.

[88] Weinberg, S. 1987. 'Anthropic Bound on the Cosmological Constant'. *Physical Review Letters*, **59**(Nov.), 2607–2610.

[89] Weinberg, S. 1989. 'The Cosmological Constant Problem'. *Reviews of Modern Physics*, **61**(Jan.), 1–23.

[90] Wheeler, J. A. 1977. 'Genesis and Observership'. in R. Butts and J. Hintikka, eds. *Foundational Problems in the Special Sciences*. Reidel. Pages 3–33.

[91] Wilczek, F. 2007a. 'A Model of Anthropic Reasoning: The Dark to Ordinary Matter Ratio'. in B. J. Carr, ed. *Universe or Multiverse?*, Cambridge University Press. Pages 151–162.

[92] Wilczek, F. 2007b. 'Enlightenment, Knowledge, Ignorance, Temptation'. in B. J. Carr, ed. *Universe or Multiverse?*, Cambridge University Press. Pages 43–54.

[93] Witten, E. 1995. 'String Theory Dynamics in Various Dimensions'. *Nuclear Physics B*, **443**, 85–126.

Part II
Cosmological Fine-Tunings

3

Naturalness, Fine-tuning, and Observer Selection in Cosmology

JOHN A. PEACOCK

Abstract

The intention of this article is to review those areas of cosmology where we may need to consider explicitly our role as observers in conditioning or biasing the properties of the Universe that we observe. What we will do is begin by attempting to define naturalness and fine-tuning, initially using the viewpoint of particle physics before setting up a more general Bayesian framework within which such issues can be discussed. With this background, we will then give a list of unexplained cosmological problems, focusing on strange parameter coincidences and, in particular, the challenge of explaining the observed level of the vacuum density. This will lead on to a survey of approaches to the observed properties of 'dark energy', especially the modelling of dark energy by considering dynamical signatures that go beyond a simple cosmological constant or by modifying the theory of gravity. The conclusion will be that there are problems with most of these approaches, leading to a focus on observer selection as a possible solution. This leads to consideration of the possible existence of a physical ensemble of causally distinct universes, as can arise in some models of inflation. Within such an ensemble, many elements of physics can, in practice, be different in different members of the ensemble. We begin with the possible physics of a variable cosmological constant, moving on to evidence that other pieces of 'fundamental' physics may also not be immutable. In all these cases, there are two practical difficulties with converting these general ideas into quantitative testable science. One is the 'measure problem' – the appropriate Bayesian prior for a given parameter – but the other is the observer weighting, where astronomers risk being dragged into biological issues far beyond their sphere of competence. One way to avoid this peril is to consider the impact on the history and efficiency of star formation as pieces of fundamental physics are altered.

3.1 Overview

The intention of this chapter is to review those areas of cosmology where we may need to consider explicitly our role as observers in conditioning or biasing the properties of the Universe that we observe. A shorthand term for such considerations is 'anthropic', and it is well known that many physicists (and not a few astronomers) show a strong allergy to 'the A word'. To some extent, one can sympathise with this attitude, since the tone of some older discussions of this topic was not always helpful. Therefore, we will proceed at first with some broader terminology, attempting to define naturalness and fine-tuning initially using the viewpoint of particle physics, before setting up a more general Bayesian framework within which such issues can be discussed. With this background, we will then give a list of unexplained cosmological problems, focusing on strange parameter coincidences and, in particular, the challenge of explaining the observed level of the vacuum density.

This will lead to a survey of approaches to the observed properties of 'dark energy', especially the modelling of dark energy by considering dynamical signatures that go beyond a simple cosmological constant, or by modifying the theory of gravity. The conclusion will be that there are problems with most of these approaches, leading to a focus on observer selection as a possible solution. This leads to consideration of the possible existence of a physical ensemble of causally distinct universes, as can arise in some models of inflation (see Chapter 4 for a detailed discussion). Within such an ensemble, many elements of physics can, in practice, be different in different members of the ensemble. We begin with the possible physics of a variable cosmological constant, moving on to evidence that other pieces of 'fundamental' physics may also not be immutable. In all these cases, there are two practical difficulties with converting these general ideas into quantitative testable science. One is the 'measure problem' – the appropriate Bayesian prior for a given parameter – but the other is the observer weighting, where astronomers risk being dragged into biological issues far beyond their sphere of competence. One way to avoid this peril is to consider the impact on the history and efficiency of star formation as pieces of fundamental physics are altered.

3.2 Defining Fine-Tuning

3.2.1 Naturalness in Particle Physics

When considering the credibility of a given cosmological theory, the field borrows from particle physics. One of the most powerful principles in that subject is naturalness, which amounts to a statement that all dimensionless parameters in a theory should be of order unity. Indeed, it is common to use dimensional analysis in all areas of physics to guess the form of physical laws up to some dimensionless factor, which is then assumed to be of order unity – provided this term is interpreted generously enough to cover factors such as $(2\pi)^3$. An exception to this is that a parameter that might be expected to be of order unity can be forced to zero by means of some symmetry and a small non-zero value possibly restored by means of breaking that symmetry. For example, the fact that the electron mass is tiny compared to the > 100-GeV scale of the weak interaction can be understood via an approximate chiral symmetry (see, e.g., [59]). In non-dimensionless terms, this hypothesis implies that all particles should have roughly the same mass (or zero, like the photon) – whatever fundamental value sets the scale. It is common to assert that this should be the

Planck scale, and so the relative lightness of all known particles makes particle physics uniformly unnatural. This is known as the hierarchy problem.

A related but deeper definition of naturalness concerns quantum corrections: particle masses are not fixed at some bare value but will run according to the energy scale of observation. The interesting question is whether the resulting value is dominated by the bare value or by the correction. Consider the self-energy of the electron, and contrast the corrections to the mass up to some cut-off Λ as calculated using non-relativistic or relativistic quantum theory:

$$\frac{\delta m_e}{m_e} = \frac{4\alpha}{3\pi} \left(\frac{\Lambda}{m_e} \right) \quad (\text{non} - \text{relativistic}) \tag{3.1}$$

$$= \frac{3\alpha}{2\pi} \ln \left(\frac{\Lambda}{m_e} \right) \quad (\text{relativistic}), \tag{3.2}$$

where α is the fine-structure constant. The latter correction is natural, so the measured electron mass is never very different from its bare value. But this is not always true: for example, the Higgs mass has a correction that is quadratic in Λ, meaning that the hierarchy problem is particularly severe in this case. The need to keep the Higgs at its observed light value is one of the arguments often advanced for new physics such as supersymmetry.

Barring new physics, unnaturally small observed masses can only be accounted for by a distasteful fine-tuning, in which a hypothetical bare mass is adjusted so that it almost cancels the correction, leaving a residual small mass that is much less than either term. Cosmology has a similar problem with the value of the cosmological constant. In both subjects, fine-tuning formally deals with the observed data – but in a way that convinces no one, and which cries out for an explanation.

3.2.2 Unnatural Aspects of Cosmology

Cosmology contains a number of puzzles that disturb us at the same level as the issues in particle physics, but sometimes phrased rather differently. As discussed more fully in this section, there is undoubtedly a huge fine-tuning issue with the cosmological constant. But there are also issues that are best phrased as coincidences, and these come in two kinds:

Coincidences of Value

Here, certain parameters of the Universe take very similar values, even though they are apparently physically unrelated. The most striking puzzles here are the rough equality in density of baryons and dark matter, and the fact that recombination occurs roughly when the Universe has just become matter dominated.

Regarding the matter budget, it is normally assumed that the baryon density arises via some CP-violating process so that the numbers of baryons and photons reflect an asymmetry factor: $n_B/n_\gamma = \epsilon \sim 10^{-9}$. If the dark matter consists of incompletely annihilated particles and antiparticles, then the relic density can be written as

$$\frac{n_{DM}}{n_\gamma} \simeq \left(\sigma \frac{mc}{\hbar} \frac{m_P c}{\hbar} \right)^{-1}, \tag{3.3}$$

where σ is the annihilation cross section, m is the particle mass, and m_P is the Planck mass. Thus, the ratio of baryon and dark matter densities is

$$\frac{\rho_B}{\rho_{DM}} \simeq \epsilon \left(\sigma \frac{m_p c}{\hbar} \frac{m_P c}{\hbar} \right)^{-1}, \tag{3.4}$$

where m_p is the proton mass. This ratio is known to be close to 0.2 but is composed of physical quantities that have no known relation to each other.

As for recombination, this occurs roughly when thermal photons can ionise hydrogen: $kT \sim \alpha^2 m_e c^2$, where α is the fine-structure constant. Matter-radiation equality requires $n_\gamma kT \sim \rho_B c^2 = n_B m_p c^2$. These two temperatures will coincide if

$$\frac{n_B}{n_\gamma} \sim \alpha^2 (m_e/m_p) = 3 \times 10^{-8}. \tag{3.5}$$

So the observed near equality of the two eras arises because of the numerological coincidence between this critical value and the actual baryon-to-photon ratio. But no known physics explains why the matter-antimatter asymmetry should take this particular value (see Chapter 5).

Coincidences of Time

There are also certain puzzles that are more closely tied to our existence as observers. Although the ratio between the matter density and vacuum density varies hugely with time,

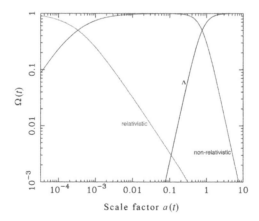

Figure 3.1 The density fraction in various components of the Universe as a function of time. The Universe is only strongly matter dominated for a little more than a single decade of expansion.

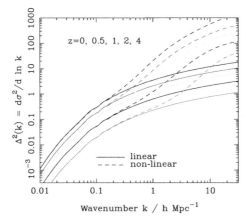

Figure 3.2 The evolution of the dimensionless mass power spectrum in the standard ΛCDM model. As the Universe expands, the high-k plateau value predicted in linear perturbation theory becomes close to unity, so non-linear effects largely erase the initial conditions up to the break in the spectrum. But as the Universe becomes Λ dominated, this evolution switches off. A small reduction in the normalisation of the spectrum would have preserved the linear information forever, but this did not happen. The curves correspond to the indicated redshifts, from left to right, respectively.

it is presently observed to be about unity – even though the history of the Universe is partitioned into many decades of expansion during which it was mostly dominated by either radiation (at early times) or vacuum energy (at late times); see Figure 3.1. A similar puzzle exists in large-scale structure, where the linear power spectrum has a break on 100-Mpc scales, to a nearly flat plateau in the dimensionless power $d\sigma^2/d\ln k \sim k^3 P(k)$ at large wave numbers. On very small scales, non-linear evolution erases the signature of the linear initial conditions, but in practice, this erasure has proceeded close to the point where all cosmological information below the break scale has been removed (see Figure 3.2). Neither of these coincidences applied in the distant past, so we have a *'why now'* problem.

3.3 Probabilistic Framework

3.3.1 Bayesian Approach

In grappling with the aforementioned puzzles, we are almost certain to use probabilistic language: 'it is very unlikely that the bare Higgs mass and its quantum corrections could cancel so precisely'; 'the vacuum density has a very large natural level, so the small observed value seems highly improbable'. Is this legitimate? The Higgs mass is what it is, so how can we talk about there being any probability that it could ever take any different value? The answer comes in the Bayesian approach to statistics. Rather than the simple view of probabilities as representing relative frequencies in an ensemble of trials, Bayesians define probability simply as a degree of belief in a proposition. Consider a theory that contains some parameter θ: if this theory is a correct description of reality, then θ has some definite value and is not a stochastic quantity. But the value of θ is unknown, and we can use the probability of different possible values as a way of quantifying our ignorance. This is done through Bayes's theorem (almost universally written as Bayes' theorem, although this is not correct English):

$$P(\theta|D) = \frac{P(\theta)\,P(D|\theta)}{P(D)}, \tag{3.6}$$

i.e., our belief about θ given data D (the posterior probability) is proportional to the prior on θ times the likelihood, $L = P(D|\theta)$. If $P(\theta|D)$ is to be normalised, then integrating over θ should give unity, requiring $P(D) = \int P(\theta)\,P(D|\theta)\,d\theta$. All of this generalises in an obvious way when θ becomes a vector of multiple parameters.

This approach has many appealing features. It focuses directly on what we care about, which is the probability of various hypotheses given the data to hand. This is in contrast to the frequentist approach, which calculates the probability of various possible outcomes on a given hypothesis, but clearly it is better to focus on what we learn from events that actually happened rather than having to think about all the other possible outcomes to the experiment that might have occurred. In particular, there is no need to have an ensemble of repeated experiments, and the Bayesian approach can happily handle unique events. The need to have a prior is a difficult and controversial part of the apparatus, however, and many are uneasy at the idea of admitting an individual's subjective degree of belief into scientific discussions.

One answer to this is that one of the main applications of the Bayesian formula is to perform inference; i.e., to estimate parameter(s) θ, together with a measure of uncertainty on them. As the quantity of data increases, the posterior probability for θ becomes concentrated around the true value, and the form of the prior becomes unimportant, provided it can be treated as constant over some small range of θ. More deeply, the prior can be considered to result from previous experimental knowledge. Consider what happens when we take two sets of data, D_1 and D_2. The posterior distribution in the face of the totality of data is

$$p(\theta|D_1, D_2) \propto p(D_1, D_2|\theta)p(\theta). \tag{3.7}$$

But the likelihood will factor into the likelihoods for the two data sets separately: $p(D_1, D_2|\theta) = p(D_1|\theta)p(D_2|\theta)$, in which case the posterior can be written as

$$p(\theta|D_1, D_2) \propto p(D_2|\theta)\left[p(D_1|\theta)p(\theta)\right]. \tag{3.8}$$

Now we see that the posterior for experiment 1 functions as the prior for experiment 2, so that Bayes's theorem neatly expresses the process of updating our belief about θ in the light of new information – although it does not remove the basic worry of how we should set a prior before we have any data at all. For further food for Bayesian thought, see, e.g., [36].

For inference – i.e., evaluating the relative probabilities of various different values of θ – the normalisation constant $P(D)$ is irrelevant, but sometimes it can be useful. This can be seen by realising that we might want to consider different models, M_i, with different parameter sets θ_i. In that case, Bayes's theorem would be written in a fuller form:

$$P(M_i, \theta_i | D) \propto P(M_i, \theta_i) \; P(D | M_i, \theta_i) \propto P(M_i) P(\theta_i | M_i) \; P(D | M_i, \theta_i), \qquad (3.9)$$

where the latter case expands the prior into a prior probability for the model itself, times a prior for the parameters in the case where that model applies. So if we want to know the probability that the *class* of model M_i applies without caring about the exact value of its parameters, then we would integrate to obtain $P(M_i | D) \propto P(M_i) \int P(\theta_i | M_i) P(D | M_i, \theta_i) \, d\theta_i$. Apart from the prior probability of the model, $P(M_i)$, this is just proportional to the earlier normalisation constant, $P(D)$, known as the Bayesian evidence. Thus, we can assess the plausibility of various models by calculating the evidence ratio between them, which allows the data to update whatever previous ratio of $P(M_i)$ values we had assigned. This is a powerful methodology, which has received considerable application in cosmology (e.g., [54]). But it is not without its difficulties, stemming from the choice of prior. This is always an issue at some level in Bayesian analyses but is often not so important in inference applications. Suppose we have a single parameter, where the prior is uniform over some range Δ: $P(\theta) = 1/\Delta$. When we compute the ratio of probabilities of two values of θ, this constant divides out as long as both values are in the allowed range, and we simply have the likelihood ratio. But in evidence ratios, the prior is more prominent. Suppose we have two models, each with a single parameter, with allowed ranges Δ_1 and Δ_2. If the likelihood has a Gaussian form, then the integral over the Gaussian is $L^{\max} \sqrt{2\pi} \sigma$, where σ is the effective precision with which θ is measured. In this case, the evidence ratio is

$$\frac{E_2}{E_1} = \left(\frac{L_2^{\max}}{L_1^{\max}} \right) \left(\frac{\sqrt{2\pi} \sigma_2}{\Delta_2} \right) \left(\frac{\sqrt{2\pi} \sigma_1}{\Delta_1} \right)^{-1}. \qquad (3.10)$$

Here, we can have the strange situation that the likelihood ratio might strongly favour model 2, but overall model 2 could be disfavoured if its prior is too wide so that the small value of σ_2/Δ_2 dominates. Things can become worse as the data improve, since σ will shrink. For example, imagine an experiment where we try to test if the vacuum energy is a cosmological constant; as discussed later, this is equivalent to introducing a parameter, w, and asking if it is consistent with -1. Suppose we measure -1.003 ± 0.001 – a 'three-sigma' detection. This corresponds to a likelihood ratio of $\exp(9/2) \simeq 90$ in favour of the more complex hypothesis. But if we had been willing to contemplate values of w between -1.5 and -0.5 ($\Delta = 1$), then the overall evidence ratio would be $1000 : 90\sqrt{2\pi} \simeq 4.4 : 1$ in favour of $w = -1$, despite the 'detection' of a deviation (the last $\sqrt{2\pi} \sigma_1 / \Delta_1$ factor is absent as the cosmological constant case has no free parameter).

This is in contrast to inference, where the prior becomes progressively less important as the data improve and the likelihood becomes narrower. Indeed, studies in parameter inference often deliberately choose uninformative priors that are much broader than any reasonable value so that the conclusions are driven by the data. This is fine so far as it goes, but such priors would be completely inappropriate for evidence calculations. Here, we really do have to believe our prior: the failure to reject $w = -1$ in our example came because the prior tells us that, e.g., $w = -1.49$ is genuinely just as likely a model as $w = -1.003$, so our model had a large list of parameter choices that we claimed to find plausible a priori, most of which failed to match the data. This is why that model framework is disfavoured. Inference requires less commitment: we can choose priors that extend to ridiculous values ($H_0 = 10$ or $200 \ \mathrm{km\,s^{-1}Mpc^{-1}}$, say), and the answers hardly alter. But this means that the term 'prior' is being used to mean two rather different things in inference, and in model testing, and this is unfortunate.

3.3.2 Observer Selection

The Bayesian framework can be used to address some of the preceding cosmological puzzles. Suppose we have some prior belief in the probabilities of different cosmological parameters, $p(\theta)$. This is not the probability that a given set of parameter values will actually be observed, since it is inevitable that the cosmological parameters will influence the number of observers that are produced. When we consider the likelihood part of Bayes's theorem, the 'data' D amount to the statement 'I am an observer, and I experience the Universe to have parameters θ'. Thus, we need to consider the probability of producing observers as a function of cosmological parameters. Furthermore, we need some assumptions about the typicality of observers since the cosmological puzzles refer to the parameter values experienced by a single observer rather than a democratic average over a whole population. To take a simple example, consider the temperature of the cosmic microwave background, currently about 2.725 K. Observers in the past would have seen a larger value, and observers in the far future will see a colder CMB, so to assess the value 2.725 K, we need to have the relative probabilities for the times when observers live. This sounds difficult at the outset: what is the prior probability of a given time interval δt? One might assert that all interval of time should be equally probable; but t is a quantity that is purely positive in a standard Big Bang universe, so it would be common to adopt a $1/t$ Jeffreys prior so that all intervals of $\ln t$ are equally probable. Fortunately, in this case, it does not matter which we pick since the likelihood requires us to calculate the number of observers living in each time interval. With the additional assumption that all observers are equally probable (i.e., that I am a randomly selected typical observer from the sequence of all observers who will ever exist), the posterior probability for the CMB temperature can be written down immediately without explicit reference to a prior on time: the probability that I measure T in some interval δT around $T = 2.725$ K is just the number of observers born in the time interval corresponding to δT divided by the total number that will ever exist.

   This appeal to the 'typical' observer sounds reasonable enough but generates problems when examined in detail – see, e.g., Bostrom [6] and Carr [9]. What counts as an observer? People, yes. Cats? Hypothetical non-carbon life forms? How should we weight for lifetime? The preceding argument assumes a time interval that is still long compared to an observer's lifetime, but what if medical science should enable immortality – is one observer who survives for 100,000 years worth 1,000+ present-day humans? Cosmologists should prefer to stay clear of such biological questions, but they cannot be evaded entirely and will inevitably leave some imprecision in our discussion. Nevertheless, the range of possible cosmological conditions is so vast that we can argue that these non-biological factors will dominate. So, for example, we may feel unsurprised that we live in an era later than recombination: it is hard to envisage that complex structures supporting intelligence could exist while the Universe is in a plasma state. But even this assertion can be challenged by the concept of the Boltzmann brain: in an infinite universe, random fluctuations in homogeneous matter of any temperature could assemble a self-aware state (e.g., [33, 40, 55]). The probability of this happening is exponentially small, but non-zero. Such considerations should be kept in mind, although they can be evaded by a little more conditional information: I am a carbon-based non-Boltzmann brain, and it is legitimate to ask what constraints that piece of data sets on the cosmological parameters that I will measure. Although the precise calculation of those limits is difficult, there is a growing acceptance in modern cosmology that such observer selection has to be taken into account when discussing issues of cosmological naturalness.

3.3.3 Ensemble Reasoning and the Multiverse

Having said that probabilities can be discussed perfectly happily without the need for an ensemble of experiments, it must be admitted that there are some advantages when such an ensemble exists. We have just seen an example of this when considering the sequence of observers in the Universe. The Bayesian approach applied to a single observer can happily reach the conclusion that the temperature of the Universe must be $< 1,000\,\mathrm{K}$, given that this observer exists, but this says nothing about whether it was inevitable that such an observer exists at all. Or consider the Earth: if it lay at a markedly different distance from the Sun (outside the 'habitable zone'), then life as we know it would not be possible. So the existence of life allows us to infer the approximate Earth-Sun distance, but why is the Earth just at this convenient distance? A universe containing just one planet far outside the habitable zone would be devoid of observers, and we would not be there to witness this unhappy situation. But although this is logically fine, it somehow fails to satisfy as an explanation for the actual location of the Earth. This is what we might term the 'something rather than nothing' problem. An alternative is to guess that, in fact, there exist innumerable Earth-like planets at all sorts of distances from their primary stars – in which case it is not only inevitable that those hosting life will find themselves in the habitable zone, but it is guaranteed that there *will* be some in the habitable zone. With recent astronomical data on the population of exoplanets, we can be confident that this is the true situation.

Such a situation may exist in cosmology, where there is interest in a multiverse solution, in which different causally disconnected domains may be able to possess different effective cosmological constants. The most natural form of such a multiverse arises in inflationary cosmology: inflationary models that display stochastic behaviour driven by quantum fluctuations of the inflation field can seed 'bubble universes' (e.g., [27]), which potentially provide the seeds for a concrete ensemble of universes. If we can further arrange for cosmological parameters or even laws of physics to be different in different members of the ensemble, then we have the raw material for potentially both solving the 'something rather than nothing' problem and explaining the cosmological coincidences. In principle, this is not hard to arrange: many physical phenomena are controlled by scalar fields, which can set particle masses in the case of the Higgs field or which can generally contribute to an effective vacuum density as discussed later. If these fields have a potential with multiple minima, then inflationary fluctuations can give the fields different values in different members of the ensemble, leading to trapping in different minima and different low-energy physics. This makes qualitative sense, but it is not easy to turn this framework into a precise prior for the low-energy physics. Not only does this depend on the scalar-field potential; there is also the complication that the seeded bubble universes are formally of infinite volume – making it unclear how different members of the ensemble should be weighted (the measure problem – see, e.g., [26]). And if this problem can be solved, we are still not finished. Given two members of the ensemble with equal prior probability, the different physical parameters within the bubbles will almost inevitably alter the efficiency with which sentient observers arise. This leads to the fundamental principle of observer selection in a multiverse: at least if the number of observers in any member of the ensemble is finite, the probability of a given set of conditions being observed is proportional to the number of observers that are generated. This criterion is an essential element of correct reasoning in such situations, and its neglect will generate paradoxes such as the infamous doomsday argument – see, e.g., [39]. Even given these serious procedural issues, the ensemble approach is a rich and stimulating vision: perhaps the greatest open question in cosmology is whether we are really part of a multiverse and how such a theory might be tested.

The situation in cosmology is perhaps analogous to that facing Darwin, as he groped for an explanation of the match between creatures and their environment. There was a choice between the argument from design (that the world was just as God happened to make it) or that it had arisen via selection from an ensemble of varied creatures. The critical element needed for the second explanation to work was, of course, the idea of a diversity of transmittable characteristics, which provides the raw material for repeated natural selection. Nevertheless, Darwin was unaware of Mendel's contemporary work, which identifies genes as these discrete carriers of heredity. Indeed, the lack of any mechanism of this sort was seen as a major defect by reviewers of *The Origin of Species*. Although he never made a definitive statement in print, Darwin was nevertheless effectively forced by the logic of his position to invent the key idea of genetics: that characteristics do not blend over generations but are carried in an atomic way: 'I have lately been inclined to speculate ... that propagation by true fertilisation will turn out to be a sort of mixture, and not true fusion of two distinct individuals' (letter to Huxley, quoted by Dawkins [16]). In this way, the evolutionary hypothesis forced Darwin towards a dim vision of the underlying genetic mechanism: the exact opposite of any abandonment of physical explanation. In the same way, the invocation of a cosmological multiverse can be a possible route to important new physics. As with evolution, this idea only works if there is *variation*, and, thus, we conclude that there must be some mechanism that allows physical constants to freeze out at different values in different members of the ensemble. If this involves scalar fields, then those fields exist here and now in our Universe, and one could imagine doing an experiment to verify that they exist. This would be like a winner of the National Lottery inferring the existence of something like a drum holding numbered balls – finding such a machine would be convincing evidence that it was indeed once used to pick a winner at random. However, this analogy between cosmology and Darwinism should not be pushed too far. In both cases, one ends up with a peculiar universe via selection from an ensemble and infers a microscopic mechanism that permits selection to operate. But the evolutionary ensemble is a continuing time series of random trials, whereas natural selection in cosmology only happens once, and the concept of inheritance does not apply.

3.4 Anthropic Principles

So far, we have deliberately refrained from using the term 'anthropic', which is how these issues have traditionally been labelled. The debate under this heading has become quite polarised, with some authorities unable to utter the 'A word', which can be seen as an excuse to avoid the hard work of doing proper physics calculations. We have, therefore, first tried to go back to basics and emphasise that observer selection must be considered, but probably also requires *more* physics, not less. And, indeed, the tone of earlier discussions was probably not helpful to this debate, raising quasi-religious associations by enunciating 'anthropic principles' of varying degrees of strength (originally due to Carter [10]; see [4] for an early and thorough review of the subject).

The Trivial Anthropic Principle

This is not strictly part of the Anthropic Principle proper but consists of anthropic ideas on which almost everyone can agree – i.e., that we can use observations of humanity as cosmological information in the same way as we use data from telescopes. For example, we can deduce that the Universe is $\gtrsim 1$ Gyr old merely by noting that carbon-based life has formed and that there needed to be time for typical stars to go through their life cycle and distribute heavy elements in order for this to have happened. The existence of humanity thus gives us a bound on H_0. That is about as far as such trivial anthropic arguments go in cosmology; they are, in a sense, unnecessary, as we have direct dating of the Earth to set a much more precise limit on H_0, a constraint that was astronomically important in the early days of the distance scale, when values of $H_0 \simeq 500 \, \mathrm{km \, s^{-1} Mpc^{-1}}$ were suggested.

However, anthropic arguments of this type have an honourable history from the 19th century, when the Earth could not be dated directly. At that time, Lord Kelvin was advocating an age for the Earth of only $\sim 10^7$ years, based on its cooling time. Evolutionary biologists were able to argue that this was an inadequate time to allow the development of species, a conclusion that was vindicated by the discovery of radioactivity (which allowed both the dating of the Earth and showed the flaw in Kelvin's argument). Here was an excellent example of important astronomical conclusions being drawn from observations of life on Earth.

The Weak Anthropic Principle

But it is not possible to go very far in discussing the astronomical consequences of our local planetary observations before we run into the question of typicality: whether the conditions we observe can be extrapolated into properties characteristic of the Universe as whole. The danger in such reasoning was pointed out by Carter [10] in what he termed the Weak Anthropic Principle: we 'must be prepared to take account of the fact that our location in the Universe is necessarily privileged to the extent of being compatible with our existence as observers'. In other words, there may be times and locations in the Universe where life is impossible, or at least highly improbable, and the allowance for such censoring may condition cosmological observables or even allow us to predict them. The outstanding success of this reasoning concerns Dirac's large-number hypothesis. Dirac noted that very large dimensionless numbers often arise in both particle physics and cosmology. The ratio of the electrostatic and gravitational forces experienced by an electron in a hydrogen atom is

$$\frac{e^2}{4\pi\epsilon_0 G m_e m_p} \simeq 10^{39.4};$$
(3.11)

one of the problems of unifying gravity and other forces is understanding how such a vast dimensionless number can be generated naturally. In a cosmological context, the weakness of gravity manifests itself in the fact that the Hubble radius is enormously greater than the Planck length:

$$\frac{c/H_0}{\sqrt{\hbar G/c^3}} \simeq 10^{61}.$$
(3.12)

This number is very nearly the 1.5 powers of the previous large number: it is as if these large numbers were quantised in steps of 10^{20}. Dirac proposed that the coincidence must indicate a causal relation; requiring the proportionality to hold at all times then yields the radical consequence

$$G \propto t^{-1}$$
(3.13)

(because H_0 declines as $\sim t^{-1}$ as the Universe ages). See also Chapter 2.

Geological evidence shows that this prediction is not upheld in practice since the Sun would have been hotter in the past. A less radical explanation of the large-number coincidence uses the anthropic idea that life presumably requires the existence of elements other than hydrogen and helium. The Universe must therefore be old enough to allow typical stars to go through their life cycle and produce 'metals'. This condition can be expressed in terms of fundamental constants as follows. Stars have masses of the order of the Chandrasekhar mass

$$M_{\text{Chandra}} \equiv \left(\frac{\hbar c}{G m_p^2}\right)^{3/2} m_p,$$
(3.14)

where m_p is the proton mass. The luminosity of a star dominated by electron-scattering opacity is

$$L \sim \frac{G^4 m_p^5 M^3}{\hbar^3 c^2 \sigma_T}. \tag{3.15}$$

The characteristic lifetime of a star, $M_{\text{Chandra}} c^2 / L$, can thus also be expressed in fundamental constants and is

$$t_* \sim \frac{c \, \sigma_T}{G \, m_p} = 5.7 \times 10^9 \text{ years.} \tag{3.16}$$

Comparing with the preceding, we see that the large-number coincidence is just $t_* \sim H_0^{-1}$; i.e., the Universe must be old enough for the stars to age. The fact that the Universe is not very much older than this may tell us that we are privileged to be in the first generation of intelligence to arise after the Big Bang: civilisations arising in $\gg 10^{10}$ years time will probably not spend their time in cosmological enquiry, as they will be in contact with experienced older races who know the answers already. Lastly, the coincidence can be used to argue for the fundamental correctness of the Big Bang as against competitors such as the steady-state theory; if the Universe is, in reality, very much older than t_*, there is no explanation for the coincidence between t_* and H_0^{-1}.

In short, the Weak Anthropic Principle states that, because intelligent life is necessary for cosmological enquiry to take place, this already imposes strong selection effects on cosmological observations. Note that, despite the name, there is no requirement for the life to be human, or even carbon based. All we say is that certain conditions can be ruled out if they do not lead to observers, and this is one of the weaknesses of the principle: are we really sure that life based on elements less massive than carbon is impossible? The whole point of these arguments is that it only has to happen once. It is nevertheless at least plausible that the 'anthropic' term may not be a complete misnomer. It has been argued (see [11]) that intelligent life may be intrinsically an extremely unlikely phenomenon, where the mean time for development could be very long:

$$\bar{t}_{\text{Intelligence}} \gg t_*. \tag{3.17}$$

If the inequality was sufficiently great, it would be surprising to find even one intelligent system within the current horizon. The Anthropic Principle provides a means for understanding why the number is non-zero, even when the expectation is small, but there would be no reason to expect a second system; humanity would then probably be alone in the Universe. On the other hand, $\bar{t}_{\text{Intelligence}}$ may be shorter, and then other intelligences would be common. Either possibility is equally consistent with the selection effect imposed by our existence, although the fact that life on Earth took billions of years to develop is consistent with the former view; if life is a rare event, it would not appear early in the allowed span of time.

Other Weak Anthropic Deductions

What other features of the Universe might be amenable to anthropic arguments? The striking aspects to explain are that we live in a universe with roughly critical density in matter, dominating a radiation background of 2.725 K by roughly four powers of 10. Our starting point is the age argument given earlier, plus the assumption that life will only arise (1) after recombination (so that we are not cooked); (2) after matter domination (so that matter can self-gravitate into stars):

$$t_0 \sim t_* > t_r, t_{eq}. \tag{3.18}$$

A first coincidence to consider is that recombination and matter-radiation equality occur at relatively similar times; why is this? Recombination requires a temperature of roughly the ionisation potential of hydrogen:

$$kT_r \sim \frac{m_e c^2}{137^2}. \tag{3.19}$$

The mass ratio of baryons and photons at recombination is therefore $137^2 m_p/m_e \sim 10^{7.5}$, and the Universe will be matter dominated at recombination unless the ratio of photon and proton number densities exceeds this value. At high temperatures, $kT > m_p c^2$, baryon, anti-baryon, and photon numbers will be comparable; it is thought that the present situation arises from a particle-physics asymmetry between matter and antimatter, so that baryons and anti-baryons do not annihilate perfectly (see Chapter 5). Matter–radiation equality thus arises anywhere after a redshift $10^{7.5}$ times larger than that of last scattering and could be infinitely delayed if the matter/antimatter asymmetry were small enough. The approximate coincidence in epochs says that the size of the particle/antiparticle asymmetry is indeed roughly $137^2 m_p/m_e$, and it is not implausible that such a relation might arise from a complete particle physics model. At any rate, it seems to have no bearing on anthropic issues.

The anthropic argument for the age of the Universe lets us work out the time of recombination if we accept for the moment that the matter density is $\Omega_m \sim 1$ (see Section 3.5). The age for an Einstein-de Sitter model would tell us both H_0 & ρ_0 and, hence, the current number density of photons if we could obtain the photon-to-baryon ratio from fundamental arguments. Since $n_\gamma \propto T_\gamma^3$, that would give the present photon temperature and, hence, the redshift of recombination. The fact that this was relatively recent (only at $z \sim 10^3$) reflects the fact that the photon-to-baryon ratio is $\sim 10^7$ rather than a much larger number.

Finally, what do anthropic arguments have to say about the matter density parameter, Ω_m? There is an instability in the evolution of Ω_m in open matter-dominated models:

$$\left| \Omega_m^{-1}(z) - 1 \right| = (1+z)^{-1} \left| \Omega_m^{-1} - 1 \right| \tag{3.20}$$

so that $\Omega_m(z)$ has to be fine-tuned at early epochs to achieve $\Omega_m \simeq 1$ now. We can go some way towards explaining this, as follows. The formation of galaxies can only occur at a redshift z_f such that $\Omega_m(z_f) \sim 1$; otherwise, growth of density perturbations switches off when the Universe becomes dominated either by curvature or by Λ. Ω_m must be unity to within a factor $\sim (1 + z_f)$. The redshift z_f must occur after matter-radiation equality;

otherwise, radiation pressure would prevent galaxy-scale systems from collapsing. Any universe with $\Omega \ll 10^{-3}$ at $t = t_*$ would then have great difficulty in generating the non-linear systems needed for life. These structure-formation arguments are explored in more detail in the following paragraphs.

The Strong Anthropic Principle

The ultimate form of anthropic reasoning is to assert that the existence of life is more than just something that operates as a selection effect on observations: rather, the Universe *must* be such as to admit the production of intelligent life at some point. This idea is known as the strong Anthropic Principle. Is such an idea a part of testable science? The whole basis of the Weak Anthropic Principle is the argument that a life-free universe cannot be observed, and observations of such a counterexample would be required in order to falsify the Strong Anthropic Principle.

However, the motivation for strong anthropic reasoning goes beyond the simple issue of observational selection effects in space and time and focuses on puzzles concerning fundamental physics. As first pointed out in a seminal article by Dicke [18], it appears that the very possibility of carbon-based life depends on a series of striking coincidences in the laws of nature. Consider our understanding of the production of the elements. It is now thought virtually certain that the abundances of the light elements up to ^{7}Li were determined by the progress of nuclear reactions in the early stages of the Big Bang, but heavier elements were produced at a much later stage by fusion in stars. Incidentally, this division of labour represents an ironic end to a historically important debate concerning the origin of the Universe, which dominated cosmology in the 1960s. The epochal paper that became universally known just as B^{2}FH [7] was concerned with showing how the elements could be built up by nuclear reactions in stars. Although this was not the motivation for the work, these mechanisms provided a vital defence for the steady-state model (which never passes through a hot phase) against the belief of Gamow and co-workers that all elements could be synthesised in the Big Bang (Gamow 1946; Alpher, Bethe & Gamow 1948). Although the steady-state model passed away, the arguments of B^{2}FH have become part of current orthodoxy, leaving only the lightest elements adhering to Gamow's vision.

Now, the fascinating aspect of all this is that synthesis of the higher elements is rather difficult, owing to the non-existence of stable elements with atomic weights $A = 5$ or $A = 8$. This makes it hard to build up heavy nuclei by collisions of ^{1}H, ^{2}D, ^{3}He, and ^{4}He nuclei. The only reason that heavier elements are produced at all is the reaction

$$3 \, {}^4\text{He} \rightarrow {}^{12}\text{C}. \tag{3.21}$$

A three-body process like this will only proceed at a reasonable rate if the cross section for the process is resonant: i.e., if there is an excited energy level of the carbon nucleus that matches the typical energy of three alpha particles in a stellar interior. The lack of such a level would lead to negligible production of heavy elements – and no carbon-based life. Using these arguments, Hoyle made a breathtaking leap of the imagination to predict that carbon would display such a resonance, which was duly found to be the case [28].

In a sense, this is just trivial anthropic reasoning: we see carbon on Earth, and nuclear physics gives us an inevitable conclusion to draw. And yet, one is struck by the *coincidence*: if the energy levels of carbon had been only slightly different, then it is reasonable to assume that the development of life anywhere in the Universe would never have occurred. Does this mean that some controlling agent designed nuclear physics specifically to produce life? This is a possible explanation, but it is interesting to ask if the appearance of design could have arisen without such interference. A tuning of nuclear physics can arise quite naturally in the context of a multiverse ensemble, provided the different members of the ensemble have some mechanism that allows fundamental physics to vary. In that case, all variants of nuclear physics will be explored: in most cases, the parameters will be such that Hoyle's coincidence does not operate, and those members of the ensemble will be sterile and devoid of life. But given a sufficiently large ensemble and the ability to vary parameters continuously, Hoyle's coincidence is bound to arise. Indeed, strong anthropic reasoning *requires* the existence of an ensemble of universes. From this point of view, the Strong Anthropic Principle really just splits weak anthropic reasoning into one-universe anthropics (which predicts that any observers will see their universe at an age of around 10 Gyr) and many-universe anthropics (which predicts that any observers will witness fundamental physics laws that are compatible with life).

What remains unresolved here is the question of why there are observers at all: just because we may be guaranteed to find some nuclear physics compatible with life, why should life take the opportunity of coming into existence? One could just accept the ultimate censoring effect of weak anthropic reasoning and admit that observers only observe cases where observers happened to be created. But we are here, and we inevitably wonder if this was inevitable. A multiverse that explores all kinds of nuclear physics creates the chance for life, and even if this is highly improbable in any given case, a non-zero probability is bound to be converted into a reality given sufficiently many trials – so our existence may provide evidence in favour of a multiverse with an extremely large number of members. But a completely different class of explanation is to be found in the interpretation of quantum mechanics, where the role of the observer is critical in determining how the Universe evolves. Chapter 7 of Barrow and Tipler [4] gives a full discussion of the relation between quantum and anthropic ideas. In the Copenhagen interpretation, the critical events in time are the moments of wave-function collapse when the act of observation singles out a concrete state from undetermined possibilities (e.g., spin up or down?). In this sense, the act of the observer may be necessary in order to bring the Universe into being at all.

3.5 The Puzzle of Dark Energy

3.5.1 *Cosmological Effects of the Vacuum*

One of the most radical conclusions of recent cosmological research has been the necessity for a non-zero vacuum density. This was detected by being open to the assumption that Einstein's cosmological constant, Λ, might contribute to the energy budget of the Universe.

We can note in passing that Einstein's 1917 paper [21] is a masterpiece of clarity, which can be read in English in, e.g., [5], and the basic argument is one that Newton might almost have generated. Consider an infinite uniform sea of matter, which we want to be static (an interesting question is whether Einstein was influenced by data in imposing this criterion or whether he took it to be self-evident): we want zero gravitational force, so both the gravitational potential, Φ, and the density, ρ, have to be constant. The trouble is this is inconsistent with Poisson's equation, $\nabla^2\Phi = 4\pi G\rho$. The 'obvious' solution (argues Einstein) is that the equation must be wrong, and he proposes instead

$$\nabla^2\Phi + \lambda\Phi = 4\pi G\rho, \tag{3.22}$$

where λ has the same logical role as the Λ term he then introduces into the field equations. In fact, this is not the correct static Newtonian limit of the field equations, which is $\nabla^2\Phi + \Lambda = 4\pi G\rho$. But either equation solves the question posed to Newton by Richard Bentley concerning the fate of an infinite mass distribution; Newton opted for a static model despite the inconsistency analysed earlier.

If we adopt the second variant of the Newtonian field equation, or the GR field equation itself, it is clear the Λ term can be taken from the left-hand side (where it represents the curvature of empty space) to the right-hand side, where it represents an additional source term for the gravitational field:

$$G^{\mu\nu} = -\frac{8\pi G}{c^4}\left(T^{\mu\nu}_{\text{matter}} + T^{\mu\nu}_{\Lambda}\right); \quad T^{\mu\nu}_{\Lambda} = \frac{\Lambda c^4}{8\pi G}g^{\mu\nu}. \tag{3.23}$$

Thus, if there is no matter, then $(\Lambda c^4/8\pi G)\,g^{\mu\nu}$ is the energy-momentum tensor of the vacuum. However, we will see later that there are good reasons to expect the vacuum to have a non-zero density on the grounds of quantum mechanics. Because such a vacuum energy would have to be invariant for all observers within Special Relativity, its energy-momentum tensor must be proportional to the metric (the only rank-2 tensor that is unchanged by Lorentz transformations): $T^{\mu\nu}_{\text{vac}} = \rho_{\text{vac}}c^2\,g^{\mu\nu}$. Thus, Einstein's classical geometrical Λ combines with any physical vacuum density into a single effective value:

$$\Lambda_{\text{eff}} = \Lambda + 8\pi G\rho_{\text{vac}}/c^2. \tag{3.24}$$

This combination into a single effective vacuum density was first noted by Sakharov [44] and Zel'dovich [60].

But if this ingredient is a reality, it raises many questions about the physical origin of the vacuum energy; as we will see, a variety of models may lead to something similar in effect to Λ, and the general term dark energy is used to describe these. The properties of dark energy can be probed by the same means that we used to deduce its existence in the first place: via its effect on the expansion history of the Universe. The vacuum density is included in the Friedmann equation, independent of the equation of state

$$\dot{R}^2 - \frac{8\pi G}{3}\rho R^2 = -kc^2. \tag{3.25}$$

At the outset, we should be very clear that the deduced existence of dark energy depends on the correctness of the Friedmann equation, and this is not guaranteed. We possibly have the wrong theory of gravity, and we have to replace the Friedmann equation with something else. Alternative models do exist, particularly in the context of extra dimensions, and these must be borne in mind. Nevertheless, as a practical framework, it makes sense to stick with the Friedmann equation and see if we can get consistent results. If this programme fails, we may be led in the direction of more radical change.

To insert vacuum energy into the Friedmann equation, we need the equation of state

$$w \equiv \frac{p}{\rho c^2}. \tag{3.26}$$

If this is constant, adiabatic expansion of the vacuum gives

$$\frac{8\pi G\rho}{3H_0^2} = \Omega_v a^{-3(w+1)}. \tag{3.27}$$

More generally, we can allow w to vary; in this case, we should regard $-3(w + 1)$ as $d \ln \rho / d \ln a$, so

$$\frac{8\pi G\rho}{3H_0^2} = \Omega_v \exp\left(\int -3(w(a) + 1)\, d \ln a\right). \tag{3.28}$$

In general, we therefore need

$$H^2(a) = H_0^2 \left[\Omega_v e^{\int -3(w(a)+1)\, d \ln a} + \Omega_m a^{-3} + \Omega_r a^{-4} - (\Omega - 1)a^{-2}\right]. \tag{3.29}$$

Some complete dynamical model is needed to calculate $w(a)$. Given the lack of a unique model, a common empirical parameterisation is

$$w(a) = w_0 + w_a(1 - a). \tag{3.30}$$

It frequently is sufficient to stick with constant w; most experiments are sensitive to w at a particular redshift of order unity, and w at this redshift can be estimated with little dependence on whether we allow dw/dz to be non-zero.

If w is negative at all, this leads to models that become progressively more vacuum dominated as time goes by. When this process is complete, the scale factor should vary as a power of time. The case $w < -1$ is particularly interesting, sometimes known as phantom dark energy [8]. Here, the vacuum energy density will eventually diverge, which has two consequences: this singularity happens in a finite time rather than asymptotically; as it does so, vacuum repulsion will overcome the normal electromagnetic binding force of matter so that all objects will be torn apart in the big rip. Integrating the Friedmann equation forward, ignoring the current matter density, the time to this event is

$$t_{\text{rip}} - t_0 \simeq \frac{2}{3} H_0^{-1} |1 + w|^{-1} (1 - \Omega_m)^{-1/2}. \tag{3.31}$$

Observable Effects of the Vacuum

The co-moving distance-redshift relation is one of the chief diagnostics of w. The general definition is

$$D \equiv R_0 r = \int_0^z \frac{c}{H(z)} \, dz. \tag{3.32}$$

Perturbing this about a fiducial $\Omega_m = 0.3$, $w = -1$ model shows a *sensitivity multiplier* of about 5 – i.e., a measurement of w to 10% requires D to 2%. Also, there is a near-perfect degeneracy with Ω_m, so this parameter must be known very well before the effect of varying w becomes detectable.

The other main diagnostic of w is its effect on the growth of density perturbations. These are also sensitive to the vacuum, as may be seen from the growth equation:

$$\ddot{\delta} + 2\frac{\dot{a}}{a}\dot{\delta} = 4\pi G \rho_0 \delta. \tag{3.33}$$

The vacuum energy manifests itself in the factor of H in the 'Hubble drag' term, $2(\dot{a}/a)\dot{\delta}$. For flat models with $w = -1$, the growing mode for density perturbations is approximately $g(a) \propto a\Omega(a)^{0.23}$ or (more accurately) $d \ln g/d \ln a = \Omega(a)^{0.55}$. For greater accuracy, the following expressions are good to a maximum error of 0.1% [41]. The cases of positive and negative Λ are somewhat distinct. For the positive case,

$$\delta(a) \simeq x(1 - x^{1.91})^{0.82} + 1.437\left(1 - (1 - x^3)^{2/3}\right), \tag{3.34}$$

where x denotes $\Omega_v(a)^{1/3}$, and we choose the $a = 1$ point to correspond to equal density in matter and vacuum:

$$\Omega_v(a) = (1 + a^{-3})^{-1} \tag{3.35}$$

so that $\delta(a) \simeq a$ for small a. For the negative-Λ case, we need time as a coordinate since the scale factor is not monotonic:

$$a(t) = \left[\sin\left(\frac{3t}{2}\right)\right]^{2/3}, \tag{3.36}$$

where here we choose units such that $a = 1$ at the point of maximum expansion, and time is measured in units of $(8\pi G|\rho_v|/3)^{-1/2}$ so that Friedmann's equation is $(\dot{a}/a)^2 = a^{-3} - 1$ and $\Omega_v(a) = (1 - a^{-3})^{-1}$. Here, the approximation for the growth function is

$$\delta(t) = \frac{(3t/2)^{2/3}}{(1 + 0.37(t/t_{\text{coll}})^{2.18})(1 - (t/t_{\text{coll}})^2)}. \tag{3.37}$$

Again, the normalisation is that $\delta(a) \simeq a$ for small a. Note that the fluctuations diverge at the collapse time ($t_{\text{coll}} = 2\pi/3$) as $1/(t_{\text{coll}} - t)$; this corresponds to the decaying mode in the expanding phase.

In the real Universe, we are interested in the positive branch of the growth factor, and this is illustrated in Figure 3.3. The plot shows two alternative points of view: either a growing $\delta(t)$ that asymptotes to a constant or a potential $\Phi(t)$ that is initially independent

John A. Peacock

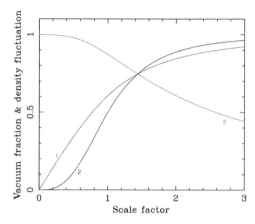

Figure 3.3 The growth of density fluctuation amplitude (1) vs dimensionless scale factor, $a(t)$, for the case of a flat universe containing matter plus a cosmological constant. The normalisation is that $a = 1$ at matter-vacuum equality. Line 2 shows $\Omega_v(a)$, and line 3 shows $\delta(a)/a$, which is proportional to the amplitude of potential fluctuations, $\Phi(a)$.

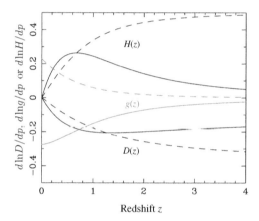

Figure 3.4 Perturbation around $\Omega_m = 0.3$ of Hubble parameter, distance-redshift, and growth-redshift relations. Solid line shows the effect of increase in w; dashed line shows the effect of increase in Ω_m.

of time, but by which these preserved initial metric fluctuations are damped away once the Universe becomes Λ dominated.

If w is made more negative, this makes the growth law closer to the Einstein-de Sitter $g(a) \propto a$ (for very large negative w, the vacuum was unimportant until very recently). Therefore, increasing w (making it less negative) has an effect in the same sense as *decreasing* Ω_m. As shown in Figure 3.4, the degeneracy between variations in Ω_m and w thus has the opposite sign to the degeneracy in $D(z)$. Ideally, one would therefore try to observe both effects.

3.5.2 Observing the Properties of Dark Energy

What are the best ways to measure w? We have seen that the two main signatures are alterations to the distance-redshift relation and the perturbation growth rate. It is possible to use both of these effects in the framework we have been discussing: observing the perturbed Universe in both the CMB and large-scale structure.

In the CMB, the main observable is the angle subtended by the horizon at last scattering

$$\theta_{\mathrm{H}} = D(z_{\mathrm{LS}})/D(z = 0). \tag{3.38}$$

This has the approximate scaling with cosmological parameters (for a flat universe)

$$\theta_{\mathrm{H}} \propto (\Omega_m h^{3.3})^{0.15} \Omega_m^{\alpha - 0.4}; \quad \alpha(w) = -2w/(1 - 3.8w). \tag{3.39}$$

The latter term comes from a convenient approximation for the current horizon size:

$$D_0 = 2 \frac{c}{H_0} \Omega_m^{-\alpha(w)}. \tag{3.40}$$

At first sight, this looks bad: the single observable of the horizon angle depends on three parameters (four, if we permit curvature). Thus, even in a flat model, we can only pin down w if we know both Ω_m and h.

However, if we have more detail on the CMB than just the main peak location, then we have seen that the $\Omega_m - h$ degeneracy is weakly broken and that this situation improves with information from large-scale structure, which yields an estimate of $\Omega_m h$. In effect, we have two constraints on the $\Omega_m - h$ plane that are consistent if $w = -1$, but this is not the case for other values of w. In this way, the current combined constraints from CMB plus alternative probes (LSS and the Supernova Hubble diagram) yield an impressive accuracy:

$$w = -1.006 \pm 0.045, \tag{3.41}$$

for a spatially flat model – see Ade [2] and Spergel *et al.* [49]. The confidence contours are plotted in detail in Figure 3.5, and it is clear that, so far, there is very good consistency with a simple cosmological constant. But as we will see, plenty of models exist in which some deviation is predicted. The next goal of the global cosmology community is therefore to push the errors on w down substantially – to about 1%. There is no guarantee that this will yield any signal, but certainly it will cut down the range of viable models for dark energy.

One of the future tools for improving the accuracy in w will be large-scale structure. We have seen how this helps pin down the parameter degeneracies inherent in a CMB-only analysis, but it also contains unique information from the acoustic horizon. Earlier, we approximated this without considering how the speed of sound would depend on the baryon density; a good approximation to the exact result is

$$D_a \simeq 55.4 \, (\Omega_m h^2)^{-0.26} (\Omega_b h^2)^{-0.13} \, \mathrm{Mpc}. \tag{3.42}$$

This forms a standard measuring rod, as seen in the 'baryon wiggles' in the galaxy power spectrum. In future galaxy surveys, the measurement of this signature as a function of redshift will be a further useful geometrical probe. Strictly, this is a slightly different

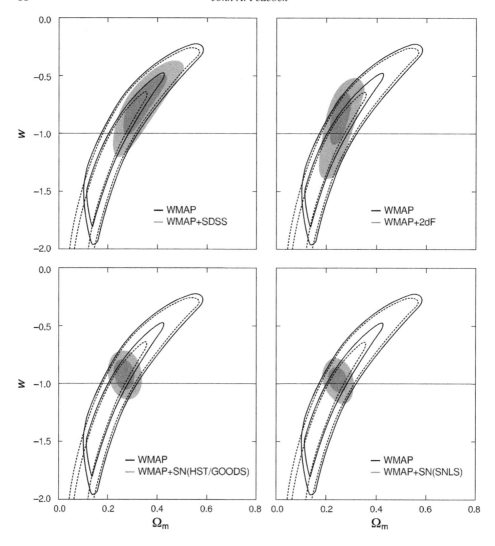

Figure 3.5 The marginalised WMAP3 confidence contours on the plane of dark-energy equation of state (w) vs Ω_m (from [49]). A flat universe is assumed, although this is not critical to the conclusions.

acoustic horizon to the one seen in the CMB, which is approximately $65.5\,(\Omega_m h^2)^{-0.25}$ $(\Omega_b h^2)^{-0.08}$ Mpc. The earlier figure is appropriate for the 'drag era' when CMB photons finally cease to interact with the baryons, and this is what sets the BAO scale in the galaxy distribution (see, e.g., [42]).

The amplitude constraint from the CMB has been harder to implement. Although CMB data provide an accurately determined temperature normalisation, this involves the uncertain optical depth due to reionisation:

$$\sigma_8(\text{CMB}) = 0.755\,(\Omega_m 0.3)^{+0.4}\exp(\tau) \pm 2\%. \tag{3.43}$$

The value of τ is constrained by large-angle polarisation data and is $\tau = 0.066 \pm 0.016$, according to Planck. This value of σ_8 is, of course, an extrapolation to $z = 0$ of the amplitude of fluctuations inferred at $z \simeq 1,100$, and so comparison with direct low-z data can, in principle, measure the growth over this period and, thus, pin down w (or deviations from Einstein gravity). Currently, this local determination is possible using gravitational lensing and yields a figure that is within 10% of the CMB determination (e.g., [1]). Currently, this is not accurate enough to give a competitive determination of w, but future constraints of this sort will be interesting.

3.5.3 Models for Dynamical Dark Energy

The simplest physical model for dynamical vacuum energy is a scalar field. We know from inflationary models that this can yield something close in properties to a cosmological constant, and so we can immediately borrow the whole apparatus for modelling vacuum energy at late times. This idea of scalar fields as a dynamical substitute for Λ was first explored by Ratra and Peebles [43]. Of course, this means yet another scalar field that is introduced without much or any motivation from fundamental physics. This hypothetical field is given the fanciful name 'quintessence', implying a new addition to the ancient Greek list of elements (fire, air, earth, water).

The Lagrangian density for a scalar field is, as usual, of the form of a kinetic minus a potential term:

$$\mathcal{L} = \tfrac{1}{2}\partial_\mu \phi\, \partial^\mu \phi - V(\phi). \tag{3.44}$$

In familiar examples of quantum fields, the potential would be

$$V(\phi) = \tfrac{1}{2} m^2 \phi^2, \tag{3.45}$$

where m is the mass of the field. However, as before, we keep the potential function general at this stage.

Suppose the Lagrangian has no explicit dependence on space-time (i.e., it depends on x^μ only implicitly through the fields and their four-derivatives). Noether's theorem then gives the energy-momentum tensor for the field as

$$T^{\mu\nu} = \partial^\mu \phi \partial^\nu \phi - g^{\mu\nu} \mathcal{L}. \tag{3.46}$$

From this, we can read off the energy density and pressure:

$$\begin{aligned}
\rho &= \tfrac{1}{2}\dot{\phi}^2 + V(\phi) + \tfrac{1}{2}(\nabla\phi)^2 \\
p &= \tfrac{1}{2}\dot{\phi}^2 - V(\phi) - \tfrac{1}{6}(\nabla\phi)^2.
\end{aligned} \tag{3.47}$$

If the field is constant both spatially and temporally, the equation of state is then $p = -\rho$, as required if the scalar field is to act as a cosmological constant; note that derivatives of the field spoil this identification.

For a homogeneous field, we have the equation of motion

$$\ddot{\phi} + 3H\dot{\phi} + dV/d\phi = 0,$$
(3.48)

which is most easily derived via energy conservation:

$$\frac{d \ln \rho}{d \ln a} = -3(1 + w) = -\frac{3\dot{\phi}^2}{(\dot{\phi}^2/2 + V)},$$
(3.49)

following which the relations $H = d \ln a/dt$ and $\dot{V} = \dot{\phi} V'$ can be used to change variables to t, and the damped oscillator equation for ϕ follows. The solution of the equation of motion becomes tractable if we make the *slow-rolling approximation* that $|\ddot{\phi}|$ is negligible in comparison with $|3H\dot{\phi}|$ and $|dV/d\phi|$ so that

$$3H\dot{\phi} = -\frac{dV}{d\phi}.$$
(3.50)

From this, we know that a sufficiently flat potential can provide a dynamical vacuum that is arbitrarily close to a cosmological constant in its equation of state. However, there are good reasons why we might want to imagine the slow-roll conditions being violated in the case of dark energy. For a detailed discussion, see Chapter 4.

Cosmic Coincidence and Quintessence

Accepting the reality of vacuum energy raises a difficult question. If the Universe contains a constant vacuum density and normal matter with $\rho \propto a^{-3}$, there is a unique epoch at which these two contributions cross over, and we seem to be living near to that time. This coincidence calls for some explanation.

We already have one coincidence, in that we live relatively close in time to the era of matter-radiation equality ($z \sim 10^3$, as opposed to $z \sim 10^{28}$ for the GUT era). This is relatively simple to understand: structure formation cannot begin until after z_{eq}, and so we would expect observers to appear before the Universe has expanded much beyond this point. The vacuum coincidence problem could therefore be solved if the vacuum density was some dynamical entity that was triggered to become Λ-like by the change in expansion history at z_{eq}. Zlatev *et al.* [61] suggested how this might happen. We have seen that the density and pressure for a quintessence field will be

$$\rho_\phi = \dot{\phi}^2/2 + V$$
$$p_\phi = \dot{\phi}^2/2 - V.$$
(3.51)

This gives us two extreme equations of state: (1) vacuum dominated, with $V \gg \dot{\phi}^2/2$, so that $p = -\rho$; (2) kinetic dominated, with $V \ll \dot{\phi}^2/2$, so that $p = \rho$. In the first case, we know that ρ does not alter as the Universe expands, so the vacuum rapidly tends to dominate over normal matter. In the second case, the equation of state is the unusual $w = +1$, so we get the rapid behaviour $\rho \propto a^{-6}$. If a quintessence-dominated universe starts off with a large kinetic term relative to the potential, it may seem that things should always evolve in the direction of being potential dominated. However, this ignores the detailed dynamics of the situation: for a suitable choice of potential, it is possible to have a tracker field, in which the kinetic and potential terms remain in a constant proportion so that we can have $\rho \propto a^{-\alpha}$, where α can be anything we choose.

Putting this condition in the equation of motion shows that the potential is required to be exponential in form. The Friedmann equation with $\rho \propto a^{-\alpha}$ requires $a \propto t^{2/\alpha}$, so we have $\rho \propto t^{-2}$, as usual. But now, both V and $\dot{\phi}^2$ must scale in the same way as ρ so that $\dot{\phi} \propto 1/t$. Both the $\ddot{\phi}$ and $3H\dot{\phi}$ terms are therefore proportional to V, so an exponential potential solves the equation of motion. More importantly, we can generalise to the case where the Universe contains scalar field and ordinary matter. Suppose the latter obeys $\rho_m \propto a^{-\alpha}$; it is then possible to have the scalar-field density obeying the same $\rho \propto a^{-\alpha}$ law, provided

$$V(\phi) = \frac{2M^4}{\lambda^2} \left(\frac{6}{\alpha} - 1 \right) \exp\left(-\frac{\lambda\phi}{M} \right),$$ (3.52)

where $M = m_P/\sqrt{8\pi}$. The scalar-field density is $\rho_\phi = (\alpha/\lambda^2)\rho_{total}$. (see, e.g., [32]). The impressive thing about this solution is that the quintessence density stays a fixed fraction of the total, whatever the overall equation of state: it automatically scales as a^{-4} at early times, switching to a^{-3} after matter-radiation equality.

This is not quite what we need, but it shows how the effect of the overall equation of state can affect the rolling field. Because of the $3H\dot{\phi}$ term in the equation of motion, ϕ 'knows' whether or not the Universe is matter dominated. This suggests that a more complicated potential than the exponential may allow the arrival of matter domination to trigger the desired Λ-like behaviour. Zlatev *et al.* [61] suggested two potentials which might achieve this:

$$V(\phi) = M^{4+\beta}\phi^{-\beta} \quad \text{or} \quad V(\phi) = M^4 \left[\exp\left(\frac{m_P}{\phi} \right) - 1 \right].$$ (3.53)

They show that these can yield an evolution in $w(t)$ so that it switches from $w \simeq 1/3$ in the radiation era to $w \simeq -1$ today.

However, a degree of fine-tuning is still required, in that the trick only works for $M \sim 1$ meV, so there is no natural reason for tracking to cease at matter-radiation equality. The idea of tracker fields thus does not remove completely the puzzle concerning the level of present-day vacuum energy. But such models are at least testable: because the Λ-like behaviour only switched on quite recently, it is hard to complete the transition, and the prediction is of something around $w \simeq -0.8$ today [61]. As we have seen, this can be firmly ruled out with current data. These ideas about the dynamical vacuum are therefore already interesting, testable science.

k-Essence

In a sense, quintessence is only half the story. We started with the usual Lagrangian for a simple massive scalar field, $\mathcal{L} = \dot{\phi}^2/2 - m^2\phi^2/2$ and generalised the quadratic mass term to an arbitrary potential, $V(\phi)$. Why not take the same liberties with the kinetic term? Even though such *k-essence* models lack the intuitive analogies of quintessence, a Lagrangian can be anything we like. The simplest models try to express things in terms of the normal kinetic expression

$$X \equiv \frac{1}{2}\partial^\mu \phi \partial_\mu \phi, \tag{3.54}$$

and one assumes that $\mathcal{L} = K(\phi) f(X)$; In the homogeneous case, $X = \dot{\phi}^2/2$. The pressure and density are

$$\rho = 2X\mathcal{L}_{,X} - \mathcal{L}$$
$$P = \mathcal{L} \tag{3.55}$$

so that the equation of state is

$$w = \frac{f}{2X\frac{df}{dX} - f}. \tag{3.56}$$

For a normal kinetic term, this gives $w = +1$ if there is no potential. The equation of motion is derived just by writing conservation of energy as for quintessence:

$$\frac{d \ln \rho}{d \ln a} = -3(1 + w). \tag{3.57}$$

What sort of k-essence Lagrangian will yield tracking? We want to fix w at the value of the dominant component, which requires

$$\frac{d \ln f}{d \ln X} = \frac{1}{2}\left(1 + \frac{1}{w}\right) \quad \Rightarrow \quad f(X) \propto X^{(1+1/w)/2}. \tag{3.58}$$

Thus, a Lagrangian proportional to the square of the usual kinetic term will produce tracking during the radiation era, but tracking in the matter era requires a step to $f(X) = 0$ to be encountered just as the Universe becomes matter dominated. This is the opposite to the case of quintessence: now fine-tuning would be required in order for tracking to be maintained. The real question is whether a simple model can achieve sufficiently strong departure from tracking to get somewhere close to $w = -1$ in the matter era in an inevitable way. This seems to be controversial: Armendariz-Picon et al. [3] claimed that it could be done, but Malquarti et al. [37] disagreed. The issue, as with quintessence, is the extent to which a tracking solution arises inevitably independent of initial conditions – i.e., whether it is an attractor. This has certainly not been demonstrated.

Perturbations in the Vacuum

In dynamical models for the vacuum, we have a peculiar kind of fluid, so it is able to respond to gravity and grow inhomogeneities. The key parameter here is the vacuum sound speed, which obeys the usual relation

$$c_s^2 = \frac{\partial p}{\partial \rho}. \tag{3.59}$$

In practice, this is evaluated as

$$c_s^2 = \frac{\partial p / \partial X}{\partial \rho / \partial X},$$ (3.60)

i.e., ignoring perturbations in the field. The justification for this is that a gauge freedom exists and that $\delta\phi = 0$ corresponds to the rest frame of the vacuum fluid.

This means that, for quintessence, the sound speed is always $c_s = c$. Even a completely flat potential with initial condition $\dot\phi = 0$ does not mimic a cosmological constant. This only happens if the Lagrangian is set up completely lacking any kinetic term. The low sound speeds in some k-essence models can have quite large effects on the CMB anisotropies, and so can be probed observationally beyond just w and its evolution (see, e.g., [13, 58]).

Scalar Fields as Dark Matter

One interesting limit of the scalar-field equation is if the 'acceleration' from the potential exceeds the Hubble drag (i.e., the Universe expands sufficiently slowly that this term can be neglected). If we further assume that the potential is mass-like (or at least parabolic near its minimum), then we have the simple oscillator equation $\ddot\phi + m^2\phi = 0$, with solution $\phi = A \sin mt$ (for a suitable origin of time). The density and pressure are

$$\rho = m^2 A^2 / 2$$
$$p = (m^2 A^2 / 2) \cos 2mt.$$ (3.61)

Therefore, averaged over many cycles, the oscillating scalar field has the equation of state of pressureless matter ($\langle p \rangle / \rho = 0$, even though there are times when $|p|$ and ρ are comparable).

It is therefore possible that the cosmological dark matter may take the form of a light scalar field rather than a supersymmetric relic WIMP (see Chapter 9). This scalar-field dark matter is normally considered to be a particle called the axion, which has some motivation in particle physics. Notice that the mass can, in principle, be anything, since the density depends on m and on the field amplitude, A. In practice, other constraints on the axion model focus attention on

$$m_{\text{axion}} \sim 10^{-5} \text{ eV}.$$ (3.62)

This is very light dark matter indeed, so shouldn't it be very hot and fail to make $\Omega \sim 1$ by a large factor? This is not so: the axion will act as cold dark matter and can have a significant relic density. The answer to the apparent paradox is that these particles should not be thought of as having been in thermal equilibrium. We are dealing with a classical field that interacts extremely weakly with ordinary matter. If this interaction was zero, there would be no prospect of detecting the axion other than via cosmology. In practice, as with WIMPs, there is some level of interaction, but the strategy for detection is completely different: the axion can interact with electromagnetic waves, and the low mass means that microwave frequencies are involved. There is therefore an active experimental programme searching for the axion using tuned microwave cavities. The problem is that, for sensitivity reasons, the bandwidth needs to be very narrow, and it takes a long time to scan an interesting frequency range: the axion model probably will be fully explored and ruled out within the next decade – or it could be detected any day now.

Despite the lack of detection of axions, the greater level of investment in WIMP searches and their lack of success has tended to focus greater attention on the axion model. Although the original QCD axion is constrained to a rather specific mass, the term 'axion' now tends to be used for a wider range of scalar fields. The most exciting possibility is the case where these are ultralight, with masses below 10^{-20}eV. In this case, the Compton wavelength $\hbar/mc$ is large enough that wave-mechanical effects may be observed on astronomical scales (see, e.g., [38, 46]).

3.5.4 The Outlook for Dark Energy

One significant problem with this line of research is the lack of a clear target. Some models, such as the $w \simeq -0.8$ from simple power-law quintessence, have been ruled out, but there is no guaranteed minimum deviation from $w = -1$. Perhaps dark energy is exactly a cosmological constant, and we are condemned to a future of ever more challenging experiments yielding increasingly precise null results around $w = -1$; at what point would we abandon the search?

Trotta [54] gave a nice Bayesian answer to this question, which applies to the general issue of asking when a theory should be expanded to include a new phenomenon. Let the effect in question be characterised by a parameter a (w in our case), such that the 'new physics' corresponds to $a \neq 0$. Let the prior probability of a be $P(a)$ so that the Bayesian odds ratio between the two hypotheses A (new physics: $a \neq 0$ and in the range da) and B (no new physics: $a = 0$ exactly) is $L(a) \times P(a)\,da$, where L is the likelihood ratio between A and B, given some data. This neglects a prior ratio of beliefs in new physics vs no new physics, which is generally taken to be of order unity (although, as usual, a sufficiently firm prejudice against new physics is not capable of being overturned, however strong the experimental evidence). To get the overall odds ratios of the two models in the face of the data, we should integrate over a to get the evidence ratio discussed earlier:

$$E = \int L(a)\,P(a)\,da. \tag{3.63}$$

Consider a simple Gaussian example, where $L(a) = \exp[-\hat{a}^2/2\sigma^2]\,/\,\exp[(a-\hat{a})^2/2\sigma^2]$ in terms of a measured value $a = \hat{a} \pm \sigma$, and let the prior on a be uniform over a range Σ. If Σ is large compared to the measuring error, σ, this gives $E = \sqrt{2\pi}\,(\sigma/\Sigma)\exp[\lambda^2/2]$, where $\lambda = \hat{a}/\sigma$ is the 'number of sigmas' at which the data 'detect' a deviation from $a = 0$. This expression is the basis of Figure 3.6 and reveals two important facts: (1) if λ is large, this need not constitute a detection if $\sigma \ll \Sigma$; (2) if we have a null result ($\lambda \lesssim 1$), then we may strongly disfavour hypothesis A if $\sigma \ll \Sigma$ – i.e., demonstrate convincingly that, in fact, $a = 0$ exactly, so there is no point in trying to measure it with improved precision. These slightly paradoxical conclusions arise because we take our prior seriously: if the prior on w

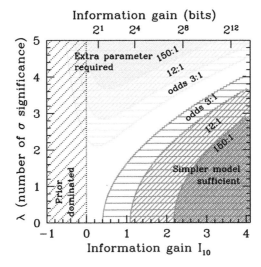

Figure 3.6 This plot, taken from [54], illustrates the interplay between the S/N of a measurement probing the existence of a new parameter (vertical axis) and the reciprocal fraction of the available parameter space ruled out by null measurements (horizontal axis). From a Bayesian point of view, improving the precision on parameter measurement sufficiently while still finding a null result can eventually yield compelling odds that the parameter will never be detected. In the context of dark energy, the initial prior range of interesting models was $|w + 1| \sim 1$; thus measuring, say, $w = -1 \pm 0.001$ would yield odds of 1,000:1 that the vacuum energy is exactly a cosmological constant.

ranged over -0.5 to -1.5, say, we are saying that the range -1.4 to -1.41 is just as likely as -1 to -1.01. But if we make a measurement with 0.01 precision, we have ruled out 99% of the possible options for a, so the original model starts to look like a poor bet.

With $\sigma/\Sigma \simeq 0.05$, dark energy is currently far from the situation envisaged in the preceding paragraph. But the next generation of experiments (Euclid; LSST) should push the measuring precision on w to below 1%, and a null result at that stage will start to be interesting from a Bayesian point of view. See, e.g., Laureijs [31].

Dark Energy vs Modified Gravity

A more radical approach to the problem of dark energy might be to wonder whether we are going entirely in the wrong direction. The acceleration of the cosmic expansion is surprising if our prejudice is that the Universe is dominated by the attractive gravitational effects of matter with normal equations of state. Introducing a new substance with $w < -1/3$ allows the acceleration to be explained and also makes new w-dependent predictions for the growth of density fluctuations. But both the expansion rate and the growth of perturbations are dictated by gravity, so is it possible that dark energy simply does not exist at all and that, in fact, Einstein's relativistic theory of gravity needs to be modified on cosmological scales?

This possibility has been explored energetically in recent cosmological research, and comprehensive reviews are given by Clifton *et al.* [12] and Joyce *et al.* [29]. There is clearly plenty of scope for building alternative models, since Einstein gravity is built on a simple scalar Lagrangian containing the Ricci scalar and the cosmological constant: $\mathcal{L} = R + 2\Lambda$. So it is only necessary to replace this with some other invariant, and we have a theory of modified gravity that is fully consistent with all the requirements of general covariance (thus, it is misleading to describe this area of research as 'testing general relativity'). The simplest such approach would be some non-linear function of R, and $f(R)$ gravity is one of the most widely explored theories of modified gravity. An alternative is to express the modification in the language of fifth forces and postulate an additional scalar field that contains explicit interaction terms with the gravitational and matter sectors (unlike the case of a simple quintessence field).

The behaviour of such models on non-linear scales is complicated, and much attention has been given to the linear regime, where we can concentrate on the behaviour of the metric perturbation potentials Ψ and Φ, which affect respectively the time and space parts of the metric. In Einstein's gravity, these potentials are both equal to the Newtonian gravitational potential, which satisfies Poisson's equation: $\nabla^2 \Phi / a^2 = 4\pi G \bar{\rho} \delta$. Empirically, modifications of gravity require us to explore a change with scale and with time of the 'slip' ($\eta \equiv \Psi/\Phi$) and the effective G on the right-hand side of Poisson's equation. The former aspect can only be probed via gravitational lensing, whereas the latter can be addressed on 10–100 Mpc scales via the growth of clustering. A common approach is to assume, as before, that the growth rate can be tied to the density parameter: $d \ln \delta / d \ln a = \Omega_m^\gamma(a)$. The parameter γ is close to 0.55 for standard relativistic gravity but can differ by around 0.1 from this value in many non-standard models. Clearly this parameterisation is incomplete, since it explicitly rejects the possibility of effects at early times ($\Omega_m(a) \to 1$ as $a \to 0$), but the recent onset of cosmological acceleration is used as a common justification for assuming that modifications of gravity only become significant at late times.

This reliance on the growth rate exposes a degeneracy, since we have seen that the growth rate is also sensitive to the equation of state of dark energy; thus, a deviation from ΛCDM growth could indicate modifications of gravity, or just that dark energy is not a simple Λ [47]. The way to break this degeneracy is by including geometrical probes that measure purely the expansion history (BAO and SNe) and, hence, measure the value of w (while being agnostic about whether this is genuinely the equation of state of a physical substance as opposed to being an effective value induced by the modification of gravity). We are then free to probe the two linear parameters that probe the effect of any modifications on the perturbations. A practical parameterisation of this is to express things in terms of factors that modify the strength of non-relativistic forces and relativistic forces (i.e., lensing) with respect to the values predicted in Einstein gravity:

$$\Psi = [1 + \mu(a,k)]\,\Psi_E$$
$$(\Psi + \Phi) = [1 + \Sigma(a,k)]\,(\Psi_E + \Phi_E), \tag{3.64}$$

where the relation to effective G and slip are $1+\mu = (G'/G)/\eta$; $1+\Sigma = (G'/2G)(1+1/\eta)$ (see, e.g., [14]). Current data are consistent with standard ΛCDM and exclude variations in slip or effective G of larger than of order 10% [48].

3.6 The Anthropic Vacuum

Whether or not one finds these approaches to dark energy compelling, there remains one big problem. All the models are constructed using Lagrangians with a particular zero level. All quintessence potentials have the field rolling down towards $V = 0$, and k-essence models lack a potential altogether. They are therefore subject to the classical dilemma of the cosmological constant: adding a pure constant to the Lagrangian has no affect on field dynamics but mimics a cosmological constant. With so many possible contributions to this vacuum energy from the zero-point energies of different fields (if nothing else), it seems contrived to force $V(\phi)$ to asymptote to zero without a reason.

To review why zero is a problematic value for the vacuum density, recall what we mean by the vacuum: $|0\rangle$, or zero occupation number for each wave mode inside a given box. But standard quantum mechanics assigns a zero-point energy of $\hbar\omega/2$ to each mode. Integrating $\hbar\omega/2c^2$ per mode over k-space (with a degeneracy of 2 for polarisation) gives a total density of

$$\rho_{\text{vac}} = \frac{\hbar}{2\pi^2 c^5} \int \omega^3 \, d\omega, \tag{3.65}$$

which diverges horribly. Is it possible that the upper limit of the integral should be finite? This would be the case if space were a lattice, which is perhaps conceivable on some unobservably small scale. However, even with a cut-off at the hardly microscopic level of $\lambda \sim 1$ mm, ρ_{vac} already exceeds the critical density of the Universe ($\sim 10^{-26}\text{kg m}^{-3}$). We can express things in terms of an energy scale E_v by writing the dimensional scaling

$$\rho_v = \frac{\hbar}{c} \left(\frac{E_v}{\hbar c} \right)^4, \tag{3.66}$$

or simply $\rho_v = E_v^4$ in natural units. if we adopt the values $\Omega_v = 0.7$ and $h = 0.7$ for the key cosmological parameters, then $E_v = 2.4$ meV is known to a tolerance of about 1%. What is a natural choice for E_v? A case can be made for E_v lying at the Planck scale, since quantum gravity effects must destroy the flat-space assumptions of quantum field theory. This would give a vacuum density 120 powers of 10 larger than observed. But this is over-dramatising the problem: one should focus on E_v rather than E_v^4. Also, the solution may lurk at much smaller energies. In unbroken supersymmetry, there would be an exact cancellation of the zero-point energy of bosonic and fermionic oscillators, and the scale of supersymmetry breaking could be as low as 10 TeV. So the vacuum problem is perhaps that the energy scale of the vacuum is 'only' 15 powers of 10 smaller than seems reasonable – a lot fewer than 120 powers of 10 but still enough to cause a problem.

The preceding argument is commonly given, but it should be taken with some caution. It is really the same argument as used to deduce black-body radiation, with a slightly different occupation number. Therefore, we would expect an equation of state $P = \rho c^2/3$: $w = +1/3$, so this is not at all a candidate calculation for the energy density of the vacuum. The problem is that the calculation is not relativistically invariant. Koksma and Prokopec [30] claim that a proper calculation changes the E_v^4 dependence to $M^4 \ln(E_v/M)$, where M is the particle rest mass associated with the field. Since we know of particles up to over $M = 100\,\mathrm{GeV}$, this makes little practical change to the magnitude of the vacuum problem.

In any case, it should be clear that this prediction is hard to make fixed, partly because of our ignorance of the field content of the Universe, and because these zero-point contributions can be supplemented by classical contributions from $V(\phi)$ of any number of scalar fields. This problem has been sharpened by recent developments in string theory, known under the heading of the landscape. For the present purpose, this can be regarded as requiring the introduction of a large number of additional scalar fields, each with an associated potential. If we assume that a vacuum state is defined by these fields sitting at the minimum of their various potentials, then the effective cosmological constant can vary. It has been estimated that there are about 10^{500} distinct minima, which divides the natural vacuum density of E_p^4 into what is almost a continuous range from the point of view of observations – so we can have almost any effective value of Λ we like.

This leads us in the direction of anthropic arguments, which are able to limit Λ to some extent: if the Universe had become vacuum dominated at $z > 1,000$, gravitational instability would have been impossible – so that galaxies, stars, and observers would not have been possible [56]. Indeed, Weinberg made the astonishingly prescient prediction on this basis that a non-zero vacuum density would be detected at Ω_v of order unity, since there was no reason for it to be much smaller.

Many Universes

At first sight, this argument seems quite appealing, but it rapidly leads us into deep waters. How can we talk about changing Λ? It has the value that it has. We are implicitly invoking an ensemble picture in which there are many universes with differing properties. This is a big step (although exciting if this turns out to be the only way to explain the vacuum level we see). In fact, the idea of an ensemble emerges inevitably from the framework of inflationary cosmology, since the fluctuations in the scalar field can affect the progress of inflation itself. We have used this idea to look at the changes in when inflation ends – but fluctuations can affect the field at all stages of its evolution. They can be thought of as adding a random-walk element to the classical rolling of the scalar field down the trough defined by $V(\phi)$. In cases where ϕ is too close to the origin for inflation to persist for sufficiently long, it is possible for the quantum fluctuations to push ϕ further out – creating further inflation in a self-sustaining process. This is the concept of stochastic eternal inflation due to Linde *et al.* [34]. Sufficiently far from the origin, the random-walk effect of fluctuations becomes more marked and can overwhelm the classical downhill rolling. This means that some regions of space can inflate for an indefinite time, and a single inflating universe automatically breaks up into different bubbles with their own histories. Some random subset of these eventually random-walk close enough to the origin that the classical end of inflation can occur, thus creating a set of 'universes' each of which can potentially host observers.

With this as a starting point, the question now becomes whether we can arrange for the different members of this ensemble to have different values of Λ. This is easily achieved. Let there be some quintessence field with a very flat potential so that it is capable of simulating Λ effectively. Quantum fluctuations during inflation can also displace this field so that each member of the multiverse would have a different Λ.

The Distribution of Λ

We are now almost in a position to calculate a probability distribution for Λ, following [20]. First, we have to set some ground rules: what will vary and what will be held fixed? We should try to change as little as possible, so we assume that all universes have the same values for

(1) the baryon fraction $f_b = \rho_b/\rho_m$
(2) the entropy per particle $S = (T/2.725)^3/\Omega_m h^2$
(3) the horizon-scale inhomogeneity $\delta_H \simeq 10^{-5}$

It is far from clear that these minimal assumptions are correct. For example, in the string theory landscape, there is no unique form for low-energy particle physics but, instead, a large number of possibilities in which numbers such as the fine-structure constant, neutrino masses, etc., are different. From the point of view of understanding Λ, we need there to be at least 10^{100} possible states so that at least some have Λ smaller than the natural m_p^4 density by a sufficient factor. The landscape hypothesis provides this variation in Λ but does not support the idea that particle physics is otherwise invariant. Still, it makes sense to start with the simplest forms of anthropic variation: if this can be ruled out, it might be taken as evidence in favour of the fuller landscape picture.

We then take a Bayesian viewpoint to the distribution of Λ, given the existence of observers:

$$P(\Lambda \mid \text{Observer}) \propto P_{\text{prior}}(\Lambda) P(\text{Observer} \mid \Lambda), \qquad (3.67)$$

where we need both the prior distribution of Λ between different members of the ensemble and how the chance of getting an observer is modified by Λ. The latter factor should be proportional to the number of stars, which is generally taken to be proportional to the fraction of the baryons that are incorporated into non-linear structures. We can estimate this using the Press-Schechter apparatus to get the collapse fraction into systems of a galaxy-scale mass. The exact definition of this is not very important since the CDM dimensionless power spectrum is very flat on small scales; any mass at all close to $10^{12} M_\odot$ gives similar answers.

The more difficult part is the prior distribution of Λ, and a common argument is to say that it has a uniform distribution – which seems reasonable enough if we are to allow it to have either sign but know that we will be interested in practice in a very small range near zero. This choice of prior is key to the results of the argument, and it is clear that different choices could completely transform the answer. For example, suppose we took a uniform prior in $\log(\Lambda)$ – or rather $\log(|\Lambda|)$, since Λ can be negative. This would multiply our flat prior by $1/|\Lambda|$, which gives a divergent spike at $\Lambda = 0$. This might have been attractive back in the days when $\Lambda = 0$ matched observations, but it actually is not an acceptable explanation. This particular prior says that there is something special about $\Lambda = 0$, whereas our whole problem is that we know of no physical argument why this should be so. As we saw earlier, the observed value of Λ is a mixture of the classical Λ and a physical vacuum density: $\Lambda_{\text{eff}} = \Lambda_{\text{bare}} + 8\pi G\rho_{\text{vac}}/c^2$. The natural value for the second term is very large, so somehow the classical term must cancel it to high precision, but there is no known reason for this to happen precisely (and, indeed, it does not). Therefore, we have to suppose that nothing violent happens to the prior as we cross from Λ being slightly positive to slightly negative – which is the basis of the uniform prior (see [57] for a detailed discussion).

We therefore have the startling proposition of the anthropic model: the effective vacuum density takes large ranges, and in almost all realisations, the values are comparable in magnitude to the natural scale m_{p}^4; such models are stupendously inimical to life. This is quantified by the simple model

$$dP(\rho_v) \propto f_c \, d\rho_v, \qquad (3.68)$$

where f_c is the collapse fraction into galaxy-scale objects. For large values of Λ, growth ceases at high redshift, and f_c is exponentially suppressed. But things are less clear-cut if $\Lambda < 0$. Here, the Universe eventually recollapses, and the high density means that the collapse fraction always tends to unity. So why do we not observe $\Lambda < 0$? The answer is that we have to cut off the calculation at late stages of recollapse: once the Universe becomes too hot, star formation may be affected, and in any case, there is little time for life to form.

With this proviso, Figure 3.7 shows the posterior distribution of Λ conditional on the existence of observers in the multiverse. Provided we consider recollapse only to a maximum temperature of about 10 K, the observed figure is matched well by the anthropic prediction: with this cut-off, most observers will see a positive Λ, and something of order 10% of observers will see Λ as big as we do, or smaller.

So is the anthropic explanation the correct one? Many people find the hypothesis too radical: why postulate an infinity of universes in order to explain a detail of one of them? Certainly, if an alternative explanation for the 'why now' problem existed in the form of, e.g., a naturally successful quintessence model, one might tend to prefer that. But so far, there is no such alternative. The longer this situation persists, the more we will be forced to accept that the Universe we see can only be understood by making proper allowance for our role as observers.

Multiverse and Curvature

To some extent, curvature presents a parallel set of problems to the vacuum. There is a scale problem, in the sense that natural initial conditions might be thought to have a total

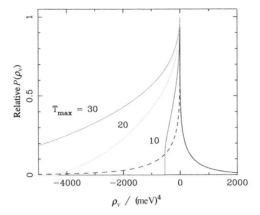

Figure 3.7 The collapse fraction as a function of the vacuum density, which is assumed to give the relative weighting of different models. The dashed line for negative density corresponds to the expanding phase only, whereas the solid lines for negative density include the recollapse phase, up to maximum temperatures of 10 K, 20 K, and 30 K.

$|\Omega - 1|$ of order unity, which would lead to a universe dominated by curvature long before today. There could also be a 'why now' problem if the present curvature was non-zero at the level of $|\Omega - 1| \sim 0.01$, which cannot currently be excluded. It is commonly assumed that inflation solves the curvature scale problem and also predicts that there is no 'why now' problem, but it is interesting to take an anthropic view of the problem. In particular, we might wonder why anthropic arguments were not applied to curvature decades ago, when many cosmologists were convinced that $\Lambda = 0$. At first sight, the issues are similar to Λ: curvature is negligible at high redshift, so we might consider a uniform prior in small early curvature values – as with Λ, appealing to the idea that zero curvature is not special. But in modern models where 'pocket' universes are formed by tunnelling, the result is an open universe, so priors on curvature might well have a discontinuity at zero (e.g., [23]). The idea of a uniform prior for curvature is therefore less well founded than it is for Λ.

In any case, the magnitude of curvature changes the situation. Decades ago, open models were seriously under consideration, and some would have argued for $\Omega_k \simeq 0.7$. This amount of curvature seriously suppresses structure formation, so an anthropic approach to explaining the density parameter in matter-only models would have yielded sensible answers. But today, we know that the Universe is flat to approximately $|\Omega_k| < 0.01$, and such a small value is suspiciously far below any anthropic upper limit. Therefore, even without the theory of inflation, we would not be attracted to anthropic explanations of the flatness problem – even though they would have worked well until perhaps the 1980s. Similarly, if we had no detection of Λ, a sufficiently strong upper limit would reject the anthropic approach, leading us to require a physical mechanism that forces $\Lambda = 0$. Anthropic reasoning is thus testable and could point to new physics. But this is not the situation we face: we have an actual detection of Λ rather than an ever-retreating upper limit, and no a priori theory predicts the observed number. An explanation in terms of anthropic selection from an ensemble matches what we see, and so far, there is no credible alternative.

More Complicated Ensembles

Weinberg's ensemble, in which all dimensionless parameters of physics are fixed at their observed values and only Λ is allowed to vary, is simple enough that we have been able to calculate its consequences in some detail. But there is no guarantee at all that variation within the multiverse is this simple, and more complex models may spoil the provisional consistency with the observed Λ. For example, Garriga and Vilenkin [24] propounded the 'Q catastrophe', in which both the normalisation of density fluctuations (Q; often also denoted δ_H) and Λ are allowed to vary. Clearly, the formation of structure is also exponentially sensitive to Q, as well as to Λ. By adopting a specific inflation model, they argued that the joint prior for Q and Λ was of the form

$$dP(\Lambda, Q) \propto Q^4 dQ \, d\Lambda \qquad (3.69)$$

and, hence, that all the anthropic weight should go to models with Q far above observation – which could, in turn, tolerate much larger values of Λ. One's response to this could be to say either that multiverse reasoning is inapplicable or that this is not the correct prior for Q (in which case we have arguably learned something about the physics of the initial conditions). But this example does emphasise the critical importance of understanding the prior, and this tends to be less easy to justify robustly than in the case of Λ.

3.7 Semi-anthropic Galaxy Formation

In Weinberg's approach to the anthropic explanation of the vacuum density, a central implicit question is the long-term efficiency of cosmic star formation. At least in simple ensembles where the ratio between the baryon and dark matter densities is held fixed, a reasonable candidate observer weighting is simply the fraction of baryons that become converted into stars. This is slightly imprecise, since some baryons may participate in star formation on multiple occasions – being recycled into new generations of stars via the process of stellar mass loss. However, Weinberg's argument hinges on an exponential suppression of structure formation when Λ is increased in value, and in this picture, one must expect that the majority of the baryon content of the Universe remains forever in a diffuse state that asymptotes towards zero density. Thus, the key question is what fraction of the baryons remain forever in this unprocessed form. This is a challenging question, given the need to account for star formation at arbitrary times in the future. There is nothing a priori special about the current time of 13.5 Gyr after the hot Big Bang phase; stars formed and could happily have hosted observers at high redshift when $t < 1\,\mathrm{Gyr}$, and similarly, it is possible to imagine a star being born a trillion years hence and hosting observers that contemplate the Universe.

Admittedly, the operation of observational cosmology will be very different in such a distant future, as the Universe becomes progressively closer to de Sitter space. There is an event horizon, beyond which causal contact with distant objects will be impossible, and the co-moving size of this is set just by the distance-redshift integral for photons that set off at the time corresponding to redshift z:

$$D_{EH} = \int_{-1}^{z} \frac{c \, dz}{H(z)} \quad (3.70)$$

(since $a \to \infty$ and $a = 1/(1+z)$, the far future corresponds to $z = -1$). At late times, H asymptotes to $H_\infty \equiv H_0/(1 - \Omega_m)^{1/2}$, so the horizon is $D_{EH} = (c/H_0)(1 - \Omega_m)^{1/2}/a$ – which, of course, has a fixed proper value. Thus, we will never lose contact with bound regions like the outer parts of the Milky Way; but galaxies in the Hubble flow will be lost to us: For $a = 2,500$, the co-moving horizon shrinks to $1 \, h^{-1}$ Mpc, excluding everything except the Local Group, so one might question whether observational cosmology would be possible then – i.e., is there an anthropic selection in time so that the very fact we are able to ask questions about the Universe as a whole makes us special observers? This is a serious question, but its importance is perhaps overstated. We are talking about the extreme long-term future: $a = 2,500$ corresponds to $t = 100 \, h^{-1}$ Gyr; also, we will be able to see objects beyond the event horizon since we are receiving light that was emitted in the past, when the event horizon was larger in co-moving terms (this applies in the present-day Universe, where every galaxy with $z \gtrsim 2.5$ is already beyond the possibility of causal contact). Quantitatively, the flux density from distant objects will be determined by the luminosity distance, $D_L = (1 + z)D$, where D is the co-moving distance and z is the redshift of the observed radiation. The redshift requires a little care since the time of observation is in the future, and so we should not use the normal formula in which $1 + z = 1/a_{emit}$; rather, we need to use the ratio of scale factors at emission and reception so that $1 + z = (1 + z_{emit})/(1 + z_{observer})$. Evaluating the integral for the co-moving distance, the result asymptotes to $D = (c/H_\infty)(z_{emit} - z_{observer})$, which itself tends to $(c/H_\infty)(1 + z_{emit})$. Thus, the observed redshift tends to $1 + z = a_{observer} D/(c/H_\infty)$, so the Universe of distant galaxies currently at $z \sim 1$ will have luminosity distances increased by a factor of roughly the future value of a compared to the present – and so would be about 10 million times fainter than at present at our illustrative $a = 2,500$. This would be challenging for future observers but not impossible, so cosmologists could operate even 100 Gyr in the future if any stars were to form then. In any case, it would be more satisfying if we could resolve the Λ question without resorting to sociological arguments, so the focus should first be on the long-term fate of star formation.

3.7.1 Direct Calculations of the Future

There is thus a strong motivation to use the most detailed modern galaxy-formation codes to estimate the long-term efficiency of star formation. The important message of the Lilly-Madau diagram is that the total co-moving density of star formation rate has declined since $z = 1$ approximately as $1/a(t)^2$, where $a(t)$ is the cosmic scale factor (e.g., [35]). The multiverse view is that this decline can be traced to the fact that we inhabit a

Λ-dominated Universe in which the gravitationally driven assembly of typical galaxies is largely suppressed by $z = 0$. This claim is to be tested by the detailed study of models for cosmological star formation in a variety of unusual contexts (see Chapter 6).

Galaxy formation requires an ability to follow the history of gas within dark matter halos, together with prescriptions for how cold gas turns into stars, followed by possible feedback of energy from the stars and from central black holes. All this must be calculated while following the hierarchical merging of dark matter halos. This can all be computed within an explicit N-body simulation, but there are disadvantages: finite resolution will mean that low-mass halos are not followed, and finite volume will limit the statistics of halos – especially at high masses. The alternative is to generate merger trees of halos via a rapid Monte Carlo algorithm: the semi-analytic approach. This is much faster than direct simulation and evades the problems of mass resolution and limited statistics – but at the price of losing the spatial relation between halos. Sudoh *et al.* [50] have applied such codes to estimate future star formation, as shown in Figure 3.8. They claim that little future star formation is permitted if Λ is raised; this is an interesting conclusion, but the calculations are only taken $\sim 10\,\mathrm{Gyr}$ beyond the present.

In principle, a more reliable alternative will be to run direct numerical simulations of 'counterfactual' universes. It is interesting to note that changing the cosmological constant at high redshift while keeping all other parameters fixed has the effect of transcribing the evolution onto another member of the ΛCDM family so that all the cosmological parameters will be different at the 'present' (i.e., when $T = 2.725\,\mathrm{K}$: we must keep this choice because the transfer function depends on the ratio of matter and radiation densities).

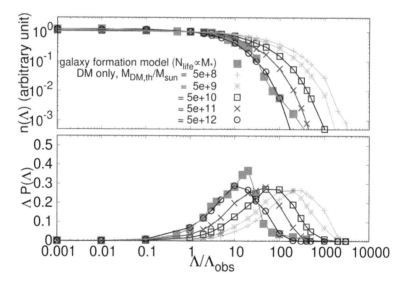

Figure 3.8 The dependence of the total stellar density on Λ, according to the semi-analytic calculations of Sudoh *et al.* [50].

Suppose we scale Λ by a factor α at some early time when the scale factor is a_i. The subsequent evolution of the Hubble parameter is $H^2 = H_i^2[(1 - \alpha\epsilon)(a_i/a)^3 + \alpha\epsilon)]$, where $\epsilon = \rho_v/\rho_{tot}$ at a_i. Note that we assume that spatial flatness is maintained, but provided ϵ is small enough, then we can neglect the $(1 - \alpha\epsilon)$ factor. For a reference standard model, $H^2 = H_{ref}^2(\Omega_{ref}a^{-3} + 1 - \Omega_{ref})$, which allows $H_i^2 a_i^3$ and $H_i^2\epsilon$ to be re-expressed in terms of the reference parameters so that the evolution of H is

$$H^2 = H_{ref}^2[\Omega_{ref}a^{-3} + \alpha(1 - \Omega_{ref})] = H_0^2[\Omega_m a^{-3} + (1 - \Omega_m)], \qquad (3.71)$$

where the last expression is ΛCDM in terms of new parameters. Thus, the new H_0 is $H_0 = H_{ref}[\Omega_{ref} + \alpha(1 - \Omega_{ref})]^{1/2}$, and the new density parameter must satisfy $(1 - 1/\Omega_m) = \alpha(1 - 1/\Omega_{ref})$. The new value of σ_8 is more complicated. If the co-moving length R is $8\,h_{ref}^{-1}$ Mpc, then $\sigma(R)$ is unchanged at high redshift, but $\sigma(R)$ is now altered at $z = 0$ because the change in Ω_m will alter the linear growth factor. Finally, because R is no longer $8\,h^{-1}$ Mpc in terms of the new h, we need to scale $\sigma(R)$ by a factor that depends on the shape of the power spectrum. The results of this exercise are shown in Figure 3.9.

The first simulation studies using this rescaling approach are now just starting to appear. Salcido *et al.* [45] used the EAGLE galaxy formation code to compute star formation $\sim 10\,$Gyr into the future and also compare ΛCDM with an Einstein-de Sitter model in which Λ is set to zero at high redshift. The results are shown in Figure 3.10, and they are intriguing. The peak in the Lilly-Madau curve is still present independent of Λ (not so surprising, as Λ is rather subdominant at the redshift of the peak); star formation may decline into the future or revive, depending on the strength of feedback. These are early days for such studies, which need to integrate for longer and consider a wider range of

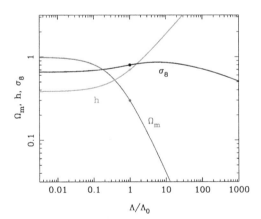

Figure 3.9 This plot shows how the principal ΛCDM parameters respond to a scaling of the cosmological constant while maintaining the high-redshift Universe otherwise unchanged and exactly flat. The 'present' is always defined as CMB temperature 2.725 K. Solid points show the default cosmology $(\Omega_m, h, \sigma_8) = (0.3, 0.7, 0.8)$, and how this scales as Λ is altered from the default value Λ_0 (which corresponds to $\Omega_v = 0.7$).

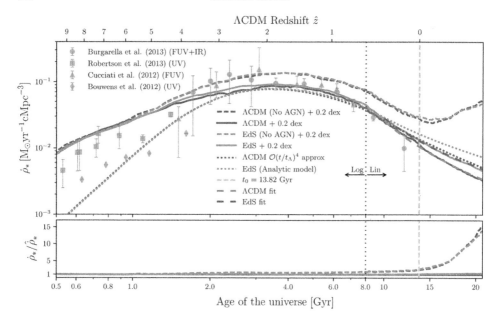

Figure 3.10 The cosmic star formation rate (Lilly-Madau diagram) as predicted from the EAGLE consortium simulations of [45].

cosmologies and possible galaxy-formation physics, but this approach has the potential to tell us much about whether observational selection really plays a part in dictating the observed value of Λ.

3.8 Outlook

The work discussed earlier represents only the first steps in exploring the practical impact of anthropic ideas in cosmology. Even the relatively simple case of Weinberg's ensemble still has much to consider: the calculations of star formation only proceed a modest distance into the future (to roughly double the present age), and a comprehensive exploration of the impact of different semi-analytic recipes remains to be performed.

More generally, the current calculations do not focus on the main issue, which is the fate of the majority of the cosmic baryons. Only about 5% of the baryons are currently in the form of stars (e.g., [22]), and the majority of the gas is perhaps not even within the virial radii of galaxy-scale halos. Based on hydrodynamical simulations, about half of the cosmic baryons are predicted to form a 'Warm-Hot Intergalactic Medium' (WHIM) that largely resides in filaments, with $T \sim 10^6$ K and over-densities of 10–30 (e.g., [15]). This diffuse gas sits in a temperature regime where it is hard to detect directly, but recently, two groups have seen this unvirialised gas via its impact on the CMB through Sunyaev-Zeldovich comptonisation [17, 52] (see Figure 3.11). Since filaments constitute a quasi-1D system, their internal density declines only as $1/a(t)$, and so the scope for this gas

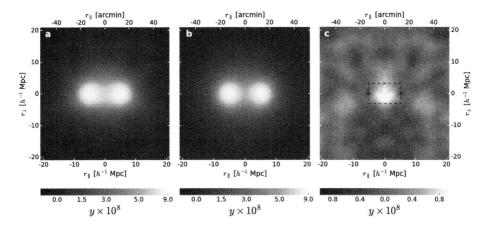

Figure 3.11 The average SZ effect of filaments, according to de Graaff *et al.* [17]. (a) the symmetrically stacked Compton y-parameter maps for 1 million close pairs of CMASS galaxies; (b) the modelled signal from the galaxy host halos only; and (c) the residual between the stacked data and model. The magnitude of the stacked signal ($y \sim 10^{-8}$) is larger than can be accounted for by gas in virialised galaxy halos, and represents the detection of the Warm-Hot Intergalactic Medium, which is where perhaps half of all baryons currently reside.

to cool and accrete onto halos in the long term needs careful exploration. Gas within the virial radius of a halo should eventually cool and form stars, provided it is not unbound by feedback, so the fraction of the WHIM that can attach itself to the deeper potential wells is the critical question in determining the asymptotic efficiency of cosmic start formation.

An interesting long-term outcome in multiverse cosmology will be the testing of specific ensembles. We do not know what physical parameters might vary, and this must be investigated empirically. The simplest ensemble, with only a variation in vacuum energy, has been claimed to predict the observed vacuum density. Once this calculation has been repeated, including realistic galaxy-formation physics for the first time, the conclusion may alter. In any case, more complex ensembles can then be investigated (see, e.g., [24, 25]), varying combinations of the dimensionless parameters that are held constant in the simplest ensemble ranging from explicitly cosmological quantities – such as the horizon-scale amplitude $\delta_H \sim 10^{-5}$, the photon-to–baryon ratio – and the baryon-to–dark matter ratio – to parameters of particle physics (coupling constants and mass ratios). In the limit, we will have the radical view is that of the string theory landscape, in which all of physics is free to vary [51]; arguably, then, there are as many as 31 dimensionless parameters that could be considered to vary [53]. But anthropic reasoning long predated the landscape [4, 10] and may outlive it. The best approach to this question is arguably an experimental one: experiment with different classes of ensemble and see which ones fail to match observation, limiting the presently uncertain physics of variation within the multiverse.

Acknowledgements

My research is currently supported by the European Research Council, under the COS-FORM grant, no. 670193.

References

[1] Abbott, T. M. C., *et al.* 2018. 'Dark Energy Survey Year 1 Results: Cosmological Constraints from Galaxy Clustering and Weak Lensing'. *Physical Review D*, **98**(4), 043526.

[2] Ade, P. A. R., *et al.*. 2016. 'Planck 2015 Results. XIII. Cosmological Parameters'. *Astronomy and Astrophysics*, **594**(Sept.), A13.

[3] Armendariz-Picon, C., Mukhanov, V., and Steinhardt, P. J. 2001. 'Essentials of *k*-Essence'. *Physical Review D*, **63**(May), 103510.

[4] Barrow, J. D., and Tipler, F. J. 1986. *The Anthropic Cosmological Principle*. Oxford University Press.

[5] Bernstein, J. and Feinberg, G. 1989. *Cosmological Constants: Papers in Modern Cosmology: A Selection of the Library Science and Astronomy Book Clubs*. Columbia University Press.

[6] Bostrom, N. 2002. *Anthropic Bias: Observation Selection Effects in Science and Philosophy*. Routledge.

[7] Burbidge, E. M., *et al.* 1957. 'Synthesis of the Elements in Stars'. *Reviews of Modern Physics*, **29**, 547–650.

[8] Caldwell, R. R., Kamionkowski, M., and Weinberg, N. N. 2003. 'Phantom Energy: Dark Energy with $w < -1$ Causes a Cosmic Doomsday'. *Physical Review Letters*, **91**(7), 071301.

[9] Carr, B. 2009. *Universe or Multiverse?* Cambridge University Press.

[10] Carter, B. 1974. 'Large Number Coincidences and the Anthropic Principle in Cosmology', in Longair, M S., ed. *Confrontation of Cosmological Theories with Observational Data*. IAU Symposium, vol. 63, Pages 291–298.

[11] Carter, B. 1983. 'The Anthropic Principle and Its Implications for Biological Evolution'. *Philosophical Transactions of the Royal Society of London Series A*, **310**(Dec.), 347–363.

[12] Clifton, T., *et al.* 2012. 'Modified Gravity and Cosmology'. *Physics Reports*, **513**(Mar.), 1–189.

[13] Copeland, E. J., Sami, M., and Tsujikawa, S. 2006. 'Dynamics of Dark Energy'. *International Journal of Modern Physics D*, **15**(Jan.), 1753–1935.

[14] Daniel, S. F., *et al.* 2010. 'Testing General Relativity with Current Cosmological Data'. *Physical Review D*, **81**(12), 123508.

[15] Davé, R., *et al.* 2001. 'Baryons in the Warm-Hot Intergalactic Medium'. *Astrophysical Journal*, **552**(May), 473–483.

[16] Dawkins, R. 2004. *A Devil's Chaplain: Selected Essays*. Phoenix.

[17] de Graaff, A., *et al.* 2017. 'Missing Baryons in the Cosmic Web Revealed by the Sunyaev-Zel'dovich Effect'. *ArXiv e-prints*, Sept., arXiv:1709.10378.

[18] Dicke, R. H., 1961. 'Dirac's Cosmology and Mach's Principle'. *Nature*, **192**(Nov.), 440–441.

[19] Dirac, P. A. M. 1937. 'The Cosmological Constants'. *Nature*, **139**(Feb.), 323.

[20] Efstathiou, G. 1995. 'An Anthropic Argument for a Cosmological Constant'. *Monthly Notices of the Royal Astronomical Society*, **274**(June), L73–L76.

[21] Einstein, A. 2006. *Kosmologische Betrachtungen zur allgemeinen Relativittstheorie*. John Wiley & Sons, Ltd. Pages 119–130.

[22] Eke, V. R., *et al.* 2005. 'Where Are the Stars?'. *Monthly Notices of the Royal Astronomical Society*, **362**(Oct.), 1233–1246.

[23] Freivogel, B., *et al.* 2006. 'Observational Consequences of a Landscape'. *Journal of High Energy Physics*, **2006**(Mar.), 039.

[24] Garriga, J., and Vilenkin, A. 2006. 'Anthropic Prediction for Λ and the Q Catastrophe'. *Progress of Theoretical Physics Supplement*, **163**(Jan.), 245–257.

[25] Garriga, J., Livio, M., and Vilenkin, A. 2000. 'Cosmological Constant and the Time of Its Dominance'. *Physical Review D*, **61**(2), 023503.

[26] Gibbons, G. W., and Turok, N. 2008. 'Measure Problem in Cosmology'. *Physical Review D*, **77**(6), 063516.

[27] Guth, A. H. 2000. 'Inflation and Eternal Inflation'. *Physics Reports*, **333**(Aug.), 555–574.

[28] Hoyle, F., *et al.* 1953. 'A State in C^{12} Predicted from Astrophysical Evidence'. *Physical Review*, **92**, 1095.

[29] Joyce, A., *et al.* 2015. 'Beyond the Cosmological Standard Model'. *Physics Reports*, **568**(Mar.), 1–98.

[30] Koksma, J. F., and Prokopec, T. 2011. 'The Cosmological Constant and Lorentz Invariance of the Vacuum State'. *ArXiv e-prints*, May, arXiv:1105.6296.

[31] Laureijs, R., *et al.* 2011. 'Euclid Definition Study Report'. *ArXiv e-prints*, Oct., arXiv:1110.3193.

[32] Liddle, A. R., and Scherrer, R. J. 1999. 'Classification of Scalar Field Potentials with Cosmological Scaling Solutions'. *Physical Review D*, **59**(2), 023509.

[33] Linde, A. 2007. 'Sinks in the Landscape, Boltzmann Brains and the Cosmological Constant Problem'. *Journal of Cosmology and Astroparticle Physics*, **1**(Jan.), 022.

[34] Linde, A., Linde, D., and Mezhlumian, A. 1994. 'From the Big Bang Theory to the Theory of a Stationary Universe'. *Physical Review D*, **49**(Feb.), 1783–1826.

[35] Mac Low, M. M., 2013. 'From Gas to Stars over Cosmic Time'. *Science*, **340**(6140).

[36] Mackay, D. J. C. 2003. *Information Theory, Inference and Learning Algorithms*. Cambridge University Press.

[37] Malquarti, M., Copeland, E. J., and Liddle, A. R. 2003. 'k-Essence and the Coincidence Problem'. *Physical Review D*, **68**(July), 023512.

[38] Marsh, D. J. E. 2016. 'Axion Cosmology'. *Physics Reports*, **643**(July), 1–79.

[39] Olum, K. D. 2002. 'The Doomsday Argument and the Number of Possible Observers'. *The Philosophical Quarterly (1950-)*, **52**(207), 164–184.

[40] Page, D. N. 2008. 'Return of the Boltzmann Brains'. *Physical Review D*, **78**(6), 063536.

[41] Peacock, J. A. 2007. 'Testing Anthropic Predictions for Λ and the Cosmic Microwave Background Temperature'. *Monthly Notices of the Royal Astronomical Society*, **379**(Aug.), 1067–1074.

[42] Percival, W. J., *et al.* 2010. 'Baryon Acoustic Oscillations in the Sloan Digital Sky Survey Data Release 7 Galaxy Sample'. *Monthly Notices of the Royal Astronomical Society*, **401**(Feb.), 2148–2168.

[43] Ratra, B., and Peebles, P. J. E. 1988. 'Cosmological Consequences of a Rolling Homogeneous Scalar Field'. *Physical Review D*, **37**(June), 3406–3427.

[44] Sakharov, A. D. 1967. 'Vacuum Quantum Fluctuations in Curved Space and the Theory of Gravitation'. *Akademiia Nauk SSSR Doklady*, **177**.

[45] Salcido, J., *et al.* 2018. 'The Impact of Dark Energy on Galaxy Formation. What Does the Future of Our Universe Hold?' *Monthly Notices of the Royal Astronomical Society*, **477**(July), 3744–3759.

[46] Schive, H. Y., Chiueh, T., and Broadhurst, T. 2014. 'Cosmic Structure as the Quantum Interference of a Coherent Dark Wave'. *Nature Physics*, **10**(July), 496–499.

[47] Simpson, F., and Peacock, J. A. 2010. 'Difficulties Distinguishing Dark Energy from Modified Gravity via Redshift Distortions'. *Physical Review D*, **81**(4), 043512.

[48] Simpson, F., *et al.* 2013. 'CFHTLenS: Testing the Laws of Gravity with Tomographic Weak Lensing and Redshift-Space Distortions'. *Monthly Notices of the Royal Astronomical Society*, **429**(Mar.), 2249–2263.

[49] Spergel, D. N., *et al.* 2007. 'Three-Year Wilkinson Microwave Anisotropy Probe (WMAP) Observations: Implications for Cosmology'. *Astrophysical Journal Supplements*, **170**(June), 377–408.

[50] Sudoh, T., *et al.* 2017. 'Testing Anthropic Reasoning for the Cosmological Constant with a Realistic Galaxy Formation Model'. *Monthly Notices of the Royal Astronomical Society*, **464**(Jan.), 1563–1568.

[51] Susskind, L. 2003 (Mar.). 'The Anthropic Landscape of String Theory', in *The Davis Meeting on Cosmic Inflation*. Page 26.

[52] Tanimura, H., *et al.* 2018. 'A Search for Warm/Hot Gas Filaments between Pairs of SDSS Luminous Red Galaxies'. *Monthly Notices of the Royal Astronomical Society*, Nov., 2970.

[53] Tegmark, M., *et al.* 2006. 'Dimensionless Constants, Cosmology, and Other Dark Matters'. *Physical Review D*, **73**(Jan.), 023505.

[54] Trotta, R. 2008. 'Bayes in the Sky: Bayesian Inference and Model Selection in Cosmology'. *Contemporary Physics*, **49**(Mar.), 71–104.

[55] Vilenkin, A. 2007. 'A Measure of the Multiverse'. *Journal of Physics A Mathematical General*, **40**(June), 6777–6785.

[56] Weinberg, S. 1989. 'The Cosmological Constant Problem'. *Reviews of Modern Physics*, **61**(Jan.), 1–23.

[57] Weinberg, S. 2000. 'A Priori Probability Distribution of the Cosmological Constant'. *Physical Review D*, **61**(10), 103505.

[58] Weller, J., and Lewis, A. M. 2003. 'Large-Scale Cosmic Microwave Background Anisotropies and Dark Energy'. *Monthly Notices of the Royal Astronomical Society*, **346**(Dec.), 987–993.

[59] Wells, J. D. 2015. 'The Utility of Naturalness, and How Its Application to Quantum Electrodynamics Envisages the Standard Model and Higgs Boson'. *Studies in History and Philosophy of Science*, **B49**, 102–108.

[60] Zel'dovich, Y. B. 1968. 'Special Issue: the Cosmological Constant and the Theory of Elementary Particles'. *Soviet Physics Uspekhi*, **11**(Mar.), 381–393.

[61] Zlatev, I., Wang, L., and Steinhardt, P. J. 1999. 'Quintessence, Cosmic Coincidence, and the Cosmological Constant'. *Physical Review Letters*, **82**(Feb.), 896–899.

4

Cosmic Inflation: Trick or Treat?

JEROME MARTIN

Abstract

Discovered almost 40 years ago, inflation has become the leading paradigm for the early Universe. Originally invented to avoid the fine-tuning puzzles of the Standard Model of cosmology, the so-called hot Big Bang phase, inflation has always been the subject of intense debates. In this chapter, we review the theoretical and observational status of inflation, discuss the criticisms that have been expressed against it, and attempt to assess whether it can be viewed as a successful solution to these issues.

4.1 Introduction

The theory of cosmic inflation was invented to solve fine-tuning problems [51, 71, 105, 106, 120–122]. Indeed, the pre-inflationary Standard Model of cosmology, the hot Big Bang model [114, 138], suffers from a number of issues, all related to a fragile adjustment of the initial conditions needed to make it work. For instance, it is well known that, in a cosmological model without inflation, when one looks at the last scattering surface (lss) where the cosmic microwave background (CMB) radiation was emitted, one looks at different causally disconnected patches of the Universe. But despite being causally disconnected, they all share approximately the same temperature. Unless one artificially fine-tunes the initial conditions, this fact is not understandable.

Soon after its advent, it was also realised that inflation provides a mechanism for structure formation [105, 106, 122]. In brief, the unavoidable vacuum quantum fluctuations of the gravitational and inflaton fields are stretched over cosmological distances by the inflationary cosmic expansion and are amplified by gravitational instability to eventually give rise to the large-scale structures observed in our Universe and to the CMB temperature anisotropy. This simple idea implies a series of remarkable predictions, among which are that the cosmological perturbations spend time outside the Hubble radius, implying the disappearance of the decaying mode and the presence of coherent oscillations in the CMB power spectrum, and that the two-point correlation function of the inflationary fluctuations should be close to scale invariance.

In 1992, the CMB anisotropies were discovered by the COsmic Background Explorer (COBE) satellite [21, 119], and this marked the beginning of a very important experimental effort by the international community to measure with a high accuracy these anisotropies in order to constrain the physics of the early Universe. This culminated recently with the publication of the Planck data, which is a cosmic variance limited experiment [1–10]. The results of these 30 years of experimental work are consistent with the predictions of single-field slow-roll inflation with a minimal kinetic term. It is worth emphasising that, in some cases, what has been confirmed are predictions and not postdictions. In particular, the prediction that the scalar spectral index should be close but not equal to 1 has been shown to be true at more than five sigmas by the Planck experiment since $n_{\rm s} = 0.9645 \pm 0.0049$ [4].

Despite these important successes and despite the fact that it has become the leading paradigm for the early Universe, inflation has always been the subject of doubts and criticisms [56, 57, 109, 110]. Soon after its invention, two main concerns were discussed: the choice of the inflationary parameters (for instance, the coupling constant in the potential) needed to match the level of CMB anisotropies, an issue related to model building and the physical nature of the inflaton field, and the question of initial conditions at the beginning of inflation. Another issue, the graceful exit or how to stop inflation, was also a hot topic but, apparently, the theory of reheating (and then preheating) gave a satisfactory answer [12, 58, 130, 133]. But the two first issues remained debated. In addition, in conjunction with the experimental efforts mentioned in the previous paragraph, various theoretical developments also took place. In particular, it was realised that single-field slow-roll models are not the only way to realise inflation, and, gradually, a large zoo of models started to appear [14, 99, 124, 137, 140]. Importantly, some of these scenarios make different predictions that single-field slow-roll inflation. For instance, the level of non-Gaussianity (NG), which is negligible for single-field slow-roll models, can be significant for a model with a non-minimal kinetic term.

Another major theoretical development is the claim that inflation can be eternal [52, 53, 70, 72, 73, 75, 129]. This is based on the fact that, due to quantum fluctuations, the various causally disconnected patches that are produced during inflation can be such that the value of the inflaton field is different from one patch to another. In particular, there can be patches where, due to quantum fluctuations, the field climbs its potential instead of rolling it down, as it does classically. And, as a consequence, this means that there are patches where inflation never stops. This idea, coupled with the concept of a string landscape, leads to the multiverse, an idea which is nowadays the subject of heated discussions.

The aims of this article are to review the present status of cosmic inflation and to assess whether it can be considered as successful given the assumptions on which it rests and given what it has achieved; for more technical details and discussion, see also Reference [34]. In particular, we discuss whether, driven out through the door, fine-tuning problems do not simply slip in again through the window under a different name. The chapter is organised as follows. In Section 4.2, we briefly present the pre-inflationary Standard Model of cosmology, namely the hot Big Bang model. We first discuss its theoretical foundations in Section 4.2.1 and then, in Section 4.2.2, how astrophysical observations can constrain it. In Section 4.3, we review the difficulties of this model – in particular, the horizon problem (see Section 4.3.1) and the flatness problem (see Section 4.3.2). In Section 4.4, we introduce inflation and discuss how it can solve the previously mentioned puzzles in Section 4.4.1. In Section 4.4.2, we study how it can be realised in practice, and we show that the presence of a scalar field dominating the energy budget of the Universe is a likely possibility. In Section 4.4.3, we present the theory of inflationary cosmological perturbations of quantum-mechanical origin, which is at the heart of the calculation of CMB anisotropy. In Section 4.4.4, we briefly review the consequences for inflation of the recently released Planck data. In Section 4.5, we discuss whether inflation is a fine-tuned scenario; in particular, we address the question of whether the choices of the parameters needed in order to have a satisfactory model of inflation is 'natural'. Then, in Section 4.6, we discuss the initial conditions at the beginning of inflation, first in a homogeneous and isotropic situation in Section 4.6.1, then in an homogeneous but anisotropic situation in Section 4.6.2, and, finally, in a general inhomogeneous situation in Section 4.6.3. We also consider the question of initial conditions for the quantum perturbations, the so-called trans-Planckian problem of inflation, in Section 4.6.4. In Section 4.7, we discuss various aspects of the multiverse question. In Section 4.7.1, we explain stochastic inflation, and in Section 4.7.2, we show how the backreaction is usually taken into account, leading to the concept of an eternal inflating Universe. In Section 4.7.3, we point out that there are models where inflation is not eternal, and in Section 4.7.4, we discuss the consequences of the possible existence of a multiverse for inflation itself. Finally, in Section 4.8, we present our conclusions.

4.2 The Standard Model of Cosmology

4.2.1 Relativistic Cosmology

Inflation is supposed to be a solution to some issues of the Standard Model of cosmology. In order to understand why this is the case, it clearly is necessary to start with a presentation of the Standard Model itself. Only after having understood its main features will it be possible to appreciate its unsatisfactory aspects.

The shape of the Universe is controlled by gravity, which – in general relativity – is described by a metric tensor $g_{\mu\nu}(x^\kappa)$. The action of the system is given by

$$S = -\frac{c^4}{16\pi G_N} \int d^4x \sqrt{-g}\,(R + 2\Lambda_B) + S_{\text{matter}}. \tag{4.1}$$

This so-called Einstein-Hilbert action involves two fundamental constants – the speed of light, $c = 3 \times 10^8$ m $\cdot$ s^{-1}, and the Newton constant, $G_N = 6.67 \times 10^{-11}$ m$^3 \cdot$ kg$^{-1} \cdot$ s^{-2}, as appropriate for a relativistic theory of the gravitational field. Quantum effects, which are controlled by the Planck constant, $\hbar = 1.05 \times 10^{-34}$ m$^2 \cdot$ kg $\cdot$ s^{-1}, are not needed to describe the dynamics of background space-time. But, as we will see, they play a fundamental role at the perturbative level. In the following, we will work in terms of natural units for which

$\hbar = c = 1$. In this system of units, everything can be expressed in terms of energy – in particular, $m_{\text{Pl}} = 1/\sqrt{G_N}$, where m_{Pl} is known as the Planck mass, $m_{\text{Pl}} \equiv \sqrt{\hbar c/G_N} = 2.17 \times 10^{-8}$kg. We will also use the reduced Planck mass, defined by $M_{\text{Pl}} \equiv m_{\text{Pl}}/\sqrt{8\pi} = 2.43 \times 10^{18}$GeV.

Let us now describe the quantities appearing in the action (4.1). g denotes the determinant of the metric tensor $g_{\mu\nu}(x^\kappa)$. $R \equiv g^{\mu\nu}R_{\mu\nu}$ is the scalar curvature, where $R_{\mu\nu} = R^\alpha{}_{\mu\alpha\nu}$ denotes the Ricci tensor, which is a contraction of the Riemann tensor. Finally, the quantity Λ_B is the bare cosmological constant. Clearly, R and, therefore, the cosmological constant Λ_B are of dimension 2, $[R] = [\Lambda_B] = 2$ (writing the natural dimension of a quantity within square brackets).

One can then obtain the equation of motion by varying the action (4.1) with respect to the metric tensor. The result reads

$$G_{\mu\nu} + \Lambda_B g_{\mu\nu} = R_{\mu\nu} - \frac{1}{2}Rg_{\mu\nu} + \Lambda_B g_{\mu\nu} = \frac{1}{M_{\text{Pl}}^2}T_{\mu\nu}, \tag{4.2}$$

where we have defined the stress-energy tensor, which describes the matter distribution responsible for the curvature of space-time by the following expression:

$$T_{\mu\nu} \equiv -\frac{2}{\sqrt{-g}}\frac{\delta S_{\text{matter}}}{\delta g^{\mu\nu}}. \tag{4.3}$$

Conservation of energy amounts to $\nabla_\alpha T^{\alpha\mu} = 0$, where ∇_α denotes the covariant derivative. Let us notice that energy conservation is compatible with the Bianchi identities, $\nabla_\alpha G^{\alpha\mu} = 0$, and that the metric tensor also has a vanishing covariant derivative. We see that the Einstein equations are a priori very complicated since they are partial, second-order, and non-linear differential equations for the metric tensor.

However, the cosmological principle states that the Universe is, on large scales, homogeneous and isotropic. Of course, this assumption is not obvious a priori and must be carefully observationally checked. We refer the reader to Reference [80], where this point is discussed in detail. Moreover, it must also be explained, rather than postulated, since it would be a bit contrived to assume that the initial state was so peculiar. We will, of course, come back to this question at length in the following sections since inflation is a scenario where this question can, in principle, be addressed. As a consequence of the cosmological principle, the metric tensor takes the Friedmann-Lemaître-Robertson-Walker (FLRW) form, namely

$$ds^2 = g_{\mu\nu}dx^\mu dx^\nu = -dt^2 + a^2(t)\gamma_{ij}^{(3)}dx^i dx^j, \tag{4.4}$$

where t is the cosmic time and x^i are space-like coordinates. The quantity $\gamma_{ij}^{(3)}$ is the metric of the three-dimensional space-like sections which have a constant scalar curvature. From (4.4), we have the relation $g_{ij} = a^2(t)\gamma_{ij}^{(3)}$. In polar coordinates, the three-dimensional metric can be written as

$$\gamma_{ij}^{(3)}dx^i dx^j = \left[\frac{dr^2}{1 - \mathcal{K}r^2} + r^2\left(d\theta^2 + \sin^2\theta d\varphi^2\right)\right], \tag{4.5}$$

while in Cartesian coordinates, it reads

$$\gamma_{ij}^{(3)} = \delta_{ij} \left[1 + \frac{\mathcal{K}}{4} \left(x^2 + y^2 + z^2 \right) \right]^{-2}.$$

(4.6)

The constant $\mathcal{K}$ describes the curvature of the space-like sections (since $^{(3)}R = 6\mathcal{K}$) and, without loss of generality, can be chosen to be $\mathcal{K} = 0, \pm 1$. As is apparent from the previous equations, there is only one unknown function left, the scale factor $a(t)$, and, moreover, this is a function of time only.

On the other hand, matter is assumed to be a collection of N perfect fluids and, as a consequence, its stress-energy tensor is given by the following expression:

$$T_{\mu\nu} = \sum_{i=1}^{i=N} T_{\mu\nu}^{(i)} = \sum_{i=1}^{i=N} \left\{ \left[\rho_i(t) + p_i(t) \right] u_\mu u_\nu + p_i(t) g_{\mu\nu} \right\},$$

(4.7)

where $\rho_i(t)$ and $p_i(t)$ are, respectively, the energy density and pressure of the fluid. The vector u_μ is the four-velocity and satisfies the relation $u_\mu u^\mu = -1$. In terms of cosmic time, this means that $u^\mu = (1,0)$ and $u_\mu = (-1,0)$. In accordance with the cosmological principle, the quantities $\rho_i(t)$ and $p_i(t)$ only depend on time. In order to close the system of equations, the relation between energy density and pressure, namely the equation of state $p_i = w_i(\rho_i)$, must also be provided.

We are now in a position to explicitly state the Einstein equations. In the case of a FLRW metric, one arrives at

$$\frac{\dot{a}^2}{a^2} + \frac{\mathcal{K}}{a^2} = \frac{1}{3M_{\rm Pl}^2} \sum_{i=1}^{N} \rho_i + \frac{\Lambda_{\rm B}}{3},$$

(4.8)

$$-\left(2\frac{\ddot{a}}{a} + \frac{\dot{a}^2}{a^2} + \frac{\mathcal{K}}{a^2} \right) = \frac{1}{M_{\rm Pl}^2} \sum_{i=1}^{N} p_i - \Lambda_{\rm B}.$$

(4.9)

We see that one has obtained an ordinary, non-linear, second-order differential equation for the scale factor $a(t)$. The fact that we now deal with ordinary differential equation is, of course, due to the cosmological principle and to the fact that the only unknown function in the metric, the scale factor, is a function of time only. Combining the two equations of motion obtained in (4.8) and (4.9), one gets an equation which gives the acceleration of the scale factor, namely

$$\frac{\ddot{a}}{a} = -\frac{1}{6M_{\rm Pl}^2} \sum_{i=1}^{N} (\rho_i + 3p_i) + \frac{1}{3}\Lambda_{\rm B}.$$

(4.10)

This equation is especially interesting because it provides the condition leading to an accelerated expansion, namely

$$\rho_{\rm T} + 3p_{\rm T} < 0,$$

(4.11)

where $\rho_T = \sum_{i=1}^{N} \rho_i$ and $p_T = \sum_{i=1}^{N} p_i$ denote the total energy density and pressure (assuming a vanishing cosmological constant or including its contribution in an extra fluid, as in (4.12) or (4.13)). Since the energy density of matter must be positive, we see that (4.11) requires a negative pressure – i.e., some exotic form of matter.

Even if the Einstein equations have been considerably simplified by the use of the cosmological principle, they remain difficult to solve analytically. However, it turns out that if the curvature term vanishes and if there is only one fluid with a constant equation of state, an exact solution to the Einstein equations is available. Of course, one can always solve these equations numerically, but exact solutions will be interesting when we discuss the puzzles of the hot Big Bang phase in the next sections. For this reason, we briefly present them. Since the equation of state is supposed to be constant, the conservation equation, which can be written as

$$\dot{\rho} + 3H(1+w)\rho = 0, \tag{4.12}$$

can be integrated exactly, and the solution reads

$$\rho(t) = \rho_f \left(\frac{a_f}{a}\right)^{3(1+w)}, \tag{4.13}$$

where ρ_f and a_f are the energy density and the scale factor expressed at a fiducial time t_f that can be chosen arbitrarily. Then, one inserts (4.13) in the Friedmann equation, namely

$$\left(\frac{1}{a}\frac{da}{dt}\right)^2 = \frac{\rho_f}{3M_{Pl}^2}\left(\frac{a_f}{a}\right)^{3(1+w)}, \tag{4.14}$$

whose solution can also be found and reads

$$\left(\frac{a}{a_f}\right)^{\frac{3(1+w)}{2}} = \frac{3(1+w)}{2}\frac{\rho_f^{1/2}}{\sqrt{3}M_{Pl}}t + C. \tag{4.15}$$

In this expression, C is an integration constant. Requiring that $a = a_f$ when $t = t_f$, one finds that $C = -3(1+w)\rho_f^{1/2}t_f/(2\sqrt{3}M_{Pl}) + 1$. Finally, noticing that $H_f = \rho_f^{1/2}/(\sqrt{3}M_{Pl})$, one arrives at

$$a(t) = a_f\left[\frac{3}{2}(1+w)H_f\left(t - t_f\right) + 1\right]^{\frac{2}{3(1+w)}}. \tag{4.16}$$

The corresponding Hubble parameter can be expressed as $H(t) = H_f/[3(1+w)H_f\left(t - t_f\right)/2 + 1]$. We notice that the scale factor vanishes when $t = t_{BB}$ with $t_{BB} = t_f - 2/[3(1+w)H_f]$. In some sense, 'time begins' at t_{BB}, and it would be meaningless to consider times such that $t < t_{BB}$. This is the famous Big Bang point where the classical analysis breaks down. This singularity is, of course, a serious problem for the hot Big Bang model. However, it is not considered to be a problem for inflation simply because inflation does not aim to address it. It could be solved if, prior to inflation, there is a bounce [19, 22] or if quantum gravitational effects take over and somehow regularise the singularity, as done, for instance, in quantum cosmology [54]. We see that the singularity problem can be treated separately and does not involve the inflationary scenario.

For future convenience, it is also interesting to rewrite the scale factor in terms of t_{BB}, and one obtains $a(t) = a_f \left[\frac{3}{2}(1+w)H_f\left(t - t_{BB}\right)\right]^{\frac{2}{3(1+w)}}$ and $H(t) = 2/[3(1+w)\left(t - t_{BB}\right)]$. If, in addition, one chooses $t_{BB} = 0$ (which can always be done), then the scale factor takes the form (with this parameterisation, $H_f = 2/\left[3(1+w)t_f\right]$)

$$a(t) = a_f \left(\frac{t}{t_f}\right)^{\frac{2}{3(1+w)}}, \tag{4.17}$$

that is to say, a power-law function. For radiation, $w = 1/3$, the scale factor behaves as $a(t) \propto t^{1/2}$ and for pressureless matter, $w = 0$, one has $a(t) \propto t^{2/3}$. We also notice that the previous expressions are ill-defined if $w = -1$. This is just because, in that case, we have an exponential solution, namely $a(t) = a_f \exp\left[H_f\left(t - t_f\right)\right]$, known as the de Sitter solution.

Putting aside the particular case $w = -1$, let us finally come back to the fact that, for $t = t_{BB}$, the scale factor vanishes. This is clearly not an artefact of the coordinate system used, as is confirmed by a calculation of the scalar curvature

$$R = \frac{4(1 - 3w)}{3(1 + w)^2} \frac{1}{\left(t - t_{BB}\right)^2}, \tag{4.18}$$

which blows up when $t \to t_{BB}$. This confirms the $t = t_{BB}$ corresponds to a real singularity.[1]

Having introduced the theoretical tools needed in order to understand the hot Big Bang model, we now discuss the parameters that describe the model and how their values can be inferred from cosmological data.

4.2.2 The Real Universe

In order to describe our Universe, we need to know its energy budget, namely the contribution of the different forms of energy density present in the Universe. Our Universe is made of photons with energy density ρ_γ, neutrinos with energy density ρ_ν, baryons with energy density ρ_b, cold dark matter with energy density ρ_c, and dark energy with energy density ρ_Λ (here assumed to be a cosmological constant). Photons and neutrinos have an equation of state $1/3$, baryons and cold dark matter have a vanishing equation of state and, finally,

---

[1] Notice also that, for radiation, R is identically zero. Of course, this does not mean that there is no singularity in a radiation-dominated epoch. This can be shown by computing another invariant – for instance, $R_{\mu\nu}R^{\mu\nu}$, which reads

$$R_{\mu\nu}R^{\mu\nu} = R_{00}R^{00} + R_{ij}R^{ij} = 9\left(\frac{\ddot{a}}{a}\right)^2 + \left(\frac{\ddot{a}}{a} + 2\frac{\dot{a}^2}{a^2}\right)^2 g_{ij}g^{ij}$$

$$= 12\left(\frac{\ddot{a}}{a}\right)^2 + 12\frac{\ddot{a}}{a}\frac{\dot{a}^2}{a^2} + 12\left(\frac{\dot{a}}{a}\right)^4$$

$$= \frac{48(3w^2 + 1)}{27(1 + w)^4}\frac{1}{\left(t - t_{BB}\right)^4}. \tag{4.19}$$

Clearly, $R_{\mu\nu}R^{\mu\nu}$ blows up as $t \to t_{BB}$ even if $w = 1/3$.

dark energy has an equation of state -1. We have, therefore, three types of fluids: radiation $\rho_{\mathrm{r}} = \rho_\gamma + \rho_\nu$, matter $\rho_{\mathrm{m}} = \rho_{\mathrm{b}} + \rho_{\mathrm{cdm}}$, and dark energy ρ_Λ. Their relative importance must be inferred from observations. In order to describe the results of those observations, it is convenient to introduce new quantities. Let us first define the critical energy density; in order to do so, we rewrite the Friedmann equation, Eq. (4.8), as

$$H^2 + \frac{\mathcal{K}}{a^2} = \frac{1}{3M_{\mathrm{Pl}}^2}\left(\rho_\Lambda + \sum_{i=1}^{i=N} \rho_i\right), \tag{4.20}$$

with $\rho_\Lambda = \Lambda_{\mathrm{B}} M_{\mathrm{Pl}}^2$ representing the vacuum energy density. We then define the critical energy density by $\rho_{\mathrm{cri}} \equiv 3H^2 M_{\mathrm{Pl}}^2$, which is clearly a time-dependent quantity. Then, the Friedmann equation can be rewritten as

$$1 + \frac{\mathcal{K}}{a^2 H^2} = \frac{\rho_{\mathrm{T}}}{\rho_{\mathrm{cri}}}, \tag{4.21}$$

where $\rho_{\mathrm{T}} = \rho_\Lambda + \sum_{i=1}^{i=N} \rho_i$ is the total energy density (compared to the definition following Eq. (4.11), we have now explicitly included the contribution of the cosmological constant in the total energy density). This means that if the spatial curvature vanishes, then $\rho_{\mathrm{T}} = \rho_{\mathrm{cri}}$, and if $\mathcal{K} > 0$ (respectively, $\mathcal{K} < 0$), then $\rho_{\mathrm{T}} > \rho_{\mathrm{cri}}$ (respectively, $\rho_{\mathrm{T}} < \rho_{\mathrm{cri}}$). One can also express the weight of a given form of matter by the quantity Ω_i defined by

$$\Omega_i \equiv \frac{\rho_i}{\rho_{\mathrm{cri}}}, \tag{4.22}$$

and, as a consequence, the Friedmann equation can be rewritten as

$$1 + \frac{\mathcal{K}}{a^2 H^2} = \Omega_\Lambda + \sum_{i=1}^{i=N} \Omega_i. \tag{4.23}$$

In particular, if the space-like sections are flat, then the sum of all the Ω_is should be 1. It follows from the previous considerations that the contributions of the different forms of energy density in our Universe are expressed through $\Omega_i^0 = \rho_i^0/\rho_{\mathrm{cri}}^0$, namely the quantity Ω_i evaluated at present time. The critical energy density today is $\rho_{\mathrm{cri}}^0 = 3H_0^2 M_{\mathrm{Pl}}^2$ with $H_0 = 100h$ km $\cdot$ s$^{-1} \cdot$ Mpc^{-1}, where h takes into account the uncertainty about H_0 (recent measurements indicate that $h \simeq 0.67$ [3]). H_0 clearly has the dimension of the inverse of a time (it is of dimension 1), and the strange units are used because of the measurement of H_0 was historically performed using the Hubble diagram [16, 61, 112, 113]. In standard units, one has $H_0 = 3.24h \times 10^{-18}$s^{-1} while, in natural units, $H_0 = 2.12h \times 10^{-42}$GeV. Therefore, we see that, by high-energy standards, the current expansion of the Universe is a low-energy phenomenon. Given the value of the reduced Planck mass, this implies that $\rho_{\mathrm{cri}}^0 \simeq 8.0990h^2 \times 10^{-47}$GeV^4.

Let us now describe the composition of our Universe. Data analysis is complicated, as it depends on which data sets are included in the analysis. For the moment, let us say that the Planck 2013 data plus the WMAP data on large-scale polarisation imply that [1, 5–7]

$$\Omega_K = -0.058^{+0.046}_{-0.026}.$$ (4.24)

If, in addition, baryonic acoustic oscillations (BAO) data are included [1, 5–7], one obtains $\Omega_K = -0.004 \pm 0.0036$. The conclusion is that everything is consistent with a vanishing spatial curvature. The photon energy density is given by $\pi^2 T_0^4/15$, where T_0 is the CMB temperature which has been measured to be $T_0 = 2.7255 \pm 0.00006$ K [40]. This implies that

$$\Omega_\gamma^0 h^2 = 2.47159 \times 10^{-5}.$$ (4.25)

In the same way, the neutrino energy density is fixed since $\rho_\nu = N_{\text{eff}}(7/8)(4/11)^{4/3}\rho_\gamma \simeq 0.68132\rho_\gamma$, with $N_{\text{eff}} = 3$. This leads to

$$\Omega_\nu^0 h^2 = 1.68394 \times 10^{-5}.$$ (4.26)

For the baryon and cold dark matter energy densities, Planck 2015 with PlanckTT, TE, EE+lowP has obtained [2–4]

$$\Omega_b^0 h^2 = 0.02225 \pm 0.00016, \quad \Omega_{\text{cdm}}^0 h^2 = 0.1198 \pm 0.0015.$$ (4.27)

Finally, since the curvature is 0, one must have $\Omega_b^0 + \Omega_{\text{cdm}}^0 + \Omega_\gamma^0 + \Omega_\nu^0 + \Omega_\Lambda^0 = 1$, from which one deduces that

$$\Omega_\Lambda h^2 = 0.306.$$ (4.28)

The previous considerations describe the current state of our Universe. The model is a six-parameter model: ρ_b, ρ_{cdm}, ρ_Λ, the optical depth τ that controls re-ionisation [49], and two parameters that describe the fluctuations, their amplitude A_s, and spectral index n_s (we discuss these two parameters in more detail in Section 4.4.3). A priori, ρ_γ and ρ_ν are also parameters, but they are usually considered as fully determined, given the precision of the measurement of the CMB temperature and given the fact that we have only three families of particles. It is impressive that with only six parameters, one can account for all the astrophysical and cosmological data.

From those numbers, using the theoretical description presented in Section 4.2.1, one can also infer the past history of the Universe. The scaling of the three different types of energy densities are given by $\rho_\gamma \propto 1/a^4$, $\rho_m \propto 1/a^3$, and ρ_Λ, is a constant. As a consequence, equality between radiation and matter occurs when

$$\left(\rho_b^0 + \rho_{\text{cdm}}^0\right)\left(\frac{a_0}{a_{\text{eq}}}\right)^3 = \left(\rho_\gamma^0 + \rho_\nu^0\right)\left(\frac{a_0}{a_{\text{eq}}}\right)^4;$$ (4.29)

that is to say,

$$1 + z_{\text{eq}} = \frac{h^2\Omega_b^0 + h^2\Omega_{\text{cdm}}^0}{h^2\Omega_\gamma^0(1 + 0.68132)} \simeq 3417,$$ (4.30)

where $z \equiv a_0/a(t) - 1$ is the redshift. In the same way, equality between pressureless matter and vacuum energy occurs at

$$1 + z_{\text{vac}} = \left(\frac{h^2 \Omega_\Lambda^0}{h^2 \Omega_b^0 + h^2 \Omega_{\text{cdm}}^0} \right)^{1/3} \simeq 1.29. \tag{4.31}$$

We thus have three different eras. In the early Universe, radiation dominated, then matter with vanishing pressure took over, and, finally, recently, the expansion of the Universe became dominated by vacuum energy. During each of these epochs, it is a good approximation to assume that the equation of state is a constant, and, therefore, the solution of the Einstein equations discussed previously – see (4.16) and (4.17) – will be very useful.

The model that we have just described – the hot Big Bang model, or, in its modern incarnation, the ΛCDM model – was the Standard Model of cosmology before the 1980s (of course, the discovery that $\Lambda_B \neq 0$ was, in fact, made later but here, we refer to the description of the Universe at very high redshifts). It is a very successful model since, with a small number of parameters, it can explain a large number of observations. Historically, three observational pillars have been the expansion of the Universe, the Big Bang nucleosynthesis (BBN) [37], and the presence of the CMB, but nowadays, the model is supported by a much larger set of observations. Nevertheless, as we are now going to explain, it possesses some undesirable features. It is not that some predictions of this model are in contradiction with the data; it is rather that the initial conditions that need to be postulated in order for the hot Big Bang model to work appear to be very unusual. In the next section, we turn to this problem.

4.3 Fine-Tuning Puzzles of the Standard Model

4.3.1 The Horizon Problem

The first puzzle that the hot Big Bang model faces is the horizon problem. As the name indicates, it is has something to do with the causality of initial conditions. A first question is when we should fix the initial conditions. A priori, this should be done at the earliest time available in the model, namely at the Big Bang. But, in practice, can we see what happens at the Big Bang? The answer is no because the Universe was opaque prior to recombination and became transparent only after ward. Recombination is the process by which free electrons and protons combine to form hydrogen atoms [108]. Before recombination, light could not propagate freely because the cross section between photons and free electrons was very large (Compton scattering). However, the cross section of photons with hydrogen atoms is much smaller, and this is the reason why the Universe became transparent after recombination. Recombination is described by the reaction $p + e^- \rightarrow H + \gamma$, which is itself controlled by the Saha equation [59]

$$\frac{1 - X_e}{X_e^2} = \frac{2\zeta(3)}{\pi^2} \eta \left(\frac{2\pi T}{m_e} \right)^{3/2} e^{B_H/T}, \tag{4.32}$$

where $X_e \equiv n_e/n_B$ with n_e representing the free electron number density and n_B representing the baryon number density. $m_e = 0.511\,\text{MeV}$ is the mass of the electron, and $B_H = m_p + m_e - m_H \simeq 13.6\,\text{eV}$, m_p being the proton mass and m_H the hydrogen atom mass, is the binding energy. Finally, $\eta \equiv n_B/n_\gamma$, where n_γ is the photons number density. If we require $X_e \simeq 0.1$, namely 90% of the free electrons have formed hydrogen atoms, then we find that $T_{\text{rec}} = 0.3\,\text{eV}$, which corresponds to $z_{\text{rec}} \simeq 1,300$. This is the furthest redshift we can reach or observe by traditional means. We see that this event takes place after equality between radiation and matter – see (4.30) – and during the matter-dominated era.

Let us now recall the definition of an horizon in cosmology. For this purpose, let us first rewrite the metric in polar coordinates; see (4.5). Assuming no spatial curvature, namely $K = 0$, one has

$$ds^2 = -dt^2 + a^2(t)\left[dr^2 + r^2\left(d\theta^2 + \sin^2\theta d\varphi^2\right)\right]. \tag{4.33}$$

The horizon problem comes from the fact that information propagates with a finite speed given by the speed of light. A photon follows a null geodesic and satisfies $ds^2 = 0$, which implies that its radial co-moving coordinate can be written as

$$r(t) = r_E - \int_{t_E}^{t} \frac{d\tau}{a(\tau)}, \tag{4.34}$$

where r_E is the co-moving radial coordinate of the source and t_E is the emission time (there is a minus sign in the equation because the 'distance' between the observer of the photon is decreasing with time as it is heading towards the telescope). Then, at time t, the proper distance is defined to be $d_p(t) = a(t)r(t)$. If, without loss of generality, we put the origin of the coordinates on Earth, then, at reception at time $t = t_R$, one has, by definition, $d_p(t_R) = 0$, which allows us to estimate the co-moving radial coordinate at emission, namely $r_E = \int_{t_E}^{t_R} d\tau/a(\tau)$. Clearly, this means that the radial coordinate of the furthest event one can, in principle, observe from Earth is obtained by taking the emission time to be the Big Bang time, namely $t_E \to 0$. This defines the size of the horizon a time t_R:

$$d_H(t_R) = a(t_R) \int_0^{t_R} \frac{d\tau}{a(\tau)}. \tag{4.35}$$

Clearly, the horizon increases as t_R increases since there is more time for light to travel, and hence, we have access to more and more remote regions of our Universe.

Then, since we have seen that recombination is the earliest event one can observe in practice, let us calculate the angular size of the horizon at that time. From the metric, we know that the apparent size D of a source is given by $D^2 = a^2(t_E)r_E^2 d\theta^2$, which implies that its angular size is given by $\delta\theta = D/[a(t_E)r_E]$. As a consequence, the angular size of the horizon at recombination (or on the lss) is given by

$$\delta\theta = \left[\int_{t_{\text{lss}}}^{t_0} \frac{d\tau}{a(\tau)}\right]^{-1} \int_0^{t_{\text{lss}}} \frac{d\tau}{a(\tau)}. \tag{4.36}$$

We see that one needs to know the behaviour of the scale factor $a(t)$ in order to carry out this calculation. Unfortunately, as was already discussed, an exact analytic solution valid at any time is not available for the hot Big Bang model. Here, a piecewise approximation, where one has several successive epochs with constant equation of state and a scale factor in each era given by (4.16), will be useful. In accordance with the preceding description of the hot Big Bang model, the first phase (phase I) is a phase dominated by radiation for which the scale factor reads $a(t) = a_i (2H_i t)^{1/2}$; see (4.17). The quantities a_i and H_i are free parameters. At $t = 0$, the scale factor vanishes, and the scalar curvature blows up; this corresponds to the Big Bang, as already discussed. The scale factor behaves according to the previous equation for times such that $0 < t < t_i$. At $t = t_i$, we assume that the behaviour of $a(t)$ changes, and for $t_i < t < t_{end}$, we assume it is given by

$$a(t) = a_i \left[\frac{3}{2} (1 + w) H_i (t - t_i) + 1 \right]^{\frac{2}{3(1+w)}} \tag{4.37}$$

(phase II), in accordance with (4.16). Notice that, here, we are using (4.16) and not (4.17). Usually, this difference is not important, but it is relevant when one considers a piecewise solution for the scale factor. The 'normalisation' of time is chosen by using $a(t) \propto t^{1/2}$ during the initial radiation-dominated era, and, then, it can no longer be modified hence the use of (4.16). The scale factor and its derivative (and, therefore, the Hubble parameter $H = \dot{a}/a$) are continuous at the transition. The quantity w is a free parameter describing the equation of state of matter during phase II. Phase II is not part of the hot Big Bang model, and we introduce it just for future convenience. If we do not want to include it in our description of the model, we just have to switch it off by taking $t_i = t_{end}$. Then, at $t = t_{end}$, phase II is over, and the radiation-dominated era starts again. This phase III has a scale factor given by $a(t) = a_{end} [2H_{end} (t - t_{end}) + 1]^{1/2}$, for times such that $t_{end} < t < t_{eq}$. The quantity a_{end} is the scale factor at $t = t_{end}$, where $a(t)$ and $H(t)$ are continuous. Again, if one switches off phase II, then there is, of course, no need to distinguish phase I and phase III. At equality between radiation and matter, at time $t = t_{eq}$, the matter-dominated era (phase IV) starts, and the scale factor can now be expressed as $a(t) = a_{eq} \left[\frac{3}{2} H_{eq} (t - t_{eq}) + 1 \right]^{2/3}$. This form is valid for times such that $t_{eq} < t < t_{de}$. Finally, at $t = t_{de}$, the phase dominated by the cosmological constant (phase V) starts, for which $a(t)$ is given by $a(t) = a_{de} e^{H_0 (t - t_{de})}$. This form is valid until the present time, for $t_{de} < t < t_0$. During this phase, the Hubble parameter is constant and given by its present value H_0. We stress again that if phase II is switched off, then the simple piecewise model exactly mimics the behaviour of $a(t)$ for the standard hot Big Bang phase.

One has then to calculate the two integrals appearing at the numerator and denominator of (4.36). This can easily be done given that the behaviour of the piecewise scale factor described previously is, during each phase, just a power-law. Straightforward manipulations lead to the following expression for the angular size of the horizon:

$$\delta\theta = \left(\frac{a_{\text{eq}}}{a_0}\right)^{1/2}\left(\frac{a_0}{a_{\text{de}}}\right)^{3/2}\left[2\left(\frac{a_{\text{lss}}}{a_{\text{eq}}}\right)^{1/2} - 1 + \frac{1-3w}{1+3w}\frac{a_{\text{end}}}{a_{\text{eq}}} - \frac{1-3w}{1+3w}\frac{a_{\text{end}}}{a_{\text{eq}}}\left(\frac{a_i}{a_{\text{end}}}\right)^{\frac{1+3w}{2}}\right]$$

$$\times\left\{2\left(\frac{a_0}{a_{\text{de}}}\right)^{3/2}\left[\left(\frac{a_{\text{de}}}{a_0}\right)^{1/2} - \left(\frac{a_{\text{lss}}}{a_0}\right)^{1/2}\right] + \frac{a_0}{a_{\text{de}}} - 1\right\}^{-1}. \tag{4.38}$$

As already emphasised, we have introduced the phase dominated by the fluid with equation of state w (i.e., the phase II) for future convenience, but in the Standard Model, this phase is absent. So we have to switch it off by assuming $a_i = a_{\text{end}}$. It is also a good approximation to take $a_0 \simeq a_{\text{de}}$ and $a_{\text{lss}} \simeq a_{\text{eq}}$. In that case, one obtains

$$\delta\theta \simeq \frac{1}{2}(1 + z_{\text{lss}})^{-1/2} \simeq 0.0138 \tag{4.39}$$

(without the simplifying assumptions $a_0 \simeq a_{\text{de}}$ and $a_{\text{lss}} \simeq a_{\text{eq}}$, one easily checks that $\delta\theta \simeq 0.0153$). This means that we should have about $40,000$ patches on the celestial sphere with completely different temperatures, meaning, a priori, with temperature fluctuations of order one. This is clearly not the case, as revealed by the impressive isotropy of the CMB; see Figure 4.1. On the Planck map, one indeed sees that the temperature anisotropy is everywhere of the order 10^{-5}.

Facing this situation, we have two options: either we say that the initial conditions were the same (meaning they were fine-tuned at the 10^{-5} level) on super-causal scales or we say that the expansion was, in the early Universe, different from that predicted by the Standard Model. The first solution corresponds to a fine-tuning (moreover on super-causal scales) while the other one corresponds to inflation. Therefore, in some sense, the concept of fine-tuning is at the heart of inflation: inflation was invented to prevent its appearance.

Figure 4.1 Map of the temperature anisotropy measured by the European Space Agency (ESA) Planck satellite. The amplitude of the anisotropy is very small, of the order of $\sim 10^{-5}$, which means that the Universe was, in fact, extremely homogeneous and isotropic on the last scattering surface. Figure taken from Reference [5].

4.3.2 The Flatness Problem

We have just discussed the horizon problem. But this problem is not the only one faced by the hot Big Bang model, and we now turn to another one, namely the flatness problem (also discussed in Reference [34]). Let us now consider (4.40) again. This equations reads

$$1 + \frac{\mathcal{K}}{a^2 H^2} = \Omega_{_{\mathrm{T}}}, \tag{4.40}$$

and we know that observations indicate that $|\Omega_{_{\mathrm{T}}}^0 - 1| < 0.01$. Clearly, this means that we live in a spatially flat Universe to a very good approximation. In the context of the Standard Model of cosmology, this is problematic. Indeed, using the Friedmann equation, one has, in general,

$$\Omega_{_{\mathrm{T}}}(t) = \frac{\sum_i \Omega_i^0 \left(\frac{a_0}{a}\right)^{3(1+w_i)}}{\sum_i \Omega_i^0 \left(\frac{a_0}{a}\right)^{3(1+w_i)} - \left(\Omega_{_{\mathrm{T}}}^0 - 1\right) \left(\frac{a_0}{a}\right)^2}. \tag{4.41}$$

In the case of the hot Big Bang model, we have seen that the Universe is made of radiation and pressureless matter. As a consequence, the preceding expression takes the form

$$\Omega_{_{\mathrm{T}}}(t) = \frac{\Omega_{\mathrm{m}}^0 \left(\frac{a_0}{a}\right)^3 + \Omega_\gamma^0 \left(\frac{a_0}{a}\right)^4}{\Omega_{\mathrm{m}}^0 \left(\frac{a_0}{a}\right)^3 + \Omega_\gamma^0 \left(\frac{a_0}{a}\right)^4 - \left(\Omega_{_{\mathrm{T}}}^0 - 1\right) \left(\frac{a_0}{a}\right)^2}. \tag{4.42}$$

Then, deep in the radiation era, this equation can be approximately expressed as

$$\Omega_{_{\mathrm{T}}}(t) \simeq 1 + \frac{\Omega_{_{\mathrm{T}}}^0 - 1}{\Omega_\gamma^0} \left(\frac{a}{a_0}\right)^2 + \cdots, \tag{4.43}$$

which implies that

$$\Omega_{_{\mathrm{T}}}^0 - 1 \simeq \Omega_\gamma^0 \left[\Omega_{_{\mathrm{T}}}(z) - 1\right] (1+z)^2 \simeq 2.47 h^{-2} \times 10^{-5} \left[\Omega_{_{\mathrm{T}}}(z) - 1\right] (1+z)^2. \tag{4.44}$$

This equation clearly shows the problem. We know as an observational fact that $|\Omega_{_{\mathrm{T}}}^0 - 1| < 0.01$. As we go backwards in time, the redshift z increases and, in order to satisfy $|\Omega_{_{\mathrm{T}}}^0 - 1| < 0.01$, $\Omega_{_{\mathrm{T}}}(z) - 1$, must be less and less. If, for instance, we evaluate $\Omega_{_{\mathrm{T}}}(z) - 1$ at BBN ($z \simeq 10^8$), we obtain $|\Omega_{_{\mathrm{T}}}^{\mathrm{BBN}} - 1| < 10^{-13} \mathcal{O} (< 0.01)$. Obviously, if we increase z (specifically consider even earlier times), this fine-tuning problem becomes even more severe. Going back all the way down to the Planck scale, one has, indeed, $|\Omega_{_{\mathrm{T}}}^{\mathrm{Pl}} - 1| < 10^{-57} \mathcal{O} (< 0.01)$. The question is then why was the Universe so flat in the early stages of its evolution? In some sense, it is like balancing a pencil on its tip for a very long time. Clearly, even a tiny fluctuation in the air (for instance) will cause the pencil to fall (more formally, this would also require us to define a measure in order to assess, in a quantitative way, how unlikely this situation is; see [34] for a full treatment of this issue). We see that, again, the question has something to do with initial conditions.

The hot Big Bang model has other puzzles, such as the presence of dangerous relics originating from phase transitions taking place in the early Universe. Rather than describing all these issues in an exhaustive way, we now turn to a possible solution, namely the theory of cosmic inflation.

4.4 Inflation

4.4.1 Solving the Standard Model Puzzles

The main idea of inflation is that the puzzles we have described in the previous sections are an indication that the dynamics of the Universe at very high redshifts was different from that implied by the hot Big Bang model. According to this model, at very high energies, the Universe was radiation dominated, with a scale factor $a(t) \propto t^{1/2}$. According to inflation, this was not the case. Let us now see how it works in practice, and let us discuss how inflation can solve the horizon problem. For this purpose, we switch on the phase dominated by the fluid with equation of state w (phase II) and rewrite Eq. (4.38) as

$$\delta\theta \simeq \frac{1}{2}(1+z_{\rm lss})^{-1/2}\left\{1+\frac{1-3w}{1+3w}\frac{a_{\rm end}}{a_{\rm lss}}\left[1-e^{-\frac{1}{2}N_{\rm T}(1+3w)}\right]\right\}, \tag{4.45}$$

where we have introduced the total number of e-folds $N_{\rm T} = \ln(a_{\rm end}/a_{\rm i})$ during phase II. The presence of phase II introduces a correction to the standard result (4.39), namely the second term in the preceding equation. If we want this correction to play a significant role, then the exponential term must be non-negligible. And this is the case if

$$1+3w < 0, \tag{4.46}$$

or, in other words, using Eq. (4.10), if the Universe was accelerating $\ddot{a} > 0$. By definition, a phase of accelerating expansion is called a phase of inflation. But having a phase of acceleration is not sufficient; we also need a phase of acceleration that lasts long enough. Indeed, requiring $\delta\theta > 2\pi$ gives $N_{\rm T} \gtrsim \ln(1+z_{\rm end})$ (here, we assume that w is not fine-tuned to $\lesssim -1/3$). If we write the energy scale at the end of inflation as $\rho_{\rm end} \simeq (10^x)^4 \, {\rm GeV}^4$, then the previous condition reduces to $N_{\rm T} \gtrsim 2.3x + 29$. For the GUT scale, namely $x = 15$, this gives $N_{\rm T} \gtrsim 63$. Therefore, one concludes that the horizon problem is solved if we have a phase of inflation and, if this phase of inflation takes place at the GUT scale, it must last more than ~ 60 e-folds. If the energy scale is lower, then we need less e-folds.

Let us now see what would be the consequence for the flatness problem. In agreement with what we have discussed before, this means that we postulate the presence of a new fluid, with an a priori unknown equation of state w. This unknown fluid dominates the energy density budget of the Universe if $t_{\rm i} < t < t_{\rm end}$, namely during phase II, and is smoothly connected to the standard Big Bang phase which takes place for $t > t_{\rm end}$. As a consequence, this implies that Eq. (4.44) can only be applied if $z < z_{\rm end}$ since $t_{\rm end}$ is the earliest time where the standard evolution is valid. In that case, one has

$$\Omega_{\rm T}^0 - 1 \simeq \Omega_\gamma^0 \left[\Omega_{\rm T}(z_{\rm end}) - 1\right](1+z_{\rm end})^2 \simeq 2.47h^{-2} \times 10^{-5}\left[\Omega_{\rm T}(z_{\rm end}) - 1\right](1+z_{\rm end})^2. \tag{4.47}$$

Now our goal is to calculate $\Omega_T(z_{end}) - 1$ in terms of $\Omega_T(z_{ini}) - 1$, namely in terms of the initial conditions at the beginning of inflation. During inflation, one has

$$\Omega_T(t) \simeq \frac{\Omega_X^{ini} \left(\frac{a_{ini}}{a}\right)^{3(1+w)}}{\Omega_X^{ini} \left(\frac{a_{ini}}{a}\right)^{3(1+w)} - \left(\Omega_T^{ini} - 1\right)\left(\frac{a_{ini}}{a}\right)^2}, \tag{4.48}$$

which implies that

$$\Omega_T(z_{end}) \simeq \frac{\Omega_X^{ini}}{\Omega_X^{ini} - \left(\Omega_T^{ini} - 1\right)\left(\frac{a_{ini}}{a_{end}}\right)^{-1-3w}}. \tag{4.49}$$

Clearly, the only way to solve the flatness problem is if inflation is such that $\Omega_T(z_{end}) \simeq 1$, and the only way to achieve it is to have $1 + 3w < 0$ – that is to say, the same condition than the one derived to solve the horizon problem; see Eq. (4.46). In that situation, the previous equation takes the form

$$\Omega_T(z_{end}) \simeq 1 - \frac{\Omega_T(z_{ini}) - 1}{\Omega_X^{ini}} e^{-N_T|1+3w|}, \tag{4.50}$$

and, as a consequence,

$$\Omega_T^0 - 1 \simeq 2.47 h^{-2} \times 10^{-5} \frac{\Omega_T(z_{ini}) - 1}{\Omega_X^{ini}} e^{-N_T|1+3w|}(1 + z_{end})^2. \tag{4.51}$$

Requiring $|\Omega_T^0 - 1| < 0.01$ without postulating that $\Omega_T(z_{ini}) - 1$ is very small, namely without postulating any fine-tuning of the initial conditions at the beginning of inflation, leads to $N_T \gtrsim \ln(1 + z_{end})$ – that is to say, again, the same condition as for the horizon problem.

We conclude that inflation can solve the fine-tuning puzzles of the Big Bang model. In addition, we mentioned before the existence of additional puzzles. One can show that inflation can also fix them. The next question is then which type of matter can produce such a phase.

4.4.2 Realising a Phase of Inflation

As explained in detail in the previous sections, a phase of accelerated expansion in the early Universe solves the puzzles of the Standard Model of cosmology. Clearly, at very high energies, the correct framework to describe matter is field theory, and its simplest version, compatible with isotropy and homogeneity, is when a scalar field dominates the energy budget of the Universe. This scalar field is called the 'inflaton'. In that case, the energy density and pressure are given by

$$\rho = \frac{\dot{\phi}^2}{2} + V(\phi), \quad p = \frac{\dot{\phi}^2}{2} - V(\phi). \tag{4.52}$$

As a consequence, if the potential energy dominates over the kinetic energy, one obtains a negative pressure and, hence, inflation. This can be achieved when the field moves slowly or, equivalently, when the potential is almost flat. From a field theory perspective, the micro-physics of inflation should therefore be described by an effective field theory characterised by a cut-off Λ. One usually assumes that the gravitational sector is described by general relativity, which itself is viewed as an effective theory with a cut-off at the Planck scale, then $\Lambda < M_{Pl}$. On the other hand, we will see that the CMB anisotropy data suggests that inflation could have taken place at energies as high as the grand unified theory (GUT) scale, and this suggests $\Lambda > 10^{15}$ GeV. Particle physics has been tested in accelerators only up to scales of $\sim$ TeV, and this implies that our freedom in building models of inflation will remain very important. A priori, without any further theoretical guidance, the effective action can therefore be written as

$$
S = \int d^4 x \sqrt{-g} \left[M_{Pl}^2 \Lambda_B + \frac{M_{Pl}^2}{2} R + a R^2 + b R_{\mu\nu} R^{\mu\nu} + \frac{c}{M_{Pl}^2} R^3 + \cdots \right.
$$
$$
- \frac{1}{2} \sum_i g^{\mu\nu} \partial_\mu \phi_i \partial_\nu \phi_i - V(\phi_1, \cdots, \phi_n) + \sum_i d_i \frac{\mathcal{O}_i}{\Lambda^{n_i - 4}} \right]
$$
$$
+ S_{int}(\phi_1, \cdots, \phi_n, A_\mu, \Psi) + \cdots \tag{4.53}
$$

In the preceding equation, the first line represents the effective Lagrangian for gravity (recall that Λ_B is the cosmological constant). In practice, we will mainly work with the Einstein-Hilbert term only. The second line represents the scalar field sector, and we have postulated that, a priori, several scalar fields are present. The first two terms represent the canonical Lagrangian while $\mathcal{O}_i$ represents a higher-order operator of dimension $n_i > 4$, the amplitude of which is determined by the coefficient d_i. Those corrections can modify the potential but also the (standard) kinetic term [32]. The last term encodes the interaction between the inflaton fields and the other fields present in nature – i.e., gauge fields A_μ and fermions Ψ. Those terms are especially important to describe how inflation ends and is connected to the Standard Model of cosmology. Finally, the dots stand for the rest of the terms such as kinetic terms of gauge bosons A_μ, fermions Ψ, etc.

Given the complexity of this Lagrangian, it is clear that it is impossible to single out a model of inflation from theoretical considerations only. However, as we will see, the CMB data have given us precious information. In particular, from the absence of non-adiabatic perturbations and from the fact that the CMB fluctuations are Gaussian, models with a single field, a minimal kinetic term and a smooth potential are favoured. This does not mean that more complicated scenarios are ruled out (as a matter of fact, they are not) but that, for the moment, they are not needed to describe the data. It is important to emphasise that we are driven to this class of models, which is clearly easier to investigate than the more complicated models mentioned before, not because we want to simplify the analysis but because this is what the CMB data suggest. Then, the Lagrangian (4.53) can be simplified to

$$
\mathcal{L} = -\frac{1}{2} g^{\mu\nu} \partial_\mu \phi \partial_\nu \phi - V(\phi) + \mathcal{L}_{int}(\phi, A_\mu, \Psi). \tag{4.54}
$$

During the accelerated phase, the interaction term is supposed to be subdominant and will be neglected. Then, only one arbitrary function remains in the Lagrangian, the potential

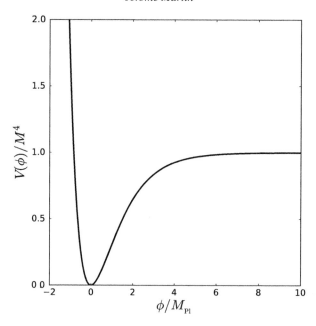

Figure 4.2 Example of a potential the Starobinsky potential (4.87) that can support inflation. Slow-roll inflation occurs along the plateau where the potential is almost flat, and the reheating phase takes place when the field oscillates around its minimum, here located at the origin.

$V(\phi)$. An example of a potential that supports inflation is given in Figure 4.2. From CMB data, one can constrain this function, and this will be discussed in the following. As already mentioned, the interaction term plays a crucial role in the process which ends inflation. Indeed, it controls how the inflaton field decays into particles describing ordinary matter. These decay products are then supposed to thermalise, and the radiation-dominated epoch starts at a temperature which is known as the reheating temperature $T_{\rm rh}$. This quantity is an important parameter of any inflationary model, and we will see that the CMB data can also say something about its value.

Following the preceding considerations, during inflation itself, the interaction term is neglected, and the evolution of the system is controlled by the Friedmann and Klein-Gordon equations, namely

$$H^2 = \frac{1}{3M_{\rm Pl}^2}\left[\frac{\dot{\phi}^2}{2} + V(\phi)\right],\tag{4.55}$$

$$\ddot{\phi} + 3H\dot{\phi} + V_\phi = 0,\tag{4.56}$$

where a subscript ϕ means a derivative with respect to the inflaton field. For an arbitrary potential, this system of equations cannot be solved analytically. This means that we have to use either numerical calculations or a perturbative method. In general, a perturbative method is based on the presence of a small parameter in the problem and on an expansion of the relevant quantities in terms of this small parameter. In the case of inflation, there exists such a small parameter which physically expresses the fact that the potential is flat. So it can be chosen as the curvature of the potential or, equivalently, as the kinetic to potential energy ratio or, given that inflation corresponds to an approximately constant Hubble parameter, as the derivative of H. Therefore, we introduce the Hubble flow functions ϵ_n defined by [67, 117]

$$\epsilon_{n+1} \equiv \frac{d \ln |\epsilon_n|}{dN}, \quad n \geq 0, \tag{4.57}$$

where $\epsilon_0 \equiv H_{\mathrm{ini}}/H$ starts the hierarchy and $N \equiv \ln(a/a_{\mathrm{ini}})$ is the number of e-folds. From the preceding expression, the first Hubble flow parameter can be written as

$$\epsilon_1 = -\frac{\dot{H}}{H^2} = 1 - \frac{\ddot{a}}{aH^2} = \frac{3\dot{\phi}^2}{2} \frac{1}{\dot{\phi}^2/2 + V(\phi)}, \tag{4.58}$$

and, therefore, inflation ($\ddot{a} > 0$) occurs if $\epsilon_1 < 1$. In terms of the Hubble flow parameters, the Friedmann and Klein-Gordon equations take the form

$$H^2 = \frac{V}{M_{\mathrm{Pl}}^2(3 - \epsilon_1)}, \tag{4.59}$$

$$\left(1 + \frac{\epsilon_2}{6 - 2\epsilon_1}\right) \frac{d\phi}{dN} = -M_{\mathrm{Pl}}^2 \frac{d \ln V}{d\phi}. \tag{4.60}$$

It is worth stressing the point that these expressions are exact. The condition $\epsilon_1 < 1$ during ~ 60 e-folds is sufficient to solve the fine-tuning problems of the Standard Model, as discussed earlier. But, if one wants to describe properly the CMB anisotropy (as in Eqs. (4.61)(4.63)), one needs $\epsilon_n \ll 1$, which is called the slow-roll regime. In this situation, the first three Hubble flow parameters can be approximated as [69]

$$\epsilon_1 \simeq \frac{M_{\mathrm{Pl}}^2}{2} \left(\frac{V_\phi}{V}\right)^2, \tag{4.61}$$

$$\epsilon_2 \simeq 2M_{\mathrm{Pl}}^2 \left[\left(\frac{V_\phi}{V}\right)^2 - \frac{V_{\phi\phi}}{V}\right], \tag{4.62}$$

$$\epsilon_2\epsilon_3 \simeq 2M_{\mathrm{Pl}}^4 \left[\frac{V_{\phi\phi\phi} V_\phi}{V^2} - 3\frac{V_{\phi\phi}}{V}\left(\frac{V_\phi}{V}\right)^2 + 2\left(\frac{V_\phi}{V}\right)^4\right]. \tag{4.63}$$

We see that the first Hubble flow parameter is also a measure of the steepness of the potential and of its first derivative. The second Hubble flow parameter is a measure of the second derivative of the potential, and so on. Therefore, if one can observationally constrain the values of the Hubble flow parameters, we can say something about the shape of the inflationary potential. The slow-roll approximation also allows us to simplify the equations of motion and to analytically integrate the inflaton trajectory. Indeed, in this regime, Eqs. (4.55) and (4.56), which control the evolution of the system, can be approximated by $H^2 \simeq V/(3M_{Pl}^2)$ and $d\phi/dN \simeq -M_{Pl}^2 d\ln V/d\phi$, from which one obtains

$$N - N_{\text{ini}} = -\frac{1}{M_{Pl}^2} \int_{\phi_{\text{ini}}}^{\phi} \frac{V(\chi)}{V_\chi(\chi)} \, d\chi, \tag{4.64}$$

ϕ_{ini} being the initial value of the inflaton. If this integral can be performed, one gets $N = N(\phi)$, and if this last equation can be inverted, one has the trajectory, $\phi = \phi(N)$.

Let us now describe the end of inflation. As already mentioned, this is the phase during which the inflaton decays into the particles of the Standard Model. During that phase, the interaction term is obviously crucial. This means that, in principle, in order to have a fair description of that process, one must specify all the interaction terms of ϕ with the other scalars, the gauge bosons, and the fermions present in the Universe together with the corresponding coupling constants. Then, one must solve the (non-linear) equations of motion of all these fields. Clearly, this is a very complicated task. However, in a cosmological context, one can proceed in a simpler way. Indeed, the reheating phase can, in fact, be described by two numbers: ρ_{reh}, the energy density at which the radiation-dominated era starts (and, therefore, at which the reheating epochs stops) and the mean equation of state $\overline{w}_{\text{reh}}$. Of course, one should also know at which energy density reheating starts, but this is not a new parameter since it is determined by the condition $\epsilon_1 = 1$. In the following, we denote this quantity ρ_{end}. Let us notice that the knowledge of ρ_{reh} is equivalent to the knowledge of the reheating temperature since

$$\rho_{\text{reh}} = g_* \frac{\pi^2}{30} T_{\text{reh}}^4, \tag{4.65}$$

where g_* encodes the number of relativistic degrees of freedom. On the other hand, the mean equation of state controls the expansion rate of the Universe during reheating. Let $\rho_T = \sum_i \rho_i$ and $p_T = \sum_i p_i$ be the total energy density and pressure, where the sum is over all the species present during reheating. Let us define the 'instantaneous' equation of state by $w_{\text{reh}} \equiv p_T/\rho_T$. Then the mean equation of state parameter, $\overline{w}_{\text{reh}}$, is given by

$$\overline{w}_{\text{reh}} \equiv \frac{1}{\Delta N} \int_{N_{\text{end}}}^{N_{\text{reh}}} w_{\text{reh}}(n) dn, \tag{4.66}$$

where $\Delta N \equiv N_{\text{reh}} - N_{\text{end}}$ is the total number of e-folds during reheating. The quantity $\overline{w}_{\text{reh}}$ allows us to determine the evolution of the total energy density since this quantity obeys

$$\rho_{\text{reh}} = \rho_{\text{end}} \, e^{-3(1+\overline{w}_{\text{reh}})\Delta N}, \tag{4.67}$$

where we recall that ρ_{end} can be determined once the model of inflation is known.

In fact, as long as the CMB is concerned, only one parameter can be constrained, and this parameter is a combination of ρ_{reh} and $\overline{w}_{\text{reh}}$. It is known as the reheating parameter and is defined by

$$R_{\text{rad}} \equiv \left(\frac{\rho_{\text{reh}}}{\rho_{\text{end}}}\right)^{(1-3\overline{w}_{\text{reh}})/(12+12\overline{w}_{\text{reh}})}. \tag{4.68}$$

The justification for this definition can be found in References [84, 92, 93, 100, 101], but a simple argument shows that it makes sense. It is clear that one cannot make the difference between a model of instantaneous reheating where $\rho_{end} = \rho_{reh}$ and a model where reheating proceeds with a mean equation of state of radiation, namely $\overline{w}_{reh} = 1/3$, since, in this last case, reheating cannot be distinguished from the subsequent radiation-dominated era. We see in the preceding definition that, in both cases, the reheating parameters have the same numerical value, $R_{rad} = 1$, which is consistent.

It may come as a surprise that a very complicated phenomenon such as reheating can be described by only one number. But one should keep in mind that this is the case only if one tries to constrain reheating from the CMB, or, to put it differently, the reheating parameter is the only quantity that can be measured if one uses CMB data. Moreover, this is not a new situation. This is indeed very similar to what happens for re-ionisation [49] for instance. Clearly, re-ionisation is, from a particle physics point of view, a very complicated process. But despite this complexity, as long as one considers CMB data only, it is described by one quantity, the optical depth τ [49].

4.4.3 Inflationary Cosmological Perturbations

So far, we have described the background space-time during inflation. We now turn to the perturbations [81–83, 107]. As is well known, this is a crucial part of the inflationary theory since it gives a convincing explanation for the origin of the large-scale structures observed in our Universe. However, in order to deal with this question, one must go beyond homogeneity and isotropy, which is a complicated task. But we know that, in the early Universe, the deviations from the cosmological principle were small, as revealed – for instance – by the magnitude of the CMB anisotropy $\delta T/T \sim 10^{-5}$. During inflation, we expect the fluctuations to be even smaller since they grow with time according to the mechanism of gravitational collapse. This means that we can treat the inhomogeneities perturbatively and, in fact, restrict ourselves to linear perturbations. Then, the idea is to write the metric tensor as $g_{\mu\nu}(\eta, \boldsymbol{x}) = g_{\mu\nu}^{\text{FLRW}}(\eta) + \delta g_{\mu\nu}(\eta, \boldsymbol{x}) + \cdots$, where $g_{\mu\nu}^{\text{FLRW}}(\eta)$ represents the metric tensor of the FLRW Universe (see Eq. (4.4)), and where $\delta g_{\mu\nu}(\eta, \boldsymbol{x}) \ll g_{\mu\nu}^{\text{FLRW}}(\eta)$. Here, η is the conformal time, related to the cosmic time by $d\eta = a dt$. In the same way, the inflaton field is expanded as $\phi(\eta, \boldsymbol{x}) = \phi^{\text{FLRW}}(\eta)z + \delta\phi(\eta, \boldsymbol{x})$, with $\delta\phi(\eta, \boldsymbol{x}) \ll \phi^{\text{FLRW}}(\eta)$. In fact, $\delta g_{\mu\nu}(\eta, \boldsymbol{x})$ can be expressed in terms of three types of perturbations – scalar, vector, and tensor. In the context of inflation, only scalar and tensor are important. Scalar perturbations are directly coupled to the perturbed scalar field $\delta\phi(\eta, \boldsymbol{x})$ while tensor fluctuations represent gravity waves. The equations of motion of each type of fluctuations are given by the perturbed Einstein equations, namely $\delta G_{\mu\nu} = \delta T_{\mu\nu}$. But we also need to specify the initial conditions. A crucial assumption of inflation is that the source of the perturbations are the unavoidable quantum vacuum fluctuations of the gravitational and scalar fields. It is clear that this has drastic implications: it means that the large-scale structures in the Universe are nothing but quantum fluctuations made classical and stretched to cosmological scales.

Let us now turn to a quantitative characterisation of the cosmological fluctuations. The amplitude of scalar perturbations is described by the curvature perturbations [18, 94] $\zeta(\eta, x) \equiv \Phi + 2(\mathcal{H}^{-1}\Phi' + \Phi)/(3 + 3w)$, with $w = p/\rho$ the equation of state during inflation and Φ the Bardeen potential [17] (not to be confused with the scalar field ϕ). The Bardeen potential is the quantity that describes scalar perturbations as revealed by writing explicitly the perturbed metric in longitudinal gauge, $ds^2 = a^2(\eta)[-(1-2\Phi)d\eta^2 + (1 - 2\Phi)\delta_{ij}dx^i dx^j]$. Since we deal with a linear theory, we can go to Fourier space and follow the time evolution of the Fourier component $\zeta_k(\eta)$. Then, the properties of the fluctuations are described by the power spectrum of scalar perturbations, which is given by

$$\mathcal{P}_\zeta(k) = \frac{k^3}{2\pi^2}|\zeta_k|^2. \tag{4.69}$$

The power spectrum depends on the model of inflation – that is to say, for the simple class of models discussed here, on the potential $V(\phi)$. Unfortunately, there exists no exact analytic calculation of $\mathcal{P}_\zeta(k)$ for an arbitrary $V(\phi)$. Therefore, one must rely on either numerical calculations or perturbative methods. Here again, the slow-roll approximation can be used, and it leads to the following result [67]

$$\mathcal{P}_\zeta(k) = \mathcal{P}_{\zeta 0}(k_{\mathrm{p}}) \left[a_0^{(\mathrm{S})} + a_1^{(\mathrm{S})} \ln\left(\frac{k}{k_{\mathrm{p}}}\right) + \frac{a_2^{(\mathrm{S})}}{2} \ln^2\left(\frac{k}{k_{\mathrm{p}}}\right) + \cdots \right], \tag{4.70}$$

where k_{p} is a pivot scale, and the overall amplitude can be written as

$$\mathcal{P}_{\zeta 0} = \frac{H_*^2}{8\pi^2 \epsilon_{1*} M_{\mathrm{Pl}}^2}. \tag{4.71}$$

In the preceding expression (and in the subsequent ones), a star means that the corresponding quantity has been evaluated at the time at which the pivot scale crossed out the Hubble radius during inflation, namely $k_{\mathrm{p}} \sim a_* H_*$. The amplitude of the spectrum depends on (the square of) the strength of the gravitational field during inflation, which is described by the expansion rate H_*. It is also inversely proportional to the first derivative of the potential through the presence of ϵ_{1*} in the denominator. The main property of $\mathcal{P}_{\zeta 0}$ is that it is does not depend on the wave number; in other words, it is scale independent. This result represents one of the main successes of inflation since a scale-invariant power spectrum was known for a long time to be in agreement with the observations. But there is even more. We see that the scale-invariant piece of the power spectrum receives scale-dependent logarithmic corrections, the amplitudes of which are controlled by the Hubble flow parameters and given by [29–31, 33, 46, 67, 78, 102, 117],

$$a_0^{(S)} = 1 - 2\,(C+1)\,\epsilon_{1*} - C\epsilon_{2*} + \left(2C^2 + 2C + \frac{\pi^2}{2} - 5\right)\epsilon_{1*}^2$$

$$+ \left(C^2 - C + \frac{7\pi^2}{12} - 7\right)\epsilon_{1*}\epsilon_{2*} + \left(\frac{1}{2}C^2 + \frac{\pi^2}{8} - 1\right)\epsilon_{2*}^2$$

$$+ \left(-\frac{1}{2}C^2 + \frac{\pi^2}{24}\right)\epsilon_{2*}\epsilon_{3*} + \cdots, \tag{4.72}$$

$$a_1^{(S)} = -2\epsilon_{1*} - \epsilon_{2*} + 2(2C+1)\epsilon_{1*}^2 + (2C-1)\epsilon_{1*}\epsilon_{2*} + C\epsilon_{2*}^2 - C\epsilon_{2*}\epsilon_{3*} + \cdots, \tag{4.73}$$

$$a_2^{(S)} = 4\epsilon_{1*}^2 + 2\epsilon_{1*}\epsilon_{2*} + \epsilon_{2*}^2 - \epsilon_{2*}\epsilon_{3*} + \cdots, \tag{4.74}$$

$$a_3^{(S)} = \mathcal{O}(\epsilon_{n*}^3), \tag{4.75}$$

where $C \equiv \gamma_E + \ln 2 - 2 \approx -0.7296$, γ_E being the Euler constant. Since the coefficients $a_1^{(S)}$, $a_2^{(S)}$ etc.... are small (being proportional to the Hubble flow parameters), this means that the inflationary power spectrum is not exactly scale invariant but, in fact, almost scale invariant. This is the main prediction of inflation, and it was confirmed recently by the CMB Planck data. We stress that this is a prediction since it was made before it was measured. In terms of spectral index, being defined as the logarithmic derivative of $\ln \mathcal{P}_\zeta(k)$, one has

$$n_s = 1 - 2\epsilon_{1*} - \epsilon_{2*}, \tag{4.76}$$

where $n_s = 1$ corresponds to exact scale invariance. We see in this expression that the small deviations from exact scale invariance carry information about the shape of the inflationary potential since ϵ_1 and ϵ_2 respectively depend on the first and second derivative of $V(\phi)$. Therefore, an accurate measurement of the power spectrum can provide information about which version of inflation was realised in the early Universe.

We have also mentioned that gravitational waves are produced during inflation. The corresponding treatment is very similar to the one we have just described. In particular, the tensor power spectrum $\mathcal{P}_h$ can be written as

$$\mathcal{P}_h(k) = \mathcal{P}_{h0}(k_p)\left[a_0^{(T)} + a_1^{(T)}\ln\left(\frac{k}{k_p}\right) + \frac{a_2^{(T)}}{2}\ln^2\left(\frac{k}{k_p}\right) + \cdots\right], \tag{4.77}$$

with a scale-invariant overall amplitude that can be expressed as

$$\mathcal{P}_{h0} = \frac{2H_*^2}{\pi^2 M_{Pl}^2}. \tag{4.78}$$

This time, and contrary to scalar perturbations, the amplitude only depends on the Hubble parameter during inflation. This has a very important implication: if one can measure the amplitude of tensor power spectrum, then one immediately determines the expansion rate during inflation or, in other words, the energy scale of inflation. Unfortunately, the inflationary gravitational waves have not yet been detected. As for scalar perturbations, the tensor power spectrum has small scale-dependent logarithmic corrections, which can be written as [67]

$$a_0^{(T)} = 1 - 2(C+1)\epsilon_{1*} + \left(2C^2 + 2C + \frac{\pi^2}{2} - 5\right)\epsilon_{1*}^2$$

$$+ \left(-C^2 - 2C + \frac{\pi^2}{12} - 2\right)\epsilon_{1*}\epsilon_{2*} + \cdots, \tag{4.79}$$

$$a_1^{(T)} = -2\epsilon_{1*} + 2(2C+1)\epsilon_{1*}^2 - 2(C+1)\epsilon_{1*}\epsilon_{2*} + \cdots, \tag{4.80}$$

$$a_2^{(T)} = 4\epsilon_{1*}^2 - 2\epsilon_{1*}\epsilon_{2*} + \cdots, \tag{4.81}$$

$$a_3^{(T)} = \mathcal{O}(\epsilon_{n*}^3), \tag{4.82}$$

corresponding to tensor spectral index given by

$$n_T = -2\epsilon_1, \tag{4.83}$$

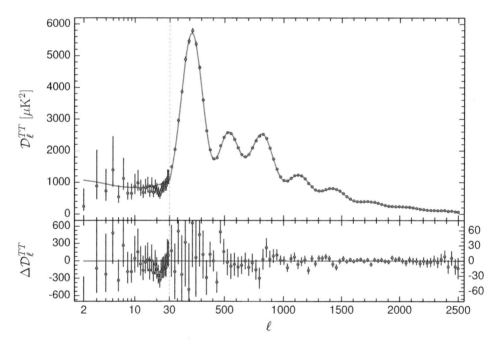

Figure 4.3 Multipole moments vs angular scale obtained from the Planck 2015 data. The multipole moments are defined from the following expression of the temperature fluctuation two-point correlation function: $\langle \delta T/T(e_1)\delta T/T(e_2)\rangle = (4\pi)^{-1}\sum_\ell (2\ell+1)C_\ell P_\ell(\cos\theta)$, where θ is the angle between the two directions e_1 and e_2. The multipole moments C_ℓ represent the power of the signal at a given spatial frequency ℓ. Notice that the quantity $\mathcal{D}_\ell$ is defined by $\mathcal{D}_\ell = \ell(\ell+1)C_\ell/(2\pi)$. The solid curve in the upper panel corresponds to the best fit in the parameter space of the ΛCDM model. This result is consistent with the predictions of inflation, for instance, because of the presence of the Doppler peaks. Figure taken from Reference [3].

an exact scale invariance corresponding, with these conventions, to $n_T = 0$ (and not one as for the scalars). Since, by the definition of inflation, one has $\epsilon_1 > 0$, this means that $n_T < 0$; i.e., we say that inflation predicts a red power spectrum (that is to say, more power on large scales) for gravitational waves. It is also interesting to measure the relative amplitude of the tensors compared to the scalars, and this is done in terms of the parameter r, defined by

$$r \equiv \frac{\mathcal{P}_h}{\mathcal{P}_\zeta} = 16\epsilon_{1*}. \qquad (4.84)$$

Clearly, since $\epsilon_{1*} \ll 1$, tensors are subdominant, which is compatible with the fact that they have not yet been detected [2, 98].

4.4.4 Constraints on Inflation

After having discussed the main features and predictions of the inflationary scenario, let us discuss what the CMB Planck data imply for inflation. The Planck data are represented in Figures 4.3, 4.4, and 4.5. As already mentioned, the most important discovery made by the

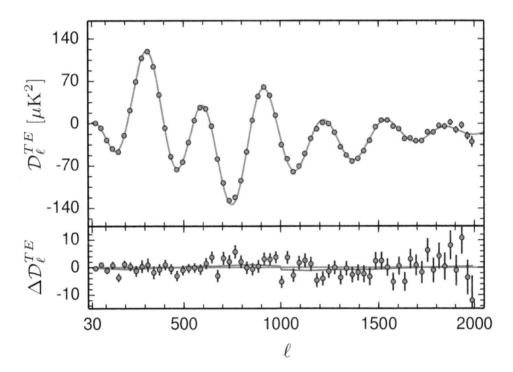

Figure 4.4 Multipole moments corresponding to the correlation between temperature and so-called E-mode polarisation anisotropies (we refer the reader to Reference [60] for definitions of polarised CMB quantities) obtained from Planck 2015. The solid curve in the top panel corresponds to prediction of the ΛCDM model obtained from the best fit in Figure 4.3 (with temperature measurements only). The lower panel shows the residual with respect to this best fit. Figure taken from Reference [3].

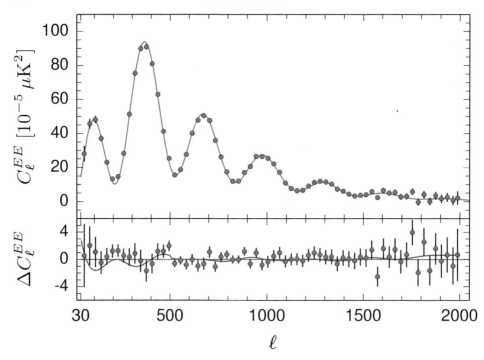

Figure 4.5 Same as in Figure 4.4 but for the E-mode power spectrum obtained from Planck 2015. Figure taken from Reference [3].

Planck satellite is probably the measurement of the scalar spectral index, which is found to be [4]

$$n_{\rm s} = 0.9645 \pm 0.0049. \tag{4.85}$$

It is a crucial result since this is the first time that a deviation from $n_{\rm s} = 1$ is measured at a statistically significant level (say, more than 5σ). It is clearly a strong point in favour of inflation. As was discussed previously, inflation also predicts the presence of a background of gravitational waves, and, unfortunately, we do not yet have a detection of those primordial gravity waves. This means that we only have an upper bound on the parameter r, namely

$$r \lesssim 0.07, \tag{4.86}$$

obtained by combining the Planck data and the BICEP/Keck data [2]. As already mentioned, the Planck data are also compatible with no non-Gaussianity [7] and no non-adiabatic perturbations [4], which is compatible with the simplest model of inflation.

One can also use the Planck data to constrain the shape of the inflationary potential. The performance of a model can be described by two numbers: the Bayesian evidence

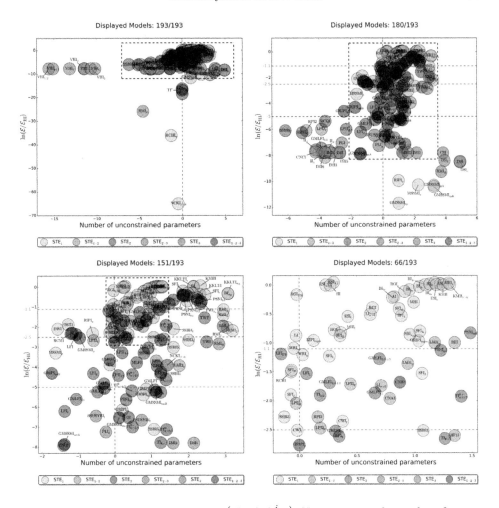

Figure 4.6 Inflationary models in the space $(N_{uc}, \ln B^i_{REF})$. N_{nuc} represents the number of unconstrained parameters of a given model while B^i_{REF} is the evidence of a given model i to evidence of a reference model ratio. Each model is represented by a circle (the radius of which has no meaning) with its acronym, taken from Reference [103], written inside. The four panels corresponds to successive zooms towards the best region (indicated by the dashed rectangles). Figures taken from Reference [103].

[131, 132], which characterises the ability of the model to fit the data in a simple way, and the Bayesian complexity [63], which is related to the number of unconstrained parameters (given a data set). A good model is a model that has a large Bayesian evidence and no unconstrained parameters. In Figure 4.6, we have represented the Bayesian evidence and complexity for nearly 200 models of inflation, given the Planck data [76, 77, 84, 92, 97–99, 103]. Based on this analysis, it is found that potentials with a plateau are favoured by the

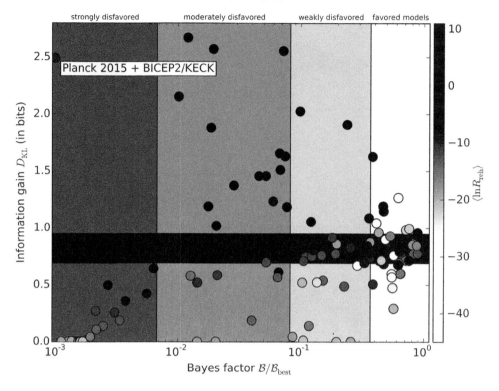

Figure 4.7 Observational constraints on reheating from the Planck and BICEP2/KECK data. The vertical axis a measure of how tight the constraint on the reheating parameter is while the horizontal axis represents the Bayesian evidence, namely the performance of a model. Each circle represents a model of inflation. The shadings give the best value of the reheating parameter.

data, the prototypical example being the Starobinsky model [121], for which the potential is given by

$$V(\phi) = M^4 \left(1 - e^{-\sqrt{2/3}\phi/M_{\mathrm{Pl}}} \right)^2 . \tag{4.87}$$

Reheating can also be constrained by means of the Planck data [84, 92, 93, 100, 101]; see Figure 4.7. We have seen that the only piece of information about the end of inflation that can be extracted from CMB data is the posterior distribution of the reheating parameter. In order to quantify whether the constraint is tight or not, one then has to compare the posterior to the prior. In technical terms, this is given by the Kullback-Leibler divergence, D_{KL}, between the prior and the posterior. In Figure 4.7, we have represented D_{KL} as a function of the Bayesian evidence for the nearly 200 models of inflation, already studied in Figure 4.6. Each model is represented by a circle. The yellow band corresponds to the one-sigma deviation around the mean value, which is given by $\langle D_{\mathrm{KL}} \rangle = 0.82 \pm 0.13$. This corresponds to the information value of almost one bit, and, therefore, this confirms that reheating is constrained by CMB data. Of course, it is not straightforward to translate these constraints into constraints on the reheating temperature unless one specifies $\overline{w}_{\mathrm{reh}}$ explicitly, in which case the reheating parameter and the reheating temperature are in one-to-one correspondence.

4.5 Is Inflation Fine-Tuned? Choosing the Free Parameters of the Inflationary Potential

In this section, we turn to the question of whether inflation, which was invented in order to solve fine-tuning problems, is itself fine-tuned. Let us discuss the first aspect of the problem, namely how the parameters of the potential must be chosen and what their numerical values are in order for the model to correctly account for the data. Let us start with a particular model, namely Large Field Model (LFI, for Large Field Inflation) for which the potential is given by

$$V(\phi) = M^4 \left(\frac{\phi}{M_{\mathrm{Pl}}}\right)^p, \tag{4.88}$$

where M and p are two free parameters. Using Eq. (4.64), one can calculate the slow-roll trajectory, and one finds

$$\phi(N) = \sqrt{\phi_{\mathrm{ini}}^2 - 2pM_{\mathrm{Pl}}^2(N - N_{\mathrm{ini}})}. \tag{4.89}$$

In order to calculate the spectral index and the scalar-to-tensor ratio, one must calculate the Hubble flow parameters. Using the expressions of ϵ_1 and ϵ_2 in the slow-roll approximation, one obtains (see Eqs. (4.61) and (4.62))

$$\epsilon_1 = \frac{p^2 M_{\mathrm{Pl}}^2}{2\phi^2}, \quad \epsilon_2 = \frac{2pM_{\mathrm{Pl}}^2}{\phi^2}. \tag{4.90}$$

This immediately leads to the vacuum expectation value at which inflation ends since the condition $\epsilon_1 = 1$ implies $\phi_{\mathrm{end}}/M_{\mathrm{Pl}} = p/\sqrt{2}$. Then, we must evaluate the Hubble flow parameters at the time that was previously denoted with a star, namely the time at which the pivot scale crossed out the Hubble radius during inflation. Using the slow-roll trajectory, it is easy to show that $\phi_*^2/M_{\mathrm{Pl}}^2 = p^2/2 + 2p\Delta N_*$, where $\Delta N_* = N_{\mathrm{end}} - N_*$ with N_{end} the number of e-folds at the end of inflation and N_* the number of e-folds at Hubble radius exit. In terms of ΔN_*, the Hubble flow parameters read

$$\epsilon_{1*} = \frac{p}{4(\Delta N_* + p/4)}, \quad \epsilon_{2*} = \frac{1}{\Delta N_* + p/4}. \tag{4.91}$$

As a consequence, one has

$$n_s - 1 = -\frac{p + 2}{2\Delta N_* + p/2}, \quad r = \frac{4p}{\Delta N_* + p/4}. \tag{4.92}$$

The measurements of n_s and the constraints on r can therefore allow us to put constraints on the parameter p. But we also see that the spectral index and the tensor-to-scalar ratio do not depend on the other free parameter, namely M. This one is, in fact, fixed by the amplitude of the fluctuations (i.e., the 'COBE normalisation') – that is to say, by the fact that $\delta T/T \sim 10^{-5}$. Using Eq. (4.71) and the slow-roll approximation for the Friedmann equation, one obtains

$$\frac{M^4}{M_{\mathrm{Pl}}^4} = 12\pi^2 p^2 \left(\frac{\phi_*}{M_{\mathrm{Pl}}}\right)^{-p-2} \mathcal{P}_{\zeta 0} = 12\pi^2 p^2 \left(\frac{p^2}{2} + 2p\Delta N_*\right)^{-p/2-1} \mathcal{P}_{\zeta 0}. \quad (4.93)$$

Knowing that the Planck 2015 data [2–4] indicate that

$$\ln\left(10^{10}\mathcal{P}_{\zeta 0}\right) = 3.094 \pm 0.0049, \quad (4.94)$$

one finds that $M/M_{\mathrm{Pl}} \simeq 1.3 \times 10^{-3}$ for $p = 2$ and $M/M_{\mathrm{Pl}} \simeq 3 \times 10^{-4}$ for $p = 4$. In order to obtain these numbers, we have assumed $\Delta N_* = 55$ and a comment is in order at this stage. In principle, one should not assume a value for ΔN_* since it is determined once the reheating temperature and the mean equation of state parameter during reheating have been chosen [84, 92, 93, 100, 101]. It can be quite dangerous to choose a 'reasonable' value blindly because, sometimes, it could imply a reheating energy density higher than the energy density at the end of inflation, which is clearly meaningless. In fact, the dependence in ΔN_* of n_s and r is precisely the reason why one can use the CMB to put constraints on the reheating epoch, as explained in the previous sections. Indeed, ΔN_* cannot take arbitrary values; otherwise, the corresponding spectral index and tensor-to-scalar ratio would be incompatible with the data. But since ΔN_* depends on T_{reh} and $\overline{w}_{\mathrm{reh}}$, this means that those quantities cannot take arbitrary values as well, or, to put it differently, they are constrained by the CMB data. Nevertheless, one can show that, for large-field inflation, ΔN_* can vary in a quite small range around the value $\Delta N_* = 55$, and this is the reason why we choose this value. Considering another value would not affect our numerical estimate much and would change nothing in the present discussion.

The estimates of the mass-scale M derived earlier show that inflation in this model takes place around the GUT scale. But let us consider the case $p = 4$ and write the potential as $V(\phi) = \lambda\phi^4$, where λ is a dimensionless coupling constant. Clearly, $\lambda = M^4/M_{\mathrm{Pl}}^4$, which implies that $\lambda \sim 10^{-13}$. This very small value can be viewed as a fine-tuning, at least if one adopts the standard lore that absence of fine-tuning means that dimensionless quantities should be 'naturally' of order one. Let us now consider the case $p = 2$ and write the corresponding potential as $V(\phi) = m^2\phi^2/2$, where m is the mass of the inflaton field. In that case, one has $m = \sqrt{2}(M/M_{\mathrm{Pl}})^2 M_{\mathrm{Pl}}$, which leads to $m \sim 2 \times 10^{-6} M_{\mathrm{Pl}}$. Is this fine-tuning? In absence of a rigorous definition of fine-tuning, this is hard to tell. But one can notice that $m/H \sim \sqrt{6}(2 + 4\Delta N_*)^{-1/2} < 1$, which may be viewed as unnatural. Indeed, we expect the mass of the inflaton to be corrected by high-energy physics according to $m^2 \rightarrow m^2 + gM^2\ln(\Lambda/\mu)$, where μ is the renormalisation scale, $M > \Lambda$ the mass of a heavy field, Λ the cut-off already discussed in Section 4.4.2, and g the coupling constant. The presence of these corrections implies $m/H \sim 1$, and keeping $m/H < 1$ may be problematic. This problem is also known as the η-problem of inflation [20]. But this at least illustrates the fact that the fine-tuning of the parameters (if any) can depend on the potential.

For this reason, it is worth studying the situation for the Starobinsky potential (4.87) since this is the favoured model.

The Starobinsky model can be derived from different assumptions. Historically, it was derived by considering R^2 corrections to the Einstein-Hilbert action. However, more recently, it was realised that it can also be viewed as a scenario in which the inflaton field is the Higgs field, this one being non-minimally coupled to gravity. In technical terms, the action of the model reads

$$S = \frac{M_{\mathrm{Pl}}^2}{2} \int d^4x \sqrt{-g} \left[\left(1 + \xi h^2\right) R - g^{\mu\nu} \partial_\mu h \partial_\nu h - 2M_{\mathrm{Pl}}^2 \frac{\lambda}{4} \left(h^2 - \frac{v^2}{M_{\mathrm{Pl}}^2}\right)^2 \right], \quad (4.95)$$

where v is the Higgs vacuum expectation value and λ the self-interacting coupling constant. The quantity ξ is a dimensionless constant which describes the non-minimal coupling. If one defines the field ϕ by $d[\phi/(\sqrt{2}M_{\mathrm{Pl}})]/dh = \sqrt{1 + \xi(1 + 6\xi)h^2}/[\sqrt{2}(1 + \xi h^2)]$ then this field has a standard Lagrangian with a potential which is exactly the potential of Eq. (4.87), the scale M being given by

$$M^4 = \frac{M_{\mathrm{Pl}}^4 \lambda}{4\xi^2}. \quad (4.96)$$

Then the COBE normalisation, which constrains the value of M, leads to

$$\xi \sim 46,000\sqrt{\lambda}, \quad (4.97)$$

where $\lambda = m_{\mathrm{H}}^2/v^2$ with $v \simeq 175\,\mathrm{GeV}$ and $m_{\mathrm{H}} \simeq 125\,\mathrm{GeV}$. We see that $\xi \gg 1$, which can imply many issues as far as the consistency of the model is concerned.

The overall picture that emerges from this section is that it is difficult to say whether the parameters of the inflationary potential are necessarily fine-tuned if one wants to account for the data. It is clear that this question is model dependent. For some potentials, the fine-tuning is present, but for others (and, in particular, those that fit the data well), it is unclear whether this is the case. The situation of the Starobinsky model is particularly interesting. The coupling between gravity and the Higgs is not small, or it is not perturbative, which may lead to technical difficulties, but this strong coupling problem is not necessarily associated with a fine-tuning problem. Here, we are just missing an objective definition of what fine-tuning is.

In fact, one could argue that such a definition exists and is nothing but the Bayesian evidence considered in Section 4.4.4. Technically, the Bayesian evidence is the integral of the likelihood over prior space, but its meaning can easily be grasped intuitively. Let us consider a model depending on, say, one free parameter. If, for all values of the parameter in the prior range, one obtains a good fit, then the Bayesian evidence is 'good'. This is, for instance, the case of the model in Figure 4.8 (left panel). Different points correspond to different values of the reheating temperature but all points are within the 1σ Planck contour. On the contrary, if one needs to tune the value of the free parameter in order to have a good fit, then the Bayesian evidence will be 'bad'. This is the case for the model

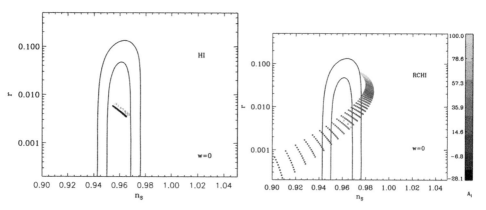

Figure 4.8 Predictions in the (n_S, r) space for two inflationary models, Higgs inflation (left panel) and Higgs inflation with quantum corrections (right panel); see Reference [99]. In both cases, a very good fit can be found, but in the case of Higgs inflation with quantum corrections, this requires a tuning of the free parameters characterising the model. As a consequence, the Bayesian evidence is smaller than that of Higgs inflation, and given the data, the model is seen as 'less good'.

in Figure 4.8 (right panel). In order to have a good compatibility with the data (i.e., points within the 1σ contour), one needs to tune the parameter A_I (which controls the amplitude of the quantum corrections), and the Bayesian evidence is 'bad'. In other words, the wasted parameter space is penalised. Obviously, the smaller the range of A_I leading to a good fit (compared to the prior), the smaller the evidence. We conclude that the evidence is a good, objective measure of fine-tuning. In this sense, the Starobinsky model is the best model because it is the less fine-tuned one.

4.6 Inflationary Initial Conditions

4.6.1 Homogeneous Initial Conditions

Let us now discuss another type of possible fine-tuning, namely the initial conditions (see also Reference [34] for a detailed discussion of this question). We have seen previously that one of the main motivations for inflation is to avoid the fine-tuning of the initial conditions that is needed in order for the Standard Model to work. If our solution to that issue were also fine-tuned, then one could wonder whether something has been gained or not. In fact, this problem has different facets. If we restrict ourselves to an homogeneous and isotropic solution, then the only question is how we should choose $\phi_{\rm ini}$ and $\dot{\phi}_{\rm ini}$. The slow-roll trajectory corresponds to $\dot{\phi}_{\rm ini} \simeq -V_\phi(\phi_{\rm ini})/[3H(\phi_{\rm ini})]$, and, therefore, there could be the worry that we have to tune the initial velocity to this value. However, this is not the case because the slow-roll trajectory is an attractor as can be seen in Figure 4.9. It is true that, for some $V(\phi)$, the corresponding basin of attraction is very small. This is, for instance, the case for Small Field Inflation (SFI). However, on the contrary, it can be very large for other models, such as Large Field Inflation (LFI) (let us also notice that the existence of

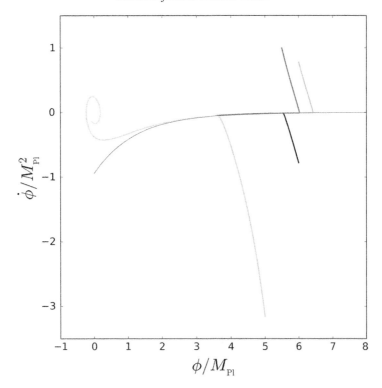

Figure 4.9 Phase space for the Starobinsky model. The central solid curve represents the slow roll trajectory while the grey lines correspond to exact trajectories with different initial conditions. It is evident from the plot that the slow-roll trajectory is an attractor in phase space.

an attractor is immune to stochastic effects; see Reference [47]). The interesting point is that it is also the case for the Starobinsky model and plateau potentials, namely the models favoured by the data. In this sense, in this restricted framework, there is no fine-tuning of the initial condition [56].

4.6.2 Anisotropic Initial Conditions

Obviously, however, the previous analysis is not entirely satisfactory. Indeed, we start from a homogeneous and isotropic situation while inflation is precisely supposed to explain why our Universe is homogeneous and isotropic. The analysis can be improved by considering that, initially, the Universe is not isotropic (but still homogeneous) [13, 127, 134]. For this purpose, let us consider the following metric (Bianchi I model):

$$ds^2 = -dt^2 + a_i^2(t)\left(dx^i\right)^2; \tag{4.98}$$

that is to say, we now have one scale factor for each space direction. This metric can also be rewritten as $ds^2 = -dt^2 + a^2(t)\gamma_{ij}dx^i dx^j$, with

$$a(t) \equiv [a_1(t)a_2(t)a_3(t)]^{1/3}, \tag{4.99}$$

and

$$\gamma_{ij} = \begin{pmatrix} e^{2\beta_1(t)} & 0 & 0 \\ 0 & e^{2\beta_2(t)} & 0 \\ 0 & 0 & e^{2\beta_3(t)} \end{pmatrix}, \tag{4.100}$$

with $\sum_{i=1}^{i=3} \beta_i = 0$. As usual, one can also introduce the conformal time η in terms of which the metric can be expressed as $ds^2 = a^2(\eta)\left(-d\eta^2 + \gamma_{ij}dx^i dx^j\right)$. Then, the next step is to introduce the shear σ_{ij}, which is defined by (as usual, a prime denotes a derivative with respect to conformal time)

$$\sigma_{ij} = \frac{1}{2}\gamma'_{ij} = \begin{pmatrix} \beta'_1 e^{2\beta_1} & 0 & 0 \\ 0 & \beta'_2 e^{2\beta_2} & 0 \\ 0 & 0 & \beta'_3 e^{2\beta_3} \end{pmatrix}. \tag{4.101}$$

Assuming that matter is described by a scalar field, it is then easy to write the Einstein equations. They read

$$3\frac{\mathcal{H}^2}{a^2} = \frac{\rho}{M_{\rm Pl}^2} + \frac{\sigma^2}{2a^2} = \frac{1}{M_{\rm Pl}^2}\left[\frac{\phi'^2}{2a^2} + V(\phi)\right] + \frac{\sigma^2}{2a^2}, \tag{4.102}$$

$$-\frac{1}{a^2}\left(\mathcal{H}^2 + 2\mathcal{H}'\right) = \frac{p}{M_{\rm Pl}^2} + \frac{\sigma^2}{2a^2} = \frac{1}{M_{\rm Pl}^2}\left[\frac{\phi'^2}{2a^2} - V(\phi)\right] + \frac{\sigma^2}{2a^2}, \tag{4.103}$$

$$\left(\sigma^i_j\right)' + 2\mathcal{H}\sigma^i_j = 0, \tag{4.104}$$

where $\sigma^2 = \sigma_{ij}\sigma^{ij} = \sum_{i=1}^{i=3} \beta_i'^2$ and $\sigma^i_j = \gamma^{ik}\sigma_{kj}$; that is to say,

$$\sigma^i_j = \begin{pmatrix} \beta'_1 & 0 & 0 \\ 0 & \beta'_2 & 0 \\ 0 & 0 & \beta'_3 \end{pmatrix}. \tag{4.105}$$

The solution for the shear can easily be found, namely $\sigma^i_j = S^i_j/a^2$, where S^i_j is a constant tensor. This implies that $\sigma^2 = S^2/a^4$, where $S^2 = S^i_j S^j_i$. As a consequence, one sees that the shear is, in fact, equivalent to a stiff fluid with an equation of state $w_\sigma \equiv p_\sigma/\rho_\sigma = 1$ and $\rho_\sigma = M_{\rm Pl}^2 S^2/(2a^6)$. Therefore, if initially the shear dominates, $\rho_\sigma \gg \rho_\phi$, the Universe will expand as $a \propto t^{1/3}$, (see Eq. (4.17)) and the expansion will not be accelerated. However, since $\rho_\sigma \propto a^{-6}$ while ρ_ϕ is approximately constant, the scalar field will eventually take over, and inflation will start. We conclude that, even if the Universe is not initially isotropic, it will become so in the presence of a scalar field whose energy density is dominated by its potential. In this sense, it is legitimate to start from an isotropic situation, as was done previously. This is clearly not a fine-tuning but, rather, an attractor of the dynamical evolution.

4.6.3 Inhomogeneous Initial Conditions

Despite the fact that taking into account the shear represents an improvement, this still does not allow us to discuss the real issue. For that, we need a framework where the initial state of the Universe is neither isotropic nor homogeneous. Technically, this is clearly very complicated since we have to solve the Einstein equations in full generality. The only way to study these questions exactly is therefore numerical relativity. However, some schemes of approximation have also been developed, and we now discuss them. Of course, the perturbative approach described before (see Section 4.4.3) is one way of taking into account the inhomogeneities. However, by definition, these fluctuations must be small while we would like to see whether inflation 'homogenises' the Universe even if it is strongly inhomogeneous initially. Another method is the so-called effective-density approximation [44, 45]. The idea is to study an inhomogeneous scalar field on a (isotropic and homogeneous) FLRW background and to add to the Friedmann equation a term which describes the backreaction of the field gradient on the geometry [44, 45]. In practice, one writes

$$\phi(t, \boldsymbol{x}) = \phi_0(t) + \Re\left[\delta\phi(t)e^{i\boldsymbol{k}\cdot\boldsymbol{x}/a(t)}\right] \tag{4.106}$$

and assumes that the corresponding Klein-Gordon equation can be split into two equations, namely

$$\ddot{\phi}_0 + 3H\dot{\phi}_0 + V_\phi(\phi_0) = 0, \tag{4.107}$$

$$\ddot{\delta\phi} + 3H\dot{\delta\phi} + \frac{k^2}{a^2}\delta\phi = 0. \tag{4.108}$$

The Friedmann equation is then written as

$$H^2 = \frac{1}{3M_{\text{Pl}}^2}\left[\frac{1}{2}\dot{\phi}_0^2 + V(\phi_0) + \frac{1}{2}\dot{\delta\phi}^2 + \frac{1}{2}\frac{k^2}{a^2}\delta\phi^2\right] - \frac{\mathcal{K}}{a^2}. \tag{4.109}$$

The wave number $\boldsymbol{k}$ should be chosen such that the wavelength of the perturbations is much smaller than the Hubble radius, namely $2\pi k/a \ll H^{-1}$. In the opposite limit, the contribution of $\delta\phi$ should just be added to the background. The energy density of the inhomogeneities $\rho_{\delta\phi} = \rho_{\dot{\delta\phi}} + \rho_\nabla$, with $\rho_{\dot{\delta\phi}} = \dot{\delta\phi}^2/2$ and $\rho_\nabla = k^2\delta\phi^2/(2a^2)$ is supposed to dominate initially (i.e., the Universe is inhomogeneous initially), $\rho_{\delta\phi} \gg \rho_{\phi_0}$. The question is whether $\rho_{\delta\phi}$ can decrease (i.e., the Universe becomes homogeneous) such that, at some point, ρ_{ϕ_0} takes over and inflation starts.

Let us now discuss the initial conditions. We take $\dot{\phi}_0$ and ϕ_0 such that, in absence of inhomogeneities, slow-roll inflation starts. Initially, the Friedmann equations can be written as

$$\frac{3a^2H^2}{k^2} \simeq \frac{1}{2}\frac{\dot{\delta\phi}^2}{M_{\text{Pl}}^2}\frac{a^2}{k^2} + \frac{1}{2}\frac{\delta\phi^2}{M_{\text{Pl}}^2}, \tag{4.110}$$

since $\rho_{\delta\phi} \gg \rho_{\phi_0}$. For simplicity, we have taken $\mathcal{K} = 0$, but it is straightforward to include the case where curvature is not vanishing. We have already mentioned that the effective density approximation is valid only if the wavelength of $\delta\phi$ is smaller than the Hubble radius.

This means that the left-hand side of Eq. (4.110) must be small. This immediately implies that $\delta\phi \ll M_{\text{Pl}}$ and $\dot{\delta\phi}^2/M_{\text{Pl}}^2 \ll k^2/a^2$ initially. Then, the corresponding solution is easily guessed: the field $\delta\phi$ oscillates and decays inversely proportional to the scale factor, namely

$$\delta\phi(t) \simeq \Re\left[\frac{\delta\phi_{\text{ini}}}{a(t)}e^{ikt/a(t)}\right]. \qquad (4.111)$$

This immediately implies that $\rho_{\delta\phi}$ behaves as radiation, namely $\rho_{\delta\phi} \propto 1/a^4$. In Figure 4.10, Eqs. (4.107), ((4.108) and (4.109) have been numerically integrated and the evolution

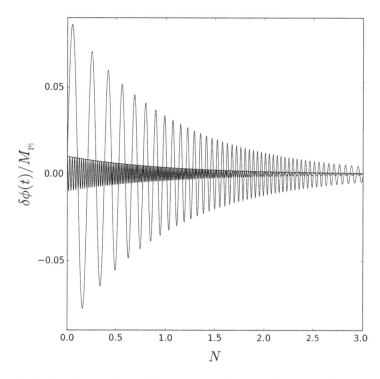

Figure 4.10 Evolution of the scalar field $\delta\phi(t)$ obtained by numerical integration of Eqs. (4.107), (4.108) and (4.109). The potential is chosen to be the Starobinsky one (see Eq. (4.87)) with a scale $M = 0.001M_{\text{Pl}}$ which, roughly speaking, matches the CMB normalisation. The initial value of the field ϕ_0 is $\phi_0 = 4M_{\text{Pl}}$ and $\dot{\phi}_0 = -V_\phi(\phi_0)/[3V(\phi_0)]$ (which is the slow-roll velocity). In absence of inhomogeneities, with these initial conditions, inflation would start and would lead to more than 60 e-folds. The initial value of $\delta\phi$ is taken to be $0.01M_{\text{Pl}}$ (and is therefore less than the Planck mass, as required; see the main text) while the initial velocity of $\delta\phi(t)$ is given by $\dot{\delta\phi}_{\text{ini}} = 0$ (central, densely oscillating curve) or $\dot{\delta\phi}_{\text{ini}} = 9 \times 10^{-5}M_{\text{Pl}}^2$ (wider oscillations). The scale k is chosen to be $k/a_{\text{ini}} = 10^{-3}$, and the curvature is given by $\mathcal{K}/a_{\text{ini}}^2 \simeq 1.36\times10^{-12}$. This implies the following Hubble parameter: $H_{\text{ini}}/M_{\text{Pl}} \simeq 3.69 \times 10^{-5}$ and $\rho_{\phi,\text{ini}} \simeq 3.33 \times 10^{-13}M_{\text{Pl}}^4$, $\rho_{\delta\phi,\text{ini}} \simeq 1.36 \times 10^{-9}M_{\text{Pl}}^4$. One easily checks that those initial conditions are such that $H_{\text{ini}}^2 a_{\text{ini}}^2/k^2 \simeq 1.36 \times 10^{-3} < 1$ and $\rho_{\phi,\text{ini}}/\rho_{\delta\phi,\text{ini}} \simeq 2.44 \times 10^{-4}$, namely the inhomogeneities largely dominate initially. Finally, the solid line represents $\delta\phi_{\text{ini}}/a = \delta\phi_{\text{ini}}e^{-N}$ for the initial conditions corresponding to the central line. We see that the envelope of the numerical solution indeed follows Eq. (4.111).

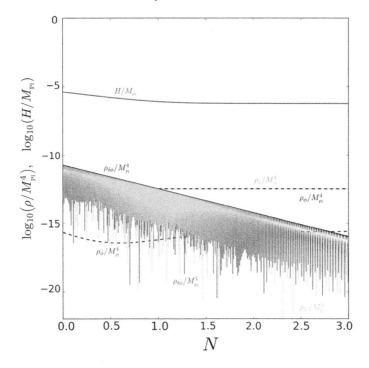

Figure 4.11 Evolution of the Hubble parameter and of the various energy densities obtained by numerical integration of Eqs. (4.107), (4.108) and (4.109). The initial conditions are those that lead to the wider curve in Figure 4.10. Initially, the Universe is strongly inhomogeneous since $\rho_{\delta\phi} \gg \rho_\phi$. However, ρ_ϕ (black line) stays approximately constant while $\rho_{\delta\phi} \propto a^{-4}$ (solid line with negative slope). As a consequence, the expansion is first radiation-dominated and then (at $N \simeq 1$ in the plot), ρ_ϕ takes over and inflation starts. Therefore, at least in this example, large inhomogeneities initially do not prevent the onset of inflation.

of $\delta\phi(t)$ is displayed and compared to Eq. (4.111)). We see that they match very well. In Figure 4.11, we have represented the corresponding energy densities. While ρ_ϕ remains constant, $\rho_{\delta\phi}$ behaves as radiation and, as a consequence, becomes very quickly subdominant. As a consequence, after a few e-folds, the Universe becomes homogeneous and inflation starts as can be seen on the evolution of the Hubble parameter (H/M_Pl) which, initially, decreases and then becomes almost constant.

The previous analysis seems to indicate that inflation does indeed homogenise the Universe. However, one should be aware of its limitations. First obviously, there is the question of the domain of validity of Eqs. (4.107), (4.108), and (4.109) and whether they can really represent a strongly inhomogeneous situation. Clearly, if $2\pi k/a \simeq H^{-1}$, this is not the case, and one has to rely on other techniques. Basically, one has two possibilities: either one obtains exact solutions [28, 111, 128], but they are very hard to find in the inhomogeneous case, or one uses numerical simulations [11, 23, 42–45, 62, 64–66]. These ones are also complicated to study since they involve full numerical relativity.

Historically, the first numerical solutions [42–45] were done under the assumption that space-time is spherically symmetric. This has the advantage to simplify the equations since they only depend on time and r, the radial coordinate. Of course, in that case, one still has to numerically solve partial differential equations. The metric considered in Reference [45] reads

$$ds^2 = -(N^2 - R^2\beta^2)dt^2 + 2R^2\beta d\chi dt + R^2(d\chi^2 + \sin^2\chi d\Omega^2),   \quad (4.112)$$

where $0 \leq \chi \leq \pi$ so that the space-like sections are closed. The lapse and shift functions N and β depend on t and r as well as the 'scale factor' R. The matter content assumed in Reference [45] is a scalar field ϕ, which is the inflaton, and another scalar field ψ without potential and playing the role of radiation. Some important technical restrictions are also postulated on the initial data. First, it is assumed that the total energy density is constant. Given an initial inhomogeneous distribution for the inflaton $\phi(\chi)$, this is achieved by choosing the initial velocity of ψ to be such that the total energy density is constant. Second, the initial momentum is taken to vanish. Based on the previous calculations (see Eqs. (4.107), (4.108), and (4.109)), it is argued in Reference [45] that, at least for large-field models, this does not restrict the significance of the results. Third, the integration is performed for values of the inflaton self-coupling that are larger than the ones necessary to CMB-normalise the model. Different initial configurations for the inflaton field are considered. In particular, the Gaussian ansatz

$$\phi_{\text{ini}}(\chi) = \phi_0 + \delta\phi \left[1 - \exp\left(-\frac{\sin^2\chi}{\Delta^2} \right) \right]   \quad (4.113)$$

was studied in details. This initial profile depends on three parameters: ϕ_0, the value of the field at the origin $\chi = 0$; $\delta\phi$, which can be viewed as the value of the field on the other side of the Universe, $\phi(\pi/2) = \phi_0 + \delta\phi(1 - e^{-1/\Delta^2})$; and Δ, which represents the width of the Gaussian.

Let us now describe the results obtained for large-field inflation. If $V(\chi = 0)$ and $V(\chi = \pi/2)$, or ϕ_0 and $\delta\phi$, are such that, in a homogeneous situation, inflation would start, then it also starts in the present case. If, on the contrary, $V(\chi = 0)$ is such that inflation would start in a homogeneous situation but not $V(\chi = \pi/2)$ (therefore, the gradients are important), then Reference [45] has shown that the outcome crucially depends on the width Δ. More precisely, the numerical simulations show that the crucial parameter is $R\Delta/H^{-1}$, which has to be large enough in order for inflation to start. Moreover, the larger the gradient, the shorter the duration of inflation. For small-field models, the sensitivity to the initial conditions is even greater.

A few years later, the analysis was improved in a significant way, and, in particular, the assumption of spherical symmetry was relaxed. Indeed, References [64–66] ran simulations of strongly inhomogeneous inflation with a three-dimensional numerical relativity code.

These simulations are such that the initial time slice has homogeneous total energy density, which means that $(\nabla\phi)^2/2 < 3M_{\text{Pl}}^2 H^2$ implying that

$$\nabla\phi < \sqrt{6}\frac{M_{\text{Pl}}}{H^{-1}}. \tag{4.114}$$

Thus, inhomogeneities that have wavelengths smaller than the Hubble radius must have a small amplitude or, to put it differently, large inhomogeneities must necessarily extend over many Hubble patches. The simulations were carried out for a quartic large field model with an initial configuration given by

$$\phi_{\text{ini}}(t_{\text{ini}}, \boldsymbol{x}) = \phi_0$$

$$+ \delta\phi \sum_{\ell,m,n=1}^{2} \frac{1}{\ell m n} \sin\left(\frac{2\pi\ell x}{L} + \theta_{x\ell}\right) \sin\left(\frac{2\pi\ell y}{L} + \theta_{ym}\right) \sin\left(\frac{2\pi\ell z}{L} + \theta_{zn}\right),$$

$$\tag{4.115}$$

where the θs are random phases. Two runs have been carried out in Reference [65], one with $L = H^{-1}$ and $\delta\phi = 0.0125m_{\text{Pl}}$ and one with $L = 32H^{-1}$ and $\delta\phi = 0.4m_{\text{Pl}}$. In both cases, one has $\phi_0 = 5m_{\text{Pl}}$ and $H_0 = 0.1m_{\text{Pl}}$. The simulations show that, in the first case, the inhomogeneities oscillate, and their amplitude is damped. At the end of the run, the inflaton field is homogeneous. But, in the second case, they do not oscillate (initially, they are larger than the Hubble radius) and are not damped.

In conclusion, it seems possible to start inflation with inhomogeneous initial conditions and to homogenise the Universe. However, admittedly, the numerical simulations that have been carried out so far all require some technical restrictions. The crucial question that emerges from the simulations is the size of the initial homogeneous patch. There is also a dependence on the model with large-field scenarios being the preferred class of scenarios. As a consequence, the Starobinsky model is (again) among the good models. Let us also notice that, even more recently, new simulations have been carried out; see References [35, 39]. These new works bring new insights into an issue that will probably be studied even more in the future.

A last comment is that we have good reasons to believe the quantum effects to play an important role at the beginning of inflation. For this reason, studying the initial conditions at only the classical level is maybe not sufficient, and even more elaborate investigations may be needed to settle this question.

4.6.4 Initial Conditions for the Perturbations

So far, we have discussed the question of the fine-tuning of the initial conditions related to the background. Obviously, there is the same question for the perturbations. We have seen that they are chosen such that the perturbations are initially placed in the vacuum state. However, if one traces back the scale of astrophysical interest today to the beginning of inflation, one notices that they correspond to physical lengths smaller than the Planck length. Clearly, in this regime, the framework used to derive the predictions of inflation, namely quantum field theory in curved space-time, is no longer valid. This is the so-called trans-Planckian problem of inflation [24–27, 68, 85, 86]. Notice that, at the same time, one has $H \ll M_{Pl}$, and, therefore, the concept of classical background is perfectly well defined. So, a priori, one could argue that the initial conditions for the perturbations are tuned in an artificial way. Then, the next question is what happens if one modifies those initial conditions: does it destroy the inflationary predictions that are so successful? To study the robustness of inflation, one can introduce ad hoc (since we do not know the theory of quantum gravity which would control the behaviour of the perturbations on scales smaller than the Planck length), but reasonable, modifications and then recompute the power spectrum of the fluctuations and see whether we obtain a result which significantly differs from the standard result. Various modifications have been proposed: a modification of the dispersion relations of the perturbations [24, 86], a modification of the commutation relations [55]. etc. However, the most general approach consists in parameterising the initial conditions of the perturbations when they emerge from the quantum foam. Let M_C be the energy scale at which the regime of quantum field theory in curved space-time breaks down (possibly the Planck scale or the string scale) [85]. A Fourier mode emerges from the quantum foam when its physical wavelength equals the length-scale associated to the scale M_C, namely

$$\lambda(\eta) = \frac{2\pi}{k} a(\eta) = \ell_c \equiv \frac{2\pi}{M_C}. \tag{4.116}$$

The initial time satisfying Eq. (4.116) is, contrary to what happens in the usual case, scale dependent. As a consequence, the corresponding power spectrum at the end of inflation is modified, and one can show that it now reads [85]

$$\mathcal{P}_\zeta(k) = \frac{H^2}{\pi \epsilon_1 m_{Pl}^2} \left\{ 1 - 2\left(C + 1\right)\epsilon_1 - C\epsilon_2 - (2\epsilon_1 + \epsilon_2)\ln\frac{k}{k_P} - 2|x|\frac{H}{M_C}\left[1 - 2(C+1)\epsilon_1\right.\right.$$

$$\left. - C\epsilon_2 - (2\epsilon_1 + \epsilon_2)\ln\frac{k}{k_P}\right]\cos\left[\frac{2M_C}{H}\left(1 + \epsilon_1 + \epsilon_1\ln\frac{k}{a_0 M_C}\right) + \varphi\right]$$

$$\left. -|x|\frac{H}{M_C}\pi\left(2\epsilon_1 + \epsilon_2\right)\sin\left[\frac{2M_C}{H}\left(1 + \epsilon_1 + \epsilon_1\ln\frac{k}{a_0 M_C}\right) + \varphi\right]\right\}. \tag{4.117}$$

This expression should be compared to Eq. (4.77). In this expression, the scale k_P is the pivot scale, and a_0 is the scale factor evaluated at the time where $k_P/a_0 = M_C$. Finally, the initial quantum state of the perturbations at the new scale-dependent initial time is characterised by a complex number x that can be written in polar form $x \equiv |x|e^{i\varphi}$, hence defining $|x|$ and φ. This power spectrum is represented in Figure 4.12.

Let us now comment on the power spectrum itself. The most obvious remark is that it is modified by the presence of super-imposed oscillations. These oscillations modify the CMB multipole moments as shown in Figure 4.13 and, therefore, have observational consequences [89–91]. The amplitude of the oscillations is, roughly speaking, given by $|x|H/M_C$, while the frequency is proportional to $(H/M_C)^{-1}$. On general grounds, we

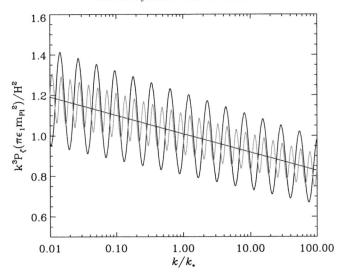

Figure 4.12 Trans-Planckian power spectra given by Eq. (4.117). The straight line corresponds to a vanilla model with $|x| = 0$ and $\epsilon_1 = 1/(2\Delta N_*)$, $\epsilon_2 = 1/\Delta N_*$ with $\Delta N_* \simeq 50$, as predicted for the $m^2\phi^2$ inflationary model. The widely oscillating line corresponds to a model with the same values for the slow-roll parameters and $H/M_C \simeq 0.002$, $|x| \simeq 50$, $\phi = 3$. Finally, the tightly oscillating line represents a model with $H/M_C \simeq 0.001$, $\phi = 2$, and the same values for the other parameters.

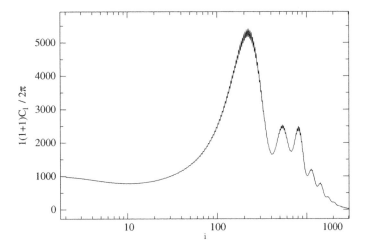

Figure 4.13 Multipole moments in presence of super-imposed trans-Planckian oscillations. Figure taken from Reference [90].

expect the ratio H/M_C to be a small number. Indeed, we know from the CMB normalisation that $H \lesssim 10^{-5} M_{Pl}$. The scale M_C is not known, but $M_C \in \left[10^{-1} M_{Pl}, 10^{-3} M_{Pl} \right]$ seems reasonable, and this implies that, at most, $H/M_C \sim 0.01$. Therefore, unless the number $|x|$ is very large, the amplitude of the oscillations is small, and one could argue that inflation is robust against trans-Planckian corrections. In this sense, assuming the vacuum state initially is not a fine-tuning. Of course, as already mentioned, $|x|$ could be large and, in this case, the modification sizable. However, the magnitude of $|x|$ is limited by the backreaction problem [26]. Physically, this is due to the fact that $|x| \neq 1$ corresponds to an excited state. But the particles present in this quantum state carry energy density, and this energy density could prevent inflation to start. Therefore, it has to be smaller than the inflationary energy density $H^2 M_{\mathrm{Pl}}^2$. One can show that this leads to an upper bound on the amplitude of the oscillations given by [89–91]

$$|x|\frac{H}{M_{\mathrm{C}}} \lesssim \frac{10^4}{\sqrt{\epsilon_1}} \left(\frac{H}{M_{\mathrm{C}}}\right)^2 \sim 4.3 \times 10^{-4}\sqrt{r}\left(\frac{M_{\mathrm{Pl}}}{M_{\mathrm{C}}}\right)^2. \tag{4.118}$$

This upper bound is not sufficient to exclude a possible detection of the oscillations in the data (although for the moment nothing has been seen). And, in this sense, one could argue that inflation is not robust to a change of the initial conditions. However, detecting the oscillations would mean opening a window on physics beyond the quantum gravity scale, clearly a fascinating possibility.

4.7 The Multiverse

4.7.1 Stochastic Inflation

The discussion of the previous section about the initial conditions misses a crucial ingredient, namely the fact that the background field is itself a quantum field. So far, the quantum effects have been taken into account but only at the perturbative level. The question is now whether they also play an important role in the evolution of the background. Classically, the inflaton field evolves according to the Klein-Gordon equation and, in the slow-roll regime, the typical variation of ϕ is then given by $\Delta\phi_{\mathrm{cl}} \simeq -V_\phi/(3H)\Delta t$. On the other hand, the amplitude of the quantum kick received by ϕ during one e-fold is, roughly speaking, of the order of the square root of the power spectrum of $\delta\phi$, namely $\Delta\phi_{\mathrm{q}} \simeq H/(2\pi)$. If $\Delta\phi_{\mathrm{q}} \gg \Delta\phi_{\mathrm{cl}}$, then quantum effects are likely to be dominant. In fact, it is easy to see that

$$\frac{\Delta\phi_{\mathrm{q}}}{\Delta\phi_{\mathrm{cl}}} = \sqrt{\mathcal{P}_{\zeta 0}}, \tag{4.119}$$

where $\mathcal{P}_{\zeta 0}$ is the amplitude of scalar perturbations; see Eq. (4.71). This equation just tells us that, when the fluctuations are of order one, quantum effects are relevant even for the background. Notice that if we want to see whether stochastic effects can modify the power spectrum of curvature perturbations, then the criterion is different; see Reference [136].

If, for instance, we consider the model $V(\phi) = M^4(\phi/M_{\mathrm{Pl}})^p$, then the condition $\Delta\phi_{\mathrm{q}} > \Delta\phi_{\mathrm{cl}}$ is equivalent to $\phi > \phi_{\mathrm{s}}$ with

$$\frac{\phi_{\mathrm{s}}}{M_{\mathrm{Pl}}} = \left[\pi p\sqrt{6}\left(\frac{M_{\mathrm{Pl}}}{M}\right)^2\right]^{\frac{2}{2+p}}. \tag{4.120}$$

Then, if one uses the expression of M given in Eq. (4.93), one arrives at

$$\frac{\phi_s}{M_{Pl}} = 2^{-\frac{1}{p+2}} \left(\frac{p^2}{2} + 2p\Delta N_*\right)^{1/2} \left(\mathcal{P}_{\zeta 0}^{Planck}\right)^{-\frac{1}{2+p}}, \qquad (4.121)$$

where $\mathcal{P}_{\zeta 0}^{Planck}$ is the amplitude of the spectrum measured by the Planck satellite; see Eq. (4.94). Using this equation, namely $\ln\left(10^{10}\mathcal{P}_{\zeta 0}\right) = 3.094 \pm 0.0049$, one obtains $\phi_s/M_{Pl} \simeq 1,743$ for the model $p = 2$ (one has taken $\Delta N_* \simeq 50$). It is also interesting to estimate the Hubble parameter for this value of the field, and one finds

$$\frac{H_s^2}{M_{Pl}^2} = 4\pi^2 p^2 \left(\frac{p^2}{2} + 2p\Delta N_*\right)^{-\frac{p}{2}-1} \mathcal{P}_{\zeta 0}^{Planck} \left(\frac{\phi_s}{M_{Pl}}\right)^p. \qquad (4.122)$$

For $p = 2$, this gives $H_s/M_{Pl} \simeq 0.005$, the important point being that we are in a regime where the quantum behaviour of the inflaton field must be taken into account but where, at the same time, the concept of a background space-time is still relevant since $H_s/M_{Pl} \ll 1$.

After these qualitative considerations, let us now try to establish more precisely the equations controlling the evolution of the system in this regime [15, 47, 79, 87, 88, 95, 123, 126, 136]. Let us first consider a quantum scalar field in a rigid de Sitter background. This means that the backreaction of the quantum scalar field is neglected or, in other words, that it is a test field living in a de Sitter space-time characterised by H. In this space-time, H^{-1} is a preferred length and can be used to distinguish between short and long wavelengths. Then one writes the scalar field according to [123, 126]:

$$\hat{\phi}(t, \boldsymbol{x}) = \hat{\phi}_{IR}(t, \boldsymbol{x}) + \frac{1}{(2\pi)^{3/2}} \int d\boldsymbol{k}\,\Theta\left(k - \sigma a H\right)\left[\mu_{\boldsymbol{k}}(t)e^{i\boldsymbol{k}\cdot\boldsymbol{x}}\hat{c}_{\boldsymbol{k}} + \mu_{\boldsymbol{k}}^*(t)e^{-i\boldsymbol{k}\cdot\boldsymbol{x}}\hat{c}_{\boldsymbol{k}}^\dagger\right],$$
$$(4.123)$$

where $\sigma \ll 1$ is a small constant. The quantity Θ is the Heaviside function, $\mu_{\boldsymbol{k}}(t)$ is the field mode function, and $\hat{c}_{\boldsymbol{k}}$ and $\hat{c}_{\boldsymbol{k}}^\dagger$ are the annihilation and creation operators satisfying the standard commutation relations $[\hat{c}_{\boldsymbol{k}}, \hat{c}_{\boldsymbol{p}}^\dagger] = \delta(\boldsymbol{k} - \boldsymbol{p})$. One can then insert this expression into the Klein-Gordon equation to find an equation of motion for the long-wavelength, infrared part of the field. In fact, one can forget that the infrared field is a quantum field and see it as a stochastic quantity obeying a Langevin equation given by [123, 126]

$$\frac{d\phi_{IR}(N, \boldsymbol{x})}{dN} = -\frac{V_\phi(\phi_{IR})}{3H^2} + \frac{H}{2\pi}\xi(N, \boldsymbol{x}), \qquad (4.124)$$

where $\xi(N)$ is a white noise due the ultraviolet part of the field with correlation function

$$\langle\xi(N, \boldsymbol{x})\xi(N', \boldsymbol{x}')\rangle = \delta(N - N')j_0\left(\sigma a H|\boldsymbol{x} - \boldsymbol{x}'|\right). \qquad (4.125)$$

Here, j_0 is a spherical Bessel function of order zero. By solving the Langevin equation, one can calculate the various correlation functions of the field and show that they coincide with the quantum correlation functions (at least in some limit). This approach, called stochastic inflation, is uncontroversial since it is a fact that the two types of correlation function perfectly match. This is another facet of the general fact that, on super-Hubble scales, the system can be described by a classical stochastic process [47, 96, 115].

4.7.2 Eternal Inflation

The next step is to relax the assumption that space-time is rigid and to take into account the backreaction of the scalar field on the geometry [52, 53, 70, 72, 73, 75, 129]. It is at this point that speculations enter the game. Since we study a regime where the inflaton field is viewed as a quantum field, it seems that there are two ways to take into account its backreaction. Either we still view the background as classical, in which case we need an equation such as $G_{\mu\nu} = \langle \hat{T}_{\mu\nu} \rangle$, or the background space-time becomes a quantum object, in which case we need an equation similar to $\hat{G}_{\mu\nu} = \hat{T}_{\mu\nu}$. In the case of eternal inflation, the second choice is made. However, since quantum objects are represented by stochastic quantities, we are, in fact, led to the concept of stochastic geometry (supposed to represent, in this approach, the behaviour of a quantum geometry). In this view, the stochastic geometry is sourced by the stochastic scalar field. Then comes the question of which equation controls the behaviour of the stochastic geometry. Here, the claim is that it is

$$H^2 = \frac{1}{3M_{\text{Pl}}^2} V\left(\phi_{\text{IR}}\right),$$ (4.126)

namely the classical equation promoted to an equation for the stochastic quantities. Here, we really deal with an equation of the type $\hat{G}_{\mu\nu} = \hat{T}_{\mu\nu}$ since ϕ_{IR} and H are now considered as stochastic quantities. We also notice that, obviously, the preceding equation is only valid in a cosmological context. Then, the Langevin equation (4.124) becomes

$$\frac{d\phi_{\text{IR}}}{dN} = -\frac{V_\phi(\phi_{\text{IR}})}{3H^2(\phi_{\text{IR}})} + \frac{H(\phi_{\text{IR}})}{2\pi} \xi(N).$$ (4.127)

Clearly, this equation is not equivalent to Eq. (4.124) and can even be ambiguous because of the second term, which is given by the product of two stochastic quantities. In Figure 4.14, we present a numerical integration of this equation for the potential $V = m^2\phi^2/2$ and for different initial conditions. It is easy to see that, in that case, the criterion (4.120) reads $\phi_s/M_{\text{Pl}} \simeq \sqrt{4\pi}\sqrt{6}(m/M_{\text{Pl}})^{-1}$. For numerical reasons, in order to clearly illustrate the effect, we choose a value of m much larger than implied by the CMB normalisation, namely $m = 0.1M_{\text{Pl}}$. This leads to $\phi_s \simeq 55 M_{\text{Pl}}$. Then, we numerically integrate Eq. (4.127) for four different initial conditions, $\phi_{\text{ini}} = 10 M_{\text{Pl}}$, $\phi_{\text{ini}} = 30 M_{\text{Pl}}$, $\phi_{\text{ini}} = 50 M_{\text{Pl}}$, and $\phi_{\text{ini}} = 70 M_{\text{Pl}}$. Using the trajectory (4.89) and the fact that $\phi_{\text{end}}/M_{\text{Pl}} = p/\sqrt{2}$, classically, these four initial conditions respectively correspond to a total of ~ 24.5, ~ 224.5, ~ 624.5, and $\sim 1,124.5$ e-folds of inflation. This plot confirms the previous analysis. When $\phi_{\text{ini}} < \phi_s$, we see that the stochastic trajectory (solid line) is very close to the classical one (dashed line). On the contrary, when $\phi_{\text{ini}} \sim \phi_s$ or $\phi_{\text{ini}} > \phi_s$, the stochastic effects dominate, the trajectory becomes 'chaotic' and strongly differs from its classical counterpart. In particular, we notice that, due to stochastic effects, the value of the field can increase. This means that the field can in fact climb its potential.

Let us now come back to Eq. (4.124), where we assume that the field is a test field living in a de Sitter space-time. If $V(\phi) = m^2\phi^2/2$, then this equation can be easily solved (since it is a linear equation), and the solution reads

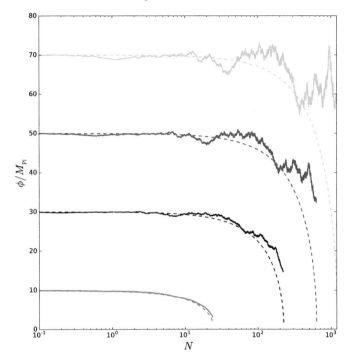

Figure 4.14 Trajectories (vacuum expectation value of the inflaton field vs number of e-folds) for the inflationary model $V(\phi) = m^2\phi^2/2$ with $m = 0.1\, M_{\rm Pl}$ and different initial conditions, $\phi_{\rm ini} = 10\, M_{\rm Pl}$ (lowest line), $\phi_{\rm ini} = 30\, M_{\rm Pl}$ (second lowest line), $\phi_{\rm ini} = 50\, M_{\rm Pl}$ (second-highest line), and $\phi_{\rm ini} = 70\, M_{\rm Pl}$ (top line). The solid lines represent the stochastic trajectories while the dashed ones correspond to the classical, slow-roll ones.

$$\phi(N,\boldsymbol{x}) = \phi_{\rm ini}(N,\boldsymbol{x})e^{-m^2(N-N_{\rm ini})/(3H^2)} + \frac{H}{2\pi}e^{-m^2 N/(3H^2)}\int_{N_{\rm ini}}^{N} e^{m^2 n/(3H^2)}\xi(n,\boldsymbol{x})dn.$$

(4.128)

Using this solution, one can then calculate the two-point correlation function at equal time. One obtains

$$\langle \phi(N,\boldsymbol{x})\, \phi(N,\boldsymbol{x}')\rangle = \left[\phi_{\rm ini}(N,\boldsymbol{x})\phi_{\rm ini}(N,\boldsymbol{x}') - \frac{3H^4}{8\pi^2 m^2}j_0\left(\sigma a H|\boldsymbol{x}-\boldsymbol{x}'|\right)\right]e^{-\frac{2m^2}{3H^2}(N-N_{\rm ini})}$$
$$+ \frac{3H^4}{8\pi^2 m^2}j_0\left(\sigma a H|\boldsymbol{x}-\boldsymbol{x}'|\right).$$

(4.129)

This expression is made of two pieces. The first one, which depends on the initial conditions, decays away exponentially for $N \gg N_{\rm ini}$ and quickly becomes subdominant. The second piece shows that the ultra-large-scale structure of the field is made of a collection of nearly homogeneous patches of size H^{-1} (i.e., the Hubble radius) since this is the distance at which the correlation function almost vanishes, thanks to the presence of the Bessel function.

Then, since inflation is an almost de Sitter expansion, what we have just described for a test field should also be true when the backreaction is taken into account, namely for the field the behaviour of which is controlled by Eq. (4.127).[2] Moreover, each patch is isolated from the others as can be seen by computing the event horizon in de Sitter space-time. Let us indeed consider a specific observer that we choose, for convenience, to be at the origin. Then, its future horizon (the part of the Universe with which the observer will be able to communicate in the future) is given by

$$d_{\mathrm{E}} = a_0 \int_{t_0}^{\infty} \frac{dt}{a(t)} = a_0 \int_{t_0}^{\infty} dt \frac{1}{a_0} e^{-H(t-t_0)} = \frac{1}{H}, \qquad (4.132)$$

namely the size of the patch itself. In other words, each patch is causally disconnected from the others and this forever. These patches are sometimes referred to as 'pocket universes'. The number of these patches is growing with time. Indeed, in one e-fold, the 'size' of the Universe increases by a factor $e^3 \sim 20$ while the 'size' of a patch is constant (since the Hubble parameter is constant). As a consequence, each e-fold, one patch gives rise to about 20 new patches, all causally disconnected.

There is also some kind of ergodic argument at play. When, see for instance Figure 4.14, we have solved the Langevin equation, each realisation of the solution of this equation was supposed to represent a specific configuration of the field over the entire homogeneous and isotropic space-time. But one can also assume that one realisation corresponds to a specific value of the field in a given patch. And, as a consequence, different realisations correspond to different values of the field in different patches. So, in this interpretation, different realisations do not represent an ensemble of different field configurations over an homogeneous and isotropic space-time but, rather, the spatial distribution of ϕ_{IR} in different patches.

The overall picture that emerges is that of an expanding space-time where the number of independent patches is increasing, the value of the field in each pocket universe being a stochastic quantity controlled by a Langevin equation. Since we have seen that, due to stochastic effects, the field can climb up its potential, there are patches where inflation will never stop. Obviously, the volume occupied by those patches, compared to the volume occupied by the patches where inflation stops, is growing, which means that patches where inflation is taking place occupy more and more regions of space-time. Globally, inflation will never stop, meaning that there are always regions of space-time undergoing inflation.

---

[2] For the potential $V(\phi) = m^2\phi^2/2$, this equation reads

$$\frac{d\phi_{\mathrm{IR}}}{dN} + \frac{2M_{\mathrm{Pl}}^2}{\phi_{\mathrm{IR}}} = \frac{m}{2\pi M_{\mathrm{Pl}}\sqrt{6}} \phi_{\mathrm{IR}} \xi. \qquad (4.130)$$

It is of the Bernouilli type and, therefore, can be solved explicitly. The solution takes the form

$$\phi_{\mathrm{IR}}^2 = e^{\frac{m}{\pi M_{\mathrm{Pl}}\sqrt{6}} \int_{N_{\mathrm{ini}}}^{N} \xi\, dn} \left[\phi_{\mathrm{ini}}^2 - 4M_{\mathrm{Pl}}^2 \int_{N_{\mathrm{ini}}}^{N} e^{-\frac{m}{\pi M_{\mathrm{Pl}}\sqrt{6}} \int_{N_{\mathrm{ini}}}^{n} \xi(\bar{n})d\bar{n}}\, dn \right]. \qquad (4.131)$$

However, it is so complicated that it is not very useful. In particular, it seems very difficult to calculate the two-point correlation function of the field from this solution.

Of course, there will also be regions of space-time where inflation stops – those where, by chance, the stochastic fluctuations do not push the field upwards. This structure is referred to as 'eternal inflation'. The stochastic effects are said to produce a 'multiverse'. Notice that the word 'multiverse' is especially awkward in the present context since we do not produce many universes as in the many-world interpretation of quantum mechanics, for instance, but just a specific spatial configuration of our single Universe made of causally independent regions, the pocket universes.

Before discussing the reliability and the implications of eternal inflation, we would like to investigate the question of whether it is unavoidable or not.

4.7.3 Avoiding Self-Replication

Before discussing the robustness of eternal inflation, it is interesting to investigate whether this is an unavoidable consequence of inflation. As recently discussed in Reference [104], it turns out that this is not the case, and in this section, we closely follow this paper although we also present some new results. We have seen that the quantum-to-classical variation of the field is given by the amplitude of the scalar power spectrum; see Eq. (4.119). If there exists a field value for which this amplitude

$$\mathcal{P}_{\zeta 0}(\phi) = \frac{H^2(\phi)}{8\pi M_{\mathrm{Pl}}^2 \epsilon_1(\phi)}, \tag{4.133}$$

is of order one, then this means that the quantum fluctuations of the field are of order one and, if the considerations presented in the previous section are correct, the regime of eternal inflation starts. Usually, this happens in the regime where $\epsilon_1(\phi) \to 0$ since $\epsilon_1(\phi)$ stands at the denominator. But this also implies that, if the shape of the potential is such that there is a field range such that $\epsilon_1 \ll 1$ (in order to have inflation!), but otherwise, $\epsilon_1(\phi)$ is large, then there could be no regime where $\mathcal{P}_{\zeta 0} > 1$. One example was found by V. Mukhanov in Reference [104]. The corresponding potential is

$$V(\phi) = M^4 \left(1 - e^{-\phi/M_{\mathrm{Pl}}}\right)^2 \left(1 - \frac{\phi}{\phi_{\mathrm{m}}}\right)^{-\alpha} \tag{4.134}$$

and is represented in Figure 4.15. It looks like the Starobinsky model corrected by a term $(1 - \phi/\phi_{\mathrm{m}})^{-\alpha}$. The model depends on three parameters: M, ϕ_{m}, and α. As usual, M is fixed by the CMB normalisation.

The first two Hubble flow parameters are given by the following expressions

$$\epsilon_1 = \frac{1}{2}\left[2\frac{e^{-\phi/M_{\mathrm{Pl}}}}{1 - e^{-\phi/M_{\mathrm{Pl}}}} + \alpha\frac{M_{\mathrm{Pl}}}{\phi_{\mathrm{m}}}\left(1 - \frac{\phi}{\phi_{\mathrm{m}}}\right)^{-1}\right]^2, \tag{4.135}$$

$$\epsilon_2 = 4\epsilon_1 + 4e^{-\phi/M_{\mathrm{Pl}}}\left(1 - e^{-\phi/M_{\mathrm{Pl}}}\right)^{-1} - 4e^{-2\phi/M_{\mathrm{Pl}}}\left(1 - e^{-\phi/M_{\mathrm{Pl}}}\right)^{-2}$$

$$- 8\alpha\frac{M_{\mathrm{Pl}}}{\phi_{\mathrm{m}}}\left(1 - e^{-\phi/M_{\mathrm{Pl}}}\right)^{-1}\left(1 - \frac{\phi}{\phi_{\mathrm{m}}}\right)^{-1} - 2(\alpha + \alpha^2)\frac{M_{\mathrm{Pl}}^2}{\phi_{\mathrm{m}}^2}\left(1 - \frac{\phi}{\phi_{\mathrm{m}}}\right)^{-2}. \tag{4.136}$$

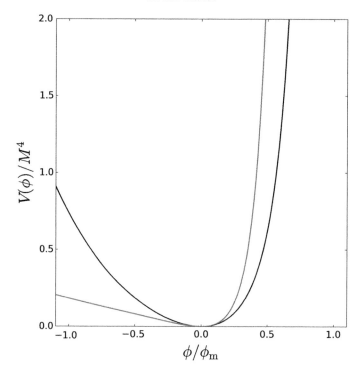

Figure 4.15 Potential given by Eq. (4.134) for two values of α, $\alpha = 2$ (black line) and $\alpha = 4$ (grey line).

The first Hubble flow parameter is represented in Figure 4.16. We see that it has exactly the expected shape. There is a field range where ϵ_1 is very small, and this is the regime during which inflation can take place. But, at large-field values, the corrections play a crucial role and $\epsilon_1 \to +, \infty$ as $\phi \to \phi_m$. As a consequence, the amplitude of the fluctuations is killed and we never reach the regime of eternal inflation.

Moreover, this model is in perfect agreement with the observations. In Figure 4.17, we have compared the predictions of the model for $\alpha = 4$ and different values of ϕ_m (indicated by the colour bar) with the CMB data (the pink contours are the WMAP7 contours while the blue contours are the Planck contours). Evidently, the model is in agreement with the data.

From the previous considerations, as we have already discussed, it should be obvious that the quantum fluctuations are suppressed. In order to check this statement explicitly, we have integrated the Langevin equation with the potential (4.134). The result is represented in Figure 4.18 and should be compared to Figue 4.14. In both plots, the value of M has been artificially increased (compared to its CMB value) in order to see the effects more clearly. It is evident that, for the model (4.134), and contrary to what happens for large-field models, the quantum fluctuations never play an important role. All the stochastic trajectories always remain close to the classical one.

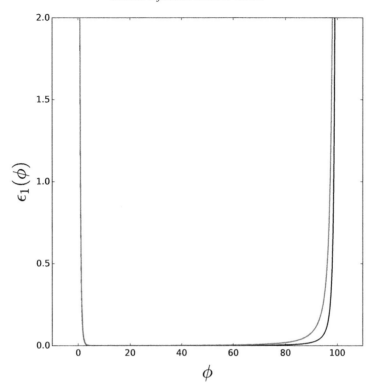

Figure 4.16 First Hubble flow parameter $\epsilon_1(\phi)$ given by Eq. (4.135) for two values of α, $\alpha = 2$ (black line) and $\alpha = 4$ (grey line).

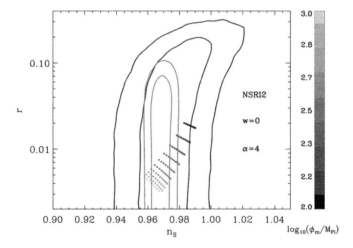

Figure 4.17 Predictions in the (r, n_S) space of the inflationary model with the potential given by Eq. (4.134). The scale M is CMB normalised, $\alpha = 4$, and $\log_{10}(\phi_m/M_{Pl}) \in [2, 3]$, its value being indicated by the shade bar. Along the same interval, different points represent different reheating temperatures. The inner contours are the 1σ and 2σ WMAP7 contours while the outer ones are the 1σ and 2σ Planck contours.

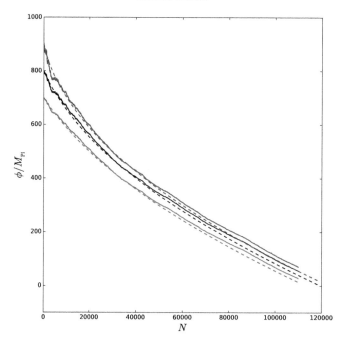

Figure 4.18 The inflaton field vacuum expectation value vs number of e-folds, for the inflationary model given by Eq. (4.134) with $\alpha = 4$ and $\phi_{\mathrm{m}} = 1{,}000$, calculated by means of the Langevin equation (solid lines) and classically (dashed lines) for different initial conditions, $\phi_{\mathrm{ini}} = 900$ (top lines), $\phi_{\mathrm{ini}} = 800$ (central lines), and $\phi_{\mathrm{ini}} = 700$ (bottom lines).

Therefore, in conclusion, the results presented here clearly indicate that eternal inflation is not mandatory and that it is perfectly possible to build a model of inflation which is in perfect agreement with the observations and where self-replication never starts. The only limitation of the previous argument is, maybe, that generic corrections are not such that self-replication is prevented. Indeed, one has $P_\zeta \sim V^3/V_\phi^2 \sim \phi^{n+2}$ if $V(\phi) \sim \phi^n$. For $n > 0$, P_ζ always grows with ϕ. So if the corrections take the form of monomials, quantum corrections will unavoidably become of order one.

4.7.4 Is the Multiverse a Threat for Inflation?

In this subsection, we discuss the implications of the previous considerations for inflation. The main point is that inflation and eternal inflation should not be put on an equal footing. The former provides a phenomenological description by means of an effective model of the early Universe which seems to be in good agreement with the observations while the latter is, at this stage, only a speculation, although definitely an interesting one. The arguments that support this point of view are the following.

First, it is important to make the distinction between stochastic inflation and eternal inflation. Stochastic inflation, which is not a model of inflation but a technique, appears to be very robust. It is just a fact that the quantum correlation functions in an expanding space-time can be recovered by focusing on the long wavelength part of the field and by requiring it to obey a Langevin equation. This has been proven beyond any doubt; see, for instance, References [123, 126]. Stochastic inflation studies test quantum fields, neglecting the backreaction of the quantum field on the geometry. In stochastic inflation, the geometry of space-time is rigid and fixed once and for all.

On the contrary, in the case of eternal inflation, one takes into account the backreaction, which means that the geometry (i.e., the gravitational field) must be viewed as a quantum (or stochastic) quantity. Clearly, this is reminiscent of quantum gravity. And, of course, the big question is which theory controls the quantum behaviour of the geometry. The theory of eternal inflation just models the coupling between the quantum field and the quantum geometry by Eq. (4.126), an equation that one could also write as

$$\hat{H}^2 = \frac{1}{3M_{\mathrm{Pl}}^2} V\left(\hat{\phi}\right), \tag{4.137}$$

where we have used hats to stress that the geometry should now be viewed as a stochastic quantity and that stochastic quantities are, in fact, quantum quantities. If this equation happened to be too simplistic, then the previous considerations about eternal inflation could be drastically modified.

Let us now discuss how the status of this equation in more detail (here, we follow the treatment of References [135, 139]). Classically, one has $\dot{H} = -(\rho + p)/(2M_{\mathrm{Pl}}^2)$. If H increases due to quantum jumps, then $\rho + p < 0$, which means that one must violate the Null Energy Condition (NEC), namely $T_{\mu\nu}n^\mu n^\nu < 0$, where n^μ is a null vector. For a scalar field, $T_{\mu\nu}n^\mu n^\nu = (n^\mu \partial_\mu \phi)^2 \geq 0$, and classically, the NEC cannot be violated. Quantum mechanically, a natural way to describe the backreaction of quantum matter on the geometry is to write the semi-classical Einstein equations, $G_{\mu\nu} = \langle \hat{T}_{\mu\nu} \rangle / M_{\mathrm{Pl}}^2$. In this approach, geometry remains classical. Then, let us introduce the NEC operator $\hat{O} \equiv \hat{T}_{\mu\nu} n^\mu n^\nu = \hat{P}^\dagger \hat{P}$, where $\hat{P} \equiv n^\mu \partial_\mu \hat{\phi}$. Generically, $\langle \hat{O} \rangle$ is infinite and must be renormalised. If this is done in a quantum state compatible with the symmetry of de Sitter, then, necessarily, $\langle \hat{T}_{\mu\nu}^{\mathrm{ren}} \rangle \propto g_{\mu\nu}$ and, therefore, $\langle \hat{O}_{\mathrm{ren}} \rangle = 0$, and the NEC cannot be violated. This means that it is necessary to go beyond semi-classical gravity if we want to treat the eternal inflation case and allow for a NEC. Notice that this is what is done in the theory of cosmological perturbations where the equations controlling the evolution of the system are $\delta \hat{G}_{\mu\nu} = \delta \hat{T}_{\mu\nu} / M_{\mathrm{Pl}}^2$ – i.e., quantum operators on both sides. In the linear regime, this has been shown to be consistent and is at the origin of the claim that inflation implies an almost scale-invariant power spectrum for cosmological perturbations. Of course, eternal inflation corresponds to a situation where the fluctuations are, by definition, not small. A possible way out is to define a smeared NEC operator [135, 139],

$$\hat{O}_W^{\mathrm{ren}} \equiv \int \mathrm{d}^4 x \sqrt{-g} W(x) \hat{O}^{\mathrm{ren}}, \tag{4.138}$$

where W is a window function which has support on a finite part of space-time. This breaks de Sitter invariance, and as a consequence, one can expect $\langle \hat{O}_W^{\text{ren}} \rangle \neq 0$. The next step would be to calculate the effects of smeared fluctuations on the metric, a framework which does not yet exist. Despite this, it is usually assumed that this effect will be described by an equation similar to Eq. (4.137). As discussed in References [135, 139], Eq. (4.137) may describe space-time before and after the fluctuation happens. But important issues are not addressed, such as the behaviour of the metric through the fluctuation or what role the conservation of energy plays in this picture. As written in Reference [50], 'An assumption is that eq. 28 is sufficient to describe this process' (where 'eq. 28' refers to Eq. (4.137) and where 'this process' refers to the response of quantum geometry to stochastic fluctuations of the field), or 'So the heuristic argument, while suggestive, is certainly not sufficient by itself to show that eternal inflation can occur'. We conclude from these considerations that Eq. (4.137), on which partially rests eternal inflation, is an assumption.

To be completely fair, we should also mention an argument which is in favour of Eq. (4.137). Let us indeed consider the Langevin equation Eq. (4.127) again. It can also be used to write a Fokker-Planck equation for $P(\phi, N)$, the probability density of having the field ϕ at time N. It reads

$$\frac{\partial}{\partial N} P(\phi, N) = \frac{\partial}{\partial \phi} \left[\frac{V_\phi}{3H^2} P(\phi, N) \right] + \frac{\partial^2}{\partial \phi^2} \left[\frac{H^2}{8\pi^2} P(\phi, N) \right]. \tag{4.139}$$

This equation can also be written as $\partial P / \partial N = \partial J / \partial \phi$, where J is a current and a stationary solution $P_{\text{sta}}(\phi)$ can be obtained by requiring that $\partial P_{\text{sta}} / \partial N = 0$. Then, the Fokker-Planck equation reduces to a first-order differential equation whose solution can be expressed as

$$P_{\text{sta}}(\phi) \propto \exp \left[\frac{24\pi^2 M_{\text{Pl}}^4}{V(\phi)} \right], \tag{4.140}$$

where we have ignored the prefactor which does not play a crucial role in our discussion. Notice that if one considers the Fokker-Planck backward equation, then one obtains the same solution but, crucially, with an overall minus sign in the argument of the exponential, namely

$$P_{\text{sta}}(\phi) \propto \exp \left[-\frac{24\pi^2 M_{\text{Pl}}^4}{V(\phi)} \right]. \tag{4.141}$$

Both Eqs. (4.140) and (4.141) are relevant for stochastic inflation. Notice that their derivation implicitly assumes Eq. (4.137).

Let us now consider the same situation but from a quantum cosmology point of view [36]. In quantum cosmology, both matter and geometry are supposed to be quantised consistently. The corresponding canonical Hamiltonian can be expressed as

$$H_c = N \left[-\frac{\pi_a^2}{48 M_{\text{Pl}}^2 v_K a} + \frac{\pi_\phi^2}{2 v_K a^3} - 12 M_{\text{Pl}}^2 k v_K a + v_K a^3 V(\phi) \right], \tag{4.142}$$

where the quantity $v_{\mathcal{K}}$ represents the volume of the space-like hypersurfaces and N is the lapse function. Carrying out Dirac quantisation leads to the Wheeler–De Witt equation for the wave function of the Universe, $\Psi(a, \phi)$, namely

$$
\frac{\partial^2}{\partial a^2} \Psi(a, \phi) + \frac{p}{a} \frac{\partial}{\partial a} \Psi(a, \phi) - 6 \frac{M_{\mathrm{Pl}}^2}{a^2} \frac{\partial^2}{\partial \phi^2} \Psi(a, \phi)
$$

$$
- 36 v_{\mathcal{K}}^2 M_{\mathrm{Pl}}^4 a_0^2 \left(\frac{a}{a_0}\right)^2 \left[\mathcal{K} - \left(\frac{a}{a_0}\right)^2\right] \Psi(a, \phi) = 0. \tag{4.143}
$$

Here, the number p takes into account the factor ordering ambiguity and $a_0 \equiv [V(\phi)/(3M_{\mathrm{Pl}}^2)]^{-1/2}$. If one neglects the second derivative with respect to ϕ and chooses $p = -1$, then the solution can be found explicitly and reads

$$
\Psi(a, \phi) = \frac{\alpha \mathrm{Ai}\,[z(a)] + \beta \mathrm{Bi}\,[z(a)]}{\alpha \mathrm{Ai}\,[z(0)] + \beta \mathrm{Bi}\,[z(0)]}, \tag{4.144}
$$

where Ai and Bi are Airy functions of first and second kinds, respectively, and $z(0) = z(a = 0)$. The quantity $z(a)$ is defined by $z(a) \equiv \left(3 v_{\mathcal{K}} M_{\mathrm{Pl}}^2 a_0^2\right)^{2/3} (\mathcal{K} - a^2/a_0^2)$, and α and β are complex numbers to be determined by boundary conditions: the tunnelling wave function corresponds to $\alpha = 1$ and $\beta = i$ and the no-boundary wave function to $\alpha = 1$ and $\beta = 0$. In order to make predictions, we need to calculate probabilities but the Wheeler–De Witt equation does not lead to positive-definite probabilities. Indeed, the associated current,

$$
j = \frac{i}{2M_{\mathrm{Pl}}^2} a^p \left(\Psi^* \partial_a \Psi - \Psi \partial_a \Psi^*\right), \tag{4.145}
$$

is not positive-definite. However, in the limit $a \gg \ell_{\mathrm{Pl}}$, the Wentzel-Kramers-Brillouin (WKB) approximation is valid and, in this regime, the probabilities are positive. For the tunnelling wave function, this gives

$$
j \simeq \frac{2}{\pi a_0^2 M_{\mathrm{Pl}}^2 |D|^2} \left(3 v_{\mathcal{K}} M_{\mathrm{Pl}}^2 a_0^2\right)^{2/3} = 6 v_{\mathcal{K}} e^{-12 v_{\mathcal{K}} M_{\mathrm{Pl}}^4 / V(\phi)}. \tag{4.146}
$$

For the no-boundary wave function, one obtains the same result except that there is no minus in the argument of the exponential. If, in addition, the space-like section are taken to be spheres, then $v_{\mathcal{K}} = 2\pi^2$, and the prediction of quantum cosmology reads

$$
j \propto \exp\left[\pm \frac{24\pi^2 M_{\mathrm{Pl}}^2}{V(\phi)}\right], \tag{4.147}
$$

which is nothing but Eqs. (4.140) and (4.141). We saw that the use of an equation $\hat{H}^2 = V(\hat{\phi})$ is questionable. The previous argument, however, seems to indicate that this could be reasonable. Indeed, as already mentioned, the stationary distribution of the Fokker-Planck equation was obtained by (implicitly) using this equation. The fact that the Wheeler–De Witt equation, which is an equation where the quantum effects of the geometry are taken into account, leads to results consistent with those obtained from the stochastic formalism retrospectively justifies the use of an equation $\hat{H}^2 = V(\hat{\phi})$. Of course, the argument is not completely conclusive since the Wheeler–De Witt equation and the minisuperspace approximation can also be questioned. We conclude that the tools used in order to model backreaction in eternal inflation are, at least for the moment, assumptions. These assumptions may be very reasonable (as seems to be suggested by this argument), but they remain assumptions.

Let us now discuss a second argument. As is clearly illustrated on the no-self-reproduction potential of Section 4.7.3, eternal inflation also rests on an extrapolation of the potential $V(\phi)$ beyond the observable window. By observing the CMB anisotropy, we probe only a limited part of $V(\phi)$ corresponding to about seven e-folds. Eternal inflation depends on another region of the potential which is not directly observed. Moreover, this part of the potential is usually relevant at energies higher than the energy scale of inflation (there are exceptions – for instance, hybrid inflation; see Reference [95]) where higher-order operators can play a crucial role. For instance, our calculation of eternal inflation in large-field models rests on the assumption that, even outside the observational window, the potential is given by $V(\phi) \propto \phi^p$. But nobody knows whether this is true since this is not directly observable. The high-energy corrections could maybe produce terms leading to the Mukhanov's potential of Section 4.7.3, in which case eternal inflation would be irrelevant. Notice that, even if one considers a plateau model, these corrections could play an important role. Indeed, it is true that, a priori, corrections in V/M_{Pl}^4 are, by construction, always negligible for plateau models. But the potential itself will generically receive corrections. For instance, if one adds a term $\propto R^3$ to the Starobinsky model, then the effective potential grows with ϕ. As a consequence, when the field is pushed upwards by the stochastic fluctuations, these corrections will be important.

Third, eternal inflation suffers from a kind of 'trans-Planckian problem'. Indeed, as discussed before, one expects the field to be pushed upwards by stochastic fluctuations. Generically, this means that the field will penetrate the region where $V(\phi) \gg M_{\mathrm{Pl}}^4$. In this regime, even the notion of a background space-time is lost. Indeed, in Reference [75], this problem was already encountered and the potential made steeper by hand in order to prevent the field to penetrate the trans-Planckian region. However, what really happens in this regime remains a matter of debate.

Fourth, the multiverse is, in fact, a combination of eternal inflation with the string landscape. A priori, string theory only depends on one parameter, the string tension. All the other parameters of high-energy physics, the masses of the particles, the coupling constant, etc., should be the vacuum expectation values of some fields appearing in string theory. Since, according to eternal inflation, the fields stochastically fluctuate from patch to patch, it should be the same for the parameters. We are thus led to a picture where what we see as fundamental parameters are, in fact, stochastic quantities fluctuating from one patch (or one 'pocket universe') to another. This is the famous multiverse. As it turns out, the concept of string landscape is not that obvious and has been discussed among string theorists [48]. At the moment, the best one could conclude is that the multiverse may pose a question, possibly justifying investigating alternatives to inflation [19]. So the multiverse problem is not only based on an an extrapolation; it relies, in fact, on a combination of extrapolations.

Based on the previous discussion, it therefore seems fair to call the multiverse 'problem' of inflation a wild speculation. Even if eternal inflation happens, it is not completely obvious that a multiverse will be present. Indeed, since the question of a string landscape remains disputed among string experts, one could imagine a situation where eternal inflation occurs but where there is no string landscape. In this case, the inflaton vacuum expectation value would still fluctuate from one patch to another, but the fundamental constants would be the same everywhere. This implies that the inflationary predictions would also be the same everywhere (for instance, Doppler peaks in the CMB would be present in each pocket universe), at least in the patches where inflation came to an end. In any case, should we reject single-field slow-roll inflation – a falsifiable, well tested, effective approach to the early Universe – in addition in perfect agreement with observations because of the multiverse? To say the least, it would be too hasty. It would be similar to rejecting the Standard Model of particle physics because (at least for the moment) it cannot be obtained from string theory.

4.8 Conclusion

In this article, we have discussed various aspects of inflation. The picture that emerges is that inflation is a very successful model of the early Universe. It has all the criterions that a good scientific theory should possess.

First, it is falsifiable. One can indeed quote two possible observations that could potentially rule out inflation. All models of inflation predict the presence of Doppler peaks in the CMB multipole moments. Therefore, if, instead of detecting them, we had obtained a bump (as predicted, for instance, if the fluctuations entirely originate from topological defects [38, 41, 116]), then inflation would have been ruled out. Another observation that could threaten the basic principles of inflation is the observation that $\Omega_{\mathcal{K}} \neq 0$. It is true that an inflationary model with $\Omega_{\mathcal{K}} \neq 0$ has been constructed in Reference [74], but this model is so peculiar that it can be viewed as a curiosity and cannot be considered as representative. Some may argue that it shows the amount of arm-twisting that needs to be done to inflation to make it predict $\Omega_{\mathcal{K}} \neq 0$. In any case, it is our point of view that $\Omega_{\mathcal{K}} \neq 0$ (beyond 10^{-5} since, of course, some curvature is present in the perturbed universe) should be considered as a fatal blow for inflation.

Second, inflation has been able to make predictions, most notably the prediction that n_s should be close to one but – and this is the crucial point – excluding one (however, see the exception [125]). As discussed at length previously, this prediction has been confirmed by the data. It is true that a scale-invariant power spectrum, the so-called Harrisson-Zeldovitch (HZ) power spectrum, was already considered before inflation. But, precisely, the HZ power spectrum has $n_s = 1$ while inflation has $n_s \sim 1$ and, crucially, $n_s - 1 \neq 0$. The prediction $n_s - 1 \neq 0$ was made by inflation, and not by any other theory, and its observational confirmation is therefore a strong argument in favour of inflation.

Third, the criticisms against inflation do not seem completely compelling (see also Reference [34], where the initial conditions problem and the measure question are dicussed in detail). The initial condition problem does not seem to be very severe, thanks to the presence of an attractor. It is true that the attractor is not present for some models (for instance, small-field inflation), but, precisely, the Planck data have singled out a model (namely the Starobinsky model) where it is present.

The multiverse question is nowadays widely debated, and there are claims that its appearance implies that standard inflation makes no prediction and, therefore, is not falsifiable. The argument is that if everything happens, there could be patches in our Universe where, for instance, the Doppler peaks are present, but there could be others where it is not the case. Or there could be patches where n_s is close to one and others where it is far from one. All that is based on the belief that the multiverse is unavoidable. However, it is, at the moment, unreasonable to put the multiverse and standard inflation on an equal footing. Indeed, at this stage, it is fair to say that the multiverse is a speculation (if it is present at all, since we have seen that it can be avoided; see Section 4.7.3), and one can argue that it would be awkward to reject a good, effective model because of a mere speculation. As already mentioned, this would be like rejecting the Standard Model of particle physics because, so far, no one has been able to derive it from string theory. To be completely fair with this analogy and the multiverse criticism, it is true that the potential modifications of the Standard Model of particle physics suggested by string theory are much less radical that what the multiverse implies for standard inflation.

It is also true that we still do not know the physical nature of the inflaton field even if the latest data raise the intriguing possibility that it could the Higgs field itself. After all, we are trying to develop a theory the typical energy scale of which could be as high as the GUT scale. So maybe this problem (if it is indeed one) is not in the inflationary scenario but, rather, in our lack of understanding of particle physics at 10^{15}GeV. In any case, with the recent discovery of the Higgs boson, a common criticism against inflation, namely that no scalar field has ever been seen, has fallen.

Of course, this does not mean that inflation has no drawback and should not be criticised. Admittedly, the question of initial conditions is clearly not completely settled. The question which is left partially unanswered is what happens when one starts from strongly inhomogeneous configurations in the most general situation; impressive numerical simulations of fully inhomogeneous situations have been performed, but they do not yet cover all the possibilities. This is technically complicated since this requires numerical relativity. But it is fair to admit that this is a remaining issue which is very important. On the other hand, it is not clear whether this question can be treated classically. Most probably, quantum effects also play an important role in this problem, which makes it even more complicated.

Another open issue is the ultraviolet (UV) sensitivity of inflation. One example is, of course, eternal inflation itself. Indeed, we have seen that it can happen or not, depending on what we assume about the shape of the potential at high energies, outside the observational window. Another example of UV dependence is the trans-Planckian problem of inflation. If the fluctuations behave in a non-standard way when their physical wavelength becomes smaller than the Planck length, and if the trans-Planckian physics is non-adiabatic, then the prediction of an almost scale-invariant power spectrum could be modified. Let us nevertheless tone down this conclusion by stressing that the corresponding modification could be very small. As was discussed earlier, we have indeed two scales in the problem: the scale M_C at which new physics pops up (typically the Planck scale) and the Hubble parameter during inflation. If the effect scales as the ratio H/M_C to some power, then the correction should be very small. Yet another example of UV dependence is the importance of higher-order operators for inflationary model building; see Reference [118].

Therefore, it is true that inflation has some UV sensitivity. But, after all, this is also the case of the Standard Model of particle physics where the Higgs mass is not stable against quantum corrections (the hierarchy problem). But no one would reject this model because of this issue. Let us also add that it is inconsistent to claim at the same time that inflation is UV dependent and that the multiverse is unavoidable: if inflation is UV dependent, then one can modify it at high energies to avoid the multiverse, and this is exactly what the calculation of Section 4.7.3 reveals. From a more general perspective concerning the IR/UV connection, it is interesting that inflation seems to provide an example in which the decoupling between physics at different scales, which is the basis of effective field theory, does not work.

In conclusion, inflation appears to be a robust and reliable scenario for the early Universe, not completely free of open issues, of course, but could it have been different for a theory which is trying to describe the first instants of the Universe, at energy scales as high as 10^{15} GeV? At this stage, admittedly, one cannot yet trust it as we trust, for example, the Standard Model of particle physics. The situation, however, could change soon if, for instance, we could check the consistency relation, $r = -n_T/8$. This is clearly a difficult task, and a first step would clearly be to detect primordial gravitational waves. After all, if the pieces of information that we have gathered so far are correct, the next generation of experiments should be able to see them. Indeed, their target is $r \sim 10^{-4}$ while our best model, the Starobinsky model, predicts $r \sim 4 \times 10^{-3}$. Then measuring n_T will be even more difficult but would be very important. The measurement of NG would also be important. The expected level, $f_{NL} \simeq 10^{-2}$, is tiny for our preferred class of models, but people are already thinking about experiments that could reach this level. In brief, inflation continues to be an inspiration for many physicists and continues to fuel new interesting works. So, inflation, trick or treat? Treat, definitively!

Acknowledgements

It is with pleasure we thank P. Peter and V. Vennin for careful reading of the manuscript.

References

[1] Ade, P. A. R., *et al.* 2014a. 'Planck 2013 Results. I. Overview of Products and Scientific Results'. *Astronomy and Astrophysics*, **571**, A1.
[2] Ade, P. A. R., *et al.* 2015. 'Joint Analysis of BICEP2/*KeckArray* and *Planck* Data'. *Physical Review Letters*, **114**, 101301.
[3] Ade, P. A. R., *et al.* 2016a. 'Planck 2015 Results. XIII. Cosmological Parameters'. *Astronomy and Astrophysics*, **594**, A13.

[4] Ade, P. A. R., *et al.* 2016b. 'Planck 2015 Results. XX. Constraints on Inflation'. *Astronomy and Astrophysics*, **594**, A20.

[5] Ade, P., *et al.* 2014b. 'Planck 2013 Results. XVI. Cosmological Parameters'. *Astronomy and Astrophysics*, **571**, A16.

[6] Ade, P., *et al.* 2014c. 'Planck 2013 Results. XXII. Constraints on Inflation'. *Astronomy and Astrophysics*, **571**, A22.

[7] Ade, P., *et al.* 2014d. 'Planck 2013 Results. XXIV. Constraints on Primordial Non-Gaussianity'. *Astronomy and Astrophysics*, **571**, A24.

[8] Aghanim, N., *et al.* 2018. 'Planck 2018 Results. VI. Cosmological Parameters'. *Astronomy and Astrophysics*, DOI: 10.1051/0004-6361/201833910.

[9] Akrami, Y., *et al.* 2018a. 'Planck 2018 Results. I. Overview and the Cosmological Legacy of Planck'. *Astronomy and Astrophysics*, DOI:10.1051/0004-6361/201833880.

[10] Akrami, Y., *et al.* 2018b. 'Planck 2018 Results. X. Constraints on Inflation'. *Astronomy and Astrophysics*, DOI:10.1051/0004-6361/201833887.

[11] Albrecht, A., Brandenberger, R. H., and Matzner, R. 1985. 'Numerical Analysis of Inflation'. *Physical Review D*, **32**, 1280.

[12] Amin, M. A., *et al.* 2014. 'Nonperturbative Dynamics of Reheating after Inflation: A Review'. *International Journal of Modern Physics D*, **24**, 1530003.

[13] Anninos, P., *et al.* 1991. 'How Does Inflation Isotropize the Universe?' *Physical Review D*, **43**, 3821–3832.

[14] Armendariz-Picon, C., Mukhanov, V. F., and Steinhardt, P. J. 2001. 'Essentials of k Essence'. *Physical Review D*, **63**, 103510.

[15] Assadullahi, H., *et al.* 2016. 'Multiple Fields in Stochastic Inflation'. *Journal of Cosmology and Astroparticle Physics*, **1606**(06), 043.

[16] Astier, P., *et al.* 2006. 'The Supernova Legacy Survey: Measurement of ωM, $\omega\lambda$ and W from the First Year Data Set'. *Astronomy and Astrophysics*, **447**, 31–48.

[17] Bardeen, J. M. 1980. 'Gauge Invariant Cosmological Perturbations'. *Physical Review D*, **22**, 1882–1905.

[18] Bardeen, J. M., Steinhardt, P. J., and Turner, M. S. 1983. 'Spontaneous Creation of Almost Scale-Free Density Perturbations in an Inflationary Universe'. *Physical Review D*, **28**, 679.

[19] Battefeld, D., and Peter, P. 2015. 'A Critical Review of Classical Bouncing Cosmologies'. *Physics Reports*, **571**, 1–66.

[20] Baumann, D., and McAllister, L. 2015. *Inflation and String Theory*. Cambridge University Press.

[21] Bennett, C. L., *et al.* 1996. 'Four year COBE DMR Cosmic Microwave Background Observations: Maps and Basic Results'. *The Astrophysical Journal Letters*, **464**, L1–L4.

[22] Brandenberger, R., and Peter, P. 2017. 'Bouncing Cosmologies: Progress and Problems'. *Foundations of Physics*, **47**(6), 797–850.

[23] Brandenberger, R. H., and Kung, J. H. 1990. 'Chaotic Inflation as an Attractor in Initial Condition Space'. *Physical Review D*, **42**, 1008–1015.

[24] Brandenberger, R. H., and Martin, J. 2001. 'The Robustness of Inflation to Changes in SuperPlanck Scale Physics'. *Modern Physics Letters A*, **16**, 999–1006.

[25] Brandenberger, R. H., and Martin, J. 2002. 'On Signatures of Short Distance Physics in the Cosmic Microwave Background'. *International Journal of Modern Physics A*, **17**, 3663–3680.

[26] Brandenberger, R. H., and Martin, J. 2005. 'Back-Reaction and the Trans-Planckian Problem of Inflation Revisited'. *Physical Review D*, **71**, 023504.

[27] Brandenberger, R. H., and Martin, J. 2013. 'Trans-Planckian Issues for Inflationary Cosmology'. *Classical and Quantum Gravity*, **30**, 113001.

[28] Calzetta, E., and Sakellariadou, M. 1992. 'Inflation in Inhomogeneous Cosmology'. *Physical Review D*, **45**, 2802–2805.

[29] Casadio, R., *et al.* 2005a. 'Higher Order Slow-Roll Predictions for Inflation'. *Physics Letters B*, **625**, 1–6.

[30] Casadio, R., *et al.* 2005b. 'Improved WKB Analysis of Cosmological Perturbations'. *Physical Review D*, **71**, 043517.

[31] Casadio, R., *et al.* 2005c. 'Improved WKB Analysis of Slow-Roll Inflation'. *Physical Review D*, **72**, 103516.

[32] Chialva, D., and Mazumdar, A. 2015. 'Cosmological Implications of Quantum Corrections and Higher-Derivative Extension'. *Modern Physics Letters A*, **30**(03n04), 1540008.

[33] Choe, J., Gong, J. O., and Stewart, E. D. 2004. 'Second Order General Slow-Roll Power Spectrum'. *Journal of Cosmology and Astroparticle Physics*, **0407**, 012.

[34] Chowdhury, D., *et al.* 2019. 'Inflation after Planck: Judgment Day'. *arXiv e-prints*, Feb, arXiv:1902.03951.

[35] Clough, K., *et al.* 2016. 'Robustness of Inflation to Inhomogeneous Initial Conditions'. arXiv:1608.04408.

[36] Coule, D. H., and Martin, J. 2000. 'Quantum Cosmology and Open Universes'. *Physical Review D*, **61**, 063501.

[37] Cyburt, R. H., *et al.* 2016. 'Big Bang Nucleosynthesis: Present Status'. *Reviews of Modern Physics*, **88**(1), 015004.

[38] Durrer, R., Gangui, A., and Sakellariadou, M. 1996. 'Doppler Peaks: A Fingerprint of Topological Defects'. *Physical Review Letters*, **76**, 579–582.

[39] East, W. E., *et al.* 2016. 'Beginning Inflation in an Inhomogeneous Universe'. *Journal of Cosmology and Astroparticle Physics*, **1609**(09), 010.

[40] Fixsen, D. J. 2009. 'The Temperature of the Cosmic Microwave Background'. *The Astrophysical Journal*, **707**(Dec.), 916–920.

[41] Gangui, A., Durrer, R., and Sakellariadou, M. 1996. 'Global Textures and the Doppler Peaks'. *Astronomical Society of the Pacific Conference Series*, **94**, 335–340.

[42] Goldwirth, D. S. 1991. 'On Inhomogeneous Initial Conditions for Inflation'. *Physical Review D*, **43**, 3204–3213.

[43] Goldwirth, D. S., and Piran, T. 1989. 'Spherical Inhomogeneous Cosmologies and Inflation. 1. Numerical Methods'. *Physical Review D*, **40**, 3263.

[44] Goldwirth, D. S., and Piran, T. 1990. 'Inhomogeneity and the Onset of Inflation'. *Physical Review Letters*, **64**, 2852–2855.

[45] Goldwirth, D. S., and Piran, T. 1992. 'Initial Conditions for Inflation'. *Physics Reports*, **214**, 223–291.

[46] Gong, J. O., and Stewart, E. D. 2001. 'The Density Perturbation Power Spectrum to Second Order Corrections in the Slow Roll Expansion'. *Physics Letters B*, **510**, 1–9.

[47] Grain, J., and Vennin, V. 2017. 'Stochastic Inflation in Phase Space: Is Slow Roll a Stochastic Attractor?' *Journal of Cosmology and Astroparticle Physics*, **1705**(05), 045.

[48] Gross, D. J., 2005. 'Where Do We Stand in Fundamental String Theory'. *Physica Scripta*, **T117**, 102–105.

[49] Gunn, J. E., and Peterson, B. A. 1965. 'On the Density of Neutral Hydrogen in Intergalactic Space'. *The Astrophysical Journal*, **142**(Nov.), 1633–1641.

[50] Guth, A., Vachaspati, T., and Winitzki, S. 'Energy Conditions in Inflation'.

[51] Guth, A. H. 1981. 'The Inflationary Universe: A Possible Solution to the Horizon and Flatness Problems'. *Physical Review D*, **23**, 347–356.

[52] Guth, A. H. 2000. 'Inflation and Eternal Inflation'. *Physics Reports*, **333**, 555–574.

[53] Guth, A. H. 2007. 'Eternal Inflation and Its Implications'. *Journal of Physics A*, **40**, 6811–6826.

[54] Hartle, J. B., and Hawking, S. W. 1983. 'Wave Function of the Universe'. *Physical Review D*, **28**, 2960–2975.

[55] Hassan, S. F., and Sloth, M. S. 2003. 'TransPlanckian Effects in Inflationary Cosmology and the Modified Uncertainty Principle'. *Nuclear Physics B*, **674**, 434–458.

[56] Ijjas, A., Steinhardt, P. J., and Loeb, A. 2013. 'Inflationary Paradigm in Trouble after Planck2013'. *Physics Letters B*, **723**, 261–266.

[57] Ijjas, A., Steinhardt, P. J., and Loeb, A. 2014. 'Inflationary Schism'. *Physics Letters B*, **736**, 142–146.

[58] Kofman, L., Linde, A. D., and Starobinsky, A. A. 1997. 'Towards the Theory of Reheating after Inflation'. *Physical Review D*, **56**, 3258–3295.

[59] Kolb, E. W., and Turner, M. S. 1990. 'The Early Universe'. *Frontiers in Physics*, **69**, 1–547.

[60] Kosowsky, A. 1996. 'Cosmic Microwave Background Polarization'. *Annals of Physics*, **246**, 49–85.

[61] Kowalski, M., *et al.* 2008. 'Improved Cosmological Constraints from New, Old and Combined Supernova Datasets'. *The Astrophysical Journal*, **686**, 749–778.

[62] Kung, J. H., and Brandenberger, R. H. 1989. 'The Initial Condition Dependence of Inflationary Universe Models'. *Physical Review D*, **40**, 2532.

[63] Kunz, M., Trotta, R., and Parkinson, D. 2006. 'Measuring the Effective Complexity of Cosmological Models'. *Physical Review D*, **74**, 023503.

[64] Kurki-Suonio, H., *et al.* 1987. 'Inflation from Inhomogeneous Initial Data in a One-Dimensional Back Reacting Cosmology'. *Physical Review D*, **35**, 435–448.

[65] Kurki-Suonio, H., Laguna, P., and Matzner, R. A. 1993. 'Inhomogeneous Inflation: Numerical Evolution'. *Physical Review D*, **48**, 3611–3624.

[66] Laguna, P., Kurki-Suonio, H., and Matzner, R. A. 1991. 'Inhomogeneous Inflation: The Initial Value Problem'. *Physical Review D*, **44**, 3077–3086.

[67] Leach, S. M., *et al.* 2002. 'Cosmological Parameter Estimation and the Inflationary Cosmology'. *Physical Review D*, **66**, 023515.

[68] Lemoine, M., *et al.* 2002. 'The Stress Energy Tensor for TransPlanckian Cosmology'. *Physical Review D*, **65**, 023510.

[69] Liddle, A. R., Parsons, P., and Barrow, J. D. 1994. 'Formalizing the Slow Roll Approximation in Inflation'. *Physical Review D*, **50**, 7222–7232.

[70] Linde, A. 2017. 'A Brief History of the Multiverse'. *Reports on Progress in Physics*, **80**(2), 022001.

[71] Linde, A. D. 1982a. 'A New Inflationary Universe Scenario: A Possible Solution of the Horizon, Flatness, Homogeneity, Isotropy and Primordial Monopole Problems'. *Physics Letters B*, **108**, 389–393.

[72] Linde, A. D. 1982b. 'Nonsingular Regenerating Inflationary Universe'. Print-82-0554, Cambridge University preprint, available at https://web.stanford.edu/~alinde/1982.pdf

[73] Linde, A. D. 1986. 'Eternally Existing Selfreproducing Chaotic Inflationary Universe'. *Physics Letters B*, **175**, 395–400.

[74] Linde, A. D., and Mezhlumian, A. 1995. 'Inflation with $\Omega \neq 1$'. *Physical Review D*, **52**, 6789–6804.

[75] Linde, A. D., Linde, D. A., and Mezhlumian, A. 1994. 'From the Big Bang Theory to the Theory of a Stationary Universe'. *Physical Review D*, **49**, 1783–1826.

[76] Lorenz, L., Martin, J., and Ringeval, C. 2008a. 'Brane Inflation and the WMAP Data: A Bayesian Analysis'. *Journal of Cosmology and Astroparticle Physics*, **0804**, 001.

[77] Lorenz, L., Martin, J., and Ringeval, C. 2008b. 'Constraints on Kinetically Modified Inflation from WMAP5'. *Physical Review D*, **78**, 063543.

[78] Lorenz, L., Martin, J., and Ringeval, C. 2008c. 'K-Inflationary Power Spectra in the Uniform Approximation'. *Physical Review D*, **78**, 083513.

[79] Lorenz, L., Martin, J., and Yokoyama, J. 2010. 'Geometrically Consistent Approach to Stochastic DBI Inflation'. *Physical Review D*, **82**, 023515.

[80] Maartens, R. 2011. 'Is the Universe Homogeneous?' *Philosophical Transactions of the Royal Society*, **369**, 5115–5137.

[81] Martin, J. 2004. 'Inflation and Precision Cosmology'. *Brazilian Journal of Physics*, **34**, 1307–1321.

[82] Martin, J. 2005. 'Inflationary Cosmological Perturbations of Quantum-Mechanical Origin'. *Lecture Notes in Physics*, **669**, 199–244.

[83] Martin, J. 2008. 'Inflationary Perturbations: The Cosmological Schwinger Effect'. *Lecture Notes in Physics*, **738**, 193–241.

[84] Martin, J. 2016. 'The Observational Status of Cosmic Inflation after Planck'. *Astrophysics and Space Science Proceedings*, **45**, 41–134.

[85] Martin, J., and Brandenberger, R. 2003. 'On the Dependence of the Spectra of Fluctuations in Inflationary Cosmology on TransPlanckian Physics'. *Physical Review D*, **68**, 063513.

[86] Martin, J., and Brandenberger, R. H. 2001. 'The TransPlanckian Problem of Inflationary Cosmology'. *Physical Review D*, **63**, 123501.

[87] Martin, J. and Musso, M. 2006a. 'On the Reliability of the Langevin Perturbative Solution in Stochastic Inflation'. *Physical Review D*, **73**, 043517.

[88] Martin, J., and Musso, M. 2006b. 'Solving Stochastic Inflation for Arbitrary Potentials'. *Physical Review D*, **73**, 043516.

[89] Martin, J., and Ringeval, C. 2004a. 'Addendum to "Superimposed oscillations in the WMAP data?"'. *Physical Review D*, **69**, 127303.

[90] Martin, J., and Ringeval, C. 2004b. 'Superimposed Oscillations in the WMAP Data?' *Physical Review D*, **69**, 083515.

[91] Martin, J., and Ringeval, C. 2005. 'Exploring the Superimposed Oscillations Parameter Space'. *Journal of Cosmology and Astroparticle Physics*, **0501**, 007.

[92] Martin, J., and Ringeval, C. 2006. 'Inflation after WMAP3: Confronting the Slow-Roll and Exact Power Spectra to CMB Data'. *Journal of Cosmology and Astroparticle Physics*, **0608**, 009.

[93] Martin, J., and Ringeval, C. 2010. 'First CMB Constraints on the Inflationary Reheating Temperature'. *Physical Review D*, **82**, 023511.

[94] Martin, J., and Schwarz, D. J. 1998. 'The Influence of Cosmological Transitions on the Evolution of Density Perturbations'. *Physical Review D*, **57**, 3302–3316.

[95] Martin, J., and Vennin, V. 2012. 'Stochastic Effects in Hybrid Inflation'. *Physical Review D*, **85**, 043525.

[96] Martin, J., and Vennin, V. 2016. 'Quantum Discord of Cosmic Inflation: Can We Show that CMB Anisotropies Are of Quantum-Mechanical Origin?' *Physical Review D*, **93**(2), 023505.

[97] Martin, J., Ringeval, C., and Trotta, R. 2011. 'Hunting Down the Best Model of Inflation with Bayesian Evidence'. *Physical Review D*, **83**, 063524.

[98] Martin, J., *et al.* 2014a. 'Compatibility of Planck and BICEP2 in the Light of Inflation'. *Physical Review D*, **90**(6), 063501.

[99] Martin, J., Ringeval, C., and Vennin, V. 2014b. 'Encyclopædia Inflationaris'. *Physics of the Dark Universe*, **5–6**, 75–235.

[100] Martin, J., Ringeval, C., and Vennin, V. 2015. 'Observing Inflationary Reheating'. *Physical Review Letters*, **114**(8), 081303.

[101] Martin, J., Ringeval, C., and Vennin, V. 2016. 'Information Gain on Reheating: The One Bit Milestone'. *Physical Review D*, **93**(10), 103532.

[102] Martin, J., Ringeval, C., and Vennin, V. 2013. 'K-Inflationary Power Spectra at Second Order'. *Journal of Cosmology and Astroparticle Physics*, **1306**, 021.

[103] Martin, J., *et al.* 2014c. 'The Best Inflationary Models after Planck'. *Journal of Cosmology and Astroparticle Physics*, **1403**, 039.

[104] Mukhanov, V. 2015. 'Inflation without Selfreproduction'. *Fortschritte der Physik*, **63**, 36–41.

[105] Mukhanov, V. F., and Chibisov, G. 1981. 'Quantum Fluctuation and Nonsingular Universe'. *Journal of Experimental and Theoretical Physics Letters*, **33**, 532–535.

[106] Mukhanov, V. F., and Chibisov, G. 1982. 'The Vacuum Energy and Large Scale Structure of the Universe'. *Journal of Experimental and Theoretical Physics*, **56**, 258–265.

[107] Mukhanov, V. F., Feldman, H., and Brandenberger, R. H. 1992. 'Theory of Cosmological Perturbations. Part 1. Classical Perturbations. Part 2. Quantum Theory of Perturbations. Part 3. Extensions'. *Physics Reports*, **215**, 203–333.

[108] Peebles, P. J. E. 1968. 'Recombination of the Primeval Plasma'. *The Astrophysical Journal*, **153**(July), 1.

[109] Peebles, P. J. E. 1999. *Summary: Inflation and Traditions of Research. Proceedings of the Pritzker Symposium on the Status of Inflationary Cosmology*, preprint available at arXiv:astro-ph/9905390.

[110] Penrose, R. 1989. 'Difficulties with Inflationary Cosmology'. *Annals of the New York Academy of Sciences*, **571**, 249–264.

[111] Perez, R. S., and Pinto-Neto, N. 2011. 'Spherically Symmetric Inflation'. *Gravitation and Cosmology*, **17**, 136–140.

[112] Perlmutter, S., *et al.* 1997. Measurements of the Cosmological Parameters Omega and Lambda from the First 7 Supernovae at z>=0.35'. *The Astrophysical Journal*, **483**, 565.

[113] Perlmutter, S., *et al.* 1998. 'Discovery of a Supernova Explosion at Half the Age of the Universe and Its Cosmological Implications'. *Nature*, **391**, 51–54.

[114] Peter, P. and Uzan, J. P. 2009. *Primordial Cosmology*. Oxford Graduate Texts. Oxford University Press.

[115] Polarski, D. and Starobinsky, A. A. 1996. 'Semiclassicality and Decoherence of Cosmological Perturbations'. *Classical and Quantum Gravity*, **13**, 377–392.

[116] Ringeval, C., Sakellariadou, M., and Bouchet, F. 2007. 'Cosmological Evolution of Cosmic String Loops'. *Journal of Cosmology and Astroparticle Physics*, **0702**, 023.

[117] Schwarz, D. J., Terrero-Escalante, C. A., and Garcia, A. A. 2001. 'Higher Order Corrections to Primordial Spectra from Cosmological Inflation'. *Physics Letters B*, **517**, 243–249.

[118] Silverstein, E. 2019. 'The Dangerous Irrelevance of String Theory'. In R. Dardashti, R. Dawid, R., and K. Thébault (eds.), Why Trust a Theory?: Epistemology of Fundamental Physics (pp. 365–376). Cambridge University Press, DOI:10.1017/9781108671224.025.

[119] Smoot, G. F., *et al.* 1992. 'Structure in the COBE Differential Microwave Radiometer First Year Maps'. *The Astrophysical Journal Letters*, **396**, L1–L5.

[120] Starobinsky, A. A. 1979. 'Spectrum of Relict Gravitational Radiation and the Early State of the Universe'. *Journal of Experimental and Theoretical Physics Letters*, **30**, 682–685.

[121] Starobinsky, A. A. 1980. 'A New Type of Isotropic Cosmological Models without Singularity'. *Physics Letters B*, **91**, 99–102.

[122] Starobinsky, A. A. 1982. 'Dynamics of Phase Transition in the New Inflationary Universe Scenario and Generation of Perturbations'. *Physics Letters B*, **117**, 175–178.

[123] Starobinsky, A. A. 1986. 'Stochastic De Sitter (Inflationary) Stage in the Early Universe'. *Lecture Notes in Physics*, **246**, 107–126.

[124] Starobinsky, A. A. 1992. 'Spectrum of Adiabatic Perturbations in the Universe When There Are Singularities in the Inflation Potential'. *JETP Letters*, **55**, 489–494.

[125] Starobinsky, A. A. 2005. 'Inflaton Field Potential Producing the Exactly Flat Spectrum of Adiabatic Perturbations'. *JETP Letters*, **82**, 169–173.

[126] Starobinsky, A. A., and Yokoyama, J. 1994. 'Equilibrium State of a Selfinteracting Scalar Field in the De Sitter Background'. *Physical Review D*, **50**, 6357–6368.

[127] Steigman, G., and Turner, M. S. 1983. 'Inflation in a Shear or Curvature Dominated Universe'. *Physics Letters B*, **128**, 295–298.

[128] Stein-Schabes, J. A. 1987. 'Inflation in Spherically Symmetric Inhomogeneous Models'. *Physical Review D*, **35**, 2345.

[129] Steinhardt, P. J. 1982. 'Natural Inflation'. In Gibbons, G,. Hawking, S., and Siklos, S. (eds.) The Very Early Universe. Cambridge University Press. Page 251.

[130] Traschen, J. H., and Brandenberger, R. H. 1990. 'Particle Production during Out-of-Equilibrium Phase Transitions'. *Physical Review D*, **42**, 2491–2504.

[131] Trotta, R. 2008. 'Bayes in the Sky: Bayesian Inference and Model Selection in Cosmology'. *Contemporary Physics*, **49**, 71–104.

[132] Trotta, R. 2017. *Bayesian Methods in Cosmology*. arXiv:1701.01467.

[133] Turner, M. S. 1983. 'Coherent Scalar Field Oscillations in an Expanding Universe'. *Physical Review D*, **28**, 1243.

[134] Turner, M. S., and Widrow, L. M. 1986. 'Homogeneous Cosmological Models and New Inflation'. *Physical Review Letters*, **57**, 2237–2240.

[135] Vachaspati, T. 2003. *Eternal Inflation and Energy Conditions in De Sitter Space-Time*.

[136] Vennin, V., and Starobinsky, A. A. 2015. 'Correlation Functions in Stochastic Inflation'. *European Physical Journal C*, **75**, 413.

[137] Wands, D. 2008. 'Multiple Field Inflation'. *Lecture Notes in Physics*, **738**, 275–304.

[138] Weinberg, S. 1972. *Gravitation and Cosmology: Principles and Applications of the General Theory of Relativity*. Wiley.

[139] Winitzki, S. 2001. *Null Energy Condition Violations in Eternal Inflation*.

[140] Ávila, S., Martin, J., and Steer, D. 2014. 'Superimposed Oscillations in Brane Inflation'. *Journal of Cosmology and Astroparticle Physics*, **1408**, 032.

5

Is the Universal Matter-Antimatter Asymmetry Fine-Tuned?

GARY STEIGMAN AND ROBERT J. SCHERRER*

Abstract

The asymmetry between matter and antimatter (baryons and anti-baryons or nucleons and anti-nucleons, along with their accompanying electrons and positrons) is key to the existence and nature of our Universe. A measure of the matter-antimatter asymmetry of the Universe is provided by the present value of the universal ratio of baryons (baryons minus anti-baryons) to photons (or the ratio of baryons to entropy). The baryon asymmetry parameter is an important physical and cosmological parameter. But how fine-tuned is it? A 'natural' value for this parameter is zero, corresponding to equal amounts of matter and antimatter. Such a universe would look nothing like ours and would be unlikely to host stars, planets, or life. Another, also possibly natural, choice for this dimensionless parameter would be of order unity, corresponding to nearly equal amounts (by number) of matter (and essentially no antimatter) and photons in every co-moving volume. However, observations suggest that in the Universe we inhabit, the value of this parameter is non-zero but smaller than this natural value by some nine to ten orders of magnitude. In this contribution we review the evidence, observational as well as theoretical, that our Universe does *not* contain equal amounts of matter and antimatter. An overview is provided of some of the theoretical proposals for extending the Standard Models of particle physics and cosmology in order to generate such an asymmetry during the early evolution of the Universe.

Any change in the magnitude of the baryon asymmetry parameter necessarily leads to a universe with physical characteristics different from those in our own. Small changes in this parameter will barely affect cosmic evolution, while large changes might alter the formation of stars and planets and affect the development of life. The degree of fine-tuning in the baryon asymmetry parameter is determined by the width of the range over which it can be varied and still allow for the existence of life. Our results suggest that the baryon asymmetry parameter can be varied over a very wide range without impacting the prospects for life; this result is *not* suggestive of fine-tuning.

---

*   Following the untimely death of Gary Steigman, the second author was brought in to complete this chapter. He has endeavoured to adhere as closely as possible to the original format and spirit of the manuscript constructed by the first author.

174

We note that, according to those extensions of the Standard Models of particle physics and cosmology that allow for a non-zero baryon number, the Universe began with zero baryon number at a time (temperature) when baryon number was conserved. As the Universe expanded and cooled, baryon number conservation was broken at some high temperature- (mass-/energy-) scale, and a non-zero baryon number was created. However, even though baryon non-conservation is strongly suppressed at late times (low temperatures), baryon number is not conserved, so matter (protons, the lightest baryons) might eventually decay, with the baryon number reverting back to zero. Ashes to ashes, dust to dust, the Universe began with zero baryon number and may well end that way.

5.1 Introduction and Overview

The asymmetry between matter and antimatter (baryons and anti-baryons or nucleons and anti-nucleons, along with their accompanying electrons and positrons) is key to the existence and nature of our Universe. Any causal Lorentz-invariant quantum theory allows for particles to come in particle-antiparticle pairs. The discovery of the antiproton [6] in 1955 quickly stimulated serious consideration of the antimatter content of the Universe [3, 18] and led to constraints on the amount of antimatter based on the astrophysical effects of interacting matter and antimatter [4]. At the time, and for many years after, the prevailing view in the physics community was that baryon number (the quantum number that distinguishes baryons and anti-baryons) was absolutely conserved, and this assumption led to two differing points of view. Either the Universe is and always has been symmetric between matter and antimatter or the Universe is and always has been asymmetric, with an excess of matter over antimatter that has remained unchanged from the beginning of the expanding Universe (the Big Bang). Those who believed the Universe to be symmetric between matter and antimatter were undeterred by the fact that that the only antimatter seen up to that time (not counting positrons) was the handful of antiprotons created in collisions at high-energy accelerators. Those who believed the Universe to be asymmetric had to come to grips with the dilemma of creating such a universe if the laws of physics dictated that particles are always created (and destroyed) in pairs and that baryon number is absolutely conserved.

Most ignored this dilemma. Andrei Sakharov [35] did not. To set the stage for Sakharov's seminal work, it is useful to recall the 1965 discovery of the cosmic microwave background (CMB) radiation [9, 30], which transformed the study of cosmology from philosophy and mathematics to physics and astronomy. It quickly became clear that the discovery of the radiation content of the Universe, along with its observed expansion, ensured that very early in its evolution, when the temperature and densities (both number and energy densities) were very high, collisions among particles would be very rapid and energetic and, at sufficiently high temperatures, particle-antiparticle pairs would be produced (and would annihilate). Sakharov explored the requirements necessary for such high-energy collisions in the early Universe to create a matter-antimatter asymmetry if none existed initially. Sakharov's recipe for cooking a universal baryon asymmetry has three ingredients. One obvious condition is that baryon number cannot be absolutely conserved; baryon number (B) conservation must be violated. Although the Standard Model of particle physics at the time did not allow for violation of baryon number conservation, the later development of grand unified theories (GUTs) did. Sakharov also noted that the discrete symmetries of parity (P) and charge conjugation (C), replacing particles with antiparticle, or of CP, would need to be broken as well. Current models, in agreement with accelerator data, do allow for P and CP violation. Sakharov's third ingredient is not from particle physics but from cosmology, relying on the expansion of the Universe. The third ingredient in the recipe requires that thermodynamic equilibrium not be maintained when the B, P, and CP violating collisions occur in the early Universe. Although at the time of Sakharov's work there was no evidence that conservation of B and CP were violated, it was already known that parity is not conserved in the weak interactions and that the expansion of the Universe could possibly provide the required departure from thermodynamic equilibrium. Sakharov set the stage for consideration of a universe with unequal amounts of matter and antimatter. We will revisit Sakharov's three conditions for baryogenesis in Section 5.4.

In the hot, dense thermal soup of the very early Universe, matter and antimatter (baryons and anti-baryons) are as abundant as all the other particles whose mass is less than the temperature. As the Universe expands and cools, particle-antiparticle pairs annihilate, leaving behind only the lightest particles, along with any particle-antiparticle pairs that evaded annihilation in the early Universe or, perhaps, in an asymmetric universe, an initial matter excess that escaped annihilation. In the late Universe, when the temperature (in energy units) is far below the masses of the unstable particles of the Standard Model (SM) of particle physics, only photons and the lightest stable (or very long lived) SM particles remain: nucleons (and possibly anti-nucleons), electrons (and possibly positrons), and the three SM neutrinos. In cosmology, it is conventional to refer to all ordinary matter consisting of nucleons and electrons (nuclei, atoms, and molecules), as 'baryons' (B)[1] to distinguish it from dark matter (DM). Electrons are not baryons, but their (very small) contribution to the present-day matter density is included in this definition of the baryon density. The photons and neutrinos are often referred to as 'radiation'. The matter-antimatter asymmetry is the difference between the numbers of baryons and anti-baryons. Since this is an extensive quantity, scaling with the size of the volume considered, it is useful to introduce the ratio (by number) of baryons to photons to quantify the size of any matter-antimatter asymmetry. The ratio of the baryon (minus the anti-baryon) and photon number densities, $\eta_B = n_B/n_\gamma$, provides a measure of the matter-antimatter asymmetry of the Universe. However, as the Universe expands and cools, the heavier, unstable SM particles annihilate and decay, increasing the number of photons N_γ in a co-moving volume V, where $N_\gamma = n_\gamma V$, while the baryon number in the same co-moving volume is unchanged (at least during those epochs when baryons are conserved). Instead, it is the entropy, $S = sV$, in the co-moving volume, not the number of photons, that is conserved as the Universe expands adiabatically. The entropy

---

[1] Throughout this article, the terms baryons, nucleons, ordinary matter, and normal matter are used interchangeably.

and the number of photons in a co-moving volume are related by $S = 1.8 g_s N_\gamma$, where the total entropy is related to the entropy in photons alone by $S \equiv (g_s/2)S_\gamma$, and $S_\gamma = 4/3$ $(\rho_\gamma/T)V = 4/3(\langle E_\gamma \rangle/T)N_\gamma$ and $\langle E_\gamma \rangle = 2.7\,T$. The quantity $g_s = g_s(T)$ counts the number of degrees of freedom contributing to the entropy at temperature T. For the SM of particle physics with three families of quarks and leptons at temperatures above the mass of the heaviest SM particle (the top quark), $g_s \approx 427/4$. For temperatures below the electron mass, after the three flavours of weakly interacting neutrinos have decoupled and the photons have been heated relative to the neutrinos by the annihilation of the $e^\pm$ pairs, $g_s \rightarrow g_{s0} \approx 43/11$. As a result, as the Universe cools from above the top quark mass to below the electron mass, the number of photons in a co-moving volume increases by a factor of ≈ 27, and the baryon to photon ratio is diluted by this same factor. In an adiabatically expanding Universe (as ours is assumed to be) the entropy in a co-moving volume is conserved, along with the net number of baryons minus anti-baryons (during those epochs when baryon number non-conservation is strongly suppressed). Therefore, the ratio of baryon number to entropy, $N_B/S = n_B/s$, provides a measure of the baryon asymmetry whose value is unchanged as the Universe expands and cools. Evaluated in the late Universe, after $e^\pm$ annihilation is complete, $s/n_\gamma \rightarrow (s/n_\gamma)_0 = 1.8\, g_{s0} \approx 7.0$, so that $n_B/s \approx n_B/7.0$. Consistent with most of the published literature, n_B is evaluated here in the late Universe, so $n_B \equiv n_{B0} \equiv (n_B/n_\gamma)_0$. In the discussion here n_B and n_B/s will both be referred to as the 'baryon asymmetry parameter'.

In a matter-antimatter *symmetric* Universe, the baryon asymmetry parameter $n_B = 0$. For a quantity that could, in principle, have any value between $-\infty$ and $+\infty$[2], zero might seem to be a 'natural' choice.

When is a physical parameter, such as the baryon asymmetry parameter, considered to be fine-tuned? The criteria for answering this question, along with a discussion of the degeneracies with other physical parameters, are discussed in Section 5.2. In Section 5.3, the overwhelming observational and theoretical evidence that our Universe is *not* matter-antimatter symmetric is reviewed, excluding the natural choice of $n_B = 0$. Faced with the necessity that a universe hosting stars, planets, and life requires $n_B \neq 0$, Section 5.4 provides an overview of the multitude of particle physics (and cosmology) models proposed to generate a non-zero baryon asymmetry during the early evolution of the Universe. These models are capable of generating a baryon asymmetry that is much smaller or much larger than that observed in our Universe, suggesting that there might be universes with almost any non-zero values of n_B. In an asymmetric universe, the quantitative value of the baryon asymmetry parameter plays an important role in primordial nucleosynthesis (Big Bang nucleosynthesis: BBN), regulating the abundances of the nuclides produced in the early Universe, before any stellar processing. BBN is reviewed for a large range of n_B in Section 5.5. The degeneracy of the baryon asymmetry parameter with other cosmological parameters is discussed in Section 5.6, and a variety of alternate cosmological models allowing

---

[2] In a universe with more 'matter' than 'antimatter', $n_B > 0$. For the opposite case, where $n_B < 0$, the definitions of matter and antimatter could be interchanged. Therefore, without loss of generality, it is assumed here that $n_B \geq 0$.

for a range of η_B values are presented in Section 5.7. The criterion used here to judge the viability of alternate cosmological models is whether their universes are capable of hosting stars, planets, and life. Our results and conclusions are summarised in Section 5.8.

5.2 Definition of Fine-Tuning of the Baryon Asymmetry Parameter

How fine-tuned is the baryon asymmetry parameter? Here we will adopt a definition of fine-tuning based on the capability of the Universe to harbour life. Clearly, small changes in the asymmetry parameter will have little effect on cosmic evolution. However, large changes in this parameter will have major effects, notably altering the production of elements in the early universe and changing the process of structure formation through the growth of primordial density perturbations. We will see that the former, even in extreme cases, is unlikely to have any effect on the development of life in the Universe, while the latter can have profound effects. In particular, if the process of galaxy and star formation is too inefficient, then there will be no planetary systems to harbour life. One must be cautious, of course, in defining the limits on environments that can support life; our argument will be based on life as we observe it, which exists on planets orbiting stars. It is always possible that more extreme environments might harbour life in ways that we have not considered; for example, Avi Loeb has pointed out that the cosmic microwave background can provide an energy source for life when the Universe was only 10 million years old and the temperature of the CMB was between the freezing and boiling points of water [27] (see Chapter 12). While we will not consider such extreme possibilities here, caution is always advised when defining the conditions needed for the existence of life.

The extent to which the value of η_B is fine-tuned will depend on how widely it can be varied while still allowing for the existence of life. The issue of the fine-tuning of η_B is not, of course, a true-false question: the best we can do is to determine an allowed range for η_B. The width of this range can then suggest the plausibility (or lack thereof) of the need for special initial conditions or special values for the underlying fundamental parameters that determine η_B. But the question, 'Is the baryon asymmetry parameter fine-tuned?' does not have a yes or no answer.

In considering the variation of one or more physical parameters, a choice must be made: do we consider the variation of the baryon asymmetry parameter alone, or do we allow other parameters to vary at the same time? In the latter case, changes in the value of one parameter may be compensated, at least in part, by changes in other parameters.

As an example, consider the way in which the relation between the baryon to entropy ratio and the baryon to photon ratio depends on the number of neutrino flavours, as well as on the neutrino decoupling temperature – which depends, in turn, on the strength of the weak interactions. In an alternate universe where there are N_ν flavours of neutrinos, instead of the SM value of $N_\nu = 3$, $g_{s0} = 43/11 + 7(N_\nu - 3)/11 = 43/11(1 + 7(N_\nu - 3)/43)$ and $g_{\rho 0} = 3.36 + 0.454(N_\nu - 3) = 3.36(1 + 0.135(N_\nu - 3))$. For these results, it has been assumed that when $N_\nu \neq 3$, the usual weak interactions are unchanged and all neutrinos decouple when $T_{dec} \gg m_e$ (but $T_{dec} \ll m_\mu$) so that $(T_\nu/T_\gamma)_0 \approx 4/11$. With these caveats, for $N_\nu \neq 3$, the late time entropy per photon is $(s/n_\gamma)_0 \approx 7.0(1 + 7(N_\nu - 3)/43)$, and the relation between η_B and n_B/s is changed,

$$\eta_B = (n_B/n_\gamma)_0 \approx 7.0(1 + 7(N_\nu - 3)/43)\,(n_B/s)\,. \tag{5.1}$$

For example, in an alternate universe with only one neutrino flavour ($N_\nu = 1$), $\eta_B \approx 4.7(n_B/s)$, while in one with eight flavours of neutrinos, ($N_\nu = 8$),[3] $\eta_B \approx 12.8(n_B/s)$.

In general, allowing multiple parameters to vary simultaneously will weaken the constraints provided when only one of them is varied, which is likely to be an issue with many of the other essays in this volume. For example, consider the atomic energy scale,

$$\epsilon \equiv \mu_H c^2 \alpha^2 = \left(\frac{m_e m_p}{m_e + m_p}\right) c^2 \left(\frac{e^2}{\hbar c}\right)^2, \tag{5.2}$$

where μ_H is the reduced mass of the proton-electron system, and the fine-structure constant is $\alpha = e^2/\hbar c \approx 1/137$ (when measured at low energies). For $m_e c^2 \approx 0.51\,\text{MeV}$ and $m_p \approx 0.94\,\text{GeV}$, $\epsilon \approx m_e c^2 \alpha^2 \approx 27\,\text{eV}$. Since ϵ is not a dimensionless parameter, perhaps it is the dimensionless parameter $\epsilon/\mu_H c^2 = \alpha^2 \approx 5.3 \times 10^{-5}$ that is fundamental. Suppose that α and $\mu_H c^2$ are allowed to change, while the atomic energy scale, $\mu_H c^2 \alpha^2$, is kept unchanged. For example, m_e and m_p might change while $m_e/m_p \ll 1$ might be (nearly) unchanged. Atomic energy levels will be largely unchanged while nuclear energies will be changed. How much freedom is there to change α along with other fundamental parameters (e.g., m_e, m_p, m_e/m_p), while leaving most of 'ordinary' atomic and nuclear physics unchanged? This issue of 'degeneracy' among physical parameters will rear its head in the subsequent discussion of the fine-tuning of the baryon asymmetry of the Universe. When exploring model universes with different values of η_B, we will keep all other parameters (e.g., α, m_e/m_p, N_ν, etc.) fixed. However, we need to remain aware that the results presented here can be considerably altered if multiple parameters are simultaneously varied.

5.3 The Case against a Symmetric Universe

Over the years, experiments at ever higher energies have confirmed that particles are created (and annihilated) in pairs and that in all collisions studied so far, baryon (and lepton) number is conserved. Perhaps only at the very highest energies, inaccessible to the current terrestrial accelerators, or in searches for proton decay, will non-conservation of baryon (and lepton) number be revealed. However, it is not unreasonable to ask how our present Universe would differ if baryon number were absolutely conserved. A complementary approach is to ask what astrophysical observations can tell us about the amount of antimatter (if any) in gas, stars, galaxies, and clusters of galaxies in the current Universe (e.g., [4]). These two approaches are explored here. The discussion here is based on several earlier

---

[3] For $N_\nu \leq 8$, QCD is asymptotically free, allowing for quark confinement and bound nuclei [20].

papers by the first author (e.g., [41–43, 45, 46]); the reader is urged to see those papers for details and for many further references.

5.3.1 The Observational Evidence against a Symmetric Universe

To paraphrase remarks by the first author in a 1976 review of the status of antimatter in the Universe [42], it is quite easy to determine if an unknown sample is made of matter or antimatter. The most rudimentary detector will suffice. Simply place your sample in the detector and wait. If the detector disappears (annihilates), your sample contained antimatter. Indeed, if you had handled your sample, you would have already known the answer. Astrophysical sources have been repeating this experiment over cosmological times. The first lunar and Venus probes confirmed that the Moon and Venus are made of matter, not antimatter. Indeed, the solar wind, sweeping past the planets of the solar system revealed by the absence of annihilation gamma rays that the Sun and the planets and other solar system bodies are all made of what we have come to define as matter. Were any of the planets made of antimatter, they would be the strongest gamma ray sources in the sky (if they had not already annihilated away). As may be inferred from the discussion in Section 5.3.2, if there were any antimatter in the material that collapsed to form the planets and other solid body objects in the solar system (the pre–solar system gas cloud) , it would have annihilated long before the solar system formed. The same is true for the stars in our galaxy. On theoretical grounds, is is highly unlikely that in a universe some 14 Gyr old, there are any non-negligible amounts of antimatter surviving in our galaxy.

In a typical nucleon – anti-nucleon annihilation, $\sim$ 5–6 pions are produced. The pions decay to muons, neutrinos, and photons, and the muons decay to electrons ($e^{\pm}$ pairs) and neutrinos. The $e^{\pm}$ pairs may annihilate in flight or, being tied to local magnetic fields, they may lose energy by Compton emission and annihilate nearly at rest (producing a characteristic 511 keV line) [4]. Photons from matter-antimatter annihilations provide the most sensitive, albeit indirect, probe of the presence of antimatter, mixed with ordinary matter, on galactic and extragalactic scales. In the galaxy, gas (clouds of atomic or molecular gas) and stars are inevitably mixed. If either contained significant amounts of antimatter, the result would be annihilation, along with the corresponding production of gamma rays. The lifetime against annihilation of an antiparticle (e.g., an antiproton) in the gas in the interstellar medium (ISM) of the galaxy is very short, $t_{ann} \approx 300$ yr [42]. It is therefore not surprising that observations of galactic gamma ray emission set very strong constraints on the antimatter fraction in the ISM, $f_{ISM} \lesssim 10^{-15}$ [42]. There can be no significant amounts of antimatter in the gas in the galaxy.

What about anti-stars? When gas collapses to form stars, the annihilation rate grows as the number density while the collapse rate increases only as the square root of the density. As a result, unless there were no normal matter in the gas that might collapse to form an anti-star, the anti-star would never form. Setting this aside, let us suppose that anti-stars had somehow formed in the galaxy. As the gas in the ISM flowed past these anti-stars, there would be annihilation, resulting in gamma rays. Using by now outdated

(40-year-old!) gamma ray data, the first author [42] determined that the absence of gamma rays indicated that the nearest anti-star in the galaxy is at least 30 pc away. This result sets an upper limit on the total number of anti-stars, N, that could be in the galaxy: $N < 10^7$, a small fraction of all the stars in the galaxy. Although more recent gamma ray data can refine these bounds, the old data were already sufficiently strong to argue against any significant amounts of antimatter in the galaxy.

Galactic cosmic rays, coming to us from outside of the solar system, provide a valuable direct probe of antimatter in the galaxy. Whatever the sources of the galactic cosmic rays, the discovery of anti-nuclei in the cosmic rays would provide direct evidence (a 'smoking gun') for the presence of antimatter in the galaxy (for more details but obsolete data, see the discussion in [42]). The antiproton would be the lightest anti-nucleus, but in high-energy collisions between cosmic rays and interstellar gas, some 'secondary' antiprotons will be produced. Indeed, antiprotons have been observed in the cosmic rays, but their numbers are consistent with a secondary origin. However, production of more complex anti-nuclei in high-energy cosmic rays – interstellar gas collisions (secondary anti-nuclei) is strongly suppressed, and to date, no anti-deuterons [16] or anti-alpha [2] particles have been detected in the cosmic rays. For example, the 1999 AMS upper bound [2] to the cosmic ray anti-helium-to-helium ratio is $< 10^{-6}$, providing a strong supplement to the gamma ray data suggesting our galaxy has no significant amounts of antimatter. The absence of primary antimatter in the cosmic rays is evidence that the sources of the galactic cosmic rays contain little, if any, antimatter (indeed, if there were some antimatter mixed with a predominant amount of ordinary matter in the cosmic ray sources, they likely would have annihilated over the lifetimes of the sources).

What of external galaxies or extragalactic high-luminosity sources such as AGNs or QSOs? If annihilations deposit their energy locally, then the gamma ray flux and the luminosity of an annihilation-powered source are connected [41]. If Φ_γ is the photon flux from annihilations (photons cm^{-2} s^{-1}) and Φ_E is the energy flux from the same source (ergs cm^{-2} s^{-1}), then $\Phi_\gamma \gtrsim 10^4 \, \Phi_E$ [41]. Although annihilation was proposed as a panacea for the energy budgets of QSOs and other high-luminosity sources [4], the detailed emission mechanisms required enormous magnetic fields, compounding the problems of an already stretched energy budget. Steigman and Strittmatter [46] explored whether observations of the annihilation neutrino flux could constrain models of annihilation-driven infrared emission in Seyfert galaxies. For individual sources, it was estimated [42] that the neutrino flux would be at least five orders of magnitude smaller than was observed at the time. The difficulty of detecting the relatively low energy ($\lesssim 500$ MeV) neutrinos, combined with improved models for the energy sources in QSOs, Seyferts, etc., have made annihilation neutrinos an unlikely probe.

Moving further away, outside our own galaxy, the strongest constraints, on the largest scales, come from observations of X-ray emitting clusters of galaxies [42, 43, 45]. Most of the baryons in clusters of galaxies are in the hot intracluster gas. The same collisions between particles in the intracluster gas responsible for producing the observed X-ray emission would result in annihilation gamma rays if some fraction of the gas consisted of antiparticles. The virtue of using X-ray emitting clusters of galaxies is that there is a direct proportionality between the X-ray emission from thermal bremsstrahlung and gamma ray emission from annihilation. This approach leads to bounds on the antimatter fraction (the fraction of antimatter mixed with ordinary matter) on the largest scales in the universe ($M \sim 10^{14} - 10^{15} M_\odot$, $R \sim$ few Mpc) [45]. Using data from 55 X-ray emitting clusters of galaxies [12] in combination with the upper bounds to the gamma ray fluxes [33], it was found that the antimatter fraction from that sample is limited to $f < 10^{-6}$ [45]. However, even stronger bounds exist for some individual clusters. For the Perseus cluster, $f < 8 \times 10^{-9}$, and for the Virgo cluster, $f < 5 \times 10^{-9}$. Perhaps the most interesting upper bound on antimatter on the largest scales comes from colliding clusters. Analysis of the Bullet Cluster gives $f < 3 \times 10^{-6}$ on the scale $M \sim 3 \times 10^{15} h^{-1} M_\odot$, where h is the Hubble parameter in units of 100 km sec^{-1} Mpc^{-1} [45].

5.3.2 The Problem of a Symmetric Universe

Very shortly after the discovery of the CMB [9, 30], Ya. B. Zeldovich [54][4] and H. Y. Chiu [7] independently considered the fate of matter and antimatter emerging from the early stages of the evolution of a hot universe. The result, whose derivation is outlined here, is easily summarised. At high temperatures, above the quark – hadron transition, there are many quark-antiquark pairs, and in a symmetric universe, there are equal numbers of quarks and antiquarks. As the Universe expands and cools, the quarks (and gluons) are confined into nucleons (neutrons and protons), which – because the strong interaction is strong – are in thermal equilibrium with the cosmic plasma (e.g., photons, neutrinos, and the light leptons and bosons). In this regime, the nucleon mass exceeds the temperature so that annihilation of nucleon–anti-nucleon pairs proceeds on a timescale short compared to the expansion rate of the Universe. But, since $m \gg T$, creation of new nucleon–anti-nucleon pairs from collisions in the background plasma is strongly (exponentially) suppressed so that up to spin-statistics factors of order unity, the ratio of nucleons (and anti-nucleons) to photons is $n_N/n_\gamma = n_{\bar{N}}/n_\gamma = n_{eq}/n_\gamma \propto (m/T)^{3/2} e^{-(m/T)} \ll 1$. Even though the abundances of nucleons and anti-nucleons (e.g., relative to photons) are very small, the strong interaction is strong, ensuring that $n_N \approx n_{eq}$ is maintained down to very low temperatures, $T \ll m_N$. However, eventually, the abundance of the nucleon–anti-nucleon pairs becomes so small that they no longer can find each other to annihilate (and the creation of new pairs is exponentially suppressed), and the abundance of nucleons (and anti-nucleons) 'freezes out' at a 'relic' abundance $(n_N/n_\gamma)_0$. The evolution of the nucleon-anti-nucleon abundances follows an evolution equation, described next, that accounts for creation, annihilation, and the expansion of the Universe. The solution, presented in the following discussion, shows that the relic abundance of the nucleon–anti-nucleon pairs in a symmetric universe is some nine orders of magnitude smaller than the nucleon abundance observed in our Universe, providing an important nail in the coffin of the symmetric Universe.

---

[4] It is interesting that Zeldovich's article was written prior to the discovery of the CMB. As a result, in his review, Zeldovich considered both hot and cold universes.

As first derived from an argument of detailed balance by Zeldovich [54] and later rediscovered and supported by many textbook derivations based on the Boltzmann equation, the evolution of the abundance of a particle (and its antiparticle) produced and annihilated in pairs, is described by the standard evolution equation (SEE); see, e.g., [7, 19, 25, 37, 41–43, 47] and references therein. For equal numbers of particles and antiparticles (no asymmetry, zero chemical potential), the SEE may be written as

$$\frac{1}{V}\left(\frac{dN}{dt}\right) = \frac{dn}{dt} + 3Hn = \langle\sigma v\rangle(n_{eq}^2 - n^2), \qquad (5.3)$$

where $N = nV$ is the number of particles (and antiparticles) in a co-moving volume V. As the Universe expands and the cosmic scale factor, a, increases, the co-moving volume grows as $V \propto a^3$. In Eq. (5.3), the number density of particles and antiparticles is n, the total annihilation cross section is $\langle\sigma v\rangle$, and $H = a^{-1}(da/dt)$ is the Hubble parameter. The SEE is a form of the Ricatti equation, for which there are no known closed-form solutions except in special cases. Although the SEE may be integrated numerically, here the approximate analytic approach first outlined by Zeldovich [54] and employed extensively in [19, 25, 37, 42, 43, 47] and elsewhere is followed.

For the approximate analytic solution to the SEE, it is convenient to write $n = (1 + \Delta)$ n_{eq} where, in the non-relativistic (NR) regime $(T < m)$, $n_{eq} = (g\,T^3/(2\pi)^{3/2})x^{3/2}$ $e^{-x} f(x)$, where $x \equiv m/T$ and $g = 2$ is the number of spin states of the proton (neutron) and of the antiproton (anti-neutron). Here, $f(x)$ is an asymptotic series in x for which $f(x) \to 1$ as $x \to \infty$. For the range of x of interest in tracking the evolution of nucleon–anti-nucleon pairs, $f(x) \approx 1$ is a very good approximation. Therefore, the evolution of the equilibrium number density (as a function of x) in the NR regime is very well described by $n_{eq} \propto T^3 x^{3/2} e^{-x} \propto x^{-3/2} e^{-x}$. Note that since the photon number density varies as T^3, $n_{eq}/n_\gamma \propto x^{3/2} e^{-x}$ in the NR regime. Instead of following the time evolution of the thermal relic abundance, it is more convenient to track its evolution as a function of x. Neglecting small logarithmic corrections involving derivatives related to the entropy and photon densities, the derivatives with respect to time and x (or T) are related by

$$dt \approx \frac{1}{H}\left(\frac{dx}{x}\right) \approx -\frac{1}{H}\left(\frac{dT}{T}\right), \qquad (5.4)$$

where $H = H(T)$ is the Hubble parameter evaluated at temperature T. Now we define the quantity $g_\rho(T)$ (in analogy to g_s) by $g_\rho/2 \equiv \rho/\rho_\gamma$, where ρ is the *total* mass/energy density and ρ_γ is the energy density in photons alone. During those epochs in the evolution of the Universe when the energy density is dominated by the contribution from relativistic particles (radiation dominated: RD), $H \propto \rho_R^{1/2} \propto g_\rho^{1/2}\rho_\gamma \propto g_\rho^{1/2}T^2$. In terms of Δ and x, the SEE may be rewritten as

$$\frac{d(\ln(1 + \Delta)N_{eq})}{d(\ln x)} = -\left(\frac{\Gamma_{eq}}{H}\right)y, \qquad (5.5)$$

where $\Gamma_{eq} \equiv \langle\sigma v\rangle n_{eq}$ and $y \equiv \Delta(2 + \Delta)/(1 + \Delta)$.

For $x \sim O(1)$ $(m \approx T)$, Δ is very small and $n = n_{eq}$ is a very good approximation. As the Universe expands and cools, x increases and Δ grows exponentially (while n_{eq} decreases exponentially), and the departure from equilibrium grows. Define x_* to be the value of x for which $\Delta(x_*) \equiv \Delta_* \sim O(1)$, so the true abundance, n_*, exceeds the equilibrium density, n_{eq*}, by factor $1 + \Delta_* > 1$. (A more precise definition of x_* is given after Eq. (5.13)). For $x \gtrsim x_*$, $\Delta \gtrsim \Delta_*$ and $n/n_{eq} > 1$ increases. In this regime, where $n \gtrsim n_{eq}$, the SEE simplifies,

$$dN/dt = \langle \sigma v \rangle (n_{eq}^2 - n^2) V \approx -\langle \sigma v \rangle n^2 V = -\langle \sigma v \rangle N^2/V . \tag{5.6}$$

This equation can be integrated directly from $t = t_*$ (when $T = T_*$ and $x = x_*$) to $t = t_0$ (when $T = T_0 \ll T_*$ and $x \gg x_*$). Replacing the evolution with time (or with x) by the evolution with temperature,

$$\frac{dN}{N^2} \approx \frac{\langle \sigma v \rangle}{VH} \frac{dT}{T} , \tag{5.7}$$

where the Hubble parameter varies as $H \approx H_*(T/T_*)^2$ and the co-moving volume increases with decreasing temperature as $V \approx V_*(T_*/T)^3$. Integrating from $T = T_*$ to $T = T_0 \ll T_*$ results in

$$N_0/N_* = [1 + (\Gamma/H)_*]^{-1}, \tag{5.8}$$

where $\Gamma_* = n_* \langle \sigma v \rangle$. For nucleon–anti-nucleon annihilation, $(\Gamma/H)_* \gg 1$, so $N_0/N_* \approx (\Gamma/H)_*^{-1} \ll 1$. When $T = T_*$ $(x = x_*)$, the number of particles (neutrons or protons) in the co-moving volume, N_*, may be compared to the number of photons in the same volume, $N_{\gamma *}$,

$$\left(\frac{N}{N_\gamma} \right)_* = \left(\frac{n}{n_\gamma} \right)_* = \left(\frac{H}{n_\gamma \langle \sigma v \rangle} \right)_* \left(\frac{\Gamma}{H} \right)_* . \tag{5.9}$$

In terms of x_*,

$$\left(\frac{H}{n_\gamma \langle \sigma v \rangle} \right)_* = \frac{6.5 \times 10^{-36} g_{\rho *}^{1/2} x_*}{m \langle \sigma v \rangle} , \tag{5.10}$$

where m is in GeV and $\langle \sigma v \rangle$ is in cm^{3}s^{-1}. As the Universe expands and cools from $T = T_*$ to $T = T_0$, the surviving nucleon (and anti-nucleon) abundance(s) decreases (decrease) to an asymptotic ('frozen out') value (ratio to photons) given by,

$$\left(\frac{N}{N_\gamma} \right)_0 = \left(\frac{N}{N_\gamma} \right)_* \left(\frac{N_0}{N_*} \right) \left(\frac{N_{\gamma *}}{N_{\gamma 0}} \right), \tag{5.11}$$

where – from entropy conservation – $N_{\gamma *}/N_0 = g_{s0}/g_{s*}$. Note that $(N/N_\gamma)_0$ is the frozen out ratio of neutrons or protons to photons (long after annihilation has ceased[5]) and is identical to the ratio of anti-neutrons or antiprotons to photons. Even though $(N/N_\gamma)_0 \neq 0$,

---

[5] Annihilations never really cease. They simply become so rare that they are unable to continue to reduce the relic abundance.

the baryon asymmetry parameter in a symmetric universe is $\eta_B = 0$. Combining the preceding equations,

$$\left(\frac{N}{N_\gamma}\right)_0 \approx \frac{2.5 \times 10^{-35}}{m \langle \sigma v \rangle} \left(\frac{g_{\rho *}^{1/2}}{g_{s *}}\right) x_*, \tag{5.12}$$

where $g_{s0} = 43/11$ – corresponding to $N_\nu = 3$ – has been adopted. For neutrons or protons (in the approximation here, they are assumed to have the same mass, $m \approx 0.94\,\text{GeV}$), the total (s-wave) annihilation cross section[6] is $\langle \sigma v \rangle \approx 1.5 \times 10^{-15}\,\text{cm}^3\,\text{s}^{-1}$ so that

$$\left(\frac{N}{N_\gamma}\right)_0 \approx 1.8 \times 10^{-20} \left(\frac{g_{\rho *}^{1/2}}{g_{s *}}\right) x_*. \tag{5.13}$$

To find x_* and $T_* = m/x_*$, in order to evaluate $g_{\rho *} = g_\rho(T_*)$ and $g_{s *} = g_s(T_*)$,[7] we impose the condition defining x_* – that is, when $x = x_*$, $\Delta(x) = \Delta(x_*) \equiv \Delta_*$. Although $\Delta_* \sim O(1)$, a specific choice needs to be made for Δ_* in order to find the corresponding value of x_* (and it needs to be checked and confirmed that the final result is insensitive to this specific choice). Here, $\Delta_* = 0.618$ (related to the 'Golden Mean') is adopted, so $y_* = \Delta_*(2 + \Delta_*)/(1 + \Delta_*) = 1$.

It may be verified that $d(\ln(1+\Delta))/d(\ln x) \ll d(\ln N_{eq})/d(\ln x)$, so Eq. (5.5) reduces to

$$-\left(\frac{\Gamma_{eq}}{H}\right) y \approx \frac{d(\ln N_{eq})}{d(\ln x)}, \tag{5.14}$$

where $N_{eq} = n_{eq} V \propto V T^3 x^{3/2} e^{-x}$. Generally, $V T^3 \propto (aT)^3 \approx$ constant, so the logarithmic derivative of $V T^3$, depending on $d(\ln g_s)/d(\ln dT)$, may be neglected, further simplifying Eq. (5.5) to an algebraic equation,

$$d(\ln N_{eq})/d(\ln x) \approx -(x - 3/2) \approx -(\Gamma_{eq}/H)\,y. \tag{5.15}$$

For $x = x_*$, $\Delta(x_*) = \Delta_* = 0.618$ and $y = y_* = 1$. As a result,

$$x_* - 3/2 = (\Gamma_{eq}/H)_* = n_{eq*} \langle \sigma v \rangle / H_* = A_* g_{\rho *}^{-1/2} x_*^{1/2} e^{-x_*}, \tag{5.16}$$

where $A_* \equiv 4 \times 10^{34} g\,m \langle \sigma v \rangle$; g is the number of neutron or proton spin states, and, as before, the mass m is in GeV and $\langle \sigma v \rangle$ is in cm^3/s. For $g = 2$, $m = 0.94$, and $\langle \sigma v \rangle = 1.5 \times 10^{-15}$, $A_* = 1.1 \times 10^{20}$. The transcendental equation for x_*, Eq. (5.16), may be solved iteratively. The solution is $x_* \approx 43.1$, corresponding to $T_* = m/x_* \approx 21.8\,\text{MeV}$, for which $g_{\rho *} \approx 11.5$ and $g_{s *} \approx 11.4$ [24]. Substituting these values into Eq. (5.13) results in the frozen-out ratios of the surviving numbers of neutrons, protons, anti-neutrons, and antiprotons to photons,

$$\left(\frac{n_n}{n_\gamma}\right)_0 = \left(\frac{n_p}{n_\gamma}\right)_0 = \left(\frac{n_{\bar{n}}}{n_\gamma}\right)_0 = \left(\frac{n_{\bar{p}}}{n_\gamma}\right)_0 \approx 2.3 \times 10^{-19}. \tag{5.17}$$

---

[6] Even though $T_* \ll m$, the nucleons are moving sufficiently rapidly that Coulomb (Sommerfeld) enhancement of the proton-antiproton annihilation cross section, relative to the neutron–anti-neutron annihilation cross section, is unimportant.
[7] For $g_\rho(T)$ and $g_s(T)$, the results of Laine and Schroeder [24] are used here.

The corresponding nucleon (neutron plus proton) and anti-nucleon–to–photon ratios are $(n_N/n_\gamma)_0 = (n_{\bar{N}}/n_\gamma)_0 \approx 4.6 \times 10^{-19}$.

Of course, for this symmetric universe, $\eta_B \equiv (n_N/n_\gamma)_0 - (n_{\bar{N}}/n_\gamma)_0 = 0$. The present mass density of matter (nucleon plus anti-nucleon) is $\rho_B = m(n_N + n_{\bar{N}}) \approx 3.5 \times 10^{-16}\,\mathrm{GeV\,cm^{-3}}$, or $\Omega_B h^2 \approx 3.3 \times 10^{-11}$. In contrast, for our observed *asymmetric* universe, where annihilation of any relic anti-nucleons is very efficient, $(n_N/n_\gamma)_0 \approx 6.1 \times 10^{-10} \gg (n_{\bar{N}}/n_\gamma)_0 \approx 0$ and $\Omega_B h^2 \approx 0.022$. In a symmetric Universe, the abundance of nucleons surviving annihilation in the early universe is smaller than the abundance of nucleons in our asymmetric Universe by some nine orders of magnitude.

Notice that when $T = T_*$, the ratio of the annihilation rate to the expansion rate is very large, $(\Gamma/H)_* \approx (1 + \Delta_*)(x_* - 3/2) \approx 67 \gg 1$. Neither annihilations nor the relic abundances freeze out when $T = T_*$. For $T < T_*$, annihilations continue to reduce the abundances of nucleons, and anti-nucleons and the ratio of the annihilation rate to the expansion rate, Γ/H, continues to decrease. Eventually, for $T \equiv T_f \approx T_*/2$, $(\Gamma/H)_f = 1$, and the relic abundances freeze out (although, depending on T_f, the number of photons in the co-moving volume may continue to increase until $T \lesssim m_e$, further reducing the relic baryon to photon ratio). For $T < T_f$, $n_N = n_{\bar{N}} = n_{Nf} (T/T_f)^3$. For temperatures even slightly below T_f, $(\Gamma/H) \approx H_f/H = (T_f/T)^2 < 1$. Thereafter, the annihilation rate scales as $n\langle\sigma v\rangle \propto T^3$ (for s-wave annihilation) while the expansion rate of the Universe scales as $H \propto T^2$ (during RD epochs in the evolution), so after freeze-out ($T \ll T_f$), $\Gamma/H \approx T/T_f \ll 1$. During matter-dominated (MD) epochs in the evolution of the universe, $H \propto T^{3/2}$, so $\Gamma/H \propto T^{3/2}$, and it is still the case that $\Gamma/H \ll 1$.

By the same argument, nuclear reactions in this universe are extremely suppressed by the very low nucleon density. There can be no primordial nucleosynthesis in a symmetric universe. After freeze-out, as the universe expands and cools, neutrons decay, and the universe is left with protons (and antiprotons) and electrons (and positrons). Note that as the protons and electrons (and antiprotons and positrons) cool and become non-relativistic, the long-range Coulomb interaction enhances, through Sommerfeld enhancement [40], the annihilation cross section, $\langle\sigma v\rangle \to 2\pi(\alpha c/v)\langle\sigma v\rangle \propto T^{-1/2}$. Even so, the ratio of the annihilation rate to the expansion rate still decreases (as $T^{1/2}$ during RD epochs and as T during MD epochs). Recombination cannot occur in such a low-baryon-density universe. In the absence of non-baryonic dark matter, it is unlikely that any collapsed structures (e.g., stars or galaxies) could form in such a low-density, ionised universe; for a more detailed discussion, see Chapter 6. The history (and future) of a symmetric universe is very bleak. The story barely changes if a symmetric universe contains non-baryonic dark matter. If, for example, the presence of DM in a symmetric universe allows collapsed DM structures to form, the relic matter and antimatter would fall into the DM potential wells, increasing their number densities, leading to renewed annihilation, further reducing their already very small abundances. A matter-antimatter symmetric universe simply bears no resemblance to our Universe.

Even in an asymmetric universe, during the very early evolution of the universe, when the temperature is very high, the *equilibrium* abundance of nucleons and anti-nucleons may be much larger than the relic abundance of nucleons in our Universe, $\eta_B = (n_N/n_\gamma)_0 \approx 6 \times 10^{-10}$. These pairs will annihilate until, at some temperature, T, $(n_N/n_\gamma)(g_s(T)/g_{s0}) \approx 6 \times 10^{-10}$. For lower temperatures, the anti-nucleons continue to be annihilated, but the nucleons, due to the asymmetry, are frozen out. For nucleons (protons plus neutrons), $g = 4$, and their equilibrium abundance relative to photons is $n_N/n_\gamma = 0.26 \, g \, x^{3/2} e^{-x} \approx x^{3/2} e^{-x}$, where $x = m_B/T$, and prior to BBN, the average mass per baryon is $m_B \approx 939$ MeV [44] so that $x \approx 939/T$, with T in MeV. Here, we have assumed that $f(x) \approx 1$. To find T, we need to solve $(939/T)^{3/2} \exp(-939/T) \approx 1.5 \times 10^{-10} g_s(T)$. Using [24] for $g_s(T)$, the solution is $T \approx 38 \, \text{MeV}$ ($x \approx 25$). To avoid the annihilation catastrophe in a symmetric universe, the baryon asymmetry must have been created when $T > 38 \, \text{MeV}$ (or when $T \gg 38 \, \text{MeV}$). Recall that $T_* \approx 22 \, \text{MeV}$, so $T > T_*$, as expected. In the extensions of the Standard Models of particle physics and cosmology that allow for a baryon asymmetry at low temperatures, the energy-/temperature-/mass-scales are orders of magnitude larger than this conservative estimate.

5.4 Particle Physics Models for Generating the Universal Matter-Antimatter Asymmetry

It is clear from the preceding two sections that a universe containing equal abundances of baryons and anti-baryons is *not* the Universe we actually observe. At some point in its evolution, the Universe must have developed an asymmetry between matter and antimatter. How did this asymmetry come about?

One possibility is that the Universe actually began in an asymmetric state, with more baryons than anti-baryons. This is, however, a very unsatisfying explanation. Furthermore, if the Universe underwent a period of inflation (i.e., very rapid expansion followed by reheating), then any pre-existing net baryon number would have been erased. A more natural explanation is that the Universe began in an initially symmetric state, with equal numbers of baryons and anti-baryons, and that it evolved later to produce a net baryon asymmetry.

As we noted in the introduction, Sakharov introduced three conditions necessary to produce a net baryon asymmetry in a universe that began with zero net baryon number. These Sakharov conditions form the basis of nearly all modern theories of baryogenesis, so we will review them in more detail here. These conditions are as follows:

1. *Baryon number violation.* This is the most obvious component needed for baryogenesis. If the Universe began with zero net baryon number, and baryon number were conserved, then it would still have zero net baryon number today.
2. *C and CP violation.* The operator C changes particles into antiparticles and vice versa, while CP also flips all three coordinate axes. A universe that is baryon–anti-baryon symmetric is unchanged when C or CP is applied, while the same is not true for a universe with a net baryon excess. Hence, the production of a baryon asymmetry requires C and CP violation.

3. *A departure from thermodynamic equilibrium.* If baryon and C/CP were violated
   while thermal equilibrium conditions prevailed, then the chemical potentials for
   baryons would be driven to zero, and the only possible difference between particle
   and antiparticle abundances would arise if there were a mass difference between them.
   But CPT invariance implies that the masses of particles and antiparticles are the same.
   Hence, Sakharov conditions 1 and 2 allow for a net baryon number to be created only
   when the particles of interest are out of thermal equilibrium.

While we know the general conditions necessary to generate a baryon asymmetry from
an initially symmetric state, we are far from having a single accepted theory of baryogenesis. Here we will outline some of the ideas that have been proposed over the years. For
some of the earliest work in this field, see References [10, 13, 28, 50, 53]. For reviews of
this topic, see References [11, 34].

Perhaps the simplest class of models (and one of the earliest to be investigated) involves
the decay of massive particles. Consider a particle-antiparticle, X and $\bar{X}$, that has dropped
out of thermal equilibrium in the early Universe, in the sense defined in Section 5.3.2.
Suppose the X can decay into two different channels, with baryon numbers B_1 and B_2,
respectively, while $\bar{X}$ decays into the corresponding 'anti'-channels, with baryon numbers
$-B_1$ and $-B_2$, respectively. Invariance under CPT guarantees that the *total* decay rate for
an antiparticle must be equal to the decay rate for the corresponding particle. However,
it says nothing about individual branching ratios. So it is possible, for instance, for the
branching ratio of X into the channel with baryon number B_1 (which we will take to be r)
to be different from the branching ratio of $\bar{X}$ into the channel with baryon number $-B_1$,
which we will call $\bar{r}$. The possibility of such a difference is the key idea underlying this
mechanism for baryogenesis. Note that $r \neq \bar{r}$ is only possible if C and CP are violated.

With the previously defined branching ratios and baryon numbers, the net baryon number produced from each pair of X and $\bar{X}$ decays is

$$B = B_1 r + B_2(1 - r) - B_1 \bar{r} - B_2(1 - \bar{r}),$$
$$= (B_1 - B_2)(r - \bar{r}). \qquad (5.18)$$

Eq. (5.18) illustrates the necessity of the three Sakharov conditions. If C and CP were
not violated, we would have $r = \bar{r}$, and the right-hand side of Eq. (5.18) would be zero.
Similarly, the possibility that X can decay into two different channels with different
baryon numbers is only possible if B is not conserved; otherwise, we would have $B_1 = B_2$
and again the right-hand side of Eq. (5.18) would be zero. Finally, we assumed out-of-
equilibrium conditions in setting up this scenario; i.e., when they decay, X and $\bar{X}$ are not
in equilibrium with the thermal background, either through annihilations with each other
or through inverse decays. If this were not the case, the particles produced in the X and
$\bar{X}$ decays would simply assume thermal equilibrium abundances, which would yield equal
baryon and anti-baryon densities.

The scenario we have sketched out here is a toy model; for more detailed models see,
e.g., Reference [22]. Models of this sort were first advanced in connection with physics

at the GUT (grand-unified) scale, $T \sim 10^{15} - 10^{16}$ GeV. However, these ideas run into trouble if inflation is assumed to occur in the early Universe. The reason is that, as we have noted, inflation wipes out any pre-existing baryon asymmetry, so baryogenesis must occur after inflation, and currently favoured models of inflation do not reheat the Universe to a temperature as high as the GUT scale.

Another possibility for baryogenesis is the Affleck-Dine mechanism [1]. This model is motivated by supersymmetry, in which all of the particles of the Standard Model have corresponding super-partners with opposite spin statistics (fermions are paired with bosonic super-particles and bosons with fermionic super-partners). The Affleck-Dine mechanism invokes a scalar field that can carry a net baryon number. The field is initially frozen at early times but begins oscillating when the Hubble parameter drops below its mass. During these oscillations, the scalar field acquires a net baryon number, which is transferred at later times into Standard Model particles.

Electroweak baryogenesis [23] is based on the idea that the Universe underwent an electroweak phase transition at a temperature $T \sim 100$ GeV, when the Higgs field dropped into its vacuum state, giving masses to the quarks, leptons, and gauge bosons. If the electroweak phase transition is first order, it can temporarily drive the Universe out of thermal equilibrium as bubbles of the low-temperature vacuum nucleate, expand, and collide, ultimately occupying all of space. The production of baryons occurs in this out-of-equilibrium state near the walls of these expanding bubbles. Electroweak baryogenesis does require physics beyond the Standard Model, as the measured Higgs boson mass implies that the phase transition would not be first order in the Standard Model. This new physics would couple to the Higgs boson, altering its production and decay. Thus, the viability of these models can be tested in the laboratory.

Another proposal goes under the heading of leptogenesis [17]. These models are based on a result by 't Hooft [48], who showed that even in the Standard Model, baryon number is violated by non-perturbative electroweak processes. These processes conserve B−L but not B (the baryon number) and L (the lepton number) separately. Furthermore, while the rates for such processes are very low at low temperatures, they can be much higher in the early Universe. Leptogenesis, then, is the production of a net lepton asymmetry in the early Universe – e.g., through massive particle decay as discussed before. Then non-perturbative electroweak effects transfer some of the net lepton number into a net baryon number.

This is by no means an exhaustive list of models for baryogenesis, which remains very much an open and active field of research. At this point, we are confident of the ingredients required in any successful model (the Sakharov conditions), and we have a very accurate measure of the desired outcome (the observed baryon asymmetry), but the determination of the correct model for baryogenesis remains an ongoing effort.

5.5 The Baryon Asymmetry Parameter and Primordial Nucleosynthesis

In the Standard Model of particle physics and cosmology, the baryon asymmetry parameter plays a key role in BBN, regulating the rates of the nuclear reactions synthesizing (and destroying) the nuclides heavier than hydrogen. BBN in the Standard Model (SBBN) and in extensions of the SM when various nuclear physics and other parameters are allowed to vary is described in Uzan's contribution to this volume (Chapter 7). Here we are mainly concerned with the BBN predicted primordial (prestellar) abundances of the light nuclides, along with the CNO abundances.

5.5.1 Standard BBN

The SBBN-predicted abundances [29], the ratios by number compared to hydrogen, are shown as a function of η_B in Figure 5.1 for a factor of 1,000 range in η_B, for the SM case of $N_\nu = 3$. Agreement between the predicted and the observationally inferred deuterium abundance and the Planck observations of the CMB power spectrum imply a value of $\eta_B \sim 6 \times 10^{-10}$. This value is shown by the dashed vertical line in Figure 5.1. This value of η_B also provides good agreement with the primordial ^{4}He abundance derived from observations. However, it predicts a primordial ^{7}Li abundance roughly three times larger than the

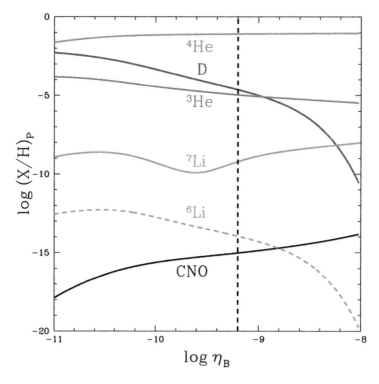

Figure 5.1 The primordial abundances predicted by SBBN [29] for a large range of the present value of the baryon to photon ratio $\eta_B = (n_B/n_\gamma)_0$. For all abundances (including ^{4}He), the ratio to hydrogen by number is shown. The dashed vertical line indicates the current SM value of $\eta_B \approx 6 \times 10^{-10}$.

observationally inferred abundance; this primordial lithium problem remains unresolved at present (see Reference [14] for a recent review).

As seen in Figure 5.1, over this large range in the baryon asymmetry parameter, the abundance trends are quite simple: as η_B increases, the ^{4}He abundance increases monotonically, but very slowly ($\sim$ logarithmically); the abundances of D, ^{3}He, and ^{6}Li are all monotonically decreasing, while the abundances of the CNO nuclides increase. In contrast, the evolution of the abundance of ^{7}Li is non-monotonic. Starting from very small values of η_B, as η_B increases, the ^{7}Li abundance first increases (until $\eta_B \sim 3 \times 10^{-11}$), then decreases (until $\eta_B \sim 3 \times 10^{-10}$), and finally increases again (eventually, for even larger values of η_B, the ^{7}Li abundance will decrease, being replaced by CNO and heavier nuclides). As η_B increases from 10^{-11} to 10^{-8}, ^{4}He/H increases by a factor of ~ 4, from ^{4}He/H ~ 0.024 ($Y_P \sim 0.09$) to ^{4}He/H ~ 0.093 ($Y_P \sim 0.27$), and the deuterium abundance decreases dramatically, from D/H $\sim 5 \times 10^{-3}$ to D/H $\sim 3 \times 10^{-11}$. Over the same range in η_B, the ^{3}He abundance decreases more slowly, from $\sim 10^{-4}$ to $\sim 3 \times 10^{-6}$, and the ^{7}Li abundance ranges from $\gtrsim 10^{-10}$ to $\lesssim 10^{-8}$, while the abundance of the CNO nuclides increases from $\sim 10^{-18}$ to $\sim 10^{-14}$.

For the value of the baryon asymmetry parameter inferred for the observed Universe, $\eta_B \sim 6 \times 10^{-10}$, ^{4}He/H ~ 0.082 ($Y_P \sim 0.25$), D/H $\sim 2.5 \times 10^{-5}$, ^{3}He/H $\sim 1.1 \times 10^{-5}$, ^{7}Li/H $\sim 5.4 \times 10^{-10}$, and the abundances of all the other primordial nuclides are $\lesssim 10^{-14}$. For a very wide range in the baryon asymmetry parameter, the gas that will become the first stars in the Universe consists mainly of hydrogen and helium (^{4}He), with only trace amounts of any other, heavier nuclides. Note, however, that for $\eta_B \gg 10^{-8}$, the primordial abundances of the CNO and heavier nuclides may become non-negligible (see Section 5.5.2). In the absence of significant CNO (or D) abundances, it is the hydrogen and helium content of the primordial gas that will most influence the formation, structure, and evolution of the first stars.

5.5.2 BBN for a Larger Range of Baryon Asymmetries

In the seminal BBN paper of Wagoner, Fowler, and Hoyle (WFH) [52], and in several follow up papers by Wagoner [51], Schramm and Wagoner [39], and Schramm [38], a much larger range in the baryon asymmetry parameter was explored than is shown here in Figure 5.1. In the WFH paper, a range of some 8.5 orders of magnitude was considered, $-12 \lesssim \log \eta_B \lesssim -3.5$, while in the other cited papers, the range is five orders of magnitude, $-11 \lesssim \log \eta_B \lesssim -6$. Although the quantitative BBN yields in those papers, based on what are now outdated nuclear and weak interaction rates (especially the much revised neutron lifetime), should be taken with a large grain of salt, the trends of the yields with η_B revealed in those papers are likely robust.

For example, over the entire range explored, the helium mass fraction increases (and the hydrogen mass fraction decreases) monotonically with η_B. Over the same range in η_B, the D and ^{3}He mass fractions decrease monotonically, with the deuterium abundance falling much more rapidly than the ^{3}He abundance. The evolution of the ^{7}Li mass fraction, X_7, is more interesting. At the lowest baryon asymmetries, X_7 increases from being negligible at $\log \eta_B \sim -12$ to a local maximum, a hint of which may be seen in Figure 5.1, when $\log \eta_B \sim -10.5$. Then, as η_B continues to increase, X_7 decreases to a local minimum at $\log \eta_B \sim -9.5$, as may be seen in Figure 5.1. For $\eta_B \gtrsim 3 \times 10^{-10}$, X_7 increases to another local maximum when $\log \eta_B \sim -6$, after which X_7 decreases monotonically for all larger values of η_B. For $\log \eta_B \lesssim -8$, the abundances of the CNO and heavier nuclides are negligible. As η_B continues to increase, so, too, do the CNO abundances, surpassing the ^{3}He and ^{7}Li abundances for $\log \eta_B \gtrsim -6$. However, almost as soon as the CNO nuclides become large enough to be of possible interest, they decrease as η_B continues to increase, being replaced by even heavier nuclides. The trend seen at the very highest values of the baryon asymmetry parameter in the WFH paper suggests that at sufficiently high values of η_B, the iron peak elements might be produced during primordial nucleosynthesis. It is interesting to speculate if even larger baryon to photon ratios might lead to the r-process elements.

As discussed in Section 5.2, in determining if the baryon asymmetry parameter is fined tuned, we are asking if stars, planets, and life could exist in alternate universes with different values of η_B. In this case, we need to check if the primordial abundances in alternate universes allow for the cooling and collapse of primordial gas clouds to form the first stars and if, in the course of evolution of those stars, the elements required for life can be synthesised.

5.6 Relation between the Baryon Asymmetry Parameter and the Observable Cosmological Parameters

In our present-day Universe, the parameter η_B is not a directly observable quantity. Instead, we measure quantities such as the baryon density or the CMB temperature, from which η_B can be inferred. In this section, we examine the relation between η_B and the observable cosmological quantities.

In a matter-antimatter asymmetric universe such as ours, the baryon asymmetry parameter is related to the contribution of baryons (normal matter) to the total mass density. As a result, the magnitude of the baryon asymmetry plays a role in the evolution of the Universe and in the growth and evolution of structure in it. For the discussion here, it is assumed that the Universe is, on average, homogeneous and is expanding isotropically so that its evolution is described by the 'Friedman equation',

$$\left(\frac{H}{H_0}\right)^2 = \Omega_R \left(\frac{a_0}{a}\right)^4 + \Omega_B \left(\frac{a_0}{a}\right)^3 + \Omega_{DM} \left(\frac{a_0}{a}\right)^3 + \Omega_k \left(\frac{a_0}{a}\right)^2 + \Omega_\Lambda. \qquad (5.19)$$

In Eq. (5.19), the subscript 0 indicates the present ($t = t_0$) value of the parameters, and $H = a^{-1}(da/dt)$ is the Hubble parameter, quantifying the expansion rate of the Universe, where $a = a(t)$ is the cosmic scale factor. The subscripts, R, B, DM, k, and Λ stand, respectively, for the contributions to the total mass/energy density from 'radiation' – i.e., massless particles or particles whose total energy (rest mass plus kinetic) far exceeds the rest mass energy), baryons ('normal' or 'ordinary' matter), dark matter (non-baryonic matter),[8] curvature, and a cosmological constant. For simplicity, we assume here that the observed accelerated expansion of the Universe is driven by a cosmological constant rather than a time-varying dark-energy component. Since the mass densities of baryonic and dark matter evolve the same way (e.g., $\rho \propto a^{-3}$), it is convenient to introduce a parameter describing the 'matter density', the total mass density in non-relativistic particles, $\Omega_M \equiv \Omega_B + \Omega_{DM}$. At the present epoch ($t = t_0$), a 'critical density' of the Universe may be identified, $\rho_{crit\,0} \equiv 3H_0^2/8\pi G = 1.05 \times 10^{-5} h^2 \, \text{GeV cm}^{-3}$, where $H_0 \equiv 100\,h \, \text{km s}^{-1} \, \text{Mpc}^{-1}$ and G is Newton's gravitational constant. Here and elsewhere, we will often set $c = 1$ and express masses in energy units. The parameters Ω_i that appear in Eq. (5.19) are the ratios of the various contributions to the present energy densities, normalised to the present critical density: $\Omega_i \equiv (\rho_i/\rho_{crit})_0$.

Consider the relation between the baryon asymmetry parameter (η_B), the baryon mass density parameter (Ω_B), and the Hubble constant (H_0). The present mass/energy density in ordinary (baryonic) matter is $\rho_{B0} = m_B n_{B0} = \Omega_B \, \rho_{crit\,0} \approx 1.05 \times 10^{-5} \, \Omega_B h^2 \, \text{GeV cm}^{-3}$. For an average mass per baryon of $m_B \approx 0.938 \, \text{MeV}$ [44][9], the present baryon number density is $n_{B0} \approx 1.12 \times 10^{-5} \, \Omega_B h^2 \, \text{cm}^{-3}$. If the present temperature of the CMB photons is T_0 (in degrees Kelvin), then the present photon number density is $n_{\gamma 0} \approx 20.3 \, T_0^3 \, \text{cm}^{-3}$, and the present baryon-to-photon ratio is $\eta_B \approx 5.54 \times 10^{-7} \, (\Omega_B h^2/T_0^3)$ so that $\eta_B \approx 5.54 \times 10^{-7} \, (\Omega_B h^2/T_0^3)$ and $n_B/s \approx 7.87 \times 10^{-8} \, (\Omega_B h^2/T_0^3)$ (for three flavours of SM neutrinos). Note that the connection between the baryon asymmetry parameter (η_B or n_B/s) and the present mass density in ordinary matter ($\propto \Omega_B h^2$) depends on the present ($t = t_0$) value of the photon temperature (T_0). If the baryon asymmetry parameter were to change by some factor, the combination $\Omega_B h^2/T_0^3$ would change by the same factor, resulting in changes to the other universal observables (e.g., Ω_B, h, T_0), separately or in combination. The baryon asymmetry parameter is degenerate with these other cosmological parameters. In particular, changes in Ω_B alone would change the expansion history of the Universe, as may be seen from the Friedman equation. The interconnections (degeneracies) among the cosmological observables complicate any discussion of the effect on the history and evolution of the Universe resulting from changes to any one of them (e.g., the baryon asymmetry parameter).

If the Friedman equation, Eq. (5.19), is evaluated at present ($t = t_0$), when $H = H_0$, there is one condition on the five parameters,

$$1 = \Omega_R + \Omega_B + \Omega_{DM} + \Omega_k + \Omega_\Lambda = \Omega_R + \Omega_M + \Omega_k + \Omega_\Lambda, \tag{5.20}$$

leaving four free parameters. For our observed Universe, $\Omega_k \ll 1$ and $\Omega_R \ll 1$ so that $\Omega_B + \Omega_{DM} + \Omega_\Lambda \approx 1$. There are still three parameters and only one constraint, leaving

---

[8] For agreement with observations of structure formation and its growth as the Universe evolves, it is assumed that the DM is 'cold', in the sense that for those epochs when deviations from homogeneity occur, the DM particles are moving slowly ($v \ll c$).

[9] In the post-BBN universe, when the baryons are mainly protons and alpha particles (hydrogen and helium), the average mass per baryon depends on the helium abundance (mass fraction, Y_P). For $Y_P \approx 0.25$, $m_B \approx 938.112 + 6.683$ $(Y_P - 0.250) \, \text{MeV}$ [44].

two free parameters. By writing $\Omega_M = \Omega_B + \Omega_{DM}$, it might appear that there are only two parameters and one constraint, $\Omega_M + \Omega_\Lambda \approx 1$. However, the ratio Ω_B / Ω_{DM} remains a free parameter, so there are still three parameters with one constraint among them.

In the next section, we will consider how the evolution of the Universe changes when η_B differs from its observed value. While our intention is to keep all of the other cosmological parameters constant, there remains an ambiguity in the way we treat them. Note that η_B is a dimensionless ratio of two quantities, the baryon and photon number densities. When we alter this quantity, we can consider two different possibilities: (1) changing n_B relative to the other cosmological parameters while leaving n_γ unchanged relative to these parameters or (2) keeping n_B fixed while changing n_γ relative to the other cosmological parameters. While each of these possibilities produces a change in η_B, they differ in their treatment of the way that n_B and n_γ change relative to the other cosmological quantities of interest. (Of course, these are only the two simplest possibilities; one could consider allowing the ratios of *both* n_B and n_γ relative to the other cosmological parameters to change, but by different amounts, thus changing η_B as well).

Which of the two approaches spelled out in the previous paragraph is the correct one? Absent a particular model for a different universe with a different value of η_B, it is impossible to say. However, the first possibility seems to be the more natural one. If we assume that baryogenesis is independent of the processes that led to dark matter or dark energy, then tweaking the model for baryogenesis will alter n_B by the same factor relative to all of the other cosmological parameters of interest. This is the case we will consider in detail.

Let F be the ratio of the value of η_B in some hypothetical universe relative to its value in our Universe; our goal will be to understand what constraints, if any, can be placed on F. We will use a tilde to denote physical quantities in a hypothetical universe in which η_B has changed, and quantities without a tilde will denote the corresponding values of these quantities in our Universe, so

$$\tilde{\eta}_B = F\eta_B. \tag{5.21}$$

In case (1) discussed earlier, the ratios ρ_B / ρ_{DM}, ρ_B / ρ_Λ, and n_B / n_ν change in proportion to the change in η_B, while n_γ / ρ_{DM}, n_γ / ρ_Λ, and n_ν / n_γ remain the same. Thus, we have

$$\tilde{\rho}_B / \tilde{\rho}_{DM} = F\rho_B / \rho_{DM}, \tag{5.22}$$

$$\tilde{\rho}_B / \tilde{\rho}_\Lambda = F\rho_B / \rho_\Lambda, \tag{5.23}$$

$$\tilde{n}_B / \tilde{n}_\nu = Fn_B / n_\nu. \tag{5.24}$$

Of course, there are other possibilities that we will not explore here. In an alternate universe with a late production of entropy, n_B would remain unchanged while the ratio of n_γ to all of the other cosmological parameters would be altered. Alternately, if baryogenesis were linked to the process that produced dark matter (as it is in some models), one might consider the possibility of changing η_B while leaving ρ_B / ρ_{DM} fixed. Nonetheless, we feel that the model spelled out in Eqs. (5.22)–(5.24) is the most natural way in which to modify η_B, and this is the case we will now attempt to constrain.

5.7 Alternate Universes with Different Baryon Asymmetry Parameters

Changing η_B alters the evolution of the Universe in two ways: it changes BBN, and it alters the processes that give rise to structure formation and ultimately yield stars and planets. We will consider both effects in turn.

First, consider our Universe at present. Our Universe is very well described by a ΛCDM cosmological model with $\Omega_k \approx 0$, $\Omega_R \ll 1$, and $\Omega_B < \Omega_{DM} < \Omega_\Lambda$ ($\Omega_B + \Omega_{DM} + \Omega_\Lambda \approx 1$). For our observed Universe, a good approximation to the 2015 Planck CMB observations [31] is $\Omega_\Lambda \approx 0.7$, $\Omega_M \approx 0.3$, $\Omega_B \approx 0.05$, $\Omega_{DM} \approx 0.25$. For a ΛCDM cosmology with $\Omega_\Lambda \approx 0.7$, $H_0 t_0 \approx 0.96$, so for $H_0 \approx 68\,\mathrm{km\,s^{-1}\,Mpc^{-1}}$, $t_0 \approx 13.8\,\mathrm{Gyr}$. For the present CMB temperature, the Fixsen *et al.* [15] result may be approximated by $T_0 \approx 2.7\,\mathrm{K}$, corresponding to a CMB photon number density $n_{\gamma 0} \approx 400\,\mathrm{cm^{-3}}$ (compared to the more accurate results, $T_0 = 2.7255\,\mathrm{K}$ and $n_{\gamma 0} \approx 411\,\mathrm{cm^{-3}}$).

5.7.1 Effect on BBN

What happens to BBN when we allow for extreme variations in η_B? As noted earlier, the most important effect of increasing η_B is to increase the primordial ^{4}He mass fraction at the expense of hydrogen. One might imagine that a universe in which stellar evolution begins with almost pure ^{4}He might be less hospitable to life. For example, Hall *et al.* [21] pointed out that in such a universe, halo cooling takes longer, stellar lifetimes are reduced, and there is less hydrogen to support organic chemistry. (The calculations in Reference [21] are focused on variations in the weak scale rather than the magnitude of the baryon asymmetry). However, even extreme increases in η_B do not produce primordial ^{4}He mass fractions close to 100%. For example, a value of η_B as large as 10^{-3} (more than six orders of magnitude larger than the observed value) yields a ^{4}He mass fraction of only 0.4 [36].

Large values of η_B also open up the possibility of producing heavier elements in BBN. Consider first the CNO elements. In standard BBN, these are produced in very small amounts, with abundances relative to hydrogen of CNO/H $\sim 10^{-15} - 10^{-14}$ [8]. However, the abundances of these elements are an increasing function of η_B, peaking at CNO/H $\sim 10^{-8}$ for $\eta_B \sim 10^{-5}$ and decreasing for larger values of η_B [52]. Even a small primordial abundance of CNO/H could affect the evolution of the first generation of stars, as noted in Reference [5]; this evolution begins to change when CNO/H increases above 10^{-11}. Nonetheless, it seems unlikely that such a change would affect the ability of the Universe to harbour life. For $\eta_B > 10^{-5}$, the abundance of the CNO elements begins to decrease as the nuclei are converted into even heavier elements [36, 52]. However, even extreme increases in the value of η_B result in only trace amounts of such heavy elements. In terms of models that can support life, it does not appear that BBN provides a useful upper bound on η_B, and it certainly does not provide a bound competitive with arguments from structure/galaxy/star formation.

Now consider BBN in the limit of very low values for η_B. In this limit, the ^{4}He abundance becomes negligible while ^{2}H increases, reaching a peak abundance of order

$D/H \sim 10^{-2}$ when $\eta_B \sim 2 \times 10^{-12}$. For smaller values of η_B, even the deuterium abundance decreases as η_B is reduced, yielding – in the limit $\eta_B \rightarrow 0$ – a primordial universe consisting essentially of pure hydrogen. The reduction in primordial helium for small values of η_B is likely to reduce the cooling of galaxies that results from the collisional excitation of ionised helium, but this is unlikely to have a major impact [21]. On the other hand, a significantly larger abundance of deuterium would lead to enhanced molecular cooling through an increase in the HD abundance [26]. While interesting, this is also unlikely to affect the prospects for a life-bearing universe.

Our conclusion, then, is that BBN provides essentially *no* constraints on universes with different values of η_B. The formation of stars and planets and the development of life is nearly completely insensitive to variations in the primordial element abundances, at least within the ranges of η_B that we have considered here.

5.7.2 Effect on Large-Scale Structure: The Linear Regime

In the Standard Model for structure formation, small initial fluctuations in the density are imprinted on the matter and radiation by inflation or some other process early in the evolution of the Universe. When the Universe is radiation dominated, these fluctuations cannot grow inside of the horizon; subhorizon fluctuations begin to grow once matter dominates the radiation. If $\delta\rho/\rho$ represents the magnitude of the fluctuation in the matter density relative to the mean matter density, then after matter domination begins, $\delta\rho/\rho$ grows proportional to the scale factor a,

$$\delta\rho/\rho \propto a. \tag{5.25}$$

Eq. (5.25) applies only as long as $\delta\rho/\rho \ll 1$; in this case, the density fluctuations are said to be in the *linear* regime. Once $\delta\rho/\rho > 1$, the Universe enters the non-linear regime, and the analytic solution given by Eq. (5.25) no longer applies. Numerical simulations are necessary to evolve the density field further forward in time. In the non-linear regime, the fluctuations in the matter density grow much more rapidly, and the dark matter ultimately collapses into halos.

This process applies in a straightforward way only to dark matter, which is collisionless. The baryons evolve in a more complicated way. At high temperatures ($T \gg 10^3$ K), the matter is ionised, and the cross section for scattering off of photons is very high. Thus, the baryons are frozen to the radiation background, and baryonic density perturbations cannot grow. As the temperature drops, the electrons become bound to the protons and to the primordial helium nuclei in a process known as recombination.[10] At this point, the density perturbations in the baryons can begin to grow along with the dark matter perturbations. A further complication is that in the non-linear regime, the baryonic matter, unlike the dark matter, is not pressureless and can also radiate away energy in the form of photons. Thus, at late times, the baryons evolve very differently than the dark matter. The end result is

---

[10] Note that this term is a bit misleading, as the electrons and atomic nuclei were never 'combined' to begin with.

that the baryons ultimately bind into fairly compact disks or ellipsoids (galaxies), fragment into stars, and form planets, while the dark matter remains in the form of diffuse halos surrounding the galaxies.

In considering the effect of changing η_B, we must therefore consider the change in two key parameters: the redshift of equal matter and radiation and the redshift at which recombination occurs. However, redshifts are defined relative to the present day, so they are not particularly useful in determining whether a modified universe can support life because we are not restricting life to form at redshift zero as it does in our Universe. Instead, we should examine the temperature of equal matter and radiation and the temperature of recombination. In our Universe, the temperature of equal matter and radiation, T_{eq} is given in terms of the present-day temperature, T_0, by $T_{eq} = T_0(\rho_M/\rho_\gamma)_0$. For the parameter values given at the beginning of this section, we obtain $T_{eq} = 9,000$ K. How does this change when η_B is altered? To determine this, note that the redshift of equal matter and radiation is given by this ratio of present-day densities:

$$1 + z_{eq} = \left(\frac{\rho_{DM} + \rho_B}{\rho_\gamma + \rho_\nu}\right)_0. \tag{5.26}$$

Here we are ignoring the fact that the neutrinos can become non-relativistic at very late times. Then we have

$$1 + \tilde{z}_{eq} = \left(\frac{\rho_{DM} + F\rho_B}{\rho_\gamma + \rho_\nu}\right)_0. \tag{5.27}$$

Using the values for the preceding cosmological parameters, we can trace out the effect of F on T_{eq}. We have $\rho_{DM}/\rho_B \approx 5$. Thus, z_{eq} changes little for $F \lesssim 5$ while, for $F \gtrsim 5$, we have $(1 + \tilde{z}_{eq}) = (F/5)(1 + z_{eq})$. Then we have

$$\tilde{T}_{eq} \approx T_{eq} \quad (F \lesssim 5), \tag{5.28}$$

$$\tilde{T}_{eq} \approx \frac{F}{5}T_{eq} \quad (F \gtrsim 5). \tag{5.29}$$

Now consider the effect of altering η_B on the recombination temperature T_{rec}. While recombination is a gradual process and does not occur suddenly at a single temperature, for the purposes of this study, it will be sufficient to take $T_{rec} \approx 3,000$ K. The process of recombination depends primarily on the ratio of the photon temperature to the binding energy of hydrogen, but there is also a residual dependence on η_B. This dependence comes about because η_B^{-1} determines the number of photons per hydrogen atom; an increase in this number makes it easier for photons to ionise the hydrogen, delaying recombination, while the reverse is true if the number of photons per hydrogen atom decreases. However, the temperature at which a given ionisation fraction is reached varies roughly logarithmically with η_B. This is a much smaller effect than the change in T_{eq} with η_B, so we will ignore it in what follows and take the recombination temperature to be roughly insensitive to changes in η_B.

Now we can investigate the effect of changing η_B on large-scale structure in the linear regime. We will not consider any possible changes in the magnitude of the primordial density fluctuations; we will assume that these are unaltered. We see that neither of the parameters affecting large-scale structure are modified if $F \ll 1$, so the process of structure formation, at least in the linear regime, proceeds in the same way as in our Universe. The density of baryons relative to dark matter will be much lower, leading to fewer galaxies per dark matter halo, but this by itself does not seem to be a barrier to the formation of stars and planets. In the opposite limit ($F \gg 1$), the Universe will be become matter dominated early on, but baryonic structure formation will not occur until the temperature drops down to T_{rec}, which is essentially unchanged from its current value. So in this case, too, we expect little change to the process of structure formation.

5.7.3 Effect on Large-Scale Structure: The Non-linear Regime

Linear perturbation growth allows density perturbations to grow until $\delta \rho / \rho \sim 1$, but it is the subsequent non-linear perturbation growth that directly produces galaxies, stars, and planets. Unfortunately, non-linear perturbation growth is more difficult to characterise for two reasons. First, it cannot be solved analytically and requires quite detailed numerical simulations. Second, non-linear baryonic physics is quite a bit more complex than the behaviour of collisionless dark matter and can be difficult to simulate, even numerically. In the absence of large-scale computer simulations of alternate universes with different values of η_B, the limits discussed here should be treated with some skepticism.

Tegmark et al. [49] have examined systematically the effects on structure formation of altering the baryon-to–dark matter density ratio, which, by assumption, is the same as the change in the baryon-to-photon ratio. Consider first the lower bound on Ω_B / Ω_M. Tegmark et al. argued that one can derive a lower bound based on the requirement that the collapsing baryon discs be able to fragment and form stars. If the baryon-to–dark matter ratio becomes too small, then the baryonic matter is insufficiently self-gravitating to allow fragmentation to occur. The limit derived in Reference [49] is $\Omega_B / \Omega_M \gtrsim 1/300$, which corresponds to the lower bound, $\widetilde{\eta}_B > 1 \times 10^{-11}$.

In the absence of detailed numerical simulations, this bound should be treated with caution. More conservative lower bounds on η_B were derived by Rahvar [32]. Star formation is significantly suppressed at very low η_B simply because there are not enough baryons around to form stars. The requirement that at least one star forms per galactic-sized halo mass gives $\widetilde{\eta}_B > 10^{-22}$. One can be even more conservative and require at least one star in the observable Universe; this requires $\widetilde{\eta}_B > 10^{-34}$ [32].

Tegmark et al. also derived an upper bound on $\widetilde{\eta}_B$ from Silk damping (also called diffusion damping). Silk damping arises near the epoch of recombination from the diffusion of photons out of over-dense (hotter) regions near the epoch of recombination. As the photons diffuse, they scatter off of charged particles and drag the baryons along with them, which tends to erase the baryonic density perturbations. Tegmark et al. argue that if the dark matter density were lower than the baryon density at recombination, Silk damping would tend to erase all fluctuations on galaxy-sized scales. Thus, they derive the limit [49]

$\Omega_B/\Omega_{DM} \lesssim 1$, corresponding to $\eta_B < 3 \times 10^{-9}$. Again, this limit should be treated with some caution; before the discovery of dark matter, cosmologists did not consider purely baryonic models to be ruled out by an absence of structure formation!

In summary, our results in this section do not point toward significant fine-tuning of the baryon asymmetry parameter, η_B. Element production in the early Universe provides essentially no limits on changes to η_B from the point of view of the habitability of the Universe, while limits from structure formation are either very weak, very speculative, or both.

5.8 Summary and Conclusions

For a dimensionless physical parameter such as the baryon asymmetry parameter, η_B, that could take on any value from $-\infty$ to $+\infty$ (or, allowing for a swap in the definition of matter and antimatter, from 0 to ∞), zero might seem to be the most natural choice. However, the value of $\eta_B = 0$ corresponds to a symmetric universe, a universe with equal amounts of matter (baryons) and antimatter (anti-baryons), which is inconsistent with what we actually observe. An overview of the problem was provided in Section 5.1, where η_B was defined and its relation to the baryon-to-entropy ratio was discussed. To address the question of whether a non-zero value for η_B is or is not, fine-tuned, some ground rules are required. These were outlined in Section 5.2. We evaluate fine-tuning in terms of the ability of the Universe to produce stars, planets, and, ultimately, life. As reviewed in Section 5.3, our Universe cannot be symmetric; observations strongly indicate that $\eta_B \neq 0$.

An overview of the models that have been proposed to account for $\eta_B \neq 0$ was offered in Section 5.4. The variety of models in the literature suggests that virtually any value of η_B, including the other 'natural' value of $\eta_B \approx O(1)$, could be 'predicted'. The observations most sensitive to η_B are the abundances of the elements produced during BBN. The dependence of BBN on η_B was reviewed in Section 5.5, revealing that while the precise abundances vary significantly with η_B, over a very large range in η_B, only hydrogen and helium (^{4}He) emerge from the early evolution of the Universe with significant abundances. The connection between η_B and a variety of other cosmological parameters was discussed in Section 5.6, and the effect of changing η_B on the evolution of the Universe was examined in Section 5.7. While large changes in η_B affect both primordial element production and the formation of galaxies and stars, it is only the latter that allows us to suggest limits on the allowed range for η_B. Our results indicate that universes with values of the baryon asymmetry parameter that differ significantly from our own can form galaxies and stars (whose evolution can produce the heavy elements necessary for life) and planets, capable of hosting life. Thus, the value of η_B can be varied by many orders of magnitude without strongly affecting the habitability of the Universe, a result that is *not* suggestive of fine-tuning.

It is likely that our Universe began with no baryon asymmetry (equal amounts of matter and antimatter), so that the initial baryon asymmetry parameter had its 'natural' value of zero. For a universe like our own, conservation of baryon number, an exact symmetry at very high temperatures, needed to be violated at some mass-/energy-scale in the very early Universe. Processes such as those described in Section 5.4, which must include baryon number non-conservation, resulted in the baryon asymmetry observed in our Universe and in those alternate universes discussed here. However, the baryon non-conservation required at high-mass/-energy scales might also lead to non-zero (even if exponentially suppressed) baryon non-conservation at very late times in the evolution of the Universe. If this were the case, then, eventually, in a universe that lives long enough, protons might decay (diamonds are not forever!), so the baryon number of the Universe (as well as the lepton number) would revert back to its natural value of zero. Ashes to ashes, dust to dust.

Acknowledgements

G. S. thanks K. M. Nollet, P. J. E. Peebles, and S. Raby for helpful and enlightening discussions. R. J. S. thanks J. F. Beacom for collecting and providing Gary Steigman's notes for this chapter.

References

[1] Affleck, I., and Dine, M. 1985. 'A New Mechanism for Baryogenesis'. *Nuclear Physics B*, **249**, 361–380.

[2] Alcaraz, J., *et al.* 1999. 'Search for Antihelium in Cosmic Rays'. *Physics Letters B*, **461**(Sept.), 387–396.

[3] Alpher, R. A., and Herman, R. 1958. 'On Nucleon-Antinucleon Symmetry in Cosmology'. *Science*, **128**(Oct.), 904.

[4] Burbidge, G. E., and Hoyle, F. 1956. 'Matter and Anti-matter'. *Il Nuovo Cimento*, **4**(Sept.), 558–564.

[5] Cassisi, S., and Castellani, V. 1993. 'An Evolutionary Scenario for Primeval Stellar Populations'. *Astrophysical Journal Supplements*, **88**(Oct.), 509–527.

[6] Chamberlain, O., *et al.* 1955. 'Observation of Anti-protons'. *Physical Review*, **100**, 947–950.

[7] Chiu, H. Y. 1966. 'Symmetry between Particle and Antiparticle Populations in the Universe'. *Physical Review Letters*, **17**(Sept.), 712–714.

[8] Coc, A., Uzan, J. P., and Vangioni, E. 2014. 'Standard Big Bang Nucleosynthesis and Primordial CNO Abundances after Planck'. *Journal of Cosmology and Astroparticle Physics*, **10**(Oct.), 050.

[9] Dicke, R. H., *et al.* 1965. 'Cosmic Black-Body Radiation'. *Astrophysical Journal*, **142**, 414–419.

[10] Dimopoulos, S., and Susskind, L. 1978. 'Baryon Number of the Universe'. *Physical Review D*, **18**(Dec.), 4500–4509.

[11] Dine, M., and Kusenko, A. 2003. 'Origin of the Matter-Antimatter Asymmetry'. *Reviews of Modern Physics*, **76**(Dec.), 1–30.

[12] Edge, A. C., *et al.* 1990. 'An X-Ray Flux-Limited Sample of Clusters of Galaxies: Evidence for Evolution of the Luminosity Function'. *Monthly Notices of the Royal Astronomical Society*, **245**(July), 559.

[13] Ellis, J., Gaillard, M. K., and Nanopoulos, D. V. 1979. 'Baryon Number Generation in Grand Unified Theories'. *Physics Letters B*, **80**(Jan.), 360–364.

[14] Fields, B. D. 2011. 'The Primordial Lithium Problem'. *Annual Review of Nuclear and Particle Science*, **61**(Nov.), 47–68.

[15] Fixsen, D. J., *et al.* 1996. 'The Cosmic Microwave Background Spectrum from the Full COBE FIRAS Data Set'. *Astrophysical Journal*, **473**(Dec.), 576.

[16] Fuke, H., *et al.* 2005. 'Search for Cosmic-Ray Antideuterons'. *Physical Review Letters*, **95**(8), 081101.

[17] Fukugita, M., and Yanagida, T. 1986. 'Barygenesis without Grand Unification'. *Physics Letters B*, **174**(June), 45–47.

[18] Goldhaber, M. 1956. 'Speculations on Cosmogony'. *Science*, **124**(Aug.), 218–219.

[19] Gondolo, P., and Gelmini, G. 1991. 'Cosmic Abundances of Stable Particles: Improved Analysis'. *Nuclear Physics B*, **360**(Aug.), 145–179.

[20] Gross, D. J., and Wilczek, F. 1973. 'Asymptotically Free G'auge Theories. I. *Physical Review*, **D8**, 3633–3652.

[21] Hall, L. J., Pinner, D., and Ruderman, J. T. 2014. 'The Weak Scale from BBN'. *Journal of High Energy Physics*, **12**(Dec.), 134.

[22] Kolb, E. W., and Turner, M. S. 1990. *The Early Universe.* Addison-Wesley.

[23] Kuzmin, V. A., Rubakov, V. A., and Shaposhnikov, M. E. 1985. 'On Anomalous Electroweak Baryon-Number Non-conservation in the Early Universe'. *Physics Letters B*, **155**(May), 36–42.

[24] Laine, M., and Schröder, Y. 2006. 'Quark Mass Thresholds in QCD Thermodynamics'. *Physical Review D*, **73**(8), 085009.

[25] Lee, B. W., and Weinberg, S. 1977. 'Cosmological Lower Bound on Heavy-Neutrino Masses'. *Physical Review Letters*, **39**(July), 165–168.

[26] Lepp, S., and Shull, J. M. 1984. 'Molecules in the Early Universe'. *Astrophysical Journal*, **280**(May), 465–469.

[27] Loeb, A. 2014. 'The Habitable Epoch of the Early Universe'. *International Journal of Astrobiology*, **13**(Sept.), 337–339.

[28] Nanopoulos, D. V., and Weinberg, S. 1979. 'Mechanisms for Cosmological Baryon Production'. *Physical Review D*, **20**(Nov.), 2484–2493.

[29] Nollett, K. M. 2016. Private communication.

[30] Penzias, A. A., and Wilson, R. W. 1965. 'A Measurement of Excess Antenna Temperature at 4080-Mc/s'. *Astrophys. J.*, **142**, 419–421.

[31] Planck Collaboration, *et al.* 2016. 'Planck 2015 Results. XIII. Cosmological Parameters'. *Astronomy and Astrophysics*, **594**(Sept.), A13.

[32] Rahvar, S. 2017. 'Cosmic Initial Conditions for a Habitable Universe'. *Monthly Notices of the Royal Astronomical Society*, **470**(Sept.), 3095–3102.

[33] Reimer, O., *et al.* 2003. 'EGRET Upper Limits on the High-Energy Gamma-Ray Emission of Galaxy Clusters'. *Astrophysical Journal*, **588**(May), 155–164.

[34] Riotto, A., and Trodden, M. 1999. 'Recent Progress in Baryogenesis'. *Annual Review of Nuclear and Particle Science*, **49**, 35–75.

[35] Sakharov, A. D. 1967. Quark-Muonic Currents and Violation of CP Invariance'. *JETP Letters*, **5**, 27–30.

[36] Scherrer, R. J. 1983. 'Primordial Element Production in Universes with Large Lepton-Baryon Ratio'. *Monthly Notices of the Royal Astronomical Society*, **205**(Nov.), 683–690.

[37] Scherrer, R. J., and Turner, M. S. 1986. 'On the Relic, Cosmic Abundance of Stable, Weakly Interacting Massive Particles'. *Physical Review D*, **33**(Mar.), 1585–1589.

[38] Schramm, D. N. 1998. 'Primordial Nucleosynthesis'. *Proceedings of the National Academy of Science*, **95**(Jan.), 42–46.

[39] Schramm, D. N., and Wagoner, R. V. 1977. 'Element Production in the Early Universe'. *Annual Review of Nuclear and Particle Science*, **27**, 37–74.

[40] Sommerfeld, A. 1931. 'Über die Beugung und Bremsung der Elektronen'. *Annalen der Physik*, **403**, 257–330.

[41] Steigman, G. 1969. 'Antimatter, Galactic Nuclei and Theories of the Universe: Antimatter and Cosmology'. *Nature*, **224**(Nov.), 477–481.

[42] Steigman, G. 1976. 'Observational Tests of Antimatter Cosmologies'. *Annual Review of Astronomy and Astrophysics*, **14**, 339–372.

[43] Steigman, G. 1979. 'Cosmology Confronts Particle Physics'. *Annual Review of Nuclear and Particle Science*, **29**, 313–338.

[44] Steigman, G. 2006. 'The Cosmological Evolution of the Average Mass per Baryon'. *Journal of Cosmology and Astroparticle Physics*, **10**(Oct.), 016.

[45] Steigman, G. 2008. 'When Clusters Collide: Constraints on Antimatter on the Largest Scales'. *Journal of Cosmology and Astroparticle Physics*, **10**(Oct.), 001.

[46] Steigman, G. and Strittmatter, P. A. 1971. 'Neutrino Limits on Antimatter Sources of Energy in Seyfert Galaxies'. *Astronomy and Astrophysics*, **11**(Mar.), 279.

[47] Steigman, G., Dasgupta, B., and Beacom, J. F. 2012. 'Precise Relic WIMP Abundance and Its Impact on Searches for Dark Matter Annihilation'. *Physical Review D*, **86**(2), 023506.

[48] 't Hooft, G. 1976. 'Symmetry Breaking through Bell-Jackiw Anomalies'. *Physical Review Letters*, **37**(July), 8–11.

[49] Tegmark, M., *et al.* 2006. 'Dimensionless Constants, Cosmology, and Other Dark Matters'. *Physical Review D*, **73**(2), 023505.

[50] Toussaint, D., *et al.* 1979. 'Matter-Antimatter Accounting, Thermodynamics, and Black-Hole Radiation'. *Physical Review D*, **19**(Feb.), 1036–1045.

[51] Wagoner, R. V. 1973. 'Big-Bang Nucleosynthesis Revisited'. *Astrophysical Journal*, **179**(Jan.), 343–360.

[52] Wagoner, R. V., Fowler, W. A., and Hoyle, F. 1967. 'On the Synthesis of Elements at Very High Temperatures'. *Astrophysical Journal*, **148**(Apr.), 3.

[53] Yoshimura, M. 1979. 'Unified Gauge Theories and the Baryon Number of the Universe'. *Physical Review Letters*, **42**(Mar.), 746.

[54] Zeldovich, Y. 1965. 'Survey of Modern Cosmology'. *Advances in Astronomy and Astrophysics*, vol. 3. Elsevier.

6

Structure Formation

A D R I A N N E S L Y Z

Abstract

In this chapter, I will describe how the galaxies populating the present-day Universe are believed to have originated from extremely small fluctuations in baryons, dark matter, and radiation in the early Universe. I will very briefly outline the currently favoured explanation as to the origin of these fluctuations before discussing their growth and evolution via linear perturbation theory. Finally, to complete the picture, I will sketch how the virialised dark matter halos hosting the galaxies we observe emerged from the cosmological density field, by combining linear theory and a simplified spherically symmetric model to follow the evolution of these perturbations once they have detached from the Universe's expansion and entered the non-linear regime.

6.1 The Emergence of Structure in the Universe

6.1.1 Introduction

Maps of the positions of galaxies in the sky reveal a complex, inhomogeneous distribution. Galaxies are organised in what is called a cosmic web, characterised by sheets, filaments, and voids (see Figure 6.1). The largest among the hierarchy of over-dense structures seen in such cosmic maps are superclusters containing one or more clusters of galaxies undergoing collapse under their own self-gravity. On the opposite end are voids, the most under-dense regions of the Universe. The development of this inhomogeneous matter distribution is called structure formation.

In the framework of the standard cosmological model, we possess a remarkably successful theory for the formation of the observed large-scale structures in the Universe. According to the Big Bang model, the Universe emerged from an initially singular state of infinite density, from which space progressively expanded. This theory is underpinned by three main pieces of observational evidence. First, support for an expanding universe came in 1930 when astronomer Edwin Hubble discovered that galaxies move away from each other at a speed proportional to the distance separating them. Second, the theory of primordial nucleosynthesis stemming from the thermal history of the Big Bang predicts a universal helium abundance consistent with the measured one (about 25% in mass). Finally,

Adrianne Slyz

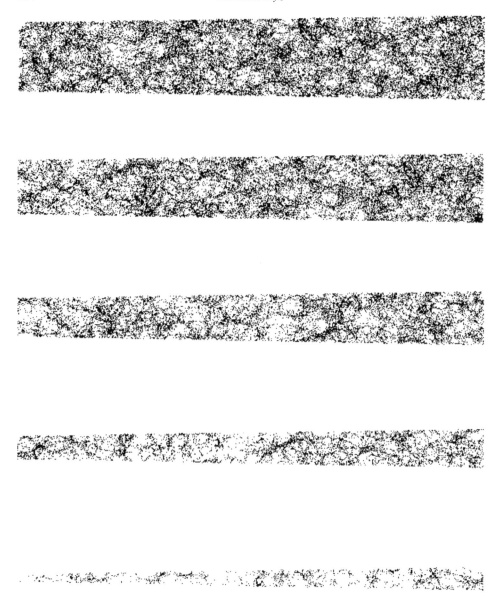

Figure 6.1 Spatial distribution of galaxies with stellar masses in excess of $2 \times 10^{10} M_\odot$ in the Horizon-AGN simulated Universe between redshifts 0 (bottom-left corner of the image) and 1 (top-right corner of the image); see www.horizon-simulation.org for detail. The observer, looking to the right and located at the apex of the bottom-most wedge, thus, sees the left hand side of the wedge situated immediately above when the bottommost wedge ends on the right-hand side and so on and so forth, all the way to the end of the uppermost wedge. In other words, the image represents the projection of a unique wedge which subtends a 2 square degree angle on the sky (about 10 times the area covered by the Moon as seen from Earth). Note how the galaxies are not distributed at random but organised in a cosmic web.

the uniform and isotropic black-body radiation prophesied to permeate the entire sky was detected by Penzias and Wilson in 1964, lending further support to the high densities and temperatures reigning in the primordial Universe.

This extreme isotropy and homogeneity of the microwave background radiation immediately raises a puzzle concerning the formation of galaxies. How can galaxies populating the present-day Universe originate from such a ball of incredibly smooth cosmic fire? The answer to this question is believed to rest with the theory of inflation, a very brief phase of exponential expansion of space itself, which occurred immediately after the Big Bang. During this phase, quantum fluctuations in density were stretched to macroscopic scales, while the observable Universe maintained a constant size. However, even though inflation extended these quantum fluctuations to sizes larger than the cosmological horizon, their amplitude, as imprinted in the black-body spectrum of the cosmic microwave background radiation remained very weak, at a level of only 1 part in 10^5. Following inflation, gravitational instability is thus invoked to amplify primordial fluctuations.

Detailed analysis of the cosmic microwave background radiation rules out structure formation in a universe dominated by baryonic matter because the time required for gravitational instability to magnify the density contrasts to present-day values from the weak level of anisotropies at the moment of photon-matter decoupling is too large. In contrast, should matter exist which interacts only weakly with photons, it could decouple at an earlier epoch and could thereby accelerate the later growth of baryonic fluctuations, up to the level of the characteristic density contrast of galaxies.

There are other observational reasons which suggest that a more exotic form of matter dominates the matter density of the Universe. For example, the rotation curves of galaxies remain flat beyond the distances where visible matter can be detected. The mass of clusters of galaxies, deduced from dynamical measurements or gravitational lensing, is higher than their luminous mass comprised of galaxies and hot gas. From the point of view of structure formation, the different possibilities for the nature of this dark matter can roughly be divided into three categories which depend on particle mass. Cold dark matter is characterised by particles with masses ~ 1 GeV or greater, warm dark matter has intermediate mass particles with masses between ~ 100 eV and ~ 100 MeV, and hot dark matter is composed of light, ultra-relativistic particles with masses smaller than 1 eV. The common characteristic of all these particles is that they interact very weakly with the rest of matter.

The nature of dark matter constitutes a very active research subject and is discussed in Chapter 9.

6.1.2 *Dynamics of the Universe*

The story of structure formation begins from the moment when Einstein's equations become a valid description of the Universe – i.e. a little after the Planck time, a few 10^{-43} seconds after the birth of the Universe. From this moment onwards, it is possible to describe the dynamical evolution of the Universe with the laws of physics as we know them. In the first instance, we will treat the distribution of matter in the Universe as completely homogeneous and isotropic. Inhomogeneities will be treated as deviations from this smooth, primordial state. We will also assume that only gravity dominates the dynamics of the Universe on large scales. This model of the Universe, the simplest one can imagine, is called the Friedmann model.

As the geometrical properties of the Universe are determined by the matter distribution according to Einstein's equations, it follows that no direction (isotropy), and no point in space (homogeneity) are privileged. In this case, one can show (see, e.g., [14]) that the most general space-time metric describing the Universe is given by the Friedmann-Lemaître-Robertson-Walker (FLRW) metric:

$$ds^2 = \mathbf{g}_{\mu\nu}dx^\mu dx^\nu = dt^2 - a^2(t)\left[\frac{dr^2}{1 - kr^2} + r^2(d\theta^2 + \sin^2\theta d\phi^2)\right], \qquad (6.1)$$

where $\mathbf{g}$ is the metric tensor, $a(t)$ is the scale factor of the Universe, which only depends on time t and is chosen to be $a(t_0) \equiv a_0 = 1$ at the present time, and k is the spatial curvature at the present epoch, equal to 0 for a flat universe, positive for a closed universe, and negative for an open universe. We also take the speed of light in vacuum to be $c = 1$ and r, θ and ϕ define a (spherical) coordinate system co-moving with the expansion of the Universe.

The same general considerations of isotropy and homogeneity allow us to jump into the co-moving coordinate frame where the matter (understood here in the general sense of the term – i.e., including all components such as radiation, baryons, and dark matter) assumed to be filling the Universe is at rest. In analogy with an ideal fluid in thermodynamic equilibrium, we can write its energy-momentum tensor $\mathbf{T}$ in the form

$$\mathbf{T}_{\mu\nu} = \mathrm{diag}(\bar\rho(t), \bar p(t), \bar p(t), \bar p(t)), \qquad (6.2)$$

where $\bar p(t)$ and $\bar\rho(t)$ are the total average pressure and energy density of the multicomponent fluid respectively and are related to one another by an effective equation of state $\bar p = \bar p(\bar\rho)$.

Finally, one uses Einstein's field equations to relate the metric tensor to the energy-momentum tensor and to thereby describe the influence of matter on space-time and vice versa:

$$\mathbf{R}_{\mu\nu} - \frac{1}{2}\mathbf{g}_{\mu\nu}R = 8\pi G\mathbf{T}_{\mu\nu} + \Lambda\mathbf{g}_{\mu\nu}, \qquad (6.3)$$

where $\mathbf{R}$ and R are functions of $\mathbf{g}$ and its first and second derivatives with respect to the coordinates, G is the (Newtonian) gravitational constant and Λ is the cosmological constant. Given our previously stated assumptions of homogeneity and isotropy for the Universe, these field equations simplify to yield the Friedmann-Lemaître equations:[1]

---

[1] To make the notation more compact, we stop writing explicit time dependence for the physical quantities when there is no ambiguity.

$$\frac{\mathrm{d}^2 a}{\mathrm{d}t^2} = -\frac{4\pi G a}{3}(\bar{\rho} + 3\bar{p}) + \frac{\Lambda a}{3} \qquad (6.4)$$

$$\frac{\mathrm{d}(\bar{\rho}a^3)}{\mathrm{d}t} = -\bar{p}\frac{\mathrm{d}a^3}{\mathrm{d}t}, \qquad (6.5)$$

where the spatial curvature k is present as a constant of integration for the second equation. The first equation states that both the energy density and the pressure cause the expansion rate of the Universe to decrease while the cosmological constant causes the expansion rate to increase. The second equation is a statement of energy conservation and can also be derived by considering an adiabatically expanding universe. Together with the equation of state, these two equations form a closed system, completely describing the dynamical evolution of an homogeneous and isotropic universe.

6.1.3 The Generation of Density Fluctuations

Having established the dynamical equations of a homogeneous and isotropic universe, the question arises of how to explain the mechanism that generates the inhomogeneities that give rise to the structures seen today. Historically, two theories were put forward for producing these inhomogeneities: inflation and topological defects. However, topological defects have difficulties explaining the power spectrum of the cosmic microwave background anisotropies as well as the existence of massive filaments, walls, and voids seen in redshift surveys on their own (see, e.g., [13]). Hence, we will focus on (very) briefly describing the basic principle behind inflation generated fluctuations.

From the perspective of structure formation by gravitational instability, the major problem that needs to be circumvented is the following: adopting standard values for the cosmological parameters, a fluctuation enclosing a galaxy as massive as our own Milky Way typically has a co-moving size ≈ 1 Mpc, which is larger than the causal distance – called the particle or cosmological horizon – that a photon can cover in ≈ 0.3 years from the original Big Bang. How, then, can a physical process possibly engender a coherent fluctuation on this scale in the primordial[2] Universe?

Inflation, in its many different guises, resolves this issue by postulating that when the Universe reaches the temperature of $T \sim 10^{14}$ GeV corresponding to the energy scale above which strong and electroweak forces are unified, a scalar field called the inflaton[3] undergoes a transition between two vacuum states, during which the energy density of the Universe – and, therefore, its expansion rate $H \equiv \frac{1}{a}\frac{\mathrm{d}a}{\mathrm{d}t}$ – remains nearly constant. This leads to an exponential growth phase during which the proper (by opposition to co-moving) Hubble radius,[4] $r_H \equiv c/H$, hardly changes. Points in space move apart faster

---

[2] Understood as the epoch where the age of the Universe is less than 0.3 years in this case.

[3] For the sake of completeness we note that inflation models with multiple scalar fields are also possible, but their dynamics generally exacerbate the fine-tuning problem, as they are more sensitive to initial conditions and couplings than single field inflation (see, e.g., [6] and references therein)

[4] Cosmologists use this scale because it defines a sphere containing all particles moving away from a given observer located at the centre with a relative velocity slower than the speed of light. Although it is *not* the particle horizon which truly defines the

than the speed of light so that by the end of inflation, any patch of the Universe is much larger than r_H. Perturbations can thus be generated by physical processes when they are within the causal sphere but then expand exponentially during inflation acquiring a size greater than r_H. In this way, perturbations are frozen until the end of the inflationary phase. Then, growing more slowly than r_H in the post-inflation Universe dominated by radiation, perturbations progressively re-enter the observable Universe.[5] For perturbations enclosing the typical galaxy mass, this re-entry corresponds to a time of ≈ 0.3 years after the Big Bang or an expansion factor $a \approx 10^{-6}$. We refer the reader to Chapter 4 for a detailed discussion of the fine-tuning issues related to the existence of such a rapid expansion phase, but note that they entail that the inflaton field must slowly roll down its potential if an inflationary phase is to occur; i.e., the potential must be flat and the mass of the inflaton small (see, e.g., [15], pp. 42–43).

In summary, if inflation is to be believed, the seed perturbations for structure formation are quantum fluctuations of a scalar field. Despite the facts that the theoretical calculations of the perturbations generated by quantum fluctuations are technically arduous and not completely conceptually understood and that many variants of inflation have been devised, they are typically predicted to be adiabatic and follow Gaussian statistics to a high level of accuracy. Thus, they are entirely specified by their power spectrum (Fourier transform of the density field autocorrelation function). This power spectrum, P, in turn, is predicted to be close to scale invariant; i.e.,

$$4\pi k^3 P(k, t_{\text{bis}}) \equiv 4\pi k^3 \left\langle |\delta(\mathbf{k}, t_{\text{bis}})|^2 \right\rangle \approx \text{constant} \approx \delta_{\text{inf}}^2, \qquad (6.6)$$

where $\mathbf{k}$ is the wave vector of the perturbation in Fourier space which corresponds to the wavelength equal to r_H at time t_{bis} when it re-enters (hence the subscript 'bis') the observable Universe, k is its amplitude, δ is the density contrast of the perturbation, $\langle\rangle$ denotes an ensemble average, and δ_{inf} is the (scale-independent) amplitude of the perturbations produced by inflation. We will come back to the evolution of the power spectrum in more detail later and, in particular, why P is only scale independent before perturbations re-enter the observable Universe but emphasise that there does not exist, as yet, a satisfactory prediction for the value of the constant in Eq. (6.6). The amplitude of density perturbations is only *measured* in the cosmic microwave background (CMB) to be around a few $\times 10^{-5}$. Some authors have argued, based on variants of inflationary cosmology, that this measured amplitude should be construed as a possible value amongst a very large ensemble of potentially vastly different ones, each associated with a different causally disconnected 'universe'. In that case, the fine-tuning question needs to be reformulated in terms of Bayesian probabilities which take into account anthropic selection effects: how

observable Universe at a given epoch, one can view it as an 'instantaneous' horizon: a particle located outside of the Hubble sphere cannot be in causal contact with the observer *at this instant*, although it may have been in the past, and may well be in the future. This is why it makes sense to compare the size of the perturbations to this length-scale at a specific moment in time.

[5]   As we will see in more detail, in the post-inflation phase, during the radiation-dominated era, r_H and the proper particle horizon are identical $(= 2ct)$, so we can use them interchangeably, and in the matter dominated-era, the proper particle horizon $(= 3ct)$ is only a factor 2 larger. They only wildly differ during the inflation-like Λ-dominated era.

likely is the observed value, given that we observe it? This is explored in some detail in [15] (and references therein), which concludes that the anthropic selection function plausibly peaks around the observed value, as values diverging from the observed one by an order of magnitude on either side already lead to radiative cooling efficiencies which radically alter structure formation and, as such, drastically reduce our chances of observing the fluctuations.

6.2 The Early Stages of Evolution: Linear Regime

6.2.1 The Linear Growth Phase

Along the lines we followed for the homogeneous and isotropic case, we will model the perturbed matter in the expanding Universe as an ideal, non-relativistic fluid evolving under Newtonian gravity (as done in, e.g., [9]), as this essentially yields the correct dynamical evolution (i.e., that obtained using General Relativity, see, e.g., [8]) provided we only consider perturbations on scales that are small compared to r_H. At the epoch when this matter is the dominant constituent, the first equation describing its behaviour is

$$\frac{\partial \rho}{\partial t} + \frac{\partial (\rho \mathbf{u})}{\partial \mathbf{r}} = 0, \tag{6.7}$$

where $\mathbf{r}$ is the position vector of a fluid element in physical coordinates and ρ and $\mathbf{u}$ are its corresponding density and velocity vector respectively. This equation is called the conservation of mass (or continuity) equation. The second fluid equation

$$\frac{\partial \mathbf{u}}{\partial t} + \left(\mathbf{u} \cdot \frac{\partial}{\partial \mathbf{r}} \right) \mathbf{u} = -\frac{1}{\rho} \frac{\partial p}{\partial \mathbf{r}} - \frac{\partial \Phi}{\partial \mathbf{r}} \tag{6.8}$$

is the Euler equation. The last equation, namely the Poisson equation for the gravitational potential Φ is

$$\left(\frac{\partial}{\partial \mathbf{r}} \right)^2 \Phi = 4\pi G \rho - \Lambda. \tag{6.9}$$

One can rewrite these equations in terms of the co-moving coordinates previously introduced for the homogeneous and isotropic Universe – i.e., coordinates such that $\mathbf{x} = \mathbf{r}/a(t)$. The velocity $\mathbf{u}$ can be expressed as $\mathbf{u} = \frac{da}{dt}\mathbf{x} + \mathbf{v}$ in this case, where the $\frac{da}{dt}\mathbf{x}$ term is called the Hubble flow and $\mathbf{v}$ is the peculiar velocity of a fluid element – i.e., its velocity relative to the Hubble flow. Using these co-moving coordinates, the fluid equations become:[6]

$$\frac{\partial \rho}{\partial t} + \frac{3}{a} \frac{da}{dt} \rho + \frac{1}{a} \frac{\partial (\rho \mathbf{v})}{\partial \mathbf{x}} = 0 \tag{6.10}$$

---

[6] Note that the partial derivatives with respect to time in these equations are taken at fixed $\mathbf{x}$ instead of fixed $\mathbf{r}$, using the relation $\left(\frac{\partial}{\partial t} \right)_{\mathbf{r}} = \left(\frac{\partial}{\partial t} \right)_{\mathbf{x}} - \frac{1}{a} \frac{da}{dt} \left(\mathbf{x} \cdot \frac{\partial}{\partial \mathbf{x}} \right)$.

$$\frac{\partial \mathbf{v}}{\partial t} + \frac{1}{a}\left(\mathbf{v} \cdot \frac{\partial}{\partial \mathbf{x}}\right)\mathbf{v} + \frac{1}{a}\frac{da}{dt}\mathbf{v} = -\frac{1}{\rho a}\frac{\partial p}{\partial \mathbf{x}} - \frac{1}{a}\frac{\partial \Phi'}{\partial \mathbf{x}} \qquad (6.11)$$

$$\left(\frac{1}{a}\frac{\partial}{\partial \mathbf{x}}\right)^2 \Phi' = 4\pi G\rho - \Lambda + \frac{3}{a}\frac{d^2 a}{dt^2}, \qquad (6.12)$$

where we have split the gravitational potential in two, writing $\Phi = \Phi' - \frac{a}{2}\frac{d^2 a}{dt^2}\mathbf{x}^2$ to somewhat simplify the form of the Euler equation.[7]

As we wish to describe all the physical quantities appearing in the dynamical evolution equations as perturbations imprinted on the homogeneous and isotropic background Universe previously discussed, we introduce the density contrast of the perturbations, δ, in these equations, by recasting the density as $\rho(\mathbf{x}, t) \equiv \bar{\rho}(t)(1+\delta(\mathbf{x}, t))$. Note that the velocity $\mathbf{u}$ and the gravitational potential Φ have already undergone a similar procedure, as $\mathbf{v}$ is the deviation from the velocity of the homogeneous and isotropic background component by definition and Φ' is the perturbed part of the term $-\frac{a}{2}\frac{d^2 a}{dt^2}\mathbf{x}^2$, which we can identify with the gravitational potential of this very same background.

If perturbations in all the physical quantities are small compared to their corresponding background values – i.e., $\delta \ll 1$, $v^2(t_e/d)^2 \ll 1$ and $\Phi'(t_e/d^2)^2/\bar{\rho} \ll 1$, where d is the coherence length of the spatial variation of δ and t_e is the expansion time $\sim (G\bar{\rho})^{-\frac{1}{2}}$ – we can linearise the evolution equations (that is to say, only keep the terms where the small perturbations appear at first order) to obtain

$$\frac{\partial \delta}{\partial t} + \frac{1}{a}\frac{\partial \mathbf{v}}{\partial \mathbf{x}} = 0 \qquad (6.13)$$

$$\frac{\partial \mathbf{v}}{\partial t} + \frac{1}{a}\frac{da}{dt}\mathbf{v} = -\frac{c_s^2}{a}\frac{\partial \delta}{\partial \mathbf{x}} - \frac{1}{a}\frac{\partial \Phi'}{\partial \mathbf{x}} \qquad (6.14)$$

$$\left(\frac{\partial}{\partial \mathbf{x}}\right)^2 \Phi' = 4\pi G\bar{\rho}a^2\delta, \qquad (6.15)$$

where we have used Eqs. 6.4 and 6.5 (dropping the $\bar{p}$ terms as the fluid is assumed to be non-relativistic), which govern the evolution of the homogeneous and isotropic Universe, to get rid of the background zeroth order terms and defined the sound speed in the polytropic[8] fluid as $c_s^2 \equiv dp/d\rho$.

Taking the gradient of Eq. (6.14), and substituting $\mathbf{v}$ using Eq. 6.13 and Φ' using Eq. 6.15 in it, we obtain the following second-order differential equation for the density contrast

$$\frac{\partial^2 \delta}{\partial t^2} + \frac{2}{a}\frac{da}{dt}\frac{\partial \delta}{\partial t} = 4\pi G\bar{\rho}\delta + \frac{c_s^2}{a^2}\left(\frac{\partial}{\partial \mathbf{x}}\right)^2 \delta, \qquad (6.16)$$

---

[7] We are allowed to do that as the second term only introduces a total time derivative in the Poisson equation, which does not change the action.

[8] By imposing a polytropic equation of state for the fluid, we force the perturbations to be adiabatic – i.e., we neglect entropy perturbations.

which fully describes the growth of density perturbations, in the regime where they remain small compared to the background density. Before discussing the solutions to this equation, we describe it qualitatively. Recalling that the Hubble expansion rate is $H \equiv \frac{1}{a}\frac{da}{dt}$, the second term on the left-hand side can be expressed as $2H\frac{\partial\delta}{\partial t}$. Called the 'Hubble drag' term, this term captures how the expansion of the Universe suppresses perturbation growth. The terms on the right-hand side, in contrast, determine how gravity enhances and spatial gradients influence perturbation growth.

The evolution equation for the density contrast is more easily solved by decomposing $\delta(\mathbf{x}, t)$ into a set of Fourier modes:

$$\delta(\mathbf{x}, t) = \sum_k \delta(\mathbf{k}, t) e^{i\mathbf{k}\cdot\mathbf{x}} = \sum_k \delta_{\mathbf{k}}(t) e^{i\mathbf{k}\cdot\mathbf{x}}. \tag{6.17}$$

The amplitude of each Fourier mode then obeys the equation

$$\frac{d^2\delta_{\mathbf{k}}}{dt^2} + 2H\frac{d\delta_{\mathbf{k}}}{dt} = \left(4\pi G\bar{\rho} - \frac{c_s^2 k^2}{a^2}\right)\delta_{\mathbf{k}}, \tag{6.18}$$

as, in the linear regime, we do not have to worry about coupling between the different modes: each Fourier mode $\delta_{\mathbf{k}}$ evolves independently.

Now, because the Universe is composed of radiation (photons, neutrinos), different matter components (baryons, collisionless dark matter), and dark energy (cosmological constant), which dominate at different epochs of the Universe evolution, solutions for the growth of the Fourier modes depend on the cosmological model, the equation of state for the different components, and the epoch considered.

6.2.2 Perturbations in a Matter-Dominated Universe

Let us start by looking at a simple universe mainly composed of ordinary matter – i.e., baryons (by opposition to pressureless dark matter). At first, this may seem like a gross oversimplification, but it constitutes an extremely useful example to understand key features of perturbation growth. Moreover, matter (cold dark matter plus baryons in that case) dominates the energy density throughout most of the evolution of the Universe. Arguably the easiest way to see this last point is to plug Eq. (6.5) into Eq. (6.4), multiply the result by $2da/dt$, and integrate it w.r.t. time to the present day, explicitly breaking up the energy density $\bar{\rho}$ into its total matter and radiation components $\bar{\rho}_m$ and $\bar{\rho}_\gamma$ in the process. This yields the following equation for the evolution of the expansion rate of the homogeneous and isotropic Universe:

$$H^2 = H_0^2\left(\Omega_{\gamma,0}\, a^{-4} + \Omega_{m,0}\, a^{-3} + \Omega_{\Lambda,0}\right), \tag{6.19}$$

where we have omitted the curvature contribution, since the Universe is measured to be flat to a high level of accuracy, and introduced the dimensionless parameters $\Omega_\gamma = \bar{\rho}_\gamma/\rho_c$, $\Omega_m = \bar{\rho}_m/\rho_c$, $\Omega_\Lambda = \bar{\rho}_\Lambda/\rho_c$, with the critical density $\rho_c = 3H^2/(8\pi G)$ and $\bar{\rho}_\Lambda = \Lambda/(8\pi G)$. Given the present-day values of $\Omega_{\gamma,0} \simeq 9.4 \times 10^{-5}$, $\Omega_{m,0} \simeq 0.31$, and $\Omega_{\Lambda,0} \simeq 0.69$ (taken from [10]), and given that Eq. (6.19) constrains their sum to remain equal to 1 at all times, one easily sees that we switch from a cosmological constant dominated expansion rate at the present epoch to one dominated by matter at $a_{\Lambda eq} = (\Omega_{m,0}/\Omega_{\Lambda,0})^{1/3} \simeq 0.77$ – i.e., a redshift $z_{\Lambda eq} \equiv a_0/a_{\Lambda eq} - 1 \simeq 0.3$ – and then to a radiation dominated one at $a_{eq} = \Omega_{\gamma,0}/\Omega_{m,0} = 3 \times 10^{-4}$ – i.e., $z_{eq} \simeq 3,300$. Overall, the Universe is thus matter dominated for more than 10 billion years out of a total age of 13.8 billion years.

Going back to our fully baryonic Universe, a solution to Eq. (6.18) depends on the sign of the right-hand side. It is positive and leads to growing perturbations only if wave numbers satisfy the criterion

$$k < k_J = \frac{2\sqrt{\pi G \bar{\rho}}}{c_s} \tag{6.20}$$

or, in terms of wavelength, if the wavelength, λ, of a perturbation is greater than a critical wavelength, λ_J, called the Jeans length

$$\lambda_J = \frac{2\pi}{k_J} = c_s\sqrt{\frac{\pi}{G\bar{\rho}}}. \tag{6.21}$$

Physically, the Jeans length encapsulates the battle between gravity and pressure. Compressing a parcel of baryons adiabatically will cause its density and temperature to rise, creating an overpressured region relative to its surroundings. If the self-gravity of the perturbation is weak, then pressure will iron out gradients, thereby suppressing the growth of perturbations, leaving them to oscillate as acoustic waves. Only if self-gravity overwhelms pressure will the perturbation grow.

Since perturbations smaller than λ_J do not grow, we consider solutions to Eq. (6.18) in the long-wavelength, gravity-dominated limit (i.e., $k \to 0$), in which case this equation becomes[9]

$$\frac{d^2\delta_{\mathbf{k}}}{dt^2} + 2H\frac{d\delta_{\mathbf{k}}}{dt} = 4\pi G\bar{\rho}\delta_{\mathbf{k}}. \tag{6.22}$$

As previously discussed, a good approximation at $0.3 < z < 3,300$ is that $\Omega_m = 1$, in which case $\bar{\rho} = \rho_c$, and Eq. (6.22) becomes

$$\frac{d^2\delta_{\mathbf{k}}}{dt^2} + 2H\frac{d\delta_{\mathbf{k}}}{dt} = \frac{3}{2}H^2\delta_{\mathbf{k}}. \tag{6.23}$$

In a non-expanding universe with $a = 1$, $H = 0$, the solutions to Eq. (6.22) would grow exponentially. The 'Hubble drag' term causes them to grow more slowly in a manner well approximated by a power-law $\delta_{\mathbf{k}} \propto t^n$, where the power law index depends on cosmology, epoch, and equation of state. The solution to Eq. (6.23) in our $\Omega_m = 1$ Einstein–de Sitter cosmology, where $H = \frac{2}{3t}$, is a linear combination of a simple growing mode $D_+ \propto t^{\frac{2}{3}} \propto a$ and a decaying mode $D_- \propto t^{-1} \propto a^{-3/2}$. Since we are interested in forming cosmological structures, we will focus our attention on the growing mode and ignore the decaying part.

In general, for other cosmological models ($\Omega_m \neq 1, \Omega_\Lambda \neq 0$), expressions for the growing mode are more complicated (not analytic) and usually expressed in terms of a

---

[9] Note that since we neglect pressure, this equation also applies to dark matter.

linear growth factor $g(z) \equiv D_+(z) \times (1 + z)$. An approximation for the linear growth factor is given by, e.g., [4] as

$$g(z) \approx \frac{5}{2}\Omega_m \left[\Omega_m^{\frac{4}{7}} - \Omega_\Lambda + \left(1 + \frac{\Omega_m}{2}\right)\left(1 + \frac{\Omega_\Lambda}{70}\right)\right]^{-1}. \tag{6.24}$$

Perturbations in an open universe or in a universe with a non-zero cosmological constant grow even more slowly than in an Einstein–de Sitter universe because the expansion rate is larger, leading to a stronger Hubble drag.

Taking the Einstein–de Sitter growth rates as an upper limit for the growth rates of perturbations in a matter-dominated universe, one sees a compelling argument against the case of a universe composed mainly of baryons. Given that CMB observations show that the amplitude of the baryonic perturbations at redshift $\sim 1{,}000$ are a few 10^{-5}, they will have grown by, at most, a factor of 1,000 by redshift 0 according to linear theory, as $D_+ \propto (1 + z)^{-1}$, which brings their amplitude only as high as a few percent, still safely in the regime where our linear solution is valid and far from the $\delta_\mathbf{k} \sim 1$ level at which non-linear evolution should take over and form the galaxies we observe today.

A resolution to this puzzle is to invoke dark matter. Because dark matter interacts with radiation only gravitationally, rather than through the scattering processes which affect baryons, larger-amplitude perturbations in the dark matter at high redshifts can be achieved without changing the CMB power spectrum. Larger dark matter perturbations at earlier times allow them to reach larger amplitudes at later times and provide gravitational potential wells for baryons to fall into, thus accelerating the growth of baryonic perturbations. We will now discuss in detail how this proceeds.

6.2.3 Perturbations in a Radiation-Dominated Universe

Given that the seed perturbations for structure formation are believed to originate during the inflationary epoch, re-entering the observed Universe around $z \sim 10^6$ for the smallest of them – i.e., galaxy-sized fluctuations – their early growth phase occurs during the radiation-dominated epoch of the Universe at $z > 3{,}300$. The fluid equations we derived for the evolution of the density perturbations were based on a non-relativistic, Newtonian approach. In principle, to describe perturbation growth during the radiation-dominated epoch, we should solve for the evolution of a photon-baryon fluid using a full relativistic treatment for the fluid in a perturbed space-time metric. In practice, we can get a good sense of the physics using the Newtonian treatment, provided we consider fluctuations in the dark matter fluid only and approximate the photon-baryon fluid as a *smooth* background dominant component which only affects the expansion rate of the Universe,[10] that is to say, we write[11]: $\rho = \bar{\rho}_m(1 + \delta) + \bar{\rho}_\gamma$.

---

[10] This approach only works to describe density fluctuations – i.e., if we are not interested in describing CMB anisotropies.

[11] Technically speaking, to be consistent with our notations, the baryon density, ρ_b, should be removed from the matter density – i.e., ρ_m should become $\rho_m - \rho_b$ – and added to the energy density of the photons ρ_γ. However, since $\rho_b \ll \rho_m$ at all times, and $\rho_m < \rho_\gamma$ before matter radiation equality (but $\delta = \delta_m > \delta_\gamma$, as perturbations in the dark matter do not oscillate), we

The validity of such an approximation during the radiation dominated era can be understood by calculating the Jeans length for the tightly coupled photon-baryon fluid. Using the
(adiabatic) sound speed $c_s = \sqrt{dp/d\rho} = c/\sqrt{3}$ since $p = \rho c^2/3$ for a relativistic fluid,
this gives

$$\lambda_J = c\sqrt{\frac{\pi}{3G\bar{\rho}}}. \tag{6.25}$$

Comparing it to the Hubble radius $r_H \equiv c/H$, Eq. (6.19) for H yields a ratio of $\lambda_J/r_H = \sqrt{8\pi^2/9} \simeq 3$. Therefore, the Jeans length during the radiation-dominated era is on the
order of three times larger than the Hubble radius, meaning that perturbations in the
photon-baryon fluid, δ_γ, smaller than the horizon scale are suppressed, hence the 'smooth'
approximation.

As for the dark matter perturbations since they are pressureless ($p_m \simeq 0$), their growth
is governed by an evolution equation similar to Eq. (6.22) in the radiation era,

$$\frac{d^2\delta_{\mathbf{k}}}{dt^2} + 2H\frac{d\delta_{\mathbf{k}}}{dt} = 4\pi G\bar{\rho}_m\delta_{\mathbf{k}}, \tag{6.26}$$

but where the density $\bar{\rho}$ in the term on the right-hand side has been replaced by $\bar{\rho}_m$. The
first Friedmann equation (6.4), however, becomes

$$\frac{1}{a}\frac{d^2a}{dt^2} = -\frac{4\pi G}{3}(\bar{\rho}_m + 2\bar{\rho}_\gamma) = -\frac{4\pi G}{3}\bar{\rho}_\gamma(2 + y) = -\frac{2+y}{2(1+y)}H^2 \tag{6.27}$$

where we have introduced the new variable $y = \bar{\rho}_m/\bar{\rho}_\gamma = a/a_{eq}$. Noticing that $dy/dt = yH$ and using Eq. (6.27), one can rewrite Eq. (6.26) as

$$\frac{d^2\delta_{\mathbf{k}}}{dy^2} + \frac{(2+3y)}{2y(1+y)}\frac{d\delta_{\mathbf{k}}}{dy} = \frac{3}{2}\frac{\delta_{\mathbf{k}}}{y(1+y)}, \tag{6.28}$$

for which the growing mode solution is $D_+ \propto 1 + 3y/2 \propto 1 + 3a/(2a_{eq})$, resulting
in a constant value for δ during the radiation dominated era ($y \ll 1$, or $a \ll a_{eq}$) –
i.e., the perturbations stagnate. Only around the epoch when matter begins to significantly
contribute to the energy density ($a \approx a_{eq}$) and later, when $y \geq 1$ and matter dominates,
do they transition smoothly to a growth proportional to the expansion factors characteristic
of an Einstein–de Sitter universe. This stagnation during the radiation dominated epoch is
called the Mészáros effect.

To summarise, during the radiation-dominated era, neither radiation, baryon, nor dark
matter perturbations on subhorizon scales grow. The expansion rate of the Universe driven
by $\bar{\rho}_\gamma$ is much faster than the growth rate of any perturbation on subhorizon scales, effectively freezing the amplitudes of the subhorizon perturbations in all components. However,
contrarily to the baryons, which remain coupled to the photons until recombination, the

neglect the baryons altogether for simplicity. Note that as long as their free-streaming length remains shorter than the
wavelengths of interest, we could include neutrinos in the smooth radiation background, as they can then be treated as a fluid.
Finally, we have also neglected the (smooth) dark-energy density, as we know that $\rho_\Lambda \ll \rho_m$ already at the start of the
matter-dominated epoch.

perturbations in the dark matter begin to grow from matter-radiation equality onwards. This head start of dark matter fluctuation growth might appear insignificant in the grand scheme of things, but it is a dramatic example of the consequences of fine-tuning: reverse the baryon-to–dark matter ratio, or simply make it larger, and this early growth phase is suppressed. As a result, the growth of baryonic fluctuations after recombination cannot be accelerated by the presence of pre-existing dark matter potential wells, and these fluctuations never reach the density contrast required to form galaxies.

6.2.4 Before and after Recombination

As the Universe transitions to the matter-dominated phase, $\bar{\rho}_m > \bar{\rho}_\gamma$, but the baryon energy density still remains $\bar{\rho}_b < \bar{\rho}_\gamma$ until recombination. This coupling is mainly achieved by Compton scattering, which efficiently transfers momentum from photons to baryons and vice versa. Varying density and pressure adiabatically in the photon-baryon fluid, one can calculate its sound speed exactly

$$c_s = \frac{c}{\sqrt{3}} \left(\frac{3}{4} \frac{\bar{\rho}_b}{\bar{\rho}_\gamma} + 1 \right)^{-1/2} \tag{6.29}$$

and realise that it drops as $\bar{\rho}_b$ becomes comparable to $\bar{\rho}_\gamma$, but not by a large factor, so the baryon Jeans length remains comparable to r_H. As a result, we expect baryon fluctuations with scale lengths smaller than the horizon to experience acoustic oscillations between a_{eq} and a_{rec} and to not grow. This is another place where fine-tuning intervenes in structure formation: changing the baryon-to-photon ratio not only delays the matter-radiation equality epoch, slowing down dark matter fluctuation growth, but also pushes back recombination. As a result, baryon fluctuations continue to oscillate for a longer period of time and, thus, significantly hamper the early formation of stars and galaxies.[12]

Eventually, close to recombination, as the density of matter and radiation drops, photons and baryons start to become less coupled, and photons can travel further before they are scattered. They therefore undergo more diffusion. In regions of photon over-density, photons diffuse out, and in the process, they pull with them any baryons that they scatter off of, smoothing out not only photon number densities but also baryon perturbations. This effect is called Silk damping and leads to a time-dependent length-scale below which acoustic oscillations are damped. A simple derivation of the Silk damping length-scale starts from the photon mean free path $\lambda_{mfp} = 1/(n_e \sigma_T)$, where σ_T is the Thomson scattering cross section and n_e is the electron number density. From the mean free path, we can write a diffusion coefficient, $\eta = \frac{1}{3} \lambda_{mfp} c$, and therefore estimate the distance out to which photons diffuse in time t, $\lambda_\eta \sim \sqrt{\eta t}$. At recombination, $t_{rec} \sim 378,000$ years, and the electron

---

[12] Although the fine-tuning issues related to the vacuum-expected-value of the Higgs field are discussed in another chapter of this volume, it feels natural to mention at the stage of recombination, when neutral atoms are created, that if this value were not reasonably close to the measured one, complex atoms (as in more complex than hydrogen) would not be able to be synthesised (e.g. [1, 5]) either in the early Universe during the post-inflation, during the radiation-dominated phase, or by stars during the structure formation stage. Needless to say, this would significantly alter galaxy formation.

number density $n_e \sim 400\,\text{cm}^{-3}$, leading to an estimate $\lambda_\eta \sim 6$ kpc in physical units (~ 6 Mpc co-moving) of the scale below which Silk damping erases perturbations in the photon-baryon fluid.

After recombination the baryons decouple from the photons to form a neutral gas of monoatomic hydrogen and helium. For quite some time (until $z \sim 100$), there are still enough residual free electrons to maintain enough interaction to keep photons and baryons in thermal equilibrium, so $T_b \simeq T_\gamma$. However, as a result of this decoupling, the sound speed of the baryons plummets dramatically from values close to the speed of light $c/\sqrt{3} \sim 0.5c$ to $\sqrt{5k_B T_b c/(3\mu m_H c^2)} \sim 2 \times 10^{-5}c$, where k_B is the Boltzmann constant, μ is the mean molecular weight of the gas, and m_H is the mass of a hydrogen atom. With this drop in sound speed, the Jeans length of the baryons also drops by a factor $\sim 4 \times 10^{-5}$, giving the possibility for baryon perturbations – which, prior to recombination, were too small – to now grow. The problem is that below λ_η, Silk damping has erased baryon perturbations. Owing to the fact that dark matter is not subjected to Silk damping because it is not coupled to the radiation, dark matter perturbations on scales smaller than λ_η should serve to regenerate the baryon perturbations.

Since the baryon Jeans length after recombination is $\ll 1$Mpc, and we are not interested in how individual galaxies form but in much larger scales, we can consider the baryon fluid as pressureless. As in the previous section, we can consider the photons (and neutrinos and dark energy) as a smooth (but now subdominant) background, but now with perturbations in both dark matter and baryonic components – i.e., we write $\rho = \bar{\rho}_d(1+\delta_d)+\bar{\rho}_b(1+\delta_b)+\bar{\rho}_\gamma$, with $\bar{\rho}_m = \bar{\rho}_d + \bar{\rho}_b$. We thus obtain for the pressureless perturbations a system of coupled evolution equations similar to Eq. (6.22):

$$\frac{d^2\delta_d}{dt^2} + 2H\frac{d\delta_d}{dt} = 4\pi G \bar{\rho}_m \delta \tag{6.30}$$

$$\frac{d^2\delta_b}{dt^2} + 2H\frac{d\delta_b}{dt} = 4\pi G \bar{\rho}_m \delta, \tag{6.31}$$

where we have dropped the subscript **k** for clarity, and $\delta = (\bar{\rho}_d\delta_d + \bar{\rho}_b\delta_b)/\bar{\rho}_m$ is the total matter perturbation.

We can now define a baryon–dark matter entropy perturbation $S_{db} \equiv \delta_d - \delta_b$ which indicates how the two components deviate from one another. Subtracting Eq. (6.31) from Eq. (6.30), we obtain an evolution equation for this entropy perturbation:

$$\frac{d^2 S_{db}}{dt^2} + 2H\frac{dS_{db}}{dt} = 0. \tag{6.32}$$

We have assumed that the perturbations are adiabatic when they re-enter the horizon so that $\delta_d(t_{\text{bis}}) = \delta_b(t_{\text{bis}})$ and $S_{db}(t_{\text{bis}}) = 0$. Thus, for large scales that re-enter after recombination, the evolution of baryon and dark matter is identical. However, for smaller scales that re-enter before recombination, $\delta_d \gg \delta_b$ as δ_d has been growing while δ_b remained stalled. Therefore, for these perturbations, at t_{rec}, $S_{db} \simeq \delta_d$. Solving Eq. (6.32) yields $S_{db}(t) \propto 1+C_-(t/t_{\text{rec}})^{-1/3}$ in the matter-dominated era, where the first term is a constant 'growing' mode, and the second term is the decaying mode. Recall that $\delta_d(t) = \delta_d(t_{\text{rec}})(t/t_{\text{rec}})^{2/3}$ was the growing mode found for the dark matter component in the matter-dominated era.[13] Assuming there is no strong cancellation between growing and decaying mode during recombination – i.e., that $C_- \ll 1$ – the constant of proportionality in front of the solution for $S_{db}(t)$ is $\sim \delta_c(t_{rec})$, and we easily see that for $t \gg t_{rec}$, $\delta_d(t) \gg S_{db}(t) \sim \delta_c(t_{rec})$. Since $S_{db}(t) \equiv \delta_d(t) - \delta_b(t)$ by definition, this implies $\delta_b(t) \simeq \delta_d(t)$ and the baryon perturbations must eventually catch up with the dark matter ones to a high level of accuracy.

6.2.5 Perturbations in a Λ-Dominated Universe

Although dark energy – assumed to be in the simplest form of a cosmological constant here – only starts dominating the energy density at late times (around $z \simeq 0.3$, as previously discussed), it is interesting to look at how it impacts the growth of perturbations in the matter field. Using an approach that should now be familiar, we will consider fluctuations in the total matter fluid only, as we just showed that baryon fluctuations rapidly tend to track dark matter ones once recombination is over, and treat the dark-energy density as a smooth, dominant background. In other words, we will write $\rho = \bar{\rho}_m(1 + \delta) + \bar{\rho}_\Lambda$ and neglect radiation.

Once again, the matter perturbation growth is governed by an evolution equation similar to Eq. (6.22):

$$\frac{d^2\delta_\mathbf{k}}{dt^2} + 2H\frac{d\delta_\mathbf{k}}{dt} = 4\pi G\bar{\rho}_m\delta_\mathbf{k}, \tag{6.33}$$

where the density $\bar{\rho}$ in the term on the right-hand side has been replaced by $\bar{\rho}_m$. The first Friedmann equation (6.4), becomes

$$\frac{1}{a}\frac{d^2a}{dt^2} = -\frac{4\pi G}{3}(\bar{\rho}_m - 2\bar{\rho}_\Lambda) = -\frac{4\pi G}{3}\bar{\rho}_\Lambda(y - 2) = -\frac{y-2}{2(1+y)}H^2, \tag{6.34}$$

where we we have introduced the new variable $y = \bar{\rho}_m/\bar{\rho}_\Lambda = (a/a_{\Lambda eq})^{-3}$. Noticing that $dy/dt = -3yH$ and using Eq. (6.34), one can rewrite Eq. (6.33) as

$$\frac{d^2\delta_\mathbf{k}}{dy^2} + \frac{(5y+2)}{6y(1+y)}\frac{d\delta_\mathbf{k}}{dy} = \frac{1}{6}\frac{\delta_\mathbf{k}}{y(1+y)}. \tag{6.35}$$

This equation can be recast in a Gaussian hypergeometric form and solved analytically. Unfortunately, the solutions can only be expressed as hypergeometric series. However, one can study their asymptotic behaviour and recover, for $y \gg 1$, the matter-dominated growth mode $D_+ \propto y^{-1/3} \propto a$, and for $y \ll 1$, Taylor expanding the solution around $y = 0$, $D_+ \propto 1 + y/2$ $[+\mathcal{O}(y^2)] \propto 1 + 0.5(a_{\Lambda eq}/a)^3$. We therefore conclude that perturbations very rapidly stop growing as a gets larger than $a_{\Lambda eq}$.[14]

---

[13] We neglect the decaying mode since it has been decaying away from the moment the perturbation re-entered the horizon and the impact of the subdominant baryon component on the growth of dark matter fluctuations.

[14] One has to pay attention that the proper wavelength of the perturbation remains smaller than r_H as $y \to 0$ for the Newtonian approximation to remain valid, which can be a challenge as a grows exponentially when $\bar{\rho}_\Lambda$ dominates, whereas r_H remains constant. In practice, the approximation holds well to the present day – i.e., $a = 1$ or $y = 0.45$ – even for perturbations with sizes comparable to r_H.

From the point of view of structure formation, this halt of perturbation growth related to the present-day value of the cosmological constant arguably constitutes the most clear-cut case of fine-tuning. Indeed, $\bar{\rho}_\Lambda$ will quickly dominate the energy density in the near future as $\bar{\rho}_m$ decays away like a^{-3}. If the value of Λ was only 43% larger than it is measured to be[15] – which is not as large an increase as it seems, given that Λ is 122 orders of magnitude smaller than it should if we interpret it as vacuum energy! – then we would have $a_{\Lambda eq} = (0.01/0.99)^{1/3} \simeq 0.2$, which is small enough to wipe any galaxy cluster out of the night sky and, thus, force galaxies to live in isolation.

6.2.6 *The Linear Power Spectrum*

After describing the linear growth of primordial fluctuations in the matter density field through different epochs of the Universe, we now turn to the question of the evolution of the spectrum of fluctuations characterised by an initially scale-independent power spectrum widely believed to result from an early inflationary phase. As inflation generates fluctuations on all super-horizon scales, different sized fluctuations will re-enter the horizon at different epochs, which will determine the physical processes that they will be subjected to. The primordial spectrum of fluctuations $P(k, t_{bis})$ defined by Eq. (6.6), at time t_{bis} when the co-moving scale $\lambda = 2\pi/k$ re-enters the horizon, gets altered by linear perturbation theory to a spectrum $P(k, t)$, at a later time t: $P(k, t) = P(k, t_{bis})D_+^2(t)/D_+^2(t_{bis})$.

Now the horizon scale itself evolves with the expansion of the Universe: if it increases with time, an observer is able to see objects at increasingly further distances. So we first need to calculate how this evolution proceeds. We will denote this co-moving particle horizon – i.e., the co-moving distance to the furthest observable point in the Universe by s_H. How s_H evolves with the scale factor a can be derived from the FLRW metric (Eq. (6.1)). Photons travel along null geodesics $ds^2 = 0$. We can choose our coordinate system to consider radially travelling photons ($d\theta^2 = d\phi^2 = 0$), place the observer at $r = 0$, and define the proper particle horizon for the photon emitted at time t to be at r_{hor}. The FLRW metric, for a flat universe, then gives

$$\int_0^t \frac{dt}{a(t)} = \frac{1}{c}\int_0^{r_{hor}} dr. \tag{6.36}$$

We can also define the distance between two events measured in a reference frame where the two events happen at the same time ($dt = 0$). This distance, called the proper distance, for an object at $r = 0$ to r_{hor} is

$$s_H^{prop} = a(t)\int_0^{r_{hor}} dr, \tag{6.37}$$

which means that the co-moving particle horizon can be expressed as

$$s_H = \int_0^t \frac{c\,dt}{a(t)}.$$ (6.38)

Using $dt = da/\frac{da}{dt}$ and the Friedmann equation (6.19) to get an expression for $da/dt = aH$, the dependence of the co-moving particle horizon on a is

$$s_H = c\int_0^a \frac{da}{a^2 H} = \frac{c}{H_0}\int_0^a \left(\Omega_{\gamma,0} + \Omega_{m,0}a + \Omega_{\Lambda,0}a^4\right)^{-1/2} da,$$ (6.39)

which yields $s_H \propto a$ in the radiation dominated epoch, $s_H \propto a^{1/2}$ in the matter dominated epoch, and $s_H \propto 1/a$ when Λ dominates.

Consider a fluctuation mode with a wavelength λ so large[16] that it re-enters the horizon only during the matter-dominated era. At the time of re-entry, $\lambda = 2\pi/k = s_H \propto a^{1/2}(t_{\text{bis}})$ and so $P(k,t) = P(k,t_{\text{bis}})\,a^2(t)/a^2(t_{\text{bis}}) \propto k\,a^2$. In other words, scale-independent perturbations during inflation produce $P(k,t) \propto k$ – i.e., a scale-invariant power spectrum called Harrison-Zel'dovich – on scales that re-enter the Hubble radius during the matter dominated epoch. Applying the same reasoning to a mode that re-enters the horizon well into the radiation-dominated era, we get $1/k \propto a(t_{\text{bis}})$ but $P(k,t) = P(k,t_{\text{bis}}) \propto k^{-3}$ since perturbations do not grow (Mészáros effect). The transition between modes whose last wavelengths re-enter the horizon during the radiation-dominated era and modes whose first wavelengths re-enter the horizon during the matter-dominated era is smooth, so the change in shape of $P(k,t)$ from $\propto k^{-3}$ to $\propto k$ is gradual.

Cosmologists encapsulate this change in the power spectrum shape through a transfer function $T(k)$, defined such as $P(k,t) \propto kT^2(k)D_+(t)$, that depends on the energy densities of the various constituents of the Universe (dark matter, baryons, photons, neutrino, dark energy, etc.) but applies regardless of the power spectrum of initial fluctuations, inflation generated or not.[17] From our previous simple analysis of the shape of $P(k,t)$, we immediately get $T(k) \simeq 1$ on large scales – i.e., for $k \ll k_{\text{eq}}$, where k_{eq} is the wave number corresponding to $s_H(a_{\text{eq}})$ – the co-moving particle horizon size at matter-radiation equality. To get a sense of scales, $k_{\text{eq}} = 2\pi/\lambda_{\text{eq}}$, where $\lambda_{\text{eq}} \sim 100$ Mpc. At the other end of the spectrum, on small scales – i.e., for $k \gg k_{\text{eq}} - T(k) \propto 1/k^2$. The transition where $T(k)$ (and $P(k,t)$) turns over, between the two regimes of scales, happens around k_{eq}.

Importantly, the dependence of $T(k)$ on wave number encodes the non-linear mapping between the primordial power spectrum of fluctuations to the spectrum at later times. This is the main consequence of smaller perturbations re-entering the horizon at earlier times. If they do so during the radiation-dominated phase, then their growth will be suppressed.

---

[16] Note that the reason why we ignore the Λ-dominated epoch is that s_H shrinks during this epoch, which means that perturbations with larger wavelengths than those which re-entered the horizon at the end of the matter-dominated phase will always remain super-horizon.

[17] For completeness sake, even though we have only discussed adiabatic perturbations up to now, we note that $T(k)$ would differ in the case of isocurvature perturbations.

Only when matter dominates will their growth resume. A major ramification of different sized modes growing at different rates at different epochs of the Universe is that structures will form from a power spectrum that is no longer scale invariant. Furthermore, the fact that the scale k_{eq} at which the transfer function bends from unity depends on cosmological parameters indicates that the galaxy distribution can be used to determine them.

We sketched out the derivation of the transfer function for dark matter and radiation focusing on the Mészáros effect, but in practice, $T(k)$ needs to account for all types of matter and radiation and the diverse physical processes that can alter the growth of fluctuations once they penetrate the observable Universe. For example, we saw that baryonic fluctuations oscillate as sound waves during recombination and that they can be damped by photons diffusing out of density peaks after recombination (Silk damping). Fluctuations can also be smoothed out if they enter the horizon while their non-collisional component is still relativistic (free-streaming). These processes will leave imprints on the transfer function.

As a result, an approximation to the transfer function for a cold dark matter–dominated universe will look like

$$T(k) = \frac{\ln(1 + 2.34q)}{2.34q}(1 + 3.89q + (16.1q)^2 + (5.46q)^3 + (6.71q)^4)^{-\frac{1}{4}}, \qquad (6.40)$$

where

$$q = \frac{k}{\Gamma} = \frac{k}{\Omega_{m,0} h \exp\left(-\Omega_{b,0}\left[1 + (2h)^{1/2} \Omega_{m,0}^{-1}\right]\right)}, \qquad (6.41)$$

where $h = H_0/100$ [12]. This approach can be generalised to other types of matter – e.g., hot dark matter composed of very light neutrinos, warm dark matter composed of intermediate mass particles, and a mixture of cold and hot dark matter. More accurate calculations of the transfer function are now routinely done numerically using Boltzmann codes (e.g., [7]) for any set of cosmological parameters and mix of matter constituents.

6.3 The Final Stretch: Non-linear Growth

6.3.1 The Spherical Top-Hat Model

The present-day density contrast of galaxies and clusters is orders of magnitude above the regime where linear perturbation theory can describe their formation (see Figure 6.2). Knowing the evolution of the Universe background density, back-of-the-envelope calculations of the density contrast give $\delta \sim 10^6$ for galaxies and $\delta \sim 10^3$ for clusters. While in the linear regime – i.e., when $\delta \ll 1$, we evolved different perturbation modes independently with linear perturbation theory. Once the density contrast of a perturbation increases to $\delta \sim 1$, over-densities run away to very large densities via gravitational instability, only prevented from reaching infinity by the random motions of collisionless dark matter and gas pressure, and perturbation modes become correlated. Thus the most accurate description of the growth of structures in the non-linear regime is left to computer simulations (such as that presented in Figure 6.2), but important insights can be gained from simple analytic models.

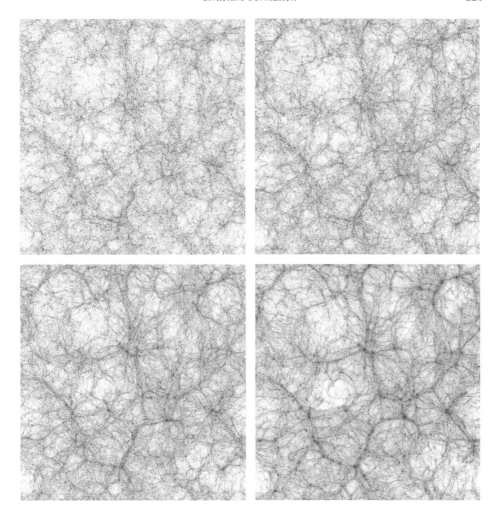

Figure 6.2 Evolution of the gas density in a 20 Mpc/h thick slice of the $(100 \text{ Mpc/h})^3$ volume of the Horizon-AGN simulated Universe between redshifts 0 and 3 (see www.horizon-simulation.org for detail). Top-left panel corresponds to $z = 3$, top-right panel to $z = 2$, bottom-left panel to $z = 1$, and bottom-right panel to $z = 0$. The scale is logarithmic to showcase the enormous range spanned by the density contrast δ, which extends well into the non-linear regime in the densest regions (in dark).

To arrive at an analytic non-linear theory, we must make drastic assumptions about the geometry and density profile of the fluctuations. The simplest model that comes to mind is that of a uniform and isolated spherical density perturbation of collisionless matter evolving in an otherwise completely homogeneous Einstein-de Sitter universe.[18] Commonly called

---

[18] This latter assumption of an $\Omega_m = 1$ universe is not necessary but yields the simplest analytic solutions while still capturing the essential physics, which is the reason why we adopt it here.

the 'top-hat' perturbation because of the form of its density profile, this very simple model is routinely used because its non-linear evolution can be completely described analytically. Moreover, its evolution provides an understanding of the key stages in the non-linear evolution of a perturbation, namely its growth at a slightly slower rate than the Hubble expansion at early times, its detachment from the Hubble flow owing to its self-gravity once it reaches a certain size, and, finally, its collapse onto itself.

Birkoff's theorem tells us that such a perturbation behaves as if it was an independent universe, with the matter exterior to it having no influence on its evolution. If we split up this perturbation into several concentric spherical shells, as long as different shells do not cross, the mass interior to each shell stays constant. We can then write the equation of motion of a shell located at a radius r inside the top hat perturbation as a function of time t as

$$\frac{d^2 r}{dt^2} = -\frac{GM}{r^2}, \tag{6.42}$$

where the mass M interior to r is

$$M \equiv M(<r) = \frac{4}{3}\pi r^3 \bar{\rho}_m (1 + \delta), \tag{6.43}$$

with δ defined as the top-hat over-density contrast and $\bar{\rho}_m$ as the mean density of the background Universe.

Integrating this equation of motion once yields the following equation of energy conservation:

$$\frac{1}{2}\left(\frac{dr}{dt}\right)^2 - \frac{GM}{r} = E, \tag{6.44}$$

where E is the specific energy the shell. For $E = 0$, r evolves at the same rate as the expansion of the Universe – i.e., $r \propto t^{2/3}$, $\propto a$ – so that, as a result, the top-hat perturbation does not grow. For $E < 0$ the shell is bound and the parametric solution for the value of its radius r at time t is given by

$$\frac{r[\eta]}{r_m} = \frac{1}{2}(1 - \cos \eta)$$

$$\frac{t[\eta]}{t_m} = \frac{1}{\pi}(\eta - \sin \eta). \tag{6.45}$$

Starting from $r[0] = 0$ at $t[0] = 0$, the shell reaches a maximum radius $r_m = r[\pi]$ at time $t_m = t[\pi]$ corresponding to turnaround. By symmetry, the shell collapses back to $r[2\pi] = 0$ and virialises at time $t[2\pi] = 2t_m$. Solutions for r_m and t_m are

$$r_m = \frac{GM}{|E|}$$

$$t_m = \frac{\pi GM}{(2|E|)^{3/2}} \tag{6.46}$$

so that $r_m^3 = (8GM/\pi^2)\, t_m^2$.

From these parametric solutions for r and t, we can deduce the solution for the evolution of δ. Indeed, the density ρ of the top-hat perturbation is given by

$$\rho = \frac{3M}{4\pi r^3} = \frac{6M}{4\pi r_m^3} (1 - \cos \eta)^{-3} \tag{6.47}$$

while that of the background Universe $\bar{\rho}_m$ follows

$$\bar{\rho}_m = \frac{1}{6\pi G t^2} = \frac{\pi}{6Gt_m^2} (\eta - \sin \eta)^{-2}. \tag{6.48}$$

Hence, using $r_m^3 = (8GM/\pi^2) t_m^2$, the density contrast of the spherical top-hat perturbation evolves as

$$1 + \delta = \frac{\rho}{\bar{\rho}_m} = \frac{9}{2} \frac{(\eta - \sin \eta)^2}{(1 - \cos \eta)^3}. \tag{6.49}$$

Small values of η correspond to the early evolution of the spherical top-hat perturbation. In that case, the solution for δ reduces to the linear regime solution. More specifically, for $\eta \ll 1$, Taylor expanding $\sin \eta$ and $\cos \eta$ around $\eta = 0$ gives

$$1 + \delta \simeq 1 + 3\frac{\eta^2}{20} \tag{6.50}$$

to second order in η, meaning $\delta \simeq 3\eta^2/20$. We can express δ in terms of t in the small η limit by first Taylor expanding t around $\eta = 0$ to get

$$t = \frac{t_m}{\pi} (\eta - \sin \eta) \simeq \frac{t_m}{\pi} \frac{\eta^3}{6}, \tag{6.51}$$

which gives

$$\eta \simeq \left(\frac{6\pi t}{t_m} \right)^{1/3}, \tag{6.52}$$

and, finally, an expression for the evolution of the density contrast in the linear regime for the Einstein-de Sitter universe:

$$\delta = \frac{3}{20} \left(\frac{6\pi t}{t_m} \right)^{2/3}. \tag{6.53}$$

We can now calculate the difference between the linear regime prediction for δ at various times and its value derived from the spherical top-hat collapse model. Key stages in the evolution of the perturbation occur at the point of maximum expansion (turnaround) and collapse. From Eq. (6.53), linear theory predicts that the density contrast at turnaround when $t = t_m$ is

$$\delta(t_m) = \frac{3}{20} (6\pi)^{2/3} \simeq 1.062 \tag{6.54}$$

and that at collapse, when $t = 2\, t_m$,

$$\delta(2\, t_m) = \frac{3}{20} (12\pi)^{2/3} \simeq 1.686. \tag{6.55}$$

Meanwhile, the spherical top-hat collapse model predicts, from Eq. (6.49), that at turnaround ($\eta = \pi$),

$$\delta(t_m) = \frac{9\pi^2}{16} - 1 \simeq 4.6, \qquad (6.56)$$

while at collapse ($\eta = 2\pi$),

$$\delta(2\,t_m) = \infty. \qquad (6.57)$$

Of course, the prediction from the spherical collapse model breaks down some point after turnaround when the shells inevitably cross because in reality the perturbation embedded in a universe with structures will not be exactly homogeneous and will experience torques from its neighbours, giving it angular momentum and causing it to deviate from perfect spherical symmetry. Hence, non-spherical motions will develop. Virialisation takes over as oscillating shells interact gravitationally and exchange energy to ultimately halt the collapse.

Neither linear perturbation theory nor the simple spherical top-hat model can give us the final end state of the perturbation, but we can determine it from energy conservation arguments. To estimate the final density of the collapsed, virialised perturbation, we use the virial theorem to estimate its radius after collapse has ended. Neglecting the surface pressure term, the perturbation will reach virial equilibrium when its final kinetic K_f and potential W_f energies obey

$$2K_f + W_f = 0. \qquad (6.58)$$

Energy conservation relates the total final energy of the perturbation, E_f, to its total energy at turnaround, E_{ta}:

$$E_f = K_f + W_f = E_{\text{ta}}. \qquad (6.59)$$

Since the velocity of the perturbation comes to a halt at turnaround, all the energy at this point is in potential form so that $E_{\text{ta}} = -GM/r_{\text{ta}}$. Combining the virial theorem and energy conservation, we can show that $E_f = W_f/2 = -GM/(2\,r_{\text{vir}})$ as $W_f = -GM/r_{\text{vir}}$, where r_{vir} is the virial radius of the perturbation. Hence, $r_{\text{vir}} = r_{\text{ta}}/2$, indicating that after it has virialised, the perturbation has collapsed to half the radius it had at turnaround.

Now we are in a position to estimate the over-density, Δ_{vir}, of the collapsed perturbation:

$$1 + \Delta_{\text{vir}} = \frac{\rho(t_{\text{coll}})}{\bar{\rho}_m(t_{\text{coll}})}, \qquad (6.60)$$

where $\rho(t_{\text{coll}})$ is the density of the top-hat perturbation at collapse and $\bar{\rho}_m(t_{\text{coll}})$ is the density of the background Universe at the same time. It is evident that the collapsed perturbation acquires its density because it increases during the collapse, while at the same time, the density of the background Universe in which it is embedded decreases due to the expansion of the Universe, thereby enhancing the density contrast of the perturbation relative to a collapse during which the background remains constant. Because the background density in a matter-dominated, Einstein–de Sitter universe scales as $a^{-3} \sim t^{-2}$ and $t_{\text{coll}} = 2\,t_{\text{ta}}$, this means that $\bar{\rho}_m(t_{\text{coll}}) = \frac{1}{4}\bar{\rho}_m(t_{\text{ta}})$. Furthermore, as the radius of the collapsed perturbation is half the radius of the perturbation at turnaround, its density at collapse is a factor 8 greater than its density at turnaround, $\rho(t_{\text{coll}}) = 8\,\rho(t_{\text{ta}})$. As a result,

$$1 + \Delta_{\text{vir}} = 32\frac{\rho(t_{\text{ta}})}{\bar{\rho}_m(t_{\text{ta}})} = 32 \times (1 + \delta(t_m)) \tag{6.61}$$

and given that we showed earlier that $1 + \delta(t_m) = 9\pi^2/16$, we obtain

$$1 + \Delta_{\text{vir}} = 18\pi^2 \simeq 178. \tag{6.62}$$

So although the spherical top-hat collapse model predicted that the density at collapse would shoot to infinity giving a singularity, virialisation intervenes when shells cross, and the spherical top-hat model breaks down, leading to an over-density of $\simeq 178$. This is two orders of magnitude higher than the prediction for the value of the over-density ($\delta_{\text{coll, lin}} \sim 1.686$) at $t_{\text{coll}} = 2\,t_{\text{ta}}$ if it had continued to evolve according to linear theory. For a low-Ω_m universe such as ours, the mean background density is lower than in an $\Omega_m = 1$, and so the virialised over-density is higher yet.

Despite its simplicity, the spherical top-hat model not only highlights the key stages in the evolution of a density perturbation (turnaround and collapse), but it also gives insight to the hierarchy of collapsing perturbations. Arguments based on energy conservation reveal that in an Einstein–de Sitter universe, all over-densities will eventually collapse. The argument goes as follows. Approximating the velocity v of a mass shell with radius r_{init} by the Hubble flow velocity $v = H\,r_{\text{init}}$, the energy of the shell can be expressed as

$$E = \frac{1}{2}H^2 r_{\text{init}}^2 - \frac{GM}{r_{\text{init}}}. \tag{6.63}$$

Using $M = 4\pi/3\, r_{\text{init}}^3 \bar{\rho}_m\,(1 + \delta(t_{\text{init}}))$ and $\bar{\rho}_m = \rho_c = 3H^2/(8\pi G)$, we have

$$E = \frac{1}{2}H^2 r_{\text{init}}^2 - \frac{1}{2}H^2 r_{\text{init}}^2(1 + \delta), \tag{6.64}$$

which gives $E < 0$ for all over-dense perturbations ($\delta > 0$) in an Einstein–de Sitter universe.

Energy conservation arguments also show that the radius at turnaround only depends on the initial over-density of the perturbation. By equating the energy at turnaround,

$$E(t_{\text{ta}}) = -\frac{GM}{r_{\text{ta}}} = -\frac{H^2 r_{\text{init}}^3}{2\,r_{\text{ta}}}(1 + \delta(t_{\text{init}})), \tag{6.65}$$

to the initial energy,

$$E(t_{\text{init}}) = -\frac{H^2 r_{\text{init}}^2}{2}\delta(t_{\text{init}}), \tag{6.66}$$

one gets

$$\frac{r_{\text{ta}}}{r_{\text{init}}} = \frac{1 + \delta(t_{\text{init}})}{\delta(t_{\text{init}})}. \tag{6.67}$$

Hence, not only does the initial over-density determine the radius at turn-around for a perturbation, but perturbations with smaller initial over-densities enclosed in the same initial radius reach larger radii at turnaround and therefore experience later collapse. In other words, for a given value of $\delta(t_{\text{init}})$, perturbations with smaller r_{init} – i.e., a smaller mass – will collapse first, leading to a hierarchical build up of structures, in a bottom-up fashion.

6.3.2 Press-Schechter Theory

Having discussed the linear evolution of perturbations and how their growth rates depend on their size and cosmic epoch, and then a simple model for their subsequent non-linear collapse which predicts different collapse times and virialised masses for different over-densities, we visit the question of whether, without performing a computer simulation, we can predict the mass spectrum of collapsed dark matter perturbations in the Universe, called halos, at different times. This mass spectrum, known as the 'halo mass function', gives the number density of halos as a function of their mass. Since galaxies are believed to form in dark matter halos, estimating the halo mass function is a key prediction from structure formation theory as it can be compared to the observed galaxy luminosity function.

The halo mass function $n(M)$ is defined as

$$dN = n(M)\,dM, \tag{6.68}$$

where dN is the number of halos per unit volume with mass between M and $M + dM$. In practice, it is measured by counting the number of halos of a given mass present in a given volume.

A simple analytic argument to derive the halo mass function was first proposed in a classic paper by Press and Schechter [16]. Despite the rather crude approximations they made, it turns out the mass function they derived yields a reasonable match to the mass function measured in numerical N-body simulations as shown in Figure 6.3.

Press and Schechter quantify how many halos form out of peaks in the matter fluctuations by assuming that fluctuations that surpass the collapse over-density predicted by linear theory ($\delta_{\text{coll, lin}} = 1.686$) will collapse to form virialised halos. Although we saw that even the simplified spherical top-hat model predicts a collapse over-density which is much higher (~ 200), taking the much lower prediction from linear theory turns out to be reasonable because gravitational instability proceeds quickly. The basic idea behind the derivation of the Press-Schechter mass function is that the fraction of mass in halos more massive than a given mass M is related to the fraction of the volume within which the smoothed density field is above threshold $\delta_{\text{coll, lin}}$. The smoothing scale R is tied to the halo mass, as we will describe below. By filtering out points in the density field that are embedded within structures of scale smaller than $R(M)$, we filter out masses less than M, leaving only structures with mass $\geq M$.

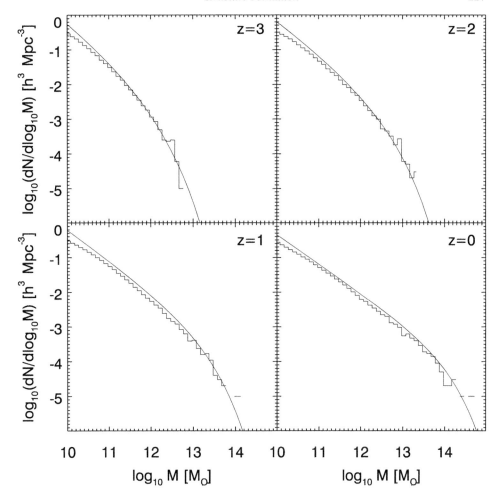

Figure 6.3 Evolution of the dark matter halo mass function (histograms) measured in the Horizon-AGN cosmological simulation between redshifts 0 and 3 (see www.horizon-simulation.org for detail). Top-left panel corresponds to $z = 3$, top-right panel to $z = 2$, bottom-left panel to $z = 1$, and bottom-right panel to $z = 0$. The solid lines on each panel indicate the Press-Schechter analytic prescription discussed in the text. The discrepancy between the two, especially noticeable at the low-mass end, is mainly caused by the various approximations used in the Press-Schechter formalism.

We now outline the Press-Schechter derivation in more detail. Consider a field of density fluctuations $\delta(\mathbf{x})$. We can filter this field on a length-scale R, where R is associated to a given mass M as the radius of a uniform-density sphere containing mass M, with the sphere density equal to the mean density of the Universe at redshift z, $\bar{\rho}_m$:

$$R(M) = \left(\frac{3M}{4\pi \bar{\rho}_m} \right)^{1/3} .$$

(6.69)

By smoothing the field on scale $R(M)$, all information about $\delta(\mathbf{x})$ below that scale is thrown out. Smoothing is achieved by convolving $\delta(\mathbf{x})$ with a window function $W_R(\mathbf{x}-\mathbf{x}')$ typically taken to be a top-hat function on scale R:

$$\delta_R(\mathbf{x}) = \int d^3x' \, \delta(\mathbf{x}) \, W_R(\mathbf{x} - \mathbf{x}'). \tag{6.70}$$

Introducing the notation $\delta_R(\mathbf{x}) = \delta_M$, and assuming that the density fluctuations $\delta(\mathbf{x})$ – and, hence, the smoothed δ_M – are a Gaussian random field, we can write the probability of finding any particular value δ_M at position $\mathbf{x}$ as

$$p(\delta_M) = \frac{1}{\sqrt{2\pi\sigma_R^2}} \exp\left(-\frac{\delta_M^2}{2\sigma_R^2}\right), \tag{6.71}$$

where σ_R is the variance of the field smoothed on scale R:

$$\sigma_R^2 = \sigma_M^2 = \int_0^\infty \frac{k^2 dk}{2\pi^2} P(k,t) \, \hat{W}_R^2(k), \tag{6.72}$$

where $\hat{W}_R$ is the Fourier transform of the window function and $P(k,t)$ is the power spectrum of fluctuations previously defined. The probability that a fluctuation exceeds $\delta_{\text{coll, lin}}$ is given by

$$
\begin{aligned}
P_{\delta > \delta_{\text{coll, lin}}}(M) &= \int_{\delta_{\text{coll, lin}}}^\infty p(\delta_M) \, d\delta_M \\
&= \frac{1}{2}\text{erfc}\left(\frac{\delta_{\text{coll, lin}}}{\sqrt{2}\sigma_M}\right),
\end{aligned} \tag{6.73}
$$

where erfc is the complementary error function.

By subtracting $P_{\delta > \delta_{\text{coll, lin}}}(M + dM)$ from $P_{\delta > \delta_{\text{coll, lin}}}(M)$, we then get the number of collapsed halos of mass M surrounded by under-dense regions. To get the mass function $n(M) \, dM$, which tells us the number of objects per unit volume, we have to multiply the probabilities by the background density divided by the mass-scale, $\bar{\rho}_m/M$. Furthermore, because we expect that most of the mass in the Universe should eventually find itself in collapsed halos, the probability should be normalised to unity. To achieve this, Press and Schechter simply multiplied the mass function by 2 to obtain:[19]

$$
\begin{aligned}
n(M) \, dM &= 2\frac{\bar{\rho}_m}{M}\left(P_{\delta > \delta_{\text{coll, lin}}}(M) - P_{\delta > \delta_{\text{coll, lin}}}(M + dM)\right) \\
&= -2\frac{\bar{\rho}_m}{M}\frac{dP_{\delta > \delta_{\text{coll, lin}}}}{dM}\, dM.
\end{aligned} \tag{6.74}
$$

---

[19] This brute force solution to the fact that under-densities ($\delta < 0$) are not treated properly by the Press-Schechter approach (the so-called cloud-in-cloud problem) was properly justified by [3] in their excursion set theory. The Press-Schechter approach of merely multiplying by 2 appears to work because, once fluctuations collapse, they are able to pull material from less dense regions, resulting in most of the mass eventually ending up in over-dense regions.

It is convenient to write

$$\frac{\mathrm{d}}{\mathrm{d}M} = \frac{\mathrm{d}\sigma_M}{\mathrm{d}M}\frac{\mathrm{d}}{\mathrm{d}\sigma_M}$$

to get

$$n(M)\,\mathrm{d}M = -2\frac{\bar{\rho}_m}{M}\frac{\mathrm{d}P_{\delta > \delta_{\mathrm{coll, lin}}}}{\mathrm{d}\sigma_M}\frac{\mathrm{d}\sigma_M}{\mathrm{d}M}\mathrm{d}M. \tag{6.75}$$

Then, using the fundamental theorem of calculus to differentiate $P_{\delta > \delta_{\mathrm{coll, lin}}}$, together with the substitution $x = \delta_{\mathrm{coll, lin}}/(\sqrt{2}\,\sigma_M)$,

$$\begin{aligned}
\frac{\mathrm{d}P_{\delta > \delta_{\mathrm{coll, lin}}}}{\mathrm{d}\sigma_M} &= \frac{\mathrm{d}}{\mathrm{d}\sigma_M}\left[\int_{\delta_{\mathrm{coll, lin}}}^{\infty}\frac{1}{\sqrt{2\pi}\sigma_M}\exp\left(-\frac{\delta_M^2}{2\sigma_M^2}\right)\mathrm{d}\delta_M\right] \\
&= \frac{1}{\sqrt{\pi}}\int_{\frac{\delta_{\mathrm{coll, lin}}}{\sqrt{2}\sigma_M}}^{\infty}\exp\left(-x^2\right)\mathrm{d}x \\
&= -\frac{1}{\sqrt{\pi}}\exp\left(-\frac{\delta_M^2}{2\sigma_R^2}\right)\frac{\mathrm{d}}{\mathrm{d}\sigma_M}\left[\frac{\delta_{\mathrm{coll, lin}}}{\sqrt{2}\sigma_M}\right] \\
&= \frac{1}{\sqrt{2\pi}}\frac{\delta_{\mathrm{coll, lin}}}{\sigma_M^2}\exp\left(-\frac{\delta_M^2}{2\sigma_R^2}\right).
\end{aligned} \tag{6.76}$$

With this expression for $\mathrm{d}P_{\delta > \delta_{\mathrm{coll, lin}}}/\mathrm{d}\sigma_M$, we finally arrive at the expression for the number density of collapsed halos of a given mass:

$$n(M)\,\mathrm{d}M = \sqrt{\frac{2}{\pi}}\frac{\bar{\rho}_m}{\sigma_M}\frac{\delta_{\mathrm{coll, lin}}\,\mathrm{d}\ln\sigma_M}{\mathrm{d}M}\exp\left(-\frac{\delta_M^2}{2\sigma_M^2}\right)\frac{\mathrm{d}M}{M}. \tag{6.77}$$

To understand the behaviour of this mass function, we can estimate σ_M by assuming a power-law function for the power spectrum, $P(k,t) \propto k^n$, and a window function that is a top-hat in k-space:

$$\hat{W}_R(k) = \begin{cases} 1, & \text{if } k \le 2\pi/R \\ 0 & \text{otherwise.} \end{cases}$$

This leads to

$$\sigma_M^2 \propto \int_0^{2\pi/R} k^{2+n}\,\mathrm{d}k \propto R^{-3+n} \propto M^{(-3+n)/3}. \tag{6.78}$$

For small scales, the argument of the exponential in the mass function goes to 1 so that

$$n(M)\,\mathrm{d}M \propto M^{(n-9)/6}\,\mathrm{d}M. \tag{6.79}$$

Earlier we estimated that the power spectrum at time t could be deduced from the scale-independent primordial power spectrum $P(k, t_{\mathrm{bis}})$ generated by inflation, and that

$P(k,t) \propto k^{-3}$ for $k \gg k_{eq}$ (corresponding to galaxy scales), meaning that $n = -3$. This gives the following scaling for the mass function at the low mass end:

$$n(M)\, dM \propto M^{-2}\, dM, \tag{6.80}$$

i.e., more low-mass halos than high mass halos. At the high-mass end, the mass function decreases exponentially so that the probability of finding a halo of arbitrarily large mass at any given redshift is extremely small.

Despite the oversimplifications it makes – for example, the fact that it assumes spherical halos and that it does not account for how many small-mass halos are subsumed into larger-mass ones – the Press-Schechter theory gives a rough estimate of the mass function across all scales from dwarf galaxies to galaxy clusters accurate to about a factor 2, as shown by Figure 6.3. Where it goes wrong the most is at the low-mass end, where it over-predicts the number of halos, and at the high-mass end, where it under-predicts the number of gravitationally bound objects compared to N-body simulations (see Figure 6.3). Nevertheless, it captures the key features of structure formation, allowing one to track when halos of a given mass first appear in the Universe.[20]

Unfortunately, galaxies do not map straightforwardly onto halos. What we measure for galaxies is their luminosity function – i.e., the number density of galaxies of a given luminosity. The faint end of the luminosity function is not as steep as the low-mass end of the halo mass function predicted by Press-Schechter, and the characteristic luminosity at which the luminosity function overturns corresponds to a much lower halo mass than the characteristic halo mass at which the Press-Schechter halo mass function overturns. Hence, to take structure formation a step further and make predictions about observed galaxies, one needs to consider more complex physics such as gas cooling, star formation, and feedback from processes such as supernovae and active galactic nuclei powered by supermassive black holes.

However, it is interesting at this point to ask how the Press-Schechter halo mass function will depend on the fine-tuning of cosmological parameters and, more specifically from an observable perspective, how much mass is locked in collapsed halos massive enough to retain baryons and form stars.[21]

This has recently been explored in some detail in Reference [2] (see their figure 3) using numerical simulations with characteristics very similar to those used in this chapter. These authors find that for the fraction of mass locked in star-forming halos to change by a factor ~ 2.5, one would need to increase Ω_Λ by a factor 100 compared to its measured value. Furthermore, the cosmic star formation rate history of the Universe, obtained when folding in the baryonic processes previously mentioned (gas cooling, star formation supernovae, and active galactic nuclei feedback) would only change by a factor eight in that case.

---

[20] It is possible to relax some of the simplifying assumptions used in the Press-Schechter theory, especially that of a spherically symmetric collapse of the perturbations, and get a better agreement with N-body simulations. We refer the interested reader to [11] for an extension of the formalism to the more general case of an ellipsoidal collapse.

[21] This can be readily deduced from Eq. (6.73) by multiplying it by two (following the original Press-Schechter argument) and setting $M = M_{min}$, where M_{min} is the minimum mass above which halos retain baryons.

The authors do point out that the environment of galaxies would be dramatically different in a larger Ω_Λ universe (disappearance of groups and clusters as argued earlier in this chapter) but, nevertheless, conclude that structure formation, as quantified by the amount of stars produced throughout the lifetime of the Universe, does not seem very sensitive to a fine-tuning of the cosmological constant.

There exists, however, a major caveat in their argument: one could easily argue that increasing the value of the measured cosmological constant by a mere two orders of magnitude is still very much within the realm of fine-tuning, given that more than 120 orders of magnitude separate it from the value expected from quantum field theory! Moreover, as the authors themselves admit, an anthropic argument calculation based on a multiverse model, whilst faring better than pure theory, remains, by and large, inconclusive in naturally predicting a low Λ value.

6.4 The Fine-Tuning of Structure Formation

Historically fine-tuning is understood as referring to the specificity of the conditions necessary for life as we know it[22] to appear in the Universe. This is generally discussed in terms of variations in the fundamental constants of physics, but from the point of view of structure formation, it seems more sensible to adopt a more prosaic point of view: that of life shelter. Obviously, the two are related: if, for instance, the gravitational coupling constant had a different value, the very presence of life shelters, such as planets, could be in jeopardy.

Science is underpinned by reproducible experimental results, so our argument shall unfold from there. Repeated CMB measurements since its discovery by Penzias and Wilson in 1965 have established that the Universe started from a highly homogeneous and isotropic state (at least to one part in 100,000). Since the 1930s and Hubble's observations that galaxies recede from one another at a rate proportional to the distance between them, we also know that the Universe is expanding. Moreover, from distant supernovae measurements performed independently by several groups in the late 1990s, we have established that this expansion occurs at an accelerated rate. The unavoidable conclusion is that the homogeneous Universe must be getting colder and more diluted as time goes by, which completely goes against the apparition of life in the Universe.

Indeed, while it is notoriously difficult to define what life is and how it came to be, there can be little doubt that it entails the formation of complex molecules. An uniform universe, whose matter density converges towards perfect vacuum and whose temperature is driven by expansion toward the absolute zero, where no movement can take place on the atomic level, does not appear as propitious conditions for complexity to emerge – hence, the concept of life shelter. If the overwhelmingly homogeneous Universe is converging to an eternal state of 'frozen in the vacuum' death, life will need to appear in small, isolated islands harbouring a more auspicious climate.

---

[22] That is to say, based on the chemistry of carbon, although to the best of my knowledge, a compelling argument that life based on another element could not exist has yet to be put forward.

Going from small to large scales, one can sketch the properties of the optimal habitat for fragile, complex molecules to be formed and survive for a significant amount of time so that they can further gain in complexity. It first needs to be dense and hot enough for various atoms to collide but not too dense or hot; otherwise, they would be too easily destroyed or not formed at all. In other words, you need a planet with a surface temperature below the energy dissociation of these complex molecules but well above absolute zero. This implies close (but not too close!) proximity with stars – first to produce the heavy elements from which complex molecules are assembled in large quantities[23] and, second, to catalyse chemical reactions and provide heat via the radiation they emit.

Now stars do not form in isolation – although there might be a few exceptions in the very early stages of structure formation – but in galaxies, which explains why galaxies are considered as the fundamental building blocks of cosmological structure formation. It also justifies why the scope of our discussion about fine-tuning in this chapter is reduced to a discussion of how fine-tuned the conditions for galaxies to exist are.

There are many places where one could see fine-tuning at work in cosmological structure formation, beginning with the initial conditions from which structures come into existence. Indeed, it may seem paradoxical that the very mechanism invoked to solve, amongst other issues, the acausal fine-tuning problem,[24] turns out to appear so fine-tuned itself. In other words, why is value of the initial fluctuation amplitudes produced by inflation so low (one part in 100,000)?

Should they be much smaller, despite the presence of dark matter, they would remain in the linear regime, and galaxies would not have enough time to grow. If, on the other hand, they were much larger, say on the order of unity, they would never grow in the linear regime to begin with, but behave like independent universes very rapidly collapsing on top of one another,[25] creating a very violent environment where galaxy cannibalism would be the rule rather than the exception. As a result, this would greatly reduce the total number of galaxies in the Universe and, hence, the number of places where life could be nurtured.

Given the measured amplitude of the fluctuations and moving on to the evolutionary stage, their growth, which will eventually give birth to galaxies, depends on two dominant 'dark' components: dark matter and dark energy (assumed in this chapter to be in the guise of a cosmological constant). Bypassing the issue of the very existence of dark matter, thanks to which, as we have seen earlier in this chapter, galaxies can form in a timely manner despite the low amplitude of initial baryonic fluctuations, one can legitimately ask the question as to why its energy density is comparable to the critical energy density today.

Once again, had it been much smaller, then hardly any galaxy would have formed by now as the epoch of matter domination during which perturbations can most efficiently grow

---

[23] We know from element abundance observations that Big Bang nucleosynthesis produces virtually no element heavier than helium.

[24] That is to say, provide an answer to the question of how perturbations on galaxy scales which are larger than the causal length-scale in the early Universe can be coherent.

[25] from Eq. (6.45) for the spherical top hat model, such large amplitude perturbations would enter the matter dominated era with a radius already greater than $r[\pi/2] = r_m/2$ at time $t[\pi/2] \simeq 0.18\, t_m$, hence rapidly reach turnaround and collapse onto themselves.

would have been considerably shortened. Conversely, had it been much larger, the Universe would have undergone a short expansion phase and recollapsed onto itself before galaxy clusters or even galaxies – depending on the exact value of $\Omega_{m,0}$ – formed. There also is the question of the mass of the dark matter particle candidate: one needs a massive dark particle with a short free-streaming length (unlike, e.g., neutrinos) in order for perturbations to exist on galaxy scales at an early stage.

Finally, as previously mentioned in the section devoted to the growth of matter fluctuations in the Λ-dominated era, comes what arguably constitutes the most clear-cut case of fine-tuning from the point of view of structure formation: the value of the cosmological constant.[26] This value is measured to be about 122 orders of magnitude lower than expected from the Standard Model of particle physics,[27] and, as discussed earlier in this chapter, should it be only marginally larger[28] than its measured value, large-scale cosmic structures and possibly galaxies would not have formed at all. Whether such a gigantic discrepancy with such extraordinary consequences simply reflects the (lack of) depth of our current knowledge of the underlying physics or possesses a more profound meaning must be regarded as the greatest mystery of modern physical cosmology.

References

[1] Agrawal, V., *et al.* 1998. 'Anthropic Considerations in Multiple-Domain Theories and the Scale of Electroweak Symmetry Breaking'. *Physical Review Letters*, **80**(Mar.), 1822–1825.

[2] Barnes, L. A., *et al.* 'Galaxy Formation Efficiency and the Multiverse Explanation of the Cosmological Constant with EAGLE Simulations'. *Monthly Notices of the Royal Astronomical Society*, **477**(3), 3727–3743.

[3] Bond, J. R., *et al.* 1991. 'Excursion Set Mass Functions for Hierarchical Gaussian Fluctuations'. *Astrophysical Journal*, **379**(Oct.), 440–460.

[4] Carroll, S. M., Press, W. H., and Turner, E. L. 1992. 'The Cosmological Constant'. *Annual Review of Astronomy and Astrophysics*, **30**, 499–542.

[5] Damour, T. and Donoghue, J. F. 2008. 'Constraints on the Variability of Quark Masses from Nuclear Binding'. *Physical Review D*, **78**(1), 014014.

[6] Easther, R. and Price, L. C. 2013. 'Initial Conditions and Sampling for Multifield Inflation'. *Journal of Cosmology and Astroparticle Physics*, **7**(July), 027.

[7] Lewis, A. and Challinor, A. 2011 (Feb.). *CAMB: Code for Anisotropies in the Microwave Background*. Astrophysics Source Code Library.

[8] Mukhanov, V. F., Feldman, H. A., and Brandenberger, R. H. 1992. 'Theory of Cosmological Perturbations'. *Physics Reports*, **215**(June), 203–333.

[9] Peebles, P. J. E. 1980. *The Large-Scale Structure of the Universe*. Princeton University Press.

---

[26] Note that if we enforce that the Universe is flat, this is the flip side of the previous question about the value of the energy density of dark matter.

[27] But no lower than that: if it were, we would not even know it exists at all!

[28] Note that even an increase by several order of magnitudes above the measured value can be considered marginal, given the amplitude of the discrepancy with theory!

[10] Planck Collaboration *et al.* 2016. 'Planck 2015 Results. XIII. Cosmological Parameters'. *Astronomy and Astrophysics*, **594**(Sept.), A13.

[11] Sheth, R. K., and Tormen, G. 2002. 'An Excursion Set Model of Hierarchical Clustering: Ellipsoidal Collapse and the Moving Barrier'. *Monthly Notices of the Royal Astronomical Society*, **329**(Jan.), 61–75.

[12] Sugiyama, N. 1995. 'Cosmic Background Anisotropies in Cold Dark Matter Cosmology'. *Astrophysical Journal Supplements*, **100**(Oct.), 281.

[13] Vilenkin, A., and Shellard, E. P. S. 2000. *Cosmic Strings and Other Topological Defects*. Cambridge University Press.

[14] Weinberg, S. 1972. *Gravitation and Cosmology: Principles and Applications of the General Theory of Relativity*. Wiley VCH.

[15] Liddle, A., and Lyth, D. 2000. *Cosmological Inflation and Large-Scale Structure*. Cambridge University Press.

[16] Press, W. H., and Schechter, P. 'Formation of Galaxies and Clusters of Galaxies by Self-Similar Gravitational Condensation'. *Astrophysical Journal*, 187, 425–438.

Part III

Fine-Tuning in Particle and Nuclear Physics

7

Nuclear Physics and Its Impact on Primordial and Stellar Nucleosynthesis

JEAN-PHILIPPE UZAN

Abstract

Nuclear physics is at work in cosmology both during Big Bang nucleosynthesis, during which light elements are formed, and during stellar evolution, where heavier elements (actually, all nuclei but hydrogen, helium, and lithium) are synthesised. It means that a successful stellar nucleosynthesis, in particular to form carbon and oxygen, is a key step in the emergence of complexity in our Universe. Besides, stars provide a long-lived, low-entropy source of energy also necessary for planetary life to emerge. Nuclear processes depend on many fundamental constants, the values of which affect cross sections, binding energies, reaction rates, and masses of particles. They also affect the stability of those nuclei. In our Universe, the Mendeleev table summarises the variety of nuclei and their isotopes that are then involved to form molecules. The goal of this chapter is to detail the connections between the Standard Model of particle physics and nuclear physics in order to understand how the fundamental constants affect the stability and production of the nuclei in the Universe and how it could limit the apparition of chemical complexity and eventually of life.

7.1 Introduction

Nature confronts us with structures at different scales, complexity, and properties, from fundamental particles to molecules and cells to planets, stars, and galaxies. Each of them can be described by a scientific theory, with its own ontology and structures. They form a hierarchy of theories organised in modules in interaction (see References [45, 97, 111] for a discussion). Indeed, those theories are not independent. Higher levels are built on lower levels and more fundamental theories that define a space of possibility for the former. And higher theories set the context in which the dynamics of the lower-level theories develop. This is related to both bottom-up action and top-down causation [45].

Hierarchy of physical theories. The fact that we can understand the Universe and its laws has a deep implication on this structure of theories. At each step in our construction of physical theories, we have been dealing with phenomena below a typical energy scale,

237

mostly for technological constraints, and it turned out (empirically) that we have always been able to design a consistent theory, valid in such a restricted regime. This is not expected a priori, and this empirical fact is deeply rooted in the mathematical structure of the theories that describe nature. In particular, they have to enjoy a *scale-decoupling principle* in the sense that there exist energy scales below which effective theories are sufficient to understand a set of physical phenomena that can be observed. *Effective theories* are then the most fundamental concepts in the scientific approach to the understanding of nature, and they always come with a domain of validity inside which they are efficient to describe all related phenomena [111]. They offer a successful explanation at a given level of complexity based on concepts of that particular level.

This implies that the structure of the theories is such that there is a kind of stability and independence of higher levels with respect to more fundamental ones. It follows that various disciplines have developed independently in almost quasi-autonomous domains, each of them having its own ontology and dynamics that are independent of our ability to formulate a theory explaining these concepts. On the one hand, we can hope to relate the concepts and constants of a given level to those of an underlying level. For instance, we understand that the proton is a composite structure of three quarks, and we may try to determine its physical characteristics (charge, mass, gyromagnetic factor, quantum numbers) in terms of these more fundamental entities [79]. We know that this reductionist approach is limited and can only be achieved for some structures since there exist emergent phenomena (information, life, consciousness) that cannot be reduced to the concepts of a lower level.

Goal of this chapter. This chapter investigates the first layer of this hierarchy, namely the one involving the properties of nuclear matter and, in particular, its stability and its production. In our Universe, the diversity of the atomic nuclei opens a large possibility space of combinations for atoms and then molecules. These properties are first roughly summarised in the Mendeleev table, and one success of the standard cosmological model and nuclear astrophysics is to understand how the Mendeleev table is populated – i.e., in which sites each nuclei is produced and with which abundance.

As can be seen from Figure 7.1, only helium is significantly synthesised during Big Bang nucleosynthesis; some lithium, beryllium, and boron and slight traces of carbon, nitrogen, and oxygen are synthesised. It is understood that most stable beryllium and boron were created in the interstellar medium when cosmic rays induced fission in heavier elements found in interstellar gas and dust. Elements lighter than iron are formed in small and large stars while heavier elements are formed mostly in explosive nucleosynthesis in supernovae. In terms of nuclear processes, one usually distinguishes different processes:

- The s-process (slow neutron-capture process). It occurs at low neutron density and intermediate temperature conditions in stars. Heavier nuclei are created by neutron capture, increasing the atomic mass of the nucleus by one, followed by a β^- decay, leading to a nucleus of higher atomic number.
- The r-process (rapid neutron-capture process). It occurs during core-collapse supernovae and is responsible for the creation of approximately half of the neutron-rich atomic nuclei

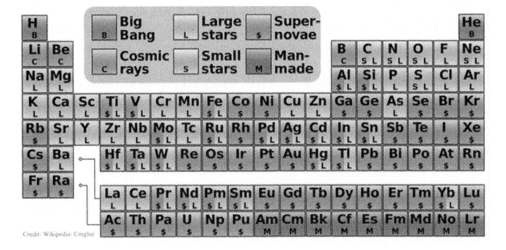

Figure 7.1 Simplified summary of the main sites of formation of the different atomic nuclei. Besides helium mostly produced during Big Bang nucleosynthesis, all nuclei are formed during the stellar evolution. It is important to note that some elements are only known to be human made and have not been found in any natural environment. A more detailed summary, included the different isotopes is presented in Figure 7.17 in Appendix A. We emphasise that the existence of isotopes, is also important, wether they are stable or not. Note that some nuclei have never been observed to be produced in natural environments. They do not seem to have any importance for the emergence of complexity but they may be a way to characterise an intelligent civilisation.

heavier than iron. The process entails a succession of rapid neutron captures (hence the name r-process) by heavy seed nuclei. It creates very neutron-rich heavy isotopes, which can then decay to the first stable isotope.

- The p-process (proton process). It refers to a proton capture process which is the source of certain naturally occurring, proton-rich isotopes of the elements

Let us also note that it is important to distinguish stable isotopes from unstable isotopes. For instance, radioactive beryllium-10 is produced in the atmosphere of the Earth by the cosmic-ray spallation of oxygen. We shall not describe all these routes here. A summary of all the sites of production, as understood today, is given in Appendix A.

Some heavy elements, for which we only have traces, are also important for the emergence of life. Our discussion mostly focuses on carbon, nitrogen, and oxygen. They are, indeed, the building blocks of life as we know them and will enable a complex chemistry. This is indeed not a sufficient condition for life to emerge. Many other elements are necessary for life as we know it to exist. But it is almost impossible to quantify their effect and whether they can be replaced by other mechanisms if they did not exist.

This raises different questions. First, one would like to determine if (and which) nuclei are stable and then if they can be produced with a large enough abundance. To answer these questions, we would need to connect quantum chromodynamics (QCD) to nuclear physics, which is a difficult task and for which many links are still not fully understood. This means that one will have to rely on different approaches: theoretical, numerical, and modelisation. It also means that we will have to embed nuclear physics in a cosmological context for Big Bang nucleosynthesis (BBN) and in an astrophysical model for stellar evolution.

Fundamental physics. To begin, let us define our starting point concerning fundamental physics. Today, gravitation is well described by general relativity, and the most fundamental (experimentally tested) theory of matter is the Standard Model of particle physics. It is a theoretical construction based on an action and many choices such as the mathematical description of the matter fields (This is not completely arbitrary. It is based on the representation of the Poincaré group, which allows one to define scalar, spinor, vector, etc., structures, but we still have to decide to identify one kind of particle with a mathematical structure [113]), symmetries, and constants. None of them can be explained by the theory at hand. In particular, it is important for the constants to be measurable. For this model, they actually are and have been measurable, as we shall describe later. There is, indeed, no way to express them in terms of more fundamental quantities, and there is no equation for them. By testing their constancy, one actually shows that at the level of accuracy of the experiments and their timescales, it is a good hypothesis. In case of a disagreement, one could promote them to a dynamical field but would have to explain why they are almost frozen today [107–110] – i.e., of a stabilisation mechanism.

It is also important to remember that any measurement is just a comparison between two physical systems, one usually defining a system of units. It follows that only dimensionless constant can be measured [46], and only the change of these constants would change the physics. We are considering only these parameters. Given a list of N constants, one can always pick up three of them to define units so that one is left with $N - 3$ fundamental parameters that affect the magnitude of any physical process.

Coming back to the Standard Model of particle physics, we assume it is our fundamental theory, even though it does not incorporate massive neutrinos and dark matter so that we know it calls for an extension. This theory offers the space of possibility for higher levels complexity to emerge. Changing the values of the fundamental constants may result in the technical impossibility for nuclei to be stable. How fine-tuned is the Universe? This is what we shall now illustrate, keeping in mind that we would like to estimate how far in the chain of physical theories of higher complexity level they propagate.

Content. This chapter focuses on the relation between QCD and nuclear physics. There, one needs to determine how cross sections, binding energies, lifetimes of unstable nuclei (or of the neutron), or simply characteristics such as the mass of the proton depend on the fundamental constants listed in Table 7.1. This has been intensively investigated, and we refer to References [107–110] for reviews. It is of huge importance for the description of Big Bang nucleosynthesis and for stellar nucleosynthesis.

This chapter will first provide some generalities on the study of the influence of fundamental constants (Section 7.2). Section 7.3 then details different approaches to relate masses, gyromagnetic factors, binding energies, and resonance energies to the parameters

Table 7.1 *List of the fundamental constants of our Standard Model. See Reference [83] for further details on the measurements.*

Constant	Symbol	Value
Speed of light	c	299,792,458 m s^{-1}
Planck constant (reduced)	$\hbar$	$1.054\,571\,628(53) \times 10^{-34}$ J s
Newton constant	G	$6.674\,28(67) \times 10^{-11}$ m^2 kg^{-1} s^{-2}
Weak coupling constant (at m_Z)	$g_2(m_Z)$	0.6520 ± 0.0001
Strong coupling constant (at m_Z)	$g_3(m_Z)$	1.221 ± 0.022
Weinberg angle	$\sin^2 \theta_W(91.2 \text{ GeV})_{\overline{MS}}$	0.23120 ± 0.00015
Electron Yukawa coupling	h_e	2.94×10^{-6}
Muon Yukawa coupling	h_μ	0.000607
Tauon Yukawa coupling	h_τ	0.0102156
Up Yukawa coupling	h_u	0.000016 ± 0.000007
Down Yukawa coupling	h_d	0.00003 ± 0.00002
Charm Yukawa coupling	h_c	0.0072 ± 0.0006
Strange Yukawa coupling	h_s	0.0006 ± 0.0002
Top Yukawa coupling	h_t	1.002 ± 0.029
Bottom Yukawa coupling	h_b	0.026 ± 0.003
Quark CKM matrix angle	$\sin \theta_{12}$	0.2243 ± 0.0016
	$\sin \theta_{23}$	0.0413 ± 0.0015
	$\sin \theta_{13}$	0.0037 ± 0.0005
Quark CKM matrix phase	δ_{CKM}	1.05 ± 0.24
Higgs potential quadratic coefficient	$\hat{\mu}^2$	$-(250.6 \pm 1.2)$ GeV^2
Higgs potential quartic coefficient	λ	1.015 ± 0.05
QCD vacuum phase	θ_{QCD}	10^{-9}

of the Standard Model of particle physics, which means that it starts with a short summary of the formulation of the Standard Model of particle physics in order for the read to understand the nature of its constants (this is complementary to Chapter 8). Masses, cross sections, and binding energies are the key quantities that appear in nuclear physics. Once we have seen how they can be related to the fundamental constants, we can give the two main astrophysical applications of these computations: the synthesis of light nuclei during Big Bang nucleosythesis in Section 7.4 (see also Chapter 5) and the production of heavier elements during stellar evolution in Section 7.5. We will pay special attention to the production of carbon-12 and oxygen-16 since they are key nuclei for life as we know it on Earth. Section 7.6 will summarise the implications for fine-tuning.

7.2 Strategy

In order to study the effects of a variation of the fundamental constants on the prediction of nuclear physics, we proceed as follows. First, we only consider a local variation; that is, we assume some changes of the fundamental constants in a domain close to the actual values observed in our Universe and, more important, we assume that the structure of the QCD theory remains unchanged. General studies on the constraints on the time variation of fundamental constants [107–110] provide upper bounds on possible variations. This means that these constraints give a measure of a universe that will appear similar to ours to observers similar to us. Our observational ability to constrain such variations provide a coarse-graining scale on the space of fundamental constants, under which two universes cannot be distinguished.

Under this hypothesis, we can define sensitivity coefficients that characterise the effect of the change of a fundamental constant. Given an observable O, the value of which depends on a set of primary parameters G_k, the sensitivity of the measured value of O to these parameters is

$$\frac{d \ln O}{d \ln G_k} = c_k. \tag{7.1}$$

The primary parameters can indeed be fundamental constants, such as the gravitational constant, the fine-structure constant, or combination of them, such as masses of composite particles, cross sections, binding energies, etc. The computation of the coefficient c_k requires a physical description of the system. Indeed, the observation O depends on both the parameters G_k and on external parameters that we shall call X, such as temperature, magnetic fields, etc.

Any observation of the system provides a value of O with some errors bars – i.e., $O_{obs} \pm \Delta O_{obs}$. Usually, the values of the primary parameters are assumed to be constant and take the value derived locally in the solar system, as tabulated in Table 7.1. Two different approaches shall be distinguished:

- One can use the observations to set constraints on the space-time variations of the primary parameters. The constraints will depend on the accuracy of the observations (ΔO_{obs}) as well as our knowledge of the effects of the external parameters X.
- One can let the constants vary in order to quantify the change that may result in the fact that the phenomena O does not happen or happens in a very different way.

The values of the parameters c_k offer a way to either choose the system O that will best constrain a set of parameters or determine the constants that play a drastic role in the phenomena O and that can thus be subject to some fine-tuning. At this stage, the approach is completely standard from a physics point of view, since it just assumes that the values of the G_k are not known a priori and that we try to measure them on the system O.

As we have already emphasised, the parameters G_k may not be fundamental constants. They need to be related to a set of fundamental constant α_i, and we define

$$\frac{d \ln G_k}{d \ln \alpha_i} = d_{ki}. \tag{7.2}$$

The computation of the coefficients d_{ki} requires one to specify the theoretical framework and depends heavily on our knowledge of nuclear physics and the assumptions of unification. This shall be discussed in the next section. In the following, a particular set of parameters d_{ki} has been singled out for the sensitivity of the mass of a body A to a variation of the fundamental constants

$$\frac{d \ln m_A}{d\alpha_i} = f_{Ai}. \tag{7.3}$$

As we have emphasised, we restrict our analysis to local variations of the fundamental constants, simply for technical reasons. In a general setting – such as in most multiverse scenarios, fundamental constants, and even the structures of the theory – can radically change. Among this range of parameters lies a subset that we shall call the *anthropic range*, which allows for the Universe to support the existence of observers (but not necessarily similar to us). The local analysis is indeed very restrictive since the mathematical form of the laws of physics may as well change so that what we are restricting to a local analysis in the neighbourhood of our observed Universe (in the space of theories). The determination of the anthropic region is not a prediction but just a characterisation of the sensitivity of 'our' Universe to a change of the fundamental constants ceteris paribus. Once this range is determined, one can ask the general question of quantifying the probability that we observe a universe as ours, hence providing a probabilistic prediction. This involves the use of the Anthropic Principle, which expresses the fact that we observe not just observations but observations made by us, and, requires us to state what an observer actually is (see References [8, 110] for a detailed discussion on fine-tuning and anthropic arguments). We shall not discuss this in detail in this chapter.

7.3 From the Standard Model of Particle Physics to Nuclear Physics

In order to detail the connection between these two theories, we start by recalling the basics of the Standard Model of particle physics. The main goal is to define the structures and the constants that appear at that level. We shall then discuss the running of the coupling constants with energy and the possibility of unification. It has deep implications on fine-tuning since it can tie the values of different constants together, hence suppressing some parts of the parameter space. We follow with a general discussion on the masses and gyromagnetic factors of composite particles. We then describe the cluster approach to model nuclei. It is an important tool that will allow us to compute binding energies and resonance energies. We finish with a summary of the tuning issues related to the stability of the nuclei.

7.3.1 The Standard Model of Particle Physics in a Nutshell

The Standard Model of non-gravitational interactions accounts for all the interactions except gravity, classified depending on the representations of the groups $SU(3)_c$ for the strong interaction, and $SU(2)_L \times U(1)_Y$ for the electroweak ones. This last symmery is broken by the Higgs mechanism along the symmetry-broken process $SU(2)_L \times U(1)_Y \rightarrow U(1)_{elec}$.

The strong interaction is based on an invariance under the transformations of the group $SU(3)$. Since this group has eight generators, its joint representation, in which the gauge bosons G_μ^a are placed, is of dimension 8. The leptons are not subject to these interactions and, thus, appear as singlets of $SU(3)$, which do not couple to the gauge boson. The quarks exist in three possible states, and are thus placed in a three-dimensional representation, denoted by **3**. The quarks are therefore represented in column vectors with these three elements,

$$Q \equiv \begin{pmatrix} u_r \\ u_g \\ u_b \end{pmatrix}, \quad \begin{pmatrix} d_r \\ d_g \\ d_b \end{pmatrix}, \quad \begin{pmatrix} s_r \\ s_g \\ s_b \end{pmatrix}, \quad \begin{pmatrix} c_r \\ c_g \\ c_b \end{pmatrix}, \quad \begin{pmatrix} t_r \\ t_g \\ t_b \end{pmatrix}, \quad \begin{pmatrix} b_r \\ b_g \\ b_b \end{pmatrix}. \tag{7.4}$$

The index is called colour. The covariant derivative of each quark Q, with respect to the strong interaction, is given by

$$D_\mu^{(3)} Q = \left(\partial_\mu - i g_3 G_{a\mu} \lambda^a \right) Q, \tag{7.5}$$

where the λ^a are eight matrices forming the algebra of $SU(3)$. We often take the Gell-Mann matrices which satisfy the appropriate commutation relations of $SU(3)$ with the structure constants of $SU(3)$, then denoted as f_c^{ab}. The coupling constant g_3 of the group measures the intensity of the interaction. The Lagrangian of QCD takes the form

$$\mathcal{L}_{QCD} = - \sum_{\text{quarks}} \overline{Q} \gamma_\mu D_{(3)}^\mu Q - \frac{1}{4} G_{a\mu\nu} G^{a\mu\nu}, \tag{7.6}$$

where $a = 1, \ldots, 8$ and $G_{a\mu\nu} \equiv \partial_\mu G_{a\nu} - \partial_\nu G_{a\mu} + i f_a^{bc} G_{b\mu} G_{c\nu}$.

The electroweak interaction combines the properties of the weak interaction, coming from the invariance under transformations of the group $SU(2)_L$, and of electromagnetism, based on $U(1)_{\text{elec}}$, brought together as a unique interaction. All particles are combined in singlets or doublets of $SU(2)_L$.

The quarks of each generation are supposed to be equivalent so that the doublets are

$$\begin{pmatrix} u_L \\ d_L \end{pmatrix}, \quad \begin{pmatrix} c_L \\ s_L \end{pmatrix}, \quad \text{and} \quad \begin{pmatrix} t_L \\ b_L \end{pmatrix}, \tag{7.7}$$

where each fermion appears as an eigenstate of chirality with the definition

$$\psi_L \equiv \left(\frac{I - \gamma^5}{2} \right) \psi, \quad \psi_R \equiv \left(\frac{I + \gamma^5}{2} \right) \psi, \tag{7.8}$$

where the matrix γ^5 is defined as $\gamma^5 \equiv i \gamma^0 \gamma^1 \gamma^2 \gamma^3$. Finally, the quarks with right chirality are defined as singlets of $SU(2)_L$:

$$u_R, d_R, \quad c_R, s_R, \quad \text{and} \quad t_R, b_R.$$

The leptons include electrons, muons, tau, and their associated neutrinos. Experimentally, neutrinos are only observed in the left state ν_L, leading to the conclusion that they should be massless. Since they only have two degrees of freedom, in order to put them in similar doublets, it is only possible to pair them with the left component of the corresponding lepton, leaving the right component in a singlet of $SU(2)_L$. The structure is, thus,

$$\begin{pmatrix} \nu_{e_L} \\ e_L \end{pmatrix}, \quad e_R, \quad \begin{pmatrix} \nu_{\mu_L} \\ \mu_L \end{pmatrix}, \quad \mu_R, \quad \begin{pmatrix} \nu_{\tau_L} \\ \tau_L \end{pmatrix}, \quad \tau_R, \quad (7.9)$$

which gathers in a similar scheme the electron, the muon, and the tau with their associated neutrinos. The covariant derivatives are defined as

$$D_\mu^{(2)} \begin{pmatrix} \nu_{e_L} \\ e_L \end{pmatrix} \equiv \left(\partial_\mu - i g_2 B_{i\mu} \sigma^i \right) \begin{pmatrix} \nu_{e_L} \\ e_L \end{pmatrix}. \quad (7.10)$$

σ^i ($i = 1, 2, 3$) are the Pauli matrices, generators of the group $SU(2)$, since they satisfy the algebra

$$\left[\sigma^i, \sigma^j \right] = 2i \epsilon^{ij}{}_k \sigma^k. \quad (7.11)$$

Similar to the strong interaction, the vector bosons $B_{i\mu}$ are in the adjoint representation of $SU(2)$. The coupling constant of the group is called g_2. The coefficients $\epsilon^{ij}{}_k$ take numerical values $+1$ (resp. -1) if (i, j, k) is an even (resp. odd) permutation of $(1, 2, 3)$, and 0 otherwise. The structure constants of the group $SU(2)$ are given by the rank 3 Levi-Civita tensor, and this group is locally equivalent to that of rotations.

Moreover, a charge with respect to the phase transformations is associated to each particle. This 'hypercharge' is denoted by Y, and we have $Y_R = -2$ for the right leptons, $Y_L = -1$ for the left leptons, $Y(q_L) = \frac{1}{3}$ for the quark doublets (7.7) of left chirality, and, for the right chirality, $Y(u_R, c_R, t_R) = \frac{4}{3}$ and $Y(d_R, s_R, b_R) = -\frac{2}{3}$. The doublet structure of Eqs. (7.7) and (7.9) reminds us of the one of spin $\frac{1}{2}$. This is why the doublet classification of $SU(2)_L$ is also called weak isospin, denoted by T. The values $T^3 = \pm\frac{1}{2}$ are respectively given to the top and bottom component of the doublet. Using this analogy and bearing in mind that the electric charges of the quarks are $Q_{u,c,t} = +\frac{2}{3}$ and $Q_{d,s,b} = -\frac{1}{3}$, that the ones for the massive leptons are $Q_{e,\mu,\tau} = -1$, and that neutrinos have no electric charge, we obtain the Gell-Mann–Nishijima relation

$$Q = T^3 + \frac{Y}{2}. \quad (7.12)$$

Note that for this relation to hold for all particles, we need to set $T = 0$ for the singlets. We can then add a covariant derivative term $U(1)_Y$ for each field,

$$D^{(1)} \psi \equiv \left(\partial_\mu + i g_1 Y C_\mu \right) \psi,$$

where g_1 is the coupling constant associated with the group $U(1)_Y$.

This provides all the building blocks to write down a complete Lagrangian for the three non-gravitational interactions and all their invariances. The kinetic terms of the Standard Model for the electroweak and strong interactions are thus given by

$$\mathcal{L}^{\text{kin}} = -i\overline{\psi}\gamma^{\mu}\left(\partial_{\mu} - ig_3 G_{a\mu}\lambda^a - ig_2 B_{i\mu}\sigma^i - ig_1 Y C_{\mu}\right)\psi$$

$$- \frac{1}{4}\left(\partial_{\mu}C_{\nu} - \partial_{\nu}C_{\mu}\right)^2$$

$$- \frac{1}{4}\left(\partial_{\mu}G_{a\nu} - \partial_{\nu}G_{a\mu} + if^{bc}_{a}G_{b\mu}G_{c\nu}\right)^2$$

$$- \frac{1}{4}\left(\partial_{\mu}B_{i\nu} - \partial_{\nu}B_{i\mu} + i\epsilon^{jk}_{i}B_{j\mu}B_{k\nu}\right)^2, \tag{7.13}$$

where the action of the covariant derivative depends on which particle it acts on. For instance, for the lepton doublet of the first family, using the hypercharges obtained earlier, this gives

$$\mathcal{L}^{\text{kin}}_e = -i\left(\overline{\nu}_{e_L}, \overline{e}_L\right)\gamma^{\mu}\left(\partial_{\mu} - ig_2 B_{i\mu}\sigma^i + ig_1 C_{\mu}\right)\begin{pmatrix}\nu_{e_L}\\e_L\end{pmatrix} - i\overline{e}_R\gamma^{\mu}\left(\partial_{\mu} + 2ig_1 C_{\mu}\right)e_R, \tag{7.14}$$

since the leptons are not subject to the strong interaction, etc. Equation (7.14) can explicitly be expanded in left and right components as

$$\mathcal{L}^{\text{kin}}_e = -i\overline{\nu}_{e_L}\gamma^{\mu}\partial_{\mu}\nu_{e_L} - i\overline{e}\gamma^{\mu}\left(\partial_{\mu} + 2ig_2\sin\theta_W A_{\mu}\right)e$$
$$- \sqrt{2}g_2\left(\overline{\nu}_{e_L}\gamma^{\mu}W^+_{\mu}e_R + \overline{e}_R\gamma^{\mu}W^-_{\mu}\nu_{e_L}\right)$$
$$- \frac{g_2}{\cos\theta_W}\left(\overline{\nu}_{e_L}\gamma^{\mu}Z_{\mu}\nu_{e_L} - \cos 2\theta_W\overline{e}_L\gamma^{\mu}Z_{\mu}e_L - 2\sin^2\theta_W\overline{e}_R\gamma^{\mu}Z_{\mu}e_R\right), \tag{7.15}$$

where the weak angle θ_W is defined by the relations

$$\sin\theta_W \equiv \frac{g_1}{\sqrt{g_1^2 + g_2^2}}, \qquad \cos\theta_W \equiv \frac{g_2}{\sqrt{g_1^2 + g_2^2}}. \tag{7.16}$$

The vector fields, are defined by

$$W^{\pm}_{\mu} = \frac{1}{\sqrt{2}}\left(B_{1\mu} \mp iB_{2\mu}\right), \tag{7.17}$$

and

$$\begin{pmatrix}Z_{\mu}\\A_{\mu}\end{pmatrix} \equiv \begin{pmatrix}\cos\theta_W & -\sin\theta_W\\\sin\theta_W & \cos\theta_W\end{pmatrix}\begin{pmatrix}B_{3\mu}\\C_{\mu}\end{pmatrix} \Longleftrightarrow \begin{pmatrix}B_{3\mu}\\C_{\mu}\end{pmatrix} \equiv \begin{pmatrix}\cos\theta_W & \sin\theta_W\\-\sin\theta_W & \cos\theta_W\end{pmatrix}\begin{pmatrix}Z_{\mu}\\A_{\mu}\end{pmatrix}, \tag{7.18}$$

A_{μ} being the electromagnetic field. The Lagrangian (7.15) can be understood in the following way. First of all, we notice that the kinetic term of the electron leads to a gauge coupling of the $U(1)$ type with the photon, the electromagnetic coupling constant, and Q the electric charge operator, with value $Q_e = -1$ for the electron. This coupling is the same for the left and right degrees of freedom of the electron. In addition to the coupling terms, an intermediate neutral boson Z_{μ} has appeared, coupling both right and left components of the electron in a different way and also adding a self-coupling term for the (left-handed) neutrino. Finally, charged intermediate bosons $W_\mu^\pm$ induce possible interactions between neutrinos and electrons (the charges of the $W^\pm$ come from the fact that each term of the final Lagrangian should preserve the electric charge, which – between a neutral neutrino and an electron – is only possible with the indicated signs for the charges).

The Lagrangian (7.13) counts almost all possible terms that satisfy the invariance of the Standard Model and brings into play the particles known experimentally. Due to its invariances and, in particular, due to the weak isospin, it is not possible to write mass terms for the leptons. As for the quarks, all elements of the same doublet of $SU(2)_L$ should have the same mass – like, for instance, both quarks u and d – which contradicts the measurements. Thus, the model can only be made compatible with experiments if the $SU(2)_L$ symmetry is broken. For this, we introduce a complex field doublet

$$\phi = \begin{pmatrix} \phi_1 \\ \phi_2 \end{pmatrix}, \tag{7.19}$$

the dynamics of which is governed by the potential

$$V(\phi) = \lambda \left(|\phi|^2 - \eta^2 \right)^2 = \lambda \left(\phi_1^\star \phi_1 + \phi_2^\star \phi_2 - \eta^2 \right)^2. \tag{7.20}$$

Around the minimum of this potential is $|\phi_1|^2 + |\phi_2|^2 = \eta^2$. Assuming that the Higgs field is not subject to the strong interaction, the kinetic term becomes

$$\mathcal{L}_\phi^{\text{cin}} = -|D\phi|^2 = -\left| \left(\partial_\mu - i g_2 B_{i\mu} \sigma^i - i g_1 Y_\phi C_\mu \right) \begin{pmatrix} \phi_1 \\ \phi_2 \end{pmatrix} \right|^2 \tag{7.21}$$

around the minimum of the potential, performing a gauge transformation to suppress any of the three redundant degrees of freedom. In the unitary gauge, for which $\phi_1 = 0$ and $\phi_2 = \eta + h/\sqrt{2}$, i.e.,

$$\phi = \phi_0 + \delta\phi, \quad \text{with} \quad \phi_0 = \begin{pmatrix} 0 \\ \eta \end{pmatrix} \quad \text{and} \quad \delta\phi = \frac{1}{\sqrt{2}} \begin{pmatrix} 0 \\ h \end{pmatrix}, \tag{7.22}$$

we are left with only the Higgs field, h, which is real. We then get

$$\mathcal{L}_\phi^{\text{cin}} = -\frac{1}{2} \partial_\mu h \partial^\mu h - \left(\eta^2 + \frac{1}{2} h^2 + \sqrt{2} \eta h \right) \left[(g_2 B_{3\mu} - g_1 Y_\phi C_\mu)^2 + 2 g_2^2 W^{+\mu} W_\mu^- \right], \tag{7.23}$$

where we recognise the intermediate vector boson Z_μ, provided that we fix the hypercharge of the Higgs scalar doublet to $Y_\phi = 1$. It is the only value that will preserve the electromagnetic invariance after the symmetry breaking, ensuring the photon defined by Eq. (7.18) remains massless. It allows one to determine the masses of the W and Z bosons:

$$M_W = \sqrt{2} g_2 \eta \quad \text{and} \quad M_Z = \frac{\sqrt{2} g_2 \eta}{\cos\theta_W} = \frac{M_W}{\cos\theta_W}. \tag{7.24}$$

All the parameters of this model can actually be measured in laboratory and accelerator experiments. Their values are summarised in Table 7.1. This sets our reference theory, concerning both its structure and the values of its free parameters.

7.3.2 Extensions

The Standard Model of particle physics briefly described before relies on the gauge group $SU(3)_c \times SU(2)_L \times U(1)_Y$. One might feel uneasy with such a structure, as the reason why this particular group determines the symmetries remains unexplained, but, indeed, we may (correctly) argue that no theory is supposed to explain its own structures. More annoying, perhaps, is the fact that the model contains three families of quarks and leptons, which again looks like an unexplained replication of the same theory but at different energy scales. These particles are placed in representations that appear to be completely arbitrary. The most severe theoretical problem from which the Standard Model suffers is probably the proliferation of free parameters, which are exclusively determined from experimental measurements and are not computable from first principles. While gravity only introduces one free parameter, the coupling constant, the electroweak and strong theory requires 19 dimensionless parameters.

Taking into account the fact that we now know that neutrinos must have a mass, then there are three additional masses, one for each generation, and as a consequence, new mixing angles and phases. This extraordinarily simple extension raises the number of free parameters to more than 25, which is considered excessive for a theory supposed to explain three of the four fundamental interactions in a unified way.

It is not the goal of this text to investigate the extensions of the Standard Model. However, we shall mention the idea of unification which relies on the fact that the value of the three gauge coupling constants run with energy. In quantum field, the calculation of scattering processes include higher-order corrections of the coupling constants related to loop corrections that introduce some integrals over internal four-momenta. Depending on the theory, these integrals may be either finite or diverging as the logarithm or power-law of a UV cut-off. In the class of theories called renormalisable, among which is the Standard Model of particle physics, the physical quantities calculated at any order do not depend on the choice of the cut-off scale. But the result may depend on $\ln E/m$ where E is the typical energy scale of the process. It follows that the values of the coupling constants of the Standard Model depend on the energy at which they are measured (or of the process in which they are involved). This running arises from the screening due to the existence of virtual particles, which are polarised by the presence of a charge. The renormalisation group allows one to compute the dependence of a coupling constants as a function of the energy E as

$$\frac{\mathrm{d}g_i(E)}{\mathrm{d}\ln E} = \beta_i(E), \tag{7.25}$$

where the beta functions, β_i, depend on the gauge group and on the matter content of the theory and may be expended in powers of g_i. For the $SU(2)$ and $U(1)$ gauge couplings of the Standard Model, they are given by

$$\beta_2(g_2) = -\frac{g_2^3}{4\pi^2}\left(\frac{11}{6} - \frac{n_g}{3}\right), \qquad \beta_1(g_1) = +\frac{g_1^3}{4\pi^2}\frac{5n_g}{9}, \tag{7.26}$$

where n_g is the number of generations for the fermions. Remember that the fine-structure constant is defined in the limit of zero momentum transfer so that cosmological variation of the fine-structure constant is independent of the issue of the renormalisation group dependence. For the $SU(3)$ sector, with fundamental Dirac fermion representations,

$$\beta_3(g_3) = -\frac{g_3^3}{4\pi^2}\left(\frac{11}{4} - \frac{n_f}{6}\right), \tag{7.27}$$

n_f being the number of quark flavours with a mass smaller than E. The negative sign implies that (1) at large momentum transfer, the coupling decreases, and loop corrections become less and less significant – QCD is said to be asymptotically free; (2) integrating the renormalisation group equation for α_3 gives

$$\alpha_3(E) = \frac{6\pi}{(33 - n_f)\ln(E/\Lambda_c)} \tag{7.28}$$

so that it diverges as the energy scale approaches Λ_c from above, Λ_c, Λ_{QCD}, which we decided to call Λ_{QCD}. This scale characterises all QCD properties, and, in particular, the masses of the hadrons are expected to be proportional to Λ_{QCD} up to corrections of order m_q/Λ_{QCD}.

It was noticed quite early that these relations imply that the weaker gauge coupling becomes stronger at high energy while the strong coupling becomes weaker, so that one can think the three non-gravitational interactions may have a single common coupling strength above a given energy. This is the driving idea of Grand Unified Theories (GUT) in which one introduces a mechanism of symmetry breaking from a higher-symmetry group at high energies. It has two important consequences for our present considerations since one gets extra relations between the free parameters of the models.

First, there may exist algebraic relations between the Yukawa couplings of the Standard Model. Second, the structure constants of the Standard Model unify at an energy scale M_U:

$$\alpha_1(M_U) = \alpha_2(M_U) = \alpha_3(M_U) \equiv \alpha_U(M_U). \tag{7.29}$$

We note that the electroweak mixing angle is fixed by the symmetry to have the value $\sin^2\theta = 3/8$ at $E = M_U$, from which we deduce that

$$\alpha^{-1}(M_Z) = \frac{5}{3}\alpha_1^{-1}(M_Z) + \alpha_2^{-1}(M_Z). \tag{7.30}$$

It follows from the renormalisation group relations that

$$\alpha_i^{-1}(E) = \alpha_i^{-1}(M_U) - \frac{b_i}{2\pi}\ln\frac{E}{M_U}, \tag{7.31}$$

where the beta-function coefficients are given by $b_i = (41/10, -19/6, 7)$ for the Standard Model and by $b_i = (33/5, 1, -3)$ for $N = 1$ supersymmetric theory. Given a field decoupling at m_{th}, one has

$$\alpha_i^{-1}(E_-) = \alpha_i^{-1}(E_+) - \frac{b_i^{(-)}}{2\pi} \ln \frac{E_-}{E_+} - \frac{b_i^{(\text{th})}}{2\pi} \ln \frac{m_{\text{th}}}{E_+}, \tag{7.32}$$

where $b_i^{(\text{th})} = b^{(+)} - b^{(-)}$ with $b^{(+/-)}$ the beta-function coefficients respectively above and below the mass threshold, with tree-level matching at m_{th}. In the case of multiple thresholds, one must sum the different contributions. The existence of these thresholds implies that the running of α_3 is complicated since it depends on the masses of heavy quarks and coloured superpartners in the case of supersymmetry. For non-supersymmetric theories, the low-energy expression of the QCD scale is

$$\Lambda_{\text{QCD}} = E \left(\frac{m_c m_b m_t}{E} \right)^{2/27} \exp \left(-\frac{2\pi}{9\alpha_3(E)} \right) \tag{7.33}$$

for $E > m_t$. This implies that the variation of Yukawa couplings, gauge couplings, Higgs vev, and Λ_{QCD}/M_P are correlated. A second set of relations arises in models in which the weak scale is determined by dimensional transmutation [47]. In such cases, the Higgs vev is related to the Yukawa constant of the top quark by [19]

$$v = M_P \exp \left(-\frac{8\pi^2 c}{h_t^2} \right), \tag{7.34}$$

where c is a constant of order unity. This would imply that $\delta \ln v = S \delta \ln h$ with $S \sim 160$ [26].

The first consequences of this unification were investigated in References [18, 19, 38, 48, 75] where the variation of the three coupling constants was reduced to the one of α_U and M_U/M_P. It was concluded that, setting

$$R \equiv \delta \ln \Lambda_{\text{QCD}}/\delta \ln \alpha, \tag{7.35}$$

$R \sim 34$ with a stated accuracy of about 20% [75] (assuming only α_U can vary), $R \sim 40.82$ in the string dilaton model (assuming grand unification) [38]. We shall not discuss all of these models here but refer to section 5.3.1 of Reference [110] for an extensive review of the models and the relations they induce between the fundamental constants.

7.3.3 Masses

Any non-fundamental particle or nuclei has a mass that can, in principle, be expressed in terms of the parameters listed in Table 7.1. When we consider 'composite' systems such as proton, neutron, nuclei, or even planets and stars, we need to compute their mass, which requires us to determine their binding energy. The electromagnetic binding energy induces a direct dependence on the fine-structure constant α of the masses of the nuclei that can be evaluated using, e.g., the Bethe–Weizäcker formula,

$$E_{\text{EM}} = 98.25 \frac{Z(Z-1)}{A^{1/3}} \alpha \text{ MeV}. \tag{7.36}$$

The dependence of the masses on the quark masses, via nuclear interactions, and the determination of the nuclear binding energy are especially difficult to estimate.

In the chiral limit of QCD in which all quark masses are negligible compared to Λ_{QCD}, all dimensionful quantities scale as some power of Λ_{QCD}. For instance, concerning the nucleon mass, $m_N = c\Lambda_{QCD}$ with $c \sim 3.9$ being computed from lattice QCD. This predicts a mass of order 860 MeV, smaller than the observed value of 940 MeV. Reference [79] shows that the mass of the proton can be expressed in terms of the masses of the light quarks as

$$\frac{\delta m_p}{m_p} = f_{T_u}\frac{\delta m_u}{m_u} + f_{T_d}\frac{\delta m_d}{m_d} + f_{T_s}\frac{\delta m_s}{m_s} + f_{T_g}\frac{\delta\Lambda_{QCD}}{\Lambda_{QCD}}, \tag{7.37}$$

where the f_{T_i} are coefficients that can be computed in different approximations. They can be found in Reference [79].

To go further and determine the sensitivity of the mass of a nucleus to the various constants,

$$m(A, Z) = Zm_p + (A - Z)m_n + Zm_e + E_S + E_{EM}, \tag{7.38}$$

one should determine the strong binding energy (see Eq. (7.39)) in function of the atomic number Z and the mass number A. If we decompose the proton and neutron masses as [60] $m_{(p,n)} = u_3 + b_{(u,d)}m_u + b_{(d,u)}m_d + B_{(p,n)}\alpha$, where u_3 is the pure QCD approximation of the nucleon mass (b_u, b_d and $B_{(n,p)}/u_3$ being pure numbers), it reduces to

$$m(A, Z) = (Au_3 + E_S) + (Zb_u + Nb_d)m_u + (Zb_d + Nb_u)m_d$$
$$+ \left(ZB_p + NB_n + 98.25\frac{Z(Z-1)}{A^{1/3}}\text{MeV}\right)\alpha, \tag{7.39}$$

with $N = A - Z$, the neutron number. For an atom, one would have to add the contribution of the electrons, Zm_e. This form depends on strong, weak, and electromagnetic quantities. The numerical coefficients $B_{(n,p)}$ are given explicitly by [60]

$$B_p\alpha = 0.63 \text{ MeV} \quad B_n\alpha = -0.13 \text{ MeV}. \tag{7.40}$$

7.3.4 Gyromagnetic Factors

Gyromagnetic factors are of importance in atomic physics, particularly when it turns to the hyperfine structure. Following Reference [79], an approximate calculation is possible in the shell model and is relatively simple for even-odd (or odd-even) nuclei where the nuclear magnetic moment is determined by the unpaired nucleon. For a single nucleon – in a particular, (l, j) state within the nucleus – we can write

$$g = \begin{cases} 2lg_l + g_s \\ \frac{j}{j+1}[2(l+1)g_l - g_s] \end{cases} \text{for} \quad \begin{cases} j = l + \frac{1}{2} \\ j = l - \frac{1}{2} \end{cases}, \tag{7.41}$$

where $g_l = 1(0)$ and $g_s = g_p(g_n)$ for a valence proton (neutron).

The differences between the shell model predicted g-factors and the experimental values can be attributed to the effects of the polarisation of the non-valence nucleons and spin-spin interaction [13, 51, 54]. Taking these effects into account, the refined formula relevant is

$$g = 2\left[g_n b \langle s_z \rangle^o + (g_p - 1)(1 - b)\langle s_z \rangle^o + j\right], \tag{7.42}$$

where $g_n = -3.826$, $\langle s_z \rangle^o$ is the spin expectation value of the single-valence proton in the shell model and is one-half of the coefficient of g_s in Eq. (7.41), and b is determined by the spin-spin interaction and appears in the expressions for the spin expectation value of the valence proton $\langle s_{z_p} \rangle = (1 - b)\langle s_z \rangle^o$ and non-valence neutrons $\langle s_{z_n} \rangle = b\langle s_z \rangle^o$. Following the preferred method in References [13, 51], it is found

$$\langle s_{z_n} \rangle = \frac{\frac{g}{2} - j - (g_p - 1)\langle s_z \rangle^o}{g_n + 1 - g_p}, \quad \text{and} \quad \langle s_{z_p} \rangle = \langle s_z \rangle^o - \langle s_{z_n} \rangle. \tag{7.43}$$

Therefore, the variation of the g-factor can be written as

$$\frac{\delta g}{g} = \frac{\delta g_p}{g_p} \frac{2g_p \langle s_{z_p} \rangle}{g} + \frac{\delta g_n}{g_n} \frac{2g_n \langle s_{z_n} \rangle}{g} + \frac{\delta b}{b} \frac{2(g_n - g_p + 1)\langle s_{z_n} \rangle}{g}. \tag{7.44}$$

The main step is, thus, the computation of the gyromagnetic factors of the proton and neutron. In Reference [79], the dependence of the g-factors was expressed as

$$\frac{\delta g_p}{g_p} = \kappa_{u_p} \frac{\delta m_u}{m_u} + \kappa_{d_p} \frac{\delta m_d}{m_d} + \kappa_{s_p} \frac{\delta m_s}{m_s} + \kappa_{QCD_p} \frac{\delta \Lambda_{QCD}}{\Lambda_{QCD}}, \tag{7.45}$$

$$\frac{\delta g_n}{g_n} = \kappa_{u_n} \frac{\delta m_u}{m_u} + \kappa_{d_n} \frac{\delta m_d}{m_d} + \kappa_{s_n} \frac{\delta m_s}{m_s} + \kappa_{QCD_n} \frac{\delta \Lambda_{QCD}}{\Lambda_{QCD}}, \tag{7.46}$$

where the coefficients κ_i have been calculated by three methods: constituent quark model, chiral perturbation theory, and based on lattice results. The results can vary according to the method by a factor 10 for some of those coefficients. This illustrates the difficulty to relate nuclear physics to QCD.

7.3.5 Effect of the Strength Nuclear Force

As just illustrated, it is sometimes difficult to relate nuclear physics parameters. In the following, we will use a microscopic model [74, 115] that proved to be useful and efficient to analyse the effect of the variation of the strength of the nuclear interaction (NN) and electromagnetic interaction on cross sections, binding energies, and energy levels. In such an approach, the wave function of a nucleus with atomic number A, spin J, and total parity π is a solution of a Schrödinger equation with a Hamiltonian given by

$$H = \sum_i^A T(\mathbf{r}_i) + \sum_{i>j=1}^A V(\mathbf{r}_{ij}). \tag{7.47}$$

$T(\mathbf{r}_i)$ is the kinetic energy of nucleon i. The nucleon-nucleon interaction $V(\mathbf{r}_{ij})$ depends only on the set of relative distances $\mathbf{r}_{ij} = \mathbf{r}_i - \mathbf{r}_j$. It can be decomposed as

$$V(\mathbf{r}_{ij}) = V_C(\mathbf{r}_{ij}) + V_N(\mathbf{r}_{ij}), \tag{7.48}$$

where the potential $V_C(\mathbf{r})$ arises from the electromagnetic interaction and $V_N(\mathbf{r})$ from the nuclear interaction. The expression for V_N is detailed in Appendix B. The eigenstates $\Psi^{JM\pi}$ with energy $E^{J\pi}$ of the system are solutions, as usual, of the Schrödinger equation associated with the Hamiltonian given in Eq. (7.47),

$$H\Psi^{JM\pi} = E^{J\pi}\Psi^{JM\pi}. \tag{7.49}$$

The total wave function $\Psi^{JM\pi}$ is a function of the $A - 1$ coordinates $\mathbf{r}_{ij}$.

When $A > 4$, no exact solutions of Eq. (7.49) can be found, and approximate solutions have to be constructed. For those cases, we use a cluster approximation in which $\Psi^{JM\pi}$ is written in terms of α-nucleus wave functions. Because the binding energy of the α particle is large, this approach has been shown to be well adapted to cluster states; and, in particular, to ^{8}Be and ^{12}C [71, 102]. In the particular case of these two nuclei, the wave functions are respectively expressed as

$$\Psi^{JM\pi}_{^8Be} = \mathcal{A}\phi_\alpha\phi_\alpha g_2^{JM\pi}(\rho)$$

$$\Psi^{JM\pi}_{^{12}C} = \mathcal{A}\phi_\alpha\phi_\alpha\phi_\alpha g_3^{JM\pi}(\rho,\mathbf{R}), \tag{7.50}$$

where ϕ_α is the α wave function, defined in the $0s$ shell model with an oscillator parameter b; $\mathcal{A}$ is the antisymmetrisation operator between the A nucleons of the system. For two-cluster systems, the wave function $g_2^{JM\pi}(\rho)$ depends on the relative coordinate ρ between the two α particles. For three-cluster systems, $\mathbf{R}$ is the relative distance between two α particles, and ρ is the relative coordinate between the third α particle and the ^{8}Be centre of mass. The relative wave functions, g_2 and g_3, are obtained by solving the Schrödinger equation (7.49).

One then needs to specify the nucleon-nucleon potential $V_N(\mathbf{r}_{ij})$. We shall use the microscopic interaction model [105], which contains one linear parameter (admixture parameter u) whose standard value is $u = 1$. It can be slightly modified to reproduce important inputs such as the resonance energy of the Hoyle state. The binding energies of the deuteron (-2.22 MeV) and the α particle (-24.28 MeV) do not depend on u. For the deuteron, the Schrödinger equation is solved exactly.

To take into account the variation of the fundamental constants, we introduce the parameters δ_α and δ_{NN} to characterise the change of the strength of the electromagnetic and nucleon-nucleon interaction respectively. This is implemented by modifying the interaction potential (7.48) so that

$$V(\mathbf{r}_{ij}) = (1 + \delta_\alpha)V_C(\mathbf{r}_{ij}) + (1 + \delta_{NN})V_N(\mathbf{r}_{ij}). \tag{7.51}$$

Such a modification will affect binding energies and resonant energies simultaneously.

7.3.6 Binding Energies

The computation of binding energies is a difficult task that can follow different paths. For light elements, several method can be used, and we shall expose that on the particular case of the deuterium binding energy, B_D. For heavier elements, one needs to rely on phenomenological models, such as the liquid drop.

Deuterium binding energy. The case of the deuterium binding energy B_D has been discussed in different ways. It plays an important role in nucleosynthesis.

- *Pion mass.* A first route is to use the dependence of the binding energy on the pion mass [12, 48], which is related to the u and d quark masses by

$$m_\pi^2 = m_q \langle \bar{u}u + \bar{d}d \rangle f_\pi^{-2} \simeq \hat{m}\Lambda_{\rm QCD}, \qquad (7.52)$$

where $m_q \equiv \frac{1}{2}(m_u + m_d)$ and assuming that the leading order of $\langle \bar{u}u + \bar{d}d \rangle f_\pi^{-2}$ depends only on $\Lambda_{\rm QCD}$, f_π being the pion decay constant. This dependence was parameterised [118] as

$$\frac{\Delta B_D}{B_D} = -r \frac{\Delta m_\pi}{m_\pi}, \qquad (7.53)$$

where r is a fitting parameter found to be between 6 [48] and 10 [12]. Prior to this result, the analysis of [52] provides two computations of this dependence, which respectively lead to $r = -3$ and $r = 18$. This shows, once more, the difficulty to get a precise prediction. Reference [82] adds an electromagnetic contribution $-0.0081\Delta\alpha/\alpha$ so that

$$\frac{\Delta B_D}{B_D} = -\frac{r}{2}\frac{\Delta m_q}{m_q} - 0.0081\frac{\Delta\alpha}{\alpha}, \qquad (7.54)$$

but this latter contribution has not been included in other works.
- *Sigma model.* In the framework of the Walecka model, where the potential for the nuclear forces keeps only the σ and ω meson exchanges,

$$V = -\frac{g_s^2}{4\pi r}\exp(-m_\sigma r) + \frac{g_v^2}{4\pi r}\exp(-m_\omega r), \qquad (7.55)$$

where g_s and g_v are two coupling constants. Describing σ as a $SU(3)$ singlet state, its mass was related to the mass of the strange quark. In this way, one can hope to take into account the effect of the strange quark, both on the nucleon mass and the binding energy. In a second step, B_D is related to the meson and nucleon mass by

$$\frac{\Delta B_D}{B_D} = -48\frac{\Delta m_\sigma}{m_\sigma} + 50\frac{\Delta m_\omega}{m_\omega} + 6\frac{\Delta m_N}{m_N} \qquad (7.56)$$

so that $\Delta B_D/B_D \simeq -17\Delta m_s/m_s$ [53]. Unfortunately, a complete treatment of all the nuclear quantities on m_s has not been performed yet.

• *Cluster model.* The analysis using the cluster model described earlier implies [44] that the deuterium binding energy scales as

$$\Delta B_D / B_D = 5.716 \times \delta_{NN}. \tag{7.57}$$

This relation allows one to relate the phenomenological parameter δ_{NN} to miscroscopic parameters. Again, this explicitly shows the difficulty of to determining the role of the QCD parameters in low-energy nuclear physics.

• *Relation to QCD parameters.* Using the expression of B_D from the sigma model and the quark matrix elements for the nucleon, variations in B_D can be related to variations in the light quark masses (particularly the strange quark) and, thus, to the corresponding quark Yukawa couplings and Higgs vev, v. In Reference [26], it was concluded that

$$\frac{\Delta B_D}{B_D} = 18 \frac{\Delta \Lambda}{\Lambda} - 17 \left(\frac{\Delta v}{v} + \frac{\Delta h_s}{h_s} \right). \tag{7.58}$$

Using the unification relations (7.33) and (7.34), one respectively gets

$$\frac{\Delta \Lambda}{\Lambda} = R \frac{\Delta \alpha}{\alpha} + \frac{2}{27} \left(3 \frac{\Delta v}{v} + \frac{\Delta h_c}{h_c} + \frac{\Delta h_b}{h_b} + \frac{\Delta h_t}{h_t} \right) \tag{7.59}$$

and

$$\frac{\Delta B_D}{B_D} = -13(1 + S) \frac{\Delta h}{h} + 18R \frac{\Delta \alpha}{\alpha}. \tag{7.60}$$

This gives expressions in which the details of the unification scheme are hidden in the two parameters (R, S).

Heavier elements. For larger nuclei, the situation is more complicated since there is no simple modelling. For large mass number A, the strong binding energy can be approximated by the liquid drop model

$$\frac{E_S}{A} = a_V - \frac{a_S}{A^{1/3}} - a_A \frac{(A - 2Z)^2}{A^2} + a_P \frac{(-1)^A + (-1)^Z}{A^{3/2}} \tag{7.61}$$

with$(a_V, a_S, a_A, a_P) = (15.7, 17.8, 23.7, 11.2)$MeV [95]. It has also been suggested [39] that the nuclear binding energy can be expressed as

$$E_S \simeq A a_3 + A^{2/3} b_3 \quad \text{with} \quad a_3 = a_3^{\text{chiral limit}} + m_\pi^2 \frac{\partial a_3}{\partial m_\pi^2}. \tag{7.62}$$

In the chiral limit, a_3 has a non-vanishing limit to which we need to add a contribution scaling like $m_\pi^2 \propto \Lambda_{\text{QCD}} m_q$. Reference [39] also pointed out that the delicate balance between attractive and repulsive nuclear interactions [100] implies that the binding energy of nuclei is expected to depend strongly on the quark masses [41]. Recently, a fitting formula derived from effective field theory and based on the semi-empirical formula derived in [58] was proposed [37] as

$$\frac{E_S}{A} = - \left(120 - \frac{97}{A^{1/3}} \right) \eta_S + \left(67 - \frac{57}{A^{1/3}} \right) \eta_V + \dots, \tag{7.63}$$

where η_S and η_V are the strength of respectively the scalar (attractive) and vector (repulsive) nuclear contact interactions normalised to their actual value. These two parameters need to be related to the QCD parameters [41]. We also refer to [55] for the study of the dependence of the binding of light ($A \leq 8$) nuclei on possible variations of hadronic masses – including meson, nucleon, and nucleon-resonance masses.

7.3.7 Resonance Energies

The formalism developed in Section 7.3.5 allows one to compute the resonance energies of different nuclear reactions. It is, indeed, out of the scope of this text to investigate all possible reactions. We shall focus of beryllium-8 and carbon-12, which are of primary importance for nucleosynthesis.

For each set of values $(\delta_\alpha, \delta_{NN})$, one can solve Eq. (7.49) with the interaction potential (7.51). We emphasise that the parameter u is determined from the experimental ^{8}Be and ^{12}C(0_2^+) energies ($u = 0.954$). We assume that δ_{NN} varies in the range $[-0.015, 0.015]$. Following Reference [44], one obtains the results depicted in Figure 7.2. The sensitivities of $E_R(^8\text{Be})$ and $E_R(^{12}\text{C})$ to δ_{NN} are scaling as

$$E_R(^8\text{Be}) \equiv -B_8 \, (0.09184 - 12.208 \times \delta_{NN}) \text{ MeV}, \tag{7.64}$$

B_8 being the ^{8}Be binding energy with respect to two-alpha break-up (with this convention, $B_8 < 0$ for unbound ^{8}Be), and

$$E_R(^{12}\text{C}) = (0.2876 - 20.412 \times \delta_{NN}) \text{ MeV}. \tag{7.65}$$

The numerical results for the sensitivities of $E_R(^8\text{Be})$ and $E_R(^{12}\text{C})$ to δ_{NN} as well as the preceding linear fits are shown in Figure 7.2. The effect of δ_α on these quantities is negligible. Note that for $\delta_{NN} \gtrsim 0.007$, $E_R(^8\text{Be})$ is negative and ^{8}Be becomes stable. Using the bijective relation (7.57) between B_D and δ_{NN}, we can also express our results as

$$E_R(^8\text{Be}) = (0.09184 - 12.208 \, \Delta B_D/B_D) \text{ MeV}, \tag{7.66}$$

$$E_R(^{12}\text{C}) = (0.2876 - 3.570 \, \Delta B_D/B_D) \text{ MeV}. \tag{7.67}$$

To estimate the effect of δ_α in Eq. (7.51), we can approximate the Coulomb energy by $(3/5)Z(Z-1)\alpha\hbar c/R_c$, where $R_c = 1.3A^{1/3}$ fm, which gives 9 MeV for ^{12}C and 0.9 MeV for ^{4}He. The variation of $Q_{\alpha\alpha\alpha}$ is thus of the order of $+6$ MeV $\times$ δ_α. The direct effect of δ_α is thus of opposite sign but considerably less important. This is in qualitative agreement with References [88, 89].

It is appropriate at this point to note that within the limits of variation of δ_{NN} that we are considering here, the effect on promoting the stability of dineutron or diproton states is negligible. Working within the context of the same nuclear model, we estimate that a value of $\delta_{NN} \geq 0.15$ (for the dineutron) or ≥ 0.35 (for the diproton) would be required in order to induce stability for the dineutron or diproton, respectively.

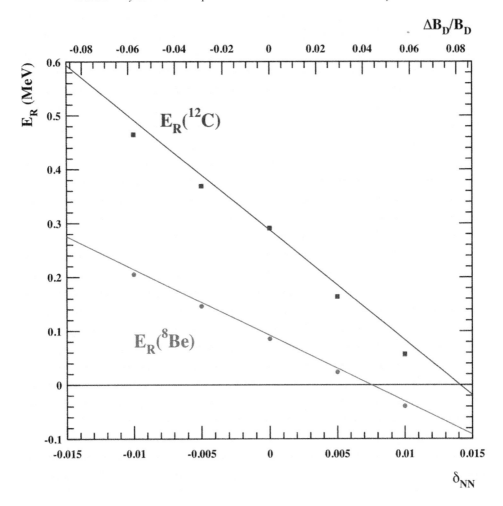

Figure 7.2 Variation of the resonance energies as a function of δ_{NN}. The symbols represent the results of the microscopic calculation while the lines correspond to the adopted linear relationship between E_R and δ_{NN}. From Reference [44].

7.3.8 *Implications for Fine-Tuning*

From the previous sections, it can be deduced that most low-energy nuclear properties will depend on the values of the gauge couplings and on the masses of the quarks. In particular, changing these constants may affect the stability of some particles (we mentioned the fact that beryliium-8 may become bound).

Several attempts to study the parameter space of the Standard Model of particle physics have been made.

They are often model dependent and restricted to a subset of parameters. For instance, Reference [9] considered a model in which up and down fermions get their masses from

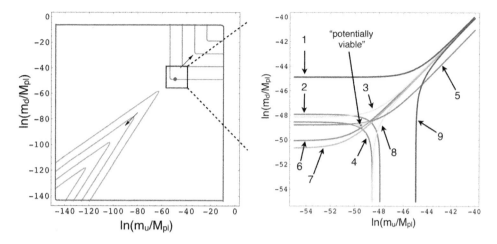

Figure 7.3 Parameter space of the masses of the up and down quarks. The lines show the limits of different life-permitting criteria, as calculated by Reference [9] and summarised in the text. The small green region marked 'potentially viable' shows where all these constraints are satisfied. From Reference [9].

different Higgs doublets. The result of their analysis in the plane of the up and down quarks masses is depicted in Figure 7.3, where they have indicated the thresholds defined by different criteria: (1) above the blue line, there exists a single stable element with chemistry similar to helium-4; (2) above the red line, the deuteron is unstable and decays via the strong force so that nucleosynthesis in hydrogen-burning stars would fail; (3) above the green line, neutrons in nuclei decay so that hydrogen is the only stable element; (4) below this red curve, the diproton is stable; two protons can fuse to helium-2 via a very fast electromagnetic reaction rather than the much slower, weak nuclear pp-chain; (5) above this red line, the production of deuterium in stars absorbs energy rather than releasing it. Also, the deuterium is unstable to weak decay; (6) below this red line, atoms are unstable since a proton in a nucleus can capture an orbiting electron to form a neutron; (7) below the orange curve, isolated protons are unstable, leaving no hydrogen left over from the early Universe to power long-lived stars and play a crucial role in organic chemistry; (8) below this green line, protons in nuclei decay so that any atoms that formed would disintegrate into a cloud of neutrons; (9) below this blue line, there is only a single stable particle, which can combine with a positron to produce an element with the chemistry of hydrogen. A handful of chemical reactions are possible, with their most complex product being (an analogue of) H_2.

Another example in the case of unification is provided by the analysis of Reference [65] in the case of the grand unified theory, in the parameter space of the mass of the electron and the difference between the masses of the down and up quarks, taking into account the stability of hydrogen and deuterium. This slice of the parameter space is depicted in Figure 7.4.

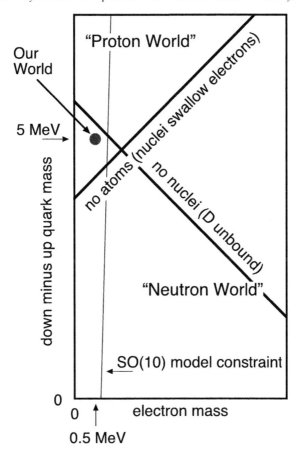

Figure 7.4 Constraints in the plane of the mass of the electron and the difference between the masses of the down and up quarks. Figure taken from Reference [65].

The effect of the QCD parameters on the strength of the interactions and masses have been extensively discussed; see, e.g., References [8, 11, 20, 21, 104]. A short summary of those constraints is the following (see Reference [8] for an extensive review on these questions).

- The existence of stable atoms requires the radius of the electron orbit to be significantly larger than the nuclear radius – i.e., $\alpha\mu/\alpha_s \ll 1$ [11].
- The existence of hydrogen – in particular, to power stars and form water – implies $m_e < m_n - m_p$. Otherwise, the electron will be captured by the proton to form a neutron [37].
- To ensure that the atomic constituents of chemical species maintain their identity in chemical reactions, one may require that the typical energy of chemical reactions is much smaller than the typical energy of nuclear reactions – i.e., $\alpha^2\mu/\alpha_s^2 \ll 1$ [11].

- Stable ordered molecular structures are not stable only if $\mu^{1/4} \ll 1$ [11].
- The stability of the proton requires $\alpha \lesssim (m_d - m_u)/141$ MeV so that the extra electromagnetic mass-energy of a proton relative to a neutron is more than counterbalanced by the bare quark masses [61, 64].
- Unless $\alpha \ll 1$, the electrons in atoms and molecules are unstable to pair creation [11].
- Unless $\alpha_s \lesssim 0.3\alpha^{1/2}$, carbon and all larger elements are unstable [11].
- Unless $\alpha_s/\alpha_{s,0} \gtrsim 0.91$ [40], the deuteron is unstable, and the main nuclear reaction in stars (pp) does not proceed.
- The grey stripe on the left of each plot shows where $\alpha < \alpha_G$, rendering electric forces weaker than gravitational ones.

Many other constraints exist [8, 11, 20, 21, 104], but they also involve cosmological or astrophysical parameters – that is, the environment in which the Standard Model of particle physics evolves. This bounds were summarised mostly for the reader to be aware that many conditions exist for the basic building blocks of nuclear physics to exist.

In the following sections of this chapter, we only consider the effect on BBN and stellar nucleosynthesis. The range of variation of the parameters will be assumed to satisfy the constraints cited earlier.

7.4 Primordial Nucleosynthesis

Big Bang nucleosynthesis describes the synthesis of light nuclei in the primordial Universe. It is considered as the *second* pillar of the Big Bang model. It is worth noting that BBN has been essential in the past, first to estimate the baryonic density of the Universe and give an upper limit on the number of neutrino families, as was later confirmed from the measurement of the Z^0 width by LEP experiments at CERN.

This section starts by describing the evolution of the Universe during the radiation era in Section 7.4.1 and then the basics of the BBN mechanism in Section 7.4.2. The observational status of the abundances of light elements is summarised in Section 7.4.3. Section 7.4.4 describes the dependence of BBN of the different constants, and Section 7.4.5 investigates the effect of a change of these parameters.

7.4.1 Description of the Universe in Its Early Phases

Cosmological Dynamics

In the standard cosmological model [92], which we shall assume here, the Universe is described by a Friedmann-Lemaître (FL) space-time with metric

$$ds^2 = -(u_\mu dx^\mu)^2 + (g_{\mu\nu} + u_\mu u_\nu)dx^\mu dx^\nu, \tag{7.68}$$

which clearly shows that the cosmic time t is the proper time measured by these fundamental observers. As a second consequence, this symmetry implies that the most general form of the stress-energy tensor is the one of a perfect fluid

$$T_{\mu\nu} = \rho u_\mu u_\nu + P(g_{\mu\nu} + u_\mu u_\nu), \tag{7.69}$$

with ρ and P being the energy density and isotropic pressure measured by the fundamental observers. The Einstein equations with the stress-energy tensor (7.69) reduce to the Friedmann equations

$$H^2 = \frac{8\pi G}{3}\rho - \frac{K}{a^2} + \frac{\Lambda}{3}, \tag{7.70}$$

$$\frac{\ddot{a}}{a} = -\frac{4\pi G}{3}(\rho + 3P) + \frac{\Lambda}{3}, \tag{7.71}$$

together with the conservation equation $(\nabla_\mu T^{\mu\nu} = 0)$

$$\dot{\rho} + 3H(\rho + P) = 0. \tag{7.72}$$

This gives two independent equations for three variables (a, ρ, P) that requires the choice of an equation of state

$$P = w\rho \tag{7.73}$$

to be integrated. It is convenient to use the conformal time defined by $dt = a(\eta)d\eta$ and the normalised density parameters

$$\Omega_i = 8\pi G\rho_i/3H_0^2, \qquad \Omega_\Lambda = \Lambda/3H_0^2, \qquad \Omega_K = -K/a_0^2 H_0^2, \tag{7.74}$$

that satisfy, from Eq. (7.70), $\sum_i \Omega_i + \Omega_\Lambda + \Omega_K = 1$, so that the Friedmann equation takes the form

$$\frac{H^2}{H_0^2} = \sum_i \Omega_i(1+z)^{3(1+w_i)} + \Omega_K(1+z)^2 + \Omega_\Lambda, \tag{7.75}$$

where the redshift z has been defined as $1 + z = a_0/a$.

Thermal History

Since for radiation $\rho \propto a^4$ and for pressureless matter $\rho \propto a^3$, it can be concluded that the Universe was dominated by radiation in its early phase. The density of radiation today is mostly determined by the temperature of the cosmic microwave background so that equality takes place at a redshift

$$z_{eq} \simeq 3612\,\Theta_{2.7}^{-4}\left(\frac{\Omega_{m0}h^2}{0.15}\right), \tag{7.76}$$

obtained by equating the matter and radiation energy densities and where $\Theta_{2.7} \equiv T_{CMB}/2.725$ mK. Since the temperature scales as $(1 + z)$, the temperature at which the matter and radiation densities were equal is $T_{eq} = T_{CMB}(1 + z_{eq})$, which is of order

$$T_{eq} \simeq 5.65\,\Theta_{2.7}^{-3}\,\Omega_{m0}h^2\,\text{eV}. \tag{7.77}$$

Above this energy, the matter content of the Universe is in a very different form from that of today. As it expands, the photon bath cools down, which implies a *thermal history*. In particular:

- When the temperature T becomes larger than twice the rest mass m of a charged particle, the energy of a photon is large enough to produce particle-antiparticle pairs. Thus, when $T \gg m_e$, both electrons and positrons were present in the Universe, so that the particle content of the Universe changes during its evolution while it cools down.
- Symmetries can be spontaneously broken.
- Some interactions may be efficient only above a temperature, typically as long as the interaction rate Γ is larger than the Hubble expansion rate.
- The freeze-out of some interaction can lead to the existence of relic particles.

Radiation Era at Thermodynamical Equilibrium

Particles interaction are mainly characterised by a reaction rate Γ. If this reaction rate is much larger than the Hubble expansion rate, then it can maintain these particles in *thermodynamic equilibrium* at a temperature T. Particles can thus be treated as perfect Fermi-Dirac and Bose-Einstein gases with distribution[1]

$$F_i(E,T) = \frac{g_i}{(2\pi)^3} \frac{1}{\exp\left[(E - \mu_i)/T_i(t)\right] \pm 1} \equiv \frac{g_i}{(2\pi)^3} f_i(E,T), \qquad (7.78)$$

where g_i is the degeneracy factor, μ_i is the chemical potential, and $E^2 = p^2 + m^2$. The normalisation of f_i is such that $f_i = 1$ for the maximum phase space density allowed by the Pauli principle for a fermion. T_i is the temperature associated with the given species, and, by symmetry, it is a function of t alone, $T_i(t)$. Interacting species have the same temperature. Among these particles, the Universe contains an electrodynamic radiation with black-body spectrum. Any species interacting with photons will, hence, have the same temperature as these photons as long as $\Gamma_i \gg H$. The photon temperature $T_\gamma = T$ will thus be called the *temperature of the Universe*.

As long as thermal equilibrium holds, one can define thermodynamical quantities such as the number density n, energy density ρ, and pressure P as

$$n_i = \int F_i(p,T)\mathrm{d}^3p \quad \rho_i = \int F_i(p,T)E(p)\mathrm{d}^3p \quad P_i = \int F_i(p,T)\frac{p^2}{3E}\mathrm{d}^3p. \qquad (7.79)$$

For ultra-relativistic particles ($m, \mu \ll T$), the density at a given temperature T is then given by

$$\rho_r(T) = g_*(T) \left(\frac{\pi^2}{30}\right) T^4. \qquad (7.80)$$

---

[1] The distribution function depends a priori on $(\boldsymbol{x}, t)$ and $(\boldsymbol{p}, E)$, but the homogeneity hypothesis implies that it does not depend on $\boldsymbol{x}$, and isotropy implies that it is a function of $p^2 = \boldsymbol{p}^2$. Thus, it follows from the cosmological principle that $f(\boldsymbol{x}, t, \boldsymbol{p}, E) = f(E, t) = f[E, T(t)]$.

g_* represents the effective number of relativistic degrees of freedom at this temperature:

$$g_*(T) = \sum_{i=\text{bosons}} g_i \left(\frac{T_i}{T}\right)^4 + \frac{7}{8} \sum_{i=\text{fermions}} g_i \left(\frac{T_i}{T}\right)^4. \qquad (7.81)$$

The factor 7/8 arises from the difference between the Fermi and Bose distributions. In the radiation era, the Friedmann equation then takes the simple form

$$H^2 = \frac{8\pi G}{3} \left(\frac{\pi^2}{30}\right) g_* T^4. \qquad (7.82)$$

Numerically, this amounts to

$$H(T) \cong 1.66 g_*^{1/2} \frac{T^2}{M_p}, \qquad t(T) \cong 0.3 g_*^{-1/2} \frac{M_p}{T^2} \sim 2.42 g_*^{-1/2} \left(\frac{T}{1\,\text{MeV}}\right)^{-2} \text{s.} \qquad (7.83)$$

Description of the Neutrinos

In order to follow the evolution of the matter content of the Universe, it is convenient to have conserved quantities such as the entropy. It can be shown [92] to be defined as $S = sa^3$ in terms of the entropy density s as

$$s \equiv \frac{\rho + P - n\mu}{T}. \qquad (7.84)$$

It satisfies $\text{d}(sa^3) = -(\mu/T)\text{d}(na^3)$ and is, hence, constant (1) as long as matter is neither destroyed nor created, since then na^3 is constant, or (2) for non-degenerate relativistic matter, $\mu/T \ll 1$. In the cases relevant for cosmology, $\text{d}(sa^3) = 0$. It can be expressed in terms of the temperature of the photon bath as

$$s = \frac{2\pi^2}{45} q_* T^3, \qquad \text{with} \qquad q_*(T) = \sum_{i=\text{bosons}} g_i \left(\frac{T_i}{T}\right)^3 + \frac{7}{8} \sum_{i=\text{fermions}} g_i \left(\frac{T_i}{T}\right)^3. \qquad (7.85)$$

If all relativistic particles are at the same temperature, $T_i = T$, then $q_* = g_*$. Note also that $s = q_* \pi^4 / 45\zeta(3) n_\gamma \sim 1.8 q_* n_\gamma$, so that the photon number density gives a measure of the entropy.

The standard example of the use of entropy is the determination of the temperature of the cosmic neutrino background. Neutrinos are in equilibrium with the cosmic plasma as long as the reactions $\nu + \bar{\nu} \longleftrightarrow e + \bar{e}$ and $\nu + e \longleftrightarrow \nu + e$ can keep them coupled. Since neutrinos are not charged, they do not interact directly with photons. The cross section of weak interactions is given by $\sigma \sim G_F^2 E^2 \propto G_F^2 T^2$ as long as the energy of the neutrinos is in the range $m_e \ll E \ll m_W$. The interaction rate is thus of the order of $\Gamma = n\langle \sigma v \rangle \simeq G_F^2 T^5$. We obtain that $\Gamma \simeq \left(\frac{T}{1\,\text{MeV}}\right)^3 H$. Thus, close to $T_D \sim 1\,\text{MeV}$, neutrinos decouple from the cosmic plasma. For $T < T_D$, the neutrino temperature decreases as $T_\nu \propto a^{-1}$ and remains equal to the photon temperature.

Slightly after decoupling, the temperature becomes smaller than m_e. Between T_D and $T = m_e$ there are four fermionic states (e^-, e^+, each having $g_e = 2$) and two bosonic states (photons with $g_\gamma = 2$) in thermal equilibrium with the photons. Thus, we have that $q_\gamma(T > m_e) = \frac{11}{2}$ while for $T < m_e$ only the photons contribute to q_γ and, hence, $q_\gamma(T < m_e) = 2$. The conservation of entropy implies that after $\bar{e} - e$ annihilation, the temperatures of the neutrinos and the photons are related by

$$T_\gamma = \left(\frac{11}{4}\right)^{1/3} T_\nu. \tag{7.86}$$

Thus, the temperature of the Universe is increased by about 40% compared to the neutrino temperature during the annihilation. Since $n_\nu = (3/11)n_\gamma$, there must exist a cosmic background of neutrinos with a density of 112 neutrinos per cubic centimetre and per family, with a temperature of around 1.95 K today.

7.4.2 Mechanism

The standard BBN scenario [15, 16, 91, 92] proceeds in three main steps:

1. For $T > 1$ MeV, $(t < 1$ s) a first stage during which the neutrons, protons, electrons, positrons, and neutrinos are kept in statistical equilibrium by the (rapid) weak interaction

$$n \longleftrightarrow p + e^- + \bar{\nu}_e, \qquad n + \nu_e \longleftrightarrow p + e^-, \qquad n + e^+ \longleftrightarrow p + \bar{\nu}_e. \tag{7.87}$$

As long as statistical equilibrium holds, the neutron-to-proton ratio is

$$\left(\frac{n}{p}\right) = \exp\left(-\frac{Q_{np}}{k_B T}\right), \tag{7.88}$$

where $Q_{np} \equiv (m_n - m_p)c^2 = 1.29$ MeV. The abundance of the other light elements is given by [92]

$$Y_A = g_A \left(\frac{\zeta(3)}{\sqrt{\pi}}\right)^{A-1} 2^{(3A-5)/2} A^{5/2} \left[\frac{k_B T}{m_N c^2}\right]^{3(A-1)/2} \eta^{A-1} Y_p^Z Y_n^{A-Z} e^{B_A/k_B T}, \tag{7.89}$$

where g_A is the number of degrees of freedom of the nucleus $^A_Z X$, m_N is the nucleon mass, η the baryon-photon ratio, and $B_A \equiv (Zm_p + (A - Z)m_n - m_A)c^2$ is the binding energy.

2. Around $T \sim 0.8$ MeV $(t \sim 2$ s), the weak interactions freeze out at a temperature T_f determined by the competition between the weak interaction rates and the expansion rate of the Universe and, thus, roughly determined by $\Gamma_w(T_f) \sim H(T_f)$; that is,

$$G_F^2(k_B T_f)^5 \sim \sqrt{G N_*}(k_B T_f)^2, \tag{7.90}$$

where G_F is the Fermi constant and N_* the number of relativistic degrees of freedom at T_f. Below T_f, the number of neutrons and protons changes only from the neutron β-decay between T_f to $T_N \sim 0.1$ MeV when $p + n$ reactions proceed faster than their inverse dissociation.

3. For 0.05 MeV $< T <$ 0.6 MeV (3 s $< t <$ 6 min), the synthesis of light elements occurs only by two-body reactions. This requires the deuteron to be synthesised ($p + n \rightarrow D$), and the photon density must be low enough for the photodissociation to be negligible. This happens roughly when

$$\frac{n_d}{n_\gamma} \sim \eta^2 \exp(-B_D/T_N) \sim 1 \tag{7.91}$$

with $\eta \sim 3 \times 10^{-10}$. The abundance of ^{4}He by mass, Y_p, is then well estimated by

$$Y_p \simeq 2 \frac{(n/p)_N}{1 + (n/p)_N} \tag{7.92}$$

with

$$(n/p)_N = (n/p)_f \exp(-t_N/\tau_n) \tag{7.93}$$

with $t_N \propto G^{-1/2} T_N^{-2}$ and $\tau_n^{-1} = 1.636 G_F^2 (1 + 3g_A^2) m_e^5/(2\pi^3)$, with $g_A \simeq 1.26$ being the axial-vector coupling of the nucleon.

4. The abundances of the light element abundances, Y_i, are then obtained by solving a series of nuclear reactions

$$\dot{Y}_i = J - \Gamma Y_i,$$

where J and Γ are time-dependent source and sink terms (see Figure 7.5).

5. Today, BBN codes include up to 424 nuclear reaction network [28] with up-to-date nuclear physics. In standard BBN, only D, ^{3}He, ^{4}He, and ^{7}Li are significantly produced as well as traces of ^{6}Li, ^{9}Be, ^{10}B, ^{11}B, and CNO. The most recent up-to-date predictions are discussed in References [30, 33].

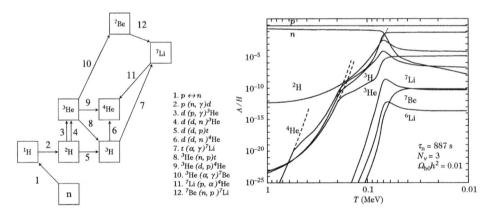

Figure 7.5 Left: The minimal 12 reactions network needed to compute the abundances up to lithium. Right: The evolution of the abundances of neutron, proton, and the lightest elements as a function of temperature (i.e., time). Below 0.01 MeV, the abundances are frozen and can be considered as the primordial abundances. Figure taken from Reference [92].

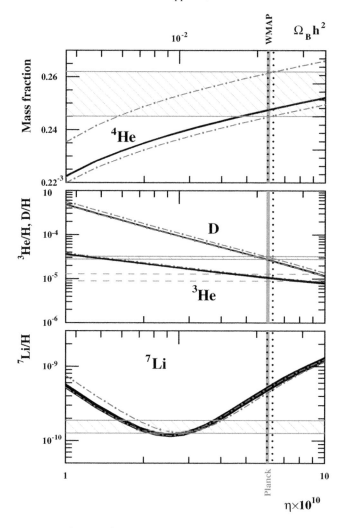

Figure 7.6 Abundances of ^{4}He D, ^{3}He, and ^{7}Li (thick solid curve) as a function of the baryon-to-photon ratio (bottom) or baryonic density (top). The vertical areas correspond to the WMAP (dot, black) and Planck (solid, grey) baryonic densities while the horizontal hatched areas represent the adopted observational abundances. The dot-dashed lines correspond to the extreme values of the *effective* neutrino families coming from CMB Planck study, $N_{eff} = (3.02, 3.70)$. Taken from Reference [33].

7.4.3 Observations

These predictions need to be compared to the observation of the abundances of the different nuclei (Figure 7.6).

Deuterium is a very fragile isotope, easily destroyed after BBN. Its most primitive abundance is determined from the observation of clouds at high redshift, on the line of sight of distant quasars. Recently, more precise observations of damped Lyman-α systems at high redshift have led to provide [35, 93] the mean value

$$D/H = (2.53 \pm 0.04) \times 10^{-5}. \tag{7.94}$$

After BBN, ^{4}He is still produced by stars, essentially during the main-sequence phase. Its primitive abundance is deduced from observations in Hıı (ionised hydrogen) regions of compact blue galaxies. The primordial ^{4}He mass fraction, Y_p, is obtained from the extrapolation to zero metallicity but is affected by systematic uncertainties. Recently, references [5, 6] have determined that

$$Y_p = 0.2465 \pm 0.0097. \tag{7.95}$$

Contrary to ^{4}He, ^{3}He is both produced and destroyed in stars all along its galactic evolution so that the evolution of its abundance as a function of time is subject to large uncertainties. Moreover, ^{3}He has been observed in our galaxy [7], and one only gets a local constraint

$$^3\text{He/H} = (1.1 \pm 0.2) \times 10^{-5}. \tag{7.96}$$

Consequently, the baryometric status of ^{3}He is not firmly established [112].

Primitive lithium abundance is deduced from observations of low-metallicity stars in the halo of our galaxy where the lithium abundance is almost independent of metallicity, displaying the so-called Spite plateau [101]. This interpretation assumes that lithium has not been depleted at the surface of these stars, so the presently observed abundance can be assumed to be equal to the primitive one. The small scatter of values around the Spite plateau is indeed an indication that depletion may not have been very efficient. However, there is a discrepancy between the value (1) deduced from these observed spectroscopic abundances and (2) the BBN theoretical predictions assuming Ω_b is determined by the CMB observations. Many studies have been devoted to the resolution of this so-called Lithium problem and many possible 'solutions', none fully satisfactory, have been proposed. For a detailed analysis, see the proceedings of the meeting 'Lithium in the Cosmos' [70]. Note that the idea according to which introducing neutrons during BBN may solve the problem has today been shown [31, 32] to be generically inconsistent with lithium and deuterium observations. Astronomical observations of these metal poor halo stars [98] have thus led to a relative primordial abundance of

$$\text{Li/H} = (1.58 \pm 0.31) \times 10^{-10}. \tag{7.97}$$

The origin of the light elements lithium, beryllium, and boron is a crossing point between optical and gamma spectroscopy, non-thermal nucleosynthesis (via spallation with galactic cosmic ray), stellar evolution, and Big Bang nucleosynthesis. We shall not discuss them in detail but just mention that, typically, ^{6}Li/H$\sim 10^{-11}$. Beryllium is a fragile nucleus formed in the vicinity of Type II supernovae by non-thermal process (spallation). The observations in metal-poor stars provide a primitive abundance at very low metallicity, of the order of Be/H $= 3. \times 10^{-14}$ at [Fe/H] $= -4$. This observation has to be compared to the typical primordial Be abundance, $Be/H = 10^{-18}$. Boron has two isotopes: ^{10}B and ^{11}B, and is also synthesised by non-thermal processes. The most recent observations give $B/H = 1.7 \times 10^{-12}$, to be compared to the typical primordial B abundance $B/H = 3 \times 10^{-16}$. For a general review of these light elements, see Reference [103].

Finally, CNO elements are observed in the lowest-metal poor stars (around [Fe/H] = -5). The observed abundance of CNO is typically [CNO/H] = -4, relatively to the solar abundance – i.e., primordial CNO/H < 10^{-7}. For a review, see Reference [56] and references therein.

7.4.4 Parameters

As can be seen in the description, the general BBN mechanism depends on

- External parameters (mostly the number of families of neutrinos or, more generally, the number of relativistic degrees of freedom, and the baryonic density)
- Several primary parameters, such as the neutron lifetime (That dictates the free neutron decay and appears in the normalisation of the proton-neutron reaction rates. It is the only weak interaction parameter, and it is related to the Higgs vev.) and the deuterium binding energies.
- Nuclear constants, such as the neutron-to-proton mass difference (it enters in the neutron-proton ratio)
- Fundamental constants, such as the Newton constant (which will affect the Hubble expansion rate at the time of nucleosynthesis in the same way as extra-relativistic degrees of freedom do so that it modifies the freeze-out time T_f), the fine-structure constant (which enters in the Coulomb barriers of the reaction rates through the Gamow factor in all the binding energies), or the Higgs vev (via the Fermi constant).

In full generality, the effect of these constants on the BBN predictions is difficult to model because of the intricate structure of QCD and its role in low-energy nuclear reactions. Thus, a solution is to proceed in *two steps*: first by determining the dependencies of the light element abundances on the BBN parameters and then by relating those parameters to the fundamental constants.

Sensitivities. In order to evaluate to these parameters, we vary them independently and compute numerically the abundances of the light elements, assuming that the cosmological parameter η is fixed to the value determined by the analysis of the cosmic microwave background anisotropies by the Planck satellite. The result is depicted on Figure 7.7 and shows that the most sensitive parameter is B_D, mostly because it controls the deuterium bottleneck.

Expression of the nuclear parameters. Q_{np} can be expressed in terms of the mass on the quarks u and d and the fine-structure constant as

$$Q_{np} = a\alpha\Lambda_{QCD} + (m_d - m_u), \tag{7.98}$$

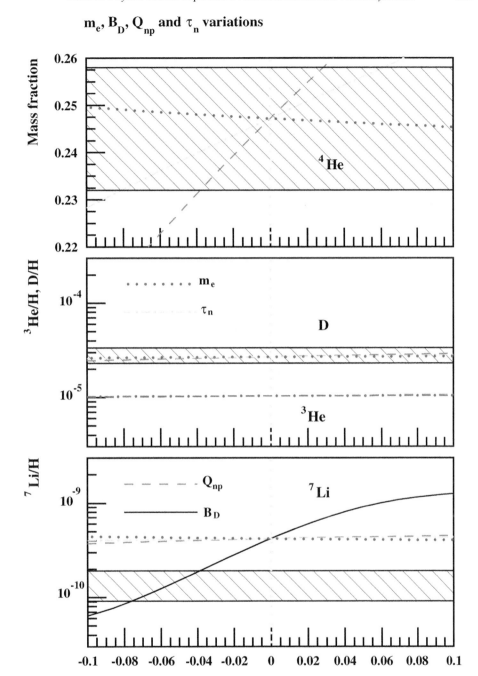

Figure 7.7 Dependence of the light element abundances on the independent variation of the main BBN parameters, assuming that η is fixed by the analysis of the Planck satellite. Figure taken from Reference [26].

where the electromagnetic contribution today is $(a\alpha\Lambda_{QCD})_0 = -0.76\text{MeV}$, and, therefore, the quark mass contribution today is $(m_d - m_u) = 2.05$ [60] so that

$$\frac{\Delta Q_{np}}{Q_{np}} = -0.59\frac{\Delta\alpha}{\alpha} + 1.59\frac{\Delta(m_d - m_u)}{(m_d - m_u)}. \tag{7.99}$$

All the preceding analyses agree on this dependence.

The neutron lifetime can be well approximated by

$$\tau_n^{-1} = \frac{1 + 3g_A^2}{120\pi^3}G_F^2 m_e^5 \left[\sqrt{q^2 - 1}(2q^4 - 9q^2 - 8) + 15\ln\left(q + \sqrt{q^2 - 1}\right)\right], \tag{7.100}$$

with $q \equiv Q_{np}/m_e$ and $G_F = 1/\sqrt{2}v^2$. Using the former expression for Q_{np}, we can express τ_n in terms of α; v; and the u, d, and electron masses. It follows:

$$\frac{\Delta\tau_n}{\tau_n} = 3.86\frac{\Delta\alpha}{\alpha} + 4\frac{\Delta v}{v} + 1.52\frac{\Delta m_e}{m_e} - 10.4\frac{\Delta(m_d - m_u)}{(m_d - m_u)}. \tag{7.101}$$

Again, all the preceding analyses agree on this dependence.

The deuterium binding energy has been extensively discussed in Section 7.3.6.

Effect of the fine-structure constant on cross sections. The fine-structure constant appears in the electromagnetic binding energy of the masses and binding energies. But it also affects all cross sections involving charged particles [14, 67].

In the non-relativistic limit, it is obtained as the thermal average of the product of the cross, the relative velocity, and the the number densities. Charged particles must tunnel through a Coulomb barrier to react. Changing α modifies these barriers and, thus, the reaction rates. Separating the Coulomb part, the low-energy cross section can be written as

$$\sigma(E) = \frac{S(E)}{E}e^{-2\pi\eta(E)}, \tag{7.102}$$

where $\eta(E)$ arises from the Coulomb barrier and is given in terms of the charges and the reduced mass M_r of the two interacting particles as

$$\eta(E) = \alpha Z_1 Z_2\sqrt{\frac{M_r c^2}{2E}}. \tag{7.103}$$

The form factor $S(E)$ has to be extrapolated from experimental nuclear data, but its α-dependence as well as the one of the reduced mass were neglected. This analysis was then extended [86] to take into account the α-dependence of the form factor to conclude that

$$\sigma(E) = \frac{2\pi\eta(E)}{\exp^{2\pi\eta(E)} - 1} \simeq 2\pi\alpha Z_1 Z_2\sqrt{\frac{M_r c^2}{2E}}\exp^{-2\pi\eta(E)}. \tag{7.104}$$

Note that Reference [86] also took into account (1) the effect that when two charged particles are produced, they must escape the Coulomb barrier. This effect is generally weak because the Q_i-values (energy release) of the different reactions are generally larger than the Coulomb barrier with the exception of two cases: $^3\text{He}(n, p)^3\text{H}$ and $^7\text{Be}(n, p)^7\text{Li}$. The rate of these reactions must be multiplied by a factor $(1+a_i \Delta\alpha/\alpha)$. (2) The radiative capture (photon-emitting processes) are proportional to α since it is the strength of the coupling of the photon and nuclear currents. All these rates need to be multiplied by $(1 + \Delta\alpha/\alpha)$.

7.4.5 Effects of a Variation of the Nuclear Parameters and Early CNO-Production

In standard BBN, the chain leading to carbon is dominated by the following reactions [92]:

$$^7\text{Li}(\alpha,\gamma)^{11}\text{B} \qquad ^7\text{Li}(n,\gamma)^8\text{Li}(\alpha,n)^{11}\text{B} \qquad (7.105)$$

followed by

$$^{11}\text{B}(p,\gamma)^{12}\text{C} \qquad ^{11}\text{B}(d,n)^{12}\text{C}, \qquad ^{11}\text{B}(d,p)^{12}\text{B} \qquad ^{11}\text{B}(n,\gamma)^{12}\text{B}, \qquad (7.106)$$

which bridge the gap between the $A \leq 7$ and $A \geq 12$. Hence, in principle, the mass gaps at $A = 5$ and $A = 8$ prevent the nucleosynthetic chain from extending beyond ^4He. The presence of these gaps is caused by the instability of ^5He, ^5Li, and ^8Be, which are respectively unbound by 0.798, 1.69, and 0.092 MeV with respect to neutron, proton, and α particle emission. CNO production in standard BBN has been investigated in References [28, 69]. While primordial CNO isotopes abundances are too low and highly unlikely with the present observational techniques, they are important for other applications. In particular, it may significantly affect the dynamics of Population III stars since hydrogen burning in low-mass Pop. III stars proceeds through the slow pp chains until enough carbon is produced, through the triple-alpha 3α reaction, to activate the CNO cycle. The minimum value of the initial CNO mass fraction that would affect Pop. III stellar evolution was estimated to be 10^{-10} [22] or even as low as 10^{-12} for less massive stars [43]. This is only two orders of magnitude above the SBBN CNO yields obtained using current nuclear reaction rates. The main difficulty in BBN calculations up to CNO is the extensive network (more than 400 reactions) needed, including n-, p-, and α- but also d-, t- and ^3He-, induced reactions on both stable and radioactive targets.

In order to quantify the effect of the previously discussed constants on the production of the light elements during BBN, at least up to carbon-12, we shall take into account their effects (1) on the lifetime and stability of ^8Be, (2) on the energy of the Hoyle state $^{12}\text{C}(0_2^+)$ for the production of carbon-12, and (3) on the $^3\text{He}(d,n)^4\text{He}$, and $^3\text{He}(d,p)^4\text{He}$ reactions, reactions involving $A = 5$ nuclei.

Beryllium-8. When the N-N interaction is modified by less than 0.75% (i.e., $\delta_{NN} < 7.52 \times 10^{-3}$), ^8Be remains unbound with respect to two α–particle emission. We can therefore take the $^4\text{He}(\alpha\alpha,\gamma)^{12}\text{C}$ rate as a function of δ_{NN}, as calculated in Reference [44]. We recall that we obtained that the energy of the ^8Be ground state with respect to the $\alpha + \alpha$ threshold was given by Eq (7.64).

Figure 7.8 depicts the reaction rate for $B_8 = 10$ and 100 keV relative to the case with $B_8 = 50$. The rate depends very little on the ^8Be binding energy for $B_8 > 0$, and the rate changes by less than $\sim 10\%$ for the three values of B_8 considered. As a result, we can safely

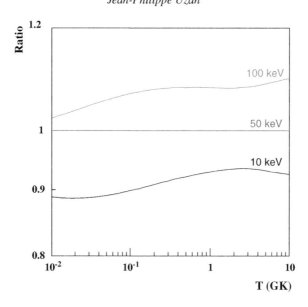

Figure 7.8 The relative variation of the $^4\text{He}(\alpha, \gamma)^8\text{Be}$ reaction rate assuming that ^8Be is bound by 10, 50, and 100 keV, relative to the 50 keV rate. Figure taken from Reference [29].

neglect the difference in the rates once $B_8 > 0$. The reaction rate is essentially given by the radiative capture cross section at the Gamow energy $E_0(T)$, which is proportional to E_γ^5, where the photon energy is $E_\gamma = E_{\text{cm}} + B_8$ and E_{cm} is the α-α centre-of-mass energy.

Hoyle state. The computation of the effect of a change of the nuclear interaction on the Hoyle state is described in Section 7.5, and the effects on the cross sections and reaction rates are detailed in Appendix B.

Taking that modifications in our BBN code, one can compute the CNO yields as a function of δ_{NN}, as displayed in Figure 7.9(a). The carbon abundance shows a maximum at $\delta_{NN} \approx 0.006$, C/H $\approx 10^{-21}$, which is *six orders of magnitude* below the carbon abundance in standard BBN. This can be understood from Figure 7.9(b), that displays the $^4\text{He}(\alpha\alpha, \gamma)^{12}\text{C}$ rate as a function of δ_{NN} for temperatures relevant to BBN – i.e., from 0.1 to 1 GK. Clearly, the variation of the rate with δ_{NN} is limited at the highest temperatures where BBN production occurs so that the amplification of ^{12}C production does not exceed a few orders of magnitudes. Indeed, while stars can process CNO at 0.1 GK over billions of years, in BBN, the optimal temperature range for producing CNO is passed through in a matter of minutes. This is not sufficient for ^{12}C (CNO) nucleosynthesis in BBN. Furthermore, the baryon density during BBN remains in the range 10^{-5}–0.1 g/cm^3 between 1.0 and 0.1 GK. This makes three-body reactions like $^4\text{He}(\alpha\alpha, \gamma)^{12}\text{C}$ much less efficient compared to two-body reactions.

The maximum of the ^{12}C production as a function of δ_{NN} in Figure 7.9 thus reflects the maxima in the $^8\text{Be}(\alpha, \gamma)^{12}\text{C}$ and $^4\text{He}(\alpha\alpha, \gamma)^{12}\text{C}$ rates. They are due to the effect of

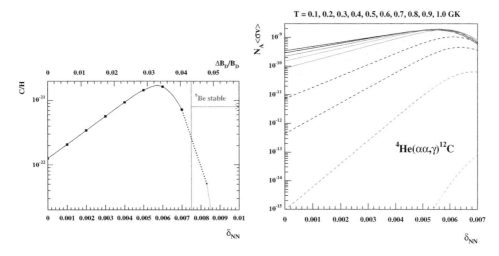

Figure 7.9 Left: ^{12}C production (in number of atoms relative to H) through the ^{4}He$(\alpha\alpha,\gamma)^{12}$C reaction (left) or via a stable ^{8}Be (right) as a function of the N-N interaction. For clarity, the rates of all other CNO producing reactions are set to zero to study these specific channels. The dotted line just connects the results of the two types of calculations: via ^{4}He$(\alpha\alpha,\gamma)^{12}$C, as in Reference [44], or via ^{4}He$(\alpha,\gamma)^8$Be$^{stable}(\alpha,\gamma)^{12}$C. Right: The ^{4}He$(\alpha\alpha,\gamma)^{12}$C rate as a function of δ_{NN} at constant temperature relevant for BBN from 0.1 GK (lower curve) to 1 GK (upper curve), in steps of 0.1 GK. Taken from Reference [29].

the sharp resonances (in both the ^{8}Be ground state and the ^{12}C 'Hoyle state' that dominate the cross section). The sensitivity of the reaction rates of a sharp resonance is detailed in Appendix D.

To summarise, for $\delta_{NN} \lesssim 0.006$, the rate decreases (increases) as a function of E_R (δ_{NN}) because of the dominating exponential factor, $\exp(-E_R/kT)$, while for $\delta_{NN} \gtrsim 0.006$, it increases (decreases) because of the penetrability. This evolution is followed by the ^{12}C production displayed in Figure 7.9. For $\delta_{NN} \geq 0.00752$, when ^{8}Be is bound, ^{12}C production drops to C/H $\approx 5 \times 10^{-23}$ for $B_8 = 10$. For larger B_8, the abundance drops sharply, as seen in Figure 7.9. For $B_8 = 50$ keV, C/H $\approx 5 \times 10^{-29}$ and is no longer in the range shown in the figure. For $B_8 = 100$ keV, corresponding to $\delta_{NN} = 0.0156$, the Hoyle state is even below threshold, and the production is vanishingly small. If ^{8}Be is bound, reactions that normally produce two α-particles could form ^{8}Be instead. We considered the following reactions

$$^7\text{Be}(n,\gamma)2\alpha \qquad ^7\text{Li}(p,\gamma)2\alpha$$
$$^7\text{Li}(n,\gamma)^8\text{Li}(\beta^+)2\alpha \qquad ^7\text{Be}(d,p)2\alpha$$
$$^7\text{Li}(d,n)2\alpha \qquad ^7\text{Be}(t,np)2\alpha$$
$$^7\text{Be}(^3\text{He},2p)2\alpha \qquad ^7\text{Be}(n,\gamma)2\alpha$$
$$^7\text{Li}(^3\text{He},d)2\alpha \qquad ^7\text{Be}(t,d)2\alpha$$
$$^7\text{Li}(t,2n)2\alpha \qquad ^7\text{Li}(^3\text{He},np)2\alpha$$

using the same rates as in Reference [28] but replacing 2α with ^{8}Be. The only significant enhancement comes from the ^{7}Li(d,n)2α reaction, but even in the most favourable case ($B_8 = 10$ keV), C/H reaches $\approx 10^{-21}$. This is still *six orders of magnitude* below the SBBN yield [28] that proceeds via the reactions listed in Eqs. (7.105) and (7.106).

^{3}H(d,n)^{4}He and ^{3}He(d,p)^{4}He reactions. Both reactions proceed through the $A = 5$ nuclei ^{5}He and ^{5}Li and are dominated by a low-energy $\frac{3}{2}^+$ resonance (at $E_R^{\text{exp}} = 0.048$ MeV for ^{5}He and $E_R^{\text{exp}} = 0.21$ MeV for ^{5}Li) and whose properties can be calculated within the same microscopic model that we used for ^{4}He($\alpha\alpha, \gamma$)^{12}C, but with ^{3}H+d and ^{3}He+d cluster structures. Unlike the case for ^{8}Be, the lifetime of the ^{5}He and ^{5}Li states is extremely short (the width of the ^{8}Be ground state is 6 eV, whereas the widths of the $\frac{3}{2}^+$ resonances in ^{5}He and ^{5}Li are of the order of 1 MeV). Therefore, the issue of producing $A = 5$ bound states, or even shifting their ground state energy down to the Gamow window, is not relevant. Even a two-step process, like the 3α reaction, where ^{5}He or ^{5}Li in thermal equilibrium would capture a subsequent nucleon to form ^{6}Li is completely negligible because they are unbound by ~ 1 MeV compared with the 92 keV of ^{8}Be. Hence, no significant equilibrium abundance of $A = 5$ nuclei can be reached.

The computation is similar as the one for the Hoyle state; i.e., (1) determine from the microscopic model the variation of the resonance energy with δ_{NN} and then (2) deduce the effect on the reaction rate.

Using the parameterisation (7.51) for the nucleon-nucleon interaction, we modify the resonance energy. Both the excitation energies of the $\frac{3}{2}^+$ resonance and of the thresholds vary. It is found that

$$\Delta E_R = -0.327 \times \delta_{NN} \qquad (7.107)$$

for ^{3}H(d,n)^{4}He and

$$\Delta E_R = -0.453 \times \delta_{NN} \qquad (7.108)$$

for ^{3}He(d,p)^{4}He (units are MeV). These energy dependences are much weaker (~ 20–30 keV for $|\delta_{NN}| \leq 0.03$) than for ^{8}Be and ^{12}C (see Eqs. (7.64) and (7.65)). This is expected for broad resonances which are weakly sensitive to the nuclear interaction [115].

The reaction rates are shown in Figure 7.10. As expected from Eqs. (7.107) and (7.108), they are only slightly affected by variations of δ_{NN} (less than 5%). From the sensitivity study of Reference [25], we deduce that ^{3}H(d,n)^{4}He rate variations have no effect on BBN while ^{3}He(d,p)^{4}He rate variations induce only very small ($\leq 4\%$) changes in the ^{7}Li and ^{3}He abundances. Because the change is so small, we can make a linear approximation to the sensitivity, as shown in Table 7.2, displaying $(\delta Y/Y)/\delta_{NN}$ values for both reactions.

Constraints on δ_{NN}. These results have been implemented in a BBN code in order to compute the primordial abundances of the light elements as a function of δ_{NN}. Reference [26] concluded that, in terms of δ_{NN}, $-0.7\% < \delta_{NN} < +0.5\%$. To test the importance

Table 7.2 *Abundance sensitivity,* $\partial \log Y / \delta_{NN}$, *to a variation of the N–N interaction at WMAP baryon density. Blank entries correspond to negligible values.*

Reaction	Y_p	D/H	^{3}He/H	^{7}Li/H
^{3}H(d,n)^{4}He				−0.015
^{3}He(d,p)^{4}He		−0.027	−1.14	−1.10

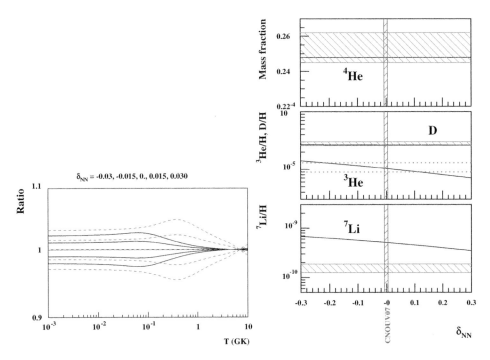

Figure 7.10 Left: Relative variation of the ^{3}H(d,n)^{4}He (solid line) and ^{3}He(d,p)^{4}He (dashed line) rates for $\delta_{NN} = -0.030$, -0.015, 0.015, and 0.030. Right: Effect of the variation of the N-N interaction induced solely by the modification of the nuclear rates of ^{3}He(d,p)^{4}He and ^{3}H(d,n)^{4}He on the primordial abundances of the light element compared to the constraints obtained in Reference [26]. Figure taken from Reference [29].

of the variations in the $A = 5$ rates, ^{3}He(d,p)^{4}He and ^{3}H(d,n)^{4}He were first considered. We emphasise that a 30% variation in δ_{NN} is unrealistic since it corresponds to a 175% variation on B_D. Given the D and ^{4}He primordial abundances, one gets the constraint

$$-0.0025 < \delta_{NN} < 0.0006. \qquad (7.109)$$

Those allowed variations in δ_{NN} are too small to reconcile ^{7}Li abundances with observations, where $\delta_{NN} \approx -0.01$ is required. We can easily extend our analysis by allowing both η_{10} (the most important external parameter) and δ_{NN} to vary. This allows one to set a joint constraint on the two parameters δ_{NN} and baryonic density, as depicted on

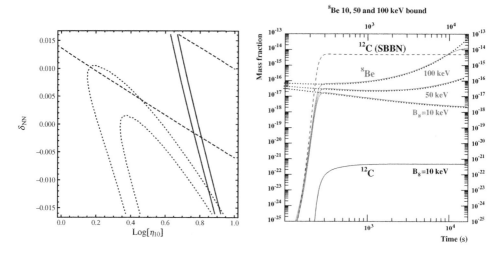

Figure 7.11 Left: Limits on of η_{10} (the number of baryons per 10^{10} photons) and δ_{NN} provided by observational constraints on D (solid) ^{4}He (dash) and ^{7}Li (dot). Right: ^{12}C and ^{8}Be mass fractions as a function of time, assuming ^{8}Be is bound by 100, 50, and 10 keV, as shown by the upper to lower solid curves, respectively [29]. (Only the ^{12}C mass fraction curve, for $B_8 = 10$ keV, is shown; others are far below the scale shown). The dotted lines correspond to the computation at thermal equilibrium and the dashed line to the SBBN [28] production. Figure taken from Reference [27].

Figure 7.11. No combination of values allow for the simultaneous fulfilment of the ^{4}He, D, and ^{7}Li observational constraints.

Note that the most influential reaction on ^{7}Li is surprisingly [26, 54] n(p,γ)d, as it affects the neutron abundance and the ^{7}Be destruction by neutron capture. The production of ^{12}C by the ^{4}He($\alpha\alpha, \gamma$)^{12}C, or the ^{4}He(α, γ)^{8}Be and ^{8}Be(α, γ)^{12}C reactions as a function of δ_{NN}, has been investigated. This is to be compared to the CNO (mostly ^{12}C) standard BBN production, CNO/H = $(0.5 - 3) \times 10^{-15}$, in number of atoms relative to hydrogen. A network of ≈ 400 reactions was used, but the main nuclear path to CNO was found to proceed from ^{7}Li(n,γ)^{8}Li(α,n)^{11}B, followed by ^{11}B(p,γ)^{12}C, ^{11}B(d,n)^{12}C, ^{11}B(d,p)^{12}B, and ^{11}B(n,γ)^{12}B reactions. To disentangle the ^{12}C production through the ^{4}He$\rightarrow^8$Be$\rightarrow^{12}$C link, from the standard ^{7}Li$\rightarrow^8$Li$\rightarrow^{11}$B$\rightarrow^{12}$C paths, we reduced the network to the reactions involved in $A < 8$ plus the ^{4}He($\alpha\alpha, \gamma$)^{12}C, or the ^{4}He(α, γ)^{8}Be and ^{8}Be(α, γ)^{12}C reactions, depending whether or not ^{8}Be would be stable for a peculiar value of δ_{NN}. The carbon abundance shows a maximum at $\delta_{NN} \approx 0.006$, C/H $\approx 10^{-21}$ [29], which is *six orders of magnitude* below the carbon abundance in SBBN [28]. This can be understood as the baryon density during BBN remains in the range 10^{-5}–0.1 g/cm^3 between 1.0 and 0.1 GK, substantially lower than in stars (e.g., 30–3,000 g/cm^3 in stars considered by Ekström *et al.* [44]). This makes three-body reactions like ^{4}He($\alpha\alpha, \gamma$)^{12}C much less efficient compared to two-body reactions. In addition, while stars can produce CNO at 0.1 GK over billions of years, in BBN, the optimal temperature range for producing CNO is passed through in a matter of minutes. Finally, in stars, ^{4}He($\alpha\alpha, \gamma$)^{12}C operates during the helium-burning phase without significant sources of ^{7}Li, d, p, and n to allow the ^{7}Li$\rightarrow$^{8}Li$\rightarrow$^{11}B$\rightarrow$^{12}C, $A = 8$, bypass process.

Note that the maximum is achieved for $\delta_{NN} \approx 0.006$ when ^{8}Be is still unbound, so contrary to a common belief, a stable ^{8}Be would not have allowed the build-up of heavy elements during BBN. This is illustrated in Figure 7.11 which displays the evolution of the ^{12}C and ^{8}Be mass fractions as a function of time when ^{8}Be is supposed to be bound by 10, 50, and 100 keV (solid lines). They both increase with time until equilibrium between two α-particle fusion and ^{8}Be photodissociation prevails, as shown by the dotted lines. For the highest values of B_8, the ^{8}Be mass fraction increases until, due to the expansion, equilibrium drops out, as shown by the late time behaviour of the upper curve ($B_8 = 100$ keV) in Figure 7.11 (right). For $B_8 \gtrsim 10$ keV, the ^{12}C production falls well below, out of the frame, because the ^{8}Be$(\alpha, \gamma)^{12}$C reaction rate decreases dramatically due to the downward shift of the Hoyle state. For comparison, the SBBN ≈ 400 reactions network (essentially the ^{7}Li$\rightarrow$^{8}Li$\rightarrow$^{11}B$\rightarrow$^{12}C chain) result [28] is plotted (dashed line) in Figure 7.11. It shows that for the ^{4}He$\rightarrow$^{8}Be$\rightarrow$^{12}C path to give a significant contribution, not only ^{8}Be should have been bound by much more than 100 keV, but also the ^{8}Be$(\alpha, \gamma)^{12}$C rate should have been much higher in order to transform most ^{8}Be in ^{12}C.

7.4.6 Summary

This section has described the dependence of the production of light nuclei during BBN on the fundamental parameters of nuclear physics. The most sensitive parameters are $(Q_{np}, B_D, \tau_n, m_e, \alpha)$.

It can be first concluded (see Figure 7.7) that to be compatible with existing observations, an independent variation of these parameters has to satisfy

$$-8.2 \times 10^{-2} \lesssim \frac{\Delta \tau_n}{\tau_n} \lesssim 6 \times 10^{-2}, \quad -4 \times 10^{-2} \lesssim \frac{\Delta Q_{np}}{Q_{np}} \lesssim 2.7 \times 10^{-2}, \quad (7.110)$$

and

$$-7.5 \times 10^{-2} \lesssim \frac{\Delta B_D}{B_D} \lesssim 6.5 \times 10^{-2}, \quad (7.111)$$

at a 2σ level. The deuterium data set the tighter constraint $-4 \times 10^{-2} \lesssim \Delta \ln B_D \lesssim 3 \times 10^{-2}$.

Similarly, the constraint on the parameter δ_{NN} has been obtained in Figure 7.11. It has also been shown that a stable ^{8}Be would not have allowed the build-up of heavy elements during BBN, and, thus, there is a large abundance of carbon-12.

There exist several public codes to compute BBN abundances: PArthENope,[2] FASTBBN [50, 77],[3] Kawano-Wagoner [72, 114], and AlterBBN.[4] No public code includes a modification of the standard physics.

---

[2] http://parthenope.na.infn.it.
[3] www-thphys.physics.ox.ac.uk/users/SubirSarkar/bbn/fastbbn.f.
[4] http://superiso.in2p3.fr/relic/alterbbn/.

7.5 Stellar Nucleosynthesis

Most nuclei are synthetised during stellar evolution (see Figure 7.1 and Appendix A). The production of carbon, oxygen, and nitrogen seems to be a central step of the complexification of the Universe – in particular, to allow for chemistry to develop in the Universe. This section describes the effect of the change of the atomic parameters on the production of these elements. It starts by a brief description of the modelling of stars in Section 7.5.1 and then investigates the production of carbon by Population III stars in Section 7.5.2.

7.5.1 Stellar Evolution Model

In the evolution of a star, the four interactions of nature are at work so that they are unique systems through which to study the impact of a modification of the laws of nature. As discussed in the introduction, only helium-4 is significantly produced during Big Bang nucleosynthesis while all other known nuclei are produced in stellar processes at different stages of the evolution of the star. It is an important issue to determine whether this process that lead to all the building blocks for atomic physics and chemistry requires some tuning.

Basics of Stellar Physics

The prediction of the abundance of the nuclei produced during the stellar evolution depends on both the microphysics (i.e., the properties of the nuclear reaction) and the stellar evolution. This means that our results will depend (in particular, in terms of external parameters) on the modelisation of the star. Thus, it is important to recall the basics of the description of stellar physics.

Naïvely thinking of a star as a gas compressed under its own gravity through a series of hydrodynamical equilibrium states, the integration of the equation for hydrostatic equilibrium,

$$\frac{\mathrm{d}P}{\mathrm{d}r} = -\frac{\rho G M_r}{r^2},$$
(7.112)

where we adopt the standard notations M_r for the mass contained inside the sphere of radius r, gives after an integration by part

$$3 \int_0^R P \mathrm{d}V = \int \rho G \frac{M_r}{r} \mathrm{d}V = \Delta \Omega_G(R).$$
(7.113)

The pressure P is always proportional to the energy density u. For instance, for a non-relativistic and non-degenerate monoatomic gas,

$$P = \frac{\rho k T}{\mu}, \qquad u = \frac{3}{2} \frac{\rho k T}{\mu}$$
(7.114)

with $\mu = A/(A + Z)$ the number of nucleons per particle, while for a relativistic gas, $P = u/3$.

It is clear that at the onset of the collapse of a cloud of gas, it is non-relativistic and non-degenerate so that Eq. (7.113) implies

$$\Delta \Omega_G(R) = 2 \int u \mathrm{d}V = 2E_{\mathrm{th}}(R). \tag{7.115}$$

This means that half of the gravitational energy is transformed in thermal energy (i.e., kinetic energy). The other half has to be radiated, and this is where electromagnetism enters the description: as soon as particles are ionized, their acceleration by gravity induces radiation (photon production). These photons are absorbed and re-emitted many times before they can escape the star, which means that we have to include a radiation component. The star shines because it is hot, and it is hot because it is composed of a gas compressed. It follows that its luminosity is the direct consequence of its mechanical equilibrium. Note that the production of energy in a star is not very efficient: it corresponds to $\sim 10^{-4}$ W/kg for the Sun, to be compared to ~ 1 W/kg for a human body!

Eq. (7.112) needs to be completed by an equation to describe the thermal transfer. For a radiative equilibrium, it takes the form

$$\frac{L_r}{4\pi r^2} = -\frac{1}{3\kappa\rho} \frac{\mathrm{d}(acT^4)}{\mathrm{d}r}, \tag{7.116}$$

where κ i the opacity, acT^4 is the energy density of the photon gas, and L_r is the flux of energy through a sphere of radius r so that the rate of photons is $\varepsilon_\gamma = \mathrm{d}L_r/\mathrm{d}M_r$.

System of Equations

In order to keep equilibrium, energy has to be extracted from some source, either from the gravitational contraction (macroscopic) or nuclear reactions (microscopic). Hence, the energy released by the nuclear reaction explains not why a star is shining but why it can shine for a long time. It also means that (1) a source term arising from the production of energy and the loss of energy by neutrino production have to be included in the evolution of L_r and (2) that the chemical composition of the star evolves with time. Concerning the transfer of the radiation and heat, one also needs to consider both radiation (in the outer part) and convection (in the central zone).

From the following discussion, it can be deduced that a stellar model can be described by the set of equations for a one-dimensional description

$$\frac{\partial P}{\partial M_r} = -\frac{\rho G M_r}{4\pi r^4}, \tag{7.117}$$

$$\frac{\partial r}{\partial M_r} = \frac{1}{4\pi r^2 \rho}, \tag{7.118}$$

$$\frac{\partial L_r}{\partial M_r} = \varepsilon_N + \varepsilon_g - \varepsilon_\nu, \tag{7.119}$$

$$\frac{\partial T}{\partial r} = \begin{cases} -\left(\frac{3}{4ac}\right)\left(\frac{\kappa\rho}{T^3}\right)\frac{L_r}{4\pi r} & \text{(radiative equilibrium)} \\ \frac{2}{5}\left(\frac{T}{\rho}\right)\frac{\mathrm{d}P}{\mathrm{d}r} & \text{(convective equilibrium)} \end{cases} \tag{7.120}$$

Table 7.3 *Typical evolution of a star given its initial mass. It gives the typical range of temperature, the total gravitational energy per nucleon emitted since the beginning of the contraction, the main nuclear reactions at work, the total nuclear energy per nucleon released since the beginning, and the minimum initial mass to reach this stage of evolution.*

	T_6 (10^6 K)	Grav. energy produced	Nuclear reactions	Nuc. energy emitted	Limit mass	Photon (%)	Neutrino (%)
Grav.	0–10	~ 1 keV/n				100	
Nuc.	10–30		$4H \rightarrow {}^4He$	6–7 MeV/n	$0.1M_\odot$	95	5
Grav.	30–100	~ 10 keV/n				100	
Nuc.	100–300		${}^4He \rightarrow {}^{12}C \rightarrow {}^{16}O$	7–8 MeV/n	$0.4M_\odot$	100	
Grav.	300–800	~ 100 keV/n				50	50
Nuc.	800–1,100		${}^{12}C \rightarrow Mb, Ne, Na, Al$	7–8 MeV/n	$0.7M_\odot$		100
Grav.	1,100–1,400	~ 150 keV/n					100
Nuc.	1,400–2,000		${}^{16}O \rightarrow S, Si, P$	8–9 MeV/n	$0.9M_\odot$		100

This gives a set of four equations for P, L_r, r and T as a function of M_r, chosen as integration variable. They involve four functions that need to be related to the microphysics: the density $\rho(P, T, \chi)$, opacity $\kappa(P, T, \chi)$, and energy sources $\varepsilon_{nuc}(P, T, \chi)$ and $\varepsilon_\nu(P, T, \chi)$. Initial conditions are set at the centre by the requirement that

$$(M, L, r, P, T)_c = (0, 0, 0, P_c, T_c). \qquad (7.121)$$

The latter two require solving the set of equations to describe the nuclear reactions

$$\frac{\partial Y_a}{\partial t} = -[a, b]Y_a Y_b + [c, d]Y_c Y_d + \dots, \qquad (7.122)$$

which also give the evolution of the chemical composition χ. Generally, the star undergoes a series of gravitational contraction and nuclear reaction energy production phases. A more massive star will go through more nuclear phases and produce heavier elements, but each phase is shorter. The typical evolution is summarised in Tables 7.3 and 7.4.

Nuclear physics enters in Eq. (7.122) through the cross sections. Indeed, they have to be integrated over temperature in order to deduce reaction rates. By changing the nuclear parameters, one affects the rate of production of nuclear energy, ε_N and, hence, the macroscopic evolution of the star (lifetime evolution in the Herzsprung-Russel diagram) and its chemical composition at the end of its evolution.

Note that the previous set of equations is extremely simplified since it does not include many important effects such as rotation, metallicity, magnetic field, binarity, etc. But in the first approximation, the evolution of the star depends on its initial mass and initial metallicity. There exist some public codes for stellar evolution such as MESA,[5] STARS,[6] or CESAM [81]. While most codes are not public, but there is a database[7] of stellar evolution

---

[5] http://mesa.sourceforge.net/.
[6] www.ast.cam.ac.uk/~stars/index.htm.
[7] https://obswww.unige.ch/Recherche/evol/Geneva-grids-of-stellar-evolution.

Table 7.4 *Evolution stages of an isolated star with solar initial metallicity. The timescales and luminosities in the sixth line are given for a star of* $15M_\odot$*. It shows that stars are more luminous in the neutrino sector. The cut in the range 9–15 is still discussed in the literature, so we do not put a clear limit between the two behaviours (see, e.g., Reference [116] for a recent discussion on this issue).*

$M/M_\odot$	Evolution	Final state
<0.08	Degenerate before H-burning	Brown dwarf
0.08–0.5	Only H-burning	
	He-core contracts and enters degenerate zone	He white dwarf
0.5–7	H-burning	
	He-burning	
	Mass loss during Asymptotic Giant Branch (AGB)	
	Ejection of envelope	C-O white dwarf
7–9	H-burning	
	He-burning	
	C-burning upto Ne-Mg	Ne-Mg white dwarf
9–X	H-burning	
	He-burning	
	C-burning	
	Electron capture triggers the SN	
	Production of ν, core collapse	Supernovae/Neutron star
X–15	H-burning (11 Myr)	
	He-burning (2 Myr)	
	C-burning (2 kyr $-L_\gamma = 72 \times 10^3 L_\odot - L_\nu = 37 \times 10^4 L_\odot$)	
	Ne-photodecay (0.7 yr $-L_\gamma = 72 \times 10^3 L_\odot - L_\nu = 14 \times 10^7 L_\odot$)	
	O-burning (2.6 yr $-L_\gamma = 72 \times 10^3 L_\odot - L_\nu = 9 \times 10^8 L_\odot$)	
	Si-burning (18 days $-L_\gamma = 72 \times 10^3 L_\odot - L_\nu = 1.3 \times 10^{11} L_\odot$)	
	Fe-burning (1 s)	
	Core collapse	Supernovae/Black hole
>150	Instability due to pair production during O-burning	

models for masses between 0.8 and 120 solar masses and metallicities from $Z = 0.001$ to 0.1.

The goal of this chapter is indeed not to discuss the whole theory of stellar evolution. We refer to standard texbooks [24, 94, 117]. The nuclear aspects will now be detailed in the particular example of carbon production in Population III stars.

7.5.2 Carbon Production in Population III Stars

Given the previous general description of the stellar dynamics, one can study the evolution of stars and of the chemical elements they produce during their evolution. The goal of this section is to focus on the production of CNO in Population III stars.

It has been argued [96, 107–110] that the synthesis of complex elements in stars (mainly the possibility of the 3α reaction as the origin of the production of ^{12}C) sets strong constraints on the values of the fine-structure and strong coupling constants. There have been several studies on the sensitivity of carbon production to the underlying nuclear rates [10, 36, 49, 78, 88, 90]. The production of ^{12}C in stars requires a triple tuning: (1) the decay lifetime of ^8Be, of order 10^{-16} s, is four orders of magnitude longer than the time for two α particles to scatter, (2) an excited state of the carbon lies just above the energy of $^8Be + \alpha$, and, finally, (3) the energy level of ^{16}O at 7.1197 MeV is non-resonant and below the energy of $^{12}C + \alpha$, at 7.1616 MeV, which ensures that most of the carbon synthesised is not destroyed by the capture of an α-particle. The existence of this excited state of ^{12}C was actually predicted by Hoyle [66] and then observed at the predicted energy by Dunbar [42] as well as its decay [34]. The variation of any constant which would modify the energy of this resonance, known as the Hoyle level, would dramatically affect the production of carbon.

Qualitative Analysis

Qualitatively, and perhaps counter-intuitively, if the energy level of the Hoyle level were increased, ^{12}C would probably be rapidly processed to ^{16}O since the star would, in fact, need to be hotter for the 3α reaction to be triggered. On the other hand, if it is decreased very little, oxygen will be produced. From the general expression of the reaction rate (see the discussion through Eq. (7.130) for details, definitions of all the quantities entering this expression, and a more accurate computation)

$$\lambda_{3\alpha} = 3^{3/2} N_\alpha^3 \left(\frac{2\pi \hbar^3}{M_\alpha k_B T} \right)^3 \frac{\Gamma}{\hbar} \exp \left[-\frac{Q_{\alpha\alpha\alpha}}{k_B T} \right], \tag{7.123}$$

where $Q_{\alpha\alpha\alpha} \sim 380$ keV is the energy of the resonance, one deduces that the sensitivity of the reaction rate to a variation of $Q_{\alpha\alpha\alpha}$ is

$$s = \frac{d \ln \lambda_{3\alpha}}{d \ln Q_{\alpha\alpha\alpha}} = -\frac{Q_{\alpha\alpha\alpha}}{k_B T} \sim \left(\frac{-4.4}{T_9} \right), \tag{7.124}$$

where $T_9 = T/10^9 K$. This effect was investigated in References [36, 88, 90] who related the variation of $Q_{\alpha\alpha\alpha}$ to a variation of the strength of the nucleon-nucleon (N-N) interaction. Focusing on the C/O ratio in red giant stars (1.3, 5, and $20 M_\odot$ with solar metallicity) up to thermally pulsing asymptotic giant branch stars (TP-AGB) [88, 90] and in low, intermediate, and high mass stars (1.3, 5, 15, and $25 M_\odot$ with solar metallicity) [99], it was estimated that outside a window of 0.5% and 4% for the values of the strong and electromagnetic forces, respectively, the stellar production of carbon or oxygen will be reduced by a factor 30–1,000.

Indeed, modifying the energy of the resonance alone is not realistic since all cross sections, reaction rates, and binding energies, etc., should be affected by the variation of the constants. One could have started by assuming independent variations of all these quantities, but it is more realistic (and, hence, more model dependent) to try to deduce their variation from a microscopic model. Our analysis can then be outlined in three main steps:

1. Relating the nuclear parameters to fundamental constants such as the Yukawa and gauge couplings and the Higgs vacuum expectation value. This is a difficult step because of the intricate structure of QCD and its role in low-energy nuclear reactions, as in the case of BBN. The nuclear parameters include the set of relevant energy levels (including the ground states), binding energies of each nucleus, and partial width of each nuclear reaction. This involves a nuclear physics model of the relevant nuclei (mainly ^{4}He, ^{8}Be, ^{12}C, and ^{16}O for our study).

2. Relating the reaction rates to the nuclear parameters, which implies an integration over energy of the cross sections.

3. Deducing the change in the stellar evolution (lifetime of the star, abundance of the nuclei, Hertzprung-Russel (H-R) diagram, etc.). This involves a stellar model.

General Description of the Computation

Such a computation involves many steps, which we briefly describe and summarise (see Figure 7.12).

The *first step* is probably the most difficult. We shall adopt a phenomenological description of the different nuclei based on a cluster model in which the wave functions of the ^{8}Be and ^{12}C nuclei are approximated by a cluster of respectively 2α and 3α wave functions. When solving the associated Schrödinger equation, we will modify the strength of the electromagnetic and nuclear N–N interaction potentials respectively by a factor $(1 + \delta_\alpha)$ and $(1 + \delta_{NN})$, where δ_α and δ_{NN} are two small dimensionless parameters that encode the variation of the fine-structure constant and other fundamental couplings. At this stage, the relation between δ_{NN} and the gauge and Yukawa couplings is not known. This will

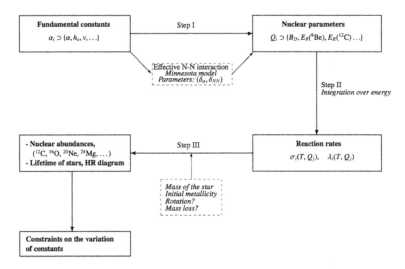

Figure 7.12 The computation of the abundance of CNO produced by Population III stars involves many steps, from nuclear physics to stellar physics, in order to constrain the effects of the fundamental constants.

allow us to obtain the energy levels, including the binding energy, of ^{2}H, ^{4}He, ^{8}Be, ^{12}C, and the first $J^\pi = 0^+$ ^{12}C excited energy level. Note that all of the relevant nuclear states are assumed to be interacting alpha clusters. In a first approximation, the variation of the α-particle mass cancels out. The partial widths (and lifetimes) of these states are scaled from their experimental laboratory values, according to their energy dependence. δ_{NN} is used as a free parameter see Section 7.3).

The *second step* requires an integration over energy to deduce the reaction rates as functions of the temperature and of the new parameters δ_α and δ_{NN}.

The *third step* involves stellar models and, in particular, some choices about the masses and initial metallicity of the stars. While theoretically uncertain, it is usually thought that the first stars were massive; however, their mass range is presently unknown (for a review, see [17]). In a hierarchical scenario of structure formation, they were formed a few $\times$ 10^8 years after the Big Bang – that is, at a redshift of $z \sim 10$–15 with zero metallicity (so that we can use the BBN abundances as initial conditions). We thus focus on Population III stars with typical masses, 15 and $60M_\odot$, assuming no rotation. Our computation is stopped at the end of core helium burning.

The *final step* would be to use these predictions to set constraints on the fundamental constants, using stellar constraints such as the C/O ratio, which is, in fact, observable in very metal-poor stars.

The 3α Reaction

To start, let us summarise the basics of the 3α process at the origin of carbon. Figure 7.13 shows the low-energy-level schemes of the nuclei participating in the ^{4}He$(\alpha\alpha, \gamma)^{12}$C reaction: ^{4}He, ^{8}Be, and ^{12}C. The 3α process begins when 2α particles fuse to produce a ^{8}Be nucleus whose lifetime is only $\sim 10^{-16}$ s but is sufficiently long so as to allow a second α capture into the second excited level of ^{12}C, at 7.65 MeV above the ground state (of ^{12}C). In the following, we shall refer to the successive α captures as first and second steps; that is, $\alpha\alpha \leftrightarrow ^8Be+\gamma$ and ^{8}Be$+\alpha \leftrightarrow ^{12}C^* \rightarrow ^{12}C+\gamma$. The excited state of ^{12}C corresponds to an $\ell = 0$ resonance, as postulated by [66] in order to increase the cross section during the helium-burning phase. This level decays to the first excited level of ^{12}C at 4.44 MeV through an $E2$ (i.e., electric with $\ell = 2$ multipolarity) radiative transition as the transition to the ground state ($0_1^+ \rightarrow 0_2^+$) is suppressed (pair emission only). At temperatures above $T_9 \approx 2$, which are not relevant for our analysis and therefore not treated, one should also consider other possible levels above the α threshold.

We define the following energies:

- $E_R(^8$Be$)$ as the energy of the ^{8}Be ground state with respect to the $\alpha + \alpha$ threshold
- $E_R(^{12}$C$)$ as the energy of the Hoyle level with respect to the ^{8}Be$+\alpha$ threshold – i.e., $E_R(^{12}$C$) \equiv ^{12}$C$(0_2^+) + Q_\alpha(^{12}$C$)$, where ^{12}C(0_2^+) is the excitation energy and $Q_\alpha(^{12}$C$)$ is the α particle separation energy
- $Q_{\alpha\alpha\alpha}$ as the energy of the Hoyle level with respect to the 3α threshold so that

$$Q_{\alpha\alpha\alpha} = E_R(^8\text{Be}) + E_R(^{12}\text{C}) \qquad (7.125)$$

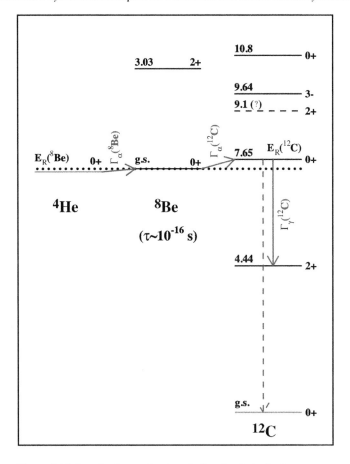

Figure 7.13 Level scheme showing the key levels in the 3α process.

- $\Gamma_\alpha(^8\text{Be})$ as the partial width of the beryllium decay ($\alpha\alpha \leftrightarrow {}^8\text{Be}+\gamma$)
- $\Gamma_{\gamma,\alpha}(^{12}\text{C})$ as the partial widths of $^8\text{Be}+\alpha \leftrightarrow {}^{12}\text{C}^* \rightarrow {}^{12}\text{C}+\gamma$

Their standard values are given in Table 7.5.

Assuming (1) thermal equilibrium between the ^4He and ^8Be nuclei so that their abundances are related by the Saha equation and (2) the sharp resonance approximation for the α capture on ^8Be, the $^4\text{He}(\alpha\alpha,\gamma)^{12}\text{C}$ rate can be expressed [2, 87] as

$$N_A^2 \langle \sigma v \rangle^{\alpha\alpha\alpha} = 3^{3/2} 6 N_A^2 \left(\frac{2\pi}{M_\alpha k_B T} \right)^3 \hbar^5 \omega\gamma \exp\left(\frac{-Q_{\alpha\alpha\alpha}}{k_B T} \right) \qquad (7.126)$$

with $\omega = 1$ (spin factor), $\gamma = \Gamma_\gamma(^{12}\text{C})\Gamma_\alpha(^{12}\text{C})/(\Gamma_\gamma(^{12}\text{C})+\Gamma_\alpha(^{12}\text{C})) \approx \Gamma_\gamma(^{12}\text{C})$ for present-day values, and M_α is the mass of the α nucleus.

During helium burning, the only other important reaction is $^{12}\text{C}(\alpha,\gamma)^{16}\text{O}$ [68], which transforms ^{12}C into ^{16}O. Its competition with the 3α reaction governs the $^{12}\text{C}/^{16}\text{O}$

Table 7.5 *Nuclear data for the two steps of the 3α reaction. See text for the definitions of the quantities [1, 4, 106].*

Nucleus	J^π	E_R (keV)	Γ_α (eV)	Γ_γ (meV)
^{8}Be	0^+	91.84 ± 0.04	5.57 ± 0.25	–
^{12}C	0_2^+	287.6 ± 0.2	8.3 ± 1.0	3.7 ± 0.5

abundance ratio at the end of the helium-burning phase. Even though the precise value of the ^{12}C$(\alpha, \gamma)^{16}$O S-factor[8] is still a matter of debate as it relies on an extrapolation of experimental data down to the astrophysical energy (≈ 300 keV), its energy dependence is much weaker than that of the 3α reaction. Indeed, as it is dominated by *broad* resonances, a shift of a few hundred keV in energy results in an S-factor variation of much less than an order of magnitude. For this reason, we can safely neglect the effect of the ^{12}C$(\alpha, \gamma)^{16}$O reaction rate variation when compared to the variation in the 3α rate. Similar considerations apply to the rate for ^{16}O$(\alpha, \gamma)^{20}$Ne.

During hydrogen burning, the pace of the CNO cycle is given by the slowest reaction, ^{14}N$(p,\gamma)^{15}$O. Its S-factor exhibits a well-known resonance at 260 keV, which is normally outside of the Gamow energy window (≈ 100 keV), but a variation of the N–N potential could shift its position downward, resulting in a higher reaction rate and more efficient CNO H– burning.

Sensitivity to the Nuclear Parameters

The microscopic analysis gave the expression of the renonance energies (7.66) and (7.65), from which one can easily deduce that the energy of the Hoyle level with respect to the 3α threshold (and not with respect to ^{8}Be$+\alpha$ threshold) is given by (see Eq. (7.125))

$$Q_{\alpha\alpha\alpha} = (0.37945 - 5.706 \times \Delta B_D / B_D) \text{ MeV}, \tag{7.127}$$
$$= (0.37945 - 32.620 \times \delta_{NN}) \text{ MeV}. \tag{7.128}$$

Then the method described earlier provides a consistent way to evaluate the sensitivity of the 3α reaction rate to a variation of the constants. This rate has been computed numerically, as explained in [3] and as described in Appendix C, where both an analytical approximation valid for sharp resonances and a numerical integration are performed.

The variation of the partial widths of both reactions have been computed in Appendix B and are depicted in Figure 7.18. Together with the results of the previous section and the details of the Appendix B, we can compute the 3α reaction rate as a function of temperature and δ_{NN}. This is summarised in Figure 7.14, which compares the rate for different values of δ_{NN} to the NACRE rate [3], which is our reference when no variation of constants is assumed (i.e., $\delta_{NN} = 0$). One can also refer to Figure 7.19, which compares

---

[8] The astrophysical S-factor is just the cross section corrected for the effect of the penetrability of the Coulomb barrier and other trivial effects.

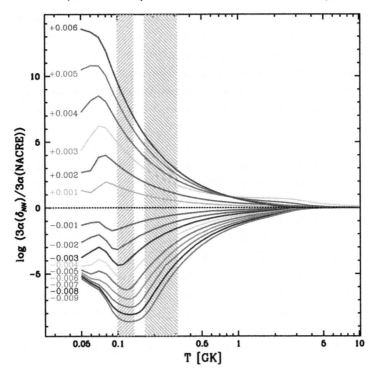

Figure 7.14 The ratio between the 3α rate obtained for $-0.009 \le \delta_{NN} \le +0.006$ and the NACRE rate, as a function of temperature. Hatched areas: values of T_c where H- and He-burning phases take place in a $15M_\odot$ model at $Z = 0$. Figure taken from Reference [44].

the full numerical integration to the analytical estimation (7.126), which turns out to be excellent in the range of temperatures of interest. As one can see, for positive values of δ_{NN}, the resonance energies are lower, so the 3α process is more efficient.

Let us compare the result of Figure 7.14, which gives $y \equiv \log[\lambda_{3\alpha}(\delta_{NN})/\lambda_{3\alpha}(0)]$, to a simple estimate. Using the analytic expression (7.126) for the reaction rate, valid only for a sharp resonance, y is simply given by

$$y = \frac{1}{\ln 10} s_{\delta_{NN}},\tag{7.129}$$

where the sensitivity $s_{\delta_{NN}} \equiv \mathrm{d}\ln\lambda_{3\alpha}/\mathrm{d}\ln\delta_{NN}$ is given, from Eq. (7.127), by $s_{\delta_{NN}} = \delta_{NN} \times$ $(32.62\mathrm{MeV})/kT$. We conclude that

$$y = 1.644 \times \left(\frac{\delta_{NN}}{10^{-3}}\right)\left(\frac{T}{10^8\,\mathrm{K}}\right)^{-1}.\tag{7.130}$$

This gives the correct order of magnitude of the curves depicted in Figure 7.14 as well as their scalings with δ_{NN} and with temperature, as long as $T_9 > 0.1$. At lower temperatures, differences arise from the fact that the analytical expression for the reaction rate is no longer accurate (see Appendix D).

The sensitivity to a variation of the intensity of the N-N interaction arises from the fact that $dQ_{\alpha\alpha\alpha}/d\delta_{NN} \sim 10^2 Q_{\alpha\alpha\alpha}$. The fact that the typical correction to the resonant energies is of order 10 MeV ($\times \delta_{NN}$) – compared to the resonant energies themselves, which are of order 0.1 MeV – allows one to put relatively strong constraints on any variation.

Implications on the Stellar Evolution

The Geneva stellar code was adapted to take into account the reaction rates computed earlier. The version of the code we use is the one described in Reference [43]. Here, we only consider models of $15M_\odot$ and $60M_\odot$ without rotation and assume an initial chemical composition given by $X = 0.7514$, $Y = 0.2486$ and $Z = 0$. This corresponds to the BBN abundance of He at the baryon density determined by WMAP [73] and at zero metallicity as is expected to be appropriate for Population III stars. For 16 values of the free parameter δ_{NN} in the range $-0.009 \leq \delta_{NN} \leq +0.006$, we computed a stellar model which was followed up to the end of core He burning (CHeB). As we will see, beyond this range in δ_{NN}, stellar nucleosynthesis is unacceptably altered. Note that for some of the most extreme cases, the set of nuclear reactions now implemented in the code should probably be adapted for a computation of the advanced evolutionary phases.

Focusing on the limited range in δ_{NN} will allow us to study the impact of a change of the fundamental constants on the production of carbon and oxygen in Population III massive stars. In this context, we recall that the observations of the most iron-poor stars in the halo offer a wonderful tool to probe the nucleosynthetic impact of the first massive stars in the Universe. Indeed, these halo stars are believed to form from material enriched by the ejecta of the first stellar generations in the Universe. Their surface chemical composition (at least on the main sequence) still bears the mark of the chemical composition of the cloud from which they formed and, thus, allows us to probe the nucleosynthetic signature of the first stellar generations. Any variation of the fundamental constants – which, for instance, would prevent the synthesis of carbon and/or oxygen – would be very hard to conciliate with present-day observations of the most iron-poor stars. For instance, the two most iron-poor stars [23, 57] both show strong overabundances of carbon and oxygen with respect to iron.

Figure 7.15 (left) shows the HR diagram for the models with δ_{NN} between -0.009 and $+0.006$ in increments of 0.001 (from left to right). Once the CNO cycle has been triggered (as we are about to discuss), the main sequence (MS) tracks are shifted towards cooler T_{eff} for increasing δ_{NN}. There is a difference of about 0.20 dex between the two extreme models. Figure 7.15 (right) shows the central temperature at the moment of the CNO-cycle ignition (lower curve). On the zero-age main sequence (ZAMS), the standard ($\delta_{NN} = 0$) model has not yet produced enough ^{12}C to be able to rely on the CNO cycle, so it starts by continuing its initial contraction until the CNO cycle ignites. In this model, CNO ignition occurs when the central H mass fraction reaches 0.724 – i.e., when less than 3% of the initial H has been burned. Models with $\delta_{NN} < 0$ (i.e., a lower 3α rate) yield a phase of

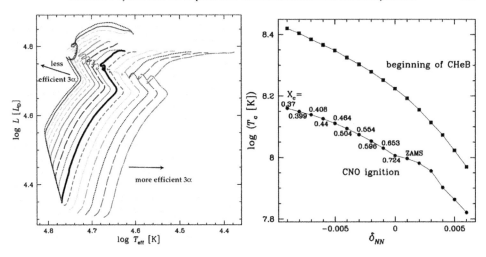

Figure 7.15 Left: HR diagrams for $15M_\odot$ models with $\delta_{NN} = 0$ (thick black) and ranges from left to right from -0.009 to $+0.006$ in steps of 0.001 (using same colour code as Figure 7.14). Right: The central temperature at CNO ignition (circles) and at the beginning of CHeB (squares) as a function of δ_{NN}. The labels on the CNO-ignition curve show the central H mass fraction at that moment. Note that above $\delta_{NN} \sim +0.001$, CNO ignition occurs on or before the ZAMS. Figure taken from Reference [44].

contraction which is longer for lower δ_{NN} (i.e., larger $|\delta_{NN}|$): in these models, the less-efficient 3α rates need a higher T_c to produce enough ^{12}C for triggering the CNO cycle. Models with $\delta_{NN} > 0$ (i.e., a higher 3α rate) are directly sustained by the CNO cycle on the ZAMS: the star can more easily counteract its own gravity and the initial contraction is stopped earlier, so H burning occurs at lower T_c and ρ_c (Figure 7.15, right), i.e., at a slower pace. The MS lifetime, τ_{MS}, is sensitive to the pace at which H is burned, so it increases with δ_{NN}. The relative difference between the Standard Model MS lifetime τ_{MS} at $\delta_{NN} = 0$ and τ_{MS} at $\delta_{NN} = -0.009$ $(+0.006)$ amounts to -17% $(+19\%)$.

Implications on the Nuclear Constants

While the differences in the 3α rates do not lead to strong effects in the evolution characteristics on the MS, the CHeB phase amplifies the differences between the models. The upper curve of Figure 7.15 (right) shows the central temperature at the beginning of CHeB. There is a factor of 2.8 in temperature between the models with $\delta_{NN} = -0.009$ and $+0.006$. To get an idea of what this difference represents, we can relate these temperatures to the grid of Population III models computed by [80]. The $15M_\odot$ model with $\delta_{NN} = -0.009$ starts its CHeB at a higher temperature than a standard 100 $M_\odot$ of the same stage. In contrast, the model with $\delta_{NN} = +0.006$ starts its CHeB phase with a lower temperature than a standard $12M_\odot$ star at CNO ignition. Table 7.6 presents the characteristics of the models for each value of δ_{NN} at the end of CHeB. From these characteristics, we distinguish four different cases:

I In the Standard Model and when δ_{NN} is very close to 0, ^{12}C is produced during He
   burning until the central temperature is high enough for the ^{12}C$(\alpha,\gamma)^{16}$O reaction to
   become efficient: during the last part of the CHeB phase, the ^{12}C is processed into ^{16}O.
   The star ends its CHeB phase with a core composed of a mixture of ^{12}C and ^{16}O.

II If the 3α rate is weakened ($-0.005 \leq \delta_{NN} \leq -0.002$), ^{12}C is produced at a slower
    pace, and T_c is high from the beginning of the CHeB phase, so the ^{12}C$(\alpha,\gamma)^{16}$O reac-
    tion becomes efficient very early: as soon as some ^{12}C is produced, it is immediately
    transformed into ^{16}O. The star ends its CHeB phase with a core composed mainly of
    ^{16}O, without any ^{12}C and with an increasing fraction of ^{24}Mg for decreasing δ_{NN}.

III For still weaker 3α rates ($\delta_{NN} \leq -0.006$), the central temperature during CHeB is
     such that the ^{16}O$(\alpha,\gamma)^{20}$Ne$(\alpha,\gamma)^{24}$Mg chain becomes efficient, reducing the final ^{16}O
     abundance. The star ends its CHeB phase with a core composed of nearly pure ^{24}Mg.
     Because the abundances of both carbon and oxygen are completely negligible, we do
     not list the irrelevant value of C/O for these cases.

IV If the 3α rate is strong ($\delta_{NN} \geq +0.003$), ^{12}C is very rapidly produced, but T_c is so low
    that the ^{12}C$(\alpha,\gamma)^{16}$O reaction can hardly enter into play: ^{12}C is not transformed into
    ^{16}O. The star ends its CHeB phase with a core almost purely composed of ^{12}C.

These results are summarised in Figure 7.16, which shows the composition of the core at the
end of the CHeB phase. One can clearly see the dramatic change in the core composition as
a function of δ_{NN} showing a nearly pure Mg core at large and negative δ_{NN}, a dominantly
O core at low but negative δ_{NN}, and a nearly pure C core at large and positive δ_{NN}. These
results are qualitatively consistent with those found by Schlattl et al. [99] for Population

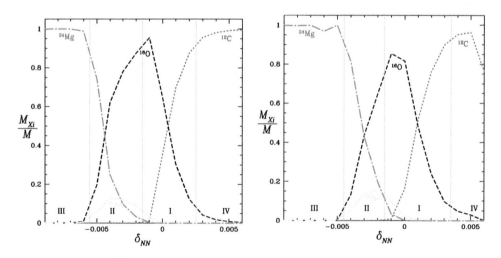

Figure 7.16 The composition of the core at the end of the central He burning in the $15M_\odot$ models as
a function of δ_{NN}. (left) for $15M_\odot$ and (right) for $60~M_\odot$. Figure taken from Reference [44].

Table 7.6 *Characteristics of the 15$M_\odot$ models with δ_{NN} ranging from -0.009 to $+0.006$ at the end of core He burning. The MS lifetime, τ_{MS}, and the core He-burning duration, τ_{CHeB}, are expressed in Myr, the CO-core mass, M_{CO}, is in $M_\odot$, $X(C)$ is the central value for the carbon mass fraction, and C/O is the ratio of the carbon to oxygen mass fractions.*

δ_{NN}	τ_{MS}	τ_{CHeB}	$M_{CO}{}^a$	$X(C)^b$	C/O	Case
-0.009	8.224	1.344	3.84	$4.4e{-}10$	–	III
-0.008	8.285	1.276	3.83	$2.9e{-}10$	–	
-0.007	8.308	1.200	3.38	$8.5e{-}10$	–	
-0.006	8.401	1.168	3.61	$4.2e{-}07$	–	
-0.005	8.480	1.130	3.59	$5.9e{-}06$	3.0e–05	II
-0.004	8.672	0.933	3.60	$3.2e{-}05$	5.2e–05	
-0.003	8.790	0.905	3.60	$1.3e{-}04$	1.7e–04	
-0.002	9.046	0.892	3.61	$5.6e{-}04$	6.4e–04	
-0.001	9.196	0.888	3.70	0.013	0.014	I
0	9.640	0.802	3.65	0.355	0.550	
$+0.001$	9.937	0.720	3.61	0.695	2.278	
$+0.002$	10.312	0.684	3.62	0.877	7.112	
$+0.003$	10.677	0.664	3.62	0.958	22.57	IV
$+0.004$	10.981	0.659	3.62	0.981	52.43	
$+0.005$	11.241	0.660	3.61	0.992	123.9	
$+0.006$	11.447	0.661	3.55	0.996	270.2	

[a] Mass coordinate where the abundance of ^{4}He drops below 10^{-3}.
[b] Here, $X(C) = X(^{12}C)$.

I–type stars. Note that their cases with $\Delta E_R = \pm 100$ keV correspond roughly to our $\delta_{NN} \approx \mp 0.005$.

Table 7.6 shows also the core size at the end of CHeB. As in Reference [63], the mass, M_{CO}, is determined as the mass coordinate where the mass fraction of ^{4}He drops below 10^{-3}. The mass of the CO core increases with decreasing δ_{NN}, the increase amounting to 8% between $\delta_{NN} = +0.006$ and -0.009. This effect is due to the higher central temperature and greater compactness at low δ_{NN}. The same effect was found by other authors [99]. As shown by these authors, this effect is expected to have an impact on the remnant mass and thus on the strength of the final explosion.

7.6 Discussion

This chapter focused on the relation between QCD and nuclear physics in order to determine the level of tuning required in order to produce a large enough variety of nuclei in the course of the evolution of the Universe in order to create a large enough space of possibility for chemistry. Several bounds on the parameters of the Standard Model of particle physics (see Table 7.1) have been discussed.

- The effect of binding energies and masses have been discussed in Section 7.3, and a sets of constraints based on the stability of matter have been summarised in Section 7.3.8.
- Light elements are formed during primordial nucleosynthesis, described in Section 7.4. While only helium-4 is significantly produced, the effects of the nuclear parameters are twofold. First, the observation allowed us the range of parameters that lead to a universe similar to ours – i.e., undistinguishable by an observer. Indeed, these bounds evolve with time and the accuracy of observations. But they allow one, at a given time, to quantify the coarse graining in the space of fundamental parameters. Second, we have shown that it was not possible to produce a large enough early abundance of carbon-12. The computation of the initial metallicity is, however, important for the evolution of Population III stars.
- Stars are the central system to understand the diversity of the Mendeleev table. Once the nuclear parameters are in the bounds that allow for the diversity and stability of these elements, one still needs them to be produce from the chemical composition after BBN, in a timescale short enough to allow for chemistry and biology to start. The main description of stellar nucleosynthesis has been described in Section 7.5. Stars are important in two respects. First, they are the source of all elements but hydrogen, helium, and lithium. Second, they provide a long-lived, low-entropy source of energy necessary for planetary life to emerge. We have shown that the production of carbon and oxygen in Population III stars requires a tuning at the level of 10^{-3} of the strength of the nuclear interaction.

This chapter has emphasised the difficulty of connecting QCD to nuclear physics. Even if, in principle, one can derive the value of the nuclear parameters (masses, binding energies, resonance energies, etc.) in terms of the constants of the Standard Model of particle physics, in practice, the accuracy of these predictions can be of one order of magnitude. This is, indeed, the main limitation of such an analysis.

Let us also emphasise that we have assumed a local variation of the constants, assuming the same structure for the theory and its symmetries. It is important to stress that there exist some attempts to evade such a limitation. Reference [62] argued for the existence of a region of parameter space which is life permitting in the absence of the weak force. It requires many tunings but provides a counterexample.

Acknowledgements

This chapter relies heavily on my work with Alain Coc, Elisabeth Vangioni, Pierre Descouvemont, Georges Meynet, and Silvia Ekström. I thank them for their collaboration over the past years, and I thank George Ellis, Cyril Pitrou for many discussions on this topic. I also thank Khalil Chamcham and Joe Silk.

References

[1] Ajzenberg-Selove, F. 1990. 'Energy Levels of Light Nuclei A = 11–12'. *Nuclear Physics A*, **506**(Jan.), 1–158.

[2] Anders, E., and Grevesse, N. 1989. 'Abundances of the Elements: Meteoritic and Solar'. *Geochimica Cosmochimica Acta*, **53**(Jan.), 197–214.

[3] Angulo, C., *et al.* 1999. 'A Compilation of Charged-Particle Induced Thermonuclear Reaction Rates'. *Nuclear Physics A*, **656**(Aug.), 3–183.

[4] Audi, G., Wapstra, A. H., and Thibault, C. 2003. 'The AME2003 Atomic Mass Evaluation: (II) Tables, Graphs and References'. *Nuclear Physics A*, **729**(Dec.), 337–676.

[5] Aver, E., Olive, K. A., and Skillman, E. D. 2012. 'An MCMC Determination of the Primordial Helium Abundance'. *Journal of Cosmology and Astroparticle Physics*, **4**(Apr.), 004.

[6] Aver, E., *et al.* 2013. 'The Primordial Helium Abundance from Updated Emissivities'. *Journal of Cosmology and Astroparticle Physics*, **11**(Nov.), 017.

[7] Bania, T. M., Rood, R. T., and Balser, D. S. 2002. 'The Cosmological Density of Baryons from Observations of ^{3}He$^+$ in the Milky Way'. *Nature*, **415**(Jan.), 54–57.

[8] Barnes, L. A. 2012. 'The Fine-Tuning of the Universe for Intelligent Life'. *Publications of the Astronomical Society of Australia*, **29**(June), 529–564.

[9] Barr, S. M., and Khan, A. 2007. 'Anthropic Tuning of the Weak Scale and of m_u/m_d in Two-Higgs-Doublet Models'. *Physical Review D*, **76**(4), 045002.

[10] Barrow, J. D. 1987. 'Observational Limits on the Time Evolution of Extra Spatial Dimensions'. *Physical Review D*, **35**(Mar.), 1805–1810.

[11] Barrow, J. D., and Tipler, F. J. 1986. *The Anthropic Cosmological Principle*. Oxford University Press.

[12] Beane, S. R., and Savage, M. J. 2003. 'Variation of Fundamental Couplings and Nuclear Forces'. *Nuclear Physics A*, **713**(Jan.), 148–164.

[13] Berengut, J. C., Flambaum, V. V., and Kava, E. M. 2011. 'Search for Variation of Fundamental Constants and Violations of Fundamental Symmetries Using Isotope Comparisons'. *Physical Review A: General Physics*, **84**(4), 042510.

[14] Bergström, L., Iguri, S., and Rubinstein, H. 1999. 'Constraints on the Cariation of the Fine Structure Constant from Big Bang Nucleosynthesis'. *Physical Review D*, **60**(4), 045005.

[15] Bernstein, J. 1988. *Kinetic Theory in the Expanding Universe*. Cambridge University Press.

[16] Bernstein, J., Brown, L. S., and Feinberg, G. 1989. 'Cosmological Helium Production Simplified'. *Reviews of Modern Physics*, **61**(Jan.), 25–39.

[17] Bromm, V., *et al.* 2009. 'The Formation of the First Stars and Galaxies'. *Nature*, **459**(May), 49–54.

[18] Calmet, X. and Fritzsch, H. 2001. 'The Cosmological Evolution of the Nucleon Mass and the Electroweak Coupling Constants'. *arXiv High Energy Physics – Phenomenology e-prints*, Dec.

[19] Campbell, B. A., and Olive, K. A. 1995. 'Nucleosynthesis and the Time Dependence of Fundamental Couplings'. *Physics Letters B*, **345**(Feb.), 429–434.

[20] Carr, B. J., and Rees, M. J. 1979. 'The Anthropic Principle and the Structure of the Physical World'. *Nature*, **278**(Apr.), 605–612.

[21] Carter, B. 1974. 'Large Number Coincidences and the Anthropic Principle in Cosmology'. Longair, M. S. ed., *Confrontation of Cosmological Theories with Observational Data*. IAU Symposium, vol. 63. Pages 291–298.

[22] Cassisi, S. and Castellani, V. 1993. 'An Evolutionary Scenario for Primeval Stellar Populations'. *Astrophysical Journal Supplements*, **88**(Oct.), 509–527.

[23] Christlieb, N., *et al.* 2004. 'HE 0107-5240, a Chemically Ancient Star: I. A Detailed Abundance Analysis'. *Astrophysical Journal*, **603**(Mar.), 708–728.

[24] Clayton, D. D. 1968. *Principles of Stellar Evolution and Nucleosynthesis*. University of Chicago Press.

[25] Coc, A. and Vangioni, E. 2010 (Jan.). 'Big-Bang Nucleosynthesis with Updated Nuclear Data'. *Journal of Physics Conference Series*. vol. 202, 012001.

[26] Coc, A., *et al.* 2007. 'Coupled Variations of Fundamental Couplings and Primordial Nucleosynthesis'. *Physical Review D*, **76**(2), 023511.

[27] Coc, A., *et al.* 2012a. 'Influence of the Cariation of Fundamental Constants on the Primordial Nucleosynthesis'. *Nuclei in the Cosmos (NIC XII)*, 73.

[28] Coc, A., *et al.* 2012b. 'Standard Big Bang Nucleosynthesis up to CNO with an Improved Extended Nuclear Network'. *Astrophysical Journal*, **744**(Jan.), 158.

[29] Coc, A., *et al.* 2012c. 'Variation of Fundamental Constants and the Role of A=5 and A=8 Nuclei on Primordial Nucleosynthesis'. *Physical Review D*, **86**(4), 043529.

[30] Coc, A., *et al.* 2013a. 'Influence of the Variation of Fundamental Constants on the Primordial Nucleosynthesis'. *arXiv e-prints*, Jan.

[31] Coc, A., Uzan, J. P., and Vangioni, E. 2013b. 'Mirror Matter Can Alleviate the Cosmological Lithium Problem'. *Physical Review D*, **87**(12), 123530.

[32] Coc, A., *et al.* 2014a. 'Modified Big Bang Nucleosynthesis with Nonstandard Neutron Sources'. *Physical Review D*, **90**(8), 085018.

[33] Coc, A., Uzan, J. P., and Vangioni, E. 2014b. 'Standard Big Bang Nucleosynthesis and Primordial CNO Abundances after Planck'. *Journal of Cosmology and Astroparticle Physics*, **10**(Oct.), 050.

[34] Cook, C. W., *et al.* 1957. 'B^{12}, C^{12}, and the Red Giants'. *Physical Review*, **107**(July), 508–515.

[35] Cooke, R. J., *et al.* 2014. 'Precision Measures of the Primordial Abundance of Deuterium'. *Astrophysical Journal*, **781**(Jan.), 31.

[36] Csótó, A., Oberhummer, H., and Schlattl, H. 2001. 'Fine-tuning the Basic Forces of Nature through the Triple-Alpha Process in Red Giant Stars'. *Nuclear Physics A*, **688**(May), 560–562.

[37] Damour, T., and Donoghue, J. F. 2008. 'Constraints on the Variability of Quark Masses from Nuclear Binding'. *Physical Review D*, **78**(1), 014014.

[38] Damour, T. and Polyakov, A. M. 1994. 'The String Dilation and a Least Coupling Principle'. *Nuclear Physics B*, **423**(July), 532–558.

[39] Damour, T. and Lilley, M. 2008. 'Course 10: String Theory, Gravity and Experiment'. Bachas, C. *et al.*, eds., *String Theory and the Real World: From Particle Physics to Astrophysics*. Les Houches, vol. 87. Elsevier. Pages 371–448.

[40] Davies, P. C. W. 1972. 'Time Variation of the Coupling Constants'. *Journal of Physics A Mathematical General*, **5**(Aug.), 1296–1304.

[41] Donoghue, J. F. 2006. 'Nuclear Central Force in the Chiral Limit'. *Physical Review C*, **74**(2), 024002.

[42] Dunbar, D. N., *et al.* 1953. 'The 7.68-Mev State in C^{12}'. *Physical Review*, **92**(Nov.), 649–650.

[43] Ekström, S., *et al.* 2008a. 'Effects of Rotation on the Evolution of Primordial Stars'. *Astronomy and Astrophysics*, **489**(Oct.), 685–698.

[44] Ekström, S., *et al.* 2010. 'Effects of the Variation of Fundamental Constants on Population III Stellar Evolution'. *Astronomy and Astrophysics*, **514**(May), A62.

[45] Ellis, G. F. R. 2005. 'Physics, Complexity and Causality'. *Nature*, **435**(June), 743.

[46] Ellis, G. F. R., and Uzan, J. P. 2005. '*c* is the Speed of Light, Isn't It?' *American Journal of Physics*, **73**(Mar.), 240–247.

[47] Ellis, J., Ibañez, L., and Ross, G. G. 1982. 'Grand Unification with Large Supersymmetry Breaking'. *Physics Letters B*, **113**(June), 283–287.

[48] Epelbaum, E., Meißner, U. G., and Glöckle, W. 2003. 'Nuclear Forces in the Chiral Limit'. *Nuclear Physics A*, **714**(Feb.), 535–574.

[49] Fairbairn, M. 1999. 'Carbon Burning in Supernovae and Evolving Physical Constants'. arXiv:astro-ph/9910328.

[50] Fiorentini, G., *et al.* 1998. 'Quantifying Uncertainties in Primordial Nucleosyn Thesis without Monte Carlo Simulations'. *Physical Review D*, **58**(6), 063506.

[51] Flambaum, V. V. 2003. 'Limits on Temporal Variation of Fine Structure Constant, Quark Masses and Strong Interaction from Atomic Clock Experiments'. *arXiv Physics e-prints*, Sept.

[52] Flambaum, V. V., and Shuryak, E. V. 2002. 'Limits on Cosmological Variation of Strong Interaction and Quark Masses from Big Bang Nucleosynthesis, Cosmic, Laboratory and Oklo Data'. *Physical Review D*, **65**(10), 103503.

[53] Flambaum, V. V., and Shuryak, E. V. 2003. 'Dependence of Hadronic Properties on Quark Masses and Constraints on Their Cosmological Variation'. *Physical Review D*, **67**(8), 083507.

[54] Flambaum, V. V., and Tedesco, A. F. 2006. Dependence of Nuclear Magnetic Moments on Quark Masses and Limits on Temporal Variation of Fundamental Constants from Atomic Clock Experiments'. *Physical Review C*, **73**(5), 055501.

[55] Flambaum, V. V., and Wiringa, R. B. 2007. 'Dependence of Nuclear Binding on Hadronic Mass Variation'. *Physical Review C*, **76**(5), 054002.

[56] Frebel, A. and Norris, J. E. 2013. *Metal-Poor Stars and the Chemical Enrichment of the Universe*. Page 55.

[57] Frebel, A., *et al.* 2008. 'HE 1327-2326, an Unevolved Star with [Fe/H]< -5.0. II. New 3D-1D Corrected Abundances from a Very Large Telescope UVES Spectrum'. *Astrophysical Journal*, **684**(Sept.), 588–602.

[58] Furnstahl, R. J., and Serot, B. D. 2000. 'Parameter Counting in Relativistic Mean-Field Models'. *Nuclear Physics A*, **671**(May), 447–460.

[59] Fynbo, H. O. U., *et al.* 2005. 'Revised Rates for the Stellar Triple-α Process from Measurement of ^{12}C Nuclear Resonances'. *Nature*, **433**(Jan.), 136–139.

[60] Gasser, J., and Leutwyler, H. 1982. 'Quark Masses'. *Physics Reports*, **87**(July), 77–169.

[61] Hall, L. J., and Nomura, Y. 2008. 'Evidence for the Multiverse in the Standard Model and Beyond'. *Physical Review D*, **78**(3), 035001.

[62] Harnik, R., Kribs, G. D., and Perez, G. 2006. 'A Universe without Weak Interactions'. *Physical Review D*, **74**(3), 035006.

[63] Heger, A., Langer, N., and Woosley, S. E. 2000. 'Presupernova Evolution of Rotating Massive Stars: I. Numerical Method and Evolution of the Internal Stellar Structure'. *Astrophysical Journal*, **528**(Jan.), 368–396.

[64] Hogan, C. J. 2000. 'Why the Universe Is Just So'. *Reviews of Modern Physics*, **72**(Oct.), 1149–1161.

[65] Hogan, C. J. 2007. *Quarks, Electrons and Atoms in Closely Related Universes.* Cambridge University Press. Page 221–230.

[66] Hoyle, F. 1954. 'On Nuclear Reactions Occuring in Very Hot Stars: I. The Synthesis of Elements from Carbon to Nickel'. *Astrophysical Journal Supplements*, **1**(Sept.), 121.

[67] Ichikawa, K., and Kawasaki, M. 2004. 'Big Bang Nucleosynthesis with a Varying Fine Structure Constant and Nonstandard Expansion Rate'. *Physical Review D*, **69**(12), 123506.

[68] Iliadis, C. 2007. *Nuclear Physics of Stars.* Wiley-VCH Verlag.

[69] Iocco, F., *et al.* 2007. 'Path to Metallicity: Synthesis of CNO Elements in Standard BBN'. *Physical Review D*, **75**(8), 087304.

[70] Iocco, F., Bonifacio, P., and Vangioni, E. 2012. 'Lithium in the Cosmos'. *Memorie della Societa Astronomica Italiana Supplementi*, **22**, 3.

[71] Kamimura, M. 1981. 'Transition Densities between the $0_1^+, 2_1^+, 4_1^+, 0_2^+, 2_2^+, 1_1^-$ and 3_1^- States in ^{12}C Derived from the Three-Alpha Resonating-Group Wave Functions'. *Nuclear Physics A*, **351**(Jan.), 456–480.

[72] Kawano, L. 1992 (Jan.). *Let's Go: Early Universe 2: Primordial Nucleosynthesis the Computer Way.* Fermilab. https://ntrs.nasa.gov/search.jsp?R=19920015920.

[73] Komatsu, E., *et al.* 2009. 'Five-Year Wilkinson Microwave Anisotropy Probe Observations: Cosmological Interpretation'. *Astrophysical Journal Supplements*, **180**(Feb.), 330–376.

[74] Korennov, S., and Descouvemont, P. 2004. 'A Microscopic Three-Cluster Model in the Hyperspherical Formalism'. *Nuclear Physics A*, **740**(Aug.), 249–267.

[75] Langacker, P., Segrè, G., and Strassler, M. J. 2002. 'Implications of Gauge Unification for Time Variation of the Fine Structure Constant'. *Physics Letters B*, **528**(Feb.), 121–128.

[76] Langanke, K., Wiescher, M., and Thielemann, F. K. 1986. 'The Triple-Alpha-Reaction at Low Temperatures'. *Zeitschrift fur Physik A Hadrons and Nuclei*, **324**, 147–152.

[77] Lisi, E., Sarkar, S., and Villante, F. L. 1999. 'Big Bang Nucleosynthesis Limit on N_ν'. *Physical Review D*, **59**(12), 123520.

[78] Livio, M., *et al.* 1989. 'The Anthropic Significance of the Existence of an Excited State of C-12'. *Nature*, **340**(July), 281–284.

[79] Luo, F., Olive, K. A., and Uzan, J. P. 2011. 'Gyromagnetic Factors and Atomic Clock Constraints on the Variation of Fundamental Constants'. *Physical Review D*, **84**(9), 096004.

[80] Marigo, P., *et al.* 2001. 'Zero-Metallicity Stars: I. Evolution at Constant Mass'. *Astronomy and Astrophysics*, **371**(May), 152–173.

[81] Morel, P., and Lebreton, Y. 2008. 'CESAM: A Free Code for Stellar Evolution Calculations'. *Astrophysics and Space Science*, **316**(Aug.), 61–73.

[82] Müller, C. M., Schäfer, G., and Wetterich, C. 2004. 'Nucleosynthesis and the Variation of Fundamental Couplings'. *Physical Review*, **70**(Oct), 083504.

[83] Nakamura, K., and Particle Data Group. 2010. 'Review of Particle Physics'. *Journal of Physics G: Nuclear Physics*, **37**(7), 075021.

[84] Navrátil, P., *et al.* 2009. 'Topical Review: Recent Developments in No-Core Shell-Model Calculations'. *Journal of Physics G: Nuclear Physics*, **36**(8), 083101.

[85] Newton, J. R., *et al.* 2007. 'Gamow peak in Thermonuclear Reactions at High Temperatures'. *Physical Review C*, **75**(4), 045801.

[86] Nollett, K. M., and Lopez, R. E. 2002. 'Primordial Nucleosynthesis with a Varying Fine Structure Constant: An Improved Estimate'. *Physical Review D*, **66**(6), 063507.

[87] Nomoto, K., Thielemann, F. K., and Miyaji, S. 1985. 'The Triple Alpha Reaction at Low Temperatures in Accreting White Dwarfs and Neutron Stars'. *Astronomy and Astrophysics*, **149**(Aug.), 239–245.

[88] Oberhummer, H., Csótó, A., and Schlattl, H. 2000. 'Stellar Production Rates of Carbon and Its Abundance in the Universe'. *Science*, **289**(July), 88–90.

[89] Oberhummer, H., Csótó, A., and Schlattl, H. 2001. 'Bridging the Mass Gaps at /A=5 and /A=8 in Nucleosynthesis'. *Nuclear Physics A*, **689**(June), 269–279.

[90] Oberhummer, H., *et al.* 2003. 'Temporal Variation of Coupling Constants and Nucleosynthesis'. *Nuclear Physics A*, **719**(May), C283–C286.

[91] Peebles, P. J. E. 1966. 'Primordial Helium Abundance and the Primordial Fireball. II'. *Astrophysical Journal*, **146**(Nov.), 542.

[92] Peter, P., and Uzan, J. 2009. *Primordial Cosmology*. Oxford Graduate Texts. Oxford University Press.

[93] Pettini, M. and Cooke, R. 2012. 'A New, Precise Measurement of the Primordial Abundance of Deuterium'. *Monthly Notices of the Royal Astronomical Society*, **425**(Oct.), 2477–2486.

[94] Reeves, H. 1968. *Stellar Evolution and Nucleosynthesis*. Cambridge University Press.

[95] Rohlf, J. 1994. *Modern Physics from [α] to Z*. Wiley.

[96] Rozental, I. L. 1988. *Big Bang, Big Bounce: How Particles and Fields Drive Cosmic Evolution*. Springer.

[97] Sánchez-Cañizares, J. 2017. 'How Can Physics Underlie the Mind? Top-Down Causation in the Human Context, by G. Ellis'. *Contemporary Physics*, **58**(Oct.), 356–356.

[98] Sbordone, L., *et al.* 2010. 'The Metal-Poor End of the Spite Plateau: I. Stellar Parameters, Metallicities, and Lithium Abundances'. *Astronomy and Astrophysics*, **522**(Nov.), A26.

[99] Schlattl, H., *et al.* 2004. 'Sensitivity of the C and O Production on the 3α Rate'. *Astrophysics and Space Science*, **291**(Apr.), 27–56.

[100] Serot, B. D., and Walecka, J. D. 1997. 'Recent Progress in Quantum Hadrodynamics'. *International Journal of Modern Physics E*, **6**, 515–631.

[101] Spite, F. and Spite, M. 1982. 'Abundance of Lithium in Unevolved Halo Stars and Old Disk Stars: Interpretation and Consequences'. *Astronomy and Astrophysics*, **115**(Nov.), 357–366.

[102] Suzuki, Y., *et al.* 2008. 'Local versus Nonlocal $\alpha\alpha$ Interactions in a 3 α Description of ^{12}C'. *Physics Letters B*, **659**(Jan.), 160–164.

[103] Symposium, I. A. U., *et al.* 2010. *Light Elements in the Universe (IAU S268)*. IAU Symposium and Colloquium Proceedings Series. Cambridge University Press.

[104] Tegmark, M. 1998. 'Is "the Theory of Everything" Merely the Ultimate Ensemble Theory?' *Annals of Physics*, **270**(Nov.), 1–51.

[105] Thompson, D. R., Lemere, M., and Tang, Y. C. 1977. 'Systematic Investigation of Scattering Problems with the Resonating-Group Method'. *Nuclear Physics A*, **286**(Aug.), 53–66.

[106] Tilley, D., *et al.* 2004. 'Energy Levels of Light Nuclei A=8,9,10'. *Nuclear Physics A*, **745**(3), 155–362.

[107] Uzan, J. P. 2003a. 'Tests of Gravity on Astrophysical Scales and Variation of the Constants'. *Annales Henri Poincaré*, **4**(Dec.), 347–369.

[108] Uzan, J. P. 2003b. 'The Fundamental Constants and Their Variation: Observational and Theoretical Status'. *Reviews of Modern Physics*, **75**(Apr.), 403–455.

[109] Uzan, J. P. 2009. 'Fundamental Constants and Tests of General Relativity: Theoretical and Cosmological Considerations'. *Space Science Reviews*, **148**(Dec.), 249–265.

[110] Uzan, J. P. 2011. 'Varying Constants, Gravitation and Cosmology'. *Living Reviews in Relativity*, **14**(Mar.), 2.

[111] Uzan, J. P. 2013. 'Models of the Cosmos and Emergence of Complexity'. In *The Causal Universe*. Copernicus Center Press. Pages 93–120.

[112] Vangioni-Flam, E., *et al.* 2003. 'On the Baryometric Status of ^{3}He'. *Astrophysical Journal*, **585**(Mar.), 611–616.

[113] Villani, C., Uzan, J., and Moncorgé, V. 2014. *La maison des mathématiques*. Beaux livres. Le Cherche Midi.

[114] Wagoner, R. V. 1973. 'Big-Bang Nucleosynthesis Revisited'. *Astrophysical Journal*, **179**(Jan.), 343–360.

[115] Wildermuth, K., and T'ang, Y. C. 1977. *A Unified Theory of the Nucleus*. Vieweg Braunschweig.

[116] Woosley, S. E., and Heger, A. 2015. 'The Remarkable Deaths of 9-11 Solar Mass Stars'. *Astrophysical Journal*, **810**(Sept.), 34.

[117] Woosley, S. E., Heger, A., and Weaver, T. A. 2002. 'The Evolution and Explosion of Massive Stars'. *Reviews of Modern Physics*, **74**(Nov.), 1015–1071.

[118] Yoo, J. J., and Scherrer, R. J. 2003. 'Big Bang Nucleosynthesis and Cosmic Microwave Background Constraints on the Time Variation of the Higgs Vacuum Expectation Value'. *Physical Review D*, **67**(4), 043517.

Appendix A: Summary of the Formation Sites of Atomic Isotopes

200   E. Anders and N. Grevesse

Table 3. Abundance of the Nuclides (Atoms/10^6 Si)

Element, A	Atom Percent	Process*	Abund.†	Element, A	Atom Percent	Process*	Abund.†	Element, A	Atom Percent	Process*	Abund.†	Element, A	Atom Percent	Process*	Abund.†
1 H 1	99.9966		2.79×10^{10}	30 Zn 64	48.63	Ex,E	613	51 Sb 121	57.362	R,s	0.177	71 Lu 175	97.41	R,s	0.0357
2	0.0034	U	9.49×10^5	66	27.90	E	352	123	42.638	R	0.132	176	2.59	S	0.000951
				67	4.10	E,S	51.7					176			0.001035
2 He 3	0.0142	U,h?	3.86×10^5	68	18.75	E,S	236	52 Te 120	0.09	P	0.0043				
4	99.9858	U,h	2.72×10^9	70	0.62	E,S	7.8	122	2.57	S	0.124	72 Hf 174	0.162	P	0.000249
								123	0.89	S	0.0428	176	5.206	S	0.00802
3 Li 6	7.5	X	4.28	31 Ga 69	60.108	S,e,r	22.7	124	4.76	S	0.229	176			0.00793
7	92.5	U,x,h	52.82	71	39.892	S,e,r	15.1	125	7.10	R,s	0.342	177	18.606	R,s	0.0287
								126	18.89	R,S	0.909	178	27.297	R,S	0.0420
4 Be 9	100	X	0.73	32 Ge 70	20.5	S,e	24.4	128	31.73	R	1.526	179	13.629	R,s	0.0210
				72	27.4	S,e,r	32.6	130	33.97	R	1.634	180	35.100	S,R	0.0541
5 B 10	19.9	X	4.22	73	7.8	e,s,r	9.28								
11	80.1	X	16.98	74	36.5	e,s,r	43.4	53 I 127	100	R	0.90	73 Ta 180	0.012	p,s,r	2.48×10^{-6}
				76	7.8	E	9.28					181	99.988	R,S	0.0207
6 C 12	98.90	He	9.99×10^6					54 Xe 124	0.121	P	0.00571				
13	1.10	H,N	1.11×10^5	33 As 75	100	R,s	6.56	126	0.108	P	0.00509	74 W 180	0.13	P	0.000173
								128	2.19	S	0.103	182	26.3	R,s	0.0350
7 N 14	99.634	H	3.12×10^6	34 Se 74	0.88	P	0.55	129	27.34	R	1.28	183	14.3	R,s	0.0190
15	0.366	H,N	1.15×10^4	76	9.0	S,p	5.6	130	4.35	S	0.205	184	30.67	R,s	0.0408
				77	7.6	R,s	4.7	131	21.69	R	1.02	186	28.6	R	0.0380
8 O 16	99.762	He	2.37×10^7	78	23.6	R,s	14.7	132	26.50	R,s	1.24				
17	0.038	N,H	9.04×10^3	80	49.7	R,s	30.9	134	9.76	R	0.459	75 Re 185	37.40	R,s	0.0193
18	0.200	He,N	4.76×10^4	82	9.2	R,S	5.7	136	7.94	R	0.373	187	62.60	R	0.0324
												187			0.0351
9 F 19	100	N	843	35 Br 79	50.69	R,s	5.98	55 Cs 133	100	R,s	0.372				
				81	49.31	R,s	5.82					76 Os 184	0.018	P	0.000122
10 Ne 20	92.99	C	3.20×10^6					56 Ba 130	0.106	P	0.00476	186	1.58	S	0.0107
21	0.226	C,Ex	7.77×10^3	36 Kr 78	0.339	P	0.153	132	0.101	P	0.00453	187	1.6	S	0.0108
22	6.79	He,N	2.34×10^5	80	2.22	S,p	0.999	134	2.417	S	0.109	187			0.00807
				82	11.45	S	5.15	135	6.592	R,s	0.296	188	13.3	R,s	0.0898
11 Na 23	100	CNeEx	5.74×10^4	83	11.47	R,s	5.16	136	7.854	S	0.353	189	16.1	R	0.109
				84	57.11	R,S	25.70	137	11.23	S,r	0.504	190	26.4	R	0.178
12 Mg 24	78.99	N,Ex	8.48×10^5	86	17.42	S,r	7.84	138	71.70	S	3.22	192	41.0	R	0.277
25	10.00	NeExC	1.07×10^5												
26	11.01	NeExC	1.18×10^5	37 Rb 85	72.165	R,s	5.12	57 La 138	0.089	P	0.000397	77 Ir 191	37.3	R	0.247
				87	27.835	S	1.97	138			0.000409	193	62.7	R	0.414
13 Al 27	100	Ne,Ex	8.49×10^4				2.11	139	99.911	S,r	0.446				
				38 Sr 84	0.56	P	0.132					78 Pt 190	0.0127	P	0.000170
14 Si 28	92.23	O,Ex	9.22×10^5	86	9.86	S	2.32	58 Ce 136	0.19	P	0.00216	192	0.78	S	0.0105
29	4.67	Ne,Ex	4.67×10^4	87	7.00	S	1.64	138	0.25	P	0.00284	194	32.9	R	0.441
30	3.10	Ne,Ex	3.10×10^4	88	82.58	S,r	1.51	138			0.00283	195	33.8	R	0.453
							19.41	140	88.48	S,r	1.005	196	25.2	R	0.338
15 P 31	100	Ne,Ex	1.04×10^4					142	11.08	R	0.126	198	7.19	R	0.0963
				39 Y 89	100	S	4.64								
16 S 32	95.02	O,Ex	4.89×10^5					59 Pr 141	100	R,S	0.167	79 Au 197	100	R	0.187
33	0.75	Ex	3.86×10^3	40 Zr 90	51.45	S	5.87								
34	4.21	O,Ex	2.17×10^4	91	11.22	S	1.28	60 Nd 142	27.13	S	0.225	80 Hg 196	0.1534	P	0.00052
36	0.02	Ex,N,e,S	1.03×10^2	92	17.15	S	1.96	143	12.18	R,s	0.101	198	9.968	S	0.0339
				94	17.38	S	1.98	143			0.100	199	16.873	R,S	0.0574
17 Cl 35	75.77	Ex	2860	96	2.80	R	0.320	144	23.80	S,R	0.197	200	23.096	S,r	0.0785
37	24.23	Ex,C,S	913					145	8.30	R,s	0.0687	201	13.181	S,r	0.0448
				41 Nb 93	100	S	0.698	146	17.19	R,S	0.142	202	29.863	S,r	0.1015
18 Ar 36	84.2	Ex	8.50×10^4					148	5.76	R	0.0477	204	6.865	R	0.0233
38	15.8	O,Ex	1.60×10^4	42 Mo 92	14.84	P	0.378	150	5.64	R	0.0467				
40		S,Ne	26	94	9.25	P	0.236					81 Tl 203	29.524	R,S	0.0543
40			25 ± 14	95	15.92	R,s	0.406	62 Sm 144	3.1	P	0.00800	205	70.476	S,R	0.1297
				96	16.68	S	0.425	147	15.0	R,s	0.0387				
19 K 39	93.2581	Ex	3516	97	9.55	R,s	0.244	147			0.0399	82 Pb 204	1.94	S	0.0611
40	0.01167	S,Ex,Nep	0.440	98	24.13	R,s	0.615	148	11.3	S	0.0292	206	19.12	R,s	0.602
40			5.48	100	9.63	R	0.246	149	13.8	R,S	0.0356	206			0.593
41	6.7302	Ex	253.7					150	7.4	S	0.0191	207	20.62	R,S	0.650
				44 Ru 96	5.52	P	0.103	152	26.7	R,S	0.0689	207			0.644
20 Ca 40	96.941	Ex	5.92×10^4	98	1.88	P	0.0350	154	22.7	R	0.0586	208	58.31	R,s	1.837
42	0.647	Ex,O	395	99	12.7	R,s	0.236					208			1.828
43	0.135	Ex,C,S	82.5	100	12.6	S	0.234	63 Eu 151	47.8	R,s	0.0465				
44	2.086	Ex,S	1275	101	17.0	R,s	0.316	153	52.2	R,s	0.0508	83 Bi 209	100	R,s	0.144
46	0.004	ExCNe	2.4	102	31.6	R,S	0.588								
48	0.187	E,Ex	114	104	18.7	R	0.348	64 Gd 152	0.20	P,s	0.00066	90 Th 232	100	RA	0.0335
								154	2.18	S	0.00719	232			0.0420
21 Sc 45	100	ExNeE	34.2	45 Rh 103	100	R,s	0.344	155	14.80	R,s	0.0488				
								156	20.47	R,s	0.0676	92 U 235	0.7200	RA	6.48×10^{-5}
22 Ti 46	8.0	Ex	192	46 Pd 102	1.020	P	0.0142	157	15.65	R,s	0.0516	235			0.00573
47	7.3	Ex	175	104	11.14	S	0.155	158	24.84	R,s	0.0820	238	99.2745	RA	0.00893
48	73.8	Ex	1771	105	22.33	R,S	0.310	160	21.86	R	0.0721	238			0.0181
49	5.5	Ex	132	106	27.33	R,S	0.380								
50	5.4	E	130	108	26.46	R,S	0.368	65 Tb 159	100	R	0.0603				
				110	11.72	R	0.163								
23 V 50	0.250	Ex,E	0.732					66 Dy 156	0.056	P	0.000221				
51	99.750	Ex	292	47 Ag 107	51.839	R,s	0.252	158	0.096	P	0.000378				
				109	48.161	R,s	0.234	160	2.34	S	0.00922				
24 Cr 50	4.345	Ex	587					161	18.91	R	0.0745				
52	83.789	Ex	1.131×10^4	48 Cd 106	1.25	P	0.0201	162	25.51	R,s	0.101				
53	9.501	Ex	1283	108	0.89	P	0.0143	163	24.90	R	0.0982				
54	2.365	E	319	110	12.49	S	0.201	164	28.19	R,S	0.111				
				111	12.80	R,S	0.206								
25 Mn 55	100	Ex,E	9550	112	24.13	S,R	0.388	67 Ho 165	100	R	0.0889				
				113	12.22	R,S	0.197								
26 Fe 54	5.8	Ex	5.22×10^4	114	28.73	S,R	0.463	68 Er 162	0.14	P	0.000351				
56	91.72	Ex,E	8.25×10^5	116	7.49	R	0.121	164	1.61	P,S	0.00404				
57	2.2	E,Ex	1.98×10^4					166	33.6	R,s	0.0843				
58	0.28	He,E,C	2.52×10^3	49 In 113	4.3	p,s,r	0.0079	167	22.95	R	0.0576				
				115	95.7	R,S	0.176	168	26.8	R,s	0.0672				
27 Co 59	100	E,C	2250					170	14.9	R	0.0374				
				50 Sn 112	0.973	P	0.0372								
28 Ni 58	68.27	E,Ex	3.37×10^4	114	0.659	P,s	0.0252	69 Tm 169	100	R,s	0.0378				
60	26.10	E	1.29×10^4	115	0.339	p,s,r	0.0129								
61	1.13	E,Ex,C	557	116	14.538	S,r	0.555	70 Yb 168	0.13	P	0.000322				
62	3.59	E,Ex,O	1770	117	7.672	R,S	0.293	170	3.05	S	0.00756				
64	0.91	Ex	449	118	24.217	S,r	0.925	171	14.3	R,s	0.0354				
				119	8.587	S,R	0.328	172	21.9	R,S	0.0543				
29 Cu 63	69.17	Ex,C	361	120	32.596	S,R	1.245	173	16.12	R,s	0.0400				
65	30.83	Ex	161	122	4.632	R	0.177	174	31.8	S,R	0.0788				
				124	5.787	R	0.221	176	12.7	R	0.0315				

*Assignments to nucleosynthetic processes are from Cameron (1982), Schramm (private communication, 1982), Walter et al. (1986), Woosley and Hoffman (1986, 1989), and Beer and Penzhorn (1987). Processes are listed in the order of importance, with minor processes (10–30% for r- and s-processes) shown in lower case. See above references for details.

U	=	cosmological nucleosynthesis
H	=	hydrogen burning
N	=	hot or explosive hydrogen burning
C	=	carbon burning
He	=	helium burning
O	=	oxygen burning
Ne	=	neon burning
Ex	=	explosive nucleosynthesis
E	=	nuclear statistical equilibrium
S	=	s-process
R	=	r-process
RA	=	r-process producing actinides
P	=	p-process
X	=	cosmic-ray spallation

†Italicized values refer to abundances 4.55×10^9 yr ago.

Figure 7.17 Abundance of the nucleides (atoms/10^6 Si) with the process of production in our Universe. Figure taken from Reference [2].

Appendix B: **Details of the Microscopic Model**

Here, we provide some technical details about the microscopic calculation used to determine the ^{8}Be and ^{12}C binding energies. This calculation is based on the description of the nucleon-nucleon interaction by the Minnesota (MN) force [105], adapted to low-mass systems.

The nuclear part of the interaction potential V_N between nucleons i and j is given by

$$V_{Nij}(r) = \left[V_R(r) + \frac{1}{2}(1 + P_{ij}^\sigma)V_t(r) + \frac{1}{2}(1 - P_{ij}^\sigma)V_s(r) \right] \left[\frac{1}{2}u + \frac{1}{2}(2 - u)P_{ij}^r \right],$$
(7.131)

where $r = |\mathbf{r}_i - \mathbf{r}_j|$ and P_{ij}^σ and P_{ij}^r are the spin and space exchange operators, respectively. The radial potentials $V_R(r), V_s(r), V_t(r)$ are expressed as Gaussians and have been optimized to reproduce various properties of the nucleon-nucleon system – such as the deuteron binding energy at $\delta_{NN} = 0$, or the low-energy phase shifts. They have been fit as [105]

$$V_R(r) = 200 \exp(-1.487r^2)$$
$$V_s(r) = -91.85 \exp(-0.465r^2)$$
$$V_t(r) = -178 \exp(-0.639r^2),$$
(7.132)

where energies are expressed in MeV and lengths in fm.

In Eq. (7.131), the exchange-admixture parameter u takes standard value $u = 1$ but can be slightly modified to reproduce important properties of the A-nucleon system (for example, the energy of a resonance). This does not affect the physical properties of the interaction. The MN force is an effective interaction, adapted to cluster models. It is not aimed at perfectly reproducing all nucleon-nucleon properties, as realistic forces used in ab initio models [84], where the cluster approximation is not employed. The potentials are expressed as Gaussian factors, well adapted to cluster models, where the nucleon orbitals are also Gaussians [115].

The wave functions (Eq. (7.50)) are written in the resonating group method (RGM), which clearly shows the factorisation of the system wave function in terms of individual cluster wave functions. In practice the radial wave functions are expanded over Gaussians, which provides the generator coordinate method (GCM), fully equivalent to the RGM [115] but better adapted to numerical calculations. Some details are given here for the simpler two cluster case. The radial function $g_2^{JM\pi}(\rho)$ is written as a sum over Gaussian functions centered at different values of the generator coordinate R_n. This allows us to write the ^{8}Be wave function (7.50) as

$$\Psi_{^8Be}^{JM\pi} = \sum_n f^{J\pi}(R_n)\Phi^{JM\pi}(R_n),$$
(7.133)

where $\Phi^{JM\pi}(R_n)$ is a projected Slater determinant.

This development corresponds to a standard expansion on a variational basis. The binding energies $E^{J\pi}$ of the system are obtained by diagonalisation of

$$\sum_n \left[H^{J\pi}(R_n, R_{n'}) - E^{J\pi} N^{J\pi}(R_n, R_{n'}) \right] f^{J\pi}(R_n) = 0, \tag{7.134}$$

where the overlap and Hamiltonian kernels are defined as

$$N^{J\pi}(R_n, R_{n'}) = \langle \Phi^{J\pi}(R_n) | \Phi^{J\pi}(R_{n'}) \rangle,$$
$$H^{J\pi}(R_n, R_{n'}) = \langle \Phi^{J\pi}(R_n) | H | \Phi^{J\pi}(R_{n'}) \rangle. \tag{7.135}$$

The Hamiltonian H is given by Eq. (7.47). Standard techniques exist for the evaluation of these many-body matrix element. The choice of the nucleon-nucleon interaction directly affects the calculation of the hamiltonian kernel and, therefore, of the eigen-energy $E^{J\pi}$.

For three-body wave functions, the theoretical developments are identical, but the presentation is more complicated due to the presence of two relative coordinates (ρ, R). The problem is addressed by using the hyperspherical formalism [74].

Appendix C: **Numerical Integration of the Cross Sections**

To take into account the (energy-dependent) finite widths of the two resonances involved in this two-step process, one has to perform numerical integrations as was done in NACRE following References [76, 87]. Here, the condition of thermal equilibrium is relaxed, but it is assumed that the timescale for alpha capture on ^{8}Be is negligible compared to its lifetime against alpha decay. The rate is calculated as in NACRE for the resonance of interest:

$$N_A^2 \langle \sigma v \rangle^{\alpha\alpha\alpha} = 3N_A \left(\frac{8\pi\hbar}{\mu_{\alpha\alpha}^2} \right) \left(\frac{\mu_{\alpha\alpha}}{2\pi k_B T} \right)^{3/2} \int_0^\infty \frac{\sigma_{\alpha\alpha}(E)}{\Gamma_\alpha(E)} \exp(-E/k_B T) N_A \langle \sigma v \rangle^{\alpha^8 Be} E \, dE,$$

where $\mu_{\alpha\alpha}$ is the reduced mass of the $\alpha + \alpha$ system, and E is the energy with respect to the $\alpha + \alpha$ threshold. The elastic cross section of $\alpha + \alpha$ scattering is given by a Breit-Wigner expression:

$$\sigma_{\alpha\alpha}(E) = \frac{\pi}{k^2} \omega \frac{\Gamma_\alpha^2(E)}{(E - E_R))^2 + \Gamma_\alpha^2(E)/4}, \tag{7.136}$$

where k is the wave number, $E_R \equiv E_R(^8 Be)$, $\Gamma_\alpha \equiv \Gamma_\alpha(^8 Be)$, and ω is a statistical factor (here, equal to 2 to account for identical particles with spin zero).

The $N_A \langle \sigma v \rangle^{\alpha^8 Be}$ rate assumes that ^{8}Be has been formed at an energy E different from $E_{^8 Be}$ [76]. This rate is given by

$$N_A \langle \sigma v \rangle^{\alpha^8 Be} = N_A \frac{8\pi}{\mu_{\alpha^8 Be}^2} \left(\frac{\mu_{\alpha^8 Be}}{2\pi k_B T} \right)^{3/2} \int_0^\infty \sigma_{\alpha^8 Be}(E'; E) \exp(-E'/k_B T) E' \, dE',$$

$$\tag{7.137}$$

where $\mu_{\alpha^8\text{Be}}$ is the reduced mass of the $\alpha + {}^8\text{Be}$ system, and E' is the energy with respect to its threshold (which varies with the formation energy E). As in References [76, 87], we parametrise $\sigma_{\alpha^8\text{Be}}(E'; E)$ as

$$\sigma_{\alpha^8\text{Be}}(E'; E) = \frac{\pi\hbar^2}{2\mu_{\alpha^8\text{Be}}E'} \frac{\Gamma_\alpha(E')\Gamma_\gamma(E' + E)}{[E' - E_R({}^{12}\text{C}) + E - E_R({}^8\text{Be})]^2 + \frac{1}{4}\Gamma(E'; E)^2}, \quad (7.138)$$

where the partial widths are those of the Hoyle state and, in particular, $\Gamma = \Gamma_\alpha({}^{12}\text{C}) + \Gamma_\gamma({}^{12}\text{C})$. The various integrals are calculated numerically. The experimental widths at resonance energy can be found in Table 7.5.

However, one must include (1) the energy dependence of those widths, away from the resonance energy and (2) the variation of the widths at the resonant energy when this energy changes due to a change in the nuclear interaction.

The energy dependence of the particle widths $\Gamma_\alpha(E)$ is given by

$$\Gamma_\alpha(E) = \Gamma_\alpha(E_R) \frac{P_\ell(E, R_c)}{P_\ell(E_R, R_c)}, \quad (7.139)$$

where P_ℓ is the penetration factor associated with the relative angular momentum, ℓ (0 here), and the channel radius, R_c.[9] The penetration factor is related to the Coulomb functions by

$$P_\ell(E, R) = \frac{\rho}{F_\ell^2(\eta, \rho) + G_\ell^2(\eta, \rho)}, \quad (7.140)$$

where $\rho = kR$ and

$$\eta = \frac{Z_1 Z_2 \alpha}{v/c} \quad (7.141)$$

is the Sommerfeld parameter.

For radiative capture reactions, the energy dependence of the gamma width $\Gamma_\gamma(E)$ is given by

$$\Gamma_\gamma(E) \propto \alpha E^{2\lambda+1}, \quad (7.142)$$

where λ is the multipolarity (here, 2 for $E2$) of the electromagnetic transition.

The relevant widths as a function of δ_{NN} are given in Figure 7.18. They are directly linked to the resulting change of $E_R({}^8\text{Be})$ and $E_R({}^{12}\text{C})$.

The radiative width, $\Gamma_\gamma({}^{12}\text{C})$, with its E^5 energy dependence, shows little evolution. (The energy of the final state at 4.44 MeV is assumed to be constant). In contrast, the ^8Be alpha width undergoes large variations due to the effect of Coulomb barrier penetrability. Note that compared to these variations, those induced by a change of α in the Coulomb barrier penetrability (Eqs. (7.141) and (7.140)) and Γ_γ are considerably smaller.

---

[9] We choose $R_c = 1.3\,(A_1^{1/3} + A_2^{1/3})$ fm, for nuclei A_1 and A_2.

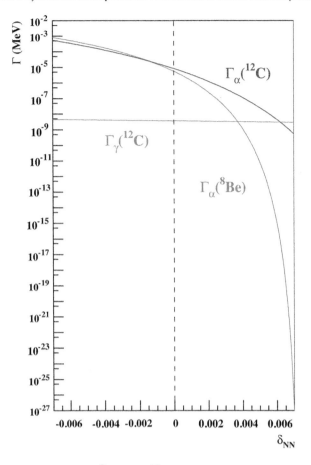

Figure 7.18 The partial widths of ^{8}Be and ^{12}C as a function of δ_{NN}. Figure taken from Reference [44].

Numerical integration is necessary at low temperatures as the reaction takes place through the low-energy wing of resonances. It takes even more relative importance at a given temperature when the resonance energy is shifted upwards. On the other hand, when δ_{NN} increases, the resonance energies decrease, and the $\Gamma_\alpha(^8Be)$ becomes so small that the numerical integration becomes useless and soon gives erroneous results because of the finite numerical resolution. For this reason, when $\Gamma_\alpha(^8Be) < 10^{-8}$ MeV, we use instead the Saha equation for the first step and the sharp resonance approximation for the second step – i.e., Eq. (7.126) when $\Gamma_\alpha(^{12}C) < 10^{-8}$ MeV. (Note that for high values of δ_{NN}, the condition $\Gamma_\gamma \ll \Gamma_\alpha$ does not hold anymore and $\gamma \neq \Gamma_\gamma$.)

At temperatures in excess of $T_9 \simeq 2$, one must include the contribution of the higher ^{12}C levels like the one observed by [59]. As this is not of importance for this study; we just

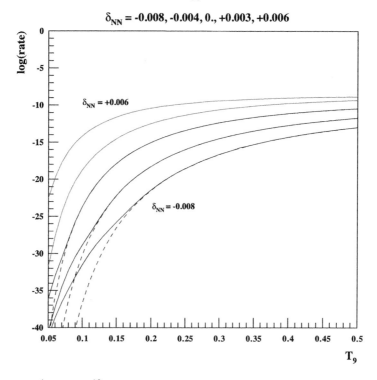

Figure 7.19 The ^{4}He$(\alpha\alpha, \gamma)^{12}$C reaction rate as a function of temperature for different values of δ_{NN}. Solid (dashed) lines represent the result of the numerical calculation (analytical approximation) with $\delta_{NN} = 0$ (centre), $\delta_{NN} > 0$ (above) and $\delta_{NN} < 0$ (below). (For the negative value of δ_{NN}, the larger difference is caused by the failure of the numerical integration and the analytical solution is preferred (see text). Figure taken from Reference [44].

added the contribution given by the last terms in the NACRE analytical approximation and neglected any induced variation.

Figure 7.19 shows the numerically integrated ^{4}He$(\alpha\alpha, \gamma)^{12}$C reaction rates for different values of δ_{NN} compared with the analytical approximation Eq. (7.126). The difference is important at low temperatures and small δ_{NN} values but becomes negligible for $\delta_{NN} \gtrsim 0$. At the highest values of δ_{NN}, we consider the numerical calculation uses the Saha equation for the first step, but the total widths of the ^{12}C level also becomes too small to be accurately numerically calculated; we use Eq. (7.126) instead.

The ^{8}Be lifetime with respect to alpha decay, $(h/\Gamma_\alpha(^8\text{Be}))$, exhibits the opposite behaviour, indicating that, for large values of δ_{NN}, it becomes stable. Before that, its lifetime is so long that the ^{4}He$(\alpha\alpha, \gamma)^{12}$C reaction should be considered as a real two-step process with ^{8}Be included in the network, as the assumption that alpha decay is much faster than alpha capture may not hold anymore. Fortunately, our network calculations show that this situation is encountered only for $\delta_{NN} \gtrsim 0.006$ for the temperatures and densities considered in our stellar evolution studies.

Appendix D: **Sensitivity of the Nuclear Reaction Rates on the Resonance Energy**

The contribution of a sharp resonance in the (α, γ) channel is given by

$$\langle \sigma v \rangle \propto \frac{\Gamma_\alpha(E_R)\Gamma_\gamma(E_R)}{\Gamma(E_R)} \exp\left(-\frac{E_R}{k_B T}\right), \tag{7.143}$$

where Γ_α is the entrance (alpha-)width, Γ_γ is the exit (gamma-)width, and Γ is the total width ($\Gamma = \Gamma_\alpha + \Gamma_\gamma$ if these are the only open channels). Figure 1 in Reference [44] displays the ^{12}C level scheme: the radiative width, Γ_γ, is associated with the decay to the first ^{12}C excited state at 4.44 MeV as the decay to the ground state proceeds only through the much less efficient electron-positron pair emission. The corresponding decay energy is then $E_\gamma(E_R) = 3.21$ MeV $+ \Delta E_R(^{12}$C$)$. Equation (7.143) shows that for a fixed E_R – i.e., δ_{NN} – the contribution increases with temperature, as seen in Figures 7.9. While the radiative width $\Gamma_\gamma(E_R) \propto E_\gamma^{2\ell+1}$ is almost insensitive to $E_R(\delta_{NN})$, $\Gamma_\alpha(E_R)$ is very sensitive to $E_R(\delta_{NN})$ variations because of the Coulomb and centrifugal barriers penetrability, $P_\ell(E)$. The reduced widths γ_x^2, defined by

$$\Gamma_x(E) = 2\gamma_x^2 P_\ell(E) \quad (x \neq \gamma), \tag{7.144}$$

are corrected for these effects so that they reflect the nuclear properties only and are, as a good approximation, independent of δ_{NN}.

Depending on whether $\Gamma_\alpha \ll \Gamma_\gamma$ or $\Gamma_\gamma \ll \Gamma_\alpha$, the sensitivity of $\langle \sigma v \rangle$ (Eq. (7.143)) to E_R or δ_{NN} variations is very different. This is due to the very different energy dependence of Γ_α and Γ_γ, as discussed in detail in Reference [85]. In the latter case, using Eq. (7.143), the sensitivity of the rate to E_R (δ_{NN}) variations is simply given by

$$\frac{\partial \ln\langle \sigma v \rangle}{\partial \ln E_R} = -\frac{E_R}{k_B T} \tag{7.145}$$

as the prefactor in Eq. (7.143) is reduced to Γ_γ, which is almost constant. Since δE_R and δ_{NN} have opposite signs (Eqs. (7.66) and (7.65)), the rate *increases* with δ_{NN}. In the former case, the same factor is reduced to the very energy-dependent Γ_α, and we have

$$\langle \sigma v \rangle \propto \gamma_\alpha^2 \exp\left(-\sqrt{\frac{E_G}{E_R}} - \frac{E_R}{k_B T}\right), \tag{7.146}$$

where the penetrability, $P_\ell(E)$, has been approximated by $\exp(-\sqrt{E_G/E})$ with Gamow energy, E_G. It is well known that the exponential in Eq. (7.146) can be well approximated (see, e.g., Reference [3]) by

$$\exp\left[-\left(\frac{E_R - E_0}{\Delta E_0/2}\right)^2\right], \tag{7.147}$$

with

$$E_0 = \left(\frac{\mu}{2}\right)^{1/3}\left(\frac{\pi e^2 Z_1 Z_2 kT}{\hbar}\right)^{2/3} = 0.1220 \, (Z_1^2 Z_2^2 A)^{1/3} \, T_9^{2/3} \text{ MeV} \tag{7.148}$$

and

$$\Delta E_0 = 4 \, (E_0 kT/3)^{1/2} = 0.2368 \, (Z_1^2 Z_2^2 A)^{1/6} \, T_9^{5/6} \text{ MeV}, \qquad (7.149)$$

which define the Gamow window. Recalling that the reduced width γ^2 only reflects the nuclear structure and is assumed to be constant, it is straightforward to calculate the sensitivity of the rate to $E_R \, (\delta_{NN})$ variations:

$$\frac{\partial \ln \langle \sigma v \rangle}{\partial \ln E_R} = 4 \left(\frac{E_0(T) - E_R(\delta_{NN})}{\Delta E_0(T)/2} \right). \qquad (7.150)$$

Since, for large δ_{NN} and T, we have $E_0 > E_R$, the rate *decreases* with δ_{NN}.

The condition $\Gamma_\alpha = \Gamma_\gamma$ that marks the boundary between these two opposite evolutions in Eq. (7.143) can be found in figure A.1 of Ekström et al. [44], at $\delta_{NN} \approx 0.006$.

8

Fine-Tunings at Particle Scales

GIULIA ZANDERIGHI

Abstract

Particle physics, like most fundamental sciences, has a number of open questions that are generally perceived as being of utmost importance. Accordingly, trying to address those questions motivates most of the experimental and theoretical activity in the field. In particle physics, these questions can be classified into two main categories. First, there are questions that arise from experimental observations – for instance, what are the constituents of dark matter, or what is the origin of the matter-antimatter asymmetry seen in our Universe? These questions have a well-defined, unique answer which must be understood before a complete understanding of the fundamental theory that governs all interactions can be claimed. Additionally, there are questions that are more related to aesthetic arguments and to our belief that the fundamental theory should obey certain criteria of symmetry and simplicity that we often refer to as the beauty or elegance of a theory. The most common queries in the second category include, for example, why are there three generations of quarks and leptons in the Standard Model (SM) of particle physics, why is the mass spectrum of fermions hierarchical, what is the origin of neutrino masses, what resolves the strong charge-parity (CP) problem, and why are the electroweak (EW) scale and the Planck scale so widely separated? These questions, some which will be discussed in detail in the following, are sometimes considered even more fundamental and interesting than those of the first type. For instance, discovering that dark matter is an axion particle would be a great achievement but would not fundamentally advance the understanding of our ultimate theory. On the other hand, it is not clear whether all questions of the second type have a well-defined answer.

8.1 Introduction

Particle physics, like most fundamental sciences, has a number of open questions that are generally perceived as being of utmost importance. Accordingly, trying to address those questions motivates most of the experimental and theoretical activity in the field. In particle physics, these questions can be classified into two main categories. First, there are questions that arise from experimental observations – for instance, what are the constituents of dark matter, or what is the origin of the matter-antimatter asymmetry seen in our Universe? These questions have a well-defined, unique answer which must be understood before a complete understanding of the fundamental theory that governs all interactions can be claimed.

Additionally, there are questions that are more related to aesthetic arguments and to our belief that the fundamental theory should obey certain criteria of symmetry and simplicity that we often refer to as the beauty or elegance of a theory. The most common queries in the second category include, for example, why are there three generations of quarks and leptons in the Standard Model (SM) of particle physics, why is the mass spectrum of fermions hierarchical, what is the origin of neutrino masses, what resolves the strong charge-parity (CP) problem, or why are the electroweak (EW) scale and the Planck scale so widely separated? These questions, some which will be discussed in detail in the following, are sometimes considered even more fundamental and interesting than those of the first type. For instance, discovering that dark matter is an axion particle would be a great achievement but would not fundamentally advance the understanding of our ultimate theory. On the other hand, it is not clear whether all questions of the second type have a well-defined answer. Hence, while a resolution is not required, if it were to happen, it would most likely lead to the discovery of new particles not present in the SM, to new symmetries of nature, and, ultimately, to the unveiling of new fundamental principles. Prominent questions of the second type address, in particular, the fine-tuning or naturalness problem in the SM that will be discussed in this chapter. In particle physics, we generally say that a parameter is fine-tuned if its observed value is much smaller than the size of corrections that the parameter receives when higher-order quantum effects are accounted for. In particular, fine-tuned parameters typically involve a cancellation between two large, apparently unrelated terms.

In the past decades, a number of situations have been encountered in particle physics when a parameter of the theory seemed fine-tuned. The apparent fine-tuning problem could be resolved by realising that the description used was only suitable at low energies and that extending the theory by including new particles and interactions could remove the alleged fine-tuning problem. A number of notable examples of apparent fine-tuning – namely the energy field of the electron, the charged and neutral pion mass difference, and the kaon transition rates – are presented in Section 8.2. Since, in the past, requiring that a theory was not fine-tuned led to formulating the correct extension of the theory, understanding how to remove the fine-tuning problems in the SM has been used as a guiding principle to construct theories of new physics, beyond the SM (BSM) over the past 40 years.

While we have some intuition about whether a parameter is fine-tuned, in order to use fine-tuning as a solid guiding principle, it is important to quantify how fine-tuned a parameter or a theory is. This is addressed in Section 8.3. The technical naturalness criterion, as formulated by 't Hooft in the 1980s [54] and discussed in more detail in Section 8.3.1, states that a parameter is allowed to be very small if setting it to zero increases the symmetry of the theory. If, however, no new symmetry is recovered and a parameter is unnaturally small, then the theory is considered to be unnatural. If a theory is found to be unnatural, it is useful to quantify the degree of fine-tuning. A number of measures that have been suggested in the past and are used today in particle physics are described in Section 8.3.2.

In contemporary particle physics, three areas face a severe fine-tuning problem: the Higgs mass related to the electroweak symmetry breaking, the strong CP-violating angle θ, and the cosmological constant Λ. The last problem is discussed extensively in Chapter 2 (by Peacock); hence, in this chapter, we will concentrate on the other two problems.

The first problem, discussed in more detail in Section 8.4, has to do with the fact that the Higgs mass takes a value that is several orders of magnitude smaller than the size of quantum corrections that it would receive in extensions of the SM. Hence, if the SM is just a low-energy effective theory, then the Higgs mass suffers from a fine-tuning problem. Section 8.4 also presents a number of possible solutions to the fine-tuning problem in the EW sector in terms of BSM theories.

Section 8.5 describes the fine-tuning of the strong CP angle – i.e., the fact that an angular parameter, usually denoted as θ, that enters the QCD Lagrangian, happens to be very small. The fine-tuning of the CP angle is quite different from the fine-tuning in the EW sector: while in the latter case the Higgs mass receives very large quantum corrections of the order of the ultraviolet (UV) cut-off of the theory, this is not the case for the strong CP angle. The problem is rather that θ is an angle that can take any value between 0 and 2π in the SM. Experimental measurements constrain θ to be very close to zero, but there is no explanation of why this angle should vanish or be so small. Section 8.5 also describes the most popular solution to the problem in terms of a new symmetry of the Lagrangian, the Peccei-Quinn symmetry. The latter leads to the existence of a new particle, the so-called QCD axion.

Finally, in Section 8.6, we will discuss anthropic arguments as possible approaches to fine-tuning problems. Anthropic arguments in their original and weaker formulation, as applied first to the cosmological constant, rely on the observation that, if the cosmological constant had a value that is much different from the one we observe, galaxies would not have formed. These arguments can be made much stronger by requiring that our life as human beings on the Earth should be possible and, hence, that light and heavy elements should have formed in the right abundance for our existence to be possible. When extended to requiring such conditions, anthropic arguments can be used to explain the fine-tuning in the EW sector, but these approaches remain controversial.

8.2 Historical Examples

It is useful to first discuss a few examples where fine-tuning and naturalness arguments provided an indication about the presence of new states and interactions. We will see that these considerations can be, and in some cases even have been, used to predict the energy scale at which new degrees of freedom are expected to appear.

8.2.1 The Energy of the Electron Field

Let us begin by considering the fine-tuning problem in the case of the energy of a static, electric field generated by an electron of negative charge e. This problem was presented first in Reference [42]. For simplicity, we will work in natural units, in which $c = \hbar = \epsilon_0 = 1$ and the inverse of a length is simply an energy. The Coulomb field $\vec{E}$ generated by a point charge e points in the radial direction and is inversely proportional to the radial distance r from the charge

$$\vec{E} = \frac{e}{4\pi r^2} \frac{\vec{r}}{r}. \tag{8.1}$$

The total self-energy W of the electric field is given by the volume integral

$$W = \frac{1}{2} \int_V d^3 r \, \vec{E}^2 \tag{8.2}$$

and is divergent at $r = 0$. If one introduces a radial cut-off $r > \Lambda^{-1}$ on the integration, one obtains

$$W_\Lambda = \frac{1}{2} \int_{r > \Lambda^{-1}} d^3 r \, \vec{E}^2 = \frac{1}{2} \alpha \Lambda, \tag{8.3}$$

where $\alpha = e^2/(4\pi) \simeq 1/137$ is the fine-structure constant. Using Einstein's equivalence between energy and mass, we add this energy to the intrinsic mass M_e of the electron to obtain the total effective, observable electron mass m_e. Accordingly, we write

$$m_e = M_e + \frac{1}{2} \alpha \Lambda. \tag{8.4}$$

The electron mass m_e has been measured very precisely via Thompson and Millikan experiments and is about 511 keV. There are also experimental limits on the charge radius of the electron, which are about 10^{-4} fm [43]. It would seem reasonable to use this bound on the radius as a cut-off Λ^{-1} on the self-energy integration. In natural units, one has 1 fm $=$ $(197\,\text{MeV})^{-1}$, so one would obtain a mass term $\frac{1}{2}\alpha\Lambda = \mathcal{O}(10^7 \text{ keV})$. This would mean that the non-electrostatic mass M_e in the right-hand side of Eq. (8.4), that seems completely unrelated to the electrostatic energy term would need to cancel it to five significant digits in order to give the experimentally observed electron mass of $m_e \simeq 511$ keV. The fact that the physical small electron mass depends on the cancellation between two very large and apparently unrelated terms is the fine-tuning problem in the case of the electron mass.

What solves the fine-tuning problem of the electron mass is quantum mechanics. In fact, the preceding is just a classical picture, probing scales of the order of 10^{-4} fm. Quantum effects allow the production of electron-positron pairs that violate energy-momentum conservation but only for short amounts of time, according to the Heisenberg uncertainty principle $\Delta E \Delta t \geq 1/2$ (in units where $\hbar = 1$). The effect of these vacuum fluctuations is to effectively reduce the electron charge, as illustrated in Figure 8.1. As one probes the electron charge to smaller and smaller distances, the electron charge is screened by the

Figure 8.1 A schematic representation of the screening of the electric field generated by a charge e provided by quantum mechanical fluctuations of the vacuum that generate pairs of positrons (e^+) and electrons (e^-).

vacuum fluctuations and seems smaller than expected. Technically, the screening effect provided by the vacuum fluctuations leads to the renormalisation of the electric charge.

An explicit calculation [61] shows that, once one consistently includes the positron contribution to the electromagnetic energy term in the electron mass, the energy does not grow as $1/r = \Lambda$ but depends just logarithmically on Λ. Through a modern field theory calculation, one easily finds that

$$m_e = M_e \left(1 - \frac{3\alpha}{2\pi} \ln \left(\frac{M_e}{\Lambda} \right) \right). \tag{8.5}$$

This contribution remains smaller than the electron mass even for scales as large as the Planck scale $M_{\text{Planck}} = 10^{19}\,\text{GeV}$ – i.e., the scale at which quantum gravity effects cannot be neglected.

The other important point to note is that the correction to the electron mass is proportional to the intrinsic electron mass, M_e, itself. The reason for this has to do with an additional symmetry that is recovered in the massless limit. In fact, consider the QED Lagrangian that governs the interaction of electrons with photons

$$\mathcal{L} = \bar{\psi} \left(i\gamma^\mu (\partial_\mu - ieA_\mu) - M_e \right) \psi, \tag{8.6}$$

where γ^μ are the standard 4×4 Dirac matrices that obey the commutation rules $\{\gamma^\mu, \gamma^\nu\} = 2g^{\mu\nu}$, A_μ is the photon field, ψ denotes the electron field, and $\bar{\psi} = \psi^\dagger \gamma^0$. If one sets $M_e = 0$, the Lagrangian becomes symmetric under a global chiral transformation

$$\psi \rightarrow e^{i\alpha\gamma^5}\psi, \tag{8.7}$$

where $\gamma^5 = i\gamma^0\gamma^1\gamma^2\gamma^3$ is the fifth gamma matrix, which anti-commutes with the four Dirac matrices γ^μ (i.e., $\{\gamma^5, \gamma^\mu\} = 0$). This transformation implies that

$$\bar{\psi} \rightarrow \left(e^{i\alpha\gamma^5}\psi\right)^\dagger \gamma^0 = \bar{\psi}e^{i\alpha\gamma^5}. \tag{8.8}$$

As a consequence, one obtains the following transformation properties:

$$\bar{\psi}\gamma^\mu\psi \rightarrow \bar{\psi}\gamma^\mu\psi \qquad M_e\bar{\psi}\psi \rightarrow e^{2i\alpha\gamma^5}M_e\bar{\psi}\psi. \tag{8.9}$$

Therefore, when $M_e = 0$, the Lagrangian remains invariant under chiral transformations. In this limit, higher-order corrections to the mass must respect the symmetry. This implies, in particular, that all corrections to the electron mass must vanish as the intrinsic mass M_e tends to zero. This is obviously the case if the correction is proportional to the mass M_e.

In conclusion, we have seen in this section that the presence of a new particle, the positron, provides a natural way to screen the electric charge and reduce the self-energy contribution. This is the first example where a quantity is only apparently fine-tuned, but the inclusion of a new state removes the fine-tuning problem. Similarly, we observe today other fine-tuned quantities in the SM and hope to use the requirement that a more complete description should be natural and not fine-tuned, as a guidance in finding new states, new interactions, and new symmetries.

8.2.2 The Charged and Neutral Pion Mass Difference

Another example of a fine-tuning problem in particle physics is provided by the mass differ ence between charged ($\pi^\pm$) and neutral (π^0) pions. This difference is due to the tadpole and self-energy diagrams where a photon is exchanged, depicted in Figure 8.2. These diagrams are present only for charged but not for neutral pions and are hence responsible for the mass difference between charged and neutral pions.

Let us try to estimate the mass difference $\Delta m^2 = m_{\pi^\pm}^2 - m_{\pi^0}^2$ that is induced by the photonic corrections to the $\pi^\pm$ propagator. The Lagrangian describing the interactions of photons and charged pions is given by

$$\mathcal{L} \supset ieA_\mu\left(\pi^+\partial^\mu\pi^- - \pi^-\partial^\mu\pi^+\right) + e^2 A_\mu A^\mu \pi^+\pi^-, \tag{8.10}$$

Figure 8.2 Photon tadpole and self-energy corrections to the $\pi^\pm$ propagator.

where A_μ denotes the photon field. The tadpole and self-energy diagram both give rise to a shift of the charged pion mass squared. From Eq. (8.10), one obtains

$$\left(\Delta m^2_{\pi^\pm}\right)_{\text{tadpole}} = -i \frac{\alpha}{4\pi^3} 4 \int d^4l \frac{1}{l^2},$$

$$\left(\Delta m^2_{\pi^\pm}\right)_{\text{self-energy}} = i \frac{\alpha}{4\pi^3} \int d^4l \frac{(l+p)^2}{(l^2 - m^2_{\pi^\pm})(l-p)^2},$$

(8.11)

where p denotes the external momentum that flows through the self-energy diagram. Notice that in the limit $l \to \infty$, both contributions involve a quadratically divergent integral. The quadratic singularities in Eq. (8.11) can be extracted by performing a Wick rotation from the Minkowskian momentum l to the Euclidean momentum l_E and introducing a UV cut-off Λ. In both cases, one has to evaluate the integral

$$\int d^4l \frac{1}{l^2} = i \int d\Omega_4 \int_0^\Lambda dl_E\, l_E = i\pi^2\Lambda^2.$$

(8.12)

Here, the factor $\int d\Omega_4$ is the surface area of a four-dimensional unit sphere, which is equal to $2\pi^2$. Combining the two contributions in Eq. (8.11) then leads to the following estimate:

$$\Delta m^2 = m^2_{\pi^\pm} - m^2_{\pi^0} \simeq \frac{3\alpha}{4\pi} \Lambda^2.$$

(8.13)

As in the case of the self-energy of the electron field discussed earlier in Section 8.2.1, one observes that the charged and neutral pion mass difference also depends sensitively on the UV cut-off scale Λ.

The fine-tuning problem related to the UV sensitivity of the charged and neutral pion mass difference has been first pointed out and addressed in [20]. In fact, by comparing the experimental determination of the mass difference [43]

$$\Delta m^2_{\text{exp}} \simeq (35.51\,\text{MeV})^2,$$

(8.14)

to Eq. (8.13), one infers that at a scale of around

$$\Lambda \simeq 850\,\text{MeV},$$

(8.15)

a new particle should arise that softens the UV behaviour of the photonic corrections to the $\pi^\pm$ propagator. An obvious candidate for this is the ρ meson which is a vector state with a mass of $m_\rho \simeq 770\,\text{MeV}$. Indeed, it turns out that this state, together with the lightest axial-vector resonance a_1 with mass $m_{a_1} \simeq 1250\,\text{MeV}$, cancels the quadratic sensitivity of the mass correction to the UV cut-off Λ. An explicit calculation of the one-loop diagrams shown in Figure 8.3 leads to [20]

$$\Delta m^2 \simeq \frac{3\alpha}{4\pi} \frac{m^2_\rho m^2_{a_1}}{m^2_{a_1} - m^2_\rho} \ln\left(\frac{m^2_{a_1}}{m^2_\rho}\right) \simeq (40\,\text{MeV})^2.$$

(8.16)

We see that, when substituting the values of the ρ and a_1 meson masses, one obtains a mass difference in reasonable agreement with the experimental measurement of Eq. (8.14).

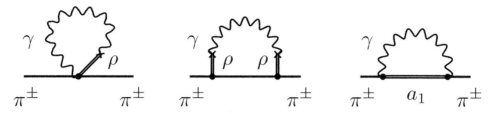

Figure 8.3 Photonic corrections to the $\pi^\pm$ propagator involving also the exchange of ρ and a_1 mesons.

Finally, notice that, in the limit $m_\rho \to m_{a_1}$, the result in Eq. (8.16) turns into $\Delta m^2 \simeq 3\alpha/(4\pi)\,m_\rho^2$, which shows clearly that the mass of the lowest-lying vector resonance acts as the UV cut-off Λ appearing in Eq. (8.13).

8.2.3 Kaon Transition Rates and Mixing

The Fermi model was proposed in 1933 to explain the β decay of the neutron $n \to p e^- \bar{\nu}_e$ [25]. According to the Fermi model, four fermions can interact directly with each other through a pointlike vertex, the interaction is universal, and its strength is proportional to the Fermi constant G_F.

Despite the large success of the Fermi model, more measurements of decay and transition rates performed in the 1960s started to reveal a number of puzzles. For instance, the measurement of the decay $\mu^- \to \nu_\mu e^- \bar{\nu}_e$ was giving a value of the Fermi constant of $G_F^\mu = (1.16632 \pm 0.00002) \cdot 10^{-5}\,\mathrm{GeV}^{-2}$, slightly larger than the one observed in β decay, namely $G_F^\beta = (1.136 \pm 0.003) \cdot 10^{-5}\,\mathrm{GeV}^{-2}$. In addition, certain hadronic decay modes were observed to be very suppressed; for instance, a universal weak interaction would predict the $\pi^- \to \mu^- \bar{\nu}_\mu$ decay rate to be equal to the $K^- \to \mu^- \bar{\nu}_\mu$ rate. However, the latter one was observed to be about 20 times smaller. This lead Cabibbo [18] in 1963 to formulate the hypothesis that weak eigenstates are different from mass eigenstates – i.e., that the one up-type quark known at the time would couple to a linear combination d' of the two down-type quarks d and s, namely

$$d' = \cos\theta_c\, d + \sin\theta_c\, s, \tag{8.17}$$

where $\sin\theta_c$ is the sine of the Cabibbo angle $\sin\theta_c \simeq 0.23$. Since the π^- meson is made up of $(\bar{u}d)$ while the K^- meson is made up to $(\bar{u}s)$, the transition rate of the latter is suppressed by a factor $\cos^2\theta_c/\sin^2\theta_c \simeq 18$, quite close to the experimental observation. Similarly, the discrepancy in the extraction of G_F from $\mu^- \to \nu_\mu e^- \bar{\nu}_e$ and $n \to p e^- \bar{\nu}_e$ could be explained by the introduction of a cosine of the Cabibbo angle in the latter decay.

Adopting the Cabibbo model, a number of new measurements were performed that still gave puzzling results. For instance, the decay of neutral K mesons to muons, $K_L \to \mu^+\mu^-$, was measured to be much smaller than predicted in the Cabibbo model. Similarly,

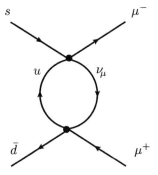

Figure 8.4 A diagram describing the decay $K_L \rightarrow \mu^+\mu^-$ in the Fermi theory. The four-fermion vertices of the Fermi theory are indicated by the black dots.

the measured K^0–$\bar{K}^0$ oscillations were considerably faster than theoretical predictions relying on the Cabibbo model. This lead Glashow, Illiopoulos, and Maiani (GIM) [30] to postulate the existence of an extra quark, the charm quark c, before its direct discovery in 1974 at SLAC [10] and Brookhaven National Laboratory [9].

The way the introduction of the charm quark solves the aforementioned discrepancies works as follows. Let us first consider the decay $K_L \rightarrow \mu^+\mu^-$ in the Fermi theory with an up quark and a muon neutrino running in the loop. The corresponding diagram is shown in Figure 8.4, and the associated decay amplitude takes the form

$$M_u = -i \left(\frac{4G_F}{\sqrt{2}}\right)^2 g_u \int \frac{d^4l}{(2\pi)^4} \frac{(\bar{s}\gamma_\mu P_L (\slashed{l} + m_u) \gamma_\nu P_L d)(\bar{\mu}\gamma^\nu P_L \slashed{l} \gamma^\mu P_L \mu)}{l^2(l^2 - m_u^2)}, \quad (8.18)$$

where

$$g_u = \sin\theta_c \cos\theta_c, \quad (8.19)$$

and $P_L = (1 - \gamma^5)/2$ projects on left-handed chiralities. In the limit $|l| \rightarrow \infty$, one can neglect the mass of the up quark in the loop, and one obtains

$$\begin{aligned} M_u &= -i \left(\frac{4G_F}{\sqrt{2}}\right)^2 g_u \int \frac{d^4l}{(2\pi)^4} \frac{l^\alpha l^\beta}{(l^2)^2} (\bar{s}\gamma_\mu \gamma^\alpha \gamma_\nu P_L d)(\bar{\mu}\gamma^\nu \gamma^\beta \gamma^\mu P_L \mu) \\ &= -i \left(\frac{4G_F}{\sqrt{2}}\right)^2 g_u \int \frac{d^4l}{(2\pi)^4} \frac{1}{4l^2} (\bar{s}\gamma_\mu \gamma^\alpha \gamma_\nu P_L d)(\bar{\mu}\gamma^\nu \gamma^\beta \gamma^\mu P_L \mu) \quad (8.20) \\ &= -i \left(\frac{4G_F}{\sqrt{2}}\right)^2 g_u \frac{i\Lambda^2}{64\pi^2} (\bar{s}\gamma_\mu \gamma_\lambda \gamma_\nu P_L d)(\bar{\mu}\gamma^\nu \gamma^\lambda \gamma^\mu P_L \mu), \end{aligned}$$

where we have used again the result in Eq. (8.12) to extract the part quadratic in Λ. Now employing

$$(\bar{s}\gamma_\mu \gamma_\lambda \gamma_\nu P_L d)(\bar{\mu}\gamma^\nu \gamma^\lambda \gamma^\mu P_L \mu) = 4 (\bar{s}\gamma_\mu P_L d)(\bar{\mu}\gamma^\mu P_L \mu), \quad (8.21)$$

one finds

$$M_u = \frac{G_F^2}{2\pi^2} g_u \Lambda^2 \left(\bar{s} \gamma_\mu P_L d \right) \left(\bar{\mu} \gamma^\mu P_L \mu \right), \tag{8.22}$$

which is – like the result in Eq. (8.13) – quadratically divergent.

However, in the diagram shown in Figure 8.4, both an up quark and a charm quark can circulate in the loop. In order to take the charm quark into account, one should replace g_u with $\sum_{q=u,c} g_q$ in Eq. (8.22). The form of the coupling g_c then follows from the observation that, in analogy to Eq. (8.17), the charm quark also couples to a linear combination of the two down-type quarks d and s, namely to

$$s' = \cos\theta_c \, s - \sin\theta_c \, d. \tag{8.23}$$

From this, it follows that

$$g_c = -\sin\theta_c \cos\theta_c, \tag{8.24}$$

and as a result, $\sum_{q=u,c} g_q = 0$. In physical terms, the latter result means that the quadratic divergence of the up-quark loop is exactly cancelled by the charm-quark contribution. This cancellation is called GIM mechanism. In the case of the up-quark and charm-quark contribution to the $K_L \rightarrow \mu^+\mu^-$ decay, the perfect GIM cancellation is broken by terms $(m_c^2 - m_u^2)/m_W^2 \simeq m_c^2/m_W^2 \ll 1$, which implies that the physical charm-quark mass acts as an effective UV cut-off.

Similarly, the presence of the charm quark can explain the observed suppression of K_0–$\bar{K}_0$ mixing. In fact, as illustrated in Figure 8.5, in this case, four diagrams contribute if we restrict ourselves to the exchange of up quarks and charm quarks. In the limit of $m_u = m_c = 0$, the individual contributions to the K_0–$\bar{K}_0$ amplitude with an exchange of a quark of flavour i and a quark of flavour j take the form

$$M_{ij} \propto G_F^2 g_i \, g_j, \tag{8.25}$$

where $i, j = u, c$. In the total amplitude $M = \sum_{i,j=u,c} M_{ij}$, hence, all Λ^2 terms cancel as a result of the GIM mechanism. Like in the case of $K_L \rightarrow \mu^+\mu^-$, finite-mass effects again spoil the complete GIM cancellation but the residual mixing effects are very small being both Cabibbo-Kobayashi-Maskawa (CKM) suppressed ($\propto \sin^2\theta_c$) and suppressed by the small charm-quark mass ($\propto m_c^2/m_W^2$). In the SM, the top quark also enters the loops that describe K_0–$\bar{K}_0$ mixing (see Figure 8.5); the discussed GIM cancellation, however, persists as a result of the unitarity of the CKM matrix [39].

Like for the mass difference between charged and neutral pions, one can also extract in the case of K_0–$\bar{K}_0$ mixing a bound on the effective cut-off scale $\Lambda \simeq m_c$. One just has to calculate the K_L–K_S mass difference $\Delta M_K = M_{K_L} - M_{K_S}$ and compare it to experiment. A calculation similar to the one leading to Eq. (8.22) gives

$$\frac{\Delta M_K}{M_K} \simeq \frac{G_F^2}{6\pi^2} f_K^2 \sin^2\theta_c \, m_c^2, \tag{8.26}$$

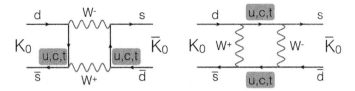

Figure 8.5 The one-loop box diagrams describing K_0–$\bar{K}_0$ mixing. To obtain the full amplitude the contributions of all internal up-type quarks have to be included.

where $M_K = (M_{K_L} + M_{K_S})/2$ and f_K denotes the kaon decay constant. Using the values $\Delta M_K / M_K \simeq 7 \cdot 10^{-15}$ and $f_K \simeq 0.1\,\text{GeV}$, one finds

$$m_c \simeq 1.4\,\text{GeV}, \tag{8.27}$$

in very good agreement with the actual mass of the charm quark as determined today by lattice QCD. In fact, the very same argument presented here has been used already more than 40 years ago by Gaillard and Lee in Reference [26] to predict that the charm quark should be lighter than 10 GeV. This is one of the first examples in the history of particle physics where the presence of a quadratic divergence in a theoretical prediction, together with an accurate measurement, has been used to predict the correct mass-scale of a new particle.

8.3 Quantifying Fine-Tuning

8.3.1 Technical Naturalness

We have seen in Section 8.2.1 that despite the fact that the electron mass is very small – in particular, much smaller than the Planck scale – this is considered natural. In fact, no fine-tuning is involved since setting the electron mass to zero one recovers an additional symmetry, the chiral symmetry. This symmetry guarantees that corrections to the electron mass remain proportional to the electron mass itself, and hence, it cannot receive large radiative corrections. This idea was formalised by 't Hooft [54] and elevated to a principle of naturalness which states that if a quantity in nature is small but a new symmetry is recovered were the quantity to vanish, then the smallness of the quantity is natural. This principle is usually referred to as technical naturalness. 't Hooft, however, made an even stronger statement, saying that a quantity in nature should be small only if the underlying theory recovers an additional symmetry when the quantity vanishes. Quantities that do not respect this principle are considered fine-tuned. Let us now consider a few examples of 'small' quantities and see if they are technically natural in the 't Hooft sense.

A first example is given by fermion masses in the SM, such as quark or neutrino masses, which exhibit a large spread in values, from less than 1 eV to more than 100 GeV. If we consider, for instance, the light quarks u, d, and s, the corresponding QCD Lagrangian contains a term of the form

$$\mathcal{L} \supset \bar{\psi} \left(i\gamma^\mu D_\mu - m \right) \psi, \tag{8.28}$$

where

$$\psi = \begin{pmatrix} u \\ d \\ s \end{pmatrix}, \qquad m = \begin{pmatrix} m_u & 0 & 0 \\ 0 & m_d & 0 \\ 0 & 0 & m_s \end{pmatrix}, \tag{8.29}$$

and

$$D_\mu = \partial_\mu - i g_s G_\mu^a t^a \tag{8.30}$$

is the covariant derivative, with g_s the strong coupling constant, G_μ^a, $a = 1,\dots,8$ the gluon fields, and t^a are the colour matrices. It is easy to see from Eq. (8.28) that if one sets the mass of the three lightest quarks to zero – i.e., $m_u = m_d = m_s = 0$, the Lagrangian has an additional $SU(3)$ flavour symmetry. Hence, the fact that the masses of the three lightest quarks are much smaller than the mass of the top quark, while curious, is technically perfectly natural.

In fact, as for the electron, in the massless limit another symmetry, the chiral symmetry is recovered, which transforms the ψ field according to Eq. (8.7). It is easy to see that in the massless limit Eq. (8.28) remains invariant under chiral transformations since γ^5 anti-commutes with the four Dirac matrices γ^μ. Similar arguments apply to all fermion masses in the SM. As a consequence of both the flavour and the chiral symmetry, while the large spread in the values of the quark and neutrino masses poses some questions, it is technically natural.

As a second example [53], one can consider an asymptotically free gauge theory with a running coupling $g(\mu)$ obeying

$$\frac{\partial g(\mu)}{\partial \ln \mu} = -\beta_0 g^3 + \cdots . \tag{8.31}$$

If the theory is asymptotically free, as is the case for QCD, $\beta_0 > 0$, the coupling becomes smaller at larger energy scales. Eq. (8.31) can be integrated between a scale Q and a scale Q_0 to give

$$\frac{1}{g^2(Q)} = \beta_0 \ln \frac{Q^2}{Q_0^2} + \frac{1}{g^2(Q_0)}, \tag{8.32}$$

where the bare coupling g_0 can be identified as $g(Q_0)$. As in QCD, the typical value of bound-state masses is of the order of Q where $g(Q)$ becomes large. Even if Eq. (8.32) becomes inaccurate in that region, it still can provide an indication about typical masses. Imposing that $g(Q)$ is very large in Eq. (8.32) leads to

$$Q = Q_0 \exp\left(-\frac{1}{2\beta_0 g_0^2}\right). \tag{8.33}$$

It is clear that, because of the exponential factor in the right-hand side, even if $Q \ll Q_0$, the bare parameter g_0 will not be very small. For instance, taking Q of the order of the proton mass, $Q = 1\,\mathrm{GeV}$, Q_0 of the order of the Planck mass, $Q_0 = 10^{19}\,\mathrm{GeV}$, and setting $2\beta_0 = 1$, one obtains $g_0 = 0.15$, which is not an unnaturally small number.

While the preceding discussion is somehow a simplification, it can be used to estimate the order of magnitude of the proton mass and explain why it is much smaller than the Planck scale. One might wonder what the symmetry is that allows the proton to remain much lighter than the Planck scale.[1] In the limit where the quark masses vanish, the classical action, $S = \int d^4 \mathcal{L}$, has a scale or conformal symmetry

$$x^\mu \to \kappa x^\mu, \qquad \partial_\mu \to \kappa^{-1}\partial_\mu, \qquad \psi \to \kappa^{-3/2}\psi. \tag{8.34}$$

If this symmetry was exact, the proton would be massless. As a consequence, the proton mass is naturally small. Note that in QCD, the conformal symmetry is broken explicitly by the renormalisation of the couplings that introduces a scale.

In summary, while most parameters in the SM are natural and not fine-tuned, according to the criterion of 't Hooft, there are also some parameters that are not natural. However, before discussing examples of fine-tuned quantities in the SM, we will introduce quantitative measures of fine-tuning in the next section.

8.3.2 Fine-Tuning Measures

We have seen that fine-tuning can be related to the presence of a delicate cancellation between apparently unrelated large terms to produce a result that is much smaller than individual contributions and that corresponds to physical observations. It is then important to provide a quantitative measure of fine-tuning, which measures how big the cancellation is. The aim is to be able to discuss questions of whether a new-physics model is fine-tuned and to compare different theories in a quantitative way in terms of how fine-tuned they are. A discussion of different fine-tuning measures can be found in Reference [34].

According to Wilson [53], the principle of naturalness requires that the physical properties of low-energy states should be stable against tiny variations of fundamental parameters. Let us illustrate a case of an unnatural scenario, where a particle receives self-energy corrections that are quadratic in a UV cut-off Λ, which has dimension of an inverse of a length-scale. For instance, let us consider a simple theory that involves a massless bare coupling g_0 and a bare mass m_0. The latter can be expressed as a dimensionless quantity by taking the ratio to the UV cut-off Λ

$$r_0 = \frac{m_0}{\Lambda}. \tag{8.35}$$

The correction Δm^2 to the mass of the particle then assumes the form

$$m^2 = m_0^2 + \Delta m^2, \qquad \Delta m^2 = g_0^2 \Lambda^2. \tag{8.36}$$

Solving for the fundamental parameter r_0, one obtains

$$r_0^2 = \frac{m^2}{\Lambda^2} - g_0^2. \tag{8.37}$$

---

[1] Note that the proton mass is not set by the light quark masses, which are more of the order of the neutron-proton mass difference.

If m is a physical mass of the order of the EW scale – i.e., $\mathcal{O}(100\,\text{GeV})$ – and Λ is taken to be the Planck scale $\mathcal{O}(10^{19}\,\text{GeV})$, then

$$r_0 = 10^{-34} - g_0^2. \tag{8.38}$$

This equation implies that r_0 must be adjusted to the 34th decimal digit, since, otherwise, the mass m will become of order $\mathcal{O}(10^{19}\,\text{GeV})$ rather than $\mathcal{O}(100\,\text{GeV})$. This high sensitivity of physical parameters to tiny variations of fundamental parameters is precisely what Wilson deemed to be unnatural.

A quantitative formalisation of naturalness, following Wilson's idea, was formulated by Barbieri and Giudice in Reference [13] in the context of quantifying the naturalness of various low-energy supersymmetry models that arise from supergravity theories. Let us consider an observable $\mathcal{O}$, which could be, for example, a mass or a mass difference, and is a function of a number of parameters a_i – for instance, the masses and couplings of new states. The fine-tuning measure Δ_{BG} of Reference [13] is defined as follows. One requires that for every a_i, the following relation holds:

$$\Delta_{\text{BG}}(\mathcal{O}; a_i) \equiv \left| \frac{a_i}{\mathcal{O}} \frac{\partial \mathcal{O}(\{a_i\})}{\partial a_i} \right| < \Delta, \tag{8.39}$$

which means that a relative variation of each of the parameters a_i does not produce a relative variation of $\mathcal{O}$ that is more than Δ times larger. So a value of Δ of 10 means that there is at most an order of magnitude cancellation between the parameters.

If we go back to the example in Eq. (8.36) and use now the measure in Eq. (8.39), we obtain

$$\Delta_{\text{BG}}(m^2, g_0) = \frac{g_0}{m^2} \frac{\partial m^2(\{g_0\})}{\partial g_0} = 2g_0^2 \frac{\Lambda^2}{m^2}. \tag{8.40}$$

This implies that unless the UV cut-off is close to the physical, observed mass m, this mass is unnatural, since $\Delta_{\text{BG}}(m^2, g_0)$ will become very large. For instance, taking $g_0 = 1$, Λ to be the Planck mass and m to be the Higgs mass, one obtains $\Delta_{\text{BG}}(m^2, g_0) = \mathcal{O}(10^{34})$.

Since in model building one typically wants to avoid models that are fine-tuned, Eq. (8.39) can be used at fixed Δ to constrain the parameter space of a theory. For instance, one can look at the allowed parameter space so that a theory is not fine-tuned by more than one or two orders of magnitude. Alternatively, given a new-physics theory, one can first exclude regions in parameter space where predictions are in conflict with experimental observations. For those regions of parameter space that remain not excluded by experimental data, one can use the Barbieri-Giudice criterion to assign a measure of fine-tuning. Originally, in Reference [13], Δ was chosen to be 10, allowing for an order of magnitude cancellation. The choice of a particular value for Δ is obviously very subjective, and after strong experimental constraints from the Large Electron Positron (LEP) and the Large Hadron Collider (LHC), larger choices of Δ became more commonly accepted.

It was also suggested in Reference [3] that the Barbieri-Giudice (BG) measure of Eq. (8.39) can be misleading, as it mixes the concept of sensitivity of a quantity to a given parameter and the notion of naturalness. In essence, the problem is related to the value of Δ that should be considered acceptable because when a quantity is more sensitive to a parameter, one should permit larger values of Δ. Hence, a new measure of fine-tuning was suggested in Reference [3] embodying the idea that 'Observable properties of a system should not be unusually unstable against minute variations of the fundamental parameters'. In essence, the new naturalness measure wants to separate the sensitivity of observables to parameters from the fine-tuning. This point can be understood by examining Eq. (8.33), which shows that changing, e.g., g_0 from, e.g., 0.15 to 0.16 changes Q from 1 GeV to 100 GeV. Hence, Q is very sensitive to g_0 despite the fact that there is no fine-tuning problem. In fact, the BG measure for Q is

$$\Delta_{BG}(Q, g_0) = \frac{1}{\beta_0 g_0^2},\tag{8.41}$$

which has no UV sensitivity to a cut-off but does depend strongly on g_0. The solution suggested in Reference [3] is to assign a probability distribution for the parameters of the theory and to calculate the fine-tuning associated with measured values after normalising away the specific sensitivity of physical quantities to parameters of the theory.

Starting from the BG measure in Eq. (8.39), Anderson and Castano (AC) define a new measure as [3]

$$\Delta_{AC}(\mathcal{O}; a_i) = \frac{\Delta_{BG}(\mathcal{O}; a_i)}{\langle \Delta_{BG}(\mathcal{O}; a_i) \rangle},\tag{8.42}$$

where

$$\frac{1}{\langle \Delta_{BG}(\mathcal{O}; a_i) \rangle} = \frac{\int da_i a_i f(a_i) \Delta_{BG}(\mathcal{O}; a_i)^{-1}}{\int da_i f(a_i)}\tag{8.43}$$

and $f(a_i)$ is a probability density. In essence, the new measure takes the BG measure and normalises it to its average over a reasonable range of parameters.

As a simplest option, one can, for instance, take a flat density $f(a_i) = 1$ or some simple form for it. One can then integrate over some minimum and maximum value of a_i, where the chosen values should correspond to a sensible range of variation for this theory parameter, with the chosen probability density. Simple forms of $f(a_i)$ give similar results in terms of the fine-tuning measure $\langle \Delta_{AC} \rangle$. In particular, the presence or absence of a fine-tuning problem is independent of the choice of $f(a_i)$, and the sensitivity to physical parameters is removed by the normalisation.

As an example, we can consider again the case of the scalar mass with quadratic corrections, Eq. (8.36). The BG measure is given in Eq. (8.40). To compute the AC measure, one needs to evaluate Eq. (8.43). For instance, for $f(g) = 1$, one obtains

$$\langle \Delta_{BG}(m^2; g_0) \rangle^{-1} = \frac{\int_{g_0^-}^{g_0^+} dg_0 g_0 f(g_0) \Delta_{BG}(m^2; g_0)^{-1}}{\int_{g_0^-}^{g_0^+} dg_0 g_0 f(g_0)} = \frac{m_0^2 \ln \frac{g_0^+}{g_0^-}}{\Lambda^2(g_0^{+2} - g_0^{-2})} + \frac{1}{2},\tag{8.44}$$

where we denoted by g_0^+ and g_0^- the upper and lower integration bounds. Similarly, for $f(g) = 1/g$, one obtains

$$\langle \Delta_{BG}(m^2; g_0) \rangle^{-1} = \frac{1}{2} - \frac{m_0^2}{2\Lambda^2 g_0^+ g_0^-}. \tag{8.45}$$

In both cases, we see that $\langle \Delta_{BG}(m^2; g_0) \rangle^{-1} = \mathcal{O}(1)$. As a consequence, $\Delta_{AC}(m; g_0) = \mathcal{O}(\Lambda^2)$, which implies that a fine-tuning is needed in order to keep the scalar mass m light.

As a second example, we can consider again the proton mass, whose value is of the order of the scale Q in Eq. (8.32). In this case, assuming a flat distribution $f(g_0) = 1$ and using Eq. (8.41), one obtains

$$\langle \Delta_{BG}(Q; g_0) \rangle^{-1} = \frac{\int_{g_0^-}^{g_0^+} dg_0 g_0 f(g_0) \Delta_{BG}(Q; g_0)^{-1}}{\int dg_0 g_0 f(g_0)} = \frac{\beta_0}{2} \left(g_0^{+2} + g_0^{-2} \right), \tag{8.46}$$

and assuming a distribution of the form $f(g_0) = 1/g_0$, one has instead

$$\langle \Delta_{BG}(m; g_0) \rangle^{-1} = \beta_0 \left(g_0^{+2} + g_0^+ g_0^- + g_0^{-2} \right). \tag{8.47}$$

In this case, regardless of the choice of the distribution $f(g_0)$, one can see that the $\Delta_{AC}(Q; g_0) = \mathcal{O}(1)$, which indicates, correctly, that there is no fine-tuning. This example illustrates the effect of the normalisation of the BG measure, since, while $\Delta_{BG}(Q; g_0) = \mathcal{O}(1/g_0^2)$ is very sensitive to g_0, the AC normalisation removes this sensitivity. On the other hand, the AC measure leads to a dependence on the integration range and to the form of the probability density, which makes it more difficult to compare different theories.

Yet another measure of fine-tuning was proposed by Athron and Miller [8]. The idea behind this measure is to generalise previous measures of fine-tuning to allow the variation of many parameters, to consider many observables, and to remove the global sensitivity by normalising to the mean value, as in the AC measure. More details can be found in the original paper, Reference [8].

In the AC measure, the particular parameters in a model are viewed as one realisation of nature over a broad distribution of possible values. This introduces the concept of likelihood of a particular choice of parameters over the possible theory distribution of allowed parameters. While Anderson and Castano do not discuss probabilities directly (because of a lack of a normalisation for the probability) they connect the notion of naturalness to that of likelihood of one model over a range of possible models. In particular, the attention smoothly shifts to having many possible realisations of a theory, with the parameters that are allowed to vary in a broad range. In this sense, the AC measure features connections to landscape theories.

While Anderson and Castano just introduce the concept of likelihood when discussing naturalness, Ciafaloni and Strumia [19] suggest to interpret directly the inverse of the BG measure as a probability

$$P \propto \frac{1}{\Delta_{BG}}; \tag{8.48}$$

hence, a large BG measure is associated to a small probability – i.e., to a large unnatural cancellation. In order to normalise the probability, they suggest requiring that the probability should be of order one in cases where no unnatural, accidental cancellation occurs. In this sense, naturalness acquires a statistical meaning of how atypical and unlikely a particular theory is. This is different from the original notion of naturalness introduced by Wilson, which merely required physics at low energy to be stable against tiny variations of fundamental parameters of the theory and did not connect the concept of naturalness to that of probability.

8.4 Fine-Tuning in the Electroweak Sector

Within the SM, the full EW symmetry $SU(2)_L \times U(1)_Y$ is broken spontaneously down to the abelian $U(1)_{em}$ subgroup describing electromagnetism. In this process, the masses of the W and Z gauge bosons as well as the fermions are generated. Since details about EW symmetry breaking (EWSB) are given in many textbooks on quantum field theory (QFT) – for instance, in References [46, 60] – we will in what follows only sketch the main ideas and steps.

The mechanism of EWSB in the SM involves a scalar field Φ that is a $SU(2)_L$ doublet. The kinetic term and the potential of Φ take the form

$$\mathcal{L} \supset |D_\mu \Phi|^2 - V(\Phi), \tag{8.49}$$

where the first term involves the covariant derivative of the $SU(2)_L \times U(1)_Y$ gauge group

$$D_\mu = \partial_\mu - ig\tau^a W_\mu^a - ig'Y B_\mu, \tag{8.50}$$

where g and g' are the $SU(2)_L$ and $U(1)_Y$ gauge coupling, respectively; $\tau^a = \sigma^a/2$ with σ^a the usual Pauli matrices; and $Y = 1/2$ represents the $U(1)_Y$ hypercharge of the field Φ. The physical gauge boson fields, i.e., the $W^\pm$, the Z, and the photon A are related to the fields W_μ^a and B_μ via

$$W_\mu^\pm = \frac{1}{\sqrt{2}} \left(W_\mu^1 \mp i W_\mu^2 \right),$$

$$Z_\mu = \cos\theta_w W_\mu^3 - \sin\theta_w B_\mu, \tag{8.51}$$

$$A_\mu = \sin\theta_w W_\mu^3 + \cos\theta_w B_\mu,$$

where $\sin\theta_w = g'/\sqrt{g^2 + g'^2} \simeq 0.22$ denotes the sine of the weak mixing angle.

The second term in Eq. (8.49) is given by

$$V(\Phi) = \mu^2 |\Phi|^2 + \lambda |\Phi|^4, \tag{8.52}$$

which is the most general potential consistent with gauge invariance and renormalisability in the case of a single scalar field. For $\mu^2 = 0$, the potential has a unique minimum at

$\Phi = 0$, while for $\mu^2 < 0$, the field value $\Phi = 0$ becomes a maximum of the potential, and the minima lie at

$$|\Phi|^2 = -\frac{\mu^2}{2\lambda} = \frac{v^2}{2},\tag{8.53}$$

with $v = \sqrt{-\mu^2/\lambda}$ the vacuum expectation value (VEV) of Φ. One can now expand the Φ field around the vacuum solutions. In the unitary gauge, one can write

$$\Phi = \frac{1}{\sqrt{2}}\begin{pmatrix} 0 \\ v + H \end{pmatrix},\tag{8.54}$$

with H a real scalar – i.e., the Higgs field. Up to an irrelevant constant, in this gauge the scalar potential reads

$$V(\Phi) = \frac{1}{2}(2\lambda v^2)H^2 + \lambda v H^3 + \frac{\lambda}{4}H^4,\tag{8.55}$$

which implies that the Higgs field H has a mass

$$m_H = \sqrt{2\lambda}\,v.\tag{8.56}$$

The term involving the covariant derivative of Φ can be also worked out easily. One obtains

$$|D_\mu \Phi|^2 = \frac{1}{2}|\partial_\mu H|^2 + \left[\frac{g^2 v^2}{4} W_\mu^+ W^{-\mu} + \frac{(g^2 + g'^2)\,v^2}{8} Z_\mu Z^\mu\right]\left(1 + \frac{H}{v}\right)^2.\tag{8.57}$$

From this expression, one can read off the values of the W and Z masses:

$$m_W = \frac{gv}{2}, \qquad m_Z = \frac{\sqrt{g^2 + g'^2}\,v}{2}.\tag{8.58}$$

The photon instead remains massless $m_A = 0$. The value of the Higgs VEV can then be obtained from precision measurements of the W and Z mass, from the Fermi constant $G_F \simeq 1.166 \cdot 10^{-5}\,\text{GeV}^{-2}$ using the relation

$$\frac{G_F}{\sqrt{2}} = \frac{g^2}{8m_W^2} = \frac{1}{2v^2},\tag{8.59}$$

which simply follows from matching the Fermi theory to the full SM. Numerically, one finds

$$v \simeq 246.22\,\text{GeV}.\tag{8.60}$$

The Higgs mass is also known very precisely today [1], to better than $\sim 1\%$ precision, and reads

$$m_H = 125.09 \pm 0.21\,(\text{stat.}) \pm (\text{syst.})\,\text{GeV}.\tag{8.61}$$

Using the second equality in Eq. (8.53) from the knowledge of the Higgs mass and the VEV, we also know the value of the quartic coupling λ in the SM. One obtains

$$\lambda \simeq 0.13.\tag{8.62}$$

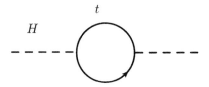

Figure 8.6 A one-loop correction to the Higgs mass coming from a top quark in the loop.

This implies that all interactions of SM particles, fermions, and bosons with the Higgs boson – in particular, also the Higgs interaction with itself through triple and quartic Higgs couplings – are fixed. A broad and detailed experimental program is ongoing at the LHC to test the validity of the Higgs mechanism. At the moment, couplings of the Higgs to fermions of the third generation and to W and Z have been measured to about 10% accuracy and seem to follow the pattern predicted in the SM. Yet couplings to lighter fermions and the Higgs self-coupling remain experimentally largely unconstrained. Better measurements of those quantities are top priorities for the upcoming LHC physics programme. In fact, although at the moment all measurements agree with the SM Higgs mechanism, the Higgs sector suffers from a fine-tuning problem, as explained in the following. Hence, it seems natural to think that the fine-tuning should be removed by the presence of BSM physics, and trying to remove or alleviate the fine-tuning in the EW sector has driven a lot of the theoretical activity in particle physics.

8.4.1 Higher-Order Corrections to the Higgs Mass

The fine-tuning problem in the Higgs sector is related to the fact that the Higgs mass receives very large quantum corrections from particles that couple directly, or even indirectly, to the Higgs boson. This fact was first observed in the context of grand unified theories (GUTs) [28, 57].

Let us consider, for example, the quantum corrections to the Higgs mass coming from a loop of fermions that couples to the Higgs through a Yukawa interaction of the form

$$\mathcal{L} \supset -\lambda_f H \bar{f} f, \tag{8.63}$$

with $\lambda_f = \sqrt{2} m_f / v$ and m_f the mass of the fermion. The diagram illustrated in Figure 8.6 gives rise to a correction to the Higgs mass of the form

$$\Delta m_H^2 = N_c i \lambda_f^2 \int \frac{d^4 l}{(2\pi)^4} \frac{\text{tr}\left[(\slashed{l} + m_t)(\slashed{l} + \slashed{p} + m_t)\right]}{(l^2 - m_t^2)\left((l + p)^2 - m_t^2\right)}, \tag{8.64}$$

which is quadratically divergent in the limit $|l| \to \infty$. The divergent contribution is given by

$$\Delta m_H^2 = N_c \frac{i\lambda_f^2}{16\pi^4} \int d^4 l \frac{4l^2}{l^4} = N_c \frac{i\lambda_f^2}{4\pi^4}\left(-i\pi^2 \Lambda^2\right) = N_c \frac{G_F m_f^2}{\sqrt{2}\pi^2} \Lambda^2, \tag{8.65}$$

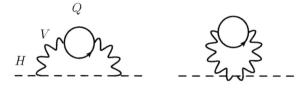

Figure 8.7 Two examples of two-loop correction to the Higgs mass coming from a heavy fermion Q that couples only indirectly to the Higgs via the exchange of a SM gauge boson V.

where we used the expression of the divergent integral in Eq. (8.12) and $N_c = 3$ is the colour factor.

Within the SM, the dominant contribution to the Higgs mass comes from the top quark because of its large Yukawa coupling. Also, loops involving massive bosons give rise to divergent corrections, and when one includes all important one-loop corrections, one obtains [55]

$$\frac{\Delta m_H^2}{m_H^2} = \frac{3G_F}{4\sqrt{2}\pi^2}\left(\frac{4m_t^2}{m_H^2} - \frac{2m_W^2}{m_H^2} - \frac{m_Z^2}{m_H^2} - 1\right)\Lambda^2 \simeq \left(\frac{\Lambda}{500\,\mathrm{GeV}}\right)^2. \tag{8.66}$$

The UV cut-off can then be interpreted as the scale at which new physics enters and modifies the high-energy behaviour of the theory. The hierarchy problem of the Higgs mass is now related to the fact that, if in Eq. (8.66) the UV cut-off Λ is taken to be the Planck mass, m_H receives corrections that are more than 15 orders of magnitude larger than the measured value of around 125 GeV. If, on the other hand, new, not-too-heavy particles exist and act as an effective UV cut-off, then this could remove the fine-tuning problem in the Higgs sector.

The interesting thing to note is that in any extension of the SM that introduces one or more heavy states, the Higgs mass squared becomes quadratically sensitive to any new degree of freedom, regardless of whether the new states interact at the tree level with the Higgs. In fact, even if the new particles do not interact directly with the Higgs boson – for instance, because their masses predominantly arise from a mechanism other than EWSB – they will generically affect the Higgs boson mass indirectly through radiative corrections. An example is shown in Figure 8.7, which displays two two-loop diagrams that involve a heavy fermion that couples only indirectly to the Higgs boson through a SM vector boson. The quadratically divergent correction to the Higgs mass squared that arises from the diagram on the right-hand side of Figure 8.7 can be written schematically as follows:

$$\Delta m_H^2 \sim g_1^2 g_2^2 C_1 C_2 \int \frac{d^4 l_1}{(2\pi)^4} \int \frac{d^4 l_2}{(2\pi)^4} \frac{\mathrm{tr}\left[(l_1 + l_2)l_2\right]}{l_1^4 l_2^2 (l_1 + l_2)^2}, \tag{8.67}$$

where the coupling of the intermediate vector boson to the Higgs is called g_1, the coupling of the vector boson to the heavy fermion is denoted by g_2, and C_1 and C_2 are the corresponding Casimir factors. The preceding expression simplifies to

$$\Delta m_H^2 \sim \frac{g_1^2}{16\pi^4}\frac{g_2^2}{16\pi^4}C_1C_2\int d^4l_1\int d^4l_2\frac{1}{l_1^4\,(l_1+l_2)^2}$$

$$\sim \frac{g_1^2}{16\pi^2}\frac{g_2^2}{16\pi^2}C_1C_2\,\Lambda^2\ln\Lambda, \tag{8.68}$$

where we used Eq. (8.12) and

$$\int d^4l\,\frac{1}{l^4}=i\int d\Omega_4\int_0^\Lambda dl_E\,\frac{1}{l_E}=i2\pi^2\ln\Lambda \tag{8.69}$$

and ignored numerical factors of $\mathcal{O}(1)$ to arrive at the final result in Eq. (8.68).

We see that like the one-loop contribution Eq. (8.65), the two-loop contribution also depends quadratically on the UV cut-off Λ. One finds a similar quadratically divergent contribution from the diagram on the left-hand side of Figure 8.7, but the two contributions do not cancel. This example illustrates that even if a new heavy particle does not interact with the Higgs boson at tree level, as long as the new state has interactions with some of the SM particles, radiative corrections will give rise to a quadratic sensitivity of m_H on the cut-off scale Λ.

Given that the technical naturalness argument is used to explain the lightness of many SM fermions, it is worth examining if an additional symmetry appears within the SM in the limit of $m_H \to 0$. A massless Higgs would imply the absence of the quadratic term in Eq. (8.52). In such a case, the Higgs VEV would be zero, the EW symmetry would be unbroken, and neither the SM gauge bosons nor the fermions would receive a mass from EWSB. In fact, at the classical level, the SM action is conformally symmetric – i.e., invariant under the transformations in Eq. (8.34) and

$$V_\mu \to \kappa^{-1}V_\mu, \qquad \Phi \to \kappa^{-1}\Phi, \tag{8.70}$$

except for the quadratic term in Eq. (8.52). In the preceding transformations, V_μ represents any of the SM gauge fields. It follows that the vanishing of the Higgs mass term $\mu^2|\Phi|^2$ would increase the symmetry of the SM by making the conformal symmetry exact. At the loop level, there is explicit breaking of the conformal symmetry reflected by the logarithmic running of the gauge and Yukawa couplings. The logarithmic running alone, however, will not generate Λ^2 terms so that in perturbation theory conformal symmetry would protect Δm_H^2 from quadratic corrections and the conventional statement of the EW fine-tuning problem. Yet the fine-tuning issue may reappear if the SM is embedded into a more complex theory, visible at short distances or high energies. Whether conformal invariance can still be used to protect the EW scale then depends on the structure and the dynamics of these more complete formulations.

Before concluding this section, it is also worth commenting on what happens if, instead of regularising the divergent UV behaviour of loop integrals by a cut-off Λ, one considers dimensional regularisation, which is the regularisation procedure usually employed in modern higher-order perturbative calculations. This point is discussed in detail, for instance, in Reference [62]. Dimensional regularisation of UV divergences proceeds by extending the four-dimensional loop integration to $D = 4 - 2\epsilon$ dimensions with $\epsilon > 0$ and infinitesimal. At the one-loop level, UV divergences appear as simple poles of the form $1/\epsilon$, and these singularities can be absorbed into the bare couplings and masses of the theory in any renormalisable QFT (such as the SM). In dimensional regularisation, the quadratic Λ dependence of the loop corrections to the Higgs mass squared is replaced by a logarithmic dependence on the so-called renormalisation scale μ, and it so might appear that there is no issue with fine-tuning of the Higgs mass (see, for instance, [14, 21, 24]). The fine-tuning problem, however, reappears in dimensional regularisation if, at a scale Λ, new degrees of freedom beyond the SM exist. These new states lead to finite threshold corrections that reintroduce a quadratic sensitivity to Λ in Δm_H^2. Irrespectively of the regularisation procedure, the EW fine-tuning problem can hence only be disregarded if the weak scale is the only relevant scale of high-energy physics, and in such a case, the SM would be natural.

Nature, however, has already revealed that there is physics beyond the SM. Dark matter and non-zero neutrino masses require new degrees of freedom and possibly new mass-scales. More indirect hints like the unification of gauge couplings, the need for a mechanism of baryogenesis, the strong CP problem, and the flavour puzzle also suggest the existence of new typically very heavy states. Ultimately, the need to include gravitation in the picture provides a concrete ultra-large energy scale close to the Planck mass, though its connection to massive particles is less clear. It, thus, seems very plausible that between the weak and the Planck scale, new degrees of freedom not present in the SM appear. Protecting the Higgs mass from the resulting large UV sensitivity then requires a new mechanism or a new symmetry that tames the size of radiative corrections. In the following, we will discuss the most popular solutions to the EW hierarchy problem.

8.4.2 Solutions of the Gauge Hierarchy Problem

The EW fine-tuning problem has been a main motivation for the construction of BSM scenarios. Most of these new-physics models predict the existence of new states at the TeV energy scale that modify the quantum corrections to the Higgs mass, thereby stabilising the weak scale. For a long time, this has been a leading argument in favour of new physics in the TeV range; however, after results from Run I and first results from Run II at the LHC, several simple and natural models to explain the lightness of the Higgs boson are not viable anymore (see, e.g., [23, 29]).

An elementary scalar can be naturally light if the mass is protected by a symmetry. The most explored possibility is supersymmetry (SUSY), an internal symmetry which relates fermions to bosons. If SUSY is unbroken, each SM particle has a SUSY partner with the same mass. The two particles are then said to be in the same supermultiplet. The mass of scalars is then related to the mass of chiral fermions, and hence, it is naturally protected from getting quadratically divergent corrections, precisely in the same way as fermion masses in the SM are protected by the chiral symmetry.

The explicit way in which the quadratic sensitivity to the UV scale is removed in SUSY models is that every SM fermionic (bosonic) quadratically divergent correction to the Higgs

Figure 8.8 One-loop correction to the Higgs mass squared from the SM top loop (left) and its SUSY partner, the stop (right).

mass comes together with a bosonic (fermionic) SUSY correction. This mechanism is illustrated by the Feynman diagrams shown in Figure 8.8. Because fermion loops involve an extra minus sign compared to bosonic loops, the divergent contribution from the two diagrams cancel. If the masses of fermionic and bosonic states were the same, the cancellation would be exact. However, if SUSY was unbroken, i.e., if SM particles and their SUSY partners had the same mass, we should have observed a plethora of SUSY particles already at LEP, if not before. Since we have not, there must be a mechanism that breaks this symmetry and that causes SUSY particles to be heavier than SM particles. A large number of models have been suggested in the past to break SUSY. For a pedagogical review, see, for instance Reference [41]. It is important to note that the heavier the SUSY particles are, the less they provide a natural a solution to the Higgs fine-tuning problem. Current bounds from the LHC on SUSY particles are already putting natural SUSY models under stress. Yet, for the moment, some regions of parameter space are not constrained – most notably, regions where SUSY spectra are compressed and where, for instance, top and stop quarks are degenerate, because, in this case, stop production can hide under the large SM top background. In general, the present situation raises the question of what level of fine-tuning one should be willing to tolerate. If, for example, one allows for a permille tuning, then SUSY particles might very well be out of reach of the LHC.

It is interesting to note that SUSY was, in fact, originally not put forward as a solution to the Higgs hierarchy problem. Rather, it was proposed for more aesthetic reasons. Hence, it is, in some sense, remarkable that SUSY provides such a simple solution to the hierarchy problem. Because of other remarkable properties of some SUSY models, such as the presence of a natural candidate for dark matter (the lightest SUSY neutral particle) and the unification of gauge couplings, despite the persisting lack of experimental evidence for SUSY particles, these models are still relatively popular.

A very different approach to solve the fine-tuning problem in the Higgs sector is due to Susskind and Weinberg, who – in 1978 – introduced the first technicolour model [53, 57, 58]. According to technicolour models, EWSB is not due to an elementary Higgs boson, but, rather, it rather thanks to a condensate of fermions of some new strong dynamics, a heavy copy of the familiar QCD interactions in the SM. If an elementary Higgs boson does not exist, then there is simply no hierarchy problem. In fact, if the Higgs boson is not an elementary particle but a composite state, then its mass is set by the binding force between the constituents. This compositeness scale Λ then becomes the cut-off of the SM. If the Higgs

boson is composite, one would expect other composite states to exist, in the same way as
a whole spectrum of mesonic and baryonic states exists in QCD. Furthermore, if the Higgs
is composite, its coupling to vector bosons and fermions would be modified compared to
the SM predictions based on an elementary Higgs boson. This imposes stringent bounds
on the scale of compositeness. While the idea behind technicolour models is appealing
and simple, these models face a number of problems. First of all, they do not account for
an explanation of the origin of fermion masses. One possibility is to introduce new gauge
interactions that break the chiral symmetry which protects fermion masses. However, these
models, besides being ad hoc, typically run into problems with flavour physics, as they give
rise to flavour-changing neutral currents that are much larger than what is observed. Hence,
again, while trying to solve one fine-tuning problem, one would introduce other ones.
Furthermore, the discovery of a very light Higgs is, in general, difficult to accommodate in
technicolour models because such models typical predict $m_H = \mathcal{O}(1\,\text{TeV})$.

   Another interesting possibility of solving the EW hierarchy is to implement the Higgs
as a pseudo-Nambu-Goldstone boson (pNGB) [7, 27, 37]. In fact, when a continuous,
global symmetry is spontaneously broken, the theory develops a massless scalar particle, the Nambu-Goldstone boson (NGB). If the symmetry is, however, not exact but only
approximate, then a pNGB appears with a small mass that depends on the symmetry
breaking in the Lagrangian. This idea applied to the Higgs boson leads to so-called little
Higgs models [4, 5, 7, 50], according to which there is a new strong interaction at a scale
$\Lambda = \mathcal{O}(1\,\text{TeV})$ with a global symmetry, which is spontaneously broken. Little Higgs
models are able to predict a naturally light Higgs particle in exactly the same way as,
in QCD, the neutral pion is much lighter than the QCD scale of about $1\,\text{GeV}$ since it is
a pNGB of chiral symmetry. Again, while the idea is appealing, explaining the origin of
fermion masses and respecting all the phenomenological constraints remains a challenge
in these models.

   An additional approach to the Higgs fine-tuning problem was suggested around 20 years
ago [6]. The resolution of the fine-tuning problem is now based on suggesting that the UV
cut-off of the SM is not the Planck scale but, rather, a much lower scale of the order of
the TeV. This is achieved by considering modifications of our four-dimensional space and,
in particular, by extending the dimensionality of space-time. In this picture, SM particles
propagate in four dimensions, while gravity propagates in additional spatial dimensions that
are much larger than $1/M_{\text{Planck}}$. If one adds n dimensions to the Einstein action, one gets

$$S \supset -\frac{M_*^{2+n}}{2} \int d^4x \, d^n y \, \sqrt{-g} R, \tag{8.71}$$

where x are our ordinary four dimensions, y are the n extra dimensions, g is the determinant
of the metric tensor $g_{\mu\nu}$, R is the scalar curvature – i.e., the trace of the Ricci curvature
tensor $R_{\mu\nu}$ – and M_* is the reduced Planck scale of the $4 + n$ – dimensional theory.

   In the model described by Eq. (8.71) the four-dimensional reduced Planck mass
$M_{\text{Planck}} = G_N/\sqrt{8\pi}$ is then related to the size of the extra dimensions R_* and M_* as
follows:

$$M_{\text{Planck}}^2 = R_*^n M_*^{2+n}. \tag{8.72}$$

Fixing M_* at around the EW scale to avoid introducing a new mass-scale in the model thus allows us to predict R_*. For $n = 1$, for instance, one obtains that the size of the extra dimension is $R_* = \mathcal{O}(10^9 \text{ km})$, a possibility that is clearly ruled out as it leads to modifications of Newton's law at solar-system distances. For $n = 2$, the extra dimensions instead are predicted to have a size of a millimetre, and as a result, modification of the gravitational law at distances smaller than that should appear. In fact, the $1/r^2$ scaling of the gravitational force and the equivalence principle – i.e., the equivalence between matter and gravity – has also been verified with very high accuracy through laboratory tests at submillimetre distances. In the case of $n = 2$, searches for deviations from Newton's law of gravitation, for example, imply $R_* > 37\,\mu\text{m}$ at 95% confidence level (CL), which translates into the lower bound $M_* > 3.6\,\text{TeV}$ [43]. These models also predict a spectrum of new particles – i.e., Kaluza-Klein (KK) excitations of all the SM fields. At the moment, no evidence for these particles has been seen, with bounds on the mass of KK modes of the order of several TeV. We finally mention that the consistency of the model introduced in Eq. (8.71) requires a stabilisation mechanism for the radii of the extra dimensions. The fact that one needs $R_* \gg 1/M_*$ leads to a new hierarchy problem, the solution of which might require further modifications such as SUSY to make the proposal both natural and phenomenologically viable.

An alternative extra-dimensional solution to the gauge hierarchy problem is provided by so-called warped extra dimensions, as proposed in the seminal papers of Randall and Sundrum (RS) [47, 48]. The simplest RS models are based on a five-dimensional theory with the extra dimension compactified in an orbifold, which is mathematically equivalent to a manifold with boundaries at $y = 0$ and $y = \pi R_*$. These boundary points are called the infrared (IR) and UV brane, respectively. Under the assumption that the five-dimensional theory has a cosmological constant Λ in the bulk, one can show that the Einstein equations in the RS model admit solutions where the metric varies like $a(y) = e^{-ky}$. Here, y denotes the extra-dimensional coordinate and $k = \sqrt{-\Lambda/(6M_*^3)}$. The factor $a(y)$ is called the 'warp' factor and determines how four-dimensional scales change as a function of y. In particular, this implies that energy scales for four-dimensional fields localised at the IR brane are redshifted by a factor $e^{-k\pi R_*}$ with respect to those on the UV brane.

Similarly to Eq. (8.72) the reduced four-dimensional Planck mass is also related in the RS framework to the five-dimensional one. Specifically, one obtains the relation

$$M_{\text{Planck}}^2 = \frac{M_*^3}{2k}\left(1 - e^{-2k\pi R_*}\right). \tag{8.73}$$

Taking $M_* \simeq k \simeq M_{\text{Planck}}$, it is now possible to generate an IR scale of $ke^{-k\pi R_*} = \mathcal{O}(\text{TeV})$ for an extra-dimensional radius $R_* \simeq 11/k$. Mechanisms to stabilise R_* to this value have been proposed [31] and do not require introducing any new small or large parameter. A natural solution to the hierarchy problem is hence achieved in the RS framework if the Higgs field is localised on (or close to) the IR brane, where the effective mass-scales are of order of a TeV. Like models with large extra dimensions, RS models also predict a rich spectrum of KK modes, which, however, have not been seen at the LHC. Furthermore, precision electroweak measurements, flavour physics, as well as Higgs measurements of both the mass and its couplings pose challenges to these models, often pushing the scale M_* into the multi-TeV range.

Recently, a new interesting idea, the cosmological relaxation [32], was suggested to explain the lightness of the Higgs boson. The reason this idea has attracted considerable attention is that it solves the fine-tuning problem without introducing any TeV-scale dynamics that could be detectable at the LHC. According to the cosmological relaxation idea, the Higgs boson is light, as its mass was driven to a value much lower than the cut-off during the dynamical evolution of the early Universe. The idea is to observe that while $m_H = 0$ is not a special point in terms of symmetries (in the sense that no new symmetry is recovered in the Lagrangian in this limit), it is a special point in terms of dynamics, since this is the point where EW symmetry is broken.

In its simplest version, this model contains, besides the SM particles, the QCD-like axion and an inflation sector. The conditions imposed on the axion are that it should have a very large (noncompact) field range and a soft symmetry-breaking coupling to the Higgs. Concretely, the Lagrangian suggested in Reference [32] to be added to the SM Lagrangian takes the form

$$\mathcal{L} = (-M^2 + g\phi)|\Phi|^2 + V(g\phi) + \frac{1}{16\pi^2}\frac{\phi}{f}\tilde{G}^a_{\mu\nu}G^{\mu\nu,a}, \tag{8.74}$$

where M is the UV cut-off of the theory, Φ denotes the Higgs doublet, $G^a_{\mu\nu}$ is the QCD field strength tensor, $\tilde{G}^a_{\mu\nu} = \epsilon_{\mu\nu\rho\sigma}G^{\rho\sigma,a}/2$ is its dual, and g is a dimensionful coupling with $g \ll M$. Notice that, in Eq. (8.74) the Higgs mass is at the UV cut-off and, thus, natural. The field ϕ has all properties of a standard QCD axion – in particular, the couplings are set by the decay constant f – however, it can take on field values much larger than f. In the limit $g \to 0$, the preceding Lagrangian has an additional shift symmetry $\phi \to \phi + 2\pi f$ (the symmetry is broken by non-perturbative QCD effects), and hence, $g \ll M$ is technically natural according to 't Hooft criterion. The coupling g can thus be treated as a spurion – i.e., an auxiliary field used to parameterise symmetry-breaking terms – and this feature allows one to expand the potential $V(g\phi)$ in powers of $g\phi/M^2$. The QCD scale Eq. (8.74) therefore effectively takes the form

$$\mathcal{L} = (-M^2 + g\phi)|\Phi|^2 + M^4\left[\frac{g\phi}{M^2} + \frac{g^2\phi^2}{M^4} + \mathcal{O}\left(\frac{g^3\phi^3}{M^6}\right)\right] + \Lambda_c^4 \cos\left(\frac{\phi}{f}\right), \tag{8.75}$$

where the $\mathcal{O}\left(g^3\phi^3/M^6\right)$ can be ignored if $\phi \lesssim M^2/g$, the periodic QCD potential has been approximated for simplicity by a cosine function, and Λ_c is of order the QCD scale Λ_{QCD}. Notice that, since Λ_c breaks a symmetry, it is technically natural to have $\Lambda_c \ll M$.

In the early Universe, one can take an initial value for ϕ such that effective mass-squared of the Higgs, $m_H^2 = -M^2 + g\phi$, is positive. The assumption on a positive initial value for m_H^2 is crucial. During inflation, ϕ will slowly roll down, scanning different values of m_H^2.

In this process, the point $m_H^2 = 0$ and the Higgs field will start to develop a VEV. This will happen for a field value $\phi = \mathcal{O}(M^2/g)$. As the VEV grows, so do the quark masses and the scale Λ. As a result, the amplitude of the cosine term in Eq. (8.75) grows, which at some point, stops the slow roll of ϕ. In fact, the rolling of ϕ stops shortly after m_H^2 crosses zero, and this sets the Higgs mass to be naturally much smaller than the cut-off M. Since it is the axion field ϕ which is responsible for the dynamical relaxation of the weak scale, the mechanism described earlier has been coined the relaxion mechanism.

While the simplest realisation of the relaxion mechanism discussed here is phenomenologically not viable, because the QCD θ angle is predicted to be of $\mathcal{O}(1)$, in stark contrast to observation (see Section 8.5), the relaxion idea is a first example of a solution to the fine-tuning in the Higgs sector that does not require new physics at the TeV scale. Hence, it opens the door to other BSM theories that solve the hierarchy problem without having any consequences at colliders. Instead, the one prediction of the model is the presence of an axion with a very small coupling which is notoriously difficult to discover in a laboratory.

8.5 Fine-Tuning in the Strong Sector

CP violation is intimately related to the baryon asymmetry of the Universe.[2] In fact, already in 1967, Sakharov [49] pointed out that three conditions are necessary to generate the baryon asymmetry: baryon number violation, C and CP violation, and interactions out of thermal equilibrium. In particular, CP violation is necessary to produce a different number of baryons of a given chirality and anti-baryons of the opposite chirality. The weak interactions of the SM are known to violate CP symmetry; however, it is also clear that the resulting amount of CP violation is not sufficient to explain the observed baryon asymmetry. For this reason, additional sources of CP violations are of considerable interest. Both the electromagnetic and the strong interaction are considered symmetric under CP transformations. However, the situation in QCD is more complicated, as there is one potential source of CP violation. The origin of the CP-violating term is intrinsically non-perturbative, a regime where calculations are notoriously difficult. As of today, there is not a full understanding of whether the potential CP-violating term in QCD is zero or not. Current experimental measurements constrain this CP-violating term to be very small. The smallness of this parameter constitutes a fine-tuning problem in the SM that is usually referred to as the strong CP problem.

8.5.1 The $U(1)_A$ and the Strong CP Problem

It is useful to first recall the $U(1)_A$ problem that puzzled physicists in the 1970s, when QCD was established as a theory of strong interactions. The QCD Lagrangian with N flavours has the form

---

[2] Under parity P, the coordinates are changed $\vec{x} \to -\vec{x}$; hence, the handedness of particles flips. Under charge conjugation particles and antiparticles are interchanged. The CP transformation is a combination of both transformations.

$$\mathcal{L} = \sum_{i=1}^{N} \bar{\psi}_i \left(i\gamma^\mu D_\mu - m_i \right) \psi_i - \frac{1}{4} G_{\mu\nu} G^{\mu\nu,a}. \tag{8.76}$$

This Lagrangian is invariant under the $SU(3)_c$ colour symmetry, which is an exact, unbroken symmetry. In the limit where the quark masses vanish – i.e., $m_i \to 0$ – the Lagrangian also has another global symmetry, namely

$$U(N)_V \times U(N)_A = SU(N)_V \times U(1)_V \times SU(N)_A \times U(1)_A, \tag{8.77}$$

where the $SU(N)_V$ factor corresponds to a global vector symmetry

$$\psi_i \to e^{i\alpha_a t^a/2} \psi_i, \tag{8.78}$$

while the $SU(A)_A$ factor is a global axial symmetry

$$\psi_i \to e^{i\alpha_a \gamma^5 t^a/2} \psi_i. \tag{8.79}$$

The latter is called axial symmetry because the associated Noether current $j_A^{a,\mu}$ is an axial vector

$$j_A^{a,\mu} = \bar{\psi} \gamma^\mu \gamma^5 t^a \psi. \tag{8.80}$$

Since two of the six quarks, the up and the down quarks, are very light, it is a good approximation to set their mass to zero. As a consequence, one expects the strong interactions to be approximately $U(2)_V \times U(2)_A$ invariant. The vector symmetry can be decomposed as $U(2)_V = SU(2)_V \times U(1)_V$, giving rise to isospin and baryon number conservation, respectively. Indeed, baryon number is an exact symmetry, and isospin is a good approximate symmetry of nature, since the proton and the neutron, or the $\pi^\pm$ or π^0, are almost degenerate multiplets. On the contrary, the corresponding $U(2)_A$ axial symmetry is not seen in the spectrum. The reason is that the $SU(2)_A$ symmetry is not preserved by the QCD vacuum but spontaneously broken. This generates light NGBs, namely the pions. However, there is no NGB for the spontaneous breaking of the global $U(1)_A$ symmetry. This is the well-known $U(1)_A$ problem that puzzled physicists in the 1970s.

The solution to the $U(1)_A$ problem came with the realisation that, while we commonly think of QCD as a theory that depends on quark masses and the strong coupling g_s only, it is, in fact, possible to add to the QCD Lagrangian in Eq. (8.76) a term of the form

$$\mathcal{L} \supset \frac{g_s^2}{16\pi^2} \theta \, \tilde{G}_{\mu\nu}^a G^{\mu\nu,a}. \tag{8.81}$$

Notice that we have already met a similar term in Eq. (8.74). While the preceding term can be added to the QCD Lagrangian, with the simple motivation that any term that is not forbidden by symmetries should be included, originally, this term was derived by studying QCD instantons – i.e., solutions to classical field equations in Euclidean space-time.

In fact, the term introduced in Eq. (8.81) is a total derivative, as can be seen, for instance, easily in the case of QED, where

$$\tilde{F}_{\mu\nu} F^{\mu\nu} = 4\vec{E} \cdot \vec{B} = \partial^\mu \left(\epsilon_{\mu\nu\rho\sigma} A^\nu F^{\rho\sigma} \right). \tag{8.82}$$

Here, $F_{\mu\nu} = \partial_\mu A_\nu - \partial_\nu A_\mu$ is the QED field strength tensor with A_μ the electromagnetic four-potential, and $\vec{E}$ and $\vec{B}$ denote the electric and magnetic three-field, respectively. Being a total derivative, this term does not contribute to the classical equations of motion. In fact, the volume integral in the action can be rewritten using Gauss's theorem as a surface term that vanishes, provided the field configuration vanishes fast enough towards infinity.

In the non-abelian case one, can also write this term as a total derivative:

$$\tilde{G}^a_{\mu\nu} G^{\mu\nu,a} = \partial^\mu K_\mu, \qquad K_\mu = \epsilon_{\mu\nu\rho\sigma} G^{\nu,a} G^{\rho\sigma,a} - \frac{2}{3} g_s f_{abc} G^{\nu,a} G^{\rho,b} G^{\sigma,c}, \qquad (8.83)$$

with f_{abc} the fully antisymmetric structure constant of QCD. One, hence, might think that this term has no physical consequence, since its contribution to the action can be written as a surface term,

$$\int d^4 x \, \tilde{G}^a_{\mu\nu} G^{\mu\nu,a} = \int_{S^3} dS_\mu \, K^\mu, \qquad (8.84)$$

which vanishes when fields vanish fast enough at infinity. In 1975, however, Belavin, Polyakov, Schwartz, and Tyupkin [16] suggested the existence of non-trivial field configurations which satisfy the vanishing boundary conditions at infinity but, nevertheless, give a non-vanishing contribution to the surface integral. These configurations are called instantons. As a result, a θ-like term appears naturally in the QCD Lagrangian once instanton configurations are considered. Instantons are associated to an integer winding number, which can be thought of as the difference between initial and final angular position in units of 2π. Different winding numbers correspond to non-equivalent instanton configurations. The classical vacuum state is the configuration which minimises the classical potential. In QCD, it turns out that an infinite number of vacuum states are possible, each labelled by a winding number, which cannot be transformed into one another by a (small) gauge transformation. Instanton configurations with winding number k, on the other hand, connect two vacuum states with winding number m and n at $t = \pm\infty$, such that $m - n = k$.

Gravitational, electromagnetic, and strong interactions without the θ term conserve CP; hence, the interactions are the same for matter and antimatter. On the other hand, if θ is not zero, the strong interaction would violate P and CP.[3] To see this, it is useful to introduce the shorthand notation $(-1)^\mu \equiv 1$ for $\mu = 0$ and $(-1)^\mu \equiv -1$ for $\mu = 1, 2, 3$. Since $G^a_{\mu\nu}$ transforms as two vectors under both P and CP, we have

$$\epsilon_{\mu\nu\rho\sigma} G^{\mu\nu,a} G^{\rho\sigma,a} \xrightarrow{P,CP} \epsilon_{\mu\nu\rho\sigma} (-1)^\mu (-1)^\nu (-1)^\rho (-1)^\sigma G^{\mu\nu,a} G^{\rho\sigma,a} = -\epsilon_{\mu\nu\rho\sigma} G^{\mu\nu,a} G^{\rho\sigma,a}, \qquad (8.85)$$

where the last equation follows from the fact that all indices must be distinct for the fully antisymmetric tensor $\epsilon_{\mu\nu\rho\sigma}$ to give a non-vanishing contribution.

---

[3] Since it is not possible to formulate a locally Lorentz-invariant QFT with a Hermitian Hamiltonian that violates CPT, CPT is believed to be a fundamental symmetry of nature (see, e.g., Reference [52]). This implies that if CP is violated, then the time reversal symmetry T is also violated by strong interactions.

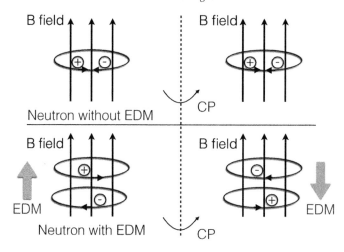

Figure 8.9 Schematic illustration of a CP transformation of a neutron in a magnetic field, without (upper figures) or with (lower figures) EDM.

Accordingly, the effect of a non-zero θ in Eq. (8.83) would be to generate an electric dipole moment (EDM) for the neutron of the form

$$d_n \simeq \frac{e}{m_n} \frac{m_q}{m_n} \theta \simeq 1.6 \cdot 10^{-16} \, \theta \, e \, \text{cm}, \tag{8.86}$$

where m_q is a typical light (up or down) quark mass and m_n is the neutron mass (see Figure 8.9 for a pictorial illustration). Experimental limits on the neutron EDM can therefore be used to constrain the θ angle. The current 95% CL upper bound on the neutron EDM is $|d_n| < 2.9 \cdot 10^{-26} \, e \, \text{cm}$ [11], which leads to

$$|\theta| \lesssim 2 \cdot 10^{-10}. \tag{8.87}$$

Since any term that is allowed by symmetry should be included in the Lagrangian, the angle θ should take a non-zero value. The question that then arises is why θ should be as small as required by Eq. (8.87) and not of order one. This is what is referred to as the fine-tuning problem in the strong sector of the SM. In fact, the problem is even more severe, as the measured θ angle is the combination of the bare one, discussed here, and a determinant of the quark mass matrix, discussed in the next section. The latter receives quantum corrections; hence, the cancellation between the two contributions has to remain intact even after radiative corrections.

Is it interesting to note that if CP was an exact symmetry of nature, a small θ angle would be technically natural according to 't Hooft criterion, since by setting θ to zero, one would recover a larger symmetry. However, we know that CP is explicitly violated in nature by the complex phase in the CKM matrix, which is of order one. Hence, no new symmetry is recovered by setting θ to zero, and its observed smallness remains just unnatural.

8.5.2 Strong CP Problem and Massless Quarks

From Eq. (8.9), one naïvely expects that the QCD Lagrangian, as given in Eq. (8.76), is invariant under the chiral transformations introduced in Eq. (8.7) if quarks are massless. This naïve expectation is, however, only correct at the classical level but not in the quantum theory. The effect of a chiral transformation in a quantised theory can be derived by studying how the measure $\mathcal{D}\psi \, \mathcal{D}\bar{\psi}$ of the functional integral

$$Z = \int \mathcal{D}\psi \, \mathcal{D}\bar{\psi} \, \exp\left[i \int d^4x \, \mathcal{L}\right] \tag{8.88}$$

of the fermion field transforms under Eq. (8.7). The derivation of the transformation property of $\mathcal{D}\psi \, \mathcal{D}\bar{\psi}$ is reviewed in great detail, for instance, in chapter 19.2 of Reference [46] or chapter 22.2 of Reference [60]. Since it is rather technical, we will simply state the final result here. It turns out that, under a chiral transformation, the functional measure transforms as

$$\mathcal{D}\psi \, \mathcal{D}\bar{\psi} \rightarrow \exp\left[i \int d^4x \, \alpha(x)\mathcal{A}(x)\right] \mathcal{D}\psi \, \mathcal{D}\bar{\psi}, \tag{8.89}$$

where the function $\mathcal{A}(x)$ is given by

$$\mathcal{A}(x) = -\frac{g_s^2}{8\pi^2} \tilde{G}^a_{\mu\nu} G^{\mu\nu,a}. \tag{8.90}$$

The QCD Lagrangian with massless quarks, hence, transforms under a global chiral transformation as

$$\mathcal{L} \rightarrow \mathcal{L} - \frac{g_s^2}{8\pi^2} \alpha \tilde{G}^a_{\mu\nu} G^{\mu\nu,a} \supset \frac{g_s^2}{16\pi^2} (\theta - 2\alpha) \tilde{G}^a_{\mu\nu} G^{\mu\nu,a}. \tag{8.91}$$

This implies that the θ angle effectively shifts $\theta \rightarrow \theta - 2\alpha$ under Eq. (8.7). The transformed θ angle can, hence, be set to zero by an appropriate global chiral transformation – i.e., by choosing $\alpha = \theta/2$. This argument leads to the conclusion that the strong CP problem would be absent for massless quarks. The up quark has, however, a non-zero mass of $m_u = (2.3^{+0.7}_{-0.5})\text{MeV}$ [43], and as a result, a different solution to the strong CP problem is required in practice.

8.5.3 The Peccei-Quinn Mechanism

The most popular solution to the strong CP problem is due to the Peccei and Quinn [44, 45], who suggested the existence of a new global chiral $U(1)_{\text{PQ}}$ symmetry (see also References [56, 63]). This symmetry is implemented at the Lagrangian level such that changes in the θ angle are equivalent to a redefinition of the fields according to a $U(1)_{\text{PQ}}$ transformation and has no physical consequences. Such a theory is then equivalent to a theory with $\theta = 0$ and, hence, has no strong CP problem. In the previous section, we have discussed how this is true in the case of massless quarks; however, Peccei and Quinn showed how this property can remain true even when all quarks are massive, provided that at least one fermion gets its entire mass from the interaction with a scalar field so that the full Lagrangian has (at least) one global $U(1)_{PQ}$ chiral symmetry.

In the simplest realisation of the Peccei-Quinn (PQ) mechanism, one adds a second Higgs doublet ϕ to the SM. One doublet then generates masses for the up-type quarks, while the other is responsible for the generation of the down-type quark masses. This fixes the $SU(2)_L \times U(1)_Y$ representation of ϕ. The Lagrangian then has an additional global chiral $U(1)_{PQ}$ symmetry. Under the corresponding transformations, the angular part of ϕ shifts. The potential of ϕ has the form

$$V(\phi) = \lambda \left(|\phi|^2 - \frac{f_a^2}{2} \right)^2 \tag{8.92}$$

and develops a VEV at $\langle \phi \rangle = f_a/\sqrt{2}$. Once the global PQ symmetry is spontaneously broken, a pNGB appears: the QCD axion a. The axion couples to SM fermions and also develops interactions to SM gauge boson through the chiral anomaly. In particular, the couplings to gluons take the form $a/f_a \tilde{G}^a_{\mu\nu} G^{\mu\nu,a}$.

It follows that the Lagrangian introduced in Eq. (8.81) is modified to include a kinetic term and an interaction term for the QCD axion field

$$\mathcal{L} \supset \frac{1}{2} (\partial_\mu a)^2 + \frac{a/f_a + \theta}{16\pi^2} \tilde{G}^a_{\mu\nu} G^{\mu\nu,a}. \tag{8.93}$$

Because, apart from $a/f_a \tilde{G}^a_{\mu\nu} G^{\mu\nu,a}$ the full theory depends only on the derivative $\partial_\mu a$ but not on the axion field a itself, the θ term appearing in Eq. (8.81) can be eliminated by the shift,

$$a \to a - \theta/f_a, \tag{8.94}$$

without changing the rest of the theory. In fact, one can prove [44, 45, 56, 63] that the axion potential has a minimum when the CP-violating term in the Lagrangian vanishes. Accordingly, the coefficient of the CP-violating term is driven to zero, and the strong CP problem is solved. By expanding the potential to second order, one can obtain the axion mass. It is approximately given by

$$m_a \simeq \frac{f_\pi}{f_a} m_\pi, \tag{8.95}$$

where m_π and f_π are the pion mass and its decay constant, respectively. The crucial property is that the axion mass as well as all its interactions to SM particles are inversely proportional to the axion decay constant f_a.

Originally, it was suggested that f_a should be of the order of the EW symmetry-breaking scale $v \simeq 246$ GeV. The corresponding pNGB is usually referred to as the Peccei-Quinn-Weinberg-Wilczek axion [44, 45, 56, 63]. However, this possibility was soon excluded by data. If, instead, f_a is very large, then the coupling of the axions becomes very small as the interaction scales as $1/f_a$ and the axion becomes very light, according to Eq. (8.95). Since these axions with $f_a \gg v$ could evade all experimental constraints, they were dubbed invisible axions. For instance, if $f_a = \mathcal{O}(10^{11}\,\text{GeV})$, then $m_a = \mathcal{O}(10^{-10}\,\text{eV})$, which is compatible with current experimental bounds.

Viable axion models include those that extend the hadronic sector of the theory to include new heavy quarks. An example is the Kim-Shifman-Vainstein-Zakharov (KSVZ) model [38, 51], which, besides the axion, introduces a weakly interacting singlet quark. Alternatively, there are models that include additional spin-0 fields in the theory. For instance, the Dine-Fischler-Srednicki-Zhitnitsky (DFSZ) model introduces an additional Higgs field besides the axion [22, 64]. In these models, the PQ symmetry-breaking mechanism is decoupled from EWSB; hence, the axion decay constant f_a can take very small values which are not excluded by current bounds.

Almost 40 years since the PQ proposal, the axion solution to the strong CP problem remains the most compelling explanation for the smallness of the θ angle. Furthermore, axions can be natural dark matter candidates since they are electrically neutral and weakly interacting. Hence, a variety of searches for axions or axion-like particles have been performed in the past years, as reviewed, for instance, in Reference [33]. Since the QCD axion has a two-photon vertex that is inherited from the mixing of the a field with neutral pions or eta mesons, photon-axion conversion experiments provide one of the main search strategies in laboratories today. Besides photon-axion conversion, many other axion searches are being pursued, as summarised in [43].

8.6 Anthropic Arguments

8.6.1 The Essence of Anthropic Arguments

Anthropic arguments [15] have been first proposed to solve the cosmological constant problem. Subsequently, they also have been applied to the fine-tuning problem in the EW sector. Anthropic arguments are based on the assumption that there are many possible vacua (possible universes); in some of these universes, some parameters appear to be fine-tuned, while in others, they do not. However, if one can argue that our existence is allowed only in those domains where the parameters seem fine-tuned, then, by construction, we have to observe fine-tuned parameters because if these parameters had different values, we would not be there to observe them. It is in this sense that anthropic arguments can justify the fact that we observe what appear to be unnatural parameters and so address fine-tuning.

In an analogy, one could think of an Earthlike planet of radius R almost completely covered with water, with only a small island on it, of radius r. If the island is inhabited by N people, one could ask what is the probability that they are all on the island, which is $(r^2/(4R^2))^N$. It is obvious that, for small r/R and large N, this probability can become extremely small. Does the observation that all people on the Earthlike planet, be it thousands or millions, live on the island pose a fine-tuning problem? The answer is, of course, no, since the island is the only place where people can survive, and hence, it is impossible to observe people far away from the island. Anthropic arguments are based, in essence, on a similar kind of reasoning.

One of the first applications of the Anthropic Principle was to explain the fine-tuning of the cosmological constant Λ [12, 17, 35, 36, 40, 59]. The argument is based on the observation that if Λ were much larger than its observed value – i.e., $\Lambda/(8\pi G_N) \simeq (10^{-3}\,\mathrm{eV})^4$ – the rapid expansion of the Universe would make it impossible for galaxies to form through gravitational collapse. The existence of galaxies – and, thus, of stars and planets, which is necessary for our existence – is only possible if Λ has a value that is of the same order of magnitude as what we observe. This is the anthropic solution of the cosmological constant problem.

In a similar way, anthropic arguments have been used to explain the observed Higgs VEV [2, 23]. In this case, it is the existence of atoms other than the hydrogen that would not be possible if v were very different from its observed value. In fact, if one keeps all parameters in the SM fixed but increases v, all masses increase. In particular, the neutron-proton mass splitting increases. Nuclear binding energies, on the other hand, decrease, as this binding can be viewed as mediated by the exchange of pions, which become heavier, and hence, their interactions become shorter range. In order for complex nuclei to form, the neutron-proton mass splitting must be smaller than the nuclear binding energy, as otherwise, all bound neutrons would decay to protons. This results in a constraint on v of about a factor of two larger than what is observed. An even stronger constraint comes from the requirement of having stable deuterium, whose existence is crucial for primordial nucleosynthesis. Since the existence of complex atoms and of deuterium is a prerequisite for living organisms to form, anthropic arguments can address the gauge hierarchy problem.

However, the discussed anthropic arguments cannot be used to explain the smallness of the θ angle (see Eq. (8.87)). If the θ parameter had any value between 0 and 2π, EDMs and binding energies would change slightly, but there would be no dramatic consequence for structure formations or for the existence of life. Hence, a simple anthropic reasoning does not solve the strong CP problem.

8.6.2 Multiverse Theories and the String Landscape

In order for the anthropic arguments to work, one needs to construct a theory that has very many ground states. This is in contrast to most theories that have just one or a small number of ground states, like the SM or SUSY theories. The need for many ground states stems from the fact that, if a parameter has a fine-tuning problem of the order of 10^{-N}, there should be of the order of 10^N ground states for there to be, on average, at least one in which anthropic arguments allow life formation. String theories are characterised by the presence of many ground states, so they are natural theories to exploit anthropic arguments. Here, one has many low-energy solutions that depend on the shape and size of the extra dimensions. Having many allowed solutions to a theory with parameters that are allowed to change is what is usually referred to as a landscape.

A criticism that anthropic arguments often face is that, by definition, it is impossible to test the existence of other vacuum states; hence, anthropic principles can never be confirmed or refuted. For this reason, anthropic principles are often rejected by physicists as being not scientific arguments. One way around this is to observe that anthropic or multiverse principles are just consequences of some fundamental theory. It is this complete theory, with a multiverse property, that should be testable and tested. The problem, however, is that, in these theories, because of the enormous landscape of solutions, it is notoriously difficult if not impossible to explore and to test the full theory space of solutions.

8.7 Final Remarks

It is still possible that nature respects the technical naturalness criterion of 't Hooft, but measurements at the LHC are challenging this criterion more and more. Still, despite the tension between 'natural' new-physics models and LHC data, the high-energy particle physics community is not yet ready to fully give up on naturalness as a guiding principle. There if still hope that the LHC will soon find signs of SUSY, extra dimensions, or other manifestations of BSM physics that will revive our belief in naturalness. Nevertheless, the unsettling possibility – that EW naturalness is not the right guiding principle to understand what physics lies beyond the SM – must be considered. If that is the case, it becomes crucial to identify what instead are the other main fundamental questions that can be addressed at the LHC and at future colliders. In fact, in the case of the LHC, there was a no-lose theorem in the sense that something new was guaranteed to show up at the LHC: either the energy of the machine was sufficient to produce a Higgs boson directly or else some sign of new physics would need to appear to prevent unitarity violations (i.e., probabilities larger than one) at the TeV scale. The LHC discovery of the Higgs boson with a mass of around 125 GeV makes the SM a consistent theory, albeit with fine-tuning problems, up to very large energy scales that are beyond the reach of even future colliders. From that point of view, there is no guarantee that the LHC or a future collider will discover physics beyond the SM. If the LHC sees no sign of new physics and natural theories are excluded or severely constrained, we find ourselves without a strong indication of what is the energy scale at which new dynamics may appear. Still, the implications of finding new particles would be so far-reaching that there is no doubt that searches for new physics at colliders must continue, irrespectively of fine-tuning arguments.

References

[1] Aad, G., *et al.* 2015. 'Combined Measurement of the Higgs Boson Mass in *pp* Collisions at $\sqrt{s} = 7$ and 8 TeV with the ATLAS and CMS Experiments'. *Physical Review Letters*, **114**, 191803.

[2] Agrawal, V., *et al.* 1998. 'Anthropic Considerations in Multiple Domain Theories and the Scale of Electroweak Symmetry Breaking'. *Physical Review Letters*, **80**, 1822–1825.

[3] Anderson, G. W., and Castano, D. J. 1995. 'Measures of Fine Tuning'. *Physics Letters B*, **347**, 300–308.

[4] Arkani-Hamed, N., *et al.* 2002a. 'The Littlest Higgs'. *Journal of High Energy Physics*, **07**, 034.

[5] Arkani-Hamed, N., *et al.* 2002b. 'The Minimal Moose for a Little Higgs'. *Journal of High Energy Physics*, **08**, 021.

[6] Arkani-Hamed, N., Dimopoulos, S., and Dvali, G. R. 1998. 'The Hierarchy Problem and New Dimensions at a Millimeter'. *Physics Letters B*, **429**, 263–272.

[7] Arkani-Hamed, N., Cohen, A. G., and Georgi, H. 2001. 'Electroweak Symmetry Breaking from Dimensional Deconstruction'. *Physics Letters B*, **513**, 232–240.

[8] Athron, P. and Miller, D. J. 2007. 'A New Measure of Fine Tuning'. *Physical Review D*, **76**, 075010.

[9] Aubert, J. J., *et al.* 1974. 'Experimental Observation of a Heavy Particle J'. *Physical Review Letters*, **33**, 1404–1406.

[10] Augustin, J. E., *et al.* 1974. 'Discovery of a Narrow Resonance in e+ e- Annihilation'. *Physical Review Letters*, **33**, 1406–1408.

[11] Baker, C. A., *et al.* 2006. 'An Improved Experimental Limit on the Electric Dipole Moment of the Neutron'. *Physical Review Letters*, **97**, 131801.

[12] Banks, T., Dine, M., and Motl, L. 2001. 'On Anthropic Solutions of the Cosmological Constant Problem'. *Journal of High Energy Physics*, **01**, 031.

[13] Barbieri, R., and Giudice, G. F. 1988. 'Upper Bounds on Supersymmetric Particle Masses'. *Nuclear Physics B*, **306**, 63–76.

[14] Bardeen, W. A. 1995. 'On Naturalness in the Standard Model'. In *Ontake Summer Institute on Particle Physics Ontake Mountain, Japan, August 27–September 2, 1995.*

[15] Barrow, J. D., and Tipler, F. J. 1988. *The Anthropic Cosmological Principle*. Oxford University Press.

[16] Belavin, A. A., *et al.* 1975. 'Pseudoparticle Solutions of the Yang-Mills Equations'. *Physics Letters B*, **59**, 85–87.

[17] Bjorken, J. D. 2001. 'Standard Model Parameters and the Cosmological Constant'. *Physical Review D*, **64**, 085008.

[18] Cabibbo, N. 1963. 'Unitary Symmetry and Leptonic Decays'. *Physical Review Letters*, **10**, 531–533.

[19] Ciafaloni, P., and Strumia, A. 1997. 'Naturalness Upper Bounds on Gauge Mediated Soft Terms'. *Nuclear Physics B*, **494**, 41–53.

[20] Das, T., *et al.* 1967. 'Electromagnetic Mass Difference of Pions'. *Physical Review Letters*, **18**, 759–761.

[21] de Gouvea, A., Hernandez, D., and Tait, T. M. P. 2014. 'Criteria for Natural Hierarchies'. *Physical Review D*, **89**(11), 115005.

[22] Dine, M., Fischler, W., and Srednicki, M. 1981. 'A Simple Solution to the Strong CP Problem with a Harmless Axion'. *Physics Letters B*, **104**, 199–202.

[23] Donoghue, J. F. 2007. 'The Fine-Tuning Problems of Particle Physics and Anthropic Mechanisms'. In Carr, B. (ed.): *Universe or Multiverse*. Cambridge University Press. 231–246.

[24] Farina, M., Perelstein, M., and Rey-Le Lorier, N. 2014. 'Higgs Couplings and Naturalness'. *Physical Review D*, **90**(1), 015014.

[25] Fermi, E. 1934. 'An Attempt of a Theory of Beta Radiation. 1'. *Zeitschrift für Physik*, **88**, 161–177.

[26] Gaillard, M. K., and Lee, B. W. 1974. 'Rare Decay Modes of the K-Mesons in Gauge Theories'. *Physical Review D*, **10**, 897.

[27] Georgi, H., and Kaplan, D. B. 1984. 'Composite Higgs and Custodial SU(2)'. *Physics Letters B*, **145**, 216–220.

[28] Gildener, E. 1976. 'Gauge Symmetry Hierarchies'. *Physical Review D*, **14**, 1667.

[29] Giudice, G. F. 2013. 'Naturalness after LHC8'. *Proceedings of Science*, **EPS-HEP2013**, 163.

[30] Glashow, S. L., Iliopoulos, J., and Maiani, L. 1970. 'Weak Interactions with Lepton-Hadron Symmetry'. *Physical Review D*, **2**, 1285–1292.

[31] Goldberger, W. D., and Wise, M. B. 1999. 'Modulus Stabilization with Bulk Fields'. *Physical Review Letters*, **83**, 4922–4925.

[32] Graham, P. W., Kaplan, D. E., and Rajendran, S. 2015a. 'Cosmological Relaxation of the Electroweak Scale'. *Physical Review Letters*, **115**(22), 221801.

[33] Graham, P. W., *et al.* 2015b. 'Experimental Searches for the Axion and Axion-Like Particles'. *Annual Review of Nuclear and Particle Science*, **65**, 485–514.

[34] Grinbaum, A. 2012. 'Which Fine-Tuning Arguments Are Fine?'. *Foundations of Physics*, **42**, 615–631.

[35] Hogan, C. J. 2000. 'Why the Universe Is Just So'. *Reviews of Modern Physics*, **72**, 1149–1161.

[36] Kallosh, R., and Linde, A. D. 2003. 'M Theory, Cosmological Constant and Anthropic Principle'. *Physical Review D*, **67**, 023510.

[37] Kaplan, D. B., and Georgi, H. 1984. 'SU(2) x U(1) Breaking by Vacuum Misalignment'. *Physics Letters B*, **136**, 183–186.

[38] Kim, J. E. 1979. 'Weak Interaction Singlet and Strong CP Invariance'. *Physical Review Letters*, **43**, 103.

[39] Kobayashi, M., and Maskawa, T. 1973. 'CP Violation in the Renormalizable Theory of Weak Interaction'. *Progress of Theoretical Physics*, **49**, 652–657.

[40] Martel, H., Shapiro, P. R. and Weinberg, S. 1998. 'Likely Values of the Cosmological Constant'. *Astrophysical Journal*, **492**, 29.

[41] Martin, S. P. 1998. *A Supersymmetry Primer*. World Scientific. Pages 1–98.

[42] Murayama, H. 2000. 'Supersymmetry Phenomenology'. *Proceedings, Summer School in Particle Physics: Trieste, Italy, June 21–July 9, 1999*. Pages 296–335.

[43] Olive, K. A. *et al.* 2014. 'Review of Particle Physics'. *Chinese Physics C*, **38**, 090001.

[44] Peccei, R. D., and Quinn, H. R. 1977a. 'Constraints Imposed by CP Conservation in the Presence of Instantons'. *Physical Review D*, **16**, 1791–1797.

[45] Peccei, R. D. and Quinn, H. R. 1977b. 'CP Conservation in the Presence of Instantons'. *Physical Review Letters*, **38**, 1440–1443.

[46] Peskin, M. E., and Schroeder, D. V. 1995. *An Introduction to Quantum Field Theory*. Avalon Publishing.

[47] Randall, L. and Sundrum, R. 1999a. 'A Large Mass Hierarchy from a Small Extra Dimension'. *Physical Review Letters*, **83**, 3370–3373.

[48] Randall, L., and Sundrum, R. 1999b. 'An Alternative to Compactification'. *Physical Review Letters*, **83**, 4690–4693.

[49] Sakharov, A. D. 1967. 'Violation of CP Invariance, c Asymmetry, and Baryon Asymmetry of the Universe'. *Pisma Zh. Eksp. Teor. Fiz.*, **5**, 32–35.

[50] Schmaltz, M. 2004. 'The Simplest Little Higgs'. *Journal of High Energy Physics*, **08**, 056.

[51] Shifman, M. A., Vainshtein, A. I., and Zakharov, V. I. 1980. 'Can Confinement Ensure Natural CP Invariance of Strong Interactions?' *Nuclear Physics B*, **B166**, 493–506.

[52] Streater, R. F., and Wightman, A. S. 1989. *PCT, Spin and Statistics, and All That*. Princeton University Press.

[53] Susskind, L. 1979. 'Dynamics of Spontaneous Symmetry Breaking in the Weinberg-Salam Theory'. *Physical Review D*, **20**, 2619–2625.

[54] 't Hooft, G., *et al.* 1980. 'Recent Developments in Gauge Theories. Proceedings, Nato Advanced Study Institute, Cargese, France, August 26–September 8, 1979'. *NATO Science Series B*, **59**, 1–438.

[55] Veltman, M. J. G. 1981. 'The Infrared: Ultraviolet Connection'. *Acta Physica Polonica B*, **12**, 437.

[56] Weinberg, S. 1978. 'A New Light Boson?' *Physical Review Letters*, **40**, 223–226.

[57] Weinberg, S. 1979a. 'Gauge Hierarchies'. *Physics Letters B*, **82**, 387–391.

[58] Weinberg, S. 1979b. 'Implications of Dynamical Symmetry Breaking: An Addendum'. *Physical Review D*, **19**, 1277–1280.

[59] Weinberg, S. 1996. 'Theories of the Cosmological Constant'. *Critical Dialogues in Cosmology. Proceedings, Celebration of the 250th Anniversary of Princeton University, Princeton, USA, June 24–27, 1996*. Pages 195–203.

[60] Weinberg, S. 2013. *The Quantum Theory of Fields. Vol. 2: Modern Applications*. Cambridge University Press.

[61] Weisskopf, V. F. 1939. 'On the Self-Energy and the Electromagnetic Field of the Electron'. *Physical Review*, **56**, 72–85.

[62] Wells, J. D. 2016. 'Higgs Naturalness and the Scalar Boson Proliferation Instability Problem'. *Synthese*, **194**, 477–490.

[63] Wilczek, F. 1978. 'Problem of Strong P and T Invariance in the Presence of Instantons'. *Physical Review Letters*, **40**, 279–282.

[64] Zhitnitsky, A. R. 1980. 'On Possible Suppression of the Axion Hadron Interactions. (In Russian)'. *Soviet Journal of Nuclear Physics*, **31**, 260.

9

Dark Matter

E D W A R D   W.   K O L B

Abstract

For more than eight decades, astronomical observations have suggested that most of the mass of the Universe is not visible at any wavelength; it is dark, hence the name *dark matter*. The nature of dark matter is one of the fundamental questions in modern cosmology. In this essay, I will briefly review the history of astronomical evidence for dark matter, discuss some of the possibilities for dark matter that have been proposed over the decades, describe the role of dark matter in structure formation, and focus on the possibility that dark matter is a yet-to-be-discovered elementary particle. Where appropriate, I will also comment on the fine-tuning aspects of dark matter.

9.1 Overview of the Current Composition of the Universe

Cosmology is the study of the origin, composition, evolution, and large-scale structure of the Universe. The subject of this essay is the composition of the Universe – in particular, the composition of one component of the present Universe: the dark matter component. This is not an insignificant issue. Roughly 20% of the total present mass-energy[1] of the Universe is in the form of dark matter, and most of the total matter density is dark matter.

We have a very good determination of the present total mass-energy density of the Universe from measurements of the temperature anisotropies in the cosmic background radiation (CBR). CBR measurements imply that on cosmological scales, the Universe has a vanishingly small *spatial* curvature. In the standard cosmological model, this implies that the present mass-energy of the Universe must be close to the critical density, $\rho_C = 3H_0^2/8\pi G_N$. Here ρ_C is the critical density, H_0 is the present value of the expansion rate of the Universe (the Hubble constant), and, of course, G_N is Newton's gravitational constant. It is traditional to express H_0 in terms of a dimensionless constant h: $H_0 = 100h$ km s^{-1} Mpc^{-1}. In terms of h, the critical density is $\rho_C = 1.88h^2 \times 10^{-29}$ g cm^{-3}.

---

[1] In order to compare mass density and energy density, I will make use of the most famous equation in twentieth-century physics: $E = mc^2$. I will almost always suppress explicit factors of the speed of light, c.

At present, there are two methods to determine the Hubble constant. The first is the traditional 'standard-candle' method pioneered by Hubble [19]. This yields a value of about $h = 0.72$ [13]. The procedure is to construct a distance ladder from a variety of 'standard candles'. Just as a chain is only as strong as its weakest link, the distance ladder is only as reliable as its weakest rung. In the last decade, tremendous advances in reducing the systematic uncertainties in the distance-ladder technique have reduced the systematic uncertainties in this approach. The second method is to infer the value of H_0 from measurements of fluctuations in the CBR. This method must assume a cosmological model for the evolution of the Universe from the time of recombination, about 380,000 years AB,[2] to today, about 14 billion years AB. The model assumed in the analysis is known as Lambda Cold Dark Matter (ΛCDM), which will be described later. Since there are model assumptions in the CBR determination of H_0, using this method provides an *indirect* determination of H_0. This yields a best-fit value of around $h = 0.68$, smaller than the standard-candle method. The two methods disagree at about three standard deviations. Whether this 'tension' between distance-ladder and CBR determinations is fundamental, or whether things will eventually sort themselves out and remove the discrepancy, remains to be seen [14]. Luckily for us, our discussion is qualitative, and the exact value of h will not matter. So we will properly regard the value of H_0 as a nuisance parameter and make the convenient choice $h = 1/\sqrt{2}$. This yields $\rho_C = 9.4 \times 10^{-30}$ g cm^{-3}.

We will not have to remember the value of the critical density because we will write the densities of all the components of the mass-energy density in terms of the fraction of the mass-energy density of a component divided by the critical density. These fractions are denoted by Ω for each component. Since we know from the CBR that the total mass-density is close to ρ_C, the sum of the omegas for all the components must be unity.

Figure 9.1 illustrates the magnitude of various components of the present mass-energy density of the Universe in the ΛCDM model. The values of the components given as a percentage correspond to the the values of omega for the component. It is illuminating to spend some time commenting on the values.

Radiation: For the first 60,000 years of cosmic history, the mass-energy density of the early Universe was dominated by radiation. But today, radiation has been redshifted by the expansion of the Universe, and it contributes only a small fraction, about 0.005%, of the total mass-energy density.

Chemical Elements: By chemical elements, I mean elements other than hydrogen and helium. Although they are crucial for our existence, they contribute only a very small percentage, about 0.025%, of the total mass-energy density. Important for us, chemistry is not so important in the composition of the Universe.

Neutrinos: Neutrinos are elementary particles. They come in three types: the electron-neutrino, the muon-neutrino, and the tau-neutrino. They have no electric charge, but they are related to charged leptons, electrons, muons, and taus by a symmetry. Very roughly,

---

[2] The notation 'AB' stands for 'After Bang'.

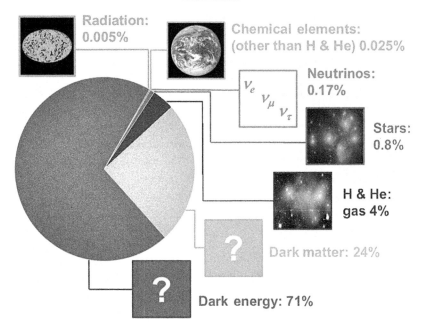

Figure 9.1 The present composition of the Universe in the ΛCDM model. See the text for explanation and elaboration.

you can imagine an electron neutrino as what would be left over if you could remove the electric charge from an electron. In the present composition of the Universe, they are even more important than the chemical elements, contributing about 0.17% of the present composition.

Hydrogen and Helium: Most of the normal matter in the Universe is in the form of a diffuse gas of hydrogen and helium that pervades large-scale structures such as galaxy clusters. Although it has a much larger fraction of the present mass-energy density than chemical elements (as defined earlier), it still contributes only about 4% to the total composition.

This exhausts the list of the components we understand. The rest of the Universe, 95%, is dark and not understood!

Dark Matter: About 24% of the present Universe is in the form of a type of matter we do not see. In the next section, I will talk about the observational evidence for dark matter. Dark matter is the subject of this essay. Using $\Omega_{DM} = 0.24$ and $h = 1/\sqrt{2}$, the useful quantity $\Omega_{DM}h^2$ takes the value $\Omega_{DM}h^2 = 0.12$.

Dark Energy: In the standard ΛCDM model, most of the mass-energy of the present Universe is contributed by a non-zero value of Einstein's cosmological constant. Dark energy is the name for the phenomenon that causes the recent expansion velocity of the Universe to increase with time. It contributes about 71% to the present composition.

Why do we care about dark matter? One reason is that it is a very significant fraction of the present Universe. If the goals of cosmology include understanding the composition of the Universe, dark matter can hardly be ignored. Another reason to care about it is that it plays a dominant role in the formation of structure in the Universe. Without dark matter, structure in the Universe would have evolved and formed in a very different way. This is not to say there would not be structure in the Universe without dark matter, but the large-scale structure of the Universe would be quite different.

Is the exact percentage of dark matter finely tuned? If it would be 14% or 34% rather than 24% would the evolution of the Universe or the formation of large-scale structure, galaxies, stars, or planets have been much affected? It does not appear so. In fact, it is hard to argue that dark matter is needed at all for our existence. I will return to some of these issues in the conclusion section.

9.2 A Brief History of the Discovery of Dark Matter

The first indications that there was more to the matter distribution than meets the eye go back at least to the 1930s. But for many decades, physicists (and astronomers as well) regarded the indications as an astronomical curiosity and not a fundamental issue. I do not think that there was any one single observation that awakened physicists and astronomers to the realisation that dark matter was real and fundamental. Rather, it was the slow accumulation of evidence that became stronger with time, as well as several influential scientists taking the matter seriously. Although there was not a single paper that jolted astronomers and physicists at the time, in hindsight, we can pick a few papers out of many that illustrate the growth of the observational evidence for dark matter.

A good starting point is the 1932 paper of Jan Oort [29], 'The Force Exerted by the Stellar System in the Direction Perpendicular to the Galactic Plane and Some Related Problems'. In this paper, Oort studied the distribution of (later-type) stars in the direction perpendicular to the galactic plane and deduced the gravitational force necessary to maintain the observed distribution. One of the purposes of his paper was 'the derivation of an accurate value for the total amount of mass, *including dark matter* [my emphasis], corresponding to a unit of luminosity surrounding of the sun'. So the phrase 'dark matter' makes an appearance at least as early as 1932. Oort also determined the mass in units of the solar mass, $M_\odot = 2 \times 10^{33}$ g, and the observed stellar luminosity in solar units, $L_\odot = 4 \times 10^{33}$ erg s^{-1}. The number he found for the mass-to-light ratio within our solar neighbourhood (in solar units) was $(\mathcal{M}/\mathcal{L})_\odot \sim 1.8$, remarkably close to the modern value. Of course, Oort found some indications about the distribution of dark matter that did not turn out to be true; for instance, Oort said 'There is an indication that the invisible mass is more strongly concentrated to the galactic plane than that of visible stars'. Characteristic of Oort, it is a complete and insightful analysis. Today it is thought that the dark matter determined by Oort has a significant contribution of non-baryonic matter.

Another significant paper on the subject of dark matter from the 1930s is the 1937 paper of Fritz Zwicky [38], 'On the Masses of Nebulae and of Clusters of Nebulae'.[3] In this paper, Zwicky proposes three methods for determining 'nebular' (galaxy) masses: the first method makes use of the virial theorem to determine the mass of galaxy clusters (in particular, the Coma cluster). Zwicky found a mass for Coma of $4.5 \times 10^{10} M_\odot$ – again, remarkably close to the presently accepted value. This led to a mass-to-light ratio of about $(\mathcal{M}/\mathcal{L})_\odot \sim 500$ for the Coma cluster. The second method is gravitational lensing. This prescient suggestion would not come to fruition for many decades. Zwicky's third proposal was to use the statistical spatial distribution of different types of nebulae. This method did not prove to be a useful technique. The modern interpretation of Zwicky's result is that some of the dark matter is baryonic (the intracluster hydrogen and helium in Figure 9.1), but most of it is non-baryonic dark matter.

By the end of the 1930s, there was evidence that a significant fraction (if not the bulk) fraction of matter associated with galaxies and clusters was dark. We now regard this as one of the fundamental unsolved issues in astronomy/physics, but it did not attract the attention of physicists (or many astronomers, for that matter) for several decades.

The significance of dark matter in astronomy and physics was only appreciated beginning in the 1970s. Both theoretical and observational advances contributed to the awareness.

The first theoretical proposal that non-baryonic matter could be dynamically important is embedded in the paper of Cowsik and McClelland [9], 'An Upper Limit on the Neutrino Rest Mass'. It was not known in 1972 that neutrinos have a small mass, but recent neutrino oscillation experiments imply that they do, and thus, they contribute a (small) fraction of the mass density of the Universe. While neutrinos are not 'the' dark matter, they provide an existence proof for a weakly interacting particle relic of the early Universe. Since neutrinos are weakly interacting and massive, they qualify as WIMPs (weakly interacting massive particles).

The most important dark matter observational program in the 1970s was the development of solid evidence for dark matter based on galactic rotation curves. Prominent among the astronomers working in this area at the time was Vera Rubin and collaborators (see, e.g., her classic paper with Kent Ford on the rotation of the Andromeda nebula [31]), although it is possible to find much earlier results along the same line (e.g., Horace Babcock's paper on the rotation of Andromeda [2], or even earlier attempts to measure the mass-to-light ratio of Andromeda by Hubble in 1929 [20]). Deducing the mass-to-light ratio from galactic rotation curves was a technique simple enough for even physicists to understand, as opposed to the astronomy-dense papers of Oort and Zwicky.

While in the 1970s, there was a growing awareness of the issue of dark matter, most physicists who knew about it still considered it just a curiosity. That dark matter merited

---

[3] Much of this paper is contained in a less well known paper of Zwicky from 1933 [37] published in German using the phrase *dunkle materie*.

the attention of particle physicists was largely due to the influential paper of Benjamin Lee and Steven Weinberg, 'Cosmological Lower Bound on Heavy Neutrino Masses' [25]. In this paper, Lee and Weinberg showed that a heavy neutrino with mass in the GeV range would be produced from weak interactions in the thermal plasma of the early Universe and survive in the correct numbers to be the dark matter.[4] Although Lee and Weinberg do not use the term 'dark matter', they concluded the article with the statement 'the gravitational field of these heavy neutrinos would provide a plausible mechanism for closing the Universe'. 'Closing the Universe' was used to mean that the heavy neutrinos would provide the unaccounted-for mass density necessary if the total mass-energy density would be the critical density. After Lee and Weinberg, it was considered legitimate for particle physics to work in cosmology, especially on the subject of dark matter. The heavy neutrino was the first example of a *cold thermal relic* – a particle species that was once in local thermodynamic equilibrium (LTE) and froze out of equilibrium when it was non-relativistic (or only semi-relativistic). Cold thermal relics as dark matter will be the main thrust of this chapter.

Since the 1970s, the observational evidence for dark matter has grown. The evidence comes from observations and theoretical considerations on a wide range of astronomical scales – from our local solar neighbourhood (essentially Oort's approach) to dwarf galaxies, galaxies, groups of galaxies, clusters, superclusters, and across the entire Universe (for a non-technical review, see Reference [30]).

9.3 Dark Matter Bestiary

Astronomy has presented physicists with the problem that the bulk of the matter density in the Universe is dark. What could be the nature of the dark matter? Theorists have not been silent on this question; they have proposed dozens (perhaps hundreds) of ideas for the nature of dark matter. It would be tedious to discuss them all (or even list them all). Here, I will just give some broad categories, delve into one particular category, and then focus on just one of the subcategories.

The three broad categories for dark matter are

1. *MOND/TeVeS:* We only deduce that unseen dark matter exists because of the effect it exerts on objects we do see. If the gravitational force law does not follow Einstein's theory of general relativity, or if the response of test particles to the gravitational force on astronomical scales of interest does not obey Newtonian dynamics, then dark matter may not exist at all. The idea of MOND (MOdified Newtonian Dynamics) was proposed by Milgrom in 1983 [27]. The relativistic version of MOND, known as TeVeS, was first proposed by Bekenstein in 2006 (see Reference [3]).

---

[4] Although the model is associated with the names Lee and Weinberg, it was discovered independently by a number of people [10, 21, 32, 35].

2. *MACHOs: (MAssive Compact Halo Objects):* The MACHO idea is that dark matter is baryonic but is contained in objects that either do not emit light (e.g., big black holes or rocky rogue planets unassociated with stellar systems) or objects that are very inefficient at emitting light (neutron stars or little dim stars like brown dwarf stars).

3. *A yet-to-be-discovered species of elementary particle:* The first 'particle' candidate for dark matter was a light (mass of a few electron volts) neutrino (see Section 9.2). The Lee-Weinberg heavy neutral lepton was the first proposal that an undiscovered species of elementary particle could be dark matter. The new particle explanation is the most studied possibility today.

Today there are still MOND/TeVeS adherents, and the Laser Interferometer Gravitational-Wave Observatory (LIGO) discovery of gravitational waves from a massive (about 30 $M_\odot$) black-hole binary system has injected new enthusiasm into MACHO fans. I will not go into the quite considerable literature arguing against MOND/TeVeS and MACHOs, but just follow my instinct that the most promising possibility is that dark matter is a new elementary particle.

9.3.1 Particle Dark Matter Taxonomy

Even within the class of particle dark matter candidates, one could fill a chapter just listing the possibilities, so some taxonomy of particle dark matter is necessary. For the sake of simplicity, I will classify relic particles from the Big Bang into two broad classes: thermal relics (or relics that are the result of the decay from or oscillation from thermal relics) and nonthermal relics.

Nonthermal relics are particles whose present abundance does not depend on whether or not they were in LTE in the thermal bath. Examples are particles produced in cosmological phase transitions. These include Bose-Einstein condensates, axions, axion miniclusters, and solitons. There are also nonthermal relics that were produced through gravitational production during inflation (WIMPzillas).

Thermal relics were once in LTE with the primordial plasma. Their present abundance is determined by when they decoupled from thermal equilibrium (when they froze out). Thermal relics include (sub-)eV mass neutrinos, sterile neutrinos, gravitini, the lightest supersymmetric particles, and the lightest Kaluza-Klein particle. Some of the thermal relics, like light neutrinos, froze out when they were relativistic. But the most promising particle dark matter candidates froze out when they were non-relativistic (or mildly relativistic). These are referred to as *cold thermal relics*. That is the possibility I will develop.

9.3.2 Cold Thermal Relics: Origin of Species

The first assumption in the cold thermal relic scenario is that there is an undiscovered particle species that is stable, or at least has a lifetime much greater than the age of the Universe. The second assumption for cold thermal relics is that there is no asymmetry between particles and antiparticle associated with the species. Finally, it is assumed that the species was in chemical equilibrium with standard model (SM) particles at temperatures greater than the mass of the dark matter (DM) particle, and it dropped out of chemical equilibrium when it was non-relativistic.

What would keep the massive cold thermal relic from decaying? Two possible schemes have been employed. The first is to assume that the particle is the lightest particle that carries some conserved additive quantum number. A familiar example of a particle that is stable because of an additive quantum number is the proton, which is stable because of conservation of baryon number. It is also possible that the particle is stable because it is the lightest state that carries a multiplicative quantum number (i.e., a conserved parity). In this case, the particle could only be created or destroyed in pairs or associated with another like-parity particle. If it was the lightest odd-parity state, it could not decay into an even-parity final state containing SM particles.

The number of baryons that survived annihilation in the early Universe is determined by the asymmetry between baryons and anti-baryons (see Chapter 5). Presumably, there is a similar asymmetry in the leptonic sector. The origin of this asymmetry between matter and antimatter is unknown. While there are dark matter scenarios where the relic abundance is determined by an asymmetry in the dark sector (for a review, see Reference [36]), the usual assumption in the cold thermal relic scenario is that there is no asymmetry.

In discussing equilibrium (or lack of it), we make the important distinction between kinetic equilibrium and chemical equilibrium. Let's denote the dark matter particle species as χ. Kinetic equilibrium is established and maintained through processes like $\chi + \gamma \longleftrightarrow \chi + \gamma$, where the net number of χ does not change in the process. Here, γ stands for any particle in the thermal bath. If such reactions are effective in establishing *kinetic equilibrium*, the phase-space occupancy of the species would be either the usual Fermi-Dirac or the Bose-Einstein distribution, depending on whether the particle has half-integer spin (Fermi-Dirac) or integer spin (Bose-Einstein).

If the particle species is in *chemical equilibrium*, then its chemical potential is related to all other chemical potentials in all processes that involve the species. For example, if χ interacts with species a, b, and c through the process $\chi + a \longleftrightarrow b + c$, then chemical equilibrium will establish $\mu_\chi + \mu_a = \mu_b + \mu_c$. Consider the process $\chi + \gamma \longleftrightarrow \chi + \gamma$, except now let γ represent a particle with zero chemical potential (like actual photons) or with a chemical potential much less than the temperature, so it can be ignored (like for quarks and leptons when relativistic). If $\mu_\gamma = 0$, then chemical equilibrium gives no information (i.e., the empty relation $\mu_\chi = \mu_\chi$). However, if the process is $\chi + \bar{\chi} \longleftrightarrow b + c$, there is information provided by the assumption of chemical equilibrium. Again, consider the case of $\mu_b = \mu_c = 0$. If this process is effective, establishing chemical equilibrium, then recalling that the lack of an asymmetry implies $\mu_\chi = \mu_{\bar{\chi}}$, we have $\mu_\chi + \mu_{\bar{\chi}} = 2\mu_\chi = 2\mu_{\bar{\chi}} = 0.^5$

---

5 In the case χ is self-conjugate and stable because of conservation of some parity, we could consider the process $\chi + \chi \longleftrightarrow b + c$ and reach the same conclusion about μ_χ.

Whether in chemical equilibrium, kinetic equilibrium, or not in equilibrium at all, the number density n of a particle with g internal degrees of freedom can be written in terms of the phase-space density of the species $f(\vec{p})$:

$$n = \frac{g}{(2\pi)^3} \int f(\vec{p}) \, d^3p. \tag{9.1}$$

Since we are assuming the Universe is homogeneous and isotropic, there is no preferred location or direction, which implies the phase-space density only depends on $|\vec{p}|$ or, equivalently, depends only on the energy E. If the particle is in kinetic equilibrium, the distribution is given by $f(E) = \exp\left[(E - \mu)/T \pm 1\right]^{-1}$, where μ is the chemical potential of the species and $+1$ obtains for Fermi-Dirac and -1 obtains for Bose-Einstein particles. Since we are assuming there is no asymmetry in the DM species, the chemical potential for particles must be equal to the chemical potential for antiparticles. If, in addition, the particle is in chemical equilibrium, the distribution is given as stated previously with zero chemical potential, and the equilibrium abundance is determined by M/T.

If a particle is relativistic ($T \gg m$) and in kinetic equilibrium, the number density for a boson with zero chemical potential is $n = g(\zeta(3)/\pi^2)T^3$ and $3/4$ of that result for a fermion. Here, $\zeta(3) = 1.202\ldots$ is the Riemann zeta function. If a particle is non-relativistic ($T \ll m$) and in kinetic equilibrium, its number density is identical for fermions and bosons and is given by $n = g(mT/2\pi)^{3/2} \exp[-(m - \mu)/T]$. If, in addition, the particle is non-relativistic and in chemical equilibrium, then $n = g(mT/2\pi)^{3/2} \exp[-m/T]$; i.e., $\mu = 0$. If we ignore factors of order unity like the number of internal degrees of freedom, $\zeta(3)$, π, etc., then the number density of a particle of mass m in chemical equilibrium relative to the equilibrium number density of a massless particle is unity for $T \gg m$ and proportional to $(m/T)^{3/2} \exp[-m/T]$ for $T \ll m$.

Processes that keep the cold thermal relic in chemical equilibrium are the production of the species from SM particles (e.g., the production process SM particle + SM particle $\rightarrow$ DM particle + DM particle, and the annihilation process DM particle + DM particle $\rightarrow$ SM particle + SM particle). Processes keeping the cold thermal relic in kinetic equilibrium are process like DM + SM particle $\longleftrightarrow$ DM + SM particle.

Here, we note the important observation that, in the non-relativistic regime, the rates for reactions establishing kinetic equilibrium are larger than the rates for reactions establishing chemical equilibrium. The rate (per χ) for the process $\chi + \gamma \rightarrow \chi + \gamma$ is proportional to $n_\gamma \sigma_{\chi\gamma \rightarrow \chi\gamma}$, while the rate (per χ) for the process $\chi + \chi \rightarrow \gamma\gamma$ is proportional to $n_\chi \sigma_{\chi\chi \rightarrow \gamma\gamma}$. (Here, again, γ represents a light particle species.) Since in the non-relativistic regime, $n_\chi \ll n_\gamma$, the first process is exponentially larger than the second process by a factor of roughly $\exp[m/T]$. Also, in the regime $T \ll m$, the rate for production of a pair of χs is suppressed because it is exponentially unlikely for a collision of two light particles to have sufficient centre-of-mass energy to produce the pair.

With the knowledge from the preceding excursion into equilibrium considerations, we can state the basic cold thermal relic scenario. The initial conditions are that at temperatures above the mass of the particle, the species is in kinetic *and* chemical equilibrium. If the interactions are sufficiently strong, the particle will remain in chemical equilibrium, even as the temperature drops somewhat below the mass. If the species is still in chemical equilibrium at temperatures below the mass, the relative abundance of the species (relative to, say, photons) becomes exponentially suppressed by the Boltzmann factor $\exp(-m/T)$. As the Universe further cools below the mass of the species, kinetic equilibrium is maintained, but the particle eventually drops out of chemical equilibrium. This is because, at temperatures much less than the mass of the species, the abundance of χ is so small it is exponentially unlikely for a χ to find another χ with which to annihilate (and it is exponentially unlikely for light particles to have sufficient collisional energy to pair-produce χs in a collision). At this point the species is said to be frozen out (or the abundance frozen in). Since the particle froze out when non-relativistic, it is a *cold* thermal relic. After freeze-out, the abundance of the species only changes because of the dilution in the number density caused by the expansion of the Universe.

So freeze-out (and, hence, the present abundance) of a cold thermal relic is governed by the interplay between the DM-SM interaction strength (set in the realm of particle physics) and the expansion rate of the Universe (gravity). These are very disparate forces that act in concert to determine the present abundance of the DM particle species and, hence, its contribution to Ω.

If one knows the properties of the DM candidate, then it is possible to calculate the cross sections for annihilation and production and find the rate for production and annihilation of DM. The rate of processes keeping the particles in chemical equilibrium should be compared to the expansion rate of the Universe. If the equilibration rates are much greater than the expansion rate, the particle will track its equilibrium chemical abundance. If the equilibration rates are much less than the expansion rate of the Universe, then the particle is frozen out, and its abundance changes only through the expansion of the Universe.

We can do quite a bit better than the preceding qualitative argument. The evolution of n, the number density of a cold thermal relic, is determined by a Boltzmann equation:

$$\dot{n} = -3Hn - \langle \sigma_A v \rangle \left(n^2 - n_{eq}^2 \right) . \tag{9.2}$$

In this equation, the overdot represents the time derivative, H is the expansion rate of the Universe, $\langle \cdots \rangle$ denotes a thermal average, $\sigma_A v$ is the total annihilation cross section times the Møller flux, and n_{eq} is the equilibrium abundance of the species with zero chemical potential. It is easy to understand the physical meaning of the terms on the right-hand side of Eq. (9.2): $-3Hn$ represents the dilution of the number density due to expansion, $-\langle \sigma_A v \rangle n^2$ represents the decrease in the number density due to annihilation, and $\langle \sigma_A v \rangle n_{eq}^2$ is the increase in the number density due to production. The final abundance of the species is calculated assuming the species is in chemical equilibrium at $T > m$ and is kept in chemical equilibrium through production and annihilation processes until the rate for the processes keeping the species in chemical equilibrium drop below the expansion rate of the Universe. The expansion rate is $H = \dot{a}/a$, where $a(t)$ is the Robertson-Walker scale factor, the only dynamical metric degree of freedom in a homogeneous and isotropic Universe. After

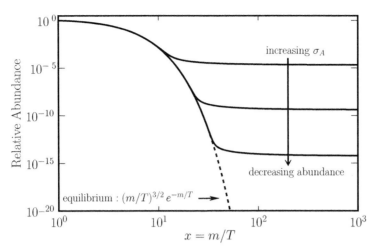

Figure 9.2 Origin of species for a cold thermal relic. Plotted is the relative abundance (say, the abundance relative to photons) as a function of the mass of the species divided by the temperature. The dashed line is the chemical equilibrium abundance. Illustrating the relative abundance, rather than the number density, removes the decrease of the number density due to expansion.

freeze-out, we can ignore interactions that change the number of χ; i.e., set $\langle \sigma_A v \rangle = 0$, and then $\dot{n}/n = -3\dot{a}/a$, which implies $n \propto a^{-3}$. Since a^3 is a volume element, the number density decreases as the volume of the Universe. So, after the species is frozen out, it decreases only due to the expansion of the Universe.

Before giving some quantitative results, it is possible to anticipate how things work. The more strongly the DM particle interacts with SM particles, the longer it will stay in equilibrium, and the lower will be its abundance when eventually it freezes out. This general feature is illustrated in Figure 9.2, where it is indicated that increasing the interaction strength, here characterised by the annihilation cross section σ_A, leads to a smaller abundance today. The more weakly interacting a particle, the larger is its contribution to the present mass density. In cosmology, at least, the weak shall dominate!

There are analytic approximations for the final abundance resulting from the Boltzmann equation. Assuming that $\sigma_A v$ is a constant in the non-relativistic limit (s-wave annihilation), the contribution to Ω from a cold thermal relic is approximately[6]

$$\Omega h^2 \simeq 0.12 \times \frac{10^{-36}\,\text{cm}^2}{\sigma_s} \qquad (\langle \sigma_A v \rangle = \sigma_s \text{ in the non-relativistic limit}) . \qquad (9.3)$$

As expected, the result decreases as the cross section increases. It is interesting to note that the result is independent of the mass of the species, or it depends on the mass of the species

---

[6] To be sure, there are many effects that can complicate this simple result: there may be a velocity dependence to the annihilation cross section, there may be resonances in the annihilation process, there is a logarithmic dependence of the mass, there might be asymmetries between particle and antiparticle, etc. The variations on the simple theme have kept many theorists (the author included) busy for many years.

only insofar as the annihilation cross section depends on the mass of the species. If $\sigma_A v$ is proportional to the velocity-squared in the non-relativistic limit (p-wave annihilation), then the estimate for Ωh^2 as a function of the annihilation cross section is modified:

$$\Omega h^2 \simeq 0.12 \times \frac{10^{-35} \, \text{cm}^2}{\sigma_p} \qquad (\langle \sigma_A v \rangle = \sigma_p v^2 \text{ in the non-relativistic limit}). \qquad (9.4)$$

Using the fact that $\hbar c = 1.97 \times 10^{-14}$ GeV-cm, we can convert the cross section in cm^2 to units of GeV^{-2}. Including the fine-structure constant α as a proxy for the square of a coupling constant, we can express $10^{-36} \, \text{cm}^2$ as approximately $\alpha^2/(150 \, \text{GeV})^2$. This suggests that 150 GeV is the mass scale for the cold thermal relic. Since 150 GeV is around the weak scale, this suggests that the cold thermal relic has weak-scale mass and interaction strength. Because the particle is weakly interacting and massive, it is called a WIMP (recall, this stands for weakly interacting massive particle). While the term WIMP is often used interchangeably with the term dark mater, we will reserve the term WIMP for a cold thermal relic with weak-scale mass and interactions. The fact that the interaction strength for a cold thermal relic is comparable to the scale of weak interactions is sometimes called the 'WIMP miracle'. Since, in principle, the cross section could have turned out to be anything, the fact that it is the magnitude associated with a known interaction is suggestive. But a miracle? Of course, apparent miracles often turn out to be mere coincidences. It is perhaps best not to use the term 'miracle' in science. Merriam-Webster's online dictionary defines miracle as 'an extraordinary event manifesting divine intervention in human affairs' – surely an inappropriate term for a scientific model. In this context, Wikipedia probably has a better description of the word miracle as something 'often used to give an impression of great and unusual value in a trivial context'.

In deriving Eqs. (9.3) and (9.4), it was assumed that the expansion rate H is given by the standard cosmology. In the standard cosmology, the early-Universe expansion rate is given by $H \sim G_N^{1/2} g_*^{1/2} T^2$, again, G_N is Newton's gravitational constant, T is the radiation temperature, and g_* counts the number of effective degrees of freedom. Here we again see the interplay between the particle physics of the DM and the overall expansion of the Universe. This is reminiscent of the interplay of the expansion rate of the Universe and nuclear cross sections that determine the abundance of the light elements produced in Big Bang nucleosynthesis.

Since the dark matter density in the vicinity of the solar system is about $0.4 \, \text{GeV cm}^{-3}$, the local number density of WIMPs is about $3 \times 10^{-3} \, \text{cm}^{-3}$ for a WIMP mass of 150 GeV (for different WIMP masses, the number density estimate should be multiplied by [150 GeV/m]. The local velocity of the WIMPs should be the galactic rotation velocity at our location in the Milky Way, about 220 km s^{-1}. This yields a local flux of about $70,000 \, \text{cm}^{-2} \, \text{s}^{-1}$. Again, if the mass differs from 150 GeV, the flux estimate should be multiplied by (150 GeV/m).

This leads to this remarkable story: while you are reading this article, invisible things are passing through you. A mysterious invisible particle species surrounds you, a relic of the first fraction of a second of the life of the Universe, and a few million are in your living room, flying around at speeds of about 800,000 kilometres per hour. About a million-million will pass through you while reading this essay, but you cannot see them, feel them, taste them, or smell them. And yet, they are the dominant form of matter in the Universe today, and they shape the large-scale structure of the Universe. That is truly a fantastical story.

Fantastical stories in science require observational or experimental verification. If the cold thermal relic idea is correct, the dark matter froze out of LTE when the Universe was about 10^{-9} s AB and the temperature of the Universe was about 10^{14} K. How can we directly observe this?

The most basic idea of the Big Bang model is that our Universe emerged from a state of high temperature and density 13.8 billion years ago and, throughout its history, has been evolving as it expanded and cooled. The Universe was different in the past and will be different in the future. How can we prove the Universe was different in the past? Astronomers can prove the Universe evolves because they literally can watch the Universe evolve. Unlike palaeontologists, astronomers can employ time machines to look back in time. These time machines are familiar telescopes. Because of the finite velocity of light, as we look out in space, we look back in time. When we look at our Sun, we see the Sun as it existed eight minutes before we observe it because it takes light eight minutes to reach us from the Sun. If we look at a bright star in the night sky, we are observing the star not as it is the night we observe it but as it existed about a decade ago because bright, nearby stars are about 10 light year distant. If we look through a telescope at a nearby galaxy about a million light years away, the light we observe left the galaxy a million years ago, before *Homo sapiens* walked the Earth. We see distant objects there and then, not there and now. The farther out in space we look, the further out in time we see. Since the light from more distant objects was emitted at an earlier time, distant objects appear younger to us.

But we cannot use telescopes to look out in space and back in time to observe the freeze-out of WIMPs because for the first 380,000 years of the history of the Universe, it was so hot and dense that it was opaque to electromagnetic radiation. We cannot see beyond the 'last-scattering surface' 380,000 years AB, when the temperature was about 0.1 eV.

We cannot see directly the freeze-out of WIMPs, but the fact that there must be some dark matter–standard model interactions is the key to observing WIMPs. Figure 9.3 illustrates how this works. In the figure is a representation of dark matter–standard model interactions necessary to establish DM in LTE and to determine its freeze-out temperature and subsequent abundance. In the simple illustration, two dark matter particles (denoted DM) interact with two standard model particles (denoted SM). The freeze-out abundance is determined by the annihilation cross section DM + DM → SM + SM[7], which is described in the diagram by the downward arrow. Freeze-out of dark matter means that annihilation of dark matter is insignificant *on average* in the Universe. Here, 'on average' means where

---

[7] For simplicity we do not differentiate between particle and antiparticle. If the dark matter species is not self-conjugate it should properly be dark matter particle + dark matter antiparticle. If the SM particle is charged under one of the SM interactions it should properly be a SM particle plus its SM antiparticle.

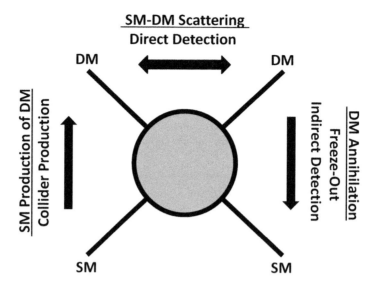

Figure 9.3 A cold thermal relic must be coupled to SM particles through some interaction, indicated by the shaded grey area. The relic abundance is determined by the WIMP annihilation cross section. The same interaction should be able to describe WIMP production from SM particles, as well as WIMP-SM scattering.

the local density of dark matter is approximately the average density of dark matter in the Universe. But if we could concentrate dark matter and increase its local density to values much greater than the average density in the Universe, the same processes responsible for the annihilation of dark matter into SM particles in the early Universe would happen today. Conveniently, nature does this concentration for us. Dark matter should accumulate in the galactic centre, in galaxy clusters, in dwarf galaxies, and in the Sun and Earth in abundances much greater than the average throughout the Universe. In these locations, DM annihilation into high-energy SM particles should be happening today. If we can detect the unambiguous signal of these processes, we would have evidence of particle dark matter. This search strategy is known as indirect detection. It will be discussed in Section 9.5.1.

The diagram in Figure 9.3 is a cartoon version of a Feynman diagram. Feynman's rules allow us to read the diagram in other directions as well. Reading the diagram top down describes annihilation of DM into standard model particles, while reading the diagram bottom up describes the process by which standard model particles collide and produce WIMPs. Since WIMPs are more massive than stable standard model particles (perhaps much more massive), there is a large threshold energy required to produce the dark matter. This large energy can be produced in particle colliders such as the CERN Large Hadron Collider (LHC), which collides protons at high energy. The production and detection of dark matter is another method to prove the WIMP hypothesis. This will be discussed in Section 9.5.2.

Finally, the diagram in Figure 9.3 can be read left to right (or right to left) to describe the scattering of a WIMP with an SM particle. Techniques that attempt to detect WIMPs through their scattering with some SM target material are known as direct detection techniques. Direct detection will be discussed in Section 9.5.3.

Before the discussion of direct detection, indirect detection, and collider production and detection of dark matter, I will make some remarks about the role of dark matter in the formation of structure in the Universe.

9.4 Role of Dark Matter in the Formation of Structure

The purpose of this section is to examine the role of dark matter in structure formation. To do so first requires us to understand the formation of structure in baryons without dark matter, then to understand the evolution of perturbations in a universe containing only dark matter, and, finally, to put things together and understand the evolution of a universe with both baryons and dark matter.

The basis of modern cosmology is the Einstein field equations, $G_{\mu\nu} = (\kappa/c^4)T_{\mu\nu}$. On the left-hand side of the Einstein equations, $G_{\mu\nu}$ is the Einstein tensor, constructed from the Ricci curvature tensor, the Ricci scalar curvature, and the metric tensor $g_{\mu\nu}$ (the Ricci tensor is a contraction of the Riemann curvature tensor, and the Ricci scalar is the trace of the Ricci tensor). On the right-hand side, $\kappa = 8\pi G_N$ and $T_{\mu\nu}$ is the stress-energy tensor which contains information about matter, energy, forces, and particles and how they are distributed in space-time. Einstein's equations are a mathematical expression of the idea that the stress-energy tensor appearing on the right-hand side informs the left-hand side and determines how space and time are curved, warped, and bent and how space expands. In turn, the curvature on the left-hand side connects to the right-hand side and tells matter and radiation how to respond to the geometry.

Note the use of plural in the phrase Einstein field equations. It may appear to be a single equation, but when the tensor nature of the expression is expanded in component form, it is, in general, 10 beastly non-linear partial differential equations. Clearly, some approximation is needed for cosmological solutions. The usual approximation is to assume that the Universe is spatially homogeneous (the same at every point) and isotropic (the same in every direction). The assumption of homogeneity and isotropy is a statement of the Cosmological Principle, which states that there is no special point or no special set of points in the Universe: the Universe is the same everywhere, and every observer sees the same Universe.

The great advantage of the Cosmological Principle is that a homogeneous/isotropic metric has only one dynamical degree of freedom that is only a function of time, the scale factor $a(t)$. The 10 beastly non-linear partial differential equations of the most general solution become two rather tame ordinary differential equations for a single degree of freedom, $a(t)$.

The scale factor represents the expansion of the Universe. In a perfectly homogeneous and isotropic Universe, the distance between any two objects scales with the scale factor due to the expansion of the Universe. If, in such a universe, the scale factor today is denoted as a_0 and the distance between any two objects is D_0, then at some earlier time t, the distance between the two objects was $D(t) = D_0[a(t)/a_0]$, where the scale factor at the earlier time is denoted as $a(t)$.

The idea of the Cosmological Principle appears in the very first paper on relativistic cosmology, Einstein's 1917 paper *Kosmologische Betrachtungen zur allgemeinen Relativitätstheorie* (*Cosmological Considerations in the General Theory of Relativity*) [12], the famous paper in which he introduced the cosmological constant. In that paper, Einstein states 'if we are concerned with the structure [of the Universe] only on large scales, we may represent matter to ourselves as being uniformally distributed over enormous spaces'.

From the very start, the Cosmological Principle was understood to be a simplification. Today, observations tell us that on the largest scales, the Universe does appear to be homogeneous and isotropic, but clearly, on small scales, it is not. Whether the 'small-scale' inhomogeneities are important for the large-scale evolution of the Universe is a matter of debate (see, e.g., References [6, 16]). For the purposes of the present chapter, we will assume the Universe is homogeneous/isotropic on large scales and adopt the standard Friedmann-Lemaître-Robertson-Walker model for the cosmological background evolution.

Taking a homogeneous universe as a starting point, how do we account for inhomogeneities such as galaxies, galaxy clusters, filaments, and other structures? The simple answer is a phenomenon known to Newton: gravitational instability. Newton realised that an infinite, perfectly uniform distribution of matter would be static in a world governed by classical Newtonian mechanics. Every mass point would be attracted gravitationally to every other mass point in the Universe, but because of the symmetry, each mass point would be pulled equally in every direction, and the Newtonian forces would cancel. Newton also realised that any departure from absolute uniformity of the mass distribution would render the system unstable: regions with higher than average density would accrete matter from regions of lower density. This is a runaway situation. The rich become richer and the poor become poorer.

Of course, to study the evolution of structure, we must have information about the initial size and nature of the departures from homogeneity. Motivated by the idea of inflation (and consistent with observations), we will assume that, in the early Universe, there were perturbations in all fluids (radiation, baryons, neutrinos, and dark matter) and that the perturbations around the time of decoupling of matter and radiation, about 400,000 years AB, were small (of order 10^{-4} or so). We will also assume, again consistent with observations as well as expectations from inflation, that the perturbations were 'adiabatic'. If the perturbations are adiabatic, there are no entropy perturbations, only pure density perturbations. A more appropriate description of these types of perturbations is 'isentropic' perturbations. All components of the mass-energy density participate in isentropic perturbations. An over-density in photons is accompanied by an over-density in baryons and dark matter.

9.4.1 Growth of Structure in a Baryon-Only Universe

The simple picture of Newtonian gravitational instability is not directly applicable to the cosmological situation because (1) there is a pressure force on baryons, (2) the Universe is expanding, and (3) in some instances, a fully general-relativistic consideration must supplant the Newtonian picture.

It is easy to extend the Newtonian analysis of gravitational instability to the situation where there is pressure support to oppose collapse. The classical analysis of Jeans studies the growth (or lack thereof) of small perturbations from homogeneity. It analyses the interplay between the force of gravity, which drives the system to collapse, and the pressure force, which opposes the collapse. The result is that if the region contains sufficient mass (mass larger than the Jeans mass), gravity wins, and it will collapse, while if the region is sufficiently small, pressure forces can support the over-density, and the system will not collapse. Furthermore, if the Jeans criterion is satisfied, the collapse is exponential in time.

Some of the same considerations are at play in an expanding universe, but there are important differences. Here, I will just list some of the differences that will be important when considering the role of dark matter in structure formation:

1. The Jeans mass in baryons, the minimum mass in baryons that will collapse, evolves in the expanding Universe. Before the decoupling of baryons and radiation at a redshift of $1 + z = a_0/a(t) \simeq 1,100$ (a_0 is the present value of the scale factor), the pressure support was provided by radiation, and the Jeans mass in baryons was $M_J(\text{baryons}) \simeq 8 \times 10^{27} M_\odot/(1+z)^3$. After recombination, the pressure support is provided by non-relativistic baryons, so the pressure is much smaller than in the radiation-dominated epoch, which leads to a precipitous drop in the Jeans mass: $M_J(\text{baryons}) \simeq 10^6 M_\odot[(1+z)/1,100]^{3/2}$. The evolution of the baryonic Jeans mass is shown in Figure 9.4.
2. We define the Hubble radius R_H as $R_H(t) = H^{-1}(t)$, where $H(t)$ is the expansion rate of the Universe. Perturbations on scales larger than the Hubble radius cannot be treated by Newtonian/Jeans considerations and require a fully general-relativistic calculation. Roughly speaking, perturbations on scales larger than R_H do not evolve; they are 'frozen'. The evolution of the mass in baryons within the Hubble-radius as a function of time is also shown in Figure 9.4.
3. Where collapse occurs in an expanding universe, the expansion of the universe impedes collapse, and the growth of small perturbations is not exponential in time but a power-law in time due to 'Hubble drag'.
4. In a matter-dominated universe, perturbations on sub-Hubble-radius scales that meet the Jeans criterion grow in time as $t^{2/3}$.
5. In a radiation-dominated universe, perturbations on sub-Hubble-radius scales that meet the Jeans criterion only grow logarithmically in time.
6. Perturbations do not grow in a spatial-curvature-dominated universe.
7. Perturbations do not grow in a universe dominated by a cosmological constant.

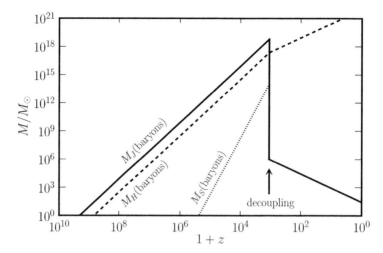

Figure 9.4 The evolution of mass scales associated with baryons as a function of $1 + z = a_0/a(t)$ (a_0 is the present scale factor). The solid curve labelled M_J is the baryonic Jeans mass. Small perturbations grow only for masses larger than the Jeans mass. Before decoupling, pressure support is provided by the coupling of baryons to photons. After decoupling pressure, support is provided by non-relativistic hydrogen atoms, hence the drop in the Jeans mass at decoupling. The dashed curve labelled M_H is the mass in baryons within the Hubble radius. Small perturbations on mass scales larger than the Hubble radius do not grow. Thus, baryonic perturbations only grow after decoupling since the Jeans mass is less than the Hubble-radius mass only after decoupling. The dotted curve labelled M_S is the Silk mass. Collisional (Silk) damping erases perturbations on mass scales less than the Silk mass. This figure is representational. Several crude assumptions have been made, like that the approximation that the Universe becomes matter dominated at decoupling.

Another important effect in a cosmological setting is collisional damping, which erases baryonic perturbations on interesting scales. For most of the history of the early Universe, it is valid to assume that the baryon/photon fluid is a 'perfect' fluid because the baryons and photons are tightly coupled, and the mean-free-paths of baryons and photons are much less than scales of interest.[8] But as decoupling of matter and radiation is approached, the mean-free-path of photons grow as the Universe passes from matter in an ionised state with free electrons and baryons to a state with matter in the form of neutral hydrogen; in other words, the assumption of a perfect fluid breaks down. We are assuming an initial condition of adiabatic (isentropic) perturbations, so the photon and baryon perturbations are correlated. As the photons begin to stream out of over-dense regions, they drag baryons along with them, smoothing the baryonic perturbations. In a cosmological context, this collisional damping is known as Silk damping [33].

---

[8] The mean free path of an electron, $\lambda_e = (n_\gamma \sigma_T)^{-1}$, is much less than the photon mean-free-path, $\lambda_\gamma = (n_e \sigma_T)^{-1}$ because the number density of photons is about 10^{10} that of electrons. Here, σ_T is the Thomson cross section.

The photon mean-free-path in the non-relativistic limit ($T \ll m_e$) is $\lambda_\gamma = (X_e n_e \sigma_T)^{-1}$, where σ_T is the Thomson cross section, X_e, is the ionisation fraction, and n_e is the total electron density (bound and free). In a time t, a photon will have t/λ_γ collisions and undergo a random walk. As it random-walks out of the perturbations, the photons will drag baryons out of the perturbations. The Silk scale reaches a maximum around decoupling; then, after the electrons and protons recombine, the photons decouple from baryons, and photons no longer can drag the baryons out of perturbations. The Silk scale is shown in Figure 9.4. After decoupling, baryon perturbations are damped for mass scales below about $10^{13} M_\odot$.

So now we can put everything together and predict what the growth of structure would be in a dark-matter-less universe where the only matter components are baryons and electrons.

1. For perturbations of a size smaller than the Hubble radius, a Newtonian/Jeans analysis is applicable.
2. In the radiation-dominated (or curvature-dominated or cosmological-constant-dominated) era perturbations do not grow. When the universe becomes matter dominated, perturbations on mass scales greater than the Jeans mass grow as $t^{2/3}$.
3. The Jeans mass becomes smaller than the horizon mass only after recombination.
4. By the time of recombination, collisional (Silk) damping strongly damped perturbations on scales less than about $10^{13} M_\odot$

In the baryon-only matter model, structure formation begins only after decoupling when the baryon Jeans mass drops and only on mass scales larger than the Silk mass scale, about $10^{13} M_\odot$. In this model universe, structure formation proceeds by a top-down fragmentation of perturbations on a mass scale larger than $10^{13} M_\odot$. This does not seem to be the way structure formed in our Universe, suggesting that we do not live in a baryon-only universe.

There is another reason to believe that we do not live in a baryon-only universe. Perturbations do not grow in a radiation-dominated universe; they only grow when the universe becomes matter dominated and then grow as $t^{2/3}$ or linearly in the scale factor a. Observationally, today, Ω in baryons is about 4.4×10^{-2}. (This value is consistent with the expectation from considerations of primordial nucleosynthesis.) The value of Ω in relativistic degrees of freedom (photons and neutrinos) today is about 4.4×10^{-5}, so today $\rho_{\text{radiation}}/\rho_{\text{baryons}} \sim 10^{-3}$. In expansion, $\rho_{\text{radiation}}$ scales as $(1 + z)^4$ and ρ_{baryons} scales as $(1 + z)^3$, so the ratio $\rho_{\text{radiation}}/\rho_{\text{baryons}} \propto (1 + z)$. Since the ratio today is 10^{-3}, the scale factor at the time of equal baryon and radiation density was $1 + z \sim 1{,}000$. If structure formation does not commence until after decoupling and when the universe becomes matter dominated (which, in a baryon-only model, happens at about the same time), perturbations grew by at most a factor of 10^3 from matter domination until today. In order for cosmological structures to form, the perturbations must grow to become non-linear, so the perturbation amplitude at decoupling must be at least 10^{-3}. This is larger than what we see in cosmic background radiation (CBR) measurements by a factor of at least 10.[9]

---

[9] For isentropic perturbations, the radiation perturbation amplitude is 1/3 as large as the matter perturbation amplitude.

In conclusion, in a baryon-only universe, structure must form top-down from fragmentation of structures of mass larger $10^{13} M_\odot$. Furthermore, for non-linear structures to form today, the amplitude of perturbations must be larger than allowed from CBR measurements.

Clearly structure as observed in our Universe did not form in a baryon-photon-neutrino universe. There must be something more in the mix. That something more is dark matter.

9.4.2 Growth of Dark-Matter Perturbations

Dark matter does not interact with radiation; in fact, it is collisionless. Rather than analyse fluid equations, in principle, one must solve the collisionless Boltzmann equation. In practice, one analyses moments of the collisionless Boltzmann equation. The analysis is similar to the Jeans analysis except the velocity dispersion of dark matter plays the role corresponding to the role played by the sound speed in the fluid equations.

Since dark matter is collisionless, it does not suffer Silk (collisional) damping. However, dark matter perturbations may suffer collisionless damping (or Landau damping). Again, in order to account for collisionless phase mixing, it is necessary to integrate the collisionless Boltzmann equation. However, it is possible to estimate the scale of collisionless damping.

If the collisionless species has a non-zero velocity dispersion with respect to the rest frame of the plasma, they will freely propagate out of over-dense regions and into under-dense regions, smoothing out inhomogeneities. Since we know how the velocity of a freely propagating particle redshifts in expansion, we can calculate the free-streaming distance.

Of course, if the dark matter is dead cold, then the free-streaming distance vanishes and the perturbations are undamped. Note that 'cold' refers to the velocity of the dark matter around the time of decoupling. Cold thermal relics are cold because the mass of the particle species is large compared to the temperature of the species around decoupling. But in general, dark-matter relics are cold because their velocity is small around decoupling. Another example of cold dark matter is the axion. Although axions are expected to have a mass smaller than the temperature of the Universe at decoupling (about 1/3 eV), they are cold because they were produced as a Bose condensate and a have small velocity dispersion.

In the other extreme, if dark matter is 'hot' (has a significant velocity around the time of decoupling), perturbations will suffer collisionless damping; perturbations will be damped on scales less than the free-streaming length. The quintessential example of hot dark matter is a light-mass neutrino. The contribution to Ω for a (two-component) neutrino of mass m_ν is $\Omega_\nu h^2 = m_\nu/91$ eV. So if the neutrino is the dark matter, $\Omega_\nu = 0.24$, and using $h^2 = 1/2$ we find $m_\nu = 11$ eV. We know the neutrino temperature and when it became non-relativistic. Calculating the free-streaming length until the time of matter-radiation equality when the perturbations can grow, we find that the free streaming length corresponds to a present distance of $\lambda_{FS} \simeq 100$ Mpc, containing, on average, a mass of about $2 \times 10^{16} M_\odot$.[10]

---

[10]   These numbers come from a more exact calculation [5] of the free-streaming length. The free-streaming length from the more exact calculation is about a factor of 2 larger than the approximate calculation.

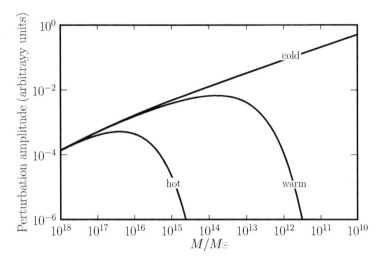

Figure 9.5 The dark matter perturbation spectrum is processed by collisionless damping. Shown are the perturbation spectra at some time after decoupling as a function of the mass in dark matter within the perturbation for three choices for dark matter: *cold*, dark matter that is non-relativistic (zero velocity) at decoupling; *hot*, dark matter that is semi-relativistic at decoupling; and *warm*, dark matter that is somewhere between hot and cold. (For aficionados, the 'perturbation spectrum' is $k^{3/2}|\delta_k|$, and for hot dark matter, I have chosen a neutrino of mass 11 eV, the appropriate mass if that species of neutrino contributes $\Omega = 0.24$.)

The dark matter perturbation amplitude processed by collisionless damping is shown in Figure 9.5. For the figure, it was assumed that dark matter today has a value of $\Omega = 0.24$, $h^2 = 1/2$, and the initial perturbation spectrum is a Harrison-Zeldovich spectrum. The spectrum shown represents the spectrum before the growth of perturbations after the Universe becomes matter dominated. For neutrino hot dark matter, the perturbations are damped for length scales today of about $2 \times 10^{16} M_\odot$. This is much larger than individual galaxy masses, so in a hot dark matter universe, structure forms top down.

Of course, there is the intermediate case of 'warm' dark matter. For warm dark matter, the dark matter is colder than neutrinos but hotter than cold dark matter. A sample warm dark matter spectrum is also shown in Figure 9.5.

Numerical simulations of structure formation highly disfavour hot dark matter. It seems that dark matter must be cold, or if it is warm, it must be very close to the spectrum for cold dark matter down to mass scales as small as astrophysically important for structure formation.

A universe with cold dark matter has another advantage over a baryon-only universe. We know from Big Bang nucleosynthesis and astronomical observations that the baryon density contributes only $\Omega_B = 4.4 \times 10^{-2}$. As mentioned before, this means that a baryon-only universe became matter dominated around $z \sim 1,000$. If we include dark matter, the total dark matter Ω plus the baryon Ω is about 0.3, or about an order of magnitude larger than a baryon-only universe. Therefore, structure formation can start earlier, with a smaller value of the perturbation amplitude, consistent with CBR measurements.

9.4.3 Fine-Tuning of Dark Matter for Structure Formation?

Dark matter is needed to explain the structure we see in the Universe today. Furthermore, the dark matter has to be cold (or very nearly cold). The model of gravitational instability in a universe with baryons and cold dark matter explains well the observed large-scale structure of the Universe. Whether this simple model is the answer or nature is more complicated remains to be seen. But the fact that we require dark matter for the Universe to turn out as observed raises the fine-tuning question.

The fine-tuning question can be stated in many ways; here, I will just address if small departures from the standard cold dark matter model would lead to a universe much different than observed. The answer, of course, depends on the meaning of the word 'small'. If small refers to changing the present value of dark matter from 24% to, say, 24% $\pm$ a few percentage points, the Universe today would be more or less as observed. If Ω_{DM} is 'much' smaller than 24%, then in a flat universe, the deficit has to be made up by something else. If that something else is a larger value of the cosmological constant, then that would spell trouble for structure formation because the Universe would have become dark-energy dominated earlier, and structure growth would have been shut down earlier.

9.5 Testing the WIMP Hypothesis

In the WIMP scenario that we are assuming here, the dark matter was in chemical and kinetic equilibrium with the primordial plasma when the dark species was relativistic, so there must be some coupling between WIMPs and standard model particles. In Section 9.3.2, we saw how the present mass density of a cold thermal relic was set by the cross section for annihilation of dark matter into standard model particles. We also saw that obtaining the desired present mass density of dark matter requires the WIMP to have an annihilation cross section of about $10^{-36}\,\text{cm}^2$. Figure 9.3 illustrates the relationship between the annihilation cross section responsible for setting the present mass density of a cold thermal relic and other processes involving dark matter (DM) and standard model (SM) particles. In this section, we exploit the SM-DM connection to explore ways of testing the WIMP hypothesis.

Unfortunately, knowledge of the DM annihilation cross section in the early Universe does not give us direct knowledge of the DM-SM interactions today. The basic reason is that freeze-out of a WIMP happened when the WIMP was mildly non-relativistic. For both s-wave and p-wave annihilation, the WIMP froze out at a temperature T_F given by $T_F \sim m/20$. For a Boltzmann distribution, this would correspond to a mean velocity of $\langle v^2 \rangle^{1/2} \sim 0.4$.[11]

---

[11] Of course, velocities are given in units of the speed of light.

But for indirect detection and direct detection, the relevant mean velocity of WIMPs is more like 10^{-3}, well into the non-relativistic regime. Thus, for indirect and direct detection searches, we expect the velocity dependence of the WIMP annihilation and scattering cross sections in the non-relativistic limit to be crucial. What we know from the criterion that WIMPs have the correct dark matter density is that $[\sigma v]_{NR} \sim 10^{-36}\,\mathrm{cm}^2$, where the notation refers to the cross section times the velocity in the non-relativistic limit, and the velocity is the velocity around freeze-out, $v \sim 0.4$. If $[\sigma v]_{NR}$ is velocity independent, then it would be the same for dark matter annihilation today, where the velocities are much smaller than the velocity at freeze out. If, however, the dark matter annihilation is velocity dependent (p-wave), then $[\sigma v]_{NR}$ in our galaxy would be a factor of $10^{-3}/0.4$ times smaller since we expect galactic WIMPs to have a velocity of about $v \sim 10^{-3}$.

The same type of considerations apply to direct detection through WIMP-nucleus scattering. Certain types of scatterings are velocity dependent, while others are velocity independent. For some models, the velocity dependence of the annihilation cross section is different than the velocity dependence of the scattering cross section. Another complication in using the annihilation cross section information from freeze-out is that it is typically about annihilation into quarks, while for direct detection, the WIMP scatters with a nucleus (which, of course, contains quarks). The typical momentum transfer in WIMP-nucleus scattering is small enough that the WIMP can see more than one nucleon. Therefore it is necessary to relate the WIMP-quark cross section to a WIMP-nucleus cross section. This involves a form factor. Furthermore, certain scatterings are spin indendent (and thus coherent) and others are spin dependent and couple to the spin of the nucleus. This means that in the spin-dependent case, some nuclear targets are more sensitive than others for WIMP detection.

It is also not straightforward to relate the non-relativistic annihilation cross section from freeze-out to the cross section for production of dark matter at colliders. To see how this issue might arise imagine that the non-relativistic WIMP annihilation at freeze-out proceeds through the WIMP pair producing some massive intermediate state that also couples to a pair of SM particles. If the mass of the intermediate state, M_I, is less than the mass of the WIMP, m, then the cross section is proportional to M_I^{-4} or M_I^{-2}, depending on whether the intermediate state is a boson or a fermion.[12] Production at colliders would also proceed through the intermediate state. However, now the total centre-of-mass energy in the process may be much larger than M_I, and the production cross section would be determined by the centre-of-mass energy and not M_I.

In conclusion, if WIMPs are the dark matter, then for velocities of about $0.4c$, $\sigma_A v \sim 10^{-36}\,\mathrm{cm}^2$. Using this to inform us about present-day annihilation rates, present-day WIMP-nucleon scattering rates, or the prospects for production and detection of WIMPs at colliders, is model dependent. Nevertheless, in spite of uncertainties, knowledge of the non-relativistic annihilation cross section is precious, and the search for WIMPs is intense: the hunt is on!

---

[12] Here, I have assumed for simplicity, I have assumed that the annihilation proceeds through the s-channel.

9.5.1 Indirect Detection

To set the scale for indirect detection, we can change units for $[\sigma v]_{NR}$ from $10^{-36}\,\mathrm{cm}^2$ by multiplying by the speed of light to obtain $[\sigma v]_{NR} = 3 \times 10^{-26}\,\mathrm{cm}^3\,\mathrm{s}^{-1}$. If locally the number density of WIMPs is n_{WIMP}, then locally the annihilation rate would be $[\sigma v]_{NR}\,n_{\mathrm{WIMP}}^2 = [\sigma v]_{NR}\,(\rho_{\mathrm{DM}}/m)^2$, where ρ_{DM} is the local dark matter density and m is the WIMP mass. It is more convenient to express the annihilation rate in terms of ρ_{DM} because that is what is inferred from astronomical measurements.

Operationally, a ground-based or space-based telescope points at some 'region of interest' to search for a signal. The region of interest is defined by some angular region on the sky described by two angles b and l.[13] Now the detector will see all particles in the line of sight s from the region of interest. As an example, let's imagine we are looking for a signal from the galactic centre. Generally, we know the dark matter mass density as a function distance r from the galactic centre. Our mastery of geometry allows us to express r in term of s, l, and b. So the analogue of $[\sigma v]_{NR}\,(\rho_{\mathrm{DM}}/m)^2$ is a factor that just depends on the mass and mass density of WIMPs as a function of r:

$$J(\text{line of sight over region of interest}) = \int \frac{\rho_{\mathrm{DM}}^2\,[r(s,l,b)]}{2m^2}\,ds\,\cos b\,db\,dl. \tag{9.5}$$

Note that the units of this expression are cm^{-5}.

There is considerable uncertainty in $\rho_{\mathrm{DM}}(r)$ near the centre of dark matter halos. To illustrate the uncertainty, I will present three density profiles often assumed to be a universal form for dark matter halos: the Navarro-Frenk-White profile $\rho_{DM}(r) = \rho_0(r/r_S)^{-1}(1 + r/r_S)^{-2}$ [28], the Einasto profile $\rho_{DM}(r) = \rho_0 \exp\{-(2/a)[(r/r_S)^a - 1]\}$ [11], and a profile with a central core $\rho_{DM}(r) = \rho_0(1+r/r_S)^{-1}(1+r^2/r_S^2)^{-2}$ [7]. In these profiles, r_S is a scale height (different for different profiles) that depends on the object, ρ_0 is the density at some value of r, and the Einasto profile has an additional parameter a. The profiles are presented in graphical form for the Milky Way in Figure 9.6. (Recall that the annihilation rate is proportional to the local density squared.) In the calculation of J, the uncertainties are the dark matter density and the dark matter mass.

Now we are interested in $d\Phi_i(E)/dE$, the differential intensity (number of particles of type i per area, per time, per solid angle, and per energy) observed from the region of interest. If the differential number of particles produced per energy in an annihilation is dN_i/dE, then

$$\frac{d\Phi_i(E)}{dE} = \frac{dN_i}{dE}\,\frac{[\sigma v]_{NR}}{4\pi}\,J, \tag{9.6}$$

where J is given in Eq. (9.5). The additional uncertainties include the annihilation cross section ($[\sigma v]_{NR}$ is here assumed to be velocity independent), the spectrum, and the

---

[13] Galactic longitude (l) measures the angular distance along the galactic equator from the galactic centre, and galactic latitude (b) measures the angle north or south of the galactic equator when viewed from Earth. Of course, any coordinate system would suffice.

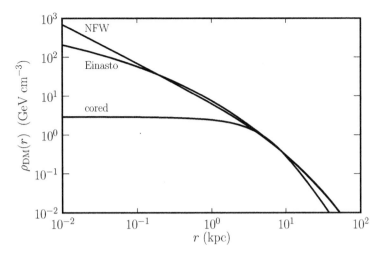

Figure 9.6 Three dark matter density profiles for the Milky Way galaxy. The curve labelled NFW is the Navarro-Frenk-White profile [28], the curve labelled Einasto is the Einasto profile [11], and the curve labelled cored is a profile with a core at the centre, in this case taken from Burkert [7].

number of particles of type i produced per annihilation. While we expect $[\sigma v]_{NR} \sim 3 \times 10^{-26}\,\mathrm{cm}^3\,\mathrm{s}^{-1}$, the spectrum of particles produced in the annihilation is model dependent.

The final step in the procedure is to convert the initial spectrum of the annihilation products into the final particle spectrum that would be detected. For instance, if quarks are produced in the annihilation, then fragmentation and hadronisation of the quark jets and subsequent decay of particles like pions and muons must be taken into account. Further environmental interactions like inverse Compton scattering, synchrotron radiation, bremsstrahlung, and scattering by magnetic fields will further affect the spectrum. Luckily, the decay, fragmentation, and hadronisation processes are well understood. The interaction of the annihilation products with the ambient astrophysical environment depends on understanding the environment.

Now I turn to the question of where to look and what to look for. First, what do we look for?

- *Charged particles, electrons, positrons, and anti-protons*: Charged particles are easy to detect, but there are astronomical backgrounds, and they are bent by magnetic fields, so it is impossible to know where they originated.

- *Continuum photons and neutrinos*: Photons are easy to detect; neutrinos are challenging to detect. There are astronomical backgrounds, but the photons and neutrinos should point back to their sources.

- *Monoenergetic photon line*: A monoenergetic photon line would come from the annihilation of dark matter into two particles, at least one of which is a photon. This is considered to be a golden detection signal since the astronomical background of line production is negligible. However, instrumental resolution makes it difficult to resolve line signals, and in most models, the annihilation channel into the two-photon final state is expected to be small.

Now, where should we look? Numerical simulations of structure formation in a cold dark matter universe yield dark matter halos on a variety of scales and roughly self-similar in form. The simulations of halos of the size and mass of the Milky Way have a smooth component modelled by one of the profiles described earlier. Since structure formed hierarchically in a cold dark matter universe, small dark matter halos collapsed first and then, through the processes of mergers and acquisitions, formed larger objects. However, some remnant of the original small halos should survive. The large end of the subhalo distribution is revealed by numerical N-body simulations, and in the small-mass end, theoretical extrapolations and arguments suggest that the halo of the Milky Way should be full of subhalos of mass in the range about 10^{-5}–$10^{-6} M_\odot$. The larger subhalos should be visible as low-surface-brightness dwarf spheroidal companions to the Milky Way such as Sculptor and Fornax, while the smaller subhalos should be very baryon poor, hence invisible to us.

With this information, we can identify the three most promising places to look:

- *The galactic centre*: The advantages of looking toward the galactic centre is that we know where to look, and it is expected to produce the largest signal of the three locations we will examine. The disadvantage is that the galactic centre is a pretty active place and has the largest backgrounds of the three places.

- *Nearby small subclumps of dark matter*: These small subhalos have essentially no baryons (therefore no stars) and, hence, should provide a very clean signal. The disadvantage of small subclumps is that we do not know where they are, and although the signal is clean, it is probably about a factor of 1,000 times smaller than the signal from the galactic centre.

- *Dwarf spheroidals*: The dwarf spheroidals are dark matter rich, with mass-to-light ratios as large as $(\mathcal{M}/\mathcal{L})_\odot \gtrsim 3,000$, so there should be a clean signal with little background. We also know where to look for them. The disadvantage is that they are far away from Earth, and the expected signal is down another factor of 1,000 or so. The signal can be enhanced by 'stacking' the signals from different dwarf spheroidals.

A large number of balloon experiments, space-based telescopes, ground-based γ-ray telescopes, and neutrino facilities have been deployed in indirect detection searches. These facilities are in remote, hostile locations like space, the upper atmosphere, Antarctica, the South Pole, Namibia, and Arizona. Every couple of years, a signal for dark matter annihilation is claimed or suggested. The issue with indirect detection is that one can have a signal that is statistically striking, but no one believes it because of background issues. Many data sets can produce a signal if the background is 'under-modelled'. On the other hand, a true signal can be removed by 'over-modelling' the background.

A variety of techniques today have reached sensitivities where one can reasonably expect to see a signal. In the meantime, there is still a lot of observational and modelling work to do: a better understanding of the galactic centre, better angular resolution that can help resolve background sources and remove emission correlated with gas, better spectral resolution that can help resolve emission lines, greater collecting area to help with lower signals, clever new techniques like dwarf stacking, and a variety of observations to inform us about backgrounds. While today there is no convincing signal, there is still discovery space, and the hunt continues.

9.5.2 Collider Production and Detection

Now let's turn to the program for producing and identifying WIMPs at colliders in particular, hadron colliders like the CERN Large Hadron Collider (LHC). If the freeze-out annihilation of WIMPs contains quarks in the final state (presumably quark + antiquark), then the collision of protons of sufficient centre-of-mass energy should produce WIMPs.[14] The velocity dependence of the non-relativistic annihilation cross section in the freeze-out calculation is irrelevant in the relativistic limit appropriate for colliders. The bigger uncertainty is whether the collision has centre-of-mass higher than any intermediate state in the production process. This, of course, is model dependent.

In discussion of the search strategy for WIMPs at the LHC, it is useful to consider a broad classification of WIMPs into two classes: social WIMPs and maverick WIMPs.

Social WIMPs are be friended by other new particles of similar mass. The quintessential example of a social WIMP is if the WIMP is the lightest supersymmetric particle. In supersymmetry (SUSY), every known elementary particle has an associated superpartner. If the particle is a boson, the superpartner is a fermion and vice versa. The superpartners are odd under a type of parity known as R-parity, while the known SM particles are even under R. This means that the lightest superparticle would be stable if R-parity is conserved because a final state containing only SM particles would have even parity. In many realisations of SUSY, the lightest superpartner can be a WIMP with the requisite annihilation cross section to be dark matter. Let's take the example that the lightest superparticle is the neutralino, a combination of the superpartners of the photino, the Z-boson, and the Higgs. If the neutralino has a mass within reach of the LHC, one would expect other superparticles to also be within reach, including squarks (superpartners of the quarks), sleptons (superpartners of charged leptons), gluinos (superpartners of gluons), etc. So if neutralinos can be produced at the LHC, one expects first to see gluinos, squarks, etc., because they are more strongly interacting and easier to produce and detect. So the search strategy would involve first searching for the supersymmetric comrades of the WIMP. In spite of huge, heroic efforts by an army of theorists, phenomenologists, and experimentalists, low-energy SUSY has

---

[14] A proton contains three 'valence' quarks (the three quarks responsible for the quantum numbers of a proton), as well as 'sea' quark-antiquark pairs and gluons. So the high-energy collision of two protons can produce quark-antiquark collisions at the constituent level.

not been discovered at the LHC. It could be that the SUSY scale is just beyond reach of the LHC, or there may be surprising twists in SUSY phenomenology. While the search box for SUSY has shrunk (and is shrinking every day), there is still some unexplored discovery space, and perhaps it is too soon to throw in the towelino.

Maverick WIMPs, on the other hand, have no like-mass friends; they are loners. Therefore, we have to somehow discover the WIMP in the debris of the collision. In the process proton + proton $\longrightarrow$ WIMP + WIMP, the final-state WIMPs would not interact in the detector, so the visible process is proton + proton $\longrightarrow$ nothing: not a promising signal. Collider experimentalists often say that looking for new particles in the debris of high-energy collisions is like looking for a needle in a haystack. Looking for maverick WIMPs is like looking for an *invisible* needle in a haystack!

But there is a strategy to detect collider-produced WIMPs by searching for missing-energy signals [4]. While the final state of two WIMPs leaves no signal, if one includes initial-state radiation, where a gluon, quark, or other gauge boson is emitted, one can have, for instance, a final state of WIMP + WIMP + quark/gluon jet. The quark/gluon jet would recoil against the WIMP pair. Since the WIMPs would not be detected, the signal in the detector would be a quark/gluon jet carrying of momentum that seems to be unbalanced. The process appears to have missing momentum in the final state since the WIMP pair is invisible.

The issue in searching for maverick WIMPs through missing energy signatures is that there are other processes that can mimic the effect of missing momentum. For instance, the final state can contain neutrinos that would be undetectable (although neutrinos are WIMPs, they are not *the* WIMP). Luckily, the sophistication of the analysis to remove standard model backgrounds from monojet signals is truly remarkable, and one can account for and subtract the signal from neutrinos and other SM processes that appear to have missing momentum.

The present situation in the search for maverick WIMPs is similar to the situation for the search for social WIMPs: heroic efforts but no signal. If dark matter is a SUSY relic, some indication of SUSY should be discovered at the LHC. Gluinos, squarks, charginos (SUSY partners of the $W^\pm$), or sleptons will be seen before a WIMP. The search strategies are well developed. If dark matter is a maverick particle, the only hope is missing-energy signals. The technique is most effective for lower-mass WIMPs, and there is no guarantee that $[\sigma v]_{NR}$ determined from freeze-out is directly applicable to collider searches. Again, there is still unexplored search regions, and the search continues.

9.5.3 Direct Detection

Finally, we consider the search for WIMPs through direct detection. The idea is that we are swimming in a sea of WIMPs, and although they have 'weak' interactions, they must have some coupling to standard model particles since they were in LTE in the early Universe. If we had sufficiently sensitive detectors free from background, we should occasionally be able to detect the scattering of a WIMP.

The experimental approach that has yielded the best limits looks for nuclear recoil from WIMP scattering with a nucleus. This technique has been used for over three decades [1]. For WIMP scattering with a nucleus, a simple estimate for the event rate in would be $R = N\, n_{\rm DM}\, v\, \sigma$, where N is the number of target nuclei in the detector, $n_{\rm DM}\, v$ is the flux of WIMPs, and σ is the scattering cross section. It is useful to take this expression for the event rate and make it a little more sophisticated and relevant for experiment. Of interest is the differential rate per recoil energy E_R expressed in terms of the differential cross section with respect to the recoil energy. This can be expressed as

$$\frac{dR}{dE_R} = N\, \frac{\rho_{\rm DM}}{m} \int_{v_{\rm MIN}} d^3v\, v\, f(v)\, \frac{d\sigma}{dE_R}\,, \tag{9.7}$$

where $\rho_{\rm DM}/m$ is the local WIMP mass density, $f(v)$ is the local WIMP velocity phase-space distribution, and $v_{\rm MIN}$ is the minimum velocity to cause a recoil energy above detection threshold.

From Eq. (9.7), we see the interplay of astrophysics, particle theory, and experimental physics. The determination of $\rho_{\rm DM}$ and $f(v)$ comes from astrophysics, particle theory provides m and $d\sigma/dR$ while a given experiment determines the number of targets, the mass and spin of the target nuclei, and the threshold energy that goes into the determination of $v_{\rm MIN}$.[15]

The usual assumption for $f(v)$ is a Maxwellian velocity distribution with $\langle v^2 \rangle^{1/2} = 220\,{\rm km\ s^{-1}}$, and the usual assumption for $\rho_{\rm DM}$ is $0.4\,{\rm GeV\ cm^{-3}}$ at our location in the Milky Way. This would result in an average recoil energy of a few to a few dozen keV for most target nuclei employed. These estimates for $f(v)$ and $\rho_{\rm DM}$ represent the average values found from numerical simulations at our location in dark matter halos the mass of the Milky Way. However, there are rare places, like in a dark matter subclump, where the values may differ.

The WIMP-nucleus cross section is crucial. It may be proportional to the velocity of the WIMPs or the momentum transfer in the scattering. This would greatly decrease the sensitivity of an experiment. For some models, the WIMP-nucleus couplings is proportional to the nuclear spin, so a target nucleus with zero spin would be insensitive. Finally, it is usually assumed that the WIMP-nucleon coupling is universal. But if WIMPs couple differently to neutrons and protons, the interpretation of experimental limits would be modified. If WIMPs are hadrophobic and only couple to leptons, the nuclear recoil technique is not appropriate.

Throughout the world, there are a couple of dozen experiments in mines and tunnels (to reduce background) using a variety of detection techniques (superheated bubbles, ionisation, phonons, light) searching for a dark matter signal. The present limits are shown in Figure 9.7 for spin-independent scattering and spin-dependent scattering. The shaded area at the bottom of the spin-independent graph denoted coherent neutrino scattering is the

---

[15] The threshold energy is related to $v_{\rm MIN}$ by $E_{\rm TH} = 2\mu^2\, v_{\rm MIN}^2/M_N$, where μ is the WIMP-nucleus reduced mass and M_N is the mass of the target nucleus.

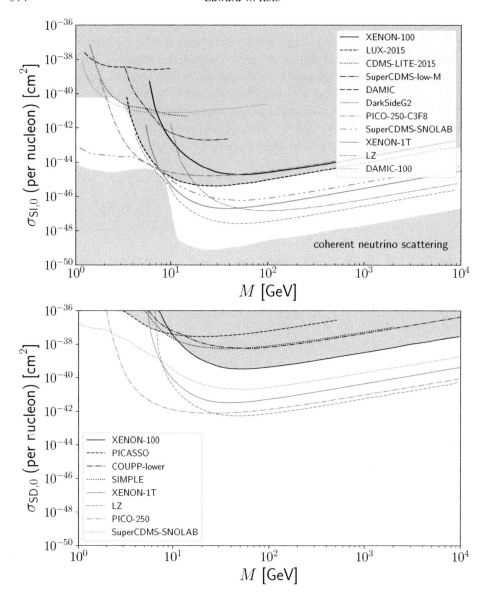

Figure 9.7 Dark matter direct detection limits. The left panel shows the limit on the WIMP scattering cross section for zero-momentum transfer from a number of experiments for spin-independent interactions, and the right panel is a similar graph for spin-dependent interactions. Completed experiments are indicated by black curves and proposed experiments by grey curves. The currently excluded area is shaded. The figures were adapted from figures by Michael Fedderke.

upper value for the cross section where the signal from neutrino scattering will have to be dealt with. There are two sources for these background neutrinos: solar neutrinos for low mass (low-momentum transfer) and the diffuse background of neutrinos from other sources at high mass (higher-momentum transfer). Of course, there is no way to shield the detector from background neutrinos.

From the figures, one can see that planned experiments will explore practically all of the region above the coherent neutrino background for spin-independent interactions. There is significant white space to be explored for spin-dependent scattering. Before leaving the figure, it is remarkable to note the size of the cross section limits. The present limits for spin-independent scattering are pushing 10^{-45} cm^2, or 1 zeptobarn! Future experiments will push 10^{-48} cm^2, which is 1 yoctobarn!

As with indirect detection efforts and collider detection efforts, the efforts to see WIMPs through direct detection proceeds unabated. There are some obvious goals. The spin-dependent limits should be pushed to the neutrino floor. There is a large amount of real estate to be explored for spin-dependent scattering. It is also desirable to push the limits to smaller mass (smaller recoil energy). That will probably require new detection techniques beyond nuclear recoil. If one does see a signal, then it should be possible to test for annual variation caused by the Sun's motion through the WIMP sea. Different targets would help resolve the question of the WIMP mass. Finally, an experiment with sufficient directional sensitivity to the incoming WIMP would add a lot.

9.5.4 The WIMP Decade

Evidence for dark matter has been around for eight decades. The idea that the dark matter is a WIMP has been around for four decades [25]. Indirect detection was first discussed in 1978 [34], and direct detection was proposed as early as 1985 [15]. The search for SUSY is decades old; the search for WIMPs through missing momentum signals is more recent [4]. But it is only in the 2010s that the three detection techniques have reached sufficient sensitivity that they can eat into expected regions. This truly was the WIMP decade!

If we are through the WIMP decade without a WIMP signal, what does that mean? While the simple WIMP hypothesis is really being squeezed (and the SUSY-WIMP hypothesis squeezed even more so), there are still regions of parameter space where the simple WIMP hypothesis can live. Let me give a simple example. Suppose the WIMP is a Majorana fermion χ that couples to quarks in a form that mimics a Fermi-type interaction:[16]

$$\Lambda^{-2} \, \chi \gamma^\mu \gamma_5 \chi \, \bar{q} \gamma_\mu (C_V - C_A \gamma_5) \, q, \tag{9.8}$$

where Λ is some mass scale. For an interaction of this form, the annihilation of WIMPs to quarks in the non-relativistic limit is proportional to v^2. While this would lead to a mild

---

[16] The Fermi interaction describing the coupling of neutrinos to quarks is $(G_F/\sqrt{2})\bar{\nu}\gamma_\mu(1-\gamma_5)\nu \, \bar{q}\gamma_\mu(C_V - C_A\gamma_5) \, q$, where G_F is the Fermi constant with units of mass^{-2} and C_V and C_A are constants of order unity. One can demonstrate that for Majorana fermions, $\chi\gamma^\mu\chi$ vanishes, so the expression in Eq. (9.8) would take a form similar to the Fermi interaction.

suppression for freeze-out, it would mean that indirect detection limits do not apply. For direct detection, the part of the operator that contains $\bar{q}\gamma_\mu\gamma_5\,q$ leads to a spin-dependent interaction, and the limits are not strong. The part of the operator proportional to $\bar{q}\gamma_\mu\,q$ when coupled to $\chi\gamma^\mu\gamma_5\chi$ is suppressed by v^2 or momentum-transfer squared. Therefore, direct-detection limits can be evaded. Furthermore, if the WIMP is massive enough, collider limits can be skirted.

While there are ways to evade present constraints, if cosmologists live by the Bayesian reasoning, they often profess, they would admit that the lack of a WIMP signal makes the simple WIMP hypothesis less likely.[17]

But an important word in the previous paragraph is 'simple'. There are scores of tweaks to the simple WIMP model described here – too many to enumerate. Also, implicitly assumed in the analysis presented here is the assumption that 100% of the dark matter abundance resides in a single particle. That may seem naïve since the visible matter we do see is comprised of a rich variety of forms.

In spite of uncertainties and lack of experimental evidence, for the WIMP hypothesis, perhaps the best thing to do is to keep calm and carry on with the experimental program!

9.6 What If Dark Matter Is Not a WIMP?

A wide range of possible candidates for dark matter was discussed in Section 9.3. In the sections following where I concentrated on the possibility that the dark matter was a cold thermal relic – i.e., a WIMP. One of the reasons the WIMP hypothesis is so attractive is that the necessity of a WIMP–standard model coupling leads to several avenues to discover WIMPs (although falsifying the WIMP hypothesis is more challenging). But what if dark matter is not the simple WIMP discussed before?

Of course, one possibility is that it is a variant of the simple WIMP hypothesis. One variant is 'freeze-in' models where the dark matter never obtains LTE but is produced due to dark matter–standard model interactions. The current techniques for detecting WIMPs may still bear fruit in this case. Another possibility is that dark matter is a WIMP, but it is leptophilic and couples to leptons rather than hadrons. In this case, direct-detection limits based on nuclear recoil would not apply.

In the three decades since theoretical physicists turned their attention to the dark matter problem, a large number of possible solutions have been proposed. As mentioned before, it would be tedious to mention them all. But it is interesting to note that the range of masses of proposed candidates cover 81 orders of magnitude, from 10^{-22} eV (10^{-56} g) Bose-Einstein condensates [18], to $10^{-8} M_\odot$ (10^{25} g) axion miniclusters [17, 22]. In this chapter, I have concentrated on weakly interacting dark matter, but candidates have been proposed with interactions ranging from strongly interacting [23] to interacting only gravitationally [8, 24]. If particle dark matter is the answer, the properties of the dark matter should fall in that wide range; one can hardly imagine it wider!

---

[17] Of course, in Bayesian reasoning, if your prior is unity, then no amount of negative evidence can change that.

9.7 Final Remarks on Dark Matter and Fine-Tuning

There are some curious aspects to dark matter. First, one might ask why dark matter exists at all. After all, we have a Standard Model of particle physics capable to accounting for (at least in principle) all laboratory experiments and interactions. Our everyday world experience does not scream out for the existence of dark matter. It is only when astronomers look to the heavens that something seems amiss. Astronomical observations do not fit the standard cosmological model without dark matter and dark energy. If not for irksome astronomers, physicists would be happy without dark matter (and dark energy). This brings to mind the famous quote attributed to I. I. Rabi in 1936 when he learned of the discovery of the muon with surprising properties: 'who ordered that?' Indeed, who ordered dark matter and dark energy? There are couple of ways to answer the question of why dark matter exists. One possible response is that nature demands it. Dark matter is part of nature, and when we understand the grander, deeper theory underpinning the current Standard Model of particle physics, dark matter will naturally be part of the theory. If so, to ask why dark matter exists is like asking why are there is a third generation of quarks and leptons. Another possible answer is to evoke the Anthropic Principle (for a review, see Reference [26]) and say that if not for dark matter, the Universe would not have turned out the way it did, and we would not be here to ask the question. Perhaps there are other universes in a multiverse without dark matter and without pesky astronomers. Some question whether the Anthropic Principle explains anything and point out that people without ideas can still have principles. But I am not sure that even a principle as accommodating as the Anthropic Principle can explain why there is dark matter. After all, one could imagine arranging the Universe in a way to accommodate galaxies, stars, planets, and people without dark matter.

Although perhaps one can imagine living in a universe without dark matter, it would not look like our Universe, at least with regard to large-scale structure. In Section 9.4, we saw that in a baryon-only universe structure would form top-down from the fragmentation of large structures. While this does not appear to be the way structure formed in our Universe, it would still eventually lead to structure.

Another curious aspect to the present composition of the Universe is the six-to-one ratio of dark matter to baryons. A priori, this could be larger or smaller. The baryon density seems to be set by the matter-antimatter asymmetry. While it is possible that such a mechanism is responsible for the dark matter density, most scenarios have dark matter arising some other way. Why would matter of such different origins end up being of the same order of magnitude? Curious indeed! There does not seem to be any anthropic reason for the six-to-one ratio. Perhaps not every coincidental numbers should be considered fine-tuning.

If dark matter is a WIMP, the presently observed dark matter density is the result of freeze-out in the early Universe. As discussed in Section 9.4, the freeze-out of WIMPs depends on the interplay of the scattering processes and the expansion rate of the Universe. There is some degree of tuning to get the necessary dark matter density, but the tuning is not excessive.

One might also ask why neutrinos contribute a small but non-negligible amount to the mass-energy budget. Whether neutrinos are massless or have perhaps 10 times the mass they do, there does not seem to be a large cosmological consequence.

Since 95% of the Universe is dark and mysterious, perhaps it is not surprising that not all fundamental cosmological questions (such as tuning) can be answered. One thing that would greatly help point us to the answers would be the discovery of dark matter, perhaps in this decade, the new decade of the WIMP!

References

[1] Ahlen, S. P., *et al.* 1987. 'Limits on Cold Dark Matter Candidates from an Ultralow Background Germanium Spectrometer'. *Physics Letters B*, **195**, 603–608.

[2] Babcock, H. W. 1939. 'The Rotation of the Andromeda Nebula'. *Lick Observatory Bulletin*, **19**, 41–51.

[3] Bekenstein, J. D., and Sanders, R. H. 2006. *A Primer to Relativistic MOND Theory*. **20**, 225–230.

[4] Beltran, M., *et al.* 2010. 'Maverick Dark Matter at Colliders'. *Journal of High Energy Physics*, **09**, 037.

[5] Bond, J. R., Efstathiou, G., and Silk, J. 1980. 'Massive Neutrinos and the Large-Scale Structure of the Universe'. *Physical Review Letters*, **45**(Dec.), 1980–1984.

[6] Buchert, T., *et al.* 2015. 'Is There Proof That Backreaction of Inhomogeneities Is Irrelevant in Cosmology?' *Classical and Quantum Gravity*, **32**, 215021.

[7] Burkert, A. 1996. 'The Structure of Dark Matter Halos in Dwarf Galaxies'. *IAU Symposium*, **171**, 175.

[8] Chung, D. J. H., Kolb, E. W., and Riotto, A. 1999. 'Superheavy Dark Matter'. *Physical Review D*, **59**, 023501.

[9] Cowsik, R. and McClelland, J. 1972. 'An Upper Limit on the Neutrino Rest Mass'. *Physical Review Letters*, **29**, 669–670.

[10] Dicus, D. A., Kolb, E. W., and Teplitz, V. L. 1977. 'Cosmological Upper Bound on Heavy Neutrino Lifetimes'. *Physical Review Letters*, **39**, 168.

[11] Einasto, J. 1965. 'On the Construction of a Composite Model for the Galaxy and on the Determination of the System of Galactic Parameters'. *Trudy Astrofizicheskogo Instituta Alma-Ata*, **5**, 87–100.

[12] Einstein, A. 1917. 'Kosmologische Betrachtungen zur allgemeinen Relativitätstheorie'. *Sitzungsberichte der Königlich Preußischen Akademie der Wissenschaften (Berlin)*, 142–152.

[13] Freedman, W. L. 2013. 'The Cosmic Distance Scale and H_0: Past, Present, and Future'. *IAU Symposium*, **289**, 3.

[14] Freedman, W. L. 2017. 'Cosmology at Crossroads: Tension with the Hubble Constant'. *Nature Astronomy*, **1**, 0169.

[15] Goodman, M. W., and Witten, E. 1985. 'Detectability of Certain Dark-Matter Candidates'. *Physical Review D*, **31**(June), 3059–3063.

[16] Green, S. R., and Wald, R. M. 2016. 'A Simple, Heuristic Derivation of Our No Backreaction Results'. *Classical and Quantum Gravity*, **33**(12), 125027.

[17] Hogan, C. J., and Rees, M. J. 1988. 'Axion Miniclusters'. *Physics Letters B*, **205**, 228–230.

[18] Hu, W., Barkana, R. and Gruzinov, A. 2000. 'Cold and Fuzzy Dark Matter'. *Physical Review Letters*, **85**, 1158–1161.

[19] Hubble, E. 1929a. 'A Relation between Distance and Radial Velocity among Extra-Galactic Nebulae'. *Proceedings of the National Academy of Science*, **15**(Mar.), 168–173.

[20] Hubble, E. 1929b. 'A Spiral Nebula as a Stellar System, Messier 31'. *Contributions from the Mount Wilson Observatory/Carnegie Institution of Washington*, **376**, 1–55.

[21] Hut, P. 1977. 'Limits on Masses and Number of Neutral Weakly Interacting Particles'. *Physics Letters B*, **69**, 85.

[22] Kolb, E. W., and Tkachev, I. I. 1993. 'Axion Miniclusters and Bose Stars'. *Physical Review Letters*, **71**, 3051–3054.

[23] Kusenko, A. 1997. 'Solitons in the Supersymmetric Extensions of the Standard Model'. *Physics Letters B*, **405**, 108.

[24] Kuzmin, V. and Tkachev, I. 1998. 'Ultrahigh-Energy Cosmic Rays, Superheavy Long Living Particles, and Matter Creation after Inflation'. *JETP Letters*, **68**, 271–275.

[25] Lee, B. W., and Weinberg, S. 1977. 'Cosmological Lower Bound on Heavy-Neutrino Masses'. *Physical Review Letters*, **39**(July), 165–168.

[26] Linde, A. 2017. 'A Brief History of the Multiverse'. *Reports on Progress in Physics*, **80**(2), 022001.

[27] Milgrom, M. 1983. 'A Modification of the Newtonian Dynamics as a Possible Alternative to the Hidden Mass Hypothesis'. *The Astrophysical Journal*, **270**(July), 365–370.

[28] Navarro, J. F., Frenk, C. S. and White, S. D. M. 1996. 'The Structure of Cold Dark Matter Halos'. *The Astrophysical Journal*, **462**, 563–575.

[29] Oort, J. H. 1932. 'The Force Exerted by the Stellar System in the Direction Perpendicular to the Galactic Plane and Some Related Problems'. *Bulletin of the Astronomical Institutes of the Netherlands*, **6**(Aug.), 249.

[30] Roos, M. 2010. 'Dark Matter: The Evidence from Astronomy, Astrophysics and Cosmology'. arXiv e-prints, Jan, arXiv:1001.0316

[31] Rubin, V. C., and Ford, Jr., W. K. 1970. 'Rotation of the Andromeda Nebula from a Spectroscopic Survey of Emission Regions'. *The Astrophysical Journal*, **159**(Feb.), 379.

[32] Sato, K. and Kobayashi, M. 1977. 'Cosmological Constraints on the Mass and the Number of Heavy Lepton Neutrinos'. *Progress in Theoretical Physics*, **58**, 1775.

[33] Silk, J. 1977. 'On the Fragmentation of Cosmic Gas Clouds. I: The Formation of Galaxies and the First Generation of Stars'. *The Astrophysical Journal*, **211**(Feb.), 638–648.

[34] Stecker, F. W. 1978. 'The Cosmic Gamma-Ray Background from the Annihilation of Primordial Stable Neutral Heavy Leptons'. *The Astrophysical Journal*, **223**(Aug.), 1032–1036.

[35] Vysotsky, M. I., Dolgov, A. D., and Zeldovich, Ya. B. 1977. 'Cosmological Restriction on Neutral Lepton Masses'. *JETP Letters*, **26**, 188–190.

[36] Zurek, K. M. 2014. 'Asymmetric Dark Matter: Theories, Signatures, and Constraints'. *Physics Reports*, **537**, 91–121.

[37] Zwicky, F. 1933. 'Die Rotverschiebung von extragalaktischen Nebeln'. *Helvetica Physica Acta*, **6**, 110–127.

[38] Zwicky, F. 1937. 'On the Masses of Nebulae and of Clusters of Nebulae'. *The Astrophysical Journal*, **86**(Oct.), 217.

Part IV

Fine-Tuning for Life

10

Fine-Tuning: From Star to Galaxies Formation

JOSEPH SILK

Abstract

Numerical simulations have had a huge impact on our visualisations of star and galaxy formation, and thereby have greatly facilitated data modelling, interpretation, and forecasting. However, little fundamental understanding has emerged. Galaxy formation involves certain types of fine-tuning, as do star formation and the growth of supermassive black holes. In this chapter, I will discuss the interplay of these diverse astrophysical phenomena and show that simple back-of-the-envelope calculations can provide insights into the origin of the fundamental scales of stars and galaxies and their fine-tuning.

10.1 Introduction

If the fundamental constants of nature differed from their measured values, life as we know it would not have emerged. Imagine a universe in which the fine-structure constant was slightly larger or smaller than the actual value that we measure. Stars like the Sun are witness to a titanic battle between the forces of electromagnetism and gravity. The solar system formed from the ashes of burnt-out stars. If this equilibrium is slightly, the very existence of nuclear-burning stars is at risk. In such a universe, stars would never have formed or might have collapsed to black holes. Galaxy formation could be obliterated or drastically modified. Supermassive black holes could be far more or far less massive. All of these issues merit some reflection in the context of fine-tuning of the fundamental constants of nature (see Chapter 1).

Massive numerical simulations have had a huge impact on our visualisations of star and galaxy formation and thereby have greatly facilitated data modelling and interpretation. The significance of simulations for phenomenological predictions and forecasting cannot be overemphasised. However, one can reasonably ask whether this is any more than a convenient, albeit powerful, tool for developing future observing proposals or designing new telescopes or whether simulations have led to the emergence of any fundamental understanding of the natural phenomena. The response seems to be depressingly brief: no!

I show here that back-of-the-envelope calculations cannot only provide insights into the findings of simulations but can also account for the fundamental scales of stars and galaxies and their fine-tuning.

It is well known that simple physical arguments can account for the fundamental scaling relations of stars and galaxies – including Larson's laws [23], the Tully-Fisher [49], Faber-Jackson [13], Kennicutt-Schmidt [21, 35], and Magorrian [14, 17, 25] relations. To be fair to simulators, the dispersion in these relations contains most of the relevant physics and can only be adequately explored via state-of-the-art numerical simulations over a vast and currently inadequate dynamical range. Here, however, I focus not on the scaling laws but on the actual scales: characteristic, minimum, and maximum, from protostars to stars, galaxies, and supermassive black holes.

Unlike the scientists of antiquity, astronomers today have two significant advantages. One consists of the huge telescopes that peer back in time to the edge of the Universe. A second is the mastery of modern physics and mathematics, which has greatly discouraged the philosopher cosmologists and their theologian counterparts from entering the fray. I do not mean to be completely discouraging; there are fundamental limits to the questions that cosmology can answer.

The fossil radiation from the Big Bang, the cosmic microwave background, contains the ultimate fossil clues. Infinitesimal fluctuations in the temperature track tiny variations in density. Slight excesses in density, seen as hotspots, have slightly more gravity and so attract the surrounding matter. The slight under-densities, viewed as cold spots, behave in the opposite way: matter leaks out. This capitalistic view of the Universe, the rich becoming richer and the poor becoming poorer, works for gravity as mass is either accumulated or lost. We end up forming either massive clouds or empty voids. We are presently 13.8 billion years after the Big Bang. The fossil radiation is a glimpse of the past: it emerges from the early Universe 380,000 years after the Big Bang. So now we have a timeline for the first stars. The fluctuations grew for about a billion years until the first star-forming clouds collapsed. These weighed a million times the mass of the Sun. We are sure of this, at least as a lower bound, because the less massive clouds are cooler. It takes a certain amount of energy between colliding hydrogen atoms in order for atomic collisions to release energy and allow the gas to radiate. If this does not happen, the gas cannot contract and become denser. If the gas cannot lose energy and fragment into dense clumps, the first stars cannot form.

Then, in the next 10 billion years or so, evolution proceeded relentlessly. The first stars exploded and polluted nearby clouds. Clouds merged together to make more massive clouds. These clouds were at first the sizes of small galaxies; those we call dwarf galaxies. And many of these merge together into our Milky Way galaxy. Massive galaxies, such as our own Milky Way, are surrounded by a cloud of leftover debris.

10.2 Stellar Basics

Eddington once famously said that a physicist on a cloud-bound planet could predict that there are stars. What is a star? A galaxy is an agglomeration of billions of stars. All devolves around knowing their masses. Why should a star like the Sun weigh in at two billion trillion trillion tons? By a Sun-like star, I mean a star with the same colour, or spectral type, and composition. All such stars weigh a solar mass – no more, no less. And all other stars, countless in number, are between a 10th of and 100 times the mass of the Sun. Once we converge on the notion of a star, we inevitably wonder where it came from. This inevitably leads us into a description of the first stars in the Universe. Although they disappeared long ago, they left traces and fossils that we try to decipher.

Nearby, in Time and Space

But let's begin with the nearest star, our Sun. First, how did it form? A cloud of interstellar gas cooled down. It collapsed under its own gravity. The cloud fragmented into dense clumps of cold gas. Each of these gave birth to a star. A forming star was surrounded by a nebula of cold gas and dust that had too much rotation to collapse. Instead, the nebula cooled and formed a dusty gaseous disc that spun around the central object, which we refer to as a protostar, destined, as it shrank further, to be a star. The protostellar disc had its own destiny: to form planets.

Most of the stars were a 10th of the mass of the Sun. Relatively few were 10 solar masses or more. Most stellar mass is in stars of around half of the mass of the Sun. And then there were the Sun-like stars. The more massive a star, the more rapidly it aged. Stars radiate by thermonuclear burning of hydrogen in their cores. A helium nucleus has atomic mass 4. It consists of two protons and two neutrons. It forms by combining four protons along with two electrons, to form a helium nucleus of mass 4 and charge 2 in atomic units. In fact, the helium nucleus weighs 7% less than four protons. This 4% is released as energy via Einstein's famous equation, $E = mc^2$. That is how the Sun battles gravity. Thermonuclear energy supplies the thermal pressure that supports the Sun. The fuel supply is good for billions of years.

When the hydrogen fuel supply in the core of the Sun, where the hydrogen is hot enough to burn, is exhausted, the Sun contracts, and the core heats up. Helium is ignited, and it burns by thermonuclear reactions into carbon. This reaction releases so much luminosity that the outer part of the Sun swells up into a red giant. Our sun is fated to become a red giant in about four billion years. At this time, the Sun's atmosphere will encompass the orbit of the Earth, burning any surface organic material into ashes. Once the helium energy supply is exhausted, the core contracts into a white hot star about the size of the Earth, but a million times denser. We call this a white dwarf. The atmosphere is ejected in the beautiful phenomenon that we see as a planetary nebula. The white dwarf gradually cools down. Our galaxy is teeming with old white dwarfs, descendants of Sun-like stars.

A star that is 10 or 30 times the mass of the Sun has a much more accelerated evolution. It burns up its nuclear fuel at a rate of the cube of its mass. This means that its lifetime as a hydrogen-burning star may be only a hundred million years or even a few million years. These are tiny timescales in the grand cosmic scheme; many such events occur over the 10 billion year time span of our galaxy. So we can use our galaxy as an astronomical zoo: it contains stars at all stages of their evolution, from birth to death. We can visualise the birth, adolescence, maturity, and old age of stars.

Eddington's Amazing Insight

Decades before we knew the energy source that powered the stars, how did famed astronomer Sir Arthur Eddington conclude that stars are inevitable? Eddington made his reputation by being one of the first to understand the significance of Einstein's theory of general relativity, published in 1915. Perhaps as a Quaker and conscientious objector, he had more time on his hands. But he realised that a key prediction, that gravity bends light, could be tested during the next total eclipse of the Sun by studying the deviations in positions of stars close to the edge of the Sun that otherwise were totally obscured by sunlight. This test occurred in 1919, when Eddington secured the possibility of the leadership of a solar eclipse expedition to Principe to substitute for the military service he refused. Viewed from this island off the west coast of Africa, totality would endure more than six minutes, one of the longest total eclipses of the century, and allowing ample time to photograph the stars visible near the position of the Sun. The measurement was a success, revealing the displacement predicted by Einstein's theory of the bending of light by gravity, twice that expected according to Newtonian gravitation.

Eddington became famous overnight, and even more so did Einstein. Eddington's major achievements, however, were in theoretical astrophysics, where he pioneered our modern understanding of stars. He reasoned that a star is a giant ball of gas supported by gravity. Here is what he wrote in 1926 (with very slight amendment):

We can imagine a physicist on a cloud-bound planet who has never heard tell of the stars calculating the ratio of radiation to gas pressure for a series of globes of gas of various sizes, stating say, with a globe of mass 10 gm., then 100 gm., 1000 gm. and so on, so that her nth globe contains 10^n gm. Regarded as a tussle between gas pressure and radiation pressure, the contest is overwhelmingly one-sided except between Nos. 33–35, where we may expect something interesting to happen. What 'happens' is the stars. We draw aside the veil of cloud beneath which our physicist has been working and have her look up at the sky. There she will find a thousand million globes of gas nearly all of mass between her 33rd and 35th globes, that is to say, between 1/2 and 50 times the sun's mass.

Eddington realised that the pressure of radiation is highly destabilising. He had studied giant globes of gas supported by the balance between their own gravity and the interior pressure of the gas. But as one turned up the mass, and consequently the gravity, the centre became so hot that the pressure of radiation exceeded that of the gas. And this was enough to blow the globe apart. And if the mass was too small, the centre of the globe was so cold that it could not resist gravity, so he reasoned. That is how he deduced, from pure thought, the mass range of the stars.

In the century that followed, astronomers measured the masses of many stars, typically by using the orbits in binary systems, and confirmed Eddington's reasoning. Most stars are similar in mass to the Sun, more typically half or a third the mass of the Sun.

10.3 Stellar Mass-Scales

Let us begin with how stars are made, by fragmentation of molecular clouds. What determines the scale of the first gravitationally bound clumps of gas, the building blocks of stars?

My analysis will be analytic and order of magnitude, designed to bring out the underlying physical scalings. All examples involve fundamental constants and illustrate the role of fine-tuning.

10.3.1 The Minimum Protostellar Fragment Mass

I first derive the collapse time of a collapsing cold gas sphere of either (1) uniform density or (2) non-uniform density. For simplicity, I take an isothermal sphere density profile $\rho(r) = v_s^2/(2\pi G r^2)$, where the sound speed is expressed in terms of the temperature $v_s^2 = kT/m_p$, and turbulence and magnetic support are neglected.

Some simple scaling relations between fragment mass M, radius R, and free-fall time t_{ff} are $R = v_s t_{ff} = M^{1/3}\rho^{-1/3} = GMv_s^{-2}$, and $\rho = M/R^3 = v_s^6 M^{-2} G^{-3}$. The luminosity of an optically thick fragment satisfies

$$L_{rad} = \sigma 4\pi T^4 R^2 = \sigma 4\pi T^4 G^2 M^2 v_s^{-4}. \tag{10.1}$$

The accretion rate onto a protostellar core that is surrounded by such a sphere controls the rate of gravitational energy release. The gravitational energy release from contraction is $L_g = GM^2 R^{-1} t_{dyn}^{-1} = GM^2 R^{-2} v_s = v_s^5 G^{-1}$. Hence, equating $L_{rad} = L_g$, $\sigma 4\pi T^4 G^2 M^2 v_s^{-4} = v_s^5 G^{-1}$ or $M = V_s^{9/2} T^{-2} G^{-3/2} (\sigma 4\pi)^{-1/2} = \text{const} T^{1/4}$. To convert to fundamental units, I use $\sigma = (2\pi^5/15)k^4 h^{-3} c^{-2}$ and write T dimensionlessly as $kT/m_p c^2$ to obtain

$$\frac{M}{m_p} = \alpha_g^{-3/2}\alpha^{1/2}\left(\frac{kT}{m_p c^2}\right)^{1/4}. \tag{10.2}$$

I also write T in units of Rydbergs, $1\text{Ry} = \alpha^2 m_e c^2/2$ to get a crude estimate of the cooling levels and energy radiated by atomic or molecular excitations. Dimensionless constants to a small power can fix this. Then one has

$$M = \alpha_g^{-3/2}\alpha\left(\frac{m_e}{m_p}\right)^{1/4}\left(\frac{kT}{1\text{Ry}}\right)^{1/4}. \tag{10.3}$$

This opacity-limited minimum fragment scale is around $0.01 M_\odot$ [30, 38] (see Figure 10.1). Note that the definition of 'gravitational coupling constant', which explicitly connects the gravitational and electromagnetic couplings between two protons, is $\alpha_g = Gm_p^2/\hbar c = 6 \times 10^{-39}$. One can also write this as $\alpha_g = Gm_p^2/\hbar c = (m_p/m_{pl})^2$, where the Planck mass is $\sqrt{\hbar c/G}$. We shall see that the characteristic mass of a star is $\alpha_g^{-3/2} = (m_{pl}/m_p)^3 = 2 \times 10^{57}$. Note that the ratio of the attractive gravitational force to the repulsive electromagnetic force between two protons is $\alpha^{-1}\alpha_g$ (see also Chapter 2).

Attainable temperatures are 10 K in nearby molecular cloud cores and 1,000 K in primordial clouds with only trace amounts of H_2 as coolants. The sound speed and accretion rate vary considerably over this temperature range. However, one always finds that the minimum mass limited by opacity is around $M \sim 0.01 M_\odot$. Remarkably, detailed numerical

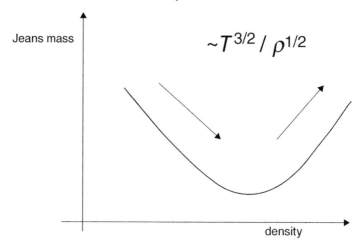

Figure 10.1 Opacity-limited fragmentation. The Jeans mass-scale decreases with increasing density until opacity intervenes, and the fragment can no longer radiate freely. This sets a minimum mass-scale of about $0.01 M_\odot$.

simulations of star formation find a similar characteristic minimum mass-scale. Fragmentation that is limited by opacity inevitably occurs down to this scale. Some physics must be added to arrive at a typical stellar mass of order a solar mass.

10.3.2 The Characteristic Mass of a Star

There is a problem with molecular cloud fragmentation. Stars are much more massive than the minimum fragment mass. The initial mass function of stars was first developed by Edwin Salpeter [33]. His great insight was to realise that counts of massive stars, being short-lived, underestimate the initial mass function by a large factor. This one insight removed any motivation for a non-stellar origin for the heavy elements as Gamow had once hoped to find in the Big Bang.

Two more pieces of astrophysics are needed to go from fragments to stars. First, the fragments grow by accretion to the stellar mass range. Second, accretion must be halted, to produce decreasing numbers of massive stars and to avoid producing overly massive stars. This occurs via feedback, from deuterium burning to magnetically suppressed accretion and magnetically driven outflows for low-mass stars to ionisation fronts and winds for massive stars.

The Role of Deuterium Burning

Deuterium burning marks an important phase in the contraction of a star to the main sequence. A minimum mass is needed for deuterium burning to occur; otherwise, one simply has a brown dwarf powered only by gravitational contraction.

I follow and reassess the arguments of Krumholz [22], in which he reevaluates the gravitational fragmentation mass for a protostar by using accretion to set the protostellar luminosity and deuterium burning to set the central temperature. He argues that approximately half of the Bonnor-Ebert mass forms the protostar – namely,

$$M_* \approx 0.6 \sqrt{\left(\frac{k_b T_e}{\mu_{H_2} M_H G}\right)^3 \frac{1}{\rho_e}}.$$ (10.4)

Here, T_e is the gas temperature at the edge of the gas accreting to form the star, and ρ_e is the local gas density, averaged over the collapsing region. He fixes T_e from the protostellar luminosity, assuming dust opacity dominates, and obtains the luminosity from the energy released by accretion onto the central protostar. The scalings are $L \propto T_c \dot{M}_* \propto M T_c \rho_e^{1/2}$, and the central temperature T_c is fixed by deuterium burning to be constant (in terms of fundamental constants, it is proportional to the Gamow energy, E_G, for the D burning reaction, itself $\propto \alpha^2 m_H c^2$). The temperature T_e is determined by the central luminosity, and for the case of fiducial Milky Way–type dust cooling in a $n = 3/2$ polytrope, scales as $(\Sigma L/M)^{1/4}$. The interstellar pressure enters via the assumption that surface density (more physically, dust opacity) is constant and, consequently, $p_{ism} \propto \Sigma^2$.

The resultant scaling for stellar mass is

$$M_* \propto \alpha_g^{-3/2} \alpha^{2/3} p^{-1/18} \Theta_c^{-4/3},$$ (10.5)

where $\Theta_c = (E_G/4kT_c)^{1/3} \sim 10$ and $T_c \approx 10^6$ K at the onset of the deuterium burning that determines the onset of the formation of the star. Krumholz, in fact, normalises the interstellar pressure to the Planck pressure $c^7/\hbar G^2$ and thereby introduces what is an unhelpful dependence of the characteristic stellar mass on the gravitational fine-structure parameter, $Gm_p^2/\hbar c$, obtaining

$$M_* = A_K m_p \left(\frac{\alpha^{41}}{\alpha_g^{25}}\right)^{1/18} \left(\frac{p}{p_{Planck}}\right)^{-1/18},$$ (10.6)

where A_k is a dimensionless constant.

In fact, it is more appropriate to normalise the temperature to natural atomic units – namely, to a Rydberg $kT_{ryd} = \alpha^2 m_e c^2/2$. After all, quantum gravity has little relevance for star formation. Only atomic or molecular processes, specifically Ly-α emission or molecular H_2 rotational excitations, are effective for primordial star formation, with $T \sim 1,000$ K, and, of course, in the present epoch interstellar medium, star formation occurs at $T \sim 10$ K. The pressure at the boundary of the protostar can be written to within dimensionless factors of order unity as

$$p = \frac{\rho k T}{m_p} = \frac{v_s^8}{G^3 M^2}.$$ (10.7)

For purposes of normalisation, I use a fiducial value $v_s = \alpha c$, equivalent to the Rydberg energy kT_{Ryd} of an electron. Now, rewriting the expression for the characteristic mass, I find that

$$M_* = A_S m_H \alpha_g^{-3/2} \alpha^{1/6} \left(\frac{kT_c}{m_p c^2} \right)^{4/9} \left(\frac{c^8 m_p^4}{e^6 p} \right), \tag{10.8}$$

where A_S is a dimensionless factor of order unity. After inserting the previous expression for the pressure in terms of v_s, this reduces to

$$M_* = A_S m_H \alpha_g^{-3/2} \alpha^{-1/2} \left(\frac{m_p}{m_e} \right)^{1/4} \left(\frac{kT_c}{m_p c^2} \right)^{1/2} \left(\frac{T_{Ryd}}{T} \right)^{1/4}. \tag{10.9}$$

Krumholz's conclusion, as corrected here, is essentially unchanged: the only dependence of M_* on astrophysical parameters is via the interstellar pressure, and the dependence is exceedingly weak. There is still an explicit dependence on the deuterium burning temperature. However. there is no longer any scaling with ambient pressure, rather with ambient temperature. Note that the temperature scaling is the inverse of that found for opacity-limited fragmentation. The scaling in fundamental units gives a mass-scale of around a solar mass.

Magnetically Driven Feedback

There is a competition between accretion and nuclear energy-driven feedback from the protostar. This limits growth of the protostar. Accretion and feedback are key to understanding the masses of subsequent generations of stars. First, there is accretion to grow the protostellar fragments. The accretion rate in the core of a collapsing isothermal cloud is $t_{acc} \sim v_s^3/G$, where v_s is the local speed of sound. At typical cold molecular cloud temperatures, $T \sim 10$ K, $v_s \sim 0.3$ km/s and $\dot{M} \sim 3 \times 10^{-6}$ $M_\odot$/yr. The protostar is powered by gravitational contraction over a Kelvin-Helmholtz time. This is of order $t_{KH} \sim GM^2/(RL) \sim 3.10^5$ yr for the Sun at a typical protostellar luminosity of $10L_\odot$ and radius of $10R_\odot$. One can form a star of around a third of a solar mass before feedback can occur once the onset of energy release by nuclear burning intervenes. This corresponds to the characteristic mass of a star.

Such phenomena as magnetically regulated contraction, controlled by dissipation of magnetorotational energy, play an important role early in the pre-main-sequence phase in determining the distribution of stellar masses. All of this culminates via feedback in the form of stellar winds from low-mass and ionisation fronts from massive young stars. The net effect is that typical stars have masses of order a third of a solar mass, with a mass function at birth that extends roughly as the inverse square of stellar mass to about 200 solar masses, above which stable stars cannot form because of Eddington's argument: they would be radiation-pressure dominated and unstable.

Derivation of the actual form of the mass function at stellar birth is a complex process, in which self-regulation by outflow-driven turbulence plays a role. Stellar self-regulation is strongly dependent on stellar mass, the more massive stars having stronger outflows that drive turbulence and progressively raise the Jeans mass, inhibiting the number of forming stars and leading to a Salpeter-like initial mass function (IMF) [39]. There are other sources of turbulence, including self-gravity, cloud collisions, spiral density waves, driving by stellar bars, and mergers with dwarf galaxies. The IMF can be derived in a general way and incorporated into a Press-Schechter-like mass function that is controlled by molecular cloud turbulence [19]. One consequence is that a top-heavy IMF can be generated in extreme situations [4].

10.3.3 The Minimum Mass of a Star

Magnetic fields play a crucial role in star formation via enabling transfer of angular momentum. This is often manifested via magnetically driven protostellar outflows, which are observed over a wide range of mass-scales. The amplification of fields via a dynamo is an essential precursor to low-mass star formation. Fields are most likely seeded by a Biermann battery. In an interstellar cloud of scale L, electron density n_e, turbulent – including thermal – velocity dispersion v_s, and pressure p, the Biermann mechanism combines differential rotation and pressure gradients to generate a seed field

$$B_{\text{Bier}} = \frac{c}{en_e} \frac{p}{Lv_s}. \tag{10.10}$$

The conventional dynamo appeals to a differentially rotating protostellar disc that develops magnetorotational instabilities (MRI) instabilities. These require a minimum field in order for the electron Larmor radius to be less than the scale of the most rapidly growing mode. The MRI disc requires a minimum field of

$$B_{\text{MRI}} = \frac{\pi^3}{2\sqrt{2}} ln\Lambda m_e^{1/2} \frac{e^2 c^2 p^{1/2}}{L(kT)^{3/2}}. \tag{10.11}$$

To initiate the MRI dynamo, we require [41] the Biermann battery to provide the initial seed $B_{\text{Bier}} > B_{\text{MRI}}$.

To reduce this further, I set

$$p = 2n_e kT; \quad kT = m_p v_s^2; \quad \rho = \frac{3v_s^6}{4\pi G^3 M^2}; \quad m_p v_s^2 = \frac{1}{2}\alpha^2 m_e c^2. \tag{10.12}$$

For MRI to be seeded, the protostellar mass cannot exceed a critical value,

$$M^{\text{MRI}} < \frac{\pi^{5/2}}{8} \sqrt{\frac{3}{2} \frac{ln\Lambda}{\alpha_g^{3/2}} \frac{c}{v_s}} m_e \sim \alpha_g^{-3/2} \alpha^{1/2} \left(\frac{T_{Ryd}}{T}\right)^{3/2} m_p. \tag{10.13}$$

Magnetically driven feedback is seen to be important for current epoch star formation.

One may compare this with the typical stellar mass: $\alpha_g^{-3/2} m_p$. We infer that $M^{\text{MRI}} \lesssim 0.1 M_\odot$, where $T \sim T_{Ryd}$. Magnetically driven feedback in current-epoch star formation is relevant for the upper limit on the mass of a brown dwarfs, a star that is too low in mass to achieve thermonuclear ignition. In the primordial situation, where much larger characteristic stellar masses are formed, magnetic feedback may still play a role, via seeding of the MRI dynamo, in allowing some lower-mass stars to form.

10.3.4 The Fate of Stars Like Our Sun: White Dwarfs

We can calculate from first principles the mass of a white dwarf. The Sun will end up as a a white dwarf, a compact star whose nuclear fuel supply is exhausted and is supported by degeneracy pressure of the electrons. This phenomenon occurs at sufficiently high density, when nuclear fuel resources are exhausted and there is no thermal pressure support. The Pauli exclusion principle comes into play, forbidding fermions from occupying the same quantum state, and manifests itself as a quantum pressure.

For a white dwarf, there are two regimes:

1. If degenerate but non-relativistic, $p_e = n\Delta p^2/m_e = n^{5/3}\hbar^2/m_e$, where $\Delta p = \hbar n^{1/3}$, $n_e = \Delta x^{-3}$, and $(\Delta x)(\Delta p) = \hbar$. This yields $GM/R = p_e/\rho \propto n^{5/3}$.
2. If degenerate and relativistic, then $p_e = n_e c\Delta p = \hbar c n^{4/3}$, with $GM/R = p_e/\rho = \hbar c(M/R^3)^{1/3}/m_p^{4/3}$. This leads to the white dwarf mass $M_{WD} = \left(\hbar c/G^{3/2}m_p^{-3}\right)$. Rewriting this expression, the white dwarf mass is

$$M_{Chandra} \approx \alpha_g^{-3/2}m_p. \tag{10.14}$$

In summary, two dimensionless fine-structure constants – combining quantum, electromagnetic, and gravitational forces – determine the future of our sun, the mass of the white dwarf to which it will collapse. The energy release en route is manifested by the formation of an ultraluminous red giant, followed by the expulsion of a planetary nebula, leaving the white-dwarf remnant in its centre. Planetary nebulae are among the most beautiful objects photographed in the sky.

10.3.5 The Most Massive Stars

Let us next see what determines the upper limit on the mass of a star. One of the greatest battles of all time occurs when the force of gravity faces up to the electromagnetic force. Gravity is an incredibly weak force. But it adds up. For stars, its main opponent, electromagnetism, provides pressure and sums over positive and negative charges. So while the repulsive electromagnetic force between a pair of protons is stronger than the attractive gravitational force by about 40 powers of 10, if we consider enough atoms with protons and electrons whose charges cancel, the forces can balance each other. When we turn on gravity, many of the atoms are crushed into bare nuclei by the immense pressure that itself opposes the force of gravity. The magic number for an equilibrium between electromagnetic and gravitational force is about the mass of the Sun, give or take Eddington's factor of 10 or 100 – which, in a nutshell, is why we are here.

One can show that the upper limit on the mass of a hydrogen-burning star is $m_{max} \approx \alpha_g^{-5/3}\alpha^{2/3}m_p$. For a massive star, we use opacity $\tau = \kappa\rho R$, where $\kappa = \sigma_T/m_p$ and σ_T is the Thomson scattering cross section. The luminosity is given by

$$L = (\kappa\rho)^{-1}\nabla(\sigma 4\pi T^4 R^2) \approx \sigma 4\pi T^4 R^2/\tau = v_s^5/G. \tag{10.15}$$

Then the electron scattering opacity can be expressed as

$$\tau_{es} = \frac{\sigma_T}{m_p} \frac{M}{R^2} = \pi \left(\frac{e^2}{m_e c^2} \right)^2 \frac{1}{m_p} \frac{1}{M} \left(\frac{kT}{Gm_p} \right)^2 = \alpha^2 \alpha_g^{-2} \frac{m_p}{M} \left(\frac{kT}{Gm_p} \right)^2. \tag{10.16}$$

The resulting expression for the maximum stellar mass is

$$M = \alpha_g^{-5/3} \alpha^{7/3} \left(\frac{m_e}{m_p} \right)^{1/6} \left(\frac{\tau}{\tau_{es}} \right)^{1/3}. \tag{10.17}$$

Here, τ is the optical depth expressed in terms of the Rosseland mean opacity. The explicit dependence on opacity is seen to be very weak.

Massive stars are short-lived because of their prolific burning of nuclear fuel. They explode after exhausting their nuclear fuel supply. Their masses are so large that there is no stable end point. Their cores collapse to form neutron stars, objects that are a thousand times more compact than white dwarfs, and supported by the quantum degeneracy pressure of neutrons. Such stars, if of initial mass less than about $25M_\odot$, end up as neutron stars. These are made of the most extreme form of matter with the nuclei so compacted together that the stars are essentially at nuclear density. The maximum mass of a neutron star is calculated to be three solar masses. The parent stars lose matter by driving winds as they evolve and end their lives as neutron stars. The radius of a solar mass neutron star is about 10 kilometres.

So much energy is released in this final collapse that a huge explosion blows off the outer layers of the star. This is a supernova explosion. The light released in this immense explosion, heralding the death throes of a massive star, is about that of the luminosity of an entire galaxy, some tens of billions of suns. The impact of such a violent explosion on the surrounding interstellar gas results in spectacular images of shock waves igniting quiescent gas clouds into glowing ribbons of hot gas that extend over hundreds of light years.

We observe objects well above the maximum mass of compact stars, produced by the death and collapse of massive hydrogen-burning stars. Stars that are initially more than about $25M_\odot$ cannot eject more than 90% of their initial mass. Any compact object that is more massive than $3M_\odot$ must be a black hole. One cannot pack matter any more densely than a neutron star without forming a black hole. The most massive stars collapse into black holes when their supply of nuclear fuel is exhausted. We measure their masses because there often is a close companion star whose atmosphere is heated by the gravity of its black hole companion and which emits prolifically in X-rays. Prior to 2016, the existence of black holes was conjecture and inference, although sound physical reasoning reinforced our belief in the existence of black holes. But the reasoning was indirect. The LIGO gravitational wave detectors provided the ultimate breakthrough via detection of merging black holes [1].

But one question leads to another. What came first, before the star clusters of Population III? Like that of the legendary old lady at a talk, often attributed to nineteenth-century philosopher William James, who had an answer as to what came first. Your cosmology is rubbish, she said – the world is supported by a giant turtle. So what supports the turtle?

It's a second turtle. And what supports that? It's turtles all the way down, she continued. This paradox of infinite regression is more commonly known as the chicken-and-egg problem: which came first? We believe that the smallest galaxies came first. Let us see what determines the mass of a galaxy today.

10.3.6 The First Stars

The most massive stars of all directly collapse into black holes. This is where our search for the first stars leads us. For we expect that the first stars were exclusively massive compared to the Sun. The early Universe was metal-free, consisting just of hydrogen and helium atoms. Now, metals control gas cooling and, hence, gas pressure. Because in today's Universe interstellar clouds are highly contaminated by metals, gas cooling is very effective. This means that the ability to support massive clouds from the compressive effects of gravity is greatly reduced. The clouds fragment into smaller clumps, and it is these that form stars today. The typical star in the Milky Way is about a third the mass of the Sun.

In the early Universe, conditions were different. There was pristine gas, hydrogen and helium. The lack of effective cooling by the metal pollutants meant that fragmentation was relatively ineffective. We refer to the first, zero metallicity, generation of stars as Population III. The gas temperature even in dense clouds is controlled by molecular hydrogen cooling. The typical temperature is determined by the lowest rotational level of H_2 and is around 1,000 K. This means that the sound speed is of order 3 km/s, and, hence, the accretion rate is around $\dot{M} \sim 3 \times 10^{-3} M_\odot \mathrm{yr}^{-1}$. Unimpeded accretion would generate typical first stars of thousands of solar masses. Such masses could accumulate in accreting gas over the typical Kelvin-Helmholtz timescale of a million years before strong energy release in the form of ionising photons occurs.

To better estimate how feedback intervenes to prevent all first stars being thousands of solar masses, consider the relation between mass and luminosity, $L \propto M^3$. Inserting this into the expression for the duration of the Kelvin-Helmholtz phase, one sees that $t_{KH} \propto 1/MR$. Let us assume v_s is controlled by molecular physics and require t_{KH} to be less than t_{acc} for effective feedback above a mass M_{cr}. Then, let's use the protostellar opacity constraint $\kappa \rho R \sim 1$ to set $R \sim (\kappa M)^{1/2}$ to deduce $M_{cr} \propto T^{-1}$, where T is the temperature of the accreting gas. I conclude that feedback is effective today in cold molecular clouds but much less effective by a factor ~ 100 in mass threshold in the primordial stellar case. Hence, the characteristic mass of the first stars is of order $100 M_\odot$.

Numerical simulations, including the effects of feedback on the accreting gas, show that there is considerable mass loss, and the typical mass of a first Population III star ranges from hundreds of solar masses at high redshift ($z > 20$), where there is intense UV radiation, to tens of solar masses at lower redshifts, where the UV background is lower and H_2 cooling is more effective [20].

These massive stars lived fast and furiously, dying in supernova explosions after millions of years. They generated elements like carbon, oxygen, and iron. The explosions polluted their environment. Once enough pollution of heavy elements occurred and enhanced cooling at low temperatures, the typical gas accretion rates onto protostellar fragments were greatly reduced. Once a metallicity $[Z] \gtrsim -4$ was reached, the first generation of exclusively massive stars gave way to formation of normal stars whose typical masses were of order a solar mass. There were fewer more-massive stars, and the mass range of Population II spanned the full range that Eddington had envisaged. Their low metallicity tells us that these are mostly old stars that formed early in the history of our Milky Way galaxy.

The spectra of the most metal-poor halo stars reveal fossil tracers from a previous generation of long-extinct stars of the first generation. The ratios of certain elements to a standard tracer such as iron are useful age indicators since iron is ejected in supernovae throughout the history of the galaxy. If we find a very iron-poor environment, we can be sure it is old. And if, in that environment, there are overabundant traces of unusual elements not normally produced in today's Universe, we can attribute such clues to the ashes from the first generation of short-lived stars, incorporated into the clouds that made all later stars. It as though we have silent witnesses to a long-vanished crime scene.

10.3.7 So What If G Was Different?

Imagine we live in a universe with a very different value of Newton's constant. If G were too small, there would not be time to make stars within the age the Universe. The lifetime of a nuclear-fuelled star is proportional to GM. Since the mass of a star is of order $\alpha_G^{-3/2}$, its lifetime is proportional to $G^{-1/2}$. So if G were smaller by a factor of 10^8, carbon would not be formed, and we would not be here. This argument assumes the age of the Universe has nothing to do with G, as it is dominated by dark energy – that is to say, the energy density associated with the cosmological constant, or some 10 billion years when it becomes dominant (see Chapter 3 for more details). That's the modern view.

Robert Dicke had a somewhat different take on this [10]. In the case of a closed Friedmann universe, considered by many at that time to be the most natural choice of a cosmological model, the age of the universe depends as $G^{-1/2}$, and so stars are inevitably fine-tuned to produce carbon. However, for an Einstein–de Sitter universe, or an open universe, we would need to be at a special epoch. This can be expressed as Dirac's coincidence [11] between the scales of cosmology, involving G and H_0, and quantum physics, involving α and $\hbar$.

10.3.8 Our Lack of Understanding of Star Formation Physics

Here is one recent example to demonstrate the fragility of our current knowledge of star formation. In our Milky Way galaxy, most dense molecular gas, where much of galactic star formation occurs in giant molecular clouds (GMCs), is in the ~ 3 kpc radial distance molecular ring. However, within ~ 0.5 kpc radial distance, the central molecular zone (CMZ) contains massive giant molecular clouds of typical mass $\sim 10^6 M_\odot$ and density

~ 100 cm^{-3}, amounting to $\sim 20\%$ of the dense molecular gas in the galaxy, but with the star formation efficiency suppressed by an order of magnitude compared to the molecular ring [2].

Compare this result with ALMA molecular mapping of the nearest starburst galaxy, NGC 253, which reveals some 10 GMCs of higher-density $n \sim 2,000$ cm^{-3} and mass 10^7 $M_\odot$ in its central molecular zone that are somewhat more extreme than those in normal galactic star–forming discs. The star formation efficiency (per free-fall time) in NGC 253 GMCs matches the standard $\sim 1\%$ for star-forming disc galaxies.

The most remarkable fact, however, is that the measured Mach number in these star-forming GMCs is ~ 90, versus ~ 10 in the star-forming GMCs in the MWG disc [24]. The turbulence driver in the Milky Way's GMCs is usually considered to be either the central bar or energy released via gravitational contraction. Perhaps the comparison between the MWG and NGC 253 indicates one should blame star formation, but the details of the turbulence driving are far from evident. Indeed, the extreme turbulence measured in the GMCs in the star-forming zone of NGC 253 and their star formation efficiency, comparable to that in most galactic disc GMCs, along with the relative longevity of the observed NGC 253 inner GMCs per disc crossing time (the cloud free-fall time is ~ 0.7 Myr, and the disc crossing time is ~ 3 Myr), seem to demand another explanation for the turbulence driver. This might possibly involve magnetic compression [3], although cloud survival remains to be demonstrated.

10.4 Galactic Scales

I now turn to larger scales. Galaxies have a characteristic mass. There is a limit to the mass of a galaxy. Essentially all galaxies host central supermassive black holes. The masses of supermassive black holes scale with the galaxy mass. About 10% of the time, these objects are fed enough gaseous or stellar fuel to be visible as active galactic nuclei or quasars. There are universal scaling relations involving galactic or SMBH masses, and these various relations involve fundamental constants and, hence, fine-tuning.

The Characteristic Mass of a Galaxy

The mass function of galaxies [29] comes from cold dark matter assembly via gravitational instability in the expanding Universe, seeded by primordial density fluctuations (see Chapter 6). The luminosity function of galaxies [34] arises from baryonic dissipation and gravity. Both functions are described by a power-law with an exponential cut-off at high masses. Both functions have a characteristic scale, in mass and in luminosity. It is possible to connect these scales and derive the characteristic mass of a galaxy in terms of fundamental constants.

To demonstrate this, I compare the free-fall time with the cooling time for an isothermal sphere. This yields a necessary condition for efficient star formation. The cooling rate is [46], with an update including time-dependent ionisation in radiatively cooling gas [18]. For a gas cloud of galactic mass, I use the cooling function

$$\Lambda_H = \alpha \sigma_T c^2 \frac{1}{v_c} E_\gamma \left(\frac{v_c}{v_{ref}}\right)^{2\beta}, \tag{10.18}$$

where v_c is the circular velocity of a test particle in the galaxy at the half-mass radius, $v_{ref} = \alpha c \left(\frac{m_e}{m_p}\right)^{1/2}$, and $E_\gamma = \alpha^2 c^2 m_e$. Here, $\beta = 0.5$ describes bremsstrahlung, and $\beta = -0.5$ approximates bound-free cooling. For a hydrogen-helium plasma at $10^5 - 10^7 \mathrm{K}$, $\beta \approx -0.5$ is a reasonable approximation. I set $t_c = 3kT/(2\Lambda_H n)$ and $t_d = GM_g/(2v_c^3)$, where M_g is the dynamical mass of a galaxy. The characteristic mass of a galaxy can now be expressed as

$$M_g = \frac{\alpha^5}{\alpha_g^2} \frac{m_p}{m_e} \frac{t_c}{t_d}. \tag{10.19}$$

Remarkably, this gives the characteristic stellar mass of a galaxy [31, 37]. If the cooling time is set equal to the free-fall time, this yields a reasonable limit on the maximum dissipation time needed to guarantee efficient star formation, which is a necessary although not a sufficient condition for star formation.

One can make the preceding result more precise by considering a more general parametrisation of the cooling function,

$$\Lambda_H = g_{bf} G^2 \frac{\alpha}{\alpha_g^2} \frac{v_s}{c^2} \frac{m_p^{9/2}}{m_e^{3/2}}, \tag{10.20}$$

where the correction to free-fall via addition of bound-free cooling is

$$g_{bf} = 1 + \frac{\alpha^2}{2} m_e/m_p \, (c/v_s)^2. \tag{10.21}$$

The characteristic galaxy mass in baryons is now

$$M_g = g_{bf} \frac{\alpha^3}{\alpha_g^2} \left(\frac{m_p}{m_e}\right)^{3/2} \left(\frac{v_s}{c}\right)^2 \frac{t_c}{t_d}. \tag{10.22}$$

This amounts to approximately $10^{11} M_\odot$. Note that, ignoring factors of order a few, the mass of a star is $\alpha_g^{-3/2}$, whereas the mass of a galaxy is $\alpha^5 \alpha_g^{-2} m_p/m_e$, since $v_s/c \sim \alpha$. Hence, the number of stars in a typical galaxy is $\alpha^5 \alpha_g^{-1/2} m_p/m_e \sim 10^{12}$.

We see explicitly that $g_{bf}(v_s/c)^2 \approx$ constant over a wide range of temperatures. This is true for H+He cooling but remains more or less the case even when metal cooling is included. One can account, from first principles, for the characteristic Schechter luminosity L_* as the mass in baryons that can cool efficiently within a dynamical time and form stars (see Figure 10.2). By abundance matching, one also infers the dark mass–to-light ratio associated with the galaxies, again in terms of fundamental constants. It does not matter whether the gaseous halo is a monolithic collapsing cloud supported by gravity or a collection of clouds contained by the same gravity field: the shock velocities are essentially the same.

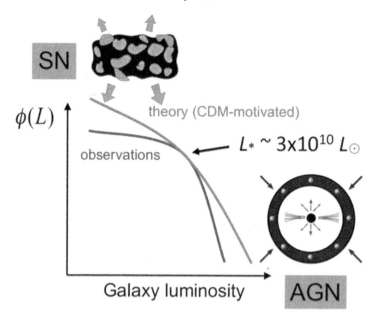

Figure 10.2 The halo mass function and the galaxy luminosity function. The normalisation of the luminosity function is from observed galaxy counts, whereas the dark matter halo function is renormalised to match at the critical, or Schechter luminosity. L_* is calculated by the cooling requirement. This gives the characteristic luminosity of a galaxy or $L_* \sim 3 \times 10^{10} L_\odot$. The renormalised ratio gives the predicted mass-to-luminosity ratio for L_* galaxies. Figure taken from Reference [45].

Next, I consider continuing accretion of gas on timescales longer than the free-fall time. This is relevant to make the most massive galaxies and would be appropriate to the massive central galaxy in a cluster. In the previous derivation, I replace the galaxy dynamical time with the age of the Universe – i.e., I set the gas cooling time equal to the Hubble time. This provides the maximum baryonic mass of a galaxy. The additional factor is

$$\frac{t_H}{t_d} = \left(\frac{12\pi G\rho}{H_0^2} \right)^{1/2} = \left(\frac{9}{2} \frac{\rho_g}{\bar{\rho}(z)} \right)^{1/2} \approx 25. \tag{10.23}$$

This nicely accounts for typical masses of the most massive galaxies, these being the brightest cluster galaxies. On group and scales, cooling is so long that star formation is inefficient. Galaxies survive as discrete objects, modulo merging, and supermassive galaxies are rare.

Confrontation of Theory and Observations

Galaxies span a range of mean densities. At the epoch of first collapse, or formation, the mean gas density, averaged over the halo virial radius, is largely determined by the epoch at which the halo initially decoupled from the expanding Universe (see Figure 10.3).

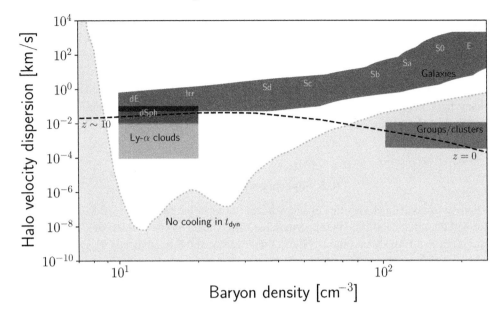

Figure 10.3 All galaxies form by dissipation in their protogalaxy gas-rich phase; clusters do not. The dotted line demarcates the region where dissipation occurs within a free-fall time. The thick dashed line shows the halo dark matter evolution, corresponding to the evolution of the virialised density with decreasing redshift. This is normalised to the CMB temperature fluctuations, for an assumed nearly scale-invariant power spectrum. All systems that dissipate and form stars lie above this line, at higher densities. Intergalactic clouds, detected as the Ly-α forest, lie below this line. Progressively more massive objects collapse later, in bottom-up evolution.

It then contracted by a factor of two to virialise at an over-density of $18\pi^2$ relative to the background at this formation epoch, in the simplest spherically symmetric, non-dissipative model. Hence, dwarf galaxies, which form earliest, have the densest halos.

Baryons dissipate and are denser than the dark halo matter. Hence, the condition for cooling within a free-fall time provides a demarcation zone in the baryon density–virial temperature diagram. The latter translates to stellar velocity dispersion in the spheroid, useful in order to make more direct contact with the galaxy data. Galaxies of all Hubble types fall within the dissipation region, from the most massive old elliptical galaxies to the younger disc star-forming galaxies, and all the way down in mass to the dwarf galaxies. On the most massive scales for collapsed systems, groups and clusters of galaxies are at sufficiently low density because of late formation that they lie outside the dissipation region, thereby accounting for their being systems of discrete galaxies.

The predicted correlation of the data in the density-temperature plane is re-expressed in more precise empirical correlations. These include the Faber-Jackson relation, the fundamental plane for early-type galaxies, and the Tully-Fisher relation for later-type galaxies. All of these are manifestations of the relations between the stellar mass or luminosity of a galaxy, its stellar velocity dispersion or rotation rate, and its size. The stellar velocity dispersion for slowly rotating early-type galaxies as well as the rotation curves for disc galaxies and rapidly rotating early-type galaxies are largely controlled by the dark matter content, at least in the outer parts of the galaxies, as is also the case for dwarf galaxies. Intergalactic clouds fall into this general arrangement. They have not collapsed and, indeed, many have been heated and expanded, so they are marginally self-gravitating. They lie below the galaxy locus.

10.5 Supermassive Black Holes

There are two distinct varieties of black holes: those in the range of 3–100 solar masses, formed by stellar death, and the supermassive variety at the centres of galaxies, observed to weight from a 100,000 to some 10 billion solar masses, and formed mostly by gas accretion. The accreting gas forms an accretion disc and fuels the black hole, and gravitational energy is released in violent explosions, outflows, or plasma jets. We observe the effects of black hole outbursts on their environment, effects for which we have no alternative explanation. These include the vast outpourings of energy from compact regions in the remote objects we call quasars, the most luminous objects in the Universe.

Essentially all galaxies contain central massive black holes. The Milky Way has a central supermassive black hole of some four million solar masses that today is quiescent. But millions of years ago, it was active. Traces of exploded debris are seen around our galactic centre that arose in a violent explosion. We do not know in any definitive detail how supermassive black holes form, whether from the mergers of many stellar black holes or possibly by swallowing large amounts of interstellar gas that accretes into the centres of galaxies. Black hole capture of stars and gas accretion are aided by the mergers of galaxies that stir up the gravity field and reduce gas angular momentum near the black hole, and facilitate fuelling by directing material inwards. Astronomers observe a correlation between central black hole mass and the total mass of stars in the spheroidal components of galaxies: the latter are the oldest stars of the galaxy. The disc component of a galaxy contains many young stars and is a recent acquisition. This tells us that massive black holes co-evolved along with the oldest components of galaxies. Massive black holes formed far away and a long time ago. These immensely luminous objects in the nuclei of galaxies were active when the Universe was young. Current data taken with the world's largest telescopes suggest that supermassive black holes formed along with the first galaxies.

We see very massive black holes in the early Universe as quasars because they accrete interstellar gas. The gas heats up in an accretion disc, and glows in X-rays, and the black hole grows as it feeds. Quasars are the most luminous objects in the Universe. The mass-doubling time for a typical massive black hole is about 50 million years. This has two implications. First, we expect to find massive black holes a few hundred million years after the Big Bang. And, indeed, black holes as massive as 10 billion solar masses are found when the Universe was a billion years old. Second, we probably need seeds, smaller black holes, to accrete gas or merge to make the massive ones. If we started from typical stellar black holes, around 10 or 20 solar masses, there would not be time enough to grow the monsters that we find at early epochs.

Quasars also fulfill another important function in the course of massive galaxy evolution at late times. Quasar eruptions continue sporadically as gas clouds are captured by and feed the central supermassive black hole. These events succeed in cleaning out the interstellar medium in the immediate environment of the host galaxy, as will be discussed later. The result is a massive galaxy that is said to be red and dead. Red because there are only old stars, and old stars are cool and hence red. Dead because young stars are not forming in the absence of molecular gas clouds.

Radio Jets

Supermassive black holes at the centres of galaxies are generally dormant giants. They become reactivated only when fresh gaseous fuel is provided. This may happen after billions of years, when a merger occurs with a nearby galaxy. This is observed when we peer into the distant Universe with our largest telescopes and observe galaxies in their youth. We occasionally see powerful jets of plasma that drive giant radio-emitting lobes. These are vigorous outflows that collide with and eject interstellar gas clouds into the surrounding circumgalactic medium, where the gas eventually cools and ends up as intergalactic clouds, to enrich new generations of galaxies.

Powerful radio jets are produced in the black hole ergosphere by release of energy arising from the rotation and winding-up of magnetic fields. The supermassive black hole acts like a gigantic flywheel, with the spin of space providing the momentum. Gripping the flywheel is presumably done with powerful magnetic fields, thought to be omnipresent. The fields initially have a dipole pattern but soon tangle up because of the differential spin and turbulence. Magnetic reconnection releases huge amounts of energy that is initially channelled along the axis of rotation, emerging as a collimated jet and continuing for thousands of parsecs.

One can estimate the jet power from the Bondi accretion rate. Suppose a fraction α of the Bondi accretion rate mass flux fuels the radio jet with efficiency ηc^2. One has $P_j = 4\pi\alpha\eta GM_{BH}^2\sigma^{-3}\rho c^2$. I evaluate the ambient density as follows. I assume the SMBH is immersed in the cooled core of a cluster or massive galaxy. I set local dynamical time equal to cooling time to obtain

$$\frac{1}{\sqrt{3\pi G\rho_c/32}} = \frac{3kTm_p}{\left(2\Lambda_0\sqrt{\frac{T}{T_0}}\rho_c\right)},$$
(10.24)

or

$$\rho \approx \sigma^2 T_0 G\pi m_p^2 k\Lambda_0^{-2}.$$
(10.25)

I infer that

$$P_j = M_{BH}^2\sigma^{-1}[4\pi\alpha\eta GT_0 G\pi m_p^2 k/\Lambda_0^2] \propto M_{BH}^{7/4} \text{ or } M_{BH}^{9/5}$$
(10.26)

for the range of the usual M_{BH}–σ scaling relation. This tells us that jet power and any associated signals, such as γ-ray signals or high-energy cosmic ray and neutrino contributions, come predominantly from the most massive galaxies, typically the brightest cluster galaxies.

Quasar energy release regulates the limiting black hole mass. This happens during the formation phase of the black hole and the massive galaxy, inferred from the observed correlation between bulge (old stellar) component and black hole mass to be coeval. Hence, we control the growth of black holes observationally, at least in constraining the growth by accretion. Growth also occurs by mergers of stars with black holes and of black holes with each other. But this stellar merger rate is inefficient and, hence, likely to be subdominant. Supermassive black holes generally formed long ago in the nuclei of forming galaxies.

SMBH Growth

I assume that the growing SMBH is fed by spherical Bondi accretion. This is most likely an upper limit on the actual accretion, which may be geometrically limited. However, filamentary flow will feed the central accretion disc, and the overall mass supply will be enhanced over the mass reservoir within the Bondi radius. Hence, Bondi accretion may give a reasonable estimate. I compare the Bondi rate $\pi G^2 M_{BH}^2 \rho / v_s^3$ with the Eddington-limited accretion rate onto the SMBH, $G M_{BH} 4\pi m_p / (\eta \sigma_T c)$. I assume that the accretion onto the SMBH is cooling limited, and expect that $v_s \sim 10$km/s, as appropriate for the Lyman-alpha-cooling primordial clouds – typically of mass $\sim 10^8$–$10^9 M_\odot$ – that are often assumed to be the sites for the first SMBHs, or at least for their seeds. In the case of a massive galaxy, the nuclear gas will be enriched and ionised. If photo-ionised, the effective value of v_s will be similar; if collisionally ionised, it will be somewhat larger. Data on nuclear emission lines generally motivate photoionisation models.

The ratio should give a measure of the Eddington ratio f_E, provided that we evaluate ρ carefully. I implement the following argument. I adopt a generalised power-law profile for the inner host galaxy, $\rho(r) = \rho_{1/2}(r_{1/2}/r)^\beta$, where $\rho_{1/2} \delta_c f_b^{-1} \rho_0 (1+z)^3$, $r_{1/2} = f_b r_{1/2}^{DM} = G M_{sph} v_c^{-2}$, and $\beta = 1$ for a NFW profile. I evaluate the density at the gravitational capture radius of the SMBH. This yields $\rho(r_{BH}) = \rho_{1/2}(M_{sph}/M_{BH})^\beta$. The Eddington ratio can now be written as

$$f_{Edd} = \frac{\dot{M}_{BH}^{accr}}{\dot{M}_{Edd}} = \left(\frac{v_c}{v_s}\right)^3 \left(\frac{M_{sph}}{M_{BH}}\right)^{\beta-1} \left(\frac{\delta_c}{f_b}\right)^{1/2} \eta n_0 \sigma_T t_0 (1+z)^{3/2}. \tag{10.27}$$

For the accretion rate, one may fit the expression given by [9], and I obtain $\dot{M}_{acc} = \frac{1}{2} M_{sph,0}(1+z)^{5/2} t_0^{-1}$. Assume a fraction f_{acc} ends up in the spheroid. I obtain

$$M_{sph,\,z} = M_{sph,0} \exp\left(-\frac{3}{4} f_{acc} z\right). \tag{10.28}$$

Now we write $M_{sph,\,z}$, evaluated at $r_{1/2}$, as

$$M_{sph,z} = 4\pi (3-\beta)^{-1} \rho_{1/2} r_{1/2}^3 \tag{10.29}$$

and obtain

$$M_{sph,0} = \frac{v_c^3}{G} \left(\frac{3 - \beta}{4\pi G\rho_0} \right)^{1/2} \left(\frac{f_b}{\delta_c(1 + z)^3} \right)^{1/2}. \tag{10.30}$$

This specifies v_c at redshift z in terms of present-day spheroid mass,

$$v_{c,z} = (GM_{sph,0})^{1/3} \left(\frac{4\pi\rho_0\delta_c}{3 - \beta} \right)^{1/6} (1 + z)^{1/2} e^{-z/4} \left(\frac{\delta_c}{f_b} \right)^{1/6}. \tag{10.31}$$

The spheroid half-mass radius is then given by

$$r_{1/2,z}^3 = \frac{3 - \beta}{4\pi\rho_0} \frac{f_b e^{-\frac{3}{4}f_{acc}z}}{\delta_c(1 + z)^3} M_{sph,0}. \tag{10.32}$$

The redshift scaling applies if z is equal to or less than the redshift of spheroid formation. Next, we estimate the redshift scaling of black hole to spheroid mass. We have

$$\frac{M_{BH}^{SR}}{M_{sph,z}} = \frac{v_c f_g \sigma_T}{((3 - \beta)4\pi G)^{1/2} m_p} \rho_{1/2}^{1/2}. \tag{10.33}$$

This gives the explicit z dependence

$$\frac{M_{BH}^{SR}}{M_{sph,z}} = f_g \frac{\sigma_T}{m_p} \left(\frac{\delta_c \rho_0}{f_b(3 - \beta)} \right)^{2/3} \left(\frac{M_{sph,0}}{4\pi} \right)^{1/3} (1 + z)^2 \exp(-f_{acc}z/4). \tag{10.34}$$

At constant accretion efficiency, black holes are expected to be more obese, relative to the associated stellar components, by an order of magnitude at $z \gtrsim 5$. This is essentially consistent with the high-redshift data [50].

The Mass of a Seed MBH

We find massive black holes a few hundred million years after the Big Bang. And, indeed, black holes as massive as 10 billion solar masses are found when the Universe was a billion years old. To explain this, we need seeds, smaller black holes, to accrete gas or merge to make the massive ones. If we started from typical stellar black holes, around 10 or 20 solar masses, there would not be time enough to grow the monsters we find. Intermediate-mass black holes undergo catastrophic accretion as dwarf-mass halos merge to eventually form supermassive black holes in massive galaxies.

How the seeds are formed is somewhat of a mystery. Presumably, they are formed from the first generation of million-solar-mass clouds formed after the Big Bang. These clouds were chemically pure; no heavy elements had been formed yet in supernovae. This means that cooling occurs by hydrogen-atom excitations. Electrons jump from one atomic orbit to a higher, more energetic level, due to a collision with another atom or electron, and then de-excite by emitting a photon. Atomic cooling provides a powerful channel for losing energy and guarantees that the clouds will undergo direct collapse to form black holes, typically of 10,000 solar masses. The only obstacle is that too much cooling may occur. Hydrogen molecules are a catalyst for cooling as they are more easily excited than hydrogen atoms.

Trace amounts of hydrogen molecules form from residual ionisation and H^- formation, and the enhanced cooling leads fragmentation into stars. This fate can be avoided if there is enough turbulence present, as this can remove the molecules in shocks or via UV radiation from nearby stars and black holes that destroy any molecules in their vicinities.

For a seed black hole, Ly-α cooling is thought to be the prevalent cooling mechanism for black hole feeding. Take v_c to correspond to $m_p v_c^2 = 0.75 \text{Ry} = 10.2 \text{eV}$, corresponding approximately to the sound speed in a medium where Ly-α cooling ($n = 2$ to $n = 1$) prevails. This is equivalent to $v_{ref} = \alpha c \left(\frac{m_e}{m_p} \right)^{1/2}$. The inferred black hole mass mass is

$$M_{\text{seed}} = \left(\frac{v_c}{v_{ref}} \right)^5 \frac{\alpha^5}{\alpha_g^2} \left(\frac{m_e}{m_p} \right)^{1/2}. \tag{10.35}$$

The ratio of seed black hole to galaxy mass is $\alpha^2 \left(\frac{m_e}{m_p} \right)^{3/2} \sim 10^{-7}$. This will provide the necessary boost for Eddington-limited growth to form supermassive black holes at early epochs.

Astrophysics of SMBH

Most massive galaxies are elliptical galaxies and today are indeed red and dead. All contain massive central black holes. Because of the paucity of interstellar gas, these black holes are not being fed and are not active. Occasionally, there is a merger, most typically with a small galaxy. If this contains gas, some of this will fall into the black hole and drive a new phase of activity.

A massive black hole lurks in the centre of our own galaxy. The Milky Way black hole is located in the constellation of Sagittarius. Studies of orbiting stars allowed a precise mass determination. The stars can be directly resolved in our galactic centre at infrared wavelengths, and their orbital motions and speeds around the black hole have been followed for more than a decade.

Our massive black hole is a bright source of radio waves, one of the brightest in the sky, hence its name Sagittarius A. Despite the radio emission, it is not very active today. The X-ray emission from Sagittarius A is very low, and we infer that it is currently lacking fuel, accreting very little gas from its surroundings. But in the past, the situation was occasionally very different, and Sagittarius A has experienced outbursts of violent activity. About 10 million years ago, a giant explosion occurred that left behind traces in the gamma ray sky. The Fermi gamma ray satellite telescope image has revealed twin bubbles of gamma ray emission hundreds of light years in extent that attest to the violence of this explosion a million years ago.

Quasars are much more vigorous manifestations of black hole activity. Their scaled-up outbursts and X-ray luminosities enable us to detect them at the far end of the Universe. Perhaps the most intriguing result of studies with the largest telescopes show that intense black hole activity and extreme rates of star formation occur in the same objects. They are also driving powerful outflows of gas. This is one of the biggest mysteries of structure formation. Why are these phenomena coinciding at early epochs, especially reaching extreme rates of star formation and mass loss that are rarely seen in the absence of supermassive black holes?

Perhaps black hole feeding, with its immense energy release, is a consequence of star formation, with the stellar debris feeding the black hole. If so, this fails to address the extreme intensity of star formation seen in many of the most distant galaxies. Both AGN activity and star formation bursts may be the collateral damage from gas-accretion events. Gas-rich galaxy mergers can provide the gas supply responsible both for star formation and supermassive black hole fuelling. In a merger, gas-cloud orbits are perturbed, and some clouds are directed into the capture zone of the supermassive black hole to refuel its activity. The increase in gas mass stimulates star formation at the same time.

An alternative pathway has the powerful outflows from the black holes compressing nearby clouds and triggering an intense burst of star formation in the surrounding clouds. This is predicted analytically [42] and in simulations [16]. Of course, the vigorous outflows show that the star formation rate is being quenched. But this may have been preceded by a phase of triggering that induced the jets and winds. Predicted signatures include star formation by positive feedback, leaving an imprint on the kinematics of stars formed via AGN ouflow-triggered star formation. Such kinematic tracers would be both long-lived and differ from the more chaotic kinematics of cloud collision and/or SN-induced star formation in the conventional scenario of gravitationally unstable disc star formation [12].

Observational evidence has recently been found for positive feedback by AGN, despite the expected short duty cycle before negative feedback kicks in. Examples include stacked star-forming galaxies [5] that show evidence of radial trajectories of hypervelocity stars, as found in the Milky Way and interpreted as being induced by Fermi-bubble-type explosions millions of years ago [44], and in AGN-driven outflows [27]. Indications of positive feedback via enhanced star formation rates are also found in green valley, X-ray detected AGN galaxies [26]; in a Seyfert galaxy [8]; along with simultaneous negative and positive feedback in an obscured radio-quiet quasar [7].

SMBH Feedback and the M_{BH}-σ Scaling Relation

Black hole feeding, with its immense energy release, may be a consequence of star formation, with the stellar debris feeding the black hole. It more likely is fed by gas streams similar to those that replenish the gas reservoir and drive star formation. Alternatively, a merger between two galaxies helps shed gas angular momentum and allows gas to pour into the vicinity of the massive central black hole. The paucity of major mergers favours the merger hypothesis only for the most luminous quasars.

The outpouring of energy from the accreting SMBH exerts strong feedback on the surrounding interstellar medium. It is conjectured that there is first a period of positive feedback during which gas clouds are compressed and a burst of star formation is triggered, followed a million years later by a phase of gas-outflow and star formation quenching because of the combined effects of AGN outflows and supernovae from the newly formed massive stars [40]. A key issue is the efficiency of the coupling of the high-energy ejecta with the surrounding ISM. This may involve momentum or energy conservation in order to explain the massive outflows of escaping gas observed from the host galaxy. Nuclear outflows are observed at $\sim 0.1c$, and these drive outflows on galactic scales at typical velocities of thousands of km/s which are responsible for depleting the gas reservoir and quenching star formation [15, 48]. Quenching must occur in order to explain the red and dead nature of early-type galaxies, and SMBHs are the principal culprit.

It is not simply momentum conservation of the SMBH ejecta that drives galactic winds. Reality, as found by simulations and observations, is more complicated. Feedback by massive black holes is controlled by the balance between gravitational attraction and radiation pressure acting on ionised accreting gas. The former is controlled by

$$\frac{L_{\text{Edd}}}{c} = \frac{GMM_{gas}}{r^2} \tag{10.36}$$

and the latter by

$$L_{\text{Edd}} = 4\pi c G M_{\text{BH}} m_p / \sigma_T. \tag{10.37}$$

When blowout occurs at a high enough rate, the gas reservoir is exhausted, and star formation terminates. This should apply for a homogeneous ISM when the $M_{\text{BH}} - \sigma$ relation saturates, and then

$$M_{\text{BH}} = 3 \times 10^9 \left(\frac{\sigma}{300 \text{ km/s}}\right)^4.$$

Mechanical feedback by SMBH is driven both by jets and winds. Jets drive bow shocks that interact much as do winds, geometrical factors aside. The Eddington luminosity and Eddington mass-outflow rate are related by $L_{\text{Edd}} = \eta c^2 \dot{M}_{\text{Edd}}$. We can infer the central wind velocity, since $L_{\text{Edd}}/c = v_w \dot{M}_{\text{Edd}}$ and the outflow speed is $v_w = \eta c \sim 0.1c$ for a nuclear wind, as observed.

There are two limiting cases that correspond to momentum or energy conserving outflows. Momentum balance gives $M_{\text{BH}} = \sigma^4 (\sigma_T / m_p)/(\pi G^2)$. Inclusion of cooling results in failure to eject enough material to quench star formation, failing to attain both the massive galaxy luminosity function and the massive black hole scaling relation.

Energy balance $1/2 M_{\text{outflow}} v_{\text{shock}}^2 = \text{const}$ means that the momentum increases as $P \propto 1/v_w$. This results in enough momentum boost to reduce excessive SMBH feedback and growth, thereby reproducing the normalisation of the Magorrian relation. The predicted scaling is given by

$$M_{\text{BH}} = \alpha \sigma^5 \frac{(\sigma_T / m_p)}{(\pi \eta c G^2)}. \tag{10.38}$$

Naïve application of this expression gives too low a normalisation for the observed relation. These results apply to highly simplified outflows. Cosmological gas accretion is ignored, as is the more realistic ISM structure. In reality, inflow is not quenched in a multiphase ISM. One needs nuclear momentum at a level $\sim 10 L_{\text{Edd}}/c$ [43]. Because the ISM is multiphase, a more realistic treatment is needed, as diffuse gas is ejected more easily, whereas dense gas clumps tend to fall in.

Simulations with cosmological accretion converge on the energy-driving case, albeit needing more effective coupling than single photon radiation scattering as in the preceding naïve formulae. Energy-driven outflows can reach momentum fluxes exceeding $10L_{Edd}/c$ within the innermost 10 kpc of the galaxy. The observed large-scale AGN-driven outflows seem to be energy driven, according to simulations. The momentum of the swept-up shell grows with decreasing shell velocity and, hence, gives effective outflows that shut off star formation and can recover the observed normalisation [6].

Observations favour energy conservation, connecting nuclear outflows from accretion discs around the AGN at $\sim 0.1c$ with kiloparsec-scale galactic-scale molecular outflows at several hundreds of km/s. The more massive the central SMBH is, the stronger the quasar outflow is found to be [32]. The spheroid mass and SMBH mass both grow with time because of the merger and gas accretion history.

There is a large variance in the scaling relations. In particular, while IMBH and SMBH generally lie on the $M_{BH}-\sigma$ relation, there is more dispersion in the $M_{BH}-M_*$ relation. For early-type massive galaxies, star formation is preferentially quenched in the quasar phase; hence, the SMBH can be overmassive relative to the stellar component. Indeed, a much broader dispersion is found of black hole mass to stellar mass [51] at high redshift relative to the nearby universe, with some extraordinarily massive SMBH outliers [36]. For late-type galaxies, continuing star formation at late times is more important, and black holes can be under-massive. Consequently, while there is expected to be little deviation in the $M_{BH}-\sigma$ relation, which primarily samples the gravitational potential, there should be a trend for SMBH at high redshift to often be obese and for IMBH, necessarily observed at low redshift, to be anorexic [28], as observed.

10.6 Conclusions

We have made immense progress in star and galaxy formation over the past several decades. Much of this is driven by data from space and terrestrial telescopes–ranging from Spitzer, Hubble, and Herschel to IRAM and ALMA. Much is also driven by numerical simulations at ever-improving resolution. It is possible to understand some of the key scales in terms of the fine-tuning of fundamental parameters. I discussed estimates for, among other characteristic scales, the typical protostellar fragment mass, stellar mass, maximum stellar mass, galaxy mass, maximum galactic mass, supermassive black hole mass, and SMBH seed mass. I showed how feedback may be characterised by ratios of fundamental constants, both for galaxies and supermassive black holes.

The essence of this discussion ultimately is star formation. The truth remains that simulations of star formation, especially in the cosmological context, cannot span the essential dynamical range between galaxy and star formation. Our models for star formation remain phenomenological rather than fundamental, and all predictions for the high-redshift universe should consequently be taken with a grain of salt. Semi-analytical models of galaxy formation have 50 or more parameters. These parameters are set phenomenologically, based on studies of the nearby universe, and one cannot be confident that similar rules for star formation apply at high redshift. Numerical simulations rely on sub-grid modelling of star formation physics. The paradigm is reasonable but, again, is based on experience deduced from nearby regions of star formation. The early Universe differs in being not only gas rich but also AGN and quasars being far more prevalent, given that their duty cycles render them almost permanently in active mode.

Galaxy formation is not well understood in our vicinity, where, for example, only half the baryon content of the Universe is accounted for. There are problems in understanding the cored nature and the abundance of dwarf galaxies, for which, admittedly, there are solutions that, however, require a certain degree of fine-tuning. Similar remarks apply to the prevalence of bulgeless galaxies and of ultra-diffuse galaxies in our vicinity.

The most direct evidence for fine-tuning is usually associated with the adopted strength of gravity. If the gravitational constant is too low, stars do not evolve sufficiently to produce the carbon needed for life. On the larger scales, galaxies provide the backdrop for star formation. Here, the most relevant parameter is the strength of the initial density fluctuations, often attributed to inflation-boosted quantum fluctuations (see Chapter 4). Over a modest range, this determines the abundance of galaxies and their epoch of formation.

If, however, one varies the fluctuation strength by an order of magnitude or more, one runs into serious problems. If the initial fluctuations are much too large, the Universe contains relatively few galaxies but predominantly supermassive black holes. If the fluctuations are much too small, there are no galaxies [47]. The fluctuation strength is a phenomenological parameter that cannot be directly related to fundamental constants; rather, it is a product of inflation and of poorly constrained inflationary physics.

It is not clear, short of a compelling theory of inflation, how one can control the initial conditions of the Universe. Inflation cannot generate isotropy and flatness of space from generic initial conditions. Nor is there a simple measure of initial conditions, although we know that our observed Universe represents a small subset of possible universes. Here, I prefer to focus on the phenomenology of galaxies and stars and to link this phenomenology to the fundamental constants of nature, motivated by the possibility that these may vary between possible universes. That is the cosmic connection that one can develop without straying too far beyond observational constraints.

References

[1] Abbott, B. P., *et al.* 2016. 'Observation of Gravitational Waves from a Binary Black Hole Merger'. *Physical Review Letters*, **116**(6), 061102.

[2] Barnes, A. T., *et al.* 2017. 'Star Formation Rates and Efficiencies in the Galactic Centre'. *Monthly Notices of the Royal Astronomical Society*, **469**(Aug.), 2263–2285.

[3] Birnboim, Y., Federrath, C., and Krumholz, M. 2018. 'Compression of Turbulent Magnetized Gas in Giant Molecular Clouds'. *Monthly Notices of the Royal Astronomical Society*, **473**(Jan.), 2144–2159.

[4] Chabrier, G., Hennebelle, P., and Charlot, S. 2014. 'Variations of the Stellar Initial Mass Function in the Progenitors of Massive Early-Type Galaxies and in Extreme Starburst Environments'. *Astrophysical Journal*, **796**(Dec.), 75.

[5] Cicone, C., Maiolino, R., and Marconi, A. 2016. 'Outflows and Complex Stellar Kinematics in SDSS Star-Forming Galaxies'. *Astronomy and Astrophysics*, **588**(Apr.), A41.

[6] Costa, T., *et al.* 2018. 'Quenching Star Formation with Quasar Outflows Launched by Trapped IR Radiation'. *Monthly Notices of the Royal Astronomical Society*, **479**(Sept.), 2079–2111.

[7] Cresci, G., *et al.* 2015a. 'Blowin' in the Wind: Both "Negative" and "Positive" Feedback in an Obscured High-*z* Quasar'. *Astrophysical Journal*, **799**(Jan.), 82.

[8] Cresci, G., *et al.* 2015b. 'The MAGNUM Survey: Positive Feedback in the Nuclear Region of NGC 5643 Suggested by MUSE'. *Astronomy and Astrophysics*, **582**(Oct.), A63.

[9] Dekel, A., and Birnboim, Y. 2006. 'Galaxy Bimodality Due to Cold Flows and Shock Heating'. *Monthly Notices of the Royal Astronomical Society*, **368**(May), 2–20.

[10] Dicke, R. H. 1961. 'Dirac's Cosmology and Mach's Principle'. *Nature*, **192**(Nov.), 440–441.

[11] Dirac, P. A. M. 1937. 'The Cosmological Constants'. *Nature*, **139**(Feb.), 323.

[12] Dugan, Z., Gaibler, V., and Silk, J. 2017. 'Feedback by AGN Jets and Wide-Angle Winds on a Galactic Scale'. *Astrophysical Journal*, **844**(July), 37.

[13] Faber, S. M., and Jackson, R. E. 1976. 'Velocity Dispersions and Mass-to-Light Ratios for Elliptical Galaxies'. *Astrophysical Journal*, **204**(Mar.), 668–683.

[14] Ferrarese, L., and Merritt, D. 2000. 'A Fundamental Relation between Supermassive Black Holes and Their Host Galaxies'. *Astrophysical Journal Letters*, **539**(Aug.), L9–L12.

[15] Feruglio, C., *et al.* 2015. 'The Multi-Phase Winds of Markarian 231: From the Hot, Nuclear, Ultra-Fast Wind to the Galaxy-Scale, Molecular Outflow'. *Astronomy and Astrophysics*, **583**(Nov.), A99.

[16] Gaibler, V., *et al.* 2012. 'Jet-Induced Star Formation in Gas-Rich Galaxies'. *Monthly Notices of the Royal Astronomical Society*, **425**(Sept.), 438–449.

[17] Gebhardt, K., *et al.* 2000. 'A Relationship between Nuclear Black Hole Mass and Galaxy Velocity Dispersion'. *Astrophysical Journal Letters*, **539**(Aug.), L13–L16.

[18] Gnat, O., and Sternberg, A. 2007. 'Time-Dependent Ionization in Radiatively Cooling Gas'. *Astrophysical Journal Supplements*, **168**(Feb.), 213–230.

[19] Hennebelle, P., and Chabrier, G. 2008. 'Analytical Theory for the Initial Mass Function: CO Clumps and Prestellar Cores'. *Astrophysical Journal*, **684**(Sept.), 395–410.

[20] Hirano, S., *et al.* 2015. 'Primordial Star Formation under the Influence of Far Ultraviolet Radiation: 1540 Cosmological Haloes and the Stellar Mass Distribution'. *Monthly Notices of the Royal Astronomical Society*, **448**(Mar.), 568–587.

[21] Kennicutt, Jr., R. C. 1989. 'The Star Formation Law in Galactic Disks'. *Astrophysical Journal*, **344**(Sept.), 685–703.

[22] Krumholz, M. R. 2011. 'On the Origin of Stellar Masses'. *Astrophysical Journal*, **743**(Dec.), 110.

[23] Larson, R. B. 1981. 'Turbulence and Star Formation in Molecular Clouds'. *Monthly Notices of the Royal Astronomical Society*, **194**(Mar.), 809–826.

[24] Leroy, A. K., *et al.* 2015. 'ALMA Reveals the Molecular Medium Fueling the Nearest Nuclear Starburst'. *Astrophysical Journal*, **801**(Mar.), 25.

[25] Magorrian, J., *et al.* 1998. 'The Demography of Massive Dark Objects in Galaxy Centers'. *Astronomical Journal*, **115**(June), 2285–2305.

[26] Mahoro, A., Pović, M., and Nkundabakura, P. 2017. 'Star Formation of Far-IR AGN and Non-AGN Galaxies in the Green Valley: Possible Implication of AGN Positive Feedback'. *Monthly Notices of the Royal Astronomical Society*, **471**(Nov.), 3226–3233.

[27] Maiolino, R., *et al.* 2017. 'Star Formation inside a Galactic Outflow'. *Nature*, **544**(Mar.), 202–206.

[28] Nguyen, D. D., *et al.* 2018. 'Nearby Early-Type Galactic Nuclei at High Resolution: Dynamical Black Hole and Nuclear Star Cluster Mass Measurements'. *Astrophysical Journal*, **858**(May), 118.

[29] Press, W. H., and Schechter, P. 1974. 'Formation of Galaxies and Clusters of Galaxies by Self-Similar Gravitational Condensation'. *Astrophysical Journal*, **187**(Feb.), 425–438.

[30] Rees, M. J. 1976. 'Opacity-Limited Hierarchical Fragmentation and the Masses of Protostars'. *Monthly Notices of the Royal Astronomical Society*, **176**(Sept.), 483–486.

[31] Rees, M. J., and Ostriker, J. P. 1977. 'Cooling, Dynamics and Fragmentation of Massive Gas Clouds: Clues to the Masses and Radii of Galaxies and Clusters'. *Monthly Notices of the Royal Astronomical Society*, **179**(June), 541–559.

[32] Rupke, D. S. N., Gültekin, K., and Veilleux, S. 2017. 'Quasar-Mode Feedback in Nearby Type 1 Quasars: Ubiquitous Kiloparsec-Scale Outflows and Correlations with Black Hole Properties'. *Astrophysical Journal*, **850**(Nov.), 40.

[33] Salpeter, E. E. 1964. 'Accretion of Interstellar Matter by Massive Objects'. *Astrophysical Journal*, **140**(Aug.), 796–800.

[34] Schechter, P. 1976. 'An Analytic Expression for the Luminosity Function for Galaxies'. *Astrophysical Journal*, **203**(Jan.), 297–306.

[35] Schmidt, M. 1959. 'The Rate of Star Formation'. *Astrophysical Journal*, **129**(Mar.), 243.

[36] Shao, Y., *et al.* 2017. 'Gas Dynamics of a Luminous $z = 6.13$ Quasar ULAS J1319+0950 Revealed by ALMA High-Resolution Observations'. *Astrophysical Journal*, **845**(Aug.), 138.

[37] Silk, J. 1977a. 'On the Fragmentation of Cosmic Gas Clouds. I. The Formation of Galaxies and the First generation of Stars'. *Astrophysical Journal*, **211**(Feb.), 638–648.

[38] Silk, J. 1977b. 'On the Fragmentation of Cosmic Gas Clouds. II. Opacity-Limited Star Formation'. *Astrophysical Journal*, **214**(May), 152–160.

[39] Silk, J. 1977c. 'On the Fragmentation of Cosmic Gas Clouds. III. The Initial Stellar Mass Function'. *Astrophysical Journal*, **214**(June), 718–724.

[40] Silk, J. 2013. 'Unleashing Positive Feedback: Linking the Rates of Star Formation, Supermassive Black Hole Accretion, and Outflows in Distant Galaxies'. *Astrophysical Journal*, **772**(Aug.), 112.

[41] Silk, J., and Langer, M. 2006. 'On the first Generation of Stars'. *Monthly Notices of the Royal Astronomical Society*, **371**(Sept.), 444–450.

[42] Silk, J. and Norman, C. 2009. 'Global Star Formation Revisited'. *Astrophysical Journal*, **700**(July), 262–275.

[43] Silk, J., and Nusser, A. 2010. 'The Massive-Black-Hole-Velocity-Dispersion Relation and the Halo Baryon Fraction: A Case for Positive Active Galactic Nucleus Feedback'. *Astrophysical Journal*, **725**(Dec.), 556–560.

[44] Silk, J., *et al.* 2012. 'Jet Interactions with a Giant Molecular Cloud in the Galactic Centre and Ejection of Hypervelocity Stars'. *Astronomy and Astrophysics*, **545**(Sept.), L11.

[45] Silk, J., and Mamon, G. A. 2012. 'The Current Status of Galaxy Formation'. *Research in Astronomy and Astrophysics*, **12**(Aug), 917–946.

[46] Sutherland, R. S., and Dopita, M. A. 1993. 'Cooling Functions for Low-Density Astrophysical Plasmas'. *Astrophysical Journal Supplements*, **88**(Sept.), 253–327.

[47] Tegmark, M., *et al.* 2006. 'Dimensionless Constants, Cosmology, and Other Dark Matters'. *Physical Review D*, **73**(2), 023505.

[48] Tombesi, F., *et al.* 2015. 'Wind from the Black-Hole Accretion Disk Driving a Molecular Outflow in an Active Galaxy'. *Nature*, **519**(Mar.), 436–438.

[49] Tully, R. B., and Fisher, J. R. 1977. 'A New Method of Determining Distances to Galaxies'. *Astronomy and Astrophysics*, **54**(Feb.), 661–673.

[50] Wang, R., *et al.* 2013. 'Star Formation and Gas Kinematics of Quasar Host Galaxies at $z \sim 6$: New Insights from ALMA'. *Astrophysical Journal*, **773**(Aug.), 44.

[51] Willott, C. J., Bergeron, J. and Omont, A. 2017. 'A Wide Dispersion in Star Formation Rate and Dynamical Mass of 10^8 Solar Mass Black Hole Host Galaxies at Redshift 6'. *Astrophysical Journal*, **850**(Nov.), 108.

11

How Special Is the Solar System?

MARIO LIVIO

Abstract

Given the fact that Earth is so far the only place in the Milky Way galaxy known to harbour life, the question arises of whether the solar system is in any way special. To address this question, I compare the solar system to the many recently discovered exoplanetary systems. I identify two main features that appear to distinguish the solar system from the majority of other systems: (1) the lack of super-Earths and (2) the absence of close-in planets. I examine models for the formation of super-Earths, as well as models for the evolution of asteroid belts, the rate of asteroid impacts on Earth, and of snow lines, all of which may have some implications for the emergence and evolution of life on a terrestrial planet. Finally, I revisit an argument by Brandon Carter on the rarity of intelligent civilisations, and I review a few of the criticisms of this argument.

11.1 Introduction

The discovery of thousands of extrasolar planets in the Milky Way galaxy in recent years (e.g., [13] and references therein) has led to the realisation that our galaxy may contain (on average) as many as 0.16 Earth-size planets and 0.12 super-Earths (planets with a mass of a few Earth masses) per every M-dwarf habitable zone [46]. The habitable zone is that relatively narrow region in orbital distances from the central star that allows for liquid water to exist on the surface of a rocky planet (see Chapter 12). Given that it is still the case that, to date, Earth is the only place known to support complex life (or any life form, for that matter), the plethora of potentially habitable extrasolar planets raises the important question of whether the solar system is, in any sense, special. This question is further motivated by the so-called Fermi Paradox – the absence of any signs for the existence of other intelligent civilisations in the Milky Way [81, 97]. It is interesting to explore, therefore, the status of the solar system in the context of the entire known population of extrasolar planets.

The solar system includes eight planets (suggestions for a ninth planet are yet to be confirmed [14]), and two belts composed of generally smaller bodies – the asteroid belt and the Kuiper belt. In an attempt to compare the solar system to other exoplanetary systems, Martin and Livio [106] first considered the statistical distributions of orbital separations

412

and eccentricities of the observed exoplanetary orbits. To this goal, and in order to allow for a more meaningful statistical analysis, they performed a transformation on the available data that makes them closer to a normal distribution [20]. Specifically, they transformed the data with the function

$$
y_\lambda(a) = \begin{cases} \frac{a^\lambda - 1}{\lambda} & \text{if } \lambda \neq 0 \\ \log a & \text{if } \lambda = 0, \end{cases} \tag{11.1}
$$

where a is the examined parameter (e.g., eccentricity or semimajor axis), and λ is a constant determined through the process described in Eq. (11.4). The maximum likelihood estimator of the mean of the transformed data is

$$
\overline{y}_\lambda = \sum_{i=1}^{n} \frac{y_{\lambda,i}}{n}, \tag{11.2}
$$

where $y_{\lambda,i} = y_\lambda(a_i)$ and a_i is the ith measurement of the total of n. Similarly, the maximum likelihood estimator of the variance of the transformed data is

$$
s_\lambda^2 = \sum_{i=1}^{n} \frac{(y_{\lambda,i} - \overline{y}_\lambda)^2}{n}. \tag{11.3}
$$

Martin and Livio [106] chose λ such that they maximise the log likelihood function

$$
\ell(\lambda) = \frac{n}{2} \log(2\pi) - \frac{n}{2} - \frac{n}{2} \log s_\lambda^2 + (\lambda - 1) \sum_{i-1}^{n} \log a_i. \tag{11.4}
$$

The new distribution, $y_\lambda(a)$, becomes an exact normal distribution if $\lambda = 0$ or $1/\lambda$ is an even integer.

Figure 11.1 shows the Box-Cox transformed eccentricities for 539 extrasolar planets with measured eccentricities. As we can see, Jupiter lies at -0.97σ from the mean, and the Earth (the unlabelled arrow) is at -1.60σ. In other words, while the eccentricities in the solar system (ranging from $e = 0.0068$ for Venus to $e = 0.21$ for Mercury) are on the low side compared to the general distribution of exoplanets, they are not altogether exceptional. Furthermore, since the mean eccentricity of a planetary system appears to be anti-correlated with the number of the planets in the system [91], the relatively low mean eccentricity of the solar system is actually the one expected for an eight-planet system (a conclusion that is further strengthened when selection biases are taken into account; e.g., [74, 118, 163]).

An examination of the transformed distribution of the semimajor axis for 5,289 candidate planets in the Kepler sample shows that Jupiter lies at 2.4σ from the mean (Figure 11.2). At first blush, this may suggest that the largest planet in the solar system is rather special, but a closer inspection reveals that this fact most likely result from selection effects. For example, if we repeat the analysis after removing planets found through transits (a method favouring planets that are close in, which is only complete in the Kepler sample for periods of up to one year [160]), we find that Jupiter's deviation from the mean is reduced to 1.44σ. This trend appears to be further strengthened by the fact that the number

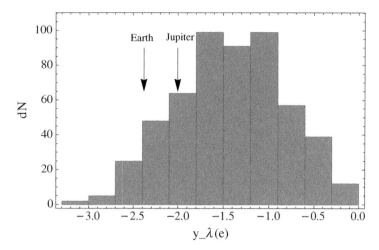

Figure 11.1 Box-Cox transformed distribution of exoplanet eccentricities. The total number of exoplanets is 539.

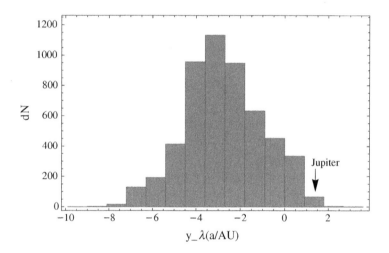

Figure 11.2 Box-Cox transformed distribution of exoplanet semimajor axis, including all planet candidates. The total number of planets is 5,289.

of detections by direct imaging is constantly increasing, suggesting that there may indeed exist an entire population of planets with semimajor axes longer than those of Jupiter, which have so far escaped detection.

Martin and Livio [106] also found that the masses of the gas giants in the solar system fit nicely within the distribution of extrasolar planets (Figure 11.3; the observations of small terrestrial planets are most likely still affected by selection effects), as do their densities (Figure 11.4). In addition, Dressing *et al.* [47] showed, that Earth and Venus can be

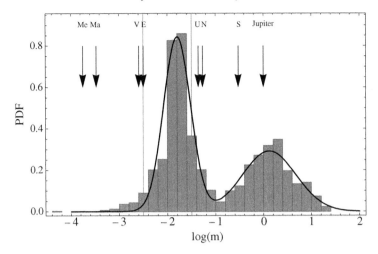

Figure 11.3 The exoplanet mass distribution. The arrows indicate the masses of the planets in the solar system. The vertical lines show the range of planets considered to be super-Earths. The total number of exoplanets is 1,516.

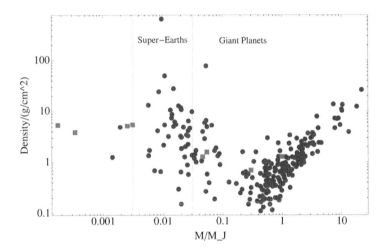

Figure 11.4 Densities of planets as a function of their mass. The dark circles show the exoplanets (a total of 287), and the lighter grey squares the planets in our solar system. The vertical lines again indicate the range of super-Earths.

modelled with a ratio of iron to magnesium silicate similar to that of the low-mass extrasolar planets.

Is there anything, then, in terms of planetary and orbital properties, that makes the solar system (even somewhat) special? Martin and Livio [106] identified two properties that are at least intriguing: (1) the lack of super-Earths and (2) the lack of very close-in planets. Let's briefly discuss each one of these characteristics.

11.2 Lack of Super-Earths

Super-Earths are planets with masses typically between 1 and 10 Earth masses. They are very common in exoplanetary systems. In fact, more than half of the observed Sun-like stars in the solar neighbourhood are orbited by one or more super-Earths with periods of days to months (e.g., [12, 23, 55, 109]). Furthermore, systems observed to contain a super-Earth usually have more than one. Most recently, observations of the star Trappist-1 found that it hosts at least seven Earth-size planets. Their orbits, six of which form a near-resonant chain (with orbital periods of 1.51, 2.42, 4.04, 6.06, 9.1, and 12.35 days), suggest that these planets formed farther from their host star and then migrated inwards (e.g., [146]). If super-Earths form via mergers of inwardly migrating cores, then having more than one super-Earth is theoretically expected (e.g., [39]).

The fact that the solar system does not contain any close-in super-Earths does set it somewhat apart from most observed exoplanetary systems. Before I examine potential *reasons* for this lack of super-Earths, it is interesting to contemplate whether not containing super-Earths (by a planetary system) can in any way be related to the emergence of life in such a system. It is interesting to note that the presence of super-Earths may affect the formation process and properties of terrestrial planets. In particular, Izidoro *et al.* [80] found that if a super-Earth migrates slowly enough through the habitable zone of its host star (the range of orbital radii that allows for liquid water to exist on a rocky planetary surface), then terrestrial planets subsequently forming in this region would be rich in volatiles – far from being an Earth twin. A super-Earth at a very small orbital radius could also (in principle, at least) disturb the dynamical stability of a terrestrial planet in the habitable zone. In contrast, the terrestrial planets in the solar system are expected to remain dynamically stable until the Sun expands to become a red giant (e.g., [86]). Finally, a super-Earth relatively close to the orbital distance of a terrestrial planet could significantly affect the rate of asteroid impacts on such a planet (see Section 11.5.3 for an extensive discussion).

11.3 Lack of Close-In Planets or Debris

The second element that appears to distinguish the solar system from most observed exoplanetary systems is the lack of very close-in planets (or, for that matter, any type of mass). Specifically, while Mercury is at 0.39 AU from the Sun, other exoplanetary systems harbour planets much closer to their host star [15]. In particular, most systems observed to have three or more planets contain a planet with a semimajor axis smaller than that of Mercury.

Again, it is not clear to what extent (if at all) the absence of close-in planets is related to the Earth's habitability. In the next section, I examine the conditions that are necessary for the formation of super-Earths and their potential implications for the two characteristics that make the solar system somewhat special.

11.4 On the Formation of Super-Earths

Super-Earths either form in situ with no significant migration through the protoplanetary disc or they form outside the snow line (the distance from the central star where the temperature is sufficiently low for water to solidify; [88]) and then migrate inwards. Chiang and Laughlin [35] used observations of super-Earths with orbital periods of $P \lesssim 100$ days to construct a minimum-mass extrasolar nebula (MMEN) with a surface density of the gas disc of (R is the distance from the star)

$$\Sigma_{\mathrm{MMEN}} = 9,900 \left(\frac{R}{1 \text{ AU}} \right)^{-1.6} \text{ g cm}^{-2}. \tag{11.5}$$

Equation (11.5) gives a somewhat higher value for the surface density than the minimum mass solar nebula (MMSN, which gives 1,700 g cm^{-2} at $R = 1$ AU, required to form the planets in our solar system [70, 155]). However, it is not clear whether the MMSN is applicable at distances smaller than 0.4 AU (inside Mercury's orbit). Hansen and Murray [67, 68] found that forming super-Earths in situ required having about 50–100 $M_\oplus$ of rocky material interior to 1 AU. Other researchers suggested that super-Earths form farther out in the disc (where solid material is more readily available) and migrate inwards (e.g., [39, 78, 147]).

Martin and Livio [103] showed that in fully turbulent disc models (i.e., discs in which the magnetorotational instability generates viscosity throughout) planets that are close to their host stars cannot form in situ since the mass of the disc interior to $R = 1$ AU is too low (Figure 11.5). Martin and Livio [103] have also shown, however, that in fully turbulent discs super-Earths can (in principle, at least) form farther out, followed by inward migration.

Still, it is generally believed that protoplanetary discs are not fully turbulent. Rather, they most likely contain an unionised region of low or no turbulence, known as a 'dead zone' (e.g., [58, 141]). The dead zone blocks the accretion and allows material to accumulate until it becomes self-gravitating. (Note though that it has been suggested that the Hall effect can 'revive' the dead zone under certain conditions, e.g., [89]).

Martin and Livio [103] considered two prescriptions for determining the surface density in the dead zone. In the first, they assumed that the disc surface layers are ionised by external sources to a maximum surface density depth of $\Sigma_{\mathrm{crit}}/2$ (see discussion in Section 11.4.1). In the second, they assumed that the surface density is determined via a critical magnetic Reynolds number (e.g., [52, 69]), $Re_{\mathrm{M,crit}}$, such that the zone is 'dead' if $Re_{\mathrm{M}} < Re_{\mathrm{M,crit}}$. By following the time-dependent evolution of the disc, they showed that at early times the disc undergoes FU Orionis-type outbursts (see also [5, 165]). These outbursts occur as the extra heating by self-gravity triggers the magnetorotational instability within the dead zone (when the infall accretion rate is still high). At later times, as the accretion rate decreases, there are no further outbursts (Figure 11.6). Planets that survive must form after the cessation of outbursts; otherwise, they are likely to be swept into the star during accretion episodes. Through time-dependent numerical simulations, Martin and Livio [103] have demonstrated that depending on the dead-zone parameters, the disc surface density and the mass inside of 1 AU can build up to several times that of the MMEN

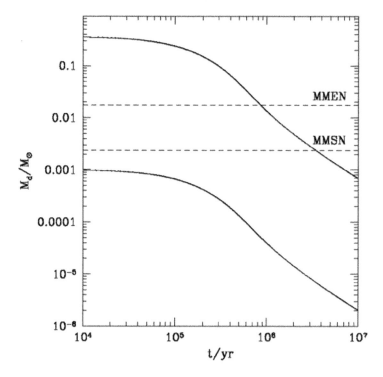

Figure 11.5 Disc mass up to a radius $R = 1$ AU as a function of time (lower curve) and the total disc mass up to 40 AU (upper curve) for a fully turbulent disc model. The infall accretion rate was assumed to decrease exponentially, from an initial $\dot{M} \simeq 10^{-5}\ M_\odot\ \mathrm{yr}^{-1}$, on a timescale of $t_{ff} = 10^5$ yr. The dashed lines show the mass at $R < 1$ AU for the MMSN and MMEN.

(Eq. (11.5)) and, therefore, that the formation of super-Earths in this region is possible (figure 11.6).

11.4.1 Application to the Solar System

There are two possible explanations for the fact that the solar system does not contain any super-Earths: (1) either the conditions in the solar nebula were such that super-Earths could not have formed or (2) super-Earths did form but were later removed by some mechanism. In the latter possibility, it is highly unlikely that the removal mechanism was that of ejection through planet-planet scattering since the average eccentricity in the solar system is low and the planets are quite coplanar. Consequently, if super-Earths had indeed formed in the solar system, they must have been 'swallowed' by the Sun.

There is another constraint that determines the likely formation site. As I noted earlier, if the super-Earths had formed outside the snow line, unless they migrated on a timescale shorter than 0.01–0.1 Myr, they would have shepherded the rocky material interior to their orbit, thereby depleting the Earth's formation zone. This would have made the Earth (and

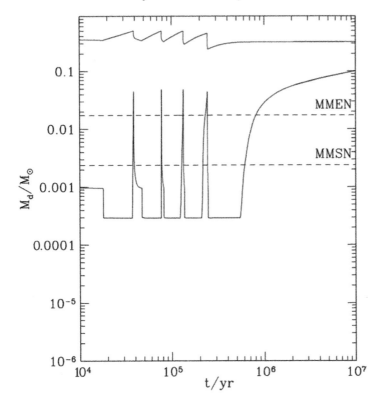

Figure 11.6 Disc mass up to a radius of $R = 1$ AU as a function of time (lower curve) and total disc mass up to a radius of 40 AU (upper curve). The disc has a dead zone defined by $Re_{M, crit} = 5 \times 10^4$. The dashed lines show the mass inside $R = 1$ AU for the MMSN (lower) and MMEN (upper).

other terrestrial planets) volatile-rich and more similar to a water world than to the current Earth. Consequently, if they formed at all, super-Earths in the solar system likely formed in the innermost regions, inside Mercury's orbit.

This suggests a way (in principle) to solve both the lack of super-Earths problem and the lack of close-in planets in the solar system through a single mechanism: super-Earths could form close to the Sun, clear the inner region of debris, and then fall into the Sun following migration through the gas disc. For this chain of events to actually happen requires some degree of *fine-tuning*. Specifically, the surface density in the active (turbulent) layer must be sufficiently large for the planets to migrate into the Sun but at the same time small enough to allow the planets to form in situ to begin with. For a super-Earth to migrate into the Sun, it must do so on a timescale shorter than the time it takes the disc to accrete (the viscous timescale). By equating the timescale for type I migration (obtained when the planet is not massive enough to open a gap in the disc) with the viscous timescale in the disc, Martin and Livio [103] found that the minimum surface density for the planet to migrate into the Sun is given by

$$\Sigma_{\min} = 940.5 \left(\frac{\alpha}{0.01}\right) \left(\frac{H/R}{0.05}\right)^4 \left(\frac{M}{M_\odot}\right)^2 \left(\frac{R}{5 \text{ AU}}\right)^{-3/2} \left(\frac{a}{1 \text{ AU}}\right)^{-1/2} \left(\frac{M_p}{5 M_\oplus}\right)^{-1} \text{ g cm}^{-2}.$$

(11.6)

Here, α is the viscosity parameter, H/R is the disc aspect ratio, M is the mass of the star, R is the radial distance from the central star, a is the orbital radius of the planet, and M_p is the planet's mass.

Generally, it is expected that the surface layers of the disc are ionised by cosmic rays or X-rays from the central star (e.g., [62, 108]), to a maximum surface density depth of $\Sigma_{\text{crit}}/2$ (on each of the disc surfaces). Therefore, if $\Sigma_{\text{crit}} > \Sigma_{\min}$, super-Earths formed in situ will migrate into the Sun at the end of the disc's lifetime. If, on the other hand, $\Sigma_{\text{crit}} < \Sigma_{\min}$, then migration may not be able to allow the super-Earth to be accreted (even if some type I migration takes place). Given the sensitive dependence of $\Sigma_{\min}$ on the disc's aspect ratio and, therefore, on its temperature (see Eq. [11.6]), Martin and Livio [103] speculated that in the solar system, super-Earths formed close to the inner boundary of the dead zone of a relatively cool disc (small H/R). In that case, there was sufficient time for the super-Earths to migrate and be accreted by the Sun. This would explain both the clearing of the region inside Mercury's orbit and the lack of super-Earths. In this scenario, Mercury and Mars would have formed from a relatively narrow annulus of rocky debris in the orbital range of 0.7–1 AU (see also [66]), where the annulus is being truncated at its inner edge by the clearing process. This would explain the relatively small masses of these planets.

I should note that an alternative mechanism, known as the 'grand tack', has been suggested for pushing the super-Earths into the Sun [15, 150]. In this scenario, Jupiter migrates inwards to 1.5 AU before it gets locked into a resonance with Saturn, and then both Jupiter and Saturn move outwards to their current positions. The migration of Jupiter causes the innermost super-Earths to be shepherded into the Sun.

To conclude this section, the lack of super-Earths in the solar system appears to be somewhat puzzling. However, the second characteristic that makes the solar system some-what special – the fact that there are no planets or debris inside of Mercury's orbit – may not be a coincidence. Super-Earths that formed in situ in the inner region of a disc that contains a dead zone could have cleared it of all solid material, with the super-Earths subsequently spiralling into the Sun. Martin and Livio [103] showed that for the dead zone to last a sufficiently long time for the super-Earths to form, the surface density in the active (turbulent) layer must satisfy $\Sigma_{\text{crit}} \lesssim 100$ g cm^{-2}. At the same time, for the super-Earths to eventually migrate into the Sun during the final accretion process, Σ_{crit} needs to be sufficiently large so that there would still be sufficient material in the disc. Satisfying both of these constraints (and for the dead zone to last throughout the entire disc lifetime) requires the disc to be sufficiently cool during late accretion. The necessary level of fine-tuning is not excessive, but it still makes the solar system somewhat special in this respect.

An additional element in the solar system that may have played a role in the emergence and evolution of life on Earth is the existence of an asteroid belt, with its associated characteristics. In the next section, I discuss asteroid belts and their possible functions in determining the habitability of terrestrial planets.

11.5 The Potential Significance of Asteroid Belts for Life

There are several ways in which the presence of an asteroid belt in a solar system can (in principle, at least) affect the habitability of a terrestrial planet:

1. Since terrestrial planets tend to form in the dry regions of protoplanetary discs [105], water must be later delivered to the rocky surface, and one of the possible mechanisms is through asteroid impacts (e.g., [120]).
2. Large moons can stabilise the rotation axis of planets against chaotic motion, thereby preventing weather extremes. The formation of such moons may again require asteroid or planet impacts (e.g., [25, 136]).
3. Life itself or its ingredients may have been delivered to Earth by asteroids (e.g., [50, 77]).
4. Since the early Earth was molten, gravitational settling resulted in the Earth's crust being depleted of heavy elements such as iron and gold. Some of those are essential for life, and they were probably brought to the crust by asteroids (e.g., [157]).
5. Finally, on a more speculative note, the dominance of mammals and the emergence of intelligent life on Earth might not have happened if it were not for the asteroid impact that brought about the extinction of the dinosaurs [3].

While all of these potential effects are somewhat uncertain, even if only one of them is operative, it makes the study of the formation and evolution of asteroid belts important for understanding life on a terrestrial planet.

The asteroid belt in our own solar system is composed of millions of irregularly shaped bodies made of rock, ices, and metals. It is located between the inner terrestrial planets and the outer giant planets, and its total current mass is about $5 \times 10^{-4} M_\oplus$, with about 80% of the mass being contained in the three largest asteroids: Ceres, Pallas, and Vesta.

Observations and models of the solar system suggest that at the time of planetesimal formation, the snow line – the radial location outside which ice forms – was located inside the asteroid belt [1, 120]. In particular, while the asteroids in the inner part of the belt are dry, those more distant than about 2.7 AU (from the Sun) are icy C-class objects. It is generally believed that the asteroid belt is the result of gravitational perturbations caused by Jupiter. Those perturbations did not allow planetesimals to merge and grow, which resulted in violent collisions producing fragmentation rather than fusion (e.g., [49]).

Since giant planets likely form outside the snow line [119] because the density of solid material there is much higher (due to water ice condensation [130]), Martin and Livio [101] proposed that *asteroid belts* (if they form at all) *should be located around the snow line*. To test their hypothesis, they calculated the expected location of the snow line in protoplanetary disc models and compared their results with observations of warm dust in exoplanetary systems, since those may indicate the location of exo-asteroid belts.

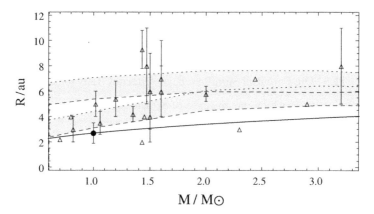

Figure 11.7 Radius of the observed warm dust. The shaded regions show the snow border found from numerical models at times $t = 10^6$ yr (upper) and $t = 10^7$ yr (lower). The solid line shows the analytic approximation to the water snow line, given by Eq. (11.7). The open triangles show the individual systems in Table 11.1. The filled circle shows the location of our solar system's water snow line and the range indicates the extent of our asteroid belt.

11.5.1 Water Snow Lines and Asteroid Belts

The water snow line marks the distance from the star exterior to which ice forms. It is thought to occur at a temperature of about 170 K (e.g., [88]). In an extended region down to about 100 K, the snow border, icy and dry planetesimals can coexist [100].

Particles that migrate through the disc accumulate near the snow line, in a region of relatively small radial extent. This allows them to grow through collisions. Consequently, the formation rate of planetesimals increases by an order of magnitude or more when crossing the snow line (since the solid surface density doubles).

Martin and Livio [101] modelled the evolution of the water snow line in a layered protoplanetary disc with a dead zone (e.g., [5, 165]). They found that the snow line moves inward over time, but that its location is only weakly dependent on the mass of the central star. Figure 11.7 shows the radius R of the inner and outer edges of the water snow border found from numerical simulations (corresponding to temperatures of $T = 170$ K and 100 K, at times $t = 10^6$ yr and 10^7 yr). The functional dependence on the stellar mass can be heuristically derived as follows: the temperature that is obtained in the disc as a result of accretion obeys $T^4 \propto M/R^3$. If we scale this to the radial location of the water snow line in the solar system, we obtain

$$R_{\text{snow}} \simeq 2.7 \left(\frac{M}{M_\odot} \right)^{1/3} \text{ AU.} \tag{11.7}$$

To compare the theoretical predictions with observations, Martin and Livio [101] used observations of debris discs that have a warm infrared component, which could be attributed

Table 11.1 *Observations of warm dust belts that may be exo-asteroid belts.*

Source ID	Name	Spectral type	M ($M_\odot$)	R_{dust} (AU)	T_{dust} (K)	Age (Myr)	Reference
HD 12039		G3/5V	1.02	4–6	109	30	[71]
HD 13246		F8V	1.06	3.5 ± 0.9	166 ± 18	30	[116]
HD 15115		F2	1.5	4 ± 2	179 ± 46	12	[117]
HD 15745		F0	(1.6)	6 ± 2	147 ± 22	12	[117]
HD 16743		F0/F2III/IV	(3.2)	8 ± 3	147 ± 24	10–50	[117]
HD 22049	ϵ Eri	K2V	0.82	3 ± 1	100–150	850	[8]
HD 30447		F3V	(1.5)	6 ± 3	159 ± 36	30	[117]
HD 38678	ζ Lep	A2 IV-V(n)	2.3	3	327	231	[114]
HD 53143		G9V/K1V	0.8	4	120 ± 60	1	[32]
HD 53842		F5V	1.20	5.4 ± 1.4	151 ± 24	30	[116]
HD 86087	HR 3927	A0V	2.44	7	80	50	[32]
HD 98800		K4/5V	(0.7)	2.2	160	10	[98]
HD 109085	η Corvi	F2V	1.43	2	180	1000	[32]
HD 113766		F3/F5V	(1.5/1.4)	4	200	16	[32]
HD 152598		F0V	1.43	9.3 ± 1.5	135 ± 11	210 ± 70	[116]
HD 169666		F5	1.35	4.2 ± 0.6	198 ± 13	21,00	[116]
HD 172555	HR 7012	A51V-V	2.0	5.8 ± 0.6	200	12	[32]
HD 181296	η Tel	A0Vn	2.9	5	115	12	[32]
HD 192758		F0V	(1.6)	7 ± 3	154 ± 31	40	[117]
HD 218396	HR 8799	A5V	1.5	8 ± 3	150 ± 30	30–160	[33, 122]
Samples from Morales *et al.* [119], median values (range)							
19 solar-type stars		G0V (K0V-F5)	(1.1 (0.8–1.4))		177 (99–220)	270 (40–900)	
50 A-type stars		A0V (B8-A7)	(2.9 (1.8–3.8))		203 (98–324)	100 (5–1000)	

Note. The masses in parentheses have been derived from the spectral type.

to an asteroid belt. Table 11.1 lists a compilation of such debris discs with their determined temperature and inferred radius. Also included in the table are the median temperature values for two samples from Morales *et al.* [119], for which radii were not determined. The radii and temperatures of these putative asteroid belts are also shown in Figures 11.7 and 11.8, respectively. As we can see from the figures, both the radii and the temperatures agree with the numerical models of the snow line. This supports the proposed scenario of Martin and Livio [101], in which the location of asteroid belts is around the water snow line.

11.5.2 Giant Planet Location

Figure 11.9 shows the periastron separation for 520 giant planets (with masses larger than $10\ M_\oplus$) as a function of the central star's mass [161]. Also shown in the figure is the lower limit of the distances of the water snow line obtained in the numerical models of Martin

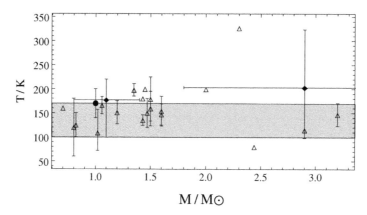

Figure 11.8 The temperature of the observed warm dust. The shaded region marks the snow border. Open triangles show the individual systems in Table 11.1. Filled diamonds show two samples from Morales *et al.* [119]. The filled circle indicates our solar system.

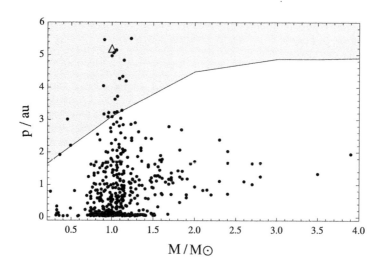

Figure 11.9 The distribution of observed giant planet periastron separation, p, against the mass of the central star, M. The open triangle shows where Jupiter lies. The shaded region shows the icy domain outside of the lower limit to the water snow line predicted by numerical models.

and Livio [101]. The region exterior to the water snow line is shaded. The planets that are observed to be close to their host star are thought to have migrated inwards through a gas-rich disc [64, 92]. Only giant planets that form when the gas is already considerably depleted can avoid migration. Simulations show that the conditions required for a gas giant around a Sun-like star to linger around Jupiter's orbital distance are obtained only in about 1%–2% of the systems [6].

In systems in which giant planet migration occurs, we do not expect to find substantial or compact asteroid belts, since the asteroids are scattered to larger distances or are accreted by the planet or the central star [53]. The observed warm dust belts listed in Table 11.1, therefore, likely remained intact because the giant planets in these systems migrated little or not at all. For this non-migration to have happened, the giant planets must have formed towards the end of the lifetime of the protoplanetary disc.

Figure 11.9 shows that only 19 out of 520 giant planets (less than 4% of these observed systems) are located outside the water snow line. These statistics suggest that only a small fraction of the observed systems contain a compact asteroid belt, making our solar system somewhat special. I should note, however, that the observed statistics are almost certainly affected by selection effects because planets with larger orbital separations are more difficult to detect (see, e.g., [40] for a discussion).

To conclude this part, in our solar system, Jupiter may have migrated only by about 0.2–0.3 AU [121]. The asteroid belt was probably much more massive initially, consisting of about one Earth mass. Due to Jupiter's migration, however slight, most of this mass has been ejected, leaving behind only about 0.001 of the original mass. This course of events may have also been important for the emergence and evolution of life on Earth, since had the asteroid belt remained very massive, the number of impacts on Earth (due to the continued perturbations from Jupiter) might have been too high to allow for the evolutionary processes to follow their course. For example, any planet around Tau Ceti (which may be orbited by five planets) would experience many more impacts than Earth, due to a much more massive debris disc in that system. It may therefore be that the time interval during which the giant planet should form is rather restricted, if it is to allow complex life to emerge and evolve. Too much migration may altogether disrupt the asteroid belt, and too little may produce far too many devastating impacts. This conclusion assumes, of course, that some asteroid impacts are indeed necessary for life. Interestingly, asteroids are presently regarded as a potential threat for humanity in the future, and NASA has been testing a computer program called Scout that is supposed to act as a celestial intruder alert system to warn against incoming asteroids. In that sense, asteroids can giveth and asteroids can taketh away.

Given that not only the individual impacts themselves but also their *rate* can have significant effects on the habitability of terrestrial planets, it is important to examine a few of the elements that determine this rate. Specifically, a high rate of asteroid impacts can render a highly cratered planet not hospitable for life. On the other hand, too low a rate could suppress the delivery of elements that are essential for the emergence and evolution of life.

11.5.3 Asteroid Impacts on Terrestrial Planets

Introduction

As I noted earlier, if asteroid belts form at all, they most likely form around the location of the water snow line, interior to giant planets. Throughout the time that the gas disc still exists, the eccentricities and inclinations of the asteroids are damped by tidal interactionsx with the protoplanetary discs (e.g., [7, 85, 153]). The lifetime of the gas disc is typically a few million years [64], after which it is being dispersed by photoevaporation (see discussion in [4]). After the gas disc is removed, gravitational perturbations clear asteroids at many resonance locations (e.g., [37, 61, 112]). This clearing creates potential Earth (and other terrestrial planets) impactors. In the solar system, Jupiter and Saturn, which are the largest planets, are the main drivers of the dynamical evolution of the asteroid belt, even though the effects that Jupiter had on the collision rate with the Earth is debated [75, 76]. Similarly, giant planets can be the driving force of evolution (if an asteroid belt exists) in exoplanetary systems.

Resonances that play a major role in the asteroid belt dynamics are *mean-motion* resonances and *secular* resonances. In a mean-motion resonance, the ratio of the orbital periods of two objects is a ratio of two integers. Several of the known mean-motion resonances in the solar system, such as the Kirkwood gaps [43, 115, 128], are found within the asteroid belt. Secular resonances arise when the apsidal precession rate of two objects orbiting a common central object are close to each other (e.g., [56, 162]). From our perspective here, the most important secular resonance in the solar system is the ν_6 resonance ([19, 113] and references therein). It relates the apsidal precessions of the asteroids and Saturn, and it sets the inner boundary (at about 2 AU) of the solar system's asteroid belt. Each resonance has a certain libration width in semimajor axis over which it is effective. Asteroids that fall within a libration width undergo perturbations that cause their eccentricities to increase to the point where they are either ejected from the system or they collide with a planet or the central star. Regions in which libration widths overlap are dynamically chaotic regions [125], and almost all asteroids are cleared from such regions. The outer edge of the asteroid belt in the solar system (at about 3.3 AU) is determined by the overlapping of Jupiter's resonances.

The fact that many of the exoplanetary systems contain a super-Earth (e.g., [82]; Figure 11.10) raises the additional question of how the presence of a super-Earth in the solar system might have affected the rate of asteroid impacts on Earth. Smallwood *et al.* [143] studied in particular multi-planet systems in which there had been no process (e.g., migration) that could have destroyed the asteroid belt or could have prevented the formation of terrestrial planets. More specifically, they investigated how the architecture of systems such as the solar system affects the rate of asteroid impacts on Earth (or a similar terrestrial planet).

The Numerical Method: N-body Simulations

Smallwood *et al.* [143] used the hybrid symplectic integrator in the orbital dynamics package MERCURY to model the structure of the asteroid belt and the rate of impacts on Earth (see [30] for details of the package). The motions of Jupiter, Saturn, a super-Earth, Earth, and a distribution of asteroids all orbiting a central object were simulated. The asteroids were assumed to gravitationally interact with the star and the planets but not with each other. The evolution of the orbit of each asteroid was followed for ten million years.

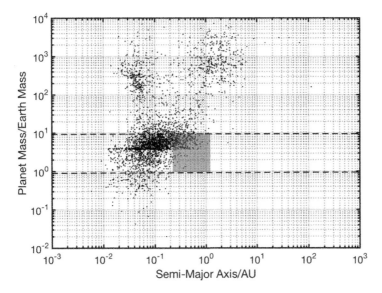

Figure 11.10 Planet masses and semimajor axes of observed exoplanets (from [65]). The area between the dashed lines contains the range of super-Earth masses used in the simulations described in Section 11.5.3. The transparent grey box highlights the observed super-Earths with a semi-major axis corresponding to the inner solar system.

In the solar system, there are more than 10,000 asteroids with high-accuracy measurements of their semimajor axes. The mean of those values is $\langle a \rangle = 2.74 \pm 0.616$ AU. The mean eccentricity is $\langle e \rangle = 0.148 \pm 0.086$, and the mean inclination $\langle i \rangle = 8.58° \pm 6.62°$ (see description in [124]). At the same time, the precise initial distribution of the asteroids within the belt (immediately following the dispersal of the protoplanetary gas disc) is not accurately known. In their simulations, Smallwood *et al.* [143] therefore assumed a uniform distribution (see also [87]), with the semimajor axis of each asteroid given by

$$a_i = (a_{\max} - a_{\min})\,\xi_r + a_{\min}, \tag{11.8}$$

where $a_{\min} = 1.558$ AU is the inner boundary of the distribution, $a_{\max} = 4.138$ AU is the outer boundary, and ξ_r is a randomly generated number between 0 and 1. The inner and outer boundaries were based on the structure of the solar system, with $a_{\min}$ being three Hill radii (the region in which a body's gravity dominates) beyond the semimajor axis of Mars, and $a_{\max}$ being three Hill radii inside Jupiter's orbit. The Hill radius is given by

$$R_H = a_p \left(\frac{M_p}{3\,M_s} \right)^{1/3}, \tag{11.9}$$

where a_p is the planet's semimajor axis, M_p is its mass, and M_s is the mass of the star. No asteroids are likely to be located within three Hill radii because this is the planet's gravitational reach (e.g., [31, 60, 123]).

The orbit of each asteroid is defined by six orbital elements: (1) the semi-major axis, a, which was taken to be distributed uniformly in the range $a_{min} < a < a_{max}$; (2) the inclination angle i, randomly chosen from the range $0°$–$10°$; (3) the eccentricity, e, randomly generated from the range 0.0–0.1; (4) the longitude of the ascending node, n; (5) the argument of perihelion (the angle from the ascending node to the object's periastion), g; (6) the mean anomaly (angular distance from pariastion), M_a. The last three elements were uniformly randomly sampled from the range $0°$–$360°$. Given that the solar system is stable over long timescales [48, 79], the current orbital parameters of the planets were also taken as initial parameters.

To test their numerical scheme, Smallwood *et al.* [143] first checked the scalability of the results with respect to the number of asteroids used and the radius used for the Earth. Based on these tests, they decided to run their simulations with 10^4 asteroids and an inflated Earth, with a radius of 2×10^6 km (otherwise, the number of impacts during the simulation is too low to allow for statistically significant conclusions). Neglecting asteroid-asteroid interactions was fully justified by the fact that the timescale for asteroid-asteroid collisions is much longer (in fact, of the order of the age of the solar system) than the timescale for action by resonance effects (of the order of 1 Myr; e.g., [44]).

The Effects of the Architecture of the Inner Solar System

Since Martin and Livio [106] identified the absence of super-Earths as perhaps the most significant characteristic that distinguishes the solar system from other exoplanetary systems, Smallwood *et al.* [143] first varied the mass and semimajor axis of an artificially added super-Earth in the inner solar system. The super-Earth initially was taken to be in a circular orbit with zero inclination. The super-Earth's mass was varied in the different simulations in the range 1–$10M_\oplus$, and its semimajor axis was taken to be in the range 0.2–1.4 AU. In each one of the simulations, Smallwood *et al.* [143] followed the dynamics for 10 Myr and determined the number of asteroid impacts on Earth (which was inflated in radius, as described earlier), impacts on Jupiter, Saturn, and the central star; the number of asteroids ejected from the system (achieving semimajor axes larger than 100 AU); and the number of asteroids remaining within the initial distribution of the asteroid belt. All the runs were then compared to that of a solar system without a super-Earth in order to evaluate the significance of the absence of a super-Earth in our solar system. Figure 11.11 gives the total number of collisions with Earth during the period of 10 Myr. Overall, the trend is that the addition of a super-Earth interior to the Earth's orbit increases the number of asteroid collisions with the Earth.

When the super-Earth is located exterior to the Earth's orbit, the total number of impacts onto the Earth is lower than in the absence of a super-Earth. For the parameters used in the simulations, a $10M_\oplus$ super-Earth located at a semimajor axis of 0.8 AU produced the largest number of impacts on Earth, whereas a $10M_\oplus$ super-Earth located at 1.20 AU caused the lowest number of impacts. At the same time, when the super-Earth was placed at 1.2 and 1.4 AU, the number of asteroids ejected from the system increased significantly. Generally,

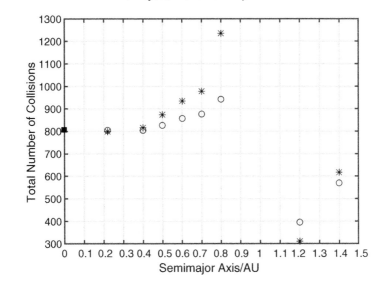

Figure 11.11 The total number of collisions with the (inflated) Earth as a function of the semimajor axis of the super-Earth. The stars represent simulations with a $10M_\oplus$ super-Earth and the circles with a $5M_\oplus$ super-Earth. The square represents the simulation without a super-Earth.

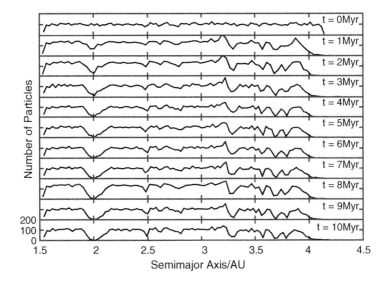

Figure 11.12 Evolution of the asteroid distribution without a super-Earth.

for *interior* (to the Earth's orbits) super-Earths, the number of impacts was found to increase with increasing distance from the Sun.

Smallwood *et al.* [143] also examined the evolution of the asteroid belt itself, and this is shown in Figures 11.12–11.15. The four panels show a simulation with no super-Earth

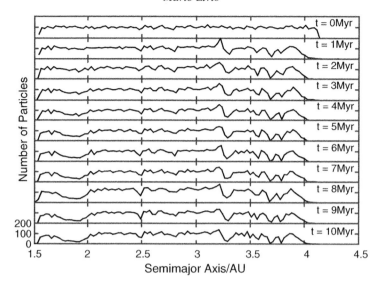

Figure 11.13 Evolution of the asteroid distribution with a $10M_\oplus$ super-Earth located at a semimajor axis of 0.8 AU.

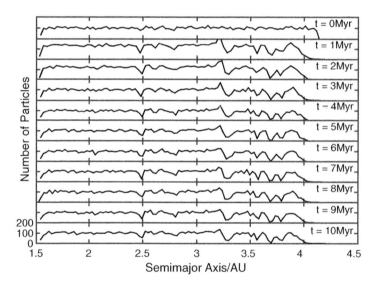

Figure 11.14 Evolution of the asteroid distribution with a $10M_\oplus$ super-Earth located at 1.2 AU.

(fig. 11.12), one with a 10 $M_\oplus$ super-Earth at $a = 0.8$ AU (Figure 11.13), one with a $10M_\oplus$ super-Earth at $a = 1.2$ AU (Figure 11.14), and one with a $10M_\oplus$ super-Earth at $a = 1.4$ AU (Figure 11.15). The distribution of the asteroids was calculated every million years for 10 Myr. As time progresses, perturbations caused by mean-motion and secular

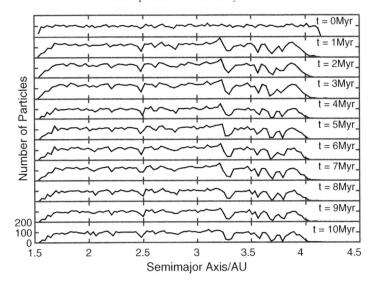

Figure 11.15 Evolution of the asteroid distribution with a $10M_\oplus$ super-Earth at 1.4 AU.

resonances with Jupiter and Saturn clear out regions in the asteroid distribution. The most important mean-motion resonances are the 3:1, 5:2, 7:3, and 2:1. These are located at 2.5 AU, 2.8 AU, 2.9 AU, and 3.3 AU, respectively. Overlapping libration widths of the mean-motion resonances produce Jupiter's chaotic region, which is located from about 3.6 AU to the outer edge of the asteroid distribution (at about 4.13 AU). The resonance that plays the most important role in determining the number of collisions with Earth is the v_6 resonance, which I will discuss in detail in the following paragraphs.

There are five potential outcomes for the fate of each asteroid during the various simulations: it can be ejected from the solar system, impact the Earth, collide with another planet, collide with the Sun, or remain in the asteroid belt. The ejections and impacts with the Earth are shown in Figures 11.16–11.19, by the black circles and grey squares, respectively. Figure 11.16 shows the results for a system with no super-Earth. Figure 11.17 contains a $10M_\oplus$ at a semimajor axis of $a = 0.8$ AU. Figure 11.18 has a $10M_\oplus$ super-Earth at $a = 1.2$ AU, and Figure 11.19 has a 10 $M_\oplus$ super-Earth at $a = 1.4$ AU. The figures show the *initial* semimajor axis for each asteroid (that is, its point of origin) as a function of the time of its final outcome. Asteroids that were originally at resonance locations are cleared out because their eccentricities increase through the action of the mean-motion and secular resonances. The locations of the mean-motion resonances between Jupiter and the super-Earth are shown on the vertical axis at the right-hand side. Placing a $10M_\oplus$ super-Earth at $a = 0.8$ AU widens the v_6 secular resonance, thereby increasing the number of asteroids perturbed into Earth-impacting orbits. When the super-Earth is placed exterior to Earth's orbit, two effects act to decrease the number of impacts on Earth: (1) the v_6 resonance is suppressed, and (2) a 2:1 mean-motion resonance is created within the asteroid belt, which acts to cause additional clearing out of asteroids. When a $10M_\oplus$ super-Earth is placed at

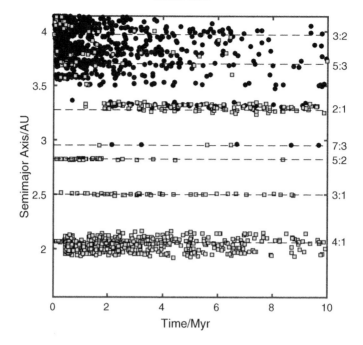

Figure 11.16 The original semimajor axis of each asteroid as a function of the time when the final outcome occurred, for a system with no super-Earth. The outcomes depicted are collisions with the Earth (grey squares) and ejections (black circles). The mean-motion resonances with Jupiter are represented by the dashed lines.

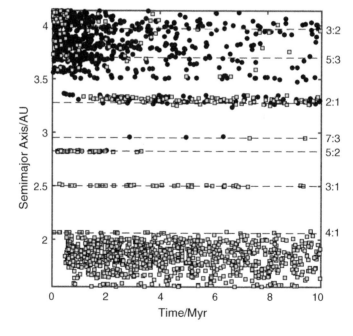

Figure 11.17 Same as Figure 11.16, with a super-Earth located at 0.8 AU.

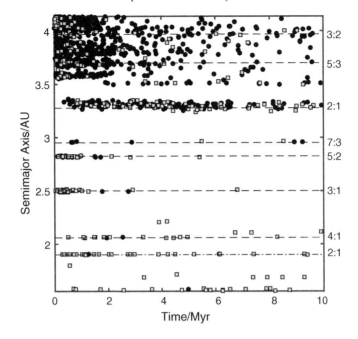

Figure 11.18 Same as Figure 11.16, with a super-Earth located at 1.2 AU.

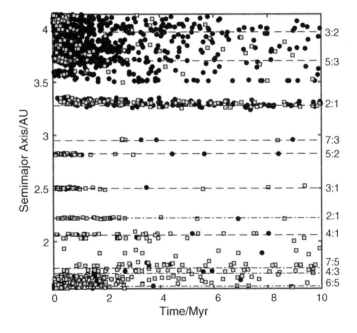

Figure 11.19 Same as Figure 11.16, with a super-Earth located at 1.4 AU.

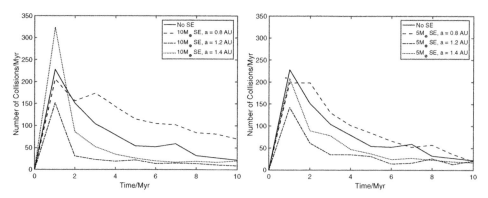

Figure 11.20 The collision rate with Earth per million years. The left panel involves a 10 $M_\oplus$ super-Earth, and the right panel a 5 $M_\oplus$ super-Earth.

1.4 AU, it creates a chaotic zone which is produced by the overlapping of the libration widths of the super-Earth's 6:5, 4:3, and 7:5 mean-motion resonances. In turn, this chaotic zone clears a large number of asteroids from the inner parts of the asteroid belt. The chaotic region can be clearly seen in Figure 11.19.

The effects of the mass of the super-Earth on the rate of asteroid impacts on Earth is shown in Figure 11.20. What is depicted is the number of impacts per million years for a super-Earth with a mass of $10M_\oplus$ (left panel) and a mass of $5M_\oplus$ (right panel), for various semimajor axis values. The initial spike in the impact rate is due to the fact that the number of asteroids in the belt is larger at the beginning of the simulations than at their end. The fact that a $10M_\oplus$ super-Earth at 1.4 AU produces the highest impact rate at 1 Myr is due to the chaotic region created by the super-Earth. The rate, however, rapidly declines as asteroids are cleared out. Over the 10 Myr duration covered by the simulation, a super-Earth located at 0.8 AU produces the highest rates, both for a $10M_\oplus$ super-Earth and a $5M_\oplus$ one.

As I have noted earlier, the ν_6 resonance is a major contributor to the rate of asteroid impacts on Earth. In fact, the majority of the asteroids colliding with the Earth originate from the location of the ν_6 resonance in the asteroid belt. Specifically, in the simulations of Smallwood *et al.* [143] without a super-Earth, the total number of asteroid impacts on Earth produced by secular resonances was about 2.5 times higher than that produced by mean-motion resonances. *If* asteroid impacts were indeed important for the emergence and/or evolution of life on Earth, then the ν_6 resonance may have played a significant role in our planet's habitability. The ν_6 resonance involves both Saturn and Jupiter. Basically, Jupiter increases the precession frequency of the asteroids so that they fall into a resonance with the apsidal precession rate of Saturn. In Figure 11.21 (taken from [143]), I show the precession rate of a test particle as a function of orbital separation. The solid horizontal line marks the eigenfrequency of Saturn (found using a generalised form of secular perturbation theory; e.g., [124]). The top-left panel of Figure 11.21 represents our solar system since it includes

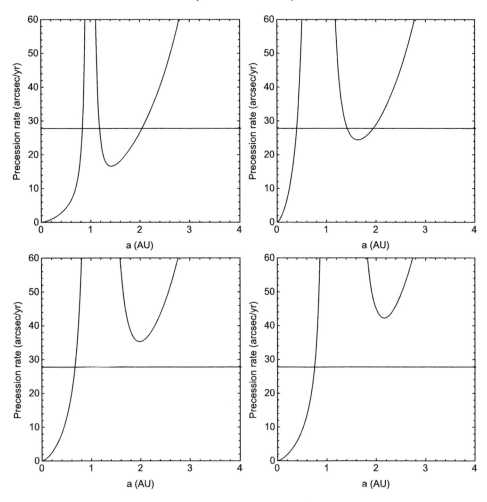

Figure 11.21 The precession rate of a test particle as a function of semimajor. The horizontal line represents the g_i eigenfrequency of Saturn. The intersection of the precession rate with the eigenfrequency denotes the location of a secular resonance. Top-left: a system with no super-Earth. Top-right: a $10M_\oplus$ super-Earth at $a = 0.8$ AU. Bottom-left: a $10M_\oplus$ super-Earth at $a = 1.2$ AU. Bottom-right: a $10M_\oplus$ super-Earth at $a = 1.4$ AU.

Earth, Jupiter, Saturn, and the asteroid belt. The intersection of the particle's precession rate with Saturn's eigenfrequency represents the location of the ν_6 resonance, at about 2 AU.

The introduction of a super-Earth can change the asteroid precession rate so as to either enhance or altogether remove the resonance with Saturn, depending on the super-Earth's location. For example, when the super-Earth has a semimajor axis of 0.8 AU, the precession rate (of the test particle) is close to Saturn's eigenfrequency for semimajor axis values in the range 1.5–2.0 AU, which enhances the ν_6 resonance (top-right panel in Figure 11.21). On the other hand, when the super-Earth is exterior to Earth's orbit, the ν_6 resonance is removed – the precession rate of the test particle does not intersect Saturn's eigenfrequency (bottom panels; the left is for a super-Earth at 1.2 AU, and the right for 1.4 AU). This behaviour agrees with that observed in Figures 11.16–11.19, where an interior super-Earth produced a widening of the libration width of the ν_6 resonance, while an exterior super-Earth led to the disappearance of the resonance. The agreement of the numerical results with the behaviour expected from the generalised form of the secular perturbation theory (represented in Figure 11.21) gives great confidence in the numerical simulations.

The Effects of the Architecture of the Outer Solar System

In the next step, Smallwood et al. [143] considered the effects of the orbital properties of the giant planets. They found that the location of the ν_6 resonance is rather insensitive to changes in Saturn's mass, moving outward only slightly as the mass is increased. The resonance was found to be much more sensitive to changes in Saturn's orbital separation from the Sun. Specifically, as Saturn is moved outward, the resonance location moves inward. However, I should note that Saturn's orbital location may not be accidental, since it is close to a 5:2 resonance with Jupiter. To further investigate the effects of the architecture of the giant planets, Smallwood et al. [143] ran additional simulations while varying Saturn's mass and semimajor axis. They found that increasing Saturn's semimajor axis (from its nominal ~9.5 AU) results in the location of the ν_6 resonance moving eventually

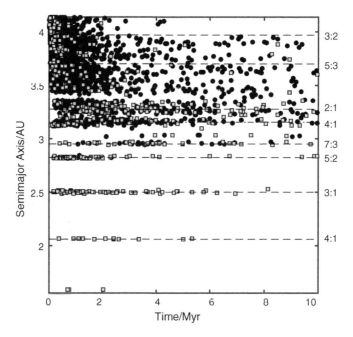

Figure 11.22 The original semimajor axis of each asteroid as a function of the time when the final outcome occurred, when Saturn is located at $a = 8$ AU. The dash-dotted line marked 4:1 represents a mean-motion resonance between the asteroids and Saturn. The dashed lines represent mean-motion resonances with Jupiter. Grey squares denote impacts on Earth and circles ejections.

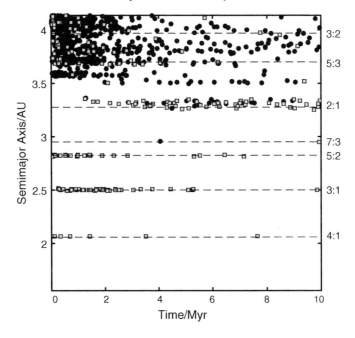

Figure 11.23 The same as Figure 11.22, with Saturn at $a = 12$ AU.

outside the boundaries of the asteroid belt. Decreasing the semimajor axis moved the ν_6 resonance towards the middle (and then outer) region of the asteroid belt. These results are also reflected in Figures 11.22–11.23. Figure 11.22 shows the initial semimajor axis for each asteroid as a function of the time of its final outcome, when Saturn's semimajor axis is decreased to be $a = 8.0$ AU. Figure 11.23 shows the results when Saturn is taken to be at $a = 12.0$ AU. For $a = 8.0$ AU, the ν_6 resonance shifts to the outer part of the asteroid belt, which decreases the number of impacts on Earth (compared to $a = 9.5$ AU) but increases the number of ejections. When Saturn's semimajor axis is taken to be 12.0 AU, the ν_6 resonance is located outside the asteroid distribution. Overall, Smallwood et al.'s simulations [143] showed that Saturn's orbital semimajor axis has a significant effect on the location of the ν_6 secular resonance and thereby on the rate of asteroid impacts on Earth.

To conclude this section, it appears that the ν_6 resonance plays an important role in producing asteroid impacts on terrestrial planets in the inner part of a planetary system. The architecture of both the inner part and of the outer part affect the location of the ν_6 resonance and thereby the rate of impacts. Super-Earths with masses larger than $5M_\oplus$ and interior to the Earth's orbit would have increased the impact rate, while super-Earths exterior to the Earth's orbit would have decreased it. The position of Saturn in the solar system also had a significant effect on the rate of asteroid impacts on Earth. Significantly changing the semimajor axis of Saturn in both directions would have generally resulted in a decrease in the number of asteroid collisions with the Earth. However, since the orbital location of Saturn is close to being in the 5:2 resonance with Jupiter, it may not be accidental.

11.5.4 The CO Snow Line

Introduction

The most abundant volatiles in a protoplanetary disc are H_2O, CO, and CO_2. A snow line marks the location in the disc where the midplane temperature is sufficiently low so that a volatile condenses out of the gas phase and becomes solid. The snow line of each volatile has its own radial distance from the central star, with the water snow line being the closest to the star. As I have noted earlier, giant planets are expected to form outside the water snow line because the density of solids there is higher (e.g., [130, 135]). In general, the composition of planets and their atmospheres is largely determined by the location of their formation relative to the snow lines [127]. While the water snow line is found at a temperature of about 170 K (see Section 11.5.1), the CO snow line occurs at $T_{CO,snow} = 17$ K [126]. Comets from the solar system's Kuiper belt show different amounts of CO, suggesting that they formed close to the CO snow line [2]. The Kuiper belt is thought to have formed in the region extending from about 27 AU (from the Sun) to about 35 AU [90]. Since the CO snow line would have been in this region at the time of planetesimal formation, it could mark the transition from the planet-forming zone to the dwarf planet/small icy body–forming zone.

While it is very difficult to detect the water snow line in exosolar systems (because of its relative proximity to the host star), the CO snow line presents an easier target because it is farther away. The best observed extrasolar snow line is in the disc around TW Hya. This star has a mass of $0.8 M_\odot$ and an age of less than 10 Myr [72]. Qi et al. [133] detected in the star's vicinity the reactive ion N_2H^+, which is only present when CO is frozen out. These authors determined the radial distance of the CO snow line to be 28–31 AU, very similar to our own solar system.

Since CO ice is needed to form methanol – a building block of more complex organic molecules – we need to understand the evolution of the CO snow line because that evolution plays a crucial role in the origin of the prebiotic molecules that had led to the emergence of life on Earth [156]. Recall that comets had intensely bombarded the young Earth, apparently delivering those ingredients that were necessary for life. I should note that debris discs – the equivalent of the solar system's Kuiper belt – are extremely common in exosolar systems and, hence, the solar system is not special in that sense. Similarly, systems that have giant planets that are the equivalent of Jupiter, may be expected to have Oort clouds.

Martin and Livio [102] studied the evolution of the CO snow line using two protoplanetary disc configurations: (1) a fully turbulent disc model and (2) the more likely model for a protoplanetary disc – a disc with a dead zone (low-turbulence region).

A Turbulent Disc Model

In fully turbulent discs, material is assumed to orbit at radius R with a Keplerian angular velocity $\Omega = \sqrt{GM_*/R^3}$, where M_* is the star's mass (e.g., [131]). The effective viscosity in a disc in which turbulence is driven by the magnetorotational instability (MRI) is usually parameterised in the form [139]

$$\nu = \alpha \frac{c_s^2}{\Omega},$$ (11.10)

where α is the viscosity parameter and c_s is the sound speed at midplane. In a steady state, mass conservation produces a surface density of

$$\Sigma = \frac{\dot{M}}{3\pi \nu},$$ (11.11)

where $\dot{M}$ is the infall (onto the disc) accretion rate. The surface temperature, T_e, in such a disc is given by energy conservation

$$\sigma T_e^4 = \frac{9}{8} \frac{\dot{M}}{3\pi} \Omega^2 + \sigma T_{irr}^4,$$ (11.12)

where T_{irr} is the irradiation (by the central star) temperature (e.g., [24]), given by

$$T_{irr} = \left(\frac{2}{3\pi}\right)^{1/4} \left(\frac{R}{R_*}\right)^{-3/4} T_*.$$ (11.13)

Here, R_* and T_* are the star's radius and temperature [36]. The midplane temperature of the disc is related to its surface temperature through $T_c^4 = \tau T_e^4$, where τ is the optical depth given by

$$\tau = \frac{3}{8} \kappa \frac{\Sigma}{2},$$ (11.14)

and the opacity is $\kappa = a T_c^b$.

While the precise values of a and b do not have a strong effect on the inferred disc's temperature, values appropriate for the low temperatures in the vicinity of the CO snow line are those obtained from absorption by dust, $a = 0.053$, $b = 0.74$ (e.g., [164]). Martin and Livio [102] solved for the CO snow line radius by equating the central temperature T_c to $T_{CO, snow}$. Their results for $R_{CO, snow}$ as a function of the accretion rate are presented by the short-dashed curve in Figure 11.24, where the assumed parameters were $M_* = 1 M_\odot$, $R_* = 3 R_\odot$, $T_* = 4,000$ K, $T_{CO, snow} = 17$ K, and $\alpha = 0.01$. Since the accretion rate drops in time, time is the implicit coordinate in this figure. The value of the viscosity parameter α is rather uncertain [84], consequently the figure also presents the results (by the long-dashed curve) obtained for $\alpha = 10^{-4}$.

The figure demonstrates that the observed location of the CO snow line in our solar system (at 27–35 AU), and in the solar system analogue TW Hydra (at 28–31 AU), *cannot* be explained by a fully turbulent disc model. Specifically, in this model, the CO snow line moves in too close to the central star during the low-accretion-rate phase towards the end of the disc's lifetime.

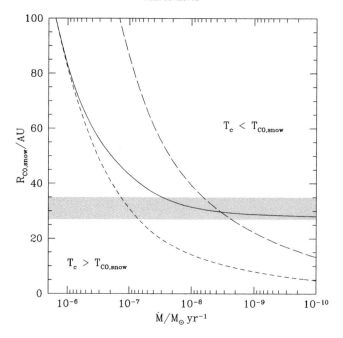

Figure 11.24 Evolution of the CO snow line as a function of the accretion rate in a steady-state disc. The dashed lines show a fully MRI turbulent disc with a viscosity parameter of 0.01 (short-dashed) and 0.0001 (long-dashed). The solid line represents a disc with a self-gravitating dead zone. The shaded region indicates the location of the CO snow line in our solar system at the time of planetesimal formation.

I should note that since irradiation is the dominant heating source (over viscous heating) on the scale of tens of AU, we can ignore the viscous heating term in Eq. (11.12), to find an approximate analytic solution that represents very closely the dashed lines in Figure 11.24 for accretion rates of up to about 10^{-8} $M_\odot$ yr^{-1}. This analytic approximation is given by

$$
\begin{aligned}
R_{\mathrm{CO,\,snow}}^{\mathrm{MRI}} \simeq 13.2 &\left(\frac{\alpha}{0.01}\right)^{-2/9} \left(\frac{M_*}{M_\odot}\right)^{1/9} \left(\frac{\dot{M}}{10^{-8} M_\odot \text{ yr}^{-1}}\right)^{2/9} \\
&\left(\frac{T_{\mathrm{CO,\,snow}}}{17 \text{ K}}\right)^{-0.95} \left(\frac{R_*}{3 R_\odot}\right)^{2/3} \left(\frac{T_*}{4000 \text{ K}}\right)^{8/9} \text{ AU.}
\end{aligned}
\tag{11.15}
$$

Since the fully turbulent disc model fails to reproduce the observed CO snow line, Martin and Livio [102] also calculated time-dependent disc models with a dead zone to follow the evolution of the CO snow line.

A Disc with a Dead Zone

A dead zone is formed when the ionisation fraction is not high enough for the MRI to drive turbulence. Typically, the hot part of the disc (close to the central star), where the midplane temperature is higher than some critical value, $T_c \simeq 800$ K [149], is thermally ionised (and, therefore, MRI turbulent.) Farther away from the central star, either cosmic rays or the X-ray flux from the star are the dominant sources of ionisation [62]. These sources can only penetrate the surface layers, down to a surface density of Σ_{crit}. Where $\Sigma > \Sigma_{crit}$, a dead zone exists at the midplane, with a surface density of $\Sigma_g = \Sigma - \Sigma_{crit}$. The precise value of Σ_{crit} depends on the ionising source, being around $\Sigma_{crit} \simeq 200$ g cm^{-2} if cosmic rays dominate [57] but much lower if X-rays dominate [108]. In the very outer parts of the disc (in terms of distance from the central star), where $\Sigma < \Sigma_{crit}$, the external ionisation sources can penetrate all the way to the midplane, and that part is again MRI active.

In the dead-zone (low-viscosity) region, material can accumulate to the point where the disc becomes self-gravitating. This is expected to occur when the Toomre [148] parameter $Q = c_s \Omega / \pi G \Sigma$ drops below a critical value Q_{crit}. Martin and Livio [102] took $Q_{crit} = 2$. This drives gravitational turbulence, with an effective viscosity parameter

$$v = \alpha_g \frac{c_s^2}{\Omega}. \tag{11.16}$$

Martin and Livio [102] adopted the functional form $\alpha_g = \alpha \exp(-Q^4)$ [104, 165].

They considered a model in which a molecular cloud collapses onto a disc, and they took the initial accretion rate to be $2 \times 10^{-6} M_\odot$ yr^{-1} and assumed that the accretion rate decreases exponentially on a timescale of 10^5 years. The initial surface density of the disc was taken to be that of a turbulent steady disc around a $1 M_\odot$ star, with an accretion rate of $2 \times 10^{-6} M_\odot$ yr^{-1}. The disc was modelled on a radial grid of 200 points evenly distributed in $\log R$ from $R = 1$ AU to $R = 200$ AU. The infalling material was added at $R = 195$ AU. To allow for a comparison, Martin and Livio [102] modelled one disc to be fully turbulent and the other to contain a dead zone, with $\Sigma_{crit} = 10$ g cm^{-2} (corresponding to ionisation by X-rays).

The disc was found to be gravo-magneto unstable for high accretion rates, which caused unsteady accretion onto the central star (for a similar behaviour, see also [5, 104]). The evolution of the CO snow line as a function of time is shown in Figure 11.25. The dashed line represents the fully turbulent disc, and the solid line, represents the disc with a dead zone. At late times, the model with the dead zone has a snow line radius that is considerably larger than that obtained for a turbulent disc and one which is consistent with the observations of the solar system and of TW Hya. The small (and brief) repetitive increases in the snow line radius at early times are caused by FU Orionis-type outbursts. Basically, with a dead zone, the small amount of self-gravity heats the more massive disc with a dead zone, and this causes the CO snow line radius to move outward, as required by observations. For low accretion rates, Martin and Livio [102] were able to find an analytic solution for the CO snow line radius in a disc with a dead zone. This is given by

$$R_{CO, snow}^{dead} \simeq 29.3 \left(\frac{M_*}{M_\odot}\right)^{1/9} \left(\frac{R_*}{3 R_\odot}\right)^{2/3} \left(\frac{T_*}{4,000 \text{ K}}\right)^{8/9} \left(\frac{T_{CO, snow}}{17 \text{ K}}\right)^{-0.61} \text{ AU.} \tag{11.17}$$

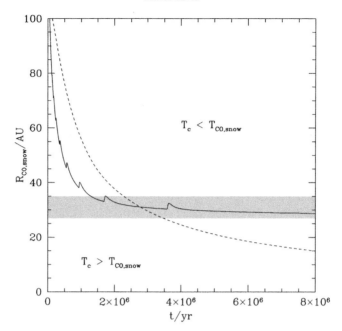

Figure 11.25 Evolution of the CO snow line in a time-dependent disc with an exponentially decreasing infall accretion rate. The dashed line shows a fully MRI turbulent disc, and the solid line, a disc with a dead zone. The shaded region indicates the location of the CO snow line in our solar system at the time of planetesimal formation.

This approximate solution agrees very well with the solid line in Figure 11.24, where the CO snow line radius is shown as a function of the accretion rate.

I should note that the CO snow line in the disc around the Herbig Ae star HD 163296 has been found to lie at a radius of 155 AU [107, 132]. For the observed parameters of this system: $M_* = 2.3M_\odot$, $R_* = 2R_\odot$, $T_* = 9,333$ K, and $\dot{M} = 7.6 \times 10^{-8} M_\odot$ yr^{-1}, the model of Martin and Livio [102] with a dead zone would predict a snow line radius of only 62 AU (a fully turbulent disc would give the even much smaller 37 AU). Disc flaring in this system could potentially account for this discrepancy, since approximations for the temperatures of flared discs would result in CO snow line radii larger than 100 AU (e.g., [36]).

In general, the analytic solution for the radius of the CO snow line could prove useful in determining the composition (and thereby maybe the habitability) of exosolar planets.

11.5.5 Conclusion of Sections 11.1–11.5

An examination of the physical properties of our solar system reveals that it is not extremely unusual when those are compared to the characteristics of the other observed exoplanetary systems. Still, there is no doubt that a few of the solar system's parameters have made it conducive to the emergence and evolution of life. For example, low eccentricity planets (as observed in the solar system) have a more stable temperature throughout the entire orbit, which may make them more likely to harbour life [158]. Planetary systems with a low mean eccentricity are also more likely to have a long-term dynamical stability.

The age of the solar system (about 4.5 billion years) may also be favourable for the emergence of complex life (which on Earth took some three billion years). However, I should note that since the current age of the Sun is about half of its total lifetime and about half the age of the Milky Way's disc, one can expect that roughly half of the stars in the disc are even older than the Sun. It is still the case, though, that if life *can* emerge and evolve around low-mass stars (that are much more numerous and live much longer), then one could expect complex life to be much more abundant in the future (about a trillion years from now; see, e.g., Chapter 12). If that is indeed the case, then the appearance of complex life on Earth could be regarded as very early. More recent work suggests, however, that planets around most M-dwarfs may experience serious atmosphere erosion and are therefore not likely to harbour life [94].

As I have already noted, the existence of terrestrial planets in the habitable zone around their host star is quite common [18, 46].

The metallicity of the host star (and of the protoplanetary disc) also does not appear too special in the solar system. The metallicity plays a role in determining the structure of the planetary system that forms [22, 151]. The metallicity was found to be correlated with the probability for the star to have a giant planet orbiting it [51, 134, 144]. However, the correlation is much less clear for lower-mass planets [21, 110]. While planets with radii smaller than four Earth radii are observed around stars with a wide range of metallicities, the average metallicity of stars hosting planets with $R_p < 1.7R_\oplus$ is close to solar [22]. It is unclear, therefore, whether having a near-solar metallicity is somehow related to forming Earth-size planets.

The variability of the Sun has also been compared to the activity of the stars in the Kepler sample, with rather ambiguous conclusions. While Basri and collaborators [10, 11] found the Sun to be quite typical, with only a quarter to a third of the sample being more active, McQuillan and his collaborators [111] found the Sun to be relatively quiet, with 60% of the stars being more active. The difference in the conclusions stems primarily from the fact that fainter stars were included in McQuillan *et al.* [111] (and also from some differences in defining the activity level of the Sun). Since 90% of M-dwarfs are more active than the Sun, the inclusion of lower-mass stars makes the Sun quiet in comparison. Compared to other Sun-like stars, however, the Sun is quite typical.

As I have noted throughout Section 11.5, the existence of a compact asteroid belt may be conducive to initiating life. Current observations do not allow us to determine categorically how common asteroid belts are. However, the list of unresolved debris disc candidates now contains hundreds of examples [34], and of these, about two-thirds are better modelled by a two-component (rather than a single) dust disc arising from two separate belts [83]. The two-belt configuration of our solar system may therefore be quite ubiquitous.

In Sections 11.2 and 11.3, I identified the lack of super-Earths and the lack of planets interior to Mercury's orbit as perhaps the two main characteristics that make the solar system somewhat special. Even though I have shown that the presence of super-Earths could affect the rate of asteroid impacts on Earth, it is not obvious that this influenced the Earth's habitability. One feature that should be further investigated is the effect of a close-in super-Earth on the dynamical stability of a terrestrial planet in the habitable zone.

The bottom line from the discussion so far is simple. There may be many factors that are necessary, but maybe not sufficient, for life to emerge and evolve on a planet. If we were to multiply the probabilities for all of these factors (in a Drake-type equation) together, we could end up with a very small probability for life in the Milky Way. Without a concrete knowledge of which ones of these factors are truly essential for life, however, such an exercise would merely represent our ignorance. If we blindly consider every single aspect of the solar system, we will obviously find it to be unique. From the parameters that I have considered here, however, I have not identified any feature that would argue for the Earth being exceptionally rare. This is what makes Section 11.6 particularly interesting, since in it I discuss a claim by astrophysicist Brandon Carter, who argues that extraterrestrial intelligent life is exceedingly rare.

11.6 How Rare Are Extrasolar Intelligent Civilisations?

11.6.1 Introduction

The existence of the so-called Fermi Paradox, the absence of any signs for the existence of other intelligent civilisations in the Milky Way galaxy, coupled with an interesting argument raised by astrophysicist Brandon Carter ([28], to be discussed later) as well as the absence of any physical law that mandates complexification, have convinced a few researchers that intelligent life may be exceedingly rare in the Milky Way (e.g., [9, 42, 152]). If true, such a reality could have implications far beyond the practical ones (i.e., the search for extraterrestrial intelligence). In fact, it would fly in the face of Copernican modesty, which argues that humanity should not be in any way 'special' in the grand cosmic scheme. Over the years, Carter's argument has generated a considerable amount of discussion, so I will briefly review it here, together with a few of the criticisms that have been raised.

11.6.2 Carter's Argument

Carter's argument [138] can be explained in very simple terms, as follows. Examine the typical timescale for biological evolution (and the emergence of intelligent life) on a planet, τ_ℓ, and the lifetime of the central star of that planetary system τ_*. If the two timescales are a priori entirely independent quantities (that is, intelligent life can develop at some random time with respect to the main-sequence lifetime of the star), then one can expect that one of the following relations holds: either $\tau_\ell \gg \tau_*$ or $\tau_\ell \ll \tau_*$. The probability that $\tau_\ell \sim \tau_*$ is very small for two truly independent quantities when each one of the two can assume a very broad range of values. If, however, *generally* $\tau_\ell \ll \tau_*$, it is very difficult to understand why in the very first system in which we found an existing intelligent civilisation (the Earth-Sun system), we find that $\tau_\ell \sim \tau_*$ (to within a factor of two; the lifetime of our Sun is about 10 billion years, and it took about 4.5 billion years for intelligent life to evolve). I take τ_ℓ to mean roughly the timescale for the appearance of land life. This means, Carter argues, that *generally* $\tau_\ell \gg \tau_*$. In that case, however, it is clear that because of observational selection, the first system found to harbour intelligent life is likely to exhibit $\tau_\ell \sim \tau_*$, since a civilisation would not have developed for $\tau_\ell \gg \tau_*$ (the evolution of life requires the star as an energy source). Consequently, Carter concludes, typically $\tau_\ell \gg \tau_*$, and intelligent civilisations do not develop. The Earth, in this case, is an extremely rare exception.

Carter's argument, if true, has very significant implications from both scientific and philosophical perspectives. In particular, it puts a very heavy burden of responsibility on humanity, as one of the very few (or only!) intelligent civilisation in the galaxy. It should come as no surprise, therefore, that several criticisms of the argument have emerged over the years since its publication.

11.6.3 Criticisms of Carter's Argument

The first significant criticism of Carter's argument, primarily on logical and methodological grounds, was launched by Wilson [159]. Wilson first explained that Carter's argument really involves three timescales: τ_* – the main-sequence lifetime of stars like the Sun ($\sim$10 Gyr), τ_ℓ – the time that biological evolution has taken on Earth ($\sim$4 Gyr), and $\bar{\tau}$ – the timescale that would intrinsically be the most likely one required for the evolution of intelligent life. Wilson then points out that we really have no idea what the value of $\bar{\tau}$ is since we do not understand all the biological processes involved in the appearance of intelligence and we have only a single observed case of biological evolution (on Earth). Wilson explains that Carter's argument states that $\bar{\tau}$ has to satisfy $\bar{\tau} \ll \tau_*$ or $\bar{\tau} \gg \tau_*$, and he rules out $\bar{\tau} \sim \tau_*$ as explained earlier, even though we are completely ignorant about $\bar{\tau}$.

In particular, Wilson points out that when Carter assumes only the three ranges of values $\bar{\tau} \gg \tau_*$, $\bar{\tau} \ll \tau_*$, and $\bar{\tau} \sim \tau_*$, he excludes the possibility, for instance, that $\bar{\tau}$ is less than τ_*, but *not much less*. Wilson also argues that since we are ignorant about the value of $\bar{\tau}$, the possibility $\bar{\tau} \sim \tau_*$ cannot truly be ruled out. In addition, Wilson explains that the claim that τ_ℓ should not differ from a given value of $\bar{\tau}$ is not equivalent to the claim that $\bar{\tau}$ should not be different from a given value of τ_ℓ, since our knowledge of the value of a statistical quantity cannot be significantly enhanced by the evidence provided by a single case.

Finally, Wilson argues that the imprecision of the coincidence between τ_ℓ and τ_* (a factor of two), also lessens the explanatory power of a very high value of $\bar{\tau}$ and thereby decreases the confirmatory power of the rough coincidence $\tau_\ell \sim \tau_*$.

Taking a different approach, I attempted to refute Carter's argument on astrophysical grounds in Livio [95]. In that paper, I used a very simple toy model for the evolution of the

Earth's atmosphere to demonstrate that Carter *could* be wrong, because of a fundamental assumption in his argument. Specifically, I showed that the two timescales τ_ℓ and τ_* could in principle, at least, be correlated, in which case the entire logical structure of the argument (which is based on their being independent) collapses. Here is a brief description of how a τ_ℓ-τ_* relation *could* arise.

On the face of it, it appears that τ_ℓ is determined entirely by biochemical reactions and Darwinian evolution, while τ_* is determined by the rate energy is produced by nuclear burning reactions, and, therefore, it appears that the two timescales are, indeed, totally independent. However, we should note the following: the absorption of UV radiation by nucleic acids peaks in the range 2,600–2,700 Å, and the absorption of UV radiation by proteins in the range 2,700–2,900 Å [29, 41, 137]. Such radiation essentially kills all cell activity [16]. The only atmospheric constituent that efficiently absorbs radiation in the 2,000–3,000 Å range is O_3 [154]. The appearance of land life, therefore, may have to await the accumulation of a protective layer of ozone in the atmosphere (e.g., [17, 99]).

The evolution of the concentration of oxygen in a planetary atmosphere is very complex, and calibrations have to rely solely on the one existing example – Earth. On Earth, dynamic rising and falling of oxygen levels started perhaps as early as three billion years ago on a background of generally rising levels from low to intermediate to high (see, e.g., [99] for a recent review). Most of the oxygen produced on Earth was of biotic origin. In the very early (non-biotic) stages (which may have lasted more than a billion years), however, oxygen was primarily released from the photodissociation of water vapor [26]. From our perspective here, the important point is that the initial (albeit very small) rise in oxygen can be produced by non-biological processes (in addition to the dissociation of water vapor, the splitting of CO_2 by intense UV radiation can also contribute; e.g., [45]). The duration of this phase is roughly inversely proportional to the intensity of the UV radiation in the range 1,000–2,000 Å (since water has significant absorption peaks in the ranges 1,100–1300 Å and 1,600–1,800 Å). Consequently, for a given planet size and orbit, the timescale for the build-up of sufficient shielding from lethal radiation (and, concomitantly, the timescale for the appearance of land life, τ_ℓ) is dependent on the spectral type of the star and thereby on τ_* (since on the main sequence the spectral type is directly related to the mass, and $\tau_* \sim 10^{10}(M/M_\odot)^{-2.45}$ yr). As an aside, I should note that the mere concept of a 'habitable zone' for the planet already introduces a correlation between the star's properties and the habitability of a planet.

Livio and Kopelman [96] used typical main sequence relations – $(L/L_\odot) = (M/M_\odot)^{3.45}$, $(R/R_\odot) = (M/M_\odot)^\beta$ (with β in the range 0.6–1, applicable at least for spectral types around that of the Sun or smaller), and empirical fractions for the UV radiation emitted in the 1,000–2,000 Å range [27, 145] – to obtain the approximate relation

$$\frac{\tau_\ell}{\tau_*} \sim 0.4 \left(\frac{\tau_*}{\tau_\odot}\right)^{1.7}. \tag{11.18}$$

As I have noted before, the potential existence of such a relation undermines a key assumption in Carter's argument. The fact that we found a relation of the form $\tau_\ell/\tau_* = f(\tau_*)$, where $f(\tau_*)$ is a monotonically increasing function (at least for a certain range of stellar spectral types), has another interesting consequence. For a Salpeter initial mass function [138], the distribution of stellar lifetimes behaves approximately as $\psi(\tau_*) \sim \tau_*$. This expresses the known fact that the number of stars increases with increasing τ_*. Since $f(\tau_*)$ also increases with τ_* (Eq. (11.18)), but complex life cannot emerge if $\tau_\ell > \tau_*$, this implies that *it is most probable that in the first place where we would encounter intelligent life, we will find that $\tau_\ell/\tau_* \sim 1$*, as in the Earth-Sun system. In other words, the observation on which Carter based his argument finds a natural explanation, and it does not have any implications for the frequency of extrasolar life. Note that this conclusion does not depend on the precise functional form of $f(\tau_*) = \tau_\ell/\tau_*$, as long as such a relation exists, and it is a monotonically increasing function of τ_*.

Since in order to obtain the τ_ℓ-τ_* relation I had to make a few extremely simplifying assumptions, I would not claim that this completely refutes Carter's argument. In particular, if instead of the one-to-one function in Eq. (11.18) there is an extremely wide 'band', this would be almost equivalent to there being no τ_ℓ-τ_* relation at all, in which case Carter's argument can be recovered. The preceding discussion does demonstrate, however, that Carter's argument *could* be wrong from an astrophysical perspective. I should emphasise that even a complete refutation of Carter's argument does not mean that extrasolar intelligent civilisations exist – only that an argument for their non-existence is wrong.

A criticism on a more fundamental level was raised by Ćirković *et al.* [38]. These authors pointed out that Carter's argument relies first of all on the assumption that *well-defined* timescales for the astrophysical and biological processes actually exist. Secondly, Carter assumes that the timescale on the astrophysical side is even fixed and approximately known. Ćirković *et al.* [38] noted, in particular, that processes other than the evolution of the host star can affect the habitability of a planet. They correctly asserted that Carter's argument regards the Earth-Sun system as a 'closed box', while concepts such as the galactic habitable zone (the region in the galaxy characterised by such physical parameters that it allows for life to emerge and evolve; e.g., [63, 93]) demonstrate a level of connectedness not envisaged by Carter. Other effects such as 'snowball Earth', glaciation episodes, and geophysical processes such as those governing the carbon-silicate cycle also demonstrate the existence of relevant timescales shorter than the main-sequence lifetime of the host star (e.g., [59, 73]). I should also point out that since some astrophysical processes and events exist (such as gamma ray bursts (GRBs); or spiral-arm crossings by the solar system) that can altogether terminate or at least strongly affect biological evolution (see, e.g., [129, 142] for recent references on GRBs and [140] on spiral-arm crossings), the governing timescales may be the ones associated with maximising the chances of life being able to survive these cosmic cataclysms rather than the lifetime of the star (the timescale for atmospheric loss for planets around M-dwarfs is also much shorter than the main-sequence lifetime [94]). These timescales are dictated primarily by the values of the Hubble constant H_0 and the value of the cosmological constant Λ (in the context of a ΛCDM cosmology). It is interesting to note, in this respect, that using cosmological N-body simulations, Piran *et al.* [129] concluded that we find ourselves in a favourable spot in the cosmological phase space, in that the exposure of the solar system to GRBs is minimised while the number of hydrogen-burning stars (around which complex life can in principle evolve) is maximised.

The bottom line is clear: Carter's argument should not be taken as a discouragement from searching for other intelligent civilisations in the Milky Way. With the realisation that punctuated equilibrium could characterise the evolution of life even on the galactic scale, even the possibility of $\tau_\ell \ll \tau_*$ cannot be convincingly rejected, since life can independently appear several times and then become extinct through catastrophic events.

Irrespective of Carter's argument, however, there are other reasons to suspect that we are not the only intelligent species to have ever existed in the Universe as a whole. For example, Frank and Sullivan [54] used the exoplanets statistics obtained by the Kepler Space Observatory to evaluate the probability that humanity is the only technological civilisation to have ever existed. They showed that for that to be true, the probability that a habitable-zone planet develops a technological species must be smaller than 10^{-24}, essentially the reciprocal of the expected number of rocky, habitable-zone planets in the observable Universe. So, unless the evolution to technology is truly extraordinarily improbable, chances are that other such civilisations existed at some point in the Universe's lifetime.

To conclude this entire chapter, I have not identified any physical parameters which convincingly demonstrate that life on Earth, or even intelligent life, is unique, either in the Universe as a whole or even in the Milky Way galaxy. Ongoing space mission such as TESS (launched in 2019) and upcoming missions such as JWST and WFIRST (to be launched in 2021 and the mid-2020s, respectively), as well as ground-based telescopes (such as a next-generation Extremely Large Telescope) will detect and start to characterise the atmospheres of super-Earth and Earth-like planets, in the search for biosignatures. Chances are that even if we do not detect extrasolar life in the next two to three decades, we will at least be able to place some meaningful limits on how rare life that dominates the planetary surface chemistry (so that it significantly alters the atmosphere) really is.

Acknowledgements

Most of the work presented in this chapter was done in collaboration with Rebecca Martin. I am also grateful to Jim Pringle, Andrew King, Joe Silk, Jeremy Smallwood, Stephen Lepp, Martin Beer, Steve Lubow, Phil Armitage, Arik Kopelman, and Lev Yungelson, who contributed to various parts of the work.

References

[1] Abe, Y., *et al.* 2000. 'Water in the Early Earth'. In Canup, R. M., and Righter, K. (eds.) *Origin of the Earth and Moon.* University of Arizona Press. Pages 413–433.

[2] A'Hearn, M. F., *et al.* 2012. Cometary Volatiles and the Origin of Comets. *Astrophysical Journal*, **758**(Oct.), 29.

[3] Alvarez, L. W., *et al.* 1980. 'Extraterrestrial Cause for the Cretaceous-Tertiary Extinction'. *Science*, **208**(June), 1095–1108.

[4] Armitage, P. J. 2013. *Astrophysics of Planet Formation* Cambridge University Press.

[5] Armitage, P. J., Livio, M., and Pringle, J. E. 2001. 'Episodic Accretion in Magnetically Layered Protoplanetary Discs'. *Monthly Notices of the Royal Astronomical Society*, **324**(June), 705–711.

[6] Armitage, P. J., *et al.* 2002. 'Predictions for the Frequency and Orbital Radii of Massive Extrasolar Planets'. *Monthly Notices of the Royal Astronomical Society*, **334**(July), 248–256.

[7] Artymowicz, P. 1993. 'Disk-Satellite Interaction via Density Waves and the Eccentricity Evolution of Bodies Embedded in Disks'. *Astrophysical Journal*, **419**(Dec.), 166.

[8] Backman, D., *et al.* 2009. 'Epsilon Eridani's Planetary Debris Disk: Structure and Dynamics Based on Spitzer and Caltech Submillimeter Observatory Observations'. *Astrophysical Journal*, **690**(Jan.), 1522–1538.

[9] Barrow, J. D., and Tipler, F. J. 1986. *The Anthropic Cosmological Principle*. Oxford University Press.

[10] Basri, G., *et al.* 2010. 'Photometric Variability in Kepler Target Stars: The Sun Among Stars: A First Look'. *Astrophysical Journal Letters*, **713**(Apr.), L155–L159.

[11] Basri, G., Walkowicz, L. M., and Reiners, A. 2013. 'Comparison of Kepler Photometric Variability with the Sun on Different Timescales'. *Astrophysical Journal*, **769**(May), 37.

[12] Batalha, N. M., *et al.* 2013. 'Planetary Candidates Observed by Kepler. III. Analysis of the First 16 Months of Data'. *Astrophysical Journal Supplements*, **204**(Feb.), 24.

[13] Batalha, N. M. 2014. 'Exploring Exoplanet Populations with NASA's Kepler Mission'. *Proceedings of the National Academy of Science*, **111**(Sept.), 12647–12654.

[14] Batygin, K., and Brown, M. E. 2016. 'Evidence for a Distant Giant Planet in the Solar System'. *Astronomical Journal*, **151**(Feb.), 22.

[15] Batygin, K., and Laughlin, G. 2015. 'Jupiter's Decisive Role in the Inner Solar System's Early Evolution'. *Proceedings of the National Academy of Sciences*.

[16] Berkner, L. V. 1952. 'Signposts to Future Ionospheric Research'. In Katz, L. and Gerson, N.C. (eds.), *Proceedings of the Conference on Ionospheric Physics*. Page 13.

[17] Berkner, L. V., and Marshall, L. C. 1965. 'On the Origin and Rise of Oxygen Concentration in the Earth's Atmosphere.' *Journal of Atmospheric Sciences*, **22**(May), 225–261.

[18] Bonfils, X., *et al.* 2013. 'The HARPS Search for Southern Extra-solar Planets. XXXI. The M-dwarf Sample'. *Astronomy and Astrophysics*, **549**(Jan.), A109.

[19] Bottke, W. F., *et al.* 2002. 'Debiased Orbital and Absolute Magnitude Distribution of the Near-Earth Objects'. *Icarus*, **156**(Apr.), 399–433.

[20] Box, G. E. P., and Cox, D. R. 1964. 'An Analysis of Transformations'. *Journal of the Royal Statistical Society. Series B (Methodological)*, **26**(2), 211–252.

[21] Buchhave, L. A., *et al.* 2012. 'An Abundance of Small Exoplanets around Stars with a Wide Range of Metallicities'. *Nature*, **486**(June), 375–377.

[22] Buchhave, L. A., *et al.* 2014. 'Three Regimes of Extrasolar Planet Radius Inferred from Host Star Metallicities'. *Nature*, **509**(May), 593–595.

[23] Burke, C. J., *et al.* 2015. 'Terrestrial Planet Occurrence Rates for the Kepler GK Dwarf Sample'. *Astrophysical Journal*, **809**(Aug.), 8.

[24] Cannizzo, J. K. 1993. 'The Accretion Disk Limit Cycle Model: Toward an Understanding of the Long-Term Behavior of SS Cygni'. *Astrophysical Journal*, **419**(Dec.), 318.

[25] Canup, R. M., and Asphaug, E. 2001. 'Origin of the Moon in a Giant Impact near the End of the Earth's Formation'. *Nature*, **412**(Aug.), 708–712.

[26] Canuto, V. M., *et al.* 1983. 'The Young Sun and the Atmosphere and PhotoChemistry of the Early Earth'. *Nature*, **305**(Sept.), 281–286.

[27] Carruthers, G. R. 1971. 'Far-Ultraviolet Spectra and Photometry of Perseus Stars'. *Astrophysical Journal*, **166**(June), 349.

[28] Carter, B. 1983. 'The Anthropic Principle and Its Implications for Biological Evolution'. *Philosophical Transactions of the Royal Society of London Series A*, **310**(Dec.), 347–363.

[29] Caspersson, T. 1950. *Cell Growth and Cell Function: A Cytochemical Study*. The Thomas William Salmon Memorial Lectures. W. W. Norton & Company.

[30] Chambers, J. E. 1999. 'A Hybrid Symplectic Integrator That Permits Close Encounters between Massive Bodies'. *Monthly Notices of the Royal Astronomical Society*, **304**(Apr.), 793–799.

[31] Chatterjee, S., *et al.* 2008. 'Dynamical Outcomes of Planet-Planet Scattering'. *Astrophysical Journal*, **686**(Oct.), 580–602.

[32] Chen, C. H., *et al.* 2006. 'Spitzer IRS Spectroscopy of IRAS-Discovered Debris Disks'. *Astrophysical Journal Supplements*, **166**(Sept.), 351–377.

[33] Chen, C. H., *et al.* 2009. 'Solar System Analogs around IRAS-Discovered Debris Disks'. *Astrophysical Journal*, **701**(Aug.), 1367–1372.

[34] Chen, C. H., *et al.* 2014. 'The Spitzer Infrared Spectrograph Debris Disk Catalog. I. Continuum Analysis of Unresolved Targets'. *Astrophysical Journal Supplements*, **211**(Apr.), 25.

[35] Chiang, E., and Laughlin, G. 2013. 'The Minimum-Mass Extrasolar Nebula: In-Situ Formation of Close-In Super-Earths'. *Monthly Notices of the Royal Astronomical Society*, **431**, 3444.

[36] Chiang, E. I., and Goldreich, P. 1997. 'Spectral Energy Distributions of T Tauri Stars with Passive Circumstellar Disks'. *Astrophysical Journal*, **490**(Nov.), 368–376.

[37] Chrenko, O., *et al.* 2015. 'The Origin of Long-Lived Asteroids in the 2:1 Mean-Motion Resonance with Jupiter'. *Monthly Notices of the Royal Astronomical Society*, **451**(Aug.), 2399–2416.

[38] Ćirković, M. M., Vukotić, B., and Dragićević, I. 2009. 'Galactic Punctuated Equilibrium: How to Undermine Carter's Anthropic Argument in Astrobiology'. *Astrobiology*, **9**(June), 491–501.

[39] Cossou, C., *et al.* 2014. 'Hot Super-Earths and Giant Planet Cores from Different Migration Histories'. *Astronomy and Astrophysics*, **569**(Sept.), A56.

[40] Cumming, A., *et al.* 2008. 'The Keck Planet Search: Detectability and the Minimum Mass and Orbital Period Distribution of Extrasolar Planets'. *Publications of the Astronomical Society of the Pacific*, **120**(May), 531.

[41] Davidson, J. N. 1960. *The Biochemistry of the Nucleic Acids*. 4th edn., rev. edn. Methuen.

[42] Davies, P. 2011. *The Eerie Silence: Renewing Our Search for Alien Intelligence*. Mariner Books.

[43] Dermott, S. F., and Murray, C. D. 1983. 'Nature of the Kirkwood Gaps in the Asteroid Belt'. *Nature*, **301**(Jan.), 201–205.

[44] Dohnanyi, J. S. 1969. 'Collisional Model of Asteroids and Their Debris'. *Journal of Geophysics Research*, **74**(May), 2531–2554.

[45] Domagal-Goldman, S. D., *et al.* 2014. 'Abiotic Ozone and Oxygen in Atmospheres Similar to Prebiotic Earth'. *Astrophysical Journal*, **792**(Sept.), 90.

[46] Dressing, C. D., and Charbonneau, D. 2015. 'The Occurrence of Potentially Habitable Planets Orbiting M Dwarfs Estimated from the Full Kepler Dataset and an Empirical Measurement of the Detection Sensitivity'. *Astrophysical Journal*, **807**(July), 45.

[47] Dressing, C. D., *et al.* 2015. 'The Mass of Kepler-93b and the Composition of Terrestrial Planets'. *Astrophysical Journal*, **800**(Feb.), 135.

[48] Duncan, M. J., and Lissauer, J. J. 1998. 'The Effects of Post-Main-Sequence Solar Mass Loss on the Stability of Our Planetary System'. *Icarus*, **134**(Aug.), 303–310.

[49] Edgar, R., and Artymowicz, P. 2004. 'Pumping of a Planetesimal Disc by a Rapidly Migrating Planet'. *Monthly Notices of the Royal Astronomical Society*, **354**(Nov.), 769–772.

[50] Ehrenfreund, P., and Charnley, S. B. 2000. 'Organic Molecules in the Interstellar Medium, Comets, and Meteorites: A Voyage from Dark Clouds to the Early Earth'. *Annual Review of Astronomy and Astrophysics*, **38**, 427–483.

[51] Fischer, D. A., and Valenti, J. 2005. 'The Planet-Metallicity Correlation'. *Astrophysical Journal*, **622**(Apr.), 1102–1117.

[52] Fleming, T. P., Stone, J. M., and Hawley, J. F. 2000. 'The Effect of Resistivity on the Nonlinear Stage of the Magnetorotational Instability in Accretion Disks'. *Astrophysical Journal*, **530**(Feb.), 464–477.

[53] Fogg, M. J., and Nelson, R. P. 2007. 'On the Formation of Terrestrial Planets in Hot-Jupiter Systems'. *Astronomy and Astrophysics*, **461**(Jan.), 1195–1208.

[54] Frank, A., and Sullivan, W. T. I. 2016. 'A New Empirical Constraint on the Prevalence of Technological Species in the Universe'. *Astrobiology*, **16**(May), 359–362.

[55] Fressin, F., *et al.* 2013. 'The False Positive Rate of Kepler and the Occurrence of Planets'. *Astrophysical Journal*, **766**(Apr.), 81.

[56] Froeschle, C., and Scholl, H. 1986. 'The Secular Resonance nu6 in the Asteroidal Belt. *Astronomy and Astrophysics*, **166**(Sept.), 326–332.

[57] Fromang, S., Terquem, C., and Balbus, S. A. 2002. 'The Ionization Fraction in α Models of Protoplanetary Discs'. *Monthly Notices of the Royal Astronomical Society*, **329**(Jan.), 18–28.

[58] Gammie, C. F., and Menou, K. 1998. 'On the Origin of Episodic Accretion in Dwarf Novae'. *Astrophysical Journal Letters*, **492**(Jan.), L75–L78.

[59] Gerstell, M. F., and Yung, Y. L. 2003. 'A Comment on Tectonics and the Future of Terrestrial Life. *Precambrian Research*, **120**(Jan.), 177–178.

[60] Gladman, B. 1993. 'Dynamics of Systems of Two Close Planets. *Icarus*, **106**(Nov.), 247.

[61] Gladman, B. J., *et al.* 1997. 'Dynamical Lifetimes of Objects Injected into Asteroid Belt Resonances'. *Science*, **277**, 197–201.

[62] Glassgold, A. E., Najita, J., and Igea, J. 2004. 'Heating Protoplanetary Disk Atmospheres'. *Astrophysical Journal*, **615**(Nov.), 972–990.

[63] Gonzalez, G. 2005. 'Habitable Zones in the Universe'. *Origins of Life and Evolution of the Biosphere*, **35**(Dec.), 555–606.

[64] Haisch, Jr., K. E., Lada, E. A., and Lada, C. J. 2001. 'Disk Frequencies and Lifetimes in Young Clusters'. *Astrophysical Journal Letters*, **553**(June), L153–L156.

[65] Han, E., *et al.* 2014. 'Exoplanet Orbit Database. II. Updates to Exoplanets.org'. *Publications of the Astronomical Society of the Pacific*, **126**(Sept.), 827.

[66] Hansen, B. M. S. 2009. 'Formation of the Terrestrial Planets from a Narrow Annulus'. *Astrophysical Journal*, **703**(Sept.), 1131–1140.

[67] Hansen, B. M. S., and Murray, N. 2012. 'Migration Then Assembly: Formation of Neptune-Mass Planets inside 1 AU'. *Astrophysical Journal*, **751**(June), 158.

[68] Hansen, B. M. S., and Murray, N. 2013. 'Testing In Situ Assembly with the Kepler Planet Candidate Sample'. *Astrophysical Journal*, **775**(Sept.), 53.

[69] Hawley, J. F., Gammie, C. F., and Balbus, S. A. 1995. 'Local Three-Dimensional Magnetohydrodynamic Simulations of Accretion Disks'. *Astrophysical Journal*, **440**(Feb.), 742.

[70] Hayashi, C. 1981. 'Structure of the Solar Nebula, Growth and Decay of Magnetic Fields and Effects of Magnetic and Turbulent Viscosities on the Nebula'. *Progress of Theoretical Physics Supplement*, **70**, 35–53.

[71] Hines, D. C., *et al.* 2006. 'The Formation and Evolution of Planetary Systems (feps): Discovery of an Unusual Debris Aystem Associated with hd 12039'. *Astrophysical Journal*, **638**, 1070–1079.

[72] Hoff, W., Henning, T., and Pfau, W. 1998. 'The Nature of Isolated T Tauri stars'. *Astronomy and Astrophysics*, **336**(Aug.), 242–250.

[73] Hoffman, P. F., *et al.* 1998. 'A Neoproterozoic Snowball Earth'. *Science*, **281**(Aug.), 1342.

[74] Hogg, D. W., Myers, A. D., and Bovy, J. 2010. 'Inferring the Eccentricity Distribution'. *Astrophysical Journal*, **725**, 2166–2175.

[75] Horner, J., and Jones, B. W. 2008. 'Jupiter Friend or Foe? I. The Asteroids'. *International Journal of Astrobiology*, **7**(Oct.), 251–261.

[76] Horner, J., and Jones, B. W. 2012. 'Jupiter: Friend or Foe? IV. The Influence of Orbital Eccentricity and Inclination'. *International Journal of Astrobiology*, **11**(July), 147–156.

[77] Houtkooper, J. M. 2011. 'Glaciopanspermia: Seeding the Terrestrial Planets with Life?' *Planetary Space Science*, **59**(Aug.), 1107–1111.

[78] Ida, S., and Lin, D. N. C. 2010. 'Toward a Deterministic Model of Planetary Formation. VI. Dynamical Interaction and Coagulation of Multiple Rocky Embryos and Super-Earth Systems around Solar-type Stars'. *Astrophysical Journal*, **719**(Aug.), 810–830.

[79] Ito, T., and Tanikawa, K. 2002. 'Long-Term Integrations and Stability of Planetary Orbits in Our Solar System'. *Monthly Notices of the Royal Astronomical Society*, **336**(Oct.), 483–500.

[80] Izidoro, A., Morbidelli, A., and Raymond, S. N. 2014. 'Terrestrial Planet Formation in the Presence of Migrating Super-Earths'. *Astrophysical Journal*, **794**(Oct.), 11.

[81] Jones, E. M. 1985. '"Where Is Everybody?" An Account of Fermi's Question', LA-10311-MS, UC-34b, CIC-14 Report Collection, March 1985. www.fas.org/sgp/othergov/doe/lanl/la-10311-ms.pdf.

[82] Jontof-Hutter, D., *et al.* 2015. 'Kepler's Low-Mass, Low Density Planets Characterized via Transit Timing'. *IAU General Assembly*, **22**(Aug.), 2257778.

[83] Kennedy, G. M., and Wyatt, M. C. 2014. 'Do Two-Temperature Debris Discs Have Multiple Belts?' *Monthly Notices of the Royal Astronomical Society*, **444**(Nov.), 3164–3182.

[84] King, A. R., Pringle, J. E., and Livio, M. 2007. 'Accretion Disc Viscosity: How Big Is Alpha?.' *Monthly Notices of the Royal Astronomical Society*, **376**(Apr.), 1740–1746.

[85] Kominami, J., and Ida, S. 2002. 'The Effect of Tidal Interaction with a Gas Disk on Formation of Terrestrial Planets'. *Icarus*, **157**(May), 43–56.

[86] Laughlin, G. 2009. 'Planetary Science: The Solar System's Extended Shelf Life'. *Nature*, **459**(June), 781–782.

[87] Lecar, M., and Franklin, F. 1997. 'The Solar Nebula, Secular Resonances, Gas Drag, and the Asteroid Belt'. *Icarus*, **129**(Sept.), 134–146.

[88] Lecar, M., *et al.* 2006. 'On the Location of the Snow Line in a Protoplanetary Disk'. *Astrophysical Journal*, **640**, 1115–1118.

[89] Lesur, G., Kunz, M. W., and Fromang, S. 2014. 'Thanatology in Protoplanetary Discs: The Combined Influence of Ohmic, Hall, and Ambipolar Diffusion on Dead Zones'. *Astronomy and Astrophysics*, **566**(June), A56.

[90] Levison, H. F., *et al.* 2008. 'Origin of the SSStructure of the Kuiper Belt during a Dynamical Instability in the Orbits of Uranus and Neptune'. *Icarus*, **196**(July), 258–273.

[91] Limbach, M. A., and Turner, E. L. 2015. 'Exoplanet Orbital Eccentricity: Multiplicity Relation and the Solar System'. *Proceedings of the National Academy of Sciences*, **112**(1), 20–24.

[92] Lin, D. N. C., and Papaloizou, J. 1986. 'On the Tidal Interaction between Protoplanets and the protoplanetary Disk: III. Orbital Migration of Protoplanets'. *Astrophysical Journal*, **309**(Oct.), 846–857.

[93] Lineweaver, C. H., Fenner, Y., and Gibson, B. K. 2004. 'The Galactic Habitable Zone and the Age Distribution of Complex Life in the Milky Way'. *Science*, **303**(Jan.), 59–62.

[94] Lingam, M., and Loeb, A. 2018. 'Is Life Most Likely around Sun-Like Stars?.' *Journal of Cosmology and Astro-Particle Physics*, **2018**(May), 020.

[95] Livio, M. 1999. 'How Rare Are Extraterrestrial Civilizations, and When Did They Emerge?.' *Astrophysical Journal*, **511**(Jan.), 429–431.

[96] Livio, M., and Kopelman, A. 1990. Life and the Sun's lifetime. *Nature*, **343**(Jan.), 25.

[97] Livio, M., and Silk, J. 2017. Where are they? *Physics Today*, **70**(3), 50–57.

[98] Low, F. J., *et al.* 2005. Exploring Terrestrial Planet Formation in the TW Hydrae Association. *Astrophysical Journal*, **631**(Oct.), 1170–1179.

[99] Lyons, T. W., Reinhard, C. T., and Planavsky, N. J. 2014. 'The Rise of Oxygen in Earth's Early Ocean and Atmosphere'. *Nature*, **506**(Feb.), 307–315.

[100] Marseille, M. G., and Cazaux, S. 2011. 'The Snow Border'. *Astronomy and Astrophysics*, **532**(Aug.), A60.

[101] Martin, R. G. and Livio, M. 2013. 'On the Evolution of the Snow Line in Protoplanetary Discs: II. Analytic Approximations'. *Monthly Notices of the Royal Astronomical Society*, **434**(Sept.), 633–638.

[102] Martin, R. G., and Livio, M. 2014. 'On the Evolution of the CO Snow Line in Protoplanetary Disks'. *Astrophysical Journal Letters*, **783**(Mar.), L28.

[103] Martin, R. G., and Livio, M. 2016. 'On the Formation of Super-Earths with Implications for the Solar System'. *Astrophysical Journal*, **822**(May), 90.

[104] Martin, R. G., and Lubow, S. H. 2013. 'Propagation of the Gravo-Magneto Disc Instability'. *Monthly Notices of the Royal Astronomical Society*, **432**(June), 1616–1622.

[105] Martin, R. G., and Livio, M. 2012. 'On the Evolution of the Snow Line in Protoplanetary Discs'. *Monthly Notices of the Royal Astronomical Society*, **425**, 6.

[106] Martin, R. G., and Livio, M. 2015. 'The Solar System as an Exoplanetary System'. *Astrophysical Journal*, **810**(Sept.), 105.

[107] Mathews, G. S., *et al.* 2013. 'ALMA Imaging of the CO Snowline of the HD 163296 Disk with DCO^+'. *Astronomy and Astrophysics*, **557**(Sept.), A132.

[108] Matsumura, S., and Pudritz, R. E. 2003. 'The Origin of Jovian Planets in Protostellar Disks: The Role of Dead Zones'. *Astrophysical Journal*, **598**(Nov.), 645–656.

[109] Mayor, M., *et al.* 2011. 'The HARPS Search for Southern Extra-solar Planets XXXIV. Occurrence, Mass Distribution and Orbital Properties of Super-Earths and Neptune-Mass Planets'. *ArXiv e-prints*, Sept., arXiv:1109.2497.

[110] Mayor, M., Lovis, C., and Santos, N. C. 2014. 'Doppler Spectroscopy as a Path to the Detection of Earth-Like Planets'. *Nature*, **513**(Sept.), 328–335.

[111] McQuillan, A., Aigrain, S., and Roberts, S. 2012. 'Statistics of Stellar Variability from Kepler.' I. Revisiting Quarter 1 with an Astrophysically Robust Systematics Correction. *Astronomy and Astrophysics*, **539**(Mar.), A137.

[112] Minton, D. A., and Malhotra, R. 2010. 'Dynamical Erosion of the Asteroid Belt and Implications for Large Impacts in the Inner Solar System'. *Icarus*, **207**(June), 744–757.

[113] Minton, D. A., and Malhotra, R. 2011. 'Secular Resonance Sweeping of the Main Asteroid Belt during Planet Migration'. *Astrophysical Journal*, **732**(May), 53.

[114] Moerchen, M. M., Telesco, C. M., and Packham, C. 2010. 'High Spatial Resolution Imaging of Thermal Emission from Debris Disks'. *Astrophysical Journal*, **723**(Nov.), 1418–1435.

[115] Moons, M. 1996. 'Review of the Dynamics in the Kirkwood Gaps'. *Celestial Mechanics and Dynamical Astronomy*, **65**(Mar.), 175–204.

[116] Moór, A., *et al.* 2009. 'The Discovery of New Warm Debris Disks around F-type Stars'. *Astrophysical Journal Letters*, **700**(July), L25–L29.

[117] Moór, A., *et al.* 2011. 'Structure and Evolution of Debris Disks around F-type Stars: I. Observations, Database, and Basic Evolutionary Aspects'. *Astrophysical Journal Supplements*, **193**(Mar.), 4.

[118] Moorhead, A. V., *et al.* 2011. 'The Distribution of Transit Durations for Kepler Planet Candidates and Implications for Their Orbital Eccentricities'. *The Astrophysical Journal Supplement Series*, **197**(Nov.), 1.

[119] Morales, F. Y., *et al.* 2011. 'Common Warm Dust Temperatures around Main-sequence Stars'. *Astrophysical Journal Letters*, **730**(Apr.), L29.

[120] Morbidelli, A., *et al.* 2000. 'Source Regions and Time Scales for the Delivery of Water to Earth'. *Meteoritics and Planetary Science*, **35**(Nov.), 1309–1320.

[121] Morbidelli, A., *et al.* 2010. 'Evidence from the Asteroid Belt for a Violent Past Evolution of Jupiter's Orbit'. *Astronomical Journal*, **140**(Nov.), 1391–1401.

[122] Moro-Martín, A., *et al.* 2010. 'Locating Planetesimal Belts in the Multiple-planet Systems HD 128311, HD 202206, HD 82943, and HR 8799'. *Astrophysical Journal*, **717**(July), 1123–1139.

[123] Morrison, S., and Malhotra, R. 2015. 'Planetary Chaotic Zone Clearing: Destinations and Timescales'. *Astrophysical Journal*, **799**(Jan.), 41.

[124] Murray, C. D., and Dermott, S. F. 2000. *Solar System Dynamics*. Cambridge University Press.

[125] Murray, N., and Holman, M. 1999. 'The Origin of Chaos in the Outer Solar System'. *Science*, **283**(Mar.), 1877.

[126] Öberg, K. I., *et al.* 2005. 'Competition between CO and N_2 Desorption from Interstellar Ices'. *Astrophysical Journal Letters*, **621**(Mar.), L33–L36.

[127] Öberg, K. I., Murray-Clay, R., and Bergin, E. A. 2011. 'The Effects of Snowlines on C/O in Planetary Atmospheres'. *Astrophysical Journal Letters*, **743**(Dec.), L16.

[128] O'Brien, D. P., Morbidelli, A., and Bottke, W. F. 2007. 'The Primordial Excitation and Clearing of the Asteroid Belt: – Revisited'. *Icarus*, **191**(Nov.), 434–452.

[129] Piran, T., *et al.* 2016. 'Cosmic Explosions, Life in the Universe, and the Cosmological Constant'. *Physical Review Letters*, **116**(8), 081301.

[130] Pollack, J. B., *et al.* 1996. 'Formation of the Giant Planets by Concurrent Accretion of Solids and Gas'. *Icarus*, **124**(Nov.), 62–85.

[131] Pringle, J. E. 1981. 'Accretion Discs in Astrophysics'. *Annual Review of Astronomy and Astrophysics*, **19**, 137–162.

[132] Qi, C., *et al.* 2011. 'Resolving the CO Snow Line in the Disk around HD 163296'. *Astrophysical Journal*, **740**(Oct.), 84.

[133] Qi, C., *et al.* 2013. 'Imaging of the CO Snow Line in a Solar Nebula Analog'. *Science*, **341**(Aug.), 630–632.

[134] Reffert, S., *et al.* 2015. 'Precise Radial Velocities of Giant Stars: VII. Occurrence Rate of Giant Extrasolar Planets as a Function of Mass and Metallicity'. *Astronomy and Astrophysics*, **574**(Feb.), A116.

[135] Ros, K., and Johansen, A. 2013. 'Ice Condensation as a Planet Formation mechanism'. *Astronomy and Astrophysics*, **552**(Apr.), A137.

[136] Rufu, R., Aharonson, O., and Perets, H. B. 2017. 'A Multiple-Impact Origin for the Moon'. *Nature Geoscience*, **10**(Jan.), 89–94.

[137] Sagan, C. 1961. 'On the Origin and Planetary Distribution of Life'. *Radiation Research*, **15**(Aug.), 174.

[138] Salpeter, E. E. 1955. 'The Luminosity Function and Stellar Evolution'. *Astrophysical Journal*, **121**(Jan.), 161.

[139] Shakura, N. I., and Sunyaev, R. A. 1973. 'Black Holes in Binary Systems: Observational Appearance'. *Astronomy and Astrophysics*, **24**, 337–355.

[140] Shaviv, N. J., 2002. 'Cosmic Ray Diffusion from the Galactic Spiral Arms, Iron Meteorites, and a Possible Climatic Connection'. *Physical Review Letters*, **89**(5), 051102.

[141] Simon, J. B., Armitage, P. J., and Beckwith, K. 2011. 'Turbulent Linewidths in Protoplanetary Disks: Predictions from Numerical Simulations'. *Astrophysical Journal*, **743**(Dec.), 17.

[142] Sloan, D., Alves Batista, R., and Loeb, A. 2017. 'The Resilience of Life to Astrophysical Events'. *Scientific Reports*, **7**(July), 5419.

[143] Smallwood, J. L., *et al.* 2018. 'Asteroid Impacts on Terrestrial Planets: The Effects of Super-Earths and the Role of the ν_6 Resonance'. *Monthly Notices of the Royal Astronomical Society*, **473**(Jan.), 295–305.

[144] Sousa, S. G., *et al.* 2011. 'Spectroscopic Stellar Parameters for 582 FGK Stars in the HARPS Volume-Limited Sample: Revising the Metallicity-Planet Correlation'. *Astronomy and Astrophysics*, **533**(Sept.), A141.

[145] Stecher, T. P. 1970. 'Stellar Spectrophotometry from a Pointed Rocket'. *Astrophysical Journal*, **159**(Feb.), 543.

[146] Tamayo, D., Rein, H., Petrovich, C., and Murray, N. 2017. 'Convergent Migration Renders TRAPPIST-1 Long-Lived'. *Astrophysical Journal*, **840**(May), L19.

[147] Terquem, C., and Papaloizou, J. C. B. 2007. 'Migration and the Formation of Systems of Hot Super-Earths and Neptunes'. *Astrophysical Journal*, **654**, 1110–1120.

[148] Toomre, A. 1964. 'On the Gravitational Stability of a Disk of Stars'. *Astrophysical Journal*, **139**(May), 1217–1238.

[149] Umebayashi, T., and Nakano, T. 1988. 'Chapter 13: Ionization State and Magnetic Fields in the Solar Nebula'. *Progress of Theoretical Physics Supplement*, **96**, 151–160.

[150] Walsh, K. J., *et al.* 2011. 'A Low Mass for Mars from Jupiter's Early Gas-Driven Migration'. *Nature*, **475**(July), 206–209.

[151] Wang, J., and Fischer, D. A. 2015. 'Revealing a Universal Planet-Metallicity Correlation for Planets of Different Sizes around Solar-Type Stars'. *Astronomical Journal*, **149**(Jan.), 14.

[152] Ward, P., and Brownlee, D. 2000. *Rare Earth : Why Complex Life Is Uncommon in the Universe*. Copernicus.

[153] Ward, W. R. 1989. 'On the Rapid Formation of Giant Planet Cores'. *Astrophysical Journal Letters*, **345**(Oct.), L99–L102.

[154] Watanabe, K. 1958. 'Ultraviolet Absorption Processes in the Upper Atmosphere'. *Advances in Geophysics*, **5**, 153–221.

[155] Weidenschilling, S. J. 1977. 'The Distribution of Mass in the Planetary System and Solar Nebula'. *Astrophysics and Space Science*, **51**(Sept.), 153–158.

[156] Wickramasinghe, J., Wickramasinghe, C., and Napier, W. 2010. *Comets and the Origin of Life*. World Scientific Publishing Co.

[157] Willbold, M., Elliott, T., and Moorbath, S. 2011. 'The Tungsten Isotopic Composition of the Earth's Mantle before the Terminal Bombardment'. *Nature*, **477**(Sept.), 195–198.

[158] Williams, D. M., and Pollard, D. 2002. 'Earth-Like Worlds on Eccentric Orbits: Excursions beyond the Habitable Zone'. *International Journal of Astrobiology*, **1**(Jan.), 61–69.

[159] Wilson, P. A. 1994. 'Carter on Anthropic Principle Predictions'. *The British Journal for the Philosophy of Science*, **45**(1), 241–253.

[160] Winn, J. N. and Fabrycky, D. C. 2015. 'The Occurrence and Architecture of Exoplanetary Systems'. *Annual Review of Astronomy and Astrophysics*, **53**(Aug.), 409–447.

[161] Wright, J. T., *et al.* 2011. 'The Exoplanet Orbit Database'. *Publications of the Astronomical Society of the Pacific*, **123**(Apr.), 412.

[162] Yoshikawa, M. 1987. 'A Simple Analytical Model for the Secular Resonance nu6 in the Asteroidal Belt'. *Celestial Mechanics*, **40**(Sept.), 233–272.

[163] Zakamska, N. L., Pan, M., and Ford, E. B. 2011. 'Observational Biases in Determining Extrasolar Planet Eccentricities in Single-Planet Systems'. *Monthly Notices of the Royal Astronomical Society*, **410**, 1895.

[164] Zhu, Z., Hartmann, L., and Gammie, C. 2009. 'Nonsteady Accretion in Protostars'. *Astrophysical Journal*, **694**(Apr.), 1045–1055.

[165] Zhu, Z., Hartmann, L., and Gammie, C. 2010. 'Long-Term Evolution of Protostellar and Protoplanetary Disks: II. Layered Accretion with Infall'. *Astrophysical Journal*, **713**(Apr.), 1143–1158.

12

On the Temporal Habitability of Our Universe

ABRAHAM LOEB

Abstract

Is life most likely to emerge at the present cosmic time near a star like the Sun? We consider the habitability of the Universe throughout cosmic history and conservatively restrict our attention to the context of 'life as we know it' and the standard cosmological model, ΛCDM. The habitable cosmic epoch started shortly after the first stars formed, about 30 Myr after the Big Bang, and will end about 10 Tyr from now, when all stars will die. We review the formation history of habitable planets and find that unless habitability around low-mass stars is suppressed or appears preferentially early, life is most likely to exist near $\sim 0.1 M_\odot$ stars 10 trillion years from now. Spectroscopic searches for biosignatures in the atmospheres of transiting Earth-mass planets around low-mass stars will determine whether present-day life is indeed premature or typical from a cosmic perspective.

12.1 Introduction

The known forms of terrestrial life involve carbon-based chemistry in liquid water [130, 131]. In the cosmological context, life could not have started earlier than 10 Myr after the Big Bang ($z \gtrsim 140$) since the entire Universe was bathed in a thermal radiation background above the boiling temperature of liquid water. Later on, however, the Universe cooled to a *habitable epoch* at a comfortable temperature of 273–373 K between 10 and 17 Myr after the Big Bang [159].

The phase diagram of water allows it to be liquid only under external pressure in an atmosphere which can be confined gravitationally on the surface of a planet. To keep the atmosphere bound against evaporation requires strong surface gravity of a rocky planet with a mass comparable to or above that of the Earth [229, 276].

The emergence of 'life as we know it' requires stars for two reasons. Stars are needed to produce the heavy elements (carbon, oxygen, and so on, up to iron) out of which rocky planets and the molecules of life are made. Stars also provide a heat source for powering the chemistry of life on the surface of their planets. Each star is surrounded by a habitable zone where the surface temperature of a planet allows liquid water to exist. The approximate distance of the habitable zone, r_{HZ}, is obtained by equating the heating rate per unit area from the stellar luminosity, L, to the cooling rate per unit area at a surface temperature of $T_{HZ} \sim 300$ K, namely $(L/4\pi r_{HZ}^2) \sim \sigma T_{HZ}^4$, where σ is the Stefan-Boltzman constant [130, 131].

Starting from the vicinity of particular stars, life could potentially spread. This process, so-called panspermia, could be mediated by the transfer of rocks between planetary systems [1]. The 'astronauts' on such rocks could be microscopic animals such as *tardigrades*, which are known to be resilient to extreme vacuum, dehydration, and exposure to radiation that characterise space travel. In 2007, dehydrated tardigrades were taken into a low Earth orbit for 10 days. After returning to Earth, most of them revived after rehydration, and many produced viable embryos [87]. Life which arose via spreading will exhibit more clustering than life which arose spontaneously, and so the existence of panspermia can be detected statistically through excess spatial correlations of life-bearing environments. Future searches for biosignatures in the atmospheres of exoplanets could test for panspermia: a smoking-gun signature would be the detection of large regions in the Milky Way where life saturates its environment, interspersed with voids where life is very uncommon. In principle, detection of as few as several tens of biologically active exoplanets could yield a highly significant detection of panspermia [151]. Once life emerges on the surface of a planet, it is difficult to extinguish it completely through astrophysical events (such as quasar activity, supernovae, gamma ray bursts, or asteroid impacts) other than the death of the host star. Life is known to be resilient to extreme environments and could be protected from harmful radiation or heat if it resides underground or on the deep ocean floor [236].

Panspermia is not limited to galactic scales and could extend over cosmological distances. Some stars and their planets are ejected from their birth galaxies at a speed approaching the speed of light, through a gravitational slingshot from pairs of supermassive black holes which are formed during galaxy mergers [104, 161]. The resulting population of relativistic stars roaming through intergalactic space could potentially transfer life between galaxies separated by vast distances across the Universe.

The spread of life could be enhanced artificially through the use of spacecrafts by advanced civilisations. Our own civilisation is currently starting to develop the technology needed to visit the nearest stars with a travel time of decades through the propulsion of lightweight sails to a fraction of the speed of light by a powerful laser.[1] The existence of advanced civilisations could be revealed through the detection of industrial pollution in the atmospheres of planets [152], the detection of powerful beacons of light used for propulsion [103] and communications [164], or through artificial lights [163]. The search for signatures of advanced civilisations is the richest interdisciplinary frontier, offering interfaces between astronomy and other disciplines, such as biology (astrobiology), chemistry (astrochemistry), statistics (astro-statistics), or engineering (astroengineering). Moreover, the prospects for communication with aliens could open new disciplines on the interface with linguistics (astro-linguistics), psychology (astro-psychology), sociology

---

[1] www.breakthroughinitiatives.org/Concept/3.

(astro-sociology), philosophy (astro-philosophy) and many other fields. 'Are we alone?' is one of the most fundamental questions in science; the answer we find to this question is likely to provide a fresh perspective on our place in the Universe. Although primitive forms of life are likely to be more abundant, intelligent civilisations could make themselves detectable out to greater distances.

The search for life was reinvigorated by the Kepler satellite which revealed that a substantial fraction of the stars in the Milky Way galaxy host habitable Earth-mass planets around them [73, 179, 207, 235]. Indeed, the nearest star to the Sun, Proxima, whose mass is only 12% of the solar mass, was recently found to host an Earth-mass planet in the habitable zone. This planet, Proxima b, is 20 times closer to its faint stellar host than the Earth is to the Sun [9]. Dwarf stars like Proxima are the most abundant stars in the Universe, and they live for trillions of years, up to a thousand times longer than the Sun. If life could form around them, it would survive long into the future. The prospects for life in the distant cosmic future can therefore be explored by searching for biosignatures around nearby dwarf stars. For example, the existence of an atmosphere around Proxima's planet could be detected relatively soon by measuring the temperature contrast between its day and night sides [140]. Speaking of the cosmic future, it is interesting to note that planets or rocky debris are known to exist around stellar remnants, such as white dwarfs [150, 259] and neutron stars [216]. The Sun is currently at the middle of its lifetime on its way to becoming a white dwarf. White dwarfs, which are billions of years old and exist in abundance comparable to that of Sun-like stars, have a surface temperature similar to that of the Sun but are a hundred times smaller in size. As a result, the habitable zone around them is a hundred times closer than the Earth is to the Sun [2]. Searches for biosignatures in the atmospheres of habitable planets which transit white dwarfs could potentially be conducted in the near future [162]. Counting all possible host stars and extrapolating to cosmological scales, there might be as many as $\sim 10^{20}$ habitable planets in the observable volume of the Universe at present [25, 275].

In the following sections of this chapter, we discuss the habitability of the Universe as a function of cosmic time, starting from the earliest habitable epoch in Section 12.2. According to the Standard Model of cosmology, the first stars in the observable Universe formed ~ 30 Myr after the Big Bang at a redshift, $z \sim 70$ [82, 159, 160, 193]. Within a few Myr, the first supernovae dispersed heavy elements into the surrounding gas, enriching the second-generation stars with heavy elements. Remnants from the second generation of stars are found in the halo of the Milky Way galaxy and may have planetary systems in the habitable zone around them [182], as discussed in Section 12.3. The related planets are likely made of carbon, and water could have been delivered to their surface by icy comets, in a similar manner to the solar system. The formation of water is expected to consume most of the oxygen in the metal poor interstellar medium of the first galaxies [29], as discussed in Section 12.4. Therefore, even if the cosmological constant was bigger than its measured value by up to a factor of $\sim 10^3$ so that galaxy formation was suppressed at redshifts $z \lesssim 10$, life could have still emerged in our Universe due to the earliest generation of galaxies [158], as discussed in Section 12.5. We conclude in Section 12.6 with a calculation of the relative likelihood per unit time for the emergence of life [165], which is of particular importance for studies attempting to gauge the level of fine-tuning required for the cosmological or fundamental physics parameters that would allow life to emerge in our Universe.

12.2 The Habitable Epoch of the Early Universe

12.2.1 Section Background

We start by pointing out that the cosmic microwave background (CMB) provided a uniform heating source at a temperature of $T_{cmb} = 272.6$ K $\times [(1+z)/100]$ [85] that could have made by itself rocky planets habitable at redshifts $(1+z) = 100$–137 in the early Universe, merely 10–17 million years after the Big Bang.

In order for rocky planets to exist at these early times, massive stars with tens to hundreds of solar masses, whose lifetimes are much shorter than the age of the Universe, had to form and enrich the primordial gas with heavy elements through winds and supernova explosions [110, 198]. Indeed, numerical simulations predict that predominantly massive stars have formed in the first halos of dark matter to collapse [41, 160]. For massive stars that are dominated by radiation pressure and shine near their Eddington luminosity $L_E = 1.3 \times 10^{40}$ erg s$^{-1}(m/100M_\odot)$, the lifetime is independent of stellar mass m and set by the 0.7% nuclear efficiency for converting rest mass to radiation, $\sim (0.007mc^2)/L_E = 3$ Myr [43, 78]. We next examine how early such stars formed within the observable volume of our Universe.

12.2.2 First Planets

In the standard cosmological model, structure forms hierarchically – starting from small spatial scales, through the gravitational growth of primordial density perturbations [160]. On any given spatial scale R, the probability distribution of fractional density fluctuations δ is assumed to have a Gaussian form, $P(\delta)d\delta = (2\pi\sigma^2)^{-1/2} \exp\{-\delta^2/2\sigma^2\}d\delta$, with a *root-mean-square* amplitude $\sigma(R)$ that is initially much smaller than unity. The initial $\sigma(R)$ is tightly constrained on large scales, $R \gtrsim 1$ Mpc, through observations of the CMB anisotropies and galaxy surveys [5, 211], and is extrapolated theoretically to smaller scales. Throughout the discussion, we normalise spatial scales to their so-called co-moving values in the present-day Universe. The assumed Gaussian shape of $P(\delta)$ has so far been tested only on scales $R \gtrsim 1$ Mpc for $\delta \lesssim 3\sigma$ [232], but was not verified in the far tail of the distribution or on small scales that are first to collapse in the early Universe.

As the density in a given region rises above the background level, the matter in it detaches from the Hubble expansion and eventually collapses, owing to its self-gravity, to make a gravitationally bound (virialised) object like a galaxy. The abundance of regions that collapse and reach virial equilibrium at any given time depends sensitively on both $P(\delta)$ and $\sigma(R)$. Each collapsing region includes a mix of dark matter and ordinary matter (often labelled as 'baryonic'). If the baryonic gas is able to cool below the virial temperature inside the dark matter halo, then it could fragment into dense clumps and make stars.

At redshifts $z \gtrsim 140$, Compton cooling on the CMB is effective on a timescale comparable to the age of the Universe, given the residual fraction of free electrons left over from cosmological recombination (see section 2.2 in Reference [160] and also Reference [215]). The thermal coupling to the CMB tends to bring the gas temperature to T_{cmb}, which at $z \sim 140$ is similar to the temperature floor associated with molecular hydrogen cooling [108, 113, 244]. In order for virialised gas in a dark matter halo to cool, condense, and fragment into stars, the halo virial temperature T_{vir} has to exceed $T_{\mathrm{min}} \approx 300$ K, corresponding to T_{cmb} at $(1 + z) \sim 110$. This implies a halo mass in excess of $M_{\mathrm{min}} = 10^4 M_{\odot}$, corresponding to a baryonic mass $M_{\mathrm{b,min}} = 1.5 \times 10^3 M_{\odot}$, a circular virial velocity $V_{\mathrm{c,min}} = 2.6$ km s^{-1}, and a virial radius $r_{\mathrm{vir,min}} = 6.3$ pc (see section 3.3 in Reference [160]). This value of M_{min} is close to the minimum halo mass to assemble baryons at that redshift (see section 3.2.1 in Reference [160] and figure 2 of Reference [250]).

The corresponding number of star-forming halos on our past light cone is given by [193],

$$N = \int_{(1+z)=100}^{(1+z)=137} n(M > M_{\mathrm{min}}, z') \frac{dV}{dz'} dz', \qquad (12.1)$$

where $n(M > M_{\mathrm{min}})$ is the co-moving number density of halos with a mass $M > M_{\mathrm{min}}$ [233], and $dV = 4\pi r^2 dr$ is the co-moving volume element with $dr = c\,dt/a(t)$. Here, $a(t) = (1 + z)^{-1}$ is the cosmological scale factor at time t, and $r(z) = c \int_0^z dz'/H(z')$ is the co-moving distance. The Hubble parameter for a flat Universe is

$$H(z) \equiv (\dot{a}/a) = H_0 \sqrt{\Omega_m (1 + z)^3 + \Omega_r (1 + z)^4 + \Omega_\Lambda}, \qquad (12.2)$$

with Ω_m, Ω_r and Ω_Λ being the present-day density parameters of matter, radiation, and vacuum, respectively. The total number of halos that existed at $(1 + z) \sim 100$ within our entire Hubble volume (not restricted to the light cone), $N_{\mathrm{tot}} \equiv n(M > M_{\mathrm{min}}, z = 99) \times (4\pi/3)(3c/H_0)^3$, is larger than N by a factor of $\sim 10^3$.

For the standard cosmological parameters [211], we find that the first star-forming halos on our past light cone reached its maximum turnaround radius[2] (with a density contrast of 5.6) at $z \sim 112$ and collapsed (with an average density contrast of 178) at $z \sim 71$. Within the entire Hubble volume, a turnaround at $z \sim 122$ resulted in the first collapse at $z \sim 77$. This result includes the delay by $\Delta z \sim 5.3$ expected from the streaming motion of baryons relative to the dark matter [82].

The preceding calculation implies that rocky planets could have formed within our Hubble volume by $(1 + z) \sim 78$ but not by $(1 + z) \sim 110$ if the initial density perturbations were perfectly Gaussian. However, the host halos of the first planets are extremely rare, representing just $\sim 2 \times 10^{-17}$ of the cosmic matter inventory. Since they lie ~ 8.5 standard deviations (σ) away on the exponential tail of the Gaussian probability distribution of initial density perturbations, $P(\delta)$, their abundance could have been significantly enhanced by

---

[2] In the spherical collapse model, the turnaround time is half the collapse time.

primordial non-Gaussianity [166, 172, 189] if the decline of $P(\delta)$ at high values of δ/σ is shallower than exponential. The needed level of deviation from Gaussianity is not ruled out by existing data sets [212]. Non-Gaussianity below the current limits is expected in generic models of cosmic inflation [173] that are commonly used to explain the initial density perturbations in the Universe.

12.2.3 Section Summary and Implications

In the discussion thus far, we highlighted a new regime of habitability made possible for ~ 6.6 Myr by the uniform CMB radiation at redshifts $(1 + z) = 100 - 137$, just when the first generation of star-forming halos (with a virial mass $\gtrsim 10^4 M_\odot$) turned around in the standard cosmological model with Gaussian initial conditions. Deviations from Gaussianity in the far (8.5σ) tail of the probability distribution of initial density perturbations, already at these redshifts, could have led to the birth of massive stars, whose heavy elements triggered the formation of rocky planets with liquid water on their surfaces.[3]

Thermal gradients are needed for life. These can be supplied by geological variations on the surface of rocky planets. Examples for sources of free energy are geothermal energy powered by the planet's gravitational binding energy at formation and radioactive energy from unstable elements produced by the earliest supernova. These internal heat sources (in addition to possible heating by a nearby star) may have kept planets warm even without the CMB, extending the habitable epoch from $z \sim 100$ to later times. The lower CMB temperature at late times may have allowed ice to form on objects that delivered water to a planet's surface and helped to maintain the cold trap of water in the planet's stratosphere. Planets could have kept a blanket of molecular hydrogen that maintained their warmth [208, 241], allowing life to persist on internally warmed planets at late cosmic times. If life persisted at $z \lesssim 100$, it could have been transported to newly formed objects through panspermia. Under the assumption that interstellar panspermia is plausible, the redshift of $z \sim 100$ can be regarded as the earliest cosmic epoch after which life was possible in our Universe.

Finally, we note that an increase in the initial amplitude of density perturbations on the mass-scale of $10^4 M_\odot$ by a modest factor of $1.4 \times [(1 + z)/110]$ would have enabled star formation within the Hubble volume at redshifts $(1 + z) > 110$ even for perfectly Gaussian initial conditions.

12.3 CEMP Stars: Possible Hosts to Carbon Planets in the Early Universe

12.3.1 Section Background

The questions of when, where, and how the first planetary systems actually formed in cosmic history remain crucial to our understanding of structure formation and the emergence

---

[3] The dynamical time of galaxies is shorter than $\sim 1/\sqrt{200} = 7\%$ of the age of the Universe at any redshift since their average density contrast is $\gtrsim 200$. After the first stars formed, the subsequent delay in producing heavy elements from the first supernovae could have been as short as a few Myr. The supernova ejecta could have produced high-metallicity islands that were not fully mixed with the surrounding primordial gas, leading to efficient formation of rocky planets within them.

of life in the early Universe [159]. The short-lived, metal-free, massive first-generation stars ultimately exploded as supernovae (SNe) and enriched the interstellar medium (ISM) with the heavy elements fused in their cores. The enrichment of gas with metals that had otherwise been absent in the early Universe enabled the formation of the first low-mass stars and perhaps marked the point at which star systems could begin to form planets [42, 57, 89]. In the core accretion model of planet formation (e.g., [123, 201]), elements heavier than hydrogen and helium are necessary not only to form the dust grains that are the building blocks of planetary cores but also to extend the lifetime of the protostellar disc long enough to allow the dust grains to grow via merging and accretion to form planetesimals [80, 126, 138, 271].

In the past four decades, a broad search has been launched for low-mass Population II stars in the form of extremely metal-poor sources within the halo of the Galaxy. The HK survey [23], the Hamburg/ESO Survey [55, 268], the Sloan Digital Sky Survey (SDSS; [274]), and the SEGUE survey [270] have all significantly enhanced the sample of metal-poor stars with [Fe/H] < −2.0. Although these iron-poor stars are often referred to in the literature as 'metal-poor' stars, it is critical to note that [Fe/H] does not necessarily reflect a stellar atmosphere's total metal content. The equivalence between 'metal-poor' and 'iron-poor' appears to fall away for stars with [Fe/H] < −3.0 since many of these stars exhibit large overabundances of elements such as C, N, and O; the total mass fractions, Z, of the elements heavier than He are therefore not much lower than the solar value in these iron-poor stars.

Carbon-enhanced metal-poor (CEMP) stars comprise one such chemically anomalous class of stars, with carbon-to-iron ratios [C/Fe] ≥ 0.7 (as defined in [11, 45, 196]). The fraction of sources that fall into this category increases from ~15%–20% for stars with [Fe/H] < −2.0, to 30% for [Fe/H] < −3.0, to ~75% for [Fe/H] < −4.0 [22, 88, 196]. Furthermore, the degree of carbon enhancement in CEMP stars has been shown to notably increase as a function of decreasing metallicity, rising from [C/Fe] ~ 1.0 at [Fe/H] = −1.5 to [C/Fe] ~ 1.7 at [Fe/H] = −2.7. [45]. Given the significant frequency and level of carbon excess in this subset of metal-poor Population II stars, the formation of carbon planets around CEMP stars in the early Universe presents itself as an intriguing possibility.

From a theoretical standpoint, the potential existence of carbon exoplanets, consisting of carbides and graphite instead of Earth-like silicates, has been suggested by Reference [143]. Using the various elemental abundances measured in planet-hosting stars, subsequent works have sought to predict the corresponding variety of terrestrial exoplanet compositions expected to exist [34, 47, 48]. Assuming that the stellar abundances are similar to those of the original circumstellar disc, related simulations yield planets with a whole range of compositions, including some that are almost exclusively C and SiC; these occur in discs with C/O > 0.8, favourable conditions for carbon condensation [144]. Observationally, there have also been indications of planets with carbon-rich atmospheres – e.g., WASP-12b [170] and carbon-rich interiors – e.g., 55 Cancri e [171].

In this section, we explore the possibility of carbon planet formation around the iron-deficient but carbon-rich subset of low-mass stars, mainly CEMP stars. Standard definitions of elemental abundances and ratios are adopted. For element X, the logarithmic absolute abundance is defined as the number of atoms of element X per 10^{12} hydrogen atoms, $\log \epsilon(X) = \log_{10}(N_X/N_Y) + 12.0$. For elements X and Y, the logarithmic abundance ratio relative to the solar ratio is defined as $[X/Y] = \log_{10}(N_X/N_Y) - \log_{10}(N_X/N_Y)_\odot$. The solar abundance set is that of Reference [14], with a solar metallicity $Z_\odot = 0.0134$.

12.3.2 Star-Forming Environment of CEMP Stars

A great deal of effort has been directed in the literature towards theoretically understanding, the origin of the most metal-poor stars and, in particular, the large fraction that is carbon-rich. These efforts have been further perturbed by the fact that CEMP stars do not form a homogenous group but, rather, can be further subdivided into two main populations [22]: carbon-rich stars that show an excess of heavy neutron-capture elements (CEMP-s, CEMP-r, and CEMP-r/s), and carbon-rich stars with a normal pattern of the heavy elements (CEMP-no). In the following sections, we focus on stars with [Fe/H] ≤ -3.0, which have been shown to fall almost exclusively in the CEMP-no subset [10].

A number of theoretical scenarios have been proposed to explain the observed elemental abundances of these stars, though there is no universally accepted hypothesis. The most extensively studied mechanism to explain the origin of CEMP-no stars is the mixing and fallback model, where a 'faint' Population III SN explodes but, due to a relatively low explosion energy, only ejects its outer layers, rich in lighter elements (up to magnesium); its innermost layers, rich in iron and heavier elements, fall back onto the remnant and are not recycled in the ISM [253, 254]. This potential link between primeval SNe and CEMP-no stars is supported by recent studies which demonstrate that the observed ratio of carbon-enriched to carbon-normal stars with [Fe/H] < -3.0 is accurately reproduced if SNe were the main source of metal enrichment in the early Universe [60, 65]. Furthermore, the observed abundance patterns of CEMP-no stars have been found to be generally well matched by the nucleosynthetic yields of primordial faint SNe [35, 121, 122, 125, 133, 177, 178, 248, 254, 272]. These findings suggest that most of the CEMP-no stars were probably born out of gas enriched by massive, first-generation stars that ended their lives as Type II SNe with low levels of mixing and a high degree of fallback.

Under such circumstances, the gas clouds which collapse and fragment to form these CEMP-no stars and their protostellar discs may contain significant amounts of carbon dust grains. Observationally, dust formation in SNe ejecta has been inferred from isotopic anomalies in meteorites where graphite, SiC, and Si_3N_4 dust grains have been identified as SNe condensates [278]. Furthermore, in situ dust formation has been unambiguously detected in the expanding ejecta of SNe such as SN 1987A [119, 167] and SN 1999em [79]. The existence of cold dust has also been verified in the supernova remnant of Cassiopeia A by SCUBA's recent submillimetre observations, and a few solar masses worth of dust is estimated to have condensed in the ejecta [75].

Theoretical calculations of dust formation in primordial core-collapsing SNe have demonstrated the condensation of a variety of grain species, starting with carbon, in the ejecta, where the mass fraction tied up in dust grains grows with increasing progenitor mass [139, 197, 247]. References [177, 178] consider, in particular, dust formation in weak Population III SNe ejecta, the type believed to have polluted the birth clouds of CEMP-no stars. Tailoring the SN explosion models to reproduce the observed elemental abundances of CEMP-no stars, they find the following. (1) For all the progenitor models investigated, amorphous carbon (AC) is the only grain species that forms in significant amounts; this is a consequence of extensive fallback, which results in a distinct, carbon-dominated ejecta composition with negligible amounts of other metals – such as Mg, Si, and Al – that can enable the condensation of alternative grain types. (2) The mass of carbon locked into AC grains increases when the ejecta composition is characterised by an initial mass of C greater than the O mass; this is particularly true in zero-metallicity supernova progenitors, which undergo less mixing than their solar metallicity counterparts [125]. In their stratified ejecta, C-grains are found only to form in layers where C/O > 1. In layers where C/O < 1, all the carbon is promptly locked in CO molecules. (3) Depending on the model, the mass fraction of dust (formed in SNe ejecta) that survives the passage of a SN reverse shock ranges between 1% and 85%. This percentage is referred to as the carbon condensation efficiency. (4) Further grain growth in the collapsing birth clouds of CEMP-no stars, due to the accretion of carbon atoms in the gas phase onto the remaining grains, occurs only if C/O > 1 and is otherwise hindered by the formation of CO molecules.

Besides the accumulation of carbon-rich grains imported from the SNe ejecta, Fischer-Tropsch-type reactions (FTTs) may also contribute to solid carbon enrichment in the protostellar discs of CEMP-no stars by enabling the conversion of nebular CO and H_2 to other forms of carbon [141]. Furthermore, in carbon-rich gas, the equilibrium condensation sequence changes signifcantly from the sequence followed in solar composition gas where metal oxides condense first. In nebular gas with C/O $\gtrsim$ 1, carbon-rich compounds such as graphite, carbides, nitrides, and sulfides are the highest-temperature condensates ($T \approx$ 1,200–1,600 K) [144]. Thus, if planet formation is to proceed in this C-rich gas, the protoplanetary discs of these CEMP-no stars may spawn many carbon planets.

12.3.3 *Orbital Radii of Potential Carbon Planets*

Given the significant abundance of carbon grains, both imported from SNe ejecta and produced by equilibrium and non-equilibrium mechanisms operating in the carbon-rich protoplanetary discs, the emerging question is, would these dust grains have enough time to potentially coagulate and form planets around their host CEMP-no stars?

In the core accretion model, terrestrial planet formation is a multistep process, starting with the aggregation and settling of dust grains in the protoplanetary disc [12, 21, 123, 155, 191, 201]. In this early stage, high densities in the disc allow particles to grow from submicron size to metre size through a variety of collisional processes including Brownian motion, settling, turbulence, and radial migration. The continual growth of such aggregates by coagulation and sticking eventually leads to the formation of kilometre-sized planetes-imals, which then begin to interact gravitationally and grow by pairwise collisions and, later, by runaway growth [155]. In order for terrestrial planets to ultimately form, these processes must all occur within the lifetime of the disc itself, a limit which is set by the relevant timescale of the physical phenomena that drive disc dissipation.

A recent study by Reference [271] of clusters in the Extreme Outer Galaxy (EOG) provides observational evidence that low-metallicity discs have shorter lifetimes (< 1 Myr) compared to solar metallicity discs ($\sim$ 5–6 Myr). This finding is consistent with models in which photoevaporation by energetic (ultraviolet or X-ray) radiation of the central star is the dominant disc dispersal mechanism. While the opacity source for EUV (extreme-ultraviolet) photons is predominantly hydrogen and is thus metallicity independent, X-ray photons are primarily absorbed by heavier elements, mainly carbon and oxygen, in the inner gas and dust shells. Therefore, in low-metallicity environments where these heavy elements are not abundant and the opacity is reduced, high-density gas at larger columns can be ionised and will experience a photoevaporative flow if heated to high enough temperatures [80, 100].

Assuming that photoevaporation is the dominant mechanism through which circumstellar discs lose mass and eventually dissipate, we adopt the metallicity-dependent disc lifetime, derived in Reference [80] using X-ray+EUV models [81],

$$t_{disc} \propto Z^{0.77(4-2p)/(5-2p)}, \tag{12.3}$$

where Z is the total metallicity of the disc and p is the power-law index of the disc surface density profile ($\Sigma \propto r^{-p}$). A mean power-law exponent of $p \sim 0.9$ is derived by modelling the spatially resolved emission morphology of young stars at (sub)millimetre wavelengths [7, 8], and the timescale is normalised such that the mean lifetime for discs of solar metallicity is 2 Myr [80]. Thus, the disc lifetime is dominated by carbon dust grains in the CEMP-no stars considered here. We adopt the carbon abundance relative to solar [C/H] as a proxy for the overall metallicity Z We adopt the carbon abundance relative to solar [C/H] as a proxy for the overall metallicity Z, since the opacity – which largely determines the photoevaporation rate and, thus, the disc lifetime – is dominated by carbon dust grains in the CEMP-no stars we consider in this work.

The timescale for planet formation is believed to be effectively set by the time it takes dust grains to settle into the disc midplane. The subsequent process of runaway planetesimal formation, possibly occurring via a series of pairwise collisions, must be quick because, otherwise, the majority of the solid disc material would radially drift towards the host star and evaporate in the hot inner regions of the circumstellar disc [12]. We adopt the one-particle model of Reference [74] to follow the mass growth of dust grains via collisions as they fall through and sweep up the small grains suspended in the disc. Balancing the gravitational force felt by a small dust particle at height z above the midplane of a disc with the aerodynamic drag (in the Epstein regime) gives a dust-settling velocity of

$$v_{sett} = \frac{dz}{dt} = \frac{3\Omega_K^2 zm}{4\rho c_s \sigma_d}, \tag{12.4}$$

where $\sigma_d = \pi a^2$ is the cross section of the dust grain with radius a and $c_s = \sqrt{k_B T(r)/\mu m_H}$ is the isothermal sound speed with m_H being the mass of a hydrogen atom and $\mu = 1.36$ being the mean molecular weight of the gas (including the contribution of helium). $\Omega_K = \sqrt{GM_*/r^3}$ is the Keplerian velocity of the disc at a distance r from the central star of mass M_*, which we take to be $M_* = 0.8\,M_\odot$ as representative of the low masses associated with CEMP-no stars [53, 88]. The disc is assumed to be in hydrostatic equilibrium with a density given by

$$\rho(z,r) = \frac{\Sigma(r)}{h\sqrt{2\pi}} \exp\left(-\frac{z^2}{2h^2}\right), \tag{12.5}$$

where the disc scale height is $h = c_s/\Omega_k$. For the disc surface density $\Sigma(r)$ and temperature $T(r)$ profiles, we adopt the radial power-law distributions fitted to (sub-)millimetre observations of circumstellar discs around young stellar objects [6–8],

$$T(r) = 200\text{ K} \left(\frac{r}{1\text{ AU}}\right)^{-0.6} \tag{12.6}$$

$$\Sigma(r) = 10^3\text{ g/cm}^2 \left(\frac{r}{1\text{ AU}}\right)^{-0.9}. \tag{12.7}$$

Although these relations were observationally inferred from discs with solar-like abundances, we choose to rely on them for our purposes given the lack of corresponding measurements for discs around stars with different abundance patterns.

The rate of grain growth, dm/dt, is determined by the rate at which grains, subject to small-scale Brownian motion, collide and stick together as they drift towards the disc midplane through a sea of smaller solid particles. If coagulation results from every collision, then the mass growth rate of a particle is effectively the amount of solid material in the volume swept out the particle's geometric cross section,

$$\frac{dm}{dt} = f_{dg}\rho\sigma_d\left(v_{rel} + \frac{dz}{dt}\right) \tag{12.8}$$

where dz/dt is the dust-settling velocity given by Equation (12.3) and

$$v_{rel} = \sqrt{\frac{8k_B T(m_1 + m_2)}{\pi m_1 m_2}} \approx \sqrt{\frac{8k_B T}{\pi m}} \tag{12.9}$$

is the relative velocity in the Brownian motion regime between grains with masses $m_1 = m_2 = m$. To calculate the dust-to-gas mass ratio in the disc f_{dg}, we follow the approach in Reference [124] and relate two expressions for the mass fraction of C: (1) the fraction of carbon in the dust, $f_{dg} M_{C,dust}/M_{dust}$, where M_{dust} is the total dust mass and $M_{C,dust}$ is the carbon dust mass ; and (2) the fraction of carbon in the gas, $\mu_C n_C/\mu n_H$, where μ_C is the molecular weight of carbon ($\sim 12m_p$) and n_C and n_H are the carbon and hydrogen number densities, respectively.

We then assume that a fraction f_{cond} (referred to from now on as the carbon condensation efficiency) of all the carbon present in the gas cloud is locked up in dust, such that

$$f_{cond} \frac{\mu_C n_C}{\mu n_H} = f_{dg} \frac{M_{C,dust}}{M_{dust}}. \tag{12.10}$$

Since faint Population III SNe are believed to have polluted the birth clouds of CEMP-no stars, and the only grain species that forms in non-negligible amounts in these ejecta is amorphous carbon [177, 178], we set $M_{dust} = M_{C,dust}$. Rewriting Eq. (12.9) in terms of abundances relative to the Sun, we obtain

$$f_{dg} = f_{cond} \frac{\mu_C}{\mu} 10^{[C/H] + \log \epsilon(C)_\odot - 12}, \tag{12.11}$$

where $\log \epsilon(C)_\odot = 8.43 \pm 0.05$ [14] is the solar carbon abundance.

For a specified metallicity [C/H] and radial distance r from the central star, we can then estimate the time it takes for dust grains to settle in the disc by integrating Eqs. (3.2) and (3.6) from an initial height of $z(t=0) = 4h$ with an initial dust grain mass of $m(t=0) = 4\pi a_{init}^3 \rho_d / 3$. The specific weight of dust is set to $\rho_d = 2.28$ g cm^{-3}, reflecting the material density of carbon grains expected to dominate the circumstellar discs of CEMP-no stars. The initial grain size a_{init} is varied between 0.01 and 1 μm to reflect the range of characteristic radii of carbon grains found when modelling CEMP-no star abundance patterns [177]. Comparing the resulting dust-settling timescale to the disc lifetime given by Eq. 12.1 for the specified metallicity, we can then determine whether there is enough time for carbon dust grains to settle in the midplane of the disc and there undergo runaway planetesimal formation before the disc is dissipated by photoevaporation. For the purposes of this simple model, we neglected possible turbulence in the disc which may counteract the effects of vertical settling, propelling particles to higher altitudes and thus preventing them from fully settling into the disc midplane [12]. We have also not accounted for the effects of radial drift, which may result in the evaporation of solid material in the hot inner regions of the circumstellar disc.

As the dust-settling timescale is dependent on the disc surface density $\Sigma(r)$ and temperature $T(r)$, we find that for a given metallicity, [C/H], there is a maximum distance r_{max} from the central star out to which planetesimal formation is possible. At larger distances from the host star, the dust-settling timescale exceeds the disc lifetime, and so carbon planets with semimajor axes $r > r_{max}$ are not expected to form. A plot of the maximum semimajor axis expected for planet formation around a CEMP-no star as a function of the carbon abundance relative to the Sun [C/H] is shown in Figure 12.1 for carbon condensation efficiencies ranging between $f_{cond} = 0.1$ and 1. As discovered in Reference [127], where the critical iron abundance for terrestrial planet formation is considered as a function of the distance from the host star, we find a linear relation between [C/H] and r_{max},

$$[C/H] = \log\left(\frac{r_{max}}{1 \text{ AU}}\right) - \alpha, \tag{12.12}$$

where $\alpha = 1.3, 1.7$, and 1.9 for $f_{cond} = 0.1, 0.5$, and 1, respectively, assuming an initial grain size of $a_{init} = 0.1\mu$m. These values for α change by less than 1% for smaller initial

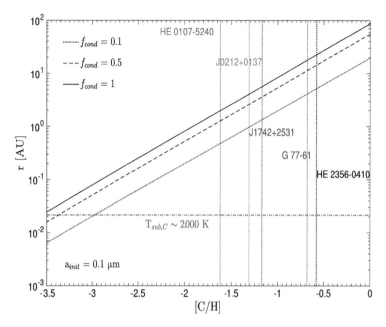

Figure 12.1 The maximum distance r_{max} from the host star out to which planetesimal formation is possible as a function of the star's metallicity, expressed as the carbon abundance relative to that of the Sun, [C/H]. The dotted, dashed, and solid black curves correspond to the results obtained assuming carbon condensation efficiencies of 10%, 50%, and 100%, respectively, and an initial grain size of $a_{init} = 0.1$ μm. The grey dash-dotted curve corresponds to the distance at which the disc temperature approaches the sublimation temperature of carbon dust grains, $T_{sub,C} \sim 2,000$ K; the formation of carbon planetesimals will therefore be suppressed at distances that fall below this line, $r \lesssim 0.02$ AU. The coloured vertical lines represent various observed CEMP stars with measured carbon abundances, [C/H].

grain sizes, $a_{init} = 0.01$ μm, and by no more than 5% for larger initial grain sizes $a_{init} = 1$ μm; given this weak dependence on a_{init}, we only show our results for a single initial grain size of $a_{init} = 0.1$ μm. The distance from the host star at which the temperature of the disc approaches the sublimation temperature of carbon dust, $T_{sub,C} \sim 2,000$ K [135], is depicted as well (dash-dotted grey curve). At distances closer to the central star than $r \simeq 0.02$ AU, temperatures well exceed the sublimation temperature of carbon grains; grain growth and subsequent carbon planetesimal formation are therefore quenched in this inner region.

Figure 12.1 shows lines representing various observed CEMP stars with measured carbon abundances – mainly HE 0107-5240 [53, 54], SDSS J0212+0137 [35], SDSS J1742+2531 [35], G 77-61 [24, 62, 213], and HE 2356-0410 [195, 223]. These stars all have iron abundances (relative to solar) [Fe/H] < −3.0, carbon abundances (relative to solar) [C/Fe] > 2.0, and carbon-to-oxygen ratios C/O > 1. This latter criteria maximises the abundance of solid carbon available for planet formation in the circumstellar discs by

Table 12.1 *Basic data^a for CEMP stars considered in this section.*

Star	log g^b	[Fe/H]	[C/Fe]	C/O^c	Source
HE 0107-5240	2.2	−5.44	3.82	14.1	[54, 58]
SDSS J0212+0137	4.0	−3.57	2.26	2.6	[35]
SDSS J1742+2531	4.0	−4.77	3.60	2.2	[35]
G 77-61	5.1	−4.03	3.35	12.0	[24, 213]
HE 2356-0410^d	2.65	−3.19	2.61	>14.1	[223]

^{a} Abundances based on one-dimensional LTE model-atmosphere analyses
^{b} Logarithm of the gravitational acceleration at the surface of stars expressed in cm s^{-2}
^{c} C/O = N_C/N_O = $10^{[C/O]+\log \epsilon (C)_\odot - \log \epsilon (O)_\odot}$
^{d} CS 22957-027

optimising carbon grain growth both in stratified SNe ejecta and later, in the collapsing molecular birth clouds of these stars. It also advances the possibility of carbon planet formation by ensuring that planet formation proceeds by a carbon-rich condensation sequence in the protoplanetary disc. SDSS J0212+0137 and HE 2356-0410 have both been classified as CEMP-no stars, with measured barium abundances [Ba/Fe] < 0 (as defined in [22]); the other three stars are barium indeterminate with only high upper limits on [Ba/Fe], but are believed to belong to the CEMP-no subclass given their light-element abundance patterns. The carbon abundance, [C/H], dominates the total metal content of the stellar atmosphere in these five CEMP objects, contributing more than 60% of the total metallicity in these stars. A summary of the relevant properties of the CEMP stars considered in this analysis can be found in Table 12.1. We find that carbon planets may be orbiting iron-deficient stars with carbon abundances [C/H] $\sim$ −0.6, such as HE 2356-0410, as far out as $\sim$ 20 AU from their host star in the case where $f_{cond} = 1$. Planets forming around stars with less carbon enhancement – i.e., HE 0107-5240 with [C/H] $\sim$ −1.6 are expected to have more compact orbits, with semimajor axes $r < 2$ AU. If the carbon condensation efficiency is only 10%, the expected orbits grow even more compact, with maximum semimajor axes of $\sim$ 5 and 0.5 AU, respectively.

12.3.4 Mass-Radius Relationship for Carbon Planets

Next we present the relationship between the mass and radius of carbon planets that we have shown may theoretically form around CEMP-no stars. These mass-radius relations have already been derived in the literature for a wide range of rocky and icy exoplanet compositions [86, 148, 230, 256, 277]. Here, we follow the approach of Reference [277] and solve the three canonical equations of internal structure for solid planets:

1. Mass conservation

$$\frac{dm(r)}{dr} = 4\pi r^2 \rho(r) \tag{12.13}$$

2. Hydrostatic equilbrium

$$\frac{dP(r)}{dr} = -\frac{Gm(r)\rho(r)}{r^2}$$ (12.14)

3. The equation of state (EOS)

$$P(r) = f(\rho(r), T(r))$$ (12.15)

$m(r)$ is the mass contained within radius r, $P(r)$ is the pressure, $\rho(r)$ is the density of the spherical planet, and f is the unique equation of state (EOS) of the material of interest – in this case, carbon.

Carbon grains in circumstellar discs most likely experience many shock events during planetesimal formation which may result in the modification of their structure. The coagulation of dust into clumps, the fragmentation of the disc into clusters of dust clumps, the merging of these clusters into ~ 1 km planetesimals, the collision of planetesimals during the accretion of meteorite parent bodies, and the subsequent collision of the parent bodies after their formation all induce strong shock waves that are expected to chemically and physically alter the materials. Subject to these high temperatures and pressures, the amorphous carbon grains polluting the protoplanetary discs around CEMP stars are expected to undergo graphitisation and may even crystallise into diamond [202, 243, 246]. In our calculations, the equation of state at low pressures, $P \leq 14$ GPa, is set to the third-order finite strain Birch-Murnagham EOS (BME; [30]) for graphite,

$$P = \frac{3}{2}K_0\left(\eta^{7/3} - \eta^{5/3}\right)\left[1 + \frac{3}{4}\left(K_0' - 4\right)\left(\eta^{2/3} - 1\right)\right],$$ (12.16)

where $\eta = \rho/\rho_0$ is the compression ratio with respect to the ambient density, ρ_0, K_0 is the bulk modulus of the material, and K_0' is the pressure derivative. Empirical fits to experimental data yield a BME EOS of graphite ($\rho_0 = 2.25$ g cm^{-3}) with parameters $K_0 = 33.8$ GPa and $K_0' = 8.9$ [109]. At 14 GPa, we incorporate the phase transition from graphite to diamond [109, 192] and adopt the Vinet EOS [262, 263],

$$P = 3K_0\eta^{2/3}\left(1 - \eta^{-1/3}\right)\exp\left[\frac{3}{2}\left(K_0' - 1\right)\left(1 - \eta^{-1/3}\right)\right]$$ (12.17)

with $K_0 = 444.5$ GPa and $K_0' = 4.18$ empirically fit for diamond, $\rho_0 = 3.51$ g cm^{-3} [66]. (As pointed out in Reference [230], the BME EOS is not fit to be extrapolated to high pressures since it is derived by expanding the elastic potential energy as a function of pressure keeping only the lowest-order terms.) Finally, at pressures $P \gtrsim 1,300$ GPa, where electron degeneracy becomes increasingly important, we use the Thomas-Fermi-Dirac (TFD) theoretical EOS ([227]; equations (40)–(49)), which intersects the diamond EOS at $P \sim 1,300$ GPa. Given that the full temperature-dependent carbon EOSs are either undetermined or dubious at best, all three EOSs adopted in this section are room-temperature EOSs for the sake of practical simplification.

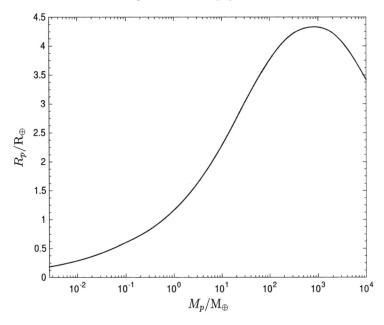

Figure 12.2 Mass-radius relation for solid homogenous, pure carbon planet.

Using a fourth-order Runge-Kutta scheme, we solve the system of equations simultaneously, numerically integrating Eqs. (12.13) and (12.14) begining at the planet's centre with the inner boundary conditions $M(r=0)=0$ and $P(r=0)=P_{central}$, where $P_{central}$ is the central pressure. The outer boundary condition $P(r=R_p)=0$ then defines the planetary radius R_p and total planetary mass $M_p=m(r=R_p)$. Integrating these equations for a range of $P_{central}$, with the appropriate EOS, $P=P(\rho)$, to close the system of equations, yields the mass-radius relationship for a given composition. We show this mass-radius relation for a purely solid carbon planet in Figure 12.2. We find that for masses $M_p \lesssim 800\ M_\oplus$, gravitational forces are small compared with electrostatic Coulomb forces in hydrostatic equilibrium, and so the planet's radius increases with increasing mass, $R_p \propto M_p^{1/3}$. However, at larger masses, the electrons are pressure ionised, and the resulting degeneracy pressure becomes significant, causing the planet radius to become constant and even decrease for increasing mass, $R_p \propto M_p^{-1/3}$ [117]. Planets which fall within the mass range $500 \lesssim M_p \lesssim 1,300\ M_\oplus$, where the competing effects of Coulomb forces and electron degeneracy pressure cancel each other out, are expected to be approximately the same size, with $R_p \simeq 4.3\ R_\oplus$, the maximum radius of a solid carbon planet. (In the case of gas giants, the planet radius can increase due to accretion of hydrogen and helium.)

Although the mass-radius relation illustrated in Figure 12.2 may alone not be enough to confidently distinguish a carbon planet from a water or silicate planet, the unique spectral features in the atmospheres of these carbon planets may provide the needed fingerprints. At high temperatures ($T \gtrsim 1,000$ K), the absorption spectra of massive ($M \sim 10$–$60\ M_\oplus$)

carbon planets are expected to be dominated by CO, in contrast with the H_2O-dominated spectra of hot massive planets with solar-composition atmospheres [143]. The atmospheres of low-mass ($M \lesssim 10$ $M_\oplus$) carbon planets are also expected to be differentiable from their solar-composition counterparts due to their abundance of CO and CH_4 and lack of oxygen-rich gases like CO_2, O_2, and O_3 [143]. Furthermore, carbon planets of all masses at low temperatures are expected to accommodate hydrocarbon synthesis in their atmospheres; stable long-chain hydrocarbons are, therefore, another signature feature that can help observers distinguish the atmospheres of cold carbon planets and more confidently determine the bulk composition of a detected planet [143].

The detection of theoretically proposed carbon planets around CEMP stars will provide us with significant clues regarding how early planet formation may have started in the Universe. While direct detection of these extrasolar planets remains difficult given the low luminosity of most planets, techniques such as the transit method are often employed to indirectly spot exoplanets and determine physical parameters of the planetary system. When a planet 'transits' in front of its host star, it partially occludes the star and causes its observed brightness to drop by a minute amount. If the host star is observed during one of these transits, the resulting dip in its measured light curve can yield information regarding the relevant sizes of the star and the planet, the orbital semimajor axis, and the orbital inclination, among other characterising properties.

12.3.5 Section Summary and Implications

We explored the possibility of carbon planet formation around the iron-deficient, carbon-rich subset of low-mass stars known as CEMP stars. The observed abundance patterns of CEMP-no stars suggest that these stellar objects were probably born out of gas enriched by massive first-generation stars that ended their lives as Type II SNe with low levels of mixing and a high degree of fallback. The formation of dust grains in the ejecta of these primordial core-collapsing SNe progenitors has been observationally confirmed and theoretically studied. In particular, amorphous carbon is the only grain species found to condense and form in non-negligible amounts in SN explosion models that are tailored to reproduce the abundance patterns measured in CEMP-no stars. Under such circumstances, the gas clouds which collapse and fragment to form CEMP-no stars and their protoplanetary discs may contain significant amounts of carbon dust grains imported from SNe ejecta. The enrichment of solid carbon in the protoplanetary discs of CEMP stars may then be further enhanced by Fischer-Tropsch-type reactions and carbon-rich condensation sequences, where the latter occurs specifically in nebular gas with C/O $\gtrsim$ 1.

For a given metallicity [C/H] of the host CEMP star, the maximum distance out to which planetesimal formation is possible can then be determined by comparing the dust-settling timescale in the protostellar disc to the expected disc lifetime. Assuming that disc dissipation is driven by a metallicity-dependent photoevaporation rate, we find a linear relation between [C/H] and the maximum semimajor axis of a carbon planet orbiting its host CEMP star. Very carbon-rich CEMP stars, such as G 77-61 and HE 2356-0410 with [C/H] $\simeq$ -0.7--0.6, can host carbon planets with semimajor axes as large $\sim$ 20 AU for 100% carbon condensation efficiencies; this maximum orbital distance reduces to $\sim$ 5 AU when the condensation efficiency drops by an order of magnitude. In the case of the observed CEMP-no stars HE 0107-5240, SDSS J0212+0137, and SDSS J1742+2531, where the carbon abundances are in the range [C/H] $\simeq$ -1.6--1.2, we expect more compact orbits, with maximum orbital distances $r_{max} \simeq$ 2, 4, and 6 AU, respectively, for $f_{cond} = 1$ and $r_{max} \simeq 0.5 - 1$ AU for $f_{cond} = 0.1$.

While the shallow transit depths of Earth-mass carbon planets around HE 0107-5240 and HE 2356-0410 may evade detection, current and future space-based transit surveys promise to achieve the precision levels ($\sim$ 0.001%) necessary to detect planetary systems around CEMP stars such as SDSS J0212+0137, SDSS J1742+2531, and G 77-61. If gas giant (Jupiter-scale) planets form around CEMP stars, their transits would be much easier to detect than rocky planets. However, they are not likely to host life as we know it. There are a number of ongoing, planned, and proposed space missions committed to this cause, including CoRot (COnvection ROtation and planetary Transits), Kepler, PLATO (PLAnetary Transits and Oscillations of stars), TESS (Transiting Exoplanet Survey Satellite), and ASTrO (All Sky Transit Observer), which are expected to achieve precisions as low as 20–30 ppm (parts per million) [26, 267].

Short orbital periods and long transit durations are also key ingredients in boosting the probability of transit detection by observers. G 77-61 is not an optimal candidate in these respects since, given its large carbon abundance ([C/Fe] $\sim$ 3.4), carbon planets may form out to very large distances and take up to a century to complete an orbit around the star for $f_{cond} = 1$ ($P_{max} \sim 10$ years for 10% carbon condensation efficiency). The small stellar radius, $R_* \sim 0.5$ $R_\odot$, also reduces chances of spotting the transit since the resulting transit duration is only $\sim$ 30 hours at most. Carbon planets around larger CEMP stars with an equally carbon-rich protoplanetary disc, such as HE 2356-0410 ($R_* \sim 7$ $R_\odot$), have a better chance of being spotted, with transit durations lasting up to $\sim$ 3 weeks. The CEMP-stars SDSS J0212+0137, and SDSS J1742+2531 are expected to host carbon planets with much shorter orbits, $P_{max} \sim 16$ years for 100% condensation efficiency ($P_{max} \sim 1$ year for $f_{cond} = 0.1$), and transit durations that last as long as $\sim$ 60 hours. If the ability to measure transit depths improves to a precision of 1 ppm, then potential carbon planets around HE 0107-5240 are the most likely to be spotted (among the group of CEMP-no stars considered here), transiting across the host star at least once every $\sim$ 5 months (10% condensation efficiency) with a transit duration of 6 days.

While our calculations place upper bounds on the distance from the host star out to which carbon planets can form, we note that orbital migration may alter a planet's location in the circumstellar disc. As implied by the existence of 'hot Jupiters', it is possible for a protoplanet that forms at radius r to migrate inward through gravitational interactions with other protoplanets, resonant interactions with planetesimals with more compact orbits, or tidal interactions with gas in the surrounding disc [201]. Since Figure 12.1 only plots r_{max}, the *maximum* distance out to which a carbon planet with [C/H] can form, our results remain consistent in the case of an inward migration. However, unless planets migrate inwards from their place of birth in the disc, we do not expect to find carbon exoplanets orbiting closer than $r \simeq 0.02$ AU from the host stars since at such close proximities, temperatures are high enough to sublimate carbon dust grains.

Protoplanets can also be gravitationally scattered into wider orbits through interactions with planetesimals in the disc [107, 260]. Such an outward migration of carbon planets may result in observations that are inconsistent with the curves in Figure 12.1. A planet that formed at radius $r \ll r_{max}$ still has room to migrate outwards without violating the 'maximum distance' depicted in Figure 12.1.

Detection of the carbon planets that we suggest may have formed around CEMP stars will provide us with significant clues regarding how planet formation may have started in the early Universe. The formation of planetary systems not only signifies an increasing degree of complexity in the young Universe, but it also carries implications for the development of life at this early junction [159]. The lowest metallicity multi-companion system detected to date is around BD+20 24 57, a K2-giant with [Fe/H] = -1.0 [194], a metallicity value once believed to yield low efficiency for planet formation [99, 209]. More recent formulations of the metallicity required for planet formation are consistent with this observation, estimating that the first Earth-like planets likely formed around stars with metallicities [Fe/H] $\lesssim -1.0$ [127]. The CEMP stars considered in this section are extremely iron deficient, with [Fe/H] $\lesssim -3.2$, and yet, given the enhanced carbon abundances which dominate the total metal content in these stars ([C/H] $\gtrsim -1.6$), the formation of solid carbon exoplanets in the protoplanetary discs of CEMP stars remains a real possibility.

An observational program aimed at searching for carbon planets around these low-mass Population II stars could therefore potentially shed light on the question of how early planets and, subsequently, life could have formed after the Big Bang.

12.4 Water Formation during the Epoch of First Metal Enrichment

12.4.1 Section Background

Water is an essential ingredient for life as we know it [130]. In the interstellar medium (ISM) of the Milky Way and also in external galaxies, water has been observed in the gas phase and as grain surface ice in a wide variety of environments. These environments include diffuse and dense molecular clouds, photon-dominated regions (PDRs), shocked gas, protostellar envelopes, and discs (see review in Reference [258]).

In diffuse and translucent clouds, H_2O is formed mainly in gas-phase reactions via ion-molecule sequences [111]. The ion-molecule reaction network is driven by cosmic-ray or X-ray ionisation of H and H_2, which leads to the formation of H^+ and H_3^+ ions. These interact with atomic oxygen and form OH^+. A series of abstractions then lead to the formation of H_3O^+, which makes OH and H_2O through dissociative recombination.

This formation mechanism is generally not very efficient, and only a small fraction of the oxygen is converted into water; the rest remains in atomic form or freezes out as water ice [114].

Reference [237] showed that the abundance of water vapor within diffuse clouds in the Milky Way galaxy is remarkably constant, with $x_{H_2O} \sim 10^{-8}$, which is $\sim 0.1\%$ of the available oxygen. Here, x_{H_2O} is the H_2O number density relative to the total hydrogen nuclei number density. Towards the galactic centre this value can be enhanced by up to a factor of ~ 3 [186, 238].

At temperatures $\gtrsim 300$ K, H_2O may form directly via the neutral-neutral reactions, $O + H_2 \rightarrow OH + H$, followed by $OH + H_2 \rightarrow H_2O + H$. This formation route is particularly important in shocks, where gas heats up to high temperatures and can drive most of the oxygen into H_2O [72, 132].

Temperatures of a few hundreds K are also expected in very-low-metallicity gas environments, with elemental oxygen and carbon abundances of $\lesssim 10^{-3}$ solar [42, 98, 199], associated with the epochs of the first enrichment of the ISM with heavy elements, in the first generation of galaxies at high redshifts [160]. At such low metallicities, cooling by fine-structure transitions of metal species such as the [CII] 158 μm line, and by rotational transitions of heavy molecules such as CO, becomes inefficient and the gas remains warm.

Could the enhanced rate of H_2O formation via the neutral-neutral sequence in such warm gas compensate for the low oxygen abundance at low metallicities?

Reference [199] studied the thermal and chemical evolution of collapsing gas clumps at low metallicities. They found that for models with gas metallicities of 10^{-3}–10^{-4} solar, x_{H_2O} may reach 10^{-8}, but only if the density, n, of the gas approaches 10^8 cm^{-3}. Photodissociation of molecules by far-ultraviolet (FUV) radiation was not included in their study. While at solar metallicity dust grains shield the interior of gas clouds from the FUV radiation, at low metallicities, photodissociation by FUV becomes a major removal process for H_2O. H_2O photodissociation produces OH, which is then itself photodissociated into atomic oxygen.

Reference [115] developed a theoretical model to study the abundances of various molecules, including H_2O, in PDRs. Their model included many important physical processes, such as freeze-out of gas species, grain surface chemistry, and also photodissociation by FUV photons. However, they focused on solar metallicity. Intriguingly, Reference [19] reports a water abundance close to 10^{-8} in the optically thick core of their single PDR model for a low metallicity of 10^{-2} (with $n = 10^3$ cm^{-3}). However, Bayet *et al.* did not investigate the effects of temperature and UV intensity variations on the water abundance in the low-metallicity regime.

Recently, a comprehensive study of molecular abundances for the bulk ISM gas as functions of the metallicity was studied in Reference [204] and Reference [28]; these models, however, focused on the 'low-temperature' ion-molecule formation chemistry.

In this section, we present results for the H_2O abundance in low-metallicity gas environments for varying temperatures, FUV intensities, and gas densities. We find that for temperatures T in the range 250–350 K, H_2O may be abundant, comparable to or higher than that found in diffuse galactic clouds, provided that the FUV intensity to density ratio is smaller than a critical value.

12.4.2 Model Ingredients

We calculate the abundance of gas-phase H_2O for low-metallicity gas parcels that are exposed to external FUV radiation and cosmic-ray and/or X-ray fluxes. Given our chemical network, we solve the steady-state rate equations using our dedicated Newton-based solver and obtain x_{H_2O} as function of T and the FUV intensity to density ratio.

We adopt a 10^5 K diluted black-body spectrum, representative of radiation produced by massive Population III stars. The photon density in the 6–13.6 eV interval, is $n_\gamma \equiv n_{\gamma,0} I_{UV}$, where $n_{\gamma,0} = 6.5 \times 10^{-3}$ photons cm^{-3} is the photon density in the interstellar radiation field [70], and I_{UV} is the 'normalised intensity'. Thus, $I_{UV} = 1$ corresponds to the FUV intensity in the Draine ISRF.

Cosmic-ray and/or X-ray ionisation drive the ion-molecule chemical network. We assume an ionisation rate per hydrogen nucleon ζ (s^{-1}). In the galaxy, [63] and References [120] showed that ζ lies within the relatively narrow range 10^{-15}–10^{-16} s^{-1}. We therefore introduce the 'normalised ionisation rate' $\zeta_{-16} \equiv (\zeta/10^{-16}$ s$^{-1})$. The ionisation rate and the FUV intensity are likely correlated, as both should scale with the formation rate of massive OB stars. We thus set $\zeta_{-16} = I_{UV}$ as our fiducial case but also consider the cases $\zeta_{-16} = 10^{-1} I_{UV}$ and $\zeta_{-16} = 10 I_{UV}$.

Dust shielding against the FUV radiation becomes ineffective at low metallicities. However, self-absorption in the H_2 lines may significantly attenuate the destructive Lyman-Werner (11.2–13.6 eV) radiation [71, 240], and high abundances of H_2 may be maintained even at low metallicity [28]. In the models presented here, we assume an H_2 shielding column of at least 5×10^{21} cm^{-2}. (For such conditions, CO is also shielded by the H_2.) The LW flux is then attenuated by a self-shielding factor of $f_{shield} \sim 10^{-8}$, and the H_2 photodissociation rate is only $5.8 \times 10^{-19} I_{UV}$ s^{-1}. With this assumption, H_2 photodissociation by LW photons is negligible compared to cosmic-ray and/or X-ray ionisation as long as $I_{UV} < 85 \zeta_{-16}$.

However, even when the Lyman-Werner band is fully blocked, OH and H_2O are photodissociated because their energy thresholds for photodissociation are 6.4 and 6 eV, respectively. For the low metallicities that we consider, photodissociation is generally the dominant removal mechanism for H_2O and OH. We adopt the calculated OH and H_2O photodissociation rates calculated by Reference [28].

We assume thermal and chemical steady states. In the Milky Way, the bulk of the ISM gas is considered to be at approximate thermal equilibrium, set by cooling and heating processes. We discuss the relevant chemical and thermal timescales in Section 12.4.4.

Given the before mentioned assumptions, the steady-state solutions for the species abundances depend on only two parameters, the temperature T and the intensity-to-density ratio I_{UV}/n_4. Here, $n_4 \equiv (n/10^4$ cm$^{-3})$ is the total number density of hydrogen nuclei normalised to typical molecular cloud densities. T and I_{UV}/n_4 form our basic parameter space in our study.

12.4.3 Chemical Model

We consider a chemical network of 503 two-body reactions, between 56 atomic, molecular, and ionic species of H, He, C, O, S, and Si. We assume cosmological elemental helium abundance of 0.1 relative to hydrogen (by number). For the metal elemental abundances, we adopt the photospheric solar abundances, multiplied by a metallicity factor Z' (i.e., $Z' = 1$ is solar metallicity). In our fiducial model, we assume $Z' = 10^{-3}$, but we also explore cases with $Z' = 10^{-2}$ and $Z' = 10^{-4}$. Since our focus here is on the very-low-metallicity regime, where dust grains play a lesser role, we neglect any depletion on dust grains and dust-grain chemistry (except for H_2, as discussed further below). Direct and induced ionisations and dissociations by the cosmic-ray/X-ray field ($\propto \zeta$) are included. For the gas-phase reactions, we adopt the rate coefficients given by the UMIST 2012 database [183].

The formation of heavy molecules relies on molecular hydrogen. We consider two scenarios for H_2 formation: (1) pure gas-phase formation and (2) gas-phase formation plus formation on dust grains. In the gas phase radiative-attachment

$$H + e \rightarrow H^- + \nu \tag{12.18}$$

followed by associative-detachment

$$H^- + H \rightarrow H_2 + e \tag{12.19}$$

is the dominant H_2 formation route.

H_2 formation on the surface of dust grains is considered to be the dominant formation channel in the Milky Way. We adopt a standard rate coefficient [49, 116, 128],

$$R \simeq 3 \times 10^{-17} \, T_2^{1/2} \, Z' \, \text{cm}^3 \, \text{s}^{-1}, \tag{12.20}$$

where $T_2 \equiv (T/100 \text{ K})$. In this expression, we assume that the dust-to-gas ratio is linearly proportional to the metallicity Z'. Thus, in scenario (b) H_2 formation on dust grains dominates even for $Z' = 10^{-3}$. Scenario (a) is the limit where the gas-phase channel dominates, as appropriate for dust-free environments or for superlinear dependence of the dust-to-gas ratio on Z'.

12.4.4 Timescales

The timescale for the system to achieve chemical steady state is dictated by the relatively long H_2 formation timescale. In the gas phase (scenario (1)) it is

$$t_{H_2} = \frac{1}{k_2 \, n \, x_e} \approx 8 \times 10^8 \, \zeta_{-16}^{-1/2} \, n_4^{-1/2} \, T_2^{-1} \text{ yr}, \tag{12.21}$$

where $x_e = 2.4 \times 10^{-5}(\zeta_{-16}/n_4)^{1/2}T_2^{0.38}$ is the electron fraction as set by ionisation recombination equilibrium, and $k_2 \approx 1.5 \times 10^{-16}T_2^{0.67}$ cm^3 s^{-1} is the rate coefficient for reaction (12.18).

For formation on dust grains ($\propto Z'$), the timescale is generally shorter, with

$$t_{H_2} = \frac{1}{R\,n} \approx 10^8 \, T_2^{-1/2} \, n_4^{-1} \left(\frac{10^{-3}}{Z'}\right) \text{ yr.} \tag{12.22}$$

Gas clouds with lifetimes $t \gg t_{H_2}$ will reach chemical steady state.

The relevant timescale for thermal equilibrium is the cooling timescale. For low-metallicity gas with $Z' = 10^{-3}$, the cooling proceeds mainly via H_2 rotational excitations [98]. If the cooling rate per H_2 molecule (in erg s^{-1}) is $W(n, T)$, then the cooling timescale is given by

$$t_{cool} = \frac{k_B \, T}{W(n, T)}. \tag{12.23}$$

Here, k_B is the Boltzmann constant. For $n = 10^4$ cm^{-3} and $T = 300$ K, $W \approx 5 \times 10^{-25}$ $(x_{H_2}/0.1)$ erg s^{-1} [147], and the cooling time is very short, $\approx 2 \times 10^3(0.1/x_{H_2})$ yr. For densities much smaller than the critical density for H_2 cooling, $W \propto n$ and $t_{cool} \propto 1/n$. In the opposite limit, W saturates and t_{cool} becomes independent of density. We see that, generally, $t_{cool} \ll t_{H_2}$.

Because the free-fall time

$$t_{ff} = \left(\frac{3\pi}{32G\rho}\right)^{1/2} = 5 \times 10^5 \, n_4^{-1/2} \text{ yr} \tag{12.24}$$

is generally much shorter than t_{H_2}, chemical steady state may be achieved only in stable, non-collapsing clouds, with lifetimes $\gg t_{ff}$. Obviously, both t_{H_2} and t_{cool} must be also shorter than the Hubble time at the redshift of interest.

12.4.5 Results

Next, we present and discuss our results for the steady-state, gas-phase H_2O fraction $x_{H_2O} \equiv n_{H_2O}/n$, as function of temperature T and the FUV intensity-to-density ratio I_{UV}/n_4.

12.4.6 x_{H_2O} as a function of T and I_{UV}/n_4

Figure 12.3 shows $\log_{10}(x_{H_2O})$ contours for the two scenarios described in Section 12.4.3. In one, H_2 forms in pure gas-phase (scenario (a) – left panel), and in the other, H_2 forms also on dust-grains (scenario (b) – right panel). Our fiducial parameters are $Z' = 10^{-3}$ and $\zeta_{-16} = I_{UV}$. In the upper-left region of the parameter space, x_{H_2O} is generally low, $\lesssim 10^{-9}$. In this regime, H_2O forms through the ion-molecule sequence, which is operative at low temperatures. In the lower-right corner, the neutral-neutral reactions become effective and x_{H_2O} rises.

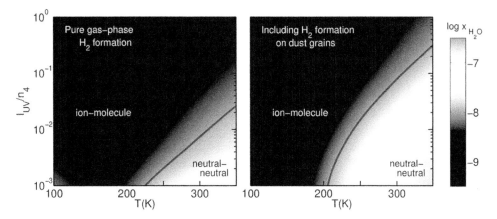

Figure 12.3 The fractional H_2O abundance x_{H_2O} as a function of T and I_{UV}/n_4, for $Z' = 10^{-3}$ and $\zeta_{-16}/I_{UV} = 1$, assuming pure gas-phase chemistry (scenario (a) – left panel), and including H_2 formation on dust grains (scenario (b) – right panels). In both panels, the solid line indicates the 10^{-8} contour, which is a characteristic value for the H_2O gas phase abundance in diffuse clouds within the Milky Way galaxy. At high temperatures (or low I_{UV}/n_4 values), the neutral-neutral reactions become effective, and x_{H_2O} rises.

In both panels, the solid line highlights the $x_{H_2O} = 10^{-8}$ contour, which resembles the H_2O gas phase-abundance in diffuse and translucent Milky Way clouds. This line also delineates the borderline between the regimes where H_2O forms via the 'cold' ion-molecule sequence and the 'warm' neutral-neutral sequence. The temperature range at which the neutral-neutral sequence kicks in is relatively narrow, ~ 250–350 K, because the neutral-neutral reactions are limited by energy barriers that introduce an exponential dependence on temperature.

The dependence on I_{UV}/n_4 is introduced because the FUV photons photodissociate OH and H_2O molecules and therefore increase the removal rate of H_2O and at the same time suppress formation via the $OH + H_2$ reaction. This gives rise to a critical value for I_{UV}/n_4, below which H_2O may become abundant.

For pure gas-phase H_2 formation (scenario (a) – left panel), the gas remains predominantly atomic, and H_2O formation is less efficient. In this case, $x_{H_2O} \gtrsim 10^{-8}$ only when I_{UV}/n_4 is smaller than a critical value of

$$(I_{UV}/n_4)^{(a)}_{crit} = 2 \times 10^{-2}. \tag{12.25}$$

However, when H_2 formation on dust is included (scenario (b) – right panel), the hydrogen becomes fully molecular, and H_2O formation is then more efficient. In this case, x_{H_2O} may reach 10^{-8} for I_{UV}/n_4 smaller than

$$(I_{UV}/n_4)^{(b)}_{crit} = 3 \times 10^{-1}, \tag{12.26}$$

an order of magnitude larger than for the pure gas phase formation scenario.

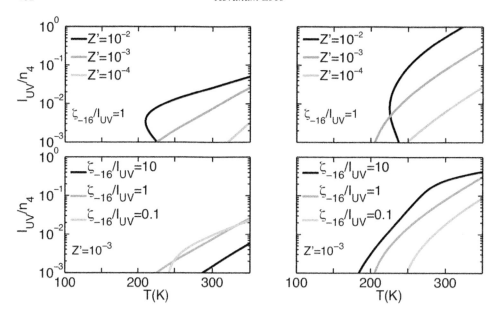

Figure 12.4 The $x_{H_2O} = 10^{-8}$ contour, for variations in Z' (upper panels) and in ζ_{-16}/I_{UV} (lower panels), assuming pure gas-phase chemistry (scenario (a) – left panels) and including H_2 formation on dust grains (scenario (b) – right panels).

12.4.7 Variations in Z' and ζ_{-16}/I_{UV}

In Figure 12.4, we investigate the effects of variations in the value of Z' and the normalisation ζ_{-16}/I_{UV}. The figure shows the $x_{H_2O} = 10^{-8}$ contours for scenarios (a) (left panels) and (b) (right panels). As discussed earlier, H_2O is generally more abundant in scenario (b) because the hydrogen is fully molecular in this case, and, therefore, the 10^{-8} contours are located at higher I_{UV}/n_4 values in both right panels.

The upper panels show the effect of variations in the metallicity value Z' for our fiducial normalisation $\zeta_{-16} = I_{UV}$. In both panels, the oxygen abundance rises, and x_{H_2O} increases with increasing Z'. Thus, at higher Z', the 10^{-8} contours shift to lower T and higher I_{UV}/n_4 and vice versa. An exception is the behaviour of the $Z' = 10^{-2}$ curve, for which the metallicity is already high enough so that reactions with metal species dominate H_2O removal for $I_{UV}/n_4 \lesssim 10^{-2}$. The increase in metallicity then results in a *decrease* of the H_2O abundance, and the 10^{-8} contour shifts to the right. For $I_{UV}/n_4 \gtrsim 10^{-2}$, removal by FUV dominates, and the behaviour is similar to that in the $Z' = 10^{-3}$ and $Z' = 10^{-4}$ cases.

The lower panels show the effects of variations in the ionisation rate normalisation ζ_{-16}/I_{UV} for our fiducial metallicity value of $Z' = 10^{-3}$. First, we consider the pure gas phase formation case (scenario (a) – lower-left panel). For the two cases $\zeta_{-16}/I_{UV} = 1$ and 10^{-1}, the H_2O removal is dominated by FUV photodissociation and, therefore, is independent of ζ. As shown by Reference [28], the H_2O formation rate is also independent of ζ when the H_2 forms in the gas phase. Therefore, x_{H_2O} is essentially independent of ζ, and the contours overlap. For the high ionisation rate $\zeta_{-16}/I_{UV} = 10$, the proton abundance becomes high, and H_2O reactions with H^+ dominate H_2O removal. In this limit, x_{H_2O} decreases with ζ, and the 10^{-8} contour moves down.

When H_2 forms on dust (scenario (b)–lower-right panel), the H_2O formation rate via the ion-molecule sequence is proportional to the H^+ and H_3^+ abundances, which rise with ζ. Since the gas is molecular, the proton fraction is low, and the removal is always dominated by FUV photodissociations (independent of ζ). Therefore, in this case, x_{H_2O} *increases* with ζ_{-16}/I_{UV}, and the curves shift up and to the left, towards lower T and higher I_{UV}/n_4.

12.4.8 Section Summary and Implications

We have demonstrated that the H_2O gas phase abundance may remain high even at very low metallicities of $Z' \sim 10^{-3}$. The onset of the efficient neutral-neutral formation sequence at $T \sim 300$ K may compensate for the low metallicity and form H_2O in abundance, similar to that found in diffuse clouds within the Milky Way.

We have considered two scenarios for H_2 formation, representing two limiting cases: one in which H_2 is formed in pure gas phase (scenario (a)), and one in which H_2 forms both in gase phase and on dust grains, assuming that the dust abundance scales linearly with Z' (scenario (b)). Recent studies by References [84, 91, 112] suggest that the dust abundance might decrease faster than then linearly with decreasing Z'. As shown by Reference [28], for $Z' = 10^{-3}$ and dust abundance that scales as Z'^β with $\beta \geq 2$, H_2 formation is dominated by the gas phase formation channel. Therefore, our scenario (a) is also applicable for models in which dust grains are present, with an abundance that scales superlinearly with Z'. For both scenarios (a) and (b), we have found that the neutral-neutral formation channel yields $x_{H_2O} \gtrsim 10^{-8}$, provided that I_{UV}/n_4 is smaller than a critical value. For the first scenario, we have found that this critical value is $(I_{UV}/n_4)_{crit} = 2 \times 10^2$. For the second scenario, $(I_{UV}/n_4)_{crit} = 3 \times 10^{-1}$.

In our analysis, we have assumed that the system had reached a chemical steady state. For initially atomic (or ionised) gas, this assumption offers the best conditions for the formation of molecules. However, chemical steady state might not always be achieved within a cloud lifetime or even within the Hubble time. The timescale to achieve chemical steady state (from an initially dissociated state) is dictated by the H_2 formation process and is generally long at low metallicities. For $Z' = 10^{-3}$ and our fiducial parameters, the timescales for both scenarios are of order of a few 10^8 years (see, e.g., Reference [27]) and are comparable to the age of the Universe at redshifts of ~ 10. The generically high water abundances we find for warm conditions and low metallicities will be maintained in dynamically evolving systems, so long as they remain H_2 shielded.

Our results might have interesting implications for the question of how early could have life originated in the Universe [159]. Our study addresses the first step of H_2O formation in early enriched, molecular gas clouds. If such a cloud is to collapse and form a protoplanetary disc, some of the H_2O may make its way to the surfaces of forming planets [258].

However, the question of to what extent the H_2O molecules that were formed in the initial molecular clouds are preserved all the way through the process of planet formation is beyond the scope of this section.

12.5 An Observational Test for the Anthropic Origin of the Cosmological Constant

12.5.1 Section Background

The distance to Type Ia supernovae [142, 206] and the statistics of the cosmic microwave background anisotropies [211] provide conclusive evidence for a finite vacuum energy density of $\rho_V = 4$ keV cm^{-3} in the present-day Universe. This value is a few times larger than the mean cosmic density of matter today. The expected exponential expansion of the Universe in the future (for a time-independent vacuum density) will halt the growth of all bound systems such as galaxies and groups of galaxies including the nearby Virgo cluster [44, 76, 190]. It will also redshift all extragalactic sources out of detectability (except for the merger remnant of the Milky Way and the Andromeda galaxies to which we are bound) – marking the end of extragalactic astronomy, as soon as the Universe ages by another factor of ten [157].

The observed vacuum density is smaller by tens of orders of magnitude than any plausible zero-point scale of the Standard Model of particle physics. Weinberg [264] first suggested that such a situation could arise in a theory that allows the cosmological constant to be a free parameter. On a scale much bigger than the observable Universe, one could then find regions in which the value of ρ_V is very different. However, if one selects those regions that give life to observers, then one would find a rather limited range of ρ_V values near its observed magnitude, since observers are most likely to appear in galaxies as massive as the Milky Way galaxy which assembled at the last moment before the cosmological constant started to dominate our Universe. Vilenkin [261] showed that this so-called 'anthropic argument' [18] can be used to calculate the probability distribution of vacuum densities with testable predictions. This notion [77, 92, 93, 180, 244, 245, 265] gained popularity when it was realised that string theory predicts the existence of an extremely large number [38, 95, 129, 175], perhaps as large as $\sim 10^{100}$–10^{500} [13], of possible vacuum states. The resulting landscape of string vacua [242] in the 'multiverse,' encompassing a volume of space far greater than our own inflationary patch, made the anthropic argument appealing to particle physicists and cosmologists alike [214, 245, 266].

The time is therefore ripe to examine the prospects for an experimental test of the anthropic argument. Any such test should be welcomed by proponents of the anthropic argument, since it would elevate the idea to the status of a falsifiable physical theory. At the same time, the test should also be welcomed by opponents of anthropic reasoning, since such a test would provide an opportunity to diminish the predictive power of the anthropic proposal and suppress discussions about it in the scientific literature.

Is it possible to dispute the anthropic argument without visiting regions of space that extend far beyond the inflationary patch of our observable Universe? The answer is *yes* if one can demonstrate that life could have emerged in our Universe even if the cosmological constant would have had values that are much larger than observed. In this section, we propose a set of astronomical observations that could critically examine this issue. We make use of the fact that dwarf galaxies formed in our Universe at redshifts as high as $z \sim 10$ when the mean matter density was larger by a factor of $\sim 10^3$ than it is today[4] [160]. If habitable planets emerged within these dwarf galaxies or their descendents (such as old globular clusters which might be the tidally truncated relics of early galaxies [184, 187]), then life would have been possible in a universe with a value of ρ_V that is a thousand times bigger than observed.

12.5.2 *Prior Probability Distribution of Vacuum Densities*

On the Planck scale of a quantum field theory which is unified with gravity (such as string theory), the vacuum energy densities under discussion represent extremely small deviations around $\rho_V = 0$. Assuming that the prior probability distribution of vacuum densities, $\mathcal{P}_*(\rho_V)$, is not divergent at $\rho_V = 0$ (since $\rho_V = 0$ is not favoured by any existing theory), it is natural to expand it in a Taylor series and keep only the leading term. Thus, in our range of interest of ρ_V values [92, 93, 265],

$$\mathcal{P}_*(\rho_V) \approx constant. \tag{12.27}$$

This implies that the probability of measuring a value equal to or smaller than the observed value of ρ_V is $\sim 10^{-3}$ if habitable planets could have formed in a Universe with a value of ρ_V that is a thousand times bigger than observed.

Numerical simulations indicate that our Universe would cease to make new bound systems in the near future [44, 76, 190]. A universe in which ρ_V is a thousand times larger would, therefore, make dwarf galaxies until $z \sim 10$ when the matter density was a thousand times larger than today. The question of whether planets can form within these dwarf galaxies can be examined observationally, as we discuss next. It is important to note that once a dwarf galaxy forms, it has an arbitrarily long time to convert its gas into stars and planets because its internal evolution is decoupled from the global expansion of the universe (as long as outflows do not carry material out of its gravitational pull).

12.5.3 *Extragalactic Planet Searches*

Gravitational microlensing is the most effective search method for planets beyond our galaxy. The planet introduces a short-term distortion to the otherwise smooth light curve produced by its parent star as that star focuses the light from a background star which happens to lie behind it [176, 203]. In an extensive search for planetary microlensing

---

[4] We note that although the cosmological constant started to dominate the mass density of our Universe at $z \sim 0.4$, its impact on the formation of bound objects became noticeable only at $z \sim 0$ or later [44, 76, 190]. For the purposes of our discussion, we therefore compare the matter density at $z \sim 10$ to that today. Coincidentally, the Milky Way galaxy formed before ρ_V dominated but it could have also formed later.

signatures, a number of collaborations named PLANET [3], μFUN [273], and RoboNET are performing follow-up observations on microlensing events which are routinely detected by the groups MOA [33] and OGLE [251]. Many 'planetary' events have been reported, including a planet of a mass of ~ 5 Earth masses at a projected separation of 2.6AU from a 0.2 $M_\odot$ M-dwarf star in the microlensing event OGLE-2005-BLG-390Lb [20] and a planet of 13 Earth masses at a projected separation of 2.3 AU from its parent star in the event OGLE-2005-BLG-169 towards the galactic bulge – in which the background star was magnified by the unusually high factor of ~ 800 [102]. Based on the statistics of these events and the search parameters, one can infer strong conclusions about the abundance of planets of various masses and orbital separations in the surveyed star population [32, 94, 102]. The technique can be easily extended to lenses outside our galaxy and out to the Andromenda galaxy (M31) using the method of pixel lensing [15, 56, 61]. For the anthropic experiment, we are particularly interested in applying this search technique to lensing of background Milky Way stars by old stars in foreground globular clusters (which may be the tidally truncated relics of $z \sim 10$ galaxies), or to lensing of background M31 stars by foreground globular clusters [118] or dwarf galaxies such as Andromeda VIII [188]. In addition, self-lensing events in which foreground stars of a dwarf galaxy lens background stars of the same galaxy are of particular interest. Such self-lensing events were observed in the form of caustic-crossing binary lens events in the Large Magellanic Cloud (LMC) and the Small Magellanic Cloud (SMC) [67]. In the observed cases, there is enough information to ascertain that the most likely lens location is in the Magellanic Clouds themselves. Yet each caustic-crossing event represents a much larger number of binary lens events from the same lens population; the majority of these may be indistinguishable from point-lens events. It is therefore possible that some of the known single-star LMC lensing events are due to self-lensing [67], as hinted at by their geometric distribution [105, 225].

As mentioned earlier, another method for finding extragalactic planets involves transit events in which the planet passes in front of its parent star and causes a slight temporary dimming of the star. Spectral modelling of the parent star allows to constrain both the size and abundance statistics of the transiting planets [52, 205]. Existing surveys reach distance scales of several kpc [174, 255] with some successful detections [37, 136, 252]. So far, a Hubble Space Telescope search for transiting Jupiters in the globular cluster 47 Tucanae resulted in no detections [96] (although a pulsar planet was discovered later by a different technique in the low-metallicity globular cluster Messier 4 [234], potentially indicating early planet formation). A future space telescope (beyond the TESS,[5] Kepler,[6] and COROT[7] missions which focus on nearby stars) or a large-aperture ground-based facility (such as the Giant Magellan Telescope [GMT],[8] the Thirty-Meter Telescope [TMT],[9] or the

---

[5] https://tess.gsfc.nasa.gov.
[6] http://kepler.nasa.gov/.
[7] http://smsc.cnes.fr/COROT/.
[8] www.gmto.org/.
[9] www.tmt.org/.

European Extremely Large Telescope [EELT][10]) could extend the transit search technique to planets at yet larger distances (but see Reference [205]). Existing searches [52] identified the need for a high signal-to-noise spectroscopy as a follow-up technique for confirming real transits out of many false events. Such follow-ups would become more challenging at large distances, making the microlensing technique more practical.

12.5.4 Observations of Dwarf Galaxies at High Redshifts

Our goal is to study stellar systems in the local Universe which are the likely descendents of the early population of $z \sim 10$ galaxies [221]. In order to refine this selection, it would be desirable to measure the characteristic size, mass, metallicity, and star formation histories of $z \sim 10$ galaxies (see Reference [160] for a review on their theoretically expected properties). As already mentioned, it is possible that the oldest globular clusters are descendents of the first galaxies [220].

Recently, a large number of faint early galaxies, born less than a billion years after the Big Bang, have been discovered (see, e.g., [134, 219]). These include starburst galaxies with star formation rates in excess of ~ 0.1 $M_\odot$ yr^{-1} and dark matter halos [239] of $\sim 10^{9-11}$ $M_\odot$ [39, 40, 218, 219, 228] at $z \sim 5$–10. Luminous Lyα emitters are routinely identified through continuum dropout and narrow-band imaging techniques [39, 40]. In order to study fainter sources which were potentially responsible for reionisation, spectroscopic searches have been undertaken near the critical curves of lensing galaxy clusters [134, 228], where gravitational magnification enhances the flux sensitivity. Because of the foreground emission and opacity of the Earth's atmosphere, it is difficult to measure spectral features other than the Lyα emission line from these feeble galaxies from ground-based telescopes.

In one example, gravitational lensing by the massive galaxy cluster A2218 allowed the detection of stellar system at $z = 5.6$ with an estimated mass of $\sim 10^6$ $M_\odot$ in stars [228]. Detection of additional low-mass systems could potentially reveal whether globular clusters formed at these high redshifts. Such a detection would be feasible with the James Webb Space Telescope.[11] Existing designs for future large-aperture (> 20 m) infrared telescopes (such as the GMT, TMT, and EELT mentioned earlier) would also enable researchers to measure the spectra of galaxies at $z \sim 10$ and infer their properties.

Based the characteristics of high-z galaxies, one would be able to identify present-day systems (such as dwarf galaxies or globular clusters) that are their likely descendents [69, 269] and search for planets within them. Since the lifetime of massive stars that explode as core-collapse supernovae is two orders of magnitude shorter than the age of the Universe at $z \sim 10$, it is possible that some of these systems would be enriched to a high metallicity despite their old age. For example, the cores of quasar host galaxies are known to possess super-solar metallicities at $z \gtrsim 6$ [68].

---

[10] www.eso.org/sci/facilities/eelt.
[11] www.jwst.nasa.gov/.

12.5.5 Section Summary and Implications

In future decades, it would be technologically feasible to search for microlensing or transit events in local dwarf galaxies or old globular clusters and to check whether planets exist in these environments. Complementary observations of early dwarf galaxies at redshifts $z \sim 10$ can be used to identify nearby galaxies or globular clusters that are their likely descendents. If planets are found in local galaxies that resemble their counterparts at $z \sim 10$, then the precise version of the anthropic argument [77, 92, 180, 261, 264] would be weakened considerably, since planets could have formed in our Universe even if the cosmological constant, ρ_V, was three orders of magnitude larger. For a flat probability distribution at these ρ_V values (which represents infinitesimal deviations from $\rho_V = 0$ relative to the Planck scale), this would imply that the probability for us to reside in a region where ρ_V obtains its observed value is lower than $\sim 10^{-3}$. The precise version of the anthropic argument [77, 92, 180, 261, 264] could then be ruled out at a confidence level of $\sim 99.9\%$, which is a satisfactory measure for an experimental test. The envisioned experiment resonates with two of the most active frontiers in astrophysics, namely the search for planets and the study of high-redshift galaxies, and if performed, it would have many side benefits to conventional astrophysics.

We note that in the hypothetical universe with a large cosmological constant, life need not form at $z \sim 10$ (merely 400 million years after the Big Bang) but rather any time later. Billions of years after a dwarf galaxy had formed, a typical astronomer within it would see the host galaxy surrounded by a void which is dominated by the cosmological constant. Of course, the volume density of life in such a case is smaller than in our Universe.

An additional factor that enters the likelihood function of ρ_V values involves the conversion efficiency of baryons into observers in the Universe. A universe in which observers only reside in galaxies that were made at $z \sim 10$ might be less effective at making observers. The fraction of baryons that have assembled into star-forming galaxies above the hydrogen cooling threshold by $z \sim 10$ is estimated to be $\sim 10\%$ [160], comparable to the final fraction of baryons that condensed into stars in the present-day Universe [90]. It is possible that more stars formed in smaller systems down to the Jeans mass of $\sim 10^{4-5}\, M_\odot$ through molecular hydrogen cooling [41]. Although today most baryons reside in a warm-hot medium of $\sim 2 \times 10^6$ K that cannot condense into stars [50, 64], most of the cosmic gas at $z \sim 10$ was sufficiently cold to fragment into stars as long as it could have cooled below the virial temperature of its host halos [160]. The star formation efficiency can be inferred [69] from dynamical measurements of the star and dark matter masses in local dwarfs or globulars that resemble their counterparts at $z \sim 10$. If only a small portion of the cosmic baryon fraction (Ω_b / Ω_m) in dwarf galaxies is converted into stars, then the probability of obtaining habitable planets would be reduced accordingly. Other physical factors, such as metallicty, may also play an important role [156]. Preliminary evidence indicates that planet formation favours environments which are abundant in heavy elements [83], although notable exceptions exist [234].

Unfortunately, it is not possible to infer the planet production efficiency for an alternative universe purely based on observations of our Universe. In our Universe, most of the baryons which were assembled into galaxies by $z \sim 10$ are later incorporated into bigger galaxies. The vast majority of the $z \sim 10$ galaxies are consumed through hierarchical mergers to make bigger galaxies; isolated descendents of $z \sim 10$ galaxies are rare among low-redshift galaxies. At any given redshift below 10, it would be difficult to separate observationally the level of planet formation in our Universe from the level that would have occurred otherwise in smaller galaxies if these were not consumed by bigger galaxies within a universe with a large vacuum density, ρ_V. In order to figure out the planet production efficiency for a large ρ_V, one must adopt a strategy that mixes observations with theory. Suppose we observe today the planet production efficiency in the descendents of $z \sim 10$ galaxies. One could then use numerical simulations to calculate the abundance that these galaxies would have had today if ρ_V was $\sim 10^3$ times bigger than its observed value. This approach implicitly takes into account the possibility that planets may form relatively late (after $\sim$10 Gyr) within these isolated descendents, irrespective of the value of ρ_V. The late time properties of gravitationally bound systems are expected to be independent of the value of ρ_V.

In our discussion, we assumed that as long as rocky planets can form at orbital radii that allow liquid water to exist on their surface (the so-called *habitable zone* [130]), life would develop over billions of years and eventually mature in intelligence. Without a better understanding of the origin of intelligent life, it is difficult to assess the physical conditions that are required for intelligence to emerge beyond the minimal requirements stated earlier. If life forms early, then civilisations might have more time to evolve to advanced levels. On the other hand, life may be disrupted more easily in early galaxies because of their higher density (making the likelihood of stellar encounters higher) [92, 245], and so it would be useful to determine the environmental density observationally. In the more distant future, it might be possible to supplement the study proposed here by the more adventurous search for radio signals from intelligent civilisations beyond the boundaries of our galaxy. Such a search would bring an extra benefit. If the anthropic argument turns out to be wrong and intelligent civilisations are common in nearby dwarf galaxies, then the older, more advanced civilisations among them might broadcast an explanation for why the cosmological constant has its observed value.

12.6 The Relative Likelihood of Life as a Function of Cosmic Time

12.6.1 Section Background

Currently, we only know of life on Earth. The Sun formed $\sim$ 4.6 Gyr ago and has a lifetime comparable to the current age of the Universe. But the lowest-mass stars near the hydrogen burning threshold at 0.08 $M_\odot$ could live a thousand times longer, up to 10 trillion years [146, 224]. Given that habitable planets may have existed in the distant past and might exist in the distant future, it is natural to ask, *what is the relative probability for the emergence of life as a function of cosmic time?* In this section, we answer this question conservatively by restricting our attention to the context of 'life as we know it' and the standard cosmological model (ΛCDM).[12] Note that since the probability distribution is normalised to have a unit integral, it only compares the relative importance of different epochs for the emergence of life but does not calibrate the overall likelihood for life in the Universe. This feature makes our results robust to uncertainties in normalisation constants associated with the likelihood for life on habitable planets.

Next, we express the relative likelihood for the appearance of life as a function of cosmic time in terms of the star formation history of the Universe, the stellar mass function, the lifetime of stars as a function of their mass, and the probability of Earth-mass planets in the habitable zone of these stars. We define this likelihood within a fixed co-moving volume which contains a fixed number of baryons. In predicting the future, we rely on an extrapolation of star formation rate until the current gas reservoir of galaxies is depleted.

12.6.2 Formalism

Master Equation

We wish to calculate the probability $dP(t)/dt$ for life to form on habitable planets per unit time within a fixed co-moving volume of the Universe [153]. This probability distribution should span the time interval between the formation time of the first stars and the maximum lifetime of all stars that were ever made (~ 10 Tyr).

The probability $dP(t)/dt$ involves a convolution of the star formation rate per co-moving volume, $\dot{\rho}_*(t')$, with the temporal window function, $g(t - t', m)$, due to the finite lifetime of stars of different masses, m, and the likelihood, $\eta_{\mathrm{Earth}}(m)$, of forming an Earth-mass rocky planet in the habitable zone (HZ) of stars of different masses, given the mass distribution of stars, $\xi(m)$, times the probability, $p(\mathrm{life}|\mathrm{HZ})$, of actually having life on a habitable planet. With all these ingredients, the relative probability per unit time for life within a fixed co-moving volume can be written in terms of the double integral,

$$\frac{dP}{dt}(t) = \frac{1}{N} \int_0^t dt' \int_{m_{\min}}^{m_{\max}} dm' \xi(m') \dot{\rho}_*(t', m') \eta_{\mathrm{Earth}}(m') p(\mathrm{life}|\mathrm{HZ}) g(t - t', m'), \quad (12.28)$$

where the prefactor $1/N$ assures that the probability distribution is normalised to a unit integral over all times. The window function, $g(t - t', m)$, determines whether a habitable planet that formed at time t' is still within a habitable zone at time t. This function is non-zero within the lifetime $\tau_*(m)$ of each star, namely $g(t - t', m) = 1$ if $0 < (t - t') < \tau_*(m)$, and zero otherwise. The quantities $m_{\min}$ and $m_{\max}$ represent the minimum and maximum masses of viable host stars for habitable planets, respectively. Next, we provide more details on each of the various components of the preceding master equation.

---

[12] We address this question from the perspective of an observer in a single co-moving Hubble volume formed after the end of inflation. As such, we do not consider issues of self-location in the multiverse nor of the measure on eternally inflating regions of space-time. We note, however, that any observers in a post-inflationary bubble will, by necessity of the eternal inflationary process, only be able to determine the age of their own bubble. We therefore restrict our attention to the question of the probability distribution of life in the history of our own inflationary bubble.

Stellar Mass Range

Life requires the existence of liquid water on the surface of Earth-mass planets during the main stage lifetime of their host star. These requirements place a lower bound on the lifetime of the host star and, thus, an upper bound on its mass.

There are several proxies for the minimum time needed for life to emerge. Certainly, the star must live long enough for the planet to form, a process which took $\sim$ 40 Myr for Earth [185]. Moreover, once the planet has formed, sufficient cooling must follow to allow the condensation of water on the planet's surface. The recent discovery of the earliest crystals, zircons, suggests that these were formed during the Archean era, as much as 160 Myr after the planet formed [257]. Thus, we arrive at a conservative minimum of 200 Myr before life could form; any star living less than this time could not host life on an Earth-like planet. At the other end of the scale, we find that the earliest evidence for life on Earth comes from around 800 Myr after the formation of the planet [185], yielding an upper bound on the minimum lifetime of the host star. For the relevant mass range of massive stars, the lifetime, τ_*, scales with stellar mass, m, roughly as $(\tau_*/\tau_\odot) = (m/M_\odot)^{-3}$, where $\tau_\odot \approx 10^{10}$ yr. Thus, we find that the maximum mass of a star capable of hosting life (m_{max}) is in the range 2.3–3.7 $M_\odot$. Due to their short lifetimes and low abundances, high-mass stars do not provide a significant contribution to the probability distribution, $dP(t)/dt$, and so the exact choice of the upper mass cut-off in the preceding range is unimportant. The lowest-mass stars above the hydrogen burning threshold have a mass $m = 0.08$ $M_\odot$.

Time Range

The stars resulted in a second generation of stars, enriched by heavy elements, merely a few Myr later. The theoretical expectation that the second generation stars should have hosted planetary systems can be tested observationally by searching for planets around metal-poor stars in the halo of the Milky Way galaxy [182].

Star formation is expected to exhaust the cold gas in galaxies as soon as the Universe ages by a factor of a few (based on the ratio between the current reservoir of cold gas in galaxies [90] and the current star formation rate), but low-mass stars would survive long after that. The lowest-mass stars near the hydrogen burning limit of 0.08 $M_\odot$ have a lifetime of order 10 trillion years [146]. The probability $dP(t)/dt$ is expected to vanish beyond that time.

Initial Mass Function

The initial mass function (IMF) of stars $\xi(m)$ is proportional to the probability that a star in the mass range between m and $m + dm$ is formed. We adopt the empirically calibrated, Chabrier functional form [51], which follows a log-normal form for masses under a solar mass, and a power-law above a solar mass, as follows:

$$\xi(m) \propto \begin{cases} \left(\dfrac{m}{M_\odot}\right)^{-2.3} & m > 1 \; M_\odot \\ a \exp\left(-\dfrac{\ln(m/m_c)^2}{2\sigma^2}\right) \dfrac{M_\odot}{m} & m \leq 1 \; M_\odot \end{cases}, \qquad (12.29)$$

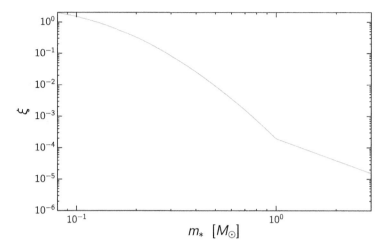

Figure 12.5 The Chabrier [51] mass function of stars, $\xi(m_*)$, plotted with a normalisation integral of unity.

where $a = 790$, $\sigma = 0.69$, and $m_c = 0.08\ M_\odot$. This IMF is plotted as a probability distribution normalised to a unit integral in Figure 12.5.

For simplicity, we ignore the evolution of the IMF with cosmic time and its dependence on galactic environment (e.g., galaxy type or metallicity [59]), as well as the uncertain dependence of the likelihood for habitable planets around these stars on metallicity [182].

Stellar Lifetime

The lifetime of stars, τ_*, as a function of their mass, m, can be modelled through a piecewise power-law form. For $m < 0.25\ M_\odot$, we follow Reference [146]. For $0.75\ M_\odot < m < 3\ M_\odot$, we adopt a scaling with an average power-law index of -2.5 and the proper normalisation for the Sun [226]. Finally, we interpolate in the range between 0.25 and 0.75 $M_\odot$ by fitting a power-law form there and enforcing continuity. In summary, we adopt,

$$\tau_*(m) = \begin{cases} 1.0 \times 10^{10}\ \text{yr}\ \left(\dfrac{m}{M_\odot}\right)^{-2.5} & 0.75\ M_\odot < m < 3\ M_\odot \\[2ex] 7.6 \times 10^{9}\ \text{yr}\ \left(\dfrac{m}{M_\odot}\right)^{-3.5} & 0.25\ M_\odot < m \le 0.75\ M_\odot \\[2ex] 5.3 \times 10^{10}\ \text{yr}\ \left(\dfrac{m}{M_\odot}\right)^{-2.1} & 0.08\ M_\odot \le m \le 0.25\ M_\odot \end{cases} \qquad (12.30)$$

This dependence is depicted in Figure 12.6.

Star Formation Rate

We adopt an empirical fit to the star formation rate per co-moving volume as a function of redshift, z [169],

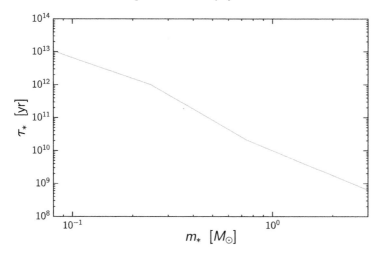

Figure 12.6 Stellar lifetime (τ_*) as a function of mass.

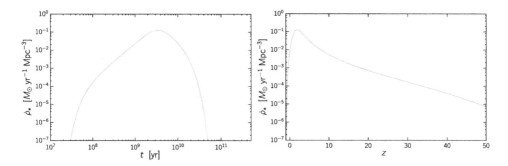

Figure 12.7 Star formation rate, $\dot{\rho}_*$, as a function of the cosmic time t (left panel) and redshift z (right panel), based on an extrapolation of a fitting function to existing data [169]. Although this extrapolation has inherent uncertainties, it provides a reasonable benchmark for theoretical expectations.

$$\dot{\rho}_*(z) = 0.015 \frac{(1+z)^{2.7}}{1 + [(1+z)/2.9]^{5.6}} \ M_\odot \mathrm{yr}^{-1}\mathrm{Mpc}^{-3}, \tag{12.31}$$

and truncate the extrapolation to early times at the expected formation time of the first stars [159]. We extrapolate the cosmic star formation history to the future or equivalently negative redshifts $-1 \leq z < 0$ (see, e.g., ref. [16]) and find that the co-moving star formation rate drops to less than 10^{-5} of the current rate at 56 Gyr into the future. We cut off the star formation at roughly the ratio between the current reservoir mass of cold gas in galaxies [90] and the current star formation rate per co-moving volume. The resulting star formation rate as a function of time and redshift is shown in Figure 12.7.

Probability of Life on a Habitable Planet

The probability for the existence of life around a star of a particular mass m can be expressed in terms of the product between the probability that there is an Earth-mass planet in the star's habitable zone (HZ) and the probability that life emerges on such a planet: $P(\text{life}|m) = P(\text{HZ}|m)P(\text{life}|\text{HZ})$. The first factor, $P(\text{HZ}|m)$, is commonly labelled η_{Earth} in the exoplanet literature [249].

Data from the NASA Kepler mission implies η_{Earth} values in the range of $6.4^{+3.4}_{-1.1}\%$ for stars of approximately a solar mass [179, 207, 235] and $24^{+18}_{-8}\%$ for lower-mass M-dwarf stars [73]. The result for solar mass stars is less robust due to lack of identified Earth-like planets at high stellar masses. Owing to the large measurement uncertainties, we assume a constant η_{Earth} within the range of stellar masses under consideration. The specific constant value of η_{Earth} drops out of the calculation due to the normalisation factor N.

There is scope for considerable refinement in the choice of the second factor $p(\text{life}|\text{HZ})$. One could suppose that the probability of life evolving on a planet increases with the amount of time that the planet exists or that increasing the surface area of the planet should increase the likelihood of life beginning. However, given our ignorance, we will set this probability factor to a constant, an assumption which can be improved upon by statistical data from future searches for biosignatures in the molecular composition of the atmospheres of habitable planets [151, 162, 222, 231]. In our simplified treatment, this constant value has no effect on $dP(t)/dt$ since its contribution is also cancelled by the normalisation factor N.

12.6.3 Results

The top and bottom panels in Figure 12.8 show the probability per log time interval $t\,dP(t)/dt = dP/d\ln t$ and the cumulative probability $P(<t) = \int_0^t [dP(t')/dt']dt'$ based on Eq. 12.28 for different choices of the low-mass cut-off in the distribution of host stars for life-hosting planets (equally spaced in $\ln m$). The upper stellar mass cut-off has a negligible influence on $dP/d\ln t$, due to the short lifetime and low abundance of massive stars. In general, $dP/d\ln t$ cuts off roughly at the lifetime of the longest-lived stars in each case, as indicated by the upper axis labels. For the full range of hydrogen-burning stars, $dP(t)/d\ln t$ peaks around the lifetime of the lowest mass stars $t \sim 10^{13}$ yr with a probability value that is a thousand times larger than for the Sun, implying that life around low-mass stars in the distant future is much more likely than terrestrial life around the Sun today.

12.6.4 Section Summary and Implications

Figure 12.8 implies that the probability for life per logarithm interval of cosmic time, $dP(t)/d\ln t$, has a broad distribution in $\ln t$ and is peaked in the future, as long as life is likely around low-mass stars. High-mass stars are shorter lived and less abundant and hence make a relatively small contribution to the probability distribution.

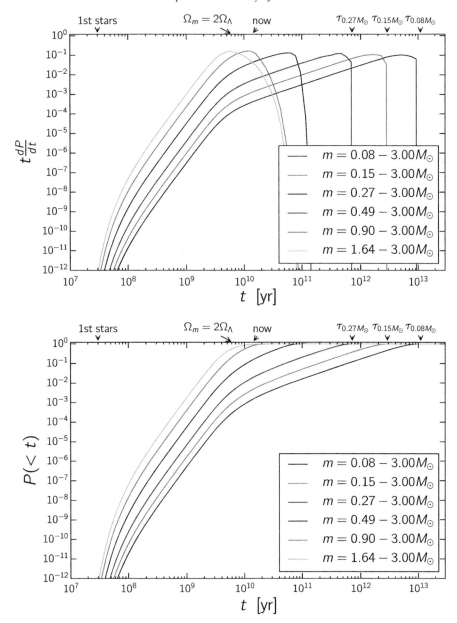

Figure 12.8 Probability distribution for the emergence of life within a fixed co-moving volume of the Universe as a function of cosmic time. We show the probability per log time, $t\,dP/dt$ (top panel) as well as the cumulative probability up to a time t, $P(<t)$ (bottom panel), for different choices of the minimum stellar mass, equally spaced in log m between 0.08 $M_\odot$ and 3 $M_\odot$. The contribution of stars above 3 $M_\odot$ to $dP(t)/dt$ is ignored due to their short lifetimes and low abundances. The labels on the top axis indicate the formation time of the first stars, the time when the cosmic expansion started accelerating (i.e., when the density parameter of matter, Ω_m, was twice that of the vacuum, Ω_Λ), the present time (now) and the lifetimes of stars with masses of 0.08 $M_\odot$, 0.15 $M_\odot$, and 0.27 $M_\odot$. The lines correspond to the masses indicated in the legend, from right to left.

Future searches for molecular biosignatures (such as O_2 combined with CH_4) in the atmospheres of planets around low-mass stars [162, 222, 231] could inform us whether life will exist at late cosmic times [137]. If we were to insist that life near the Sun is typical and not premature i.e., require that the peak in $dP(t)/d \ln t$ would coincide with the lifetime of Sun-like stars at the present time – then we must conclude that the physical environments of low-mass stars are hazardous to life. This outcome could be the result of a more conventional line of reasoning [217]). It could be a natural outcome, for example, from the enhanced UV emission and flaring activity of young low-mass stars during their extended ($\sim$ Gyr long) pre-main-sequence phase, which is capable of stripping rocky planets of their atmospheres [200]. The habitable zone is much closer in around low-mass stars, leading to tidal locking and extreme temperature contrast due between the permanent day and night sides [140] as well as an increased impact of the stellar wind and enhanced stellar variability in stripping the atmospheres of rocky planets [4, 17, 31, 168]. The recent discoveries of a habitable planet around the nearest, 0.12 $M_\odot$ star Proxima Centauri [9] and of three transiting habitable planets around the 0.08 $M_\odot$ star TRAPPIST-1 [97] offer new prospects for testing the conjecture. Empirical testing is crucial since self-selection arguments, such as the one associated with the peak in $dP(t)/d \ln t$, are known to be subject to statistical caveats [36, 101, 149]. Moreover, Figure 12.8 makes the implicit assumption that the emergence of life does not correlate with the early stages of planet formation, providing an equal probability for appearing early or late. In particular, water loss from the surface of a planet could be substantial at late times, suppressing the survival of life over trillions of years.

Values of the cosmological constant below the observed one should not affect the probability distribution, as they would introduce only mild changes to the star formation history due to the modified formation history of galaxies [44, 190]. However, much larger values of the cosmological constant would suppress galaxy formation and reduce the total number of stars per co-moving volume [158], hence limiting the overall likelihood for life altogether [264].

Our results provide a new perspective on the so-called coincidence problem, *why do we observe* $\Omega_m \sim \Omega_\Lambda$ [46]? The answer comes naturally if we consider the history of Sun-like star formation, as the number of habitable planets peaks around present time for $m \sim 1 M_\odot$. We note that for the majority of stars, this coincidence will not exist as $dP(t)/dt$ peaks in the future, where $\Omega_m \ll \Omega_\Lambda$. The question is, then, why do we find ourselves orbiting a star like the Sun now rather than a lower-mass star in the future?

We derived our numerical results based on a conservative set of assumptions and guided by the latest empirical data for the various components of Eq. (12.28). However, the emergence of life may be sensitive to additional factors that were not included in our formulation, such as the existence of a moon to stabilise the climate on an Earth-like planet [145], the existence of asteroid belts [181], the orbital structure of the host planetary system (e.g., the existence of nearby giant planets or orbital eccentricity), the effects of a binary star companion [106], the location of the planetary system within the host galaxy [154], and the detailed properties of the host galaxy (e.g., galaxy type [59] or metallicity [127, 156]), including the environmental effects of quasars, gamma ray bursts [210], or the hot gas in clusters of galaxies. These additional factors are highly uncertain and complicated to model and were ignored for simplicity in our analysis.

The probability distribution $dP(t)/d\ln t$ is of particular importance for studies attempting to gauge the level of fine-tuning required for the cosmological or fundamental physics parameters that would allow life to emerge in our Universe.

Acknowledgements

I thank my collaborators on the work described in this chapter, namely Rafael Batista, Shmuel Bialy, Laura Kreidberg, Gonzalo Gonzalez, James Guillochon, Henry Lin, Dani Maoz, Natalie Mashian, David Sloan, Amiel Sternberg, Ed Turner, and Matias Zaldarriaga.

References

[1] Adams, F. C., and Spergel, D. N. 2005. 'Lithopanspermia in Star-Forming Clusters'. *Astrobiology*, **5**(Aug.), 497–514.

[2] Agol, E. 2011. 'Transit Surveys for Earths in the Habitable Zones of White Dwarfs'. *The Astrophysical Journal Letters*, **731**(Apr.), L31.

[3] Albrow, M., *et al.* 1998. 'The 1995 Pilot Campaign of PLANET: Searching for Microlensing Anomalies through Precise, Rapid, Round-the-Clock Monitoring'. *Astrophysical Journal*, **509**(Dec.), 687–702.

[4] Alibert, Y., and Benz, W. 2017. 'Formation and Composition of Planets around Very Low Mass Stars'. *Astronomy and Astrophysics*, **598**(Jan.), L5.

[5] Anderson, L., *et al.* 2014. 'The Clustering of Galaxies in the SDSS-III Baryon Oscillation Spectroscopic Survey: Baryon Acoustic Oscillations in the Data Releases 10 and 11 Galaxy Samples'. *Monthly Notices of the Royal Astronomical Society*, **441**(June), 24–62.

[6] Andrews, S. M., and Williams, J. P. 2005. 'Circumstellar Dust Disks in Taurus-Auriga: The Submillimeter Perspective'. *Astrophysical Journal*, **631**(Oct.), 1134–1160.

[7] Andrews, S. M., *et al.* 2009. 'Protoplanetary Disk Structures in Ophiuchus'. *Astrophysical Journal*, **700**(Aug.), 1502–1523.

[8] Andrews, S. M., *et al.* 2010. 'Protoplanetary Disk Structures in Ophiuchus: II. Extension to Fainter Sources'. *Astrophysical Journal*, **723**(Nov.), 1241–1254.

[9] Anglada-Escudé, G., *et al.* 2016. 'A Terrestrial Planet Candidate in a Temperate Orbit around Proxima Centauri'. *Nature*, **536**(Aug.), 437–440.

[10] Aoki, W. 2010 (Mar.). 'Carbon-Enhanced Metal-Poor (CEMP) Stars'. Cunha, K., Spite, M., and Barbuy, B. eds. *Chemical Abundances in the Universe: Connecting First Stars to Planets*. IAU Symposium, vol. 265 Pages 111–116.

[11] Aoki, W., *et al.* 2007. 'Carbon-Enhanced Metal-Poor Stars: I. Chemical Compositions of 26 Stars'. *Astrophysical Journal*, **655**(Jan.), 492–521.

[12] Armitage, P. J. 2010. *Astrophysics of Planet Formation*. Cambridge University Press.

[13] Ashok, S. K., and Douglas, M. R. 2004. 'Counting Flux Vacua'. *Journal of High Energy Physics*, **1**(Jan.), 060.

[14] Asplund, M., *et al.* 2009. 'The Chemical Composition of the Sun'. *Annual Reviews of Astronomy and Astrophysics*, **47**(Sept.), 481–522.

[15] Baltz, E. A., and Gondolo, P. 2001. 'Binary Events and Extragalactic Planets in Pixel Microlensing'. *Astrophysical Journal*, **559**(Sept.), 41–52.

[16] Barnes, L., *et al.* 2005. 'The Influence of Evolving Dark Energy on Cosmology'. *Publications of the Astronomical Society of Australia*, **22**, 315–325.

[17] Barnes, R., *et al.* 2009. 'Tidal Limits to Planetary Habitability'. *Astrophysical Journal Letters*, **700**(July), L30–L33.

[18] Barrow, J. D., and Tipler, F. J. 1986. *The Anthropic Cosmological Principle*. Oxford University Press.

[19] Bayet, E., *et al.* 2009. 'Molecular Tracers of PDR-Dominated Galaxies'. *Astrophysical Journal*, **696**(May), 1466–1477.

[20] Beaulieu, J. P., Bennett, D. P. Fouqué, P., *et al.* 2006. 'Discovery of a Cool Planet of 5.5 Earth Masses through Gravitational Microlensing'. *Nature*, **439**(Jan.), 437–440.

[21] Beckwith, S. V. W., Henning, T., and Nakagawa, Y. 2000. 'Dust Properties and Assembly of Large Particles in Protoplanetary Disks'. *Protostars and Planets IV*, May, 533.

[22] Beers, T. C., and Christlieb, N. 2005. 'The Discovery and Analysis of Very Metal-Poor Stars in the Galaxy'. *Annual Reviews of Astronomy and Astrophysics*, **43**(Sept.), 531–580.

[23] Beers, T. C., Preston, G. W., and Shectman, S. A. 1985. 'A Search for Stars of Very Low Metal Abundance: I'. *The Astronomical Journal*, **90**(Oct.), 2089–2102.

[24] Beers, T. C., *et al.* 2007. 'Near-Infrared Spectroscopy of Carbon-Enhanced Metal-Poor Stars: I. A SOAR/OSIRIS Pilot Study'. *The Astronomical Journal*, **133**(Mar.), 1193–1203.

[25] Behroozi, P., and Peeples, M. S. 2015. 'On the History and Future of Cosmic Planet Formation'. *Monthly Notices of the Royal Astronomical Sociey*, **454**(Dec.), 1811–1817.

[26] Beichman, C. A., Greene, T., and Krist, J. 2009 (Feb.). 'Future Observations of Transits and Light Curves from Space'. Pont, F., Sasselov, D., and Holman, M. J., eds. *Transiting Planets*. IAU Symposium, vol. 253. Pages 391–328.

[27] Bell, T. A., *et al.* 2006. 'Molecular Line Intensities as Measures of Cloud Masses: I. Sensitivity of CO Emissions to Physical Parameter Variations'. *Monthly Notices of the Royal Astronomical Society*, **371**(Oct.), 1865–1872.

[28] Bialy, S., and Sternberg, A. 2015. 'CO/H_2, C/CO, OH/CO, and OH/O_2 in Dense Interstellar Gas: From High Ionization to Low Metallicity'. *Monthly Notices of the Royal Astronomical Society*, **450**(July), 4424–4445.

[29] Bialy, S., Sternberg, A., and Loeb, A. 2015. 'Water Formation during the Epoch of First Metal Enrichment'. *The Astrophysical Journal Letters*, **804**(May), L29.

[30] Birch, F. 1947. 'Finite Elastic Strain of Cubic Crystals'. *Physical Review*, **71**(June), 809–824.

[31] Bolmont, E., *et al.* 2017. 'Water Loss from Terrestrial Planets Orbiting Ultracool Dwarfs: Implications for the Planets of TRAPPIST-1'. *Monthly Notices of the Royal Astronomical Society*, **464**(Jan.), 3728–3741.

[32] Bond, I. A., *et al.* 2002a. 'Improving the Prospects for Detecting Extrasolar Planets in Gravitational Microlensing Events in 2002'. *Monthly Notices of the Royal Astronomical Society*, **331**(Mar.), L19–L23.

[33] Bond, I. A., Rattenbury, N. J., and Skuljan, J., *et al.* 2002b. 'Study by MOA of Extrasolar Planets in Gravitational Microlensing Events of High Magnification'. *Monthly Notices of the Royal Astronomical Society*, **333**(June), 71–83.

[34] Bond, J. C., O'Brien, D. P., and Lauretta, D. S. 2010. 'The Compositional Diversity of Extrasolar Terrestrial Planets: I. In Situ Simulations'. *Astrophysical Journal*, **715**(June), 1050–1070.

[35] Bonifacio, P., *et al.* 2015. 'TOPoS: II. On the Bimodality of Carbon Abundance in CEMP Stars Implications on the Early Chemical Evolution of Galaxies'. *Astronomy and Astrophysics*, **579**(July), A28.

[36] Bostrom, N. 2002. *Anthropic Bias: Observation Selection Effects in Science and Philosophy*. Routledge.

[37] Bouchy, F., *et al.* 2004. 'Two New 'Very hot Jupiters' among the OGLE Transiting Candidates. *Astronomy and Astrophysics*, **421**(July), L13–L16.

[38] Bousso, R., and Polchinski, J. 2000. 'Quantization of Four-Form Fluxes and Dynamical Neutralization of the Cosmological Constant'. *Journal of High Energy Physics*, **6**(June), 006.

[39] Bouwens, R. J., *et al.* 2015a. 'Reionization after Planck: The Derived Growth of the Cosmic Ionizing Emissivity Now Matches the Growth of the Galaxy UV Luminosity Density'. *Astrophysical Journal*, **811**(Oct.), 140.

[40] Bouwens, R. J., Illingworth, G. D., Oesch, P. A., *et al.* 2015b. 'UV Luminosity Functions at Redshifts $z = 4$ to $z = 10$: 10,000 Galaxies from HST Legacy Fields'. *Astrophysical Journal*, **803**(Apr.), 34.

[41] Bromm, V., and Larson, R. B. 2004. 'The First Stars'. *Annual Reviews of Astronomy and Astrophysics*, **42**(Sept.), 79–118.

[42] Bromm, V., and Loeb, A. 2003. 'The Formation of the First Low-Mass Stars from Gas with Low Carbon and Oxygen Abundances'. *Nature*, **425**(Oct.), 812–814.

[43] Bromm, V., Kudritzki, R. P., and Loeb, A. 2001. 'Generic Spectrum and Ionization Efficiency of a Heavy Initial Mass Function for the First Stars'. *Astrophysical Journal*, **552**(May), 464–472.

[44] Busha, M. T., Evrard, A. E., and Adams, F. C. 2007. 'The Asymptotic Form of Cosmic Structure: Small-Scale Power and Accretion History'. *Astrophysical Journal*, **665**(Aug.), 1–13.

[45] Carollo, D., *et al.* 2012. 'Carbon-Enhanced Metal-Poor Stars in the Inner and Outer Halo Components of the Milky Way'. *Astrophysical Journal*, **744**(Jan.), 195.

[46] Carter, B. 1983. 'The Anthropic Principle and Its Implications for Biological Evolution'. *Philosophical Transactions of the Royal Society of London Series A*, **310**(Dec.), 347–363.

[47] Carter-Bond, J. C., *et al.* 2012a. 'Low Mg/Si Planetary Host Stars and Their Mg-depleted Terrestrial Planets'. *Astrophysical Journal*, **747**(Mar.), L2.

[48] Carter-Bond, J. C., O'Brien, D. P., and Raymond, S. N. 2012b. 'The Compositional Diversity of Extrasolar Terrestrial Planets: II. Migration Simulations'. *Astrophysical Journal*, **760**(Nov.), 44.

[49] Cazaux, S., and Tielens, A. G. G. M. 2002. 'Molecular Hydrogen Formation in the Interstellar Medium'. *Astrophysical Journal*, **575**(Aug.), L29–L32.

[50] Cen, R., and Ostriker, J. P. 1999. 'Where Are the Baryons?'. *Astrophysical Journal*, **514**(Mar.), 1–6.

[51] Chabrier, G. 2003. 'Galactic Stellar and Substellar Initial Mass Function'. *Publications of the Astronomical Society of Australia*, **115**(July), 763–795.

[52] Charbonneau, D., *et al.* 2007. 'When Extrasolar Planets Transit Their Parent Stars'. *Protostars and Planets V*, 701–716.

[53] Christlieb, N., *et al.* 2002. 'A Stellar Relic from the Early Milky Way'. *Nature*, **419**(Oct.), 904–906.

[54] Christlieb, N., *et al.* 2004. 'HE 0107-5240, a Chemically Ancient Star: I. A Detailed Abundance Analysis'. *Astrophysical Journal*, **603**(Mar.), 708–728.

[55] Christlieb, N., *et al.* 2008. 'The Stellar Content of the Hamburg/ESO Survey: IV. Selection of Candidate Metal-Poor Stars'. *Astronomy and Astrophysics*, **484**(June), 721–732.

[56] Chung, S. J., *et al.* 2006. 'The Possibility of Detecting Planets in the Andromeda Galaxy'. *Astrophysical Journal*, **650**(Oct.), 432–437.

[57] Clark, P. C., Glover, S. C. O., and Klessen, R. S. 2008. 'The First Stellar Cluster'. *Astrophysical Journal*, **672**(Jan.), 757–764.

[58] Collet, R., Asplund, M., and Trampedach, R. 2006. 'The Chemical Compositions of the Extreme Halo Stars HE 0107-5240 and HE 1327-2326 Inferred from Three-Dimensional Hydrodynamical Model Atmospheres'. *Astrophysical Journal*, **644**(June), L121–L124.

[59] Conroy, C., *et al.* 2013. Dynamical versus Stellar Masses in Compact Early-Type Galaxies: Further Evidence for Systematic Variation in the Stellar Initial Mass Function'. *The Astrophysical Journal Letters*, **776**(Oct.), L26.

[60] Cooke, R. J., and Madau, P. 2014. 'Carbon-Enhanced Metal-Poor Stars: Relics from the Dark Ages'. *Astrophysical Journal*, **791**(Aug.), 116.

[61] Covone, G., *et al.* 2000. 'Detecting Planets around Stars in Nearby Galaxies'. *Astronomy and Astrophysics*, **357**(May), 816–822.

[62] Dahn, C. C., *et al.* 1977. 'G77-61: A Dwarf Carbon Star'. *Astrophysical Journal*, **216**(Sept.), 757–766.

[63] Dalgarno, A. 2006. 'Interstellar Chemistry Special Feature: The Galactic Cosmic Ray Ionization Rate'. *Proceedings of the National Academy of Science*, **103**(Aug.), 12269–12273.

[64] Davé, R., *et al.* 2001. 'Baryons in the Warm-Hot Intergalactic Medium'. *Astrophysical Journal*, **552**(May), 473–483.

[65] de Bennassuti, M., *et al.* 2014. 'Decoding the Stellar Fossils of the Dusty Milky Way Progenitors'. *Monthly Notices of the Royal Astronomical Society*, **445**(Dec.), 3039–3054.

[66] Dewaele, A., *et al.* 2008. 'High Pressure-High Temperature Equations of State of Neon and Diamond'. *Physical Review B*, **77**(9), 094106.

[67] Di Stefano, R., and Scalzo, R. A. 1999. 'A New Channel for the Detection of Planetary Systems through Microlensing: I. Isolated Events due to Planet Lenses'. *Astrophysical Journal*, **512**(Feb.), 564–578.

[68] Dietrich, M., *et al.* 2003. 'Fe II/Mg II Emission-Line Ratio in High-Redshift Quasars'. *Astrophysical Journal*, **596**(Oct.), 817–829.

[69] Dolphin, A. E., *et al.* 2005. 'Star Formation Histories of Local Group Dwarf Galaxies'. *ArXiv Astrophysics e-prints*, June.

[70] Draine, B. T. 2011. *Physics of the Interstellar and Intergalactic Medium*. Princeton University Press.

[71] Draine, B. T., and Bertoldi, F. 1996. 'Structure of Stationary Photodissociation Fronts'. *Astrophysical Journal*, **468**(Sept.), 269.

[72] Draine, B. T., Roberge, W. G., and Dalgarno, A. 1983. 'Magnetohydrodynamic Shock Waves in Molecular Clouds'. *Astrophysical Journal*, **264**(Jan.), 485–507.

[73] Dressing, C. D., and Charbonneau, D. 2015. 'The Occurrence of Potentially Habitable Planets Orbiting M Dwarfs Estimated from the Full Kepler Dataset and an Empirical Measurement of the Detection Sensitivity'. *Astrophysical Journal*, **807**(July), 45.

[74] Dullemond, C. P., and Dominik, C. 2005. 'Dust Coagulation in Protoplanetary Disks: A Rapid Depletion of Small Grains'. *Astronomy and Astrophysics*, **434**(May), 971–986.

[75] Dunne, L., *et al.* 2003. 'Type II Supernovae as a Significant Source of Interstellar Dust'. *Nature*, **424**(July), 285–287.

[76] Dünner, R., *et al.* 2006. 'The Limits of Bound Structures in the Accelerating Universe'. *Monthly Notices of the Royal Astronomical Society*, **366**(Mar.), 803–811.

[77] Efstathiou, G. 1995. 'An Anthropic Argument for a Cosmological Constant'. *Monthly Notices of the Royal Astronomical Society*, **274**(June), L73–L76.

[78] El Eid, M. F., Fricke, K. J., and Ober, W. W. 1983. 'Evolution of Massive Pregalactic Stars: I. Hydrogen and Helium Burning. II. Nucleosynthesis in Pair Creation Supernovae and Pregalactic Enrichment'. *Astronomy and Astrophysics*, **119**(Mar.), 54–68.

[79] Elmhamdi, A., *et al.* 2003. 'Photometry and Spectroscopy of the Type IIP SN 1999em from Outburst to Dust Formation'. *Monthly Notices of the Royal Astronomical Society*, **338**(Feb.), 939–956.

[80] Ercolano, B., and Clarke, C. J. 2010. 'Metallicity, Planet Formation and Disc Lifetimes'. *Monthly Notices of the Royal Astronomical Society*, **402**(Mar.), 2735–2743.

[81] Ercolano, B., Clarke, C. J., and Drake, J. J. 2009. 'X-Ray Irradiated Protoplanetary Disk Atmospheres: II. Predictions from Models in Hydrostatic Equilibrium'. *Astrophysical Journal*, **699**(July), 1639–1649.

[82] Fialkov, A., *et al.* 2012. Impact of the Relative Motion between the Dark Matter and Baryons on the First Stars: Semi-analytical Modelling'. *Monthly Notices of the Royal Astronomical Society*, **424**(Aug.), 1335–1345.

[83] Fischer, D. A., and Valenti, J. 2005. 'The Planet-Metallicity Correlation'. *Astrophysical Journal*, **622**(Apr.), 1102–1117.

[84] Fisher, D. B., *et al.* 2014. 'The Rarity of Dust in Metal-Poor Galaxies'. *Nature*, **505**(Jan.), 186–189.

[85] Fixsen, D. J. 2009. 'The Temperature of the Cosmic Microwave Background'. *Astrophysical Journal*, **707**(Dec.), 916–920.

[86] Fortney, J. J., Marley, M. S., and Barnes, J. W. 2007. 'Planetary Radii across Five Orders of Magnitude in Mass and Stellar Insolation: Application to Transits'. *Astrophysical Journal*, **659**(Apr.), 1661–1672.

[87] Fox-Skelly, J. *Tardigrades Return from the Dead.* www.bbc.com/earth/story/20150313-the-toughest-animals-on-earth.

[88] Frebel, A., and Norris, J. E. 2015. 'Near-Field Cosmology with Extremely Metal-Poor Stars'. *Annual Reviews of Astronomy and Astrophysics*, **53**(Aug.), 631–688.

[89] Frebel, A., Johnson, J. L., and Bromm, V. 2007. 'Probing the Formation of the First Low-Mass Stars with Stellar Aarchaeology'. *Monthly Notices of the Royal Astronomical Society*, **380**(Sept.), L40–L44.

[90] Fukugita, M., Hogan, C. J., and Peebles, P. J. E. 1998. 'The Cosmic Baryon Budget'. *Astrophysical Journal*, **503**(Aug.), 518–530.

[91] Galametz, M., *et al.* 2011. 'Probing the Dust Properties of Galaxies Up to SubMillimetre Wavelengths: II. Dust-to-Gas Mass Ratio Trends with Metallicity and the Submm Excess in Dwarf Galaxies'. *Astronomy and Astrophysics*, **532**(Aug.), A56.

[92] Garriga, J., and Vilenkin, A. 2003. 'Testable Anthropic Predictions for Dark Energy'. *Physical Review D*, **67**(4), 043503.

[93] Garriga, J., Livio, M., and Vilenkin, A. 2000. 'Cosmological Constant and the Time of its Dominance'. *Physical Review D*, **61**(2), 023503.

[94] Gaudi, B. S., and Gould, A. 1997. 'Planet Parameters in Microlensing Events'. *Astrophysical Journal*, **486**(Sept.), 85–99.

[95] Giddings, S. B., Kachru, S., and Polchinski, J. 2002. 'Hierarchies from Fluxes in String Compactifications'. *Physical Review D*, **66**(10), 106006.

[96] Gilliland, R. L., Brown, T. M., and Guhathakurta, P., *et al.* 2000. 'A Lack of Planets in 47 Tucanae from a Hubble Space Telescope Search'. *Astrophysical Journal*, **545**(Dec.), L47–L51.

[97] Gillon, M., *et al.* 2017. 'Seven Temperate Terrestrial Planets around the Nearby Ultracool Dwarf Star TRAPPIST-1'. *Nature*, **542**(Feb.), 456–460.

[98] Glover, S. C. O., and Clark, P. C. 2014. Molecular Cooling in the Diffuse Interstellar Medium'. *Monthly Notices of the Royal Astronomical Society*, **437**(Jan.), 9–20.

[99] Gonzalez, G., Brownlee, D., and Ward, P. 2001. 'The Galactic Habitable Zone: Galactic Chemical Evolution'. *Icarus*, **152**(July), 185–200.

[100] Gorti, U., and Hollenbach, D. 2009. 'Photoevaporation of Circumstellar Disks By Far-Ultraviolet, Extreme-Ultraviolet and X-Ray Radiation from the Central Star'. *Astrophysical Journal*, **690**(Jan.), 1539–1552.

[101] Gott, III, J. R. 1993. 'Implications of the Copernican Principle for Our Future Prospects'. *Nature*, **363**(May), 315–319.

[102] Gould, A., *et al.* 2006. 'Microlens OGLE-2005-BLG-169 Implies That Cool Neptune-Like Planets Are Common'. *Astrophysical Journal*, **644**(June), L37–L40.

[103] Guillochon, J., and Loeb, A. 2015a. 'SETI via Leakage from Light Sails in Exoplanetary Systems'. *The Astrophysical Journal Letters*, **811**(Oct.), L20.

[104] Guillochon, J., and Loeb, A. 2015b. 'The Fastest Unbound Stars in the Universe'. *Astrophysical Journal*, **806**(June), 124.

[105] Gyuk, G., Dalal, N., and Griest, K. 2000. 'Self-Lensing Models of the Large Magellanic Cloud'. *Astrophysical Journal*, **535**(May), 90–103.

[106] Haghighipour, N., and Kaltenegger, L. 2013. 'Calculating the Habitable Zone of Binary Star Systems: II. P-Type Binaries'. *Astrophysical Journal*, **777**(Nov.), 166.

[107] Hahn, J. M., and Malhotra, R. 1999. 'Orbital Evolution of Planets Embedded in a Planetesimal Disk'. *The Astronomical Journal*, **117**(June), 3041–3053.

[108] Haiman, Z., Thoul, A. A., and Loeb, A. 1996. 'Cosmological Formation of Low-Mass Objects'. *Astrophysical Journal*, **464**(June), 523.

[109] Hanfland, M., Beister, H., and Syassen, K. 1989. 'Graphite under Pressure: Equation of State and First-Order Raman Modes'. *Physical Review B*, **39**(June), 12598–12603.

[110] Heger, A., and Woosley, S.E. 2002. 'The Nucleosynthetic Signature of Population III'. *Astrophysical Journal*, **567**(Mar.), 532–543.

[111] Herbst, E., and Klemperer, W. 1973. 'The Formation and Depletion of Molecules in Dense Interstellar Clouds'. *Astrophysical Journal*, **185**(Oct.), 505–534.

[112] Herrera-Camus, R., *et al.* 2012. 'Dust-to-Gas Ratio in the Extremely Metal-poor Galaxy I Zw 18'. *Astrophysical Journal*, **752**(June), 112.

[113] Hirata, C. M., and Padmanabhan, N. 2006. 'Cosmological Production of H_2 before the Formation of the First Galaxies'. *Monthly Notices of the Royal Astronomical Society*, **372**(Nov.), 1175–1186.

[114] Hollenbach, D., *et al.* 2009. 'Water, O2, and Ice in Molecular Clouds'. *Astrophysical Journal*, **690**(Jan.), 1497–1521.

[115] Hollenbach, D., *et al.* 2012. 'The Chemistry of Interstellar OH^+, H_2O^+, and H_3O^+: Inferring the Cosmic-Ray Ionization Rates from Observations of Molecular Ions'. *Astrophysical Journal*, **754**(Aug.), 105.

[116] Hollenbach, D. J., Werner, M. W., and Salpeter, E. E. 1971. 'Molecular Hydrogen in H i Regions'. *Astrophysical Journal*, **163**(Jan.), 165.

[117] Hubbard, W. B. *Planetary Interiors* Van Nostrand Reinhold.

[118] Huxor, A. P., *et al.* 2005. 'A New Population of Extended, Luminous Star Clusters in the Halo of M31'. *Monthly Notices of the Royal Astronomical Society*, **360**(July), 1007–1012.

[119] Indebetouw, R., *et al.* 2014. 'Dust Production and Particle Acceleration in Supernova 1987A Revealed with ALMA'. *Astrophysical Journal*, **782**(Feb.), L2.

[120] Indriolo, N., and McCall, B. J., 2012. 'Investigating the Cosmic-Ray Ionization Rate in the Galactic Diffuse Interstellar Medium through Observations of H_3^+'. *Astrophysical Journal*, **745**(Jan.), 91.

[121] Ishigaki, M. N., *et al.* 2014. 'Faint Population III Supernovae as the Origin of the Most Iron-Poor Stars'. *Astrophysical Journal*, **792**(Sept.), L32.

[122] Iwamoto, N., *et al.* 2005. 'The First Chemical Enrichment in the Universe and the Formation of Hyper Metal-Poor Stars'. *Science*, **309**(July), 451–453.

[123] Janson, M., *et al.* 2011. 'High-Contrast Imaging Search for Planets and Brown Dwarfs around the Most Massive Stars in the Solar Neighborhood'. *Astrophysical Journal*, **736**(Aug.), 89.

[124] Ji, A. P., Frebel, A., and Bromm, V. 2014. 'The Chemical Imprint of Silicate Dust on the Most Metal-Poor Stars'. *Astrophysical Journal*, **782**(Feb.), 95.

[125] Joggerst, C. C., Woosley, S. E., and Heger, A. 2009. 'Mixing in Zero- and Solar-Metallicity Supernovae'. *Astrophysical Journal*, **693**(Mar.), 1780–1802.

[126] Johansen, A., Youdin, A., and Mac Low, M. M. 2009. 'Particle Clumping and Planetesimal Formation Depend Strongly on Metallicity'. *Astrophysical Journal*, **704**(Oct.), L75–L79.

[127] Johnson, J. L., and Li, H. 2012. 'The First Planets: The Critical Metallicity for Planet Formation'. *Astrophysical Journal*, **751**(June), 81.

[128] Jura, M. 1974. 'Formation and Destruction Rates of Interstellar H_2'. *Astrophysical Journal*, **191**(July), 375–379.

[129] Kachru, S., *et al.* 2003. 'De Sitter Vacua in String Theory'. *Physical Review D*, **68**(4), 046005.

[130] Kasting, J. 2010. *How to Find a Habitable Planet*. Princeton University Press.

[131] Kasting, J. F., Whitmire, D. P., and Reynolds, R. T. 1993. 'Habitable Zones around Main Sequence Stars'. *Icarus*, **101**(Jan.), 108–128.

[132] Kaufman, M. J., and Neufeld, D. A. 1996. 'Far-Infrared Water Emission from Magnetohydrodynamic Shock Waves'. *Astrophysical Journal*, **456**(Jan.), 611.

[133] Keller, S. C., *et al.* 2014. 'A Single Low-Energy, Iron-Poor Supernova as the Source of Metals in the Star SMSS J031300.36-670839.3'. *Nature*, **506**(Feb.), 463–466.

[134] Kneib, J. P., *et al.* 2004. 'A Probable z~7 Galaxy Strongly Lensed by the Rich Cluster A2218: Exploring the Dark Ages'. *Astrophysical Journal*, **607**(June), 697–703.

[135] Kobayashi, H., *et al.* 2011. 'Sublimation Temperature of Circumstellar Dust Particles and Its Importance for Dust Ring Formation'. *Earth, Planets, and Space*, **63**(Oct.), 1067–1075.

[136] Konacki, M., *et al.* 2003. 'An Extrasolar Planet That Transits the Disk of Its Parent Star'. *Nature*, **421**(Jan.), 507–509.

[137] Kopparapu, R. K., *et al.* 2013. 'Habitable Zones around Main-Sequence Stars: New Estimates'. *Astrophysical Journal*, **765**(Mar.), 131.

[138] Kornet, K., *et al.* 2005. 'Formation of Giant Planets in Disks with Different Metallicities'. *Astronomy and Astrophysics*, **430**(Feb.), 1133–1138.

[139] Kozasa, T., Hasegawa, H., and Nomoto, K. 1989. 'Formation of Dust Grains in the Ejecta of SN 1987A'. *Astrophysical Journal*, **344**(Sept.), 325–331.

[140] Kreidberg, L., and Loeb, A. 2016. 'Prospects for Characterizing the Atmosphere of Proxima Centauri b'. *The Astrophysical Journal Letters*, **832**(Nov.), L12.

[141] Kress, M. E., and Tielens, A. G. G. M. 2001. 'The Role of Fischer-Tropsch Catalysis in Solar Nebula Chemistry'. *Meteoritics and Planetary Science*, **36**(Jan.), 75–92.

[142] Krisciunas, K., Garnavich, P. M., Challis, P., *et al.* 2005. 'Hubble Space Telescope Observations of Nine High-Redshift ESSENCE Supernovae1'. *The Astronomical Journal*, **130**(Dec.), 2453–2472.

[143] Kuchner, M. J., and Seager, S. 2005. 'Extrasolar Carbon Planets'. *ArXiv Astrophysics e-prints*, Apr.

[144] Larimer, J. W., 1975. 'The Effect of C/O Ratio on the Condensation of Planetary Material'. *Geochimica et Cosmochimica Acta*, **39**(Mar.), 389–392.

[145] Laskar, J., Joutel, F., and Robutel, P. 1993. 'Stabilization of the Earth's Obliquity by the Moon'. *Nature*, **361**(Feb.), 615–617.

[146] Laughlin, G., Bodenheimer, P., and Adams, F. C. 1997. 'The End of the Main Sequence'. *Astrophysical Journal*, **482**(June), 420–432.

[147] Le Bourlot, J., Pineau des Forêts, G., and Flower, D. R. 1999. 'The Cooling of Astrophysical Media by H_2'. *Monthly Notices of the Royal Astronomical Society*, **305**(May), 802–810.

[148] Léger, A., *et al.* 2004. 'A New Family of Planets? "Ocean-Planets"'. *Icarus*, **169**(June), 499–504.

[149] Leslie, J. 1996. *The End of the World: The Science and Ethics of Human Extinction* Routledge.

[150] Lin, H. W., and Loeb, A. 2014. 'Finding Rocky Asteroids around White Dwarfs by Their Periodic Thermal Emission'. *The Astrophysical Journal Letters*, **793**(Oct.), L43.

[151] Lin, H. W., and Loeb, A. 2015. 'Statistical Signatures of Panspermia in Exoplanet Surveys'. *The Astrophysical Journal Letters*, **810**(Sept.), L3.

[152] Lin, H. W., Gonzalez Abad, G., and Loeb, A. 2014. 'Detecting Industrial Pollution in the Atmospheres of Earth-Like Exoplanets'. *The Astrophysical Journal Letters*, **792**(Sept.), L7.

[153] Lineweaver, C. H. 2001. 'An Estimate of the Age Distribution of Terrestrial Planets in the Universe: Quantifying Metallicity as a Selection Effect'. *Icarus*, **151**(June), 307–313.

[154] Lineweaver, C. H., Fenner, Y., and Gibson, B. K. 2004. 'The Galactic Habitable Zone and the Age Distribution of Complex Life in the Milky Way'. *Science*, **303**(Jan.), 59–62.

[155] Lissauer, J. J. 1993. 'Planet Formation'. *Annual Reviews of Astronomy and Astrophysics*, **31**, 129–174.

[156] Livio, M. 1999. 'How Rare Are Extraterrestrial Civilizations, and When Did They Emerge?' *Astrophysical Journal*, **511**(Jan.), 429–431.

[157] Loeb, A. 2002. 'Long-Term Future of Extragalactic Astronomy'. *Physical Review D*, **65**(4), 047301.

[158] Loeb, A. 2006. 'An Observational Test for the Anthropic Origin of the Cosmological Constant'. *Journal of Cosmology and Astroparticle Physics*, **5**(May), 009.

[159] Loeb, A. 2014. 'The Habitable Epoch of the Early Universe'. *International Journal of Astrobiology*, **13**(Sept.), 337–339.

[160] Loeb, A., and Furlanetto, S. R. 2013. *The First Galaxies in the Universe*. Princeton University Press.

[161] Loeb, A., and Guillochon, J. 2014. 'Observational Cosmology with Semi-Relativistic Stars'. *ArXiv e-prints*, Nov.

[162] Loeb, A., and Maoz, D. 2013. 'Detecting Biomarkers in Habitable-Zone Earths Transiting White Dwarfs'. *Monthly Notices of the Royal Astronomical Society*, **432**(May), 11–15.

[163] Loeb, A., and Turner, E. L. 2012. 'Detection Technique for Artificially Illuminated Objects in the Outer Solar System and Beyond'. *Astrobiology*, **12**(Apr.), 290–294.

[164] Loeb, A., and Zaldarriaga, M. 2007. 'Eavesdropping on Radio Broadcasts from Galactic Civilizations with Upcoming Observatories for Redshifted 21 cm Radiation'. *Journal of Cosmology and AstroParticle Physics*, **1**(Jan.), 020.

[165] Loeb, A., Batista, R. A., and Sloan, D. 2016. 'Relative Likelihood for Life as a Function of Cosmic Time'. *Journal of Cosmology and Astroparticle Physics*, **2016**(Aug.), 040.

[166] LoVerde, M., and Smith, K. M. 2011. 'The Non-Gaussian Halo Mass Function with f_{NL}, g_{NL} and τ_{NL}'. *Journal of Cosmology and Astroparticle Physics*, **8**(Aug.), 003.

[167] Lucy, L. B., *et al.* 1989. 'Dust Condensation in the Ejecta of SN 1987 A'. Tenorio-Tagle, G., Moles, M., and Melnick, J. eds., *IAU Colloquium 120: Structure and Dynamics of the Interstellar Medium*. Lecture Notes in Physics, Berlin Springer Verlag, vol. 350 Page 164.

[168] Luger, R., and Barnes, R. 2015. 'Extreme Water Loss and Abiotic O_2 Buildup on Planets throughout the Habitable Zones of M Dwarfs'. *Astrobiology*, **15**(Feb.), 119–143.

[169] Madau, P., and Dickinson, M. 2014. 'Cosmic Star-Formation History'. *Annual Reviews of Astronomy and Astrophysics*, **52**(Aug.), 415–486.

[170] Madhusudhan, N., Harrington, J., Stevenson, K. B., *et al.* 2011. 'A High C/O Ratio and Weak Thermal Inversion in the Atmosphere of Exoplanet WASP-12b'. *Nature*, **469**(Jan.), 64–67.

[171] Madhusudhan, N., Lee, K. K. M., and Mousis, O. 2012. 'A Possible Carbon-Rich Interior in Super-Earth 55 Cancri e'. *Astrophysical Journal*, **759**(Nov.), L40.

[172] Maio, U., *et al.* 2012. 'Counts of High-Redshift GRBs as Probes of Primordial Non-Gaussianities'. *Monthly Notices of the Royal Astronomical Society*, **426**(Nov.), 2078–2088.

[173] Maldacena, J. 2003. 'Non-Gaussian Features of Primordial Fluctuations in Single Field Inflationary Models'. *Journal of High Energy Physics*, **5**(May), 013.

[174] Mallén-Ornelas, G., *et al.* 2003. 'The EXPLORE Project: I. A Deep Search for Transiting Extrasolar Planets'. *Astrophysical Journal*, **582**(Jan.), 1123–1140.

[175] Maloney, A., Silverstein, E., and Strominger, A. 2002. 'De Sitter Space in Non-Critical String Theory'. *NASA STI/Recon Technical Report N*, **3**(June).

[176] Mao, S., and Paczynski, B. 1991. 'Gravitational Microlensing by Double Stars and Planetary Systems'. *Astrophysical Journal*, **374**(June), L37–L40.

[177] Marassi, S., *et al.* 2014. 'The Origin of the Most Iron-Poor Star'. *Astrophysical Journal*, **794**(Oct.), 100.

[178] Marassi, S., *et al.* 2015. 'The Metal and Dust Yields of the First Massive Stars'. *Monthly Notices of the Royal Astronomical Society*, **454**(Dec.), 4250–4266.

[179] Marcy, G. W., *et al.* 2014. 'Occurrence and Core-Envelope Structure of 1-4 Earth-Size Planets around Sun-Like Stars'. *Proceedings of the National Academy of Science*, **111**(Sept.), 12655–12660.

[180] Martel, H., Shapiro, P. R., and Weinberg, S. 1998. 'Likely Values of the Cosmological Constant'. *Astrophysical Journal*, **492**(Jan.), 29–40.

[181] Martin, R. G., and Livio, M. 2013. 'On the Formation and Evolution of Asteroid Belts and Their Potential Significance for Life'. *Monthly Notices of the Royal Astronomical Society*, **428**(Jan.), L11–L15.

[182] Mashian, N., and Loeb, A. 2016. 'CEMP Stars: Possible Hosts to Carbon Planets in the Early Universe'. *Monthly Notices of the Royal Astronomical Society*, May.

[183] McElroy, D., *et al.* 2013. 'The UMIST Database for Astrochemistry 2012'. *Astronomy and Astrophysics*, **550**(Feb.), A36.

[184] Meylan, G., and Heggie, D. C. 1997. 'Internal Dynamics of Globular Clusters'. *Astronomy and Astrophysicsr*, **8**, 1–143.

[185] Mojzsis, S. J., *et al.* 1996. 'Evidence for Life on Earth before 3,800 Million Years Ago'. *Nature*, **384**(Nov.), 55–59.

[186] Monje, R. R., *et al.* 2011. 'Herschel/HIFI Observations of Hydrogen Fluoride Toward Sagittarius B2(M)'. *Astrophysical Journal*, **734**(June), L23.

[187] Moore, B., *et al.* 2006. 'Globular Clusters, Satellite Galaxies and Stellar Haloes from Early Dark Matter Peaks'. *Monthly Notices of the Royal Astronomical Society*, **368**(May), 563–570.

[188] Morrison, H. L., *et al.* 2003. 'Andromeda VIII: A New Tidally Distorted Satellite of M31'. *Astrophysical Journal*, **596**(Oct.), L183–L186.

[189] Musso, M., and Sheth, R. K., 2014. 'The Excursion Set Approach in Non-Gaussian Random Fields'. *Monthly Notices of the Royal Astronomical Society*, **439**(Apr.), 3051–3063.

[190] Nagamine, K., and Loeb, A. 2004. 'Future Evolution of the Intergalactic Medium in a Universe Dominated by a Cosmological Constant'. *New Astronomy*, **9**(Oct.), 573–583.

[191] Nagasawa, M., *et al.* 2007. 'The Diverse Origins of Terrestrial-Planet Systems'. *Protostars and Planets V*, 639–654.

[192] Naka, S., *et al.* 1976. 'Direct Conversion of Graphite to Diamond under Static Pressure'. *Nature*, **259**(Jan.), 38–39.

[193] Naoz, S., Noter, S., and Barkana, R. 2006. 'The First Stars in the Universe'. *Monthly Notices of the Royal Astronomical Society*, **373**(Nov.), L98–L102.

[194] Niedzielski, A., *et al.* 2009. 'Substellar-Mass Companions to the K-Dwarf BD+14 4559 and the K-Giants HD 240210 and BD+20 2457'. *Astrophysical Journal*, **707**(Dec.), 768–777.

[195] Norris, J. E., Ryan, S. G., and Beers, T. C. 1997. 'Extremely Metal-Poor Stars: The Carbon-Rich, Neutron Capture Element–Poor Object CS 22957-027'. *Astrophysical Journal*, **489**(Nov.), L169.

[196] Norris, J. E., *et al.* 2013. 'The Most Metal-Poor Stars: IV. The Two Populations with [Fe/H] $<^\sim -3.0$'. *Astrophysical Journal*, **762**(Jan.), 28.

[197] Nozawa, T., *et al.* 2003. 'Dust in the Early Universe: Dust Formation in the Ejecta of Population III Supernovae'. *Astrophysical Journal*, **598**(Dec.), 785–803.

[198] Ober, W. W., El Eid, M. F., and Fricke, K. J. 1983. 'Evolution of Massive Pregalactic Stars: Part Two; Nucleosynthesis in Pair Creation Supernovae and Pregalactic Enrichment'. *Astronomy and Astrophysics*, **119**(Mar.), 61.

[199] Omukai, K., *et al.* 2005. 'Thermal and Fragmentation Properties of Star-Forming Clouds in Low-Metallicity Environments'. *Astrophysical Journal*, **626**(June), 627–643.

[200] Owen, J. E., and Mohanty, S. 2016. 'Habitability of Terrestrial-Mass Planets in the HZ of M Dwarfs: I. H/He-Dominated Atmospheres'. *Monthly Notices of the Royal Astronomical Society*, **459**(July), 4088–4108.

[201] Papaloizou, J. C. B., and Terquem, C. 2006. 'Planet Formation and Migration'. *Reports on Progress in Physics*, **69**(Jan.), 119–180.

[202] Papoular, R., *et al.* 1996. 'A Comparison of Solid-State Carbonaceous Models of Cosmic Dust'. *Astronomy and Astrophysics*, **315**(Nov.), 222–236.

[203] Park, B. G., *et al.* 2006. 'Microlensing Sensitivity to Earth-Mass Planets in the Habitable Zone'. *Astrophysical Journal*, **643**(June), 1233–1238.

[204] Penteado, E. M., Cuppen, H. M., and Rocha-Pinto, H. J. 2014. 'Modelling the Chemical Evolution of Molecular Clouds as a Function of Metallicity'. *Monthly Notices of the Royal Astronomical Society*, **439**(Apr.), 3616–3629.

[205] Pepper, J., and Gaudi, B. S. 2005. 'Searching for Transiting Planets in Stellar Systems'. *Astrophysical Journal*, **631**(Sept.), 581–596.

[206] Perlmutter, S. 2005. 'Studying Dark Energy with Supernovae: Now, Soon, and the Not-Too-Distant Future'. *Physica Scripta Volume T*, **117**(Jan.), 17–28.

[207] Petigura, E. A., Howard, A. W., and Marcy, G. W. 2013. 'Prevalence of Earth-Size Planets Orbiting Sun-Like Stars'. *Proceedings of the National Academy of Science*, **110**(Nov.), 19273–19278.

[208] Pierrehumbert, R., and Gaidos, E. 2011. 'Hydrogen Greenhouse Planets beyond the Habitable Zone'. *The Astrophysical Journal Letters*, **734**(June), L13.

[209] Pinotti, R., *et al.* 2005. 'A link between the Semimajor Axis of Extrasolar Gas Giant Planets and Stellar Metallicity'. *Monthly Notices of the Royal Astronomical Society*, **364**(Nov.), 29–36.

[210] Piran, T., and Jimenez, R. 2014. 'Possible Role of Gamma Ray Bursts on Life Extinction in the Universe'. *Physical Review Letters*, **113**(23), 231102.

[211] Planck Collaboration *et al.* 2015a. 'Planck 2015 Results: XIII. Cosmological Parameters'. *ArXiv e-prints*, Feb.

[212] Planck Collaboration *et al.* 2015b. 'Planck 2015 Results: XVII. Constraints on Primordial Non-Gaussianity'. *ArXiv e-prints*, Feb.

[213] Plez, B., and Cohen, J. G. 2005. 'Analysis of the Carbon-Rich Very Metal-Poor Dwarf G77-61'. *Astronomy and Astrophysics*, **434**(May), 1117–1124.

[214] Polchinski, J. 2006. 'The Cosmological Constant and the String Landscape'. *ArXiv High Energy Physics: Theory e-prints*, Mar.

[215] Pritchard, J. R., and Loeb, A. 2012. '21 cm Cosmology in the 21st Century'. *Reports on Progress in Physics*, **75**(8), 086901.

[216] Ray, A., and Loeb, A. 2015. 'Inferring the Composition of Super-Jupiter Mass Companions of Pulsars with Radio Line Spectroscopy'. *ArXiv e-prints*, Oct.

[217] Raymond, S. N., Scalo, J., and Meadows, V. S. 2007. 'A Decreased Probability of Habitable Planet Formation around Low-Mass Stars'. *Astrophysical Journal*, **669**(Nov.), 606–614.

[218] Rhoads, J. E., and Malhotra, S. 2001. 'Lyα Emitters at Redshift z = 5.7'. *Astrophysical Journal*, **563**(Dec.), L5–L9.

[219] Rhoads, J. E., *et al.* 2003. 'Spectroscopic Confirmation of Three Redshift z~5.7 Lyα Emitters from the Large-Area Lyman Alpha Survey'. *The Astronomical Journal*, **125**(Mar.), 1006–1013.

[220] Ricotti, M. 2002. 'Did Globular Clusters Reionize the Universe?.' *Monthly Notices of the Royal Astronomical Society*, **336**(Oct.), L33–L37.

[221] Ricotti, M., and Gnedin, N. Y., 2005. 'Formation Histories of Dwarf Galaxies in the Local Group'. *Astrophysical Journal*, **629**(Aug.), 259–267.

[222] Rodler, F., and López-Morales, M. 2014. 'Feasibility Studies for the Detection of O_2 in an Earth-Like Exoplanet'. *Astrophysical Journal*, **781**(Jan.), 54.

[223] Roederer, I. U., *et al.* 2014. 'A Search for Stars of Very Low Metal Abundance: VI. Detailed Abundances of 313 Metal-Poor Stars'. *The Astronomical Journal*, **147**(June), 136.

[224] Rushby, A. J., *et al.* 2013. 'Habitable Zone Lifetimes of Exoplanets around Main Sequence Stars'. *Astrobiology*, **13**(Sept.), 833–849.

[225] Sahu, K. C. 1994. 'Stars within the Large Magellanic Cloud as Potential Lenses for Observed Microlensing Events'. *Nature*, **370**(July), 275.

[226] Salaris, M., and Cassisi, S. 2006. *Evolution of Stars and Stellar Populations*. Wiley.

[227] Salpeter, E. E., and Zapolsky, H. S. 1967. 'Theoretical High-Pressure Equations of State Including Correlation Energy'. *Physical Review*, **158**(June), 876–886.

[228] Santos, M. R., *et al.* 2004. 'The Abundance of Low-Luminosity Lyα Emitters at High Redshift'. *Astrophysical Journal*, **606**(May), 683–701.

[229] Schaller, E. L., and Brown, M. E. 2007. 'Volatile Loss and Retention on Kuiper Belt Objects'. *The Astrophysical Journal Letters*, **659**(Apr.), L61–L64.

[230] Seager, S., *et al.* 2007. 'Mass-Radius Relationships for Solid Exoplanets'. *Astrophysical Journal*, **669**(Nov.), 1279–1297.

[231] Seager, S., Bains, W., and Hu, R. 2013. 'A Biomass-Based Model to Estimate the Plausibility of Exoplanet Biosignature Gases'. *Astrophysical Journal*, **775**(Oct.), 104.

[232] Shandera, S., *et al.* 2013. 'X-Ray Cluster Constraints on Non-Gaussianity'. *Journal of Cosmology and Astroparticle Physics*, **8**(Aug.), 004.

[233] Sheth, R. K., and Tormen, G. 1999. 'Large-Scale Bias and the Peak Background Split'. *Monthly Notices of the Royal Astronomical Society*, **308**(Sept.), 119–126.

[234] Sigurdsson, S., *et al.* 2003. 'A Young White Dwarf Companion to Pulsar B1620-26: Evidence for Early Planet Formation'. *Science*, **301**(July), 193–196.

[235] Silburt, A., Gaidos, E., and Wu, Y. 2015. 'A Statistical Reconstruction of the Planet Population around Kepler Solar-Type Stars'. *Astrophysical Journal*, **799**(Feb.), 180.

[236] Sloan, D., Alves Batista, R., and Loeb, A. 2017. 'The Resilience of Life to Astrophysical Events'. *Scientific Reports*, **7**(July), 5419.

[237] Sonnentrucker, P., Neufeld, D. A., Phillips, T. G., *et al.* 2010. 'Detection of Hydrogen Fluoride Absorption in Diffuse Molecular Clouds with Herschel/HIFI: An Ubiquitous Tracer of Molecular Gas'. *Astronomy and Astrophysics*, **521**(Oct.), L12.

[238] Sonnentrucker, P., *et al.* 2013. 'Herschel Observations Reveal Anomalous Molecular Abundances toward the Galactic Center'. *Astrophysical Journal*, **763**(Jan.), L19.

[239] Stark, D. P., and Ellis, R. S., 2006. 'Searching for the Sources Responsible for Cosmic Reionization: Probing the Redshift Range $7 < z < 10$ and Beyond'. *New Astronomy*, **50**(Mar.), 46–52.

[240] Sternberg, A., *et al.* 2014. 'H I-to-H_2 Transitions and H I Column Densities in Galaxy Star-Forming Regions'. *Astrophysical Journal*, **790**(July), 10.

[241] Stevenson, D. J. 1999. 'Life-Sustaining Planets in Interstellar Space?'. *Nature*, **400**(July), 32.

[242] Susskind, L. 2003 (Mar.). 'The Anthropic Landscape of String Theory'. In *The Davis Meeting on Cosmic Inflation*. Page 26.

[243] Takai, K., *et al.* 2003. Structure and Electronic Properties of a Nongraphitic Disordered Carbon System and Its Heat-Treatment Effects'. *Physical Review B*, **67**(21), 214202.

[244] Tegmark, M., *et al.* 1997. 'How Small Were the First Cosmological Objects?.' *Astrophysical Journal*, **474**(Jan.), 1.

[245] Tegmark, M., *et al.* 2006. 'Dimensionless Constants, Cosmology, and Other Dark Matters'. *Physical Review D*, **73**(2), 023505.

[246] Tielens, A. G. G. M., *et al.* 1987. 'Shock Processing of Interstellar Dust: Diamonds in the Sky'. *Astrophysical Journal*, **319**(Aug.), L109–L113.

[247] Todini, P., and Ferrara, A. 2001. 'Dust Formation in Primordial Type II Supernovae'. *Monthly Notices of the Royal Astronomical Society*, **325**(Aug.), 726–736.

[248] Tominaga, N., Iwamoto, N., and Nomoto, K. 2014. 'Abundance Profiling of Extremely Metal-Poor Stars and Supernova Properties in the Early Universe'. *Astrophysical Journal*, **785**(Apr.), 98.

[249] Traub, W. A. 2015. 'Steps towards Eta-Earth, from Kepler Data'. *International Journal of Astrobiology*, **14**(July), 359–363.

[250] Tseliakhovich, D., Barkana, R., and Hirata, C. M. 2011. 'Suppression and Spatial Variation of Early Galaxies and Minihaloes'. *Monthly Notices of the Royal Astronomical Society*, **418**(Dec.), 906–915.

[251] Udalski, A. 2003. 'The Optical Gravitational Lensing Experiment: Real Time Data Analysis Systems in the OGLE-III Survey'. *Acta Astronomica*, **53**(Dec.), 291–305.

[252] Udalski, A., *et al.* 2002. 'The Optical Gravitational Lensing Experiment: Search for Planetary and Low-Luminosity Object Transits in the Galactic Disk; Results of 2001 Campaign'. *Acta Astronomica*, **52**(Mar.), 1–37.

[253] Umeda, H., and Nomoto, K. 2003. 'First-Generation Black-Hole-Forming Supernovae and the Metal Abundance Pattern of a Very Iron-Poor Star'. *Nature*, **422**(Apr.), 871–873.

[254] Umeda, H., and Nomoto, K. 2005. 'Variations in the Abundance Pattern of Extremely Metal-Poor Stars and Nucleosynthesis in Population III Supernovae'. *Astrophysical Journal*, **619**(Jan.), 427–445.

[255] Urakawa, S., *et al.* 2006. 'Extrasolar Transiting Planet Search with Subaru Suprime-Cam'. *ArXiv Astrophysics e-prints*, Mar.

[256] Valencia, D., O'Connell, R. J., and Sasselov, D. 2006. 'Internal Structure of Massive Terrestrial Planets'. *Icarus*, **181**(Apr.), 545–554.

[257] Valley, J. W., *et al.* 2002. 'A Cool Early Earth'. *Geology*, **30**(Apr.), 351.

[258] van Dishoeck, E. F., Herbst, E., and Neufeld, D. A. 2013. 'Interstellar Water Chemistry: From Laboratory to Observations'. *Chemical Reviews*, **113**(Dec.), 9043–9085.

[259] Vanderburg, A., *et al.* 2015. 'A Disintegrating Minor Planet Transiting a White Dwarf'. *Nature*, **526**(Oct.), 546–549.

[260] Veras, D., Crepp, J. R., and Ford, E. B. 2009. 'Formation, Survival, and Detectability of Planets beyond 100 AU'. *Astrophysical Journal*, **696**(May), 1600–1611.

[261] Vilenkin, A. 1995. 'Predictions from Quantum Cosmology'. *Physical Review Letters*, **74**(Feb.), 846–849.

[262] Vinet, P., *et al.* 1987. 'Compressibility of Solids'. *Journal of Geophysical Research*, **92**(Aug.), 9319–9325.

[263] Vinet, P., *et al.* 1989. 'Universal Features of the Equation of State of Solids'. *Journal of Physics Condensed Matter*, **1**(Mar.), 1941–1963.

[264] Weinberg, S. 1987. 'Anthropic Bound on the Cosmological Constant'. *Physical Review Letters*, **59**(Nov.), 2607–2610.

[265] Weinberg, S. 2000. 'A Priori Probability Distribution of the Cosmological Constant'. *Physical Review D*, **61**(10), 103505.

[266] Weinberg, S. 2005. 'Living in the Multiverse'. *ArXiv High Energy Physics: Theory e-prints*, Nov.

[267] Winn, J. N., and Fabrycky, D. C. 2015. 'The Occurrence and Architecture of Exoplanetary Systems'. *Annual Reviews of Astronomy and Astrophysics*, **53**(Aug.), 409–447.

[268] Wisotzki, L., *et al.* 1996. 'The Hamburg/ESO survey for Bright QSOs: I. Survey Design and Candidate Selection Procedure'. *Astronomy and Astrophysics Supplement Series*, **115**(Feb.), 227.

[269] Wyse, R. F. G., and Gilmore, G. 2006. 'Stellar Populations with ELTs'. Whitelock, P., Dennefeld, M., and Leibundgut, B., eds. *The Scientific Requirements for Extremely Large Telescopes*. IAU Symposium, vol. 232 Pages 140–148.

[270] Yanny, B., *et al.* 2009. 'SEGUE: A Spectroscopic Survey of 240,000 Stars with g = 14–20'. *The Astronomical Journal*, **137**(May), 4377–4399.

[271] Yasui, C., *et al.* 2009. 'The Lifetime of Protoplanetary Disks in a Low-metallicity Environment'. *Astrophysical Journal*, **705**(Nov.), 54–63.

[272] Yong, D., *et al.* 2013. The Most Metal-Poor Stars: III. The Metallicity Distribution Function and Carbon-Enhanced Metal-Poor Fraction'. *Astrophysical Journal*, **762**(Jan.), 27.

[273] Yoo, J., *et al.* 2004. 'Constraints on Planetary Companions in the Magnification A = 256 Microlensing Event OGLE-2003-BLG-423'. *Astrophysical Journal*, **616**(Dec.), 1204–1214.

[274] York, D. G., Adelman, J., Anderson, J. E., *et al.* 2000. 'The Sloan Digital Sky Survey: Technical Summary'. *The Astronomical Journal*, **120**(Sept.), 1579–1587.

[275] Zackrisson, E., *et al.* 2016. 'Terrestrial Planets across Space and Time'. *ArXiv e-prints*, Feb.

[276] Zahnle, K. J., and Catling, D. C. 2017. 'The Cosmic Shoreline: The Evidence That Escape Determines Which Planets Have Atmospheres, and What This May Mean for Proxima Centauri b'. *ArXiv e-prints*, Feb.

[277] Zapolsky, H. S., and Salpeter, E. E. 1969. 'The Mass-Radius Relation for Cold Spheres of Low Mass'. *Astrophysical Journal*, **158**(Nov.), 809.

[278] Zinner, E. 1998. 'Leonard Award Address: Trends in the Study of Presolar Dust Grains from Primitive Meteorites'. *Meteoritics and Planetary Science*, **33**(July), 549–564.

13

Climbing Up the Theories of Nature: Fine-Tuning and Biological Molecules

GEORGE ELLIS, JEAN-PHILIPPE UZAN, AND DAVID SLOAN

Abstract

The understanding of the world around us requires the use of various theories representing different levels of emergent structure. While they enjoy a decoupling, they are not completely independent. It has been acknowledged that the possibility for complexity to emerge sets constraints, indeed a fine-tuning, on the fundamental constants of nature. It turns out that at each level of emergent complexity, new such constraints appear. It is not clear whether they are more stringent or obviously satisfied in view of constraints relevant for lower levels. We discuss the connection between fine-tuning of fundamental physics, atomic physics, chemistry, and biology to highlight the importance of the analysis of the connection between these different theories in relation to fine-tuning.

13.1 Introduction

13.1.1 The Basic Idea

Fine-tuning relates the existence of life to the values of physical constants. Only certain values of the fundamental constants will allow any life whatsoever to exist [2, 16, 23, 24]. This is because these parameters underlie existence of galaxies, stars, and planets with suitable heavy elements out of which organic molecules can be constructed. However, they also underlie the values of energies and angles in molecules and, hence, also govern the organic chemistry and molecular biology that underlie all living systems, so these, too, will only be possible for a restricted range of these constants. It is a fundamental problem to find out what these relations and limitations are.

Fundamental constants of physics are parameters in the theory that cannot be reduced to other parameters [30, 33]. The following are two key points:

- They are dependent on the theoretical framework used.
- They should be dimensionless to be physically meaningful.

Examples: the fine-structure constant $\alpha = e^2/(\hbar c)$ and electron-proton mass ratio $\mu = m_e/m_p$ are key dimensionless fundamental constants. The speed of light c, Planck's constant $\hbar$, and gravitational constant G are not fundamental constants because they are not dimensionless. You can make their value whatever you want by your choice of units (e.g., you can always set $c = 1$ by choice of time and length units [10]).

There is no agreement on the number of constants or what they are; a comprehensive survey has been given in Reference [27] as well as Chapter 2 of this volume. As mentioned earlier, only a limited range of values of these constants is compatible with existence of life of any kind [2, 16, 23, 24].

Fine-tuning discussions conventionally involve only issues to do with

- Existence of galaxies
- Existence of key elements out of which organic molecules can be made: C, N, O, P, S (as a result of stellar nucleosynthesis followed by explosions of first generation stars)
- Existence of second-generation stars with planets, some of which have an atmosphere and water, perhaps with a moon.

While all these are necessary, they are not sufficient. They do not touch the nature of life itself. That is what we touch on here.

Biologically Important Molecules

One needs also the existence of essential organic molecules and aqueous solutions, thus

- Water with a suitable dipole [3]
- Nucleic acids: suitable DNA [4] and RNA
- Organic molecules: suitable proteins [22]

One also needs lipids and carbohydrates, but their structure is not so crucial. Proteins are the key molecules that do the needed work in all sorts of ways [22], e.g.,

- *Catalysis:* speeding up reactions by a huge amount (enzymes)
- Controlling *gene expression* (gene transcription factors, gene corrections)
- *Logical operations*: controlling flow of ions into and out of axons (e.g., voltage gated ion channels [12])

DNA is important only because it creates proteins at the right time and place [6] (because of gene transcription networks). However, that role is, of course, crucial. RNA is, in effect, at the present time a support actor for DNA (although it may have played a different more role at earlier epochs).

The reader should note that what is presented here is very much a work in progress. We explore a variety of approaches and obtain some preliminary results, but there is much work to be done. In effect, this chapter is an outline of a research program rather than a presentation of results from a completed such program.

13.1.2 The Hierarchy of Structure

Nature confronts us with structures of different scales, complexities, and properties – from fundamental particles to molecules and cells to planets, stars, and galaxies. Each of them can be described by a scientific theory, with its own ontology and structures. The structures form a hierarchy of theories organised in modules in interaction (see References [8, 9, 11] for a discussion). Indeed, those theories are not independent. Higher levels are built on more fundamental theories. And higher-level theories set the context in which the dynamics of the lower-level theories develop. This is related to both bottom-up action and top-down causation [11].

The fact that we can understand the Universe and its laws has deep implications for this structure of theories. At each step in our construction of physical theories, we have been dealing with phenomena below a typical energy scale, mostly because of technological constraints, and it has turned out (empirically) that we always have been able to design a consistent theory that is valid in such a restricted regime. This is not expected in general and is deeply rooted in the mathematical structure of the theories that describe nature. They have to enjoy a *scale decoupling principle* in the sense that there exist energy scales below which effective theories are sufficient to understand a set of physical phenomena that can be observed. *Effective theories* are then the most fundamental concepts in the scientific approach to the understanding of nature, and they always come with a domain of validity inside which they are efficient to describe all related phenomena [28, 29]. They are a successful explanation at a given level of complexity based on the concepts of that particular level.

This implies that the structure of the theories is such that there is a kind of stability and independence of higher levels with respect to more fundamental ones. It follows that various disciplines have developed independently in almost quasi-autonomous domains, each of them having its own ontology and dynamics that are independent of our ability to formulate a theory explaining these concepts in lower-level terms. In each case, we can hope to relate the concepts and constants of a given level to those of an underlying level. For instance, we understand that the proton is a composite structure of three quarks, and we may try to determine its physical characteristic (charge, mass, gyromagnetic factor, quantum numbers) in terms of those of these more fundamental entities [19]. However, we know that this can only be achieved for some structures, since there exist emergent phenomena (information, life, consciousness) that cannot be reduced to the concepts of a lower level.

13.1.3 Fundamental Physics

Today, gravitation is well described by general relativity, and the most fundamental (experi-mentally tested) theory of matter is the Standard Model of particle physics. This is a theoret-ical construction based on an action and many choices such as the mathematical description of the matter fields (this is not completely arbitrary and is based on representations of the Poincaré group, which allows one to define scalar, spinor, vector, etc., structures, but we decide to identify each kind of particle with a particular mathematical structure [31]), symmetries (such as $SU(2)$ or $SU(3)$), and constants. None of them can be explained by the theory at hand. In particular, it is important for the constants to be measurable [21].

Table 13.1 *List of the fundamental constants of our Standard Model. See Reference [21] for further details of the measurements.*

Constant	Symbol	Value
Speed of light	c	299,792,458 m s^{-1}
Planck constant (reduced)	$\hbar$	$1.054\,571\,628(53) \times 10^{-34}$ J s
Newton constant	G	$6.674\,28(67) \times 10^{-11}$ m^2 kg^{-1} s^{-2}
Weak coupling constant (at m_Z)	$g_2(m_Z)$	0.6520 ± 0.0001
Strong coupling constant (at m_Z)	$g_3(m_Z)$	1.221 ± 0.022
Weinberg angle	$\sin^2\theta_W(91.2 \text{ GeV})_{\overline{MS}}$	0.23120 ± 0.00015
Electron Yukawa coupling	h_e	2.94×10^{-6}
Muon Yukawa coupling	h_μ	0.000607
Tauon Yukawa coupling	h_τ	0.0102156
Up Yukawa coupling	h_u	0.000016 ± 0.000007
Down Yukawa coupling	h_d	0.00003 ± 0.00002
Charm Yukawa coupling	h_c	0.0072 ± 0.0006
Strange Yukawa coupling	h_s	0.0006 ± 0.0002
Top Yukawa coupling	h_t	1.002 ± 0.029
Bottom Yukawa coupling	h_b	0.026 ± 0.003
Quark CKM matrix angle	$\sin\theta_{12}$	0.2243 ± 0.0016
	$\sin\theta_{23}$	0.0413 ± 0.0015
	$\sin\theta_{13}$	0.0037 ± 0.0005
Quark CKM matrix phase	δ_{CKM}	1.05 ± 0.24
Higgs potential quadratic coefficient	$\hat{\mu}^2$	$-(250.6 \pm 1.2)$ GeV^2
Higgs potential quartic coefficient	λ	1.015 ± 0.05
QCD vacuum phase	θ_{QCD}	10^{-9}

For this model, they actually are and have been; see Table 13.1. There is no way to express them in terms of more fundamental quantities (else they would not be fundamental constants), and there is no equation for them. Testing their constancy reveals that the hypothesis that they are constant is a good hypothesis, at least to the level of accuracy and timescale over which the experiment is conducted [27]. In case of disagreement, one can promote them to a dynamical field but then would have to explain why they are almost frozen [26, 27].

It is also important to remember that any measurement is just a comparison between two physical systems, one usually defining a system of units. It follows that only dimensionless constants can be measured [10], and only the change of such constants would change the physics. We will consider only these parameters. Given a list of N constants, one can always pick three of them to define units so that one is left with $N - 3$ fundamental parameters that affect the magnitude of any physical process.

Coming back to the Standard Model of particle physics, we have assumed it is our fundamental theory, even though it does not incorporate massive neutrinos and dark matter, so that we know it calls for an extension. This theory offers the possibility of space for higher levels of complexity to emerge. Changing the value of the fundamental constants may result in the technical impossibility for nuclei to be stable – which is an example of fine-tuning. The issue is, how fine-tuned is the Universe? This is what we shall now illustrate, keeping in mind that we would like to estimate how far up the chain of physical theories of higher complexity levels these fine-tunings propagate.

13.1.4 Towards Higher Complexity

Scaling Up to Nuclear Physics

The first level to consider above fundamental particle physics is nuclear physics. There, one needs to determine how cross sections, binding energies, lifetimes of unstable nuclei (or of the neutron), or simply characteristics such as the mass of the protons depends on the fundamental constants listed in Table 13.1. This has been intensively investigated, and we refer to References [26, 27] for reviews. They are of huge importance for the description of Big Bang nucleosynthesis and for stellar nucleosynthesis. As examples, one can compute the binding energy of the deuterium or the lifetime of the proton.

The dependence of the Hoyle state on the nuclear parameters and the fine-structure constant was studied in Reference [7]. It was shown there that the requirement that one forms both carbon-12 and oxygen-16 in population III stars set a constraint on α of the order of 10^{-3}. The dependence of the mass of the proton and gyromagnetic factors are described in Reference [19]. It is important to realise that the connection from QCD to nuclear physics is difficult and intricate so that the accuracy of these scaling is lively debated.

Scaling Up to Atomic Physics

The goal of this chapter is to describe the effect of a change in the value of the fundamental constants on the structure of molecules, where the most importat relevant atoms are H, C, N, O, P, and S. It will thus focus on the standard Schrödinger equation,

$$i\hbar \partial_t \psi(r,t) = H\psi(r,t), \tag{13.1}$$

where H is the Hamiltonian. As far as atomic physics and chemistry are concerned, this Hamiltonian depends on a few constants: the fine-structure constant α, the masses of the nuclei m_I, the mass of the electron m_e, and the gyromagnetic factors of the nuclei and electrons g_I and g_e. The observables are of two types – either dimensionless, such as geometric conformations and angles, or dimensional, such as bond lengths, energy levels, and magnetic and electric moments. These quantities affect the properties of matter. For example, a change in the electric and magnetic properties of the water molecule can impact those of water in a drastic way – e.g., if it changes the solid-liquid phase boundary in the phase diagram, which has a negative slope [3], contrary to most substances. This feature is important for the development of life as we know it.

Hypothesis: *If one changes fundamental constants, such as α and μ, chemistry will not work the same in the case of larger molecules. It will change conformation and so change biological function.*

A delicate issue is whether a change that alters all scales simultaneously would affect biological function or not or, rather, just result in the same function but rescaled in size. It would affect relevant energies, which are, of course, crucial to biology; this might or might not affect function because, for example, the photoelectric effect is not determined by the Schrödinger equation and would scale differently. Presumably, this would alter some functions – for example, those related to photosynthesis and vision – crucial ways.

The text is organised as followed. Section 13.2 describes the Schrödinger equation, focusing on the issue of units and dimensions. We then detail how to compute the effects of a variation of the constants on the properties of molecule in Section 13.3 and apply this to some simple molecules in Section 13.4. The issue of conformal scaling is considered in Section 13.5. We then draw some implications in Section 13.6 in particular, in evaluating whether the fine-tuning required from chemistry is greater or less than that arising from successful carbon production in stars.

13.2 Quantum Physics and Observables

The link from physics to chemistry is via the Schrödinger equation (13.1).

13.2.1 Schrödinger Equation in Dimensionless Form

As discussed previously, one can start by rewriting it in dimensionless form. For instance, among the constants of the problem we can pick $(m_e, \hbar, c)$ to construct our standard units in the form

$$\ell_a = \frac{\hbar}{m_e c} \qquad t_a = \frac{\hbar}{m_e c^2} \qquad E_a = m_e c^2. \tag{13.2}$$

and then perform the rescalings

$$\tau = t/t_a, \qquad \rho = \frac{r}{\ell_a}, \qquad \mathcal{E} = \frac{E}{E_a}, \qquad \mu_i \equiv \frac{m_i}{m_e}. \tag{13.3}$$

It follows that Eq. (13.1) takes the form

$$i \partial_\tau \psi(\rho, \tau) = - \left(\frac{1}{2} \frac{m_e}{m} \Delta + \mathcal{V} \right) \psi(\rho, \tau), \tag{13.4}$$

where Δ is Laplacian associated to ρ and $\mathcal{V} = V/m_e c^2$ is a dimensionless potential. For instance, the ionisation energy of hydrogen $-E_I = \frac{1}{2} m_e c^2 \alpha^2 = 13.60580\,\text{eV}$ in standard units reduces to $-E_I = \frac{1}{2}\alpha^2$ in these atomic units. What is important is that the ratios of any energies will be independent of the choice of the units. The Bohr radius $a_0 = \hbar/m_e c \alpha = 5.29 \times 10^{-11}$ m in standard units reduces to $a_0 = 1/\alpha$ in these atomic units.

This means that we are working with the Hamiltonian

$$\mathcal{H} = \sum_{\text{electrons}} \frac{1}{2} P_i^2 + \sum_{\text{nuclei}} \frac{1}{2\mu_I^2} P_I^2 + \mathcal{V} \equiv \sum_{\text{electrons}} \mathcal{T}_e + \sum_{\text{nuclei}} \mathcal{T}_N(\mu_i) + \mathcal{V}(\alpha, \mu_i, g_i) \quad (13.5)$$

so that one needs to solve

$$\mathcal{H} |\psi\rangle = \mathcal{E} |\psi\rangle. \quad (13.6)$$

13.2.2 Example: Hydrogen Atom

For the hydrogen atom including fine and hyperfine structure, we get for the electron

$$i\partial_\tau \psi_e(\rho, \tau) = -\left(\frac{1}{2} \frac{m_e}{m_p} \Delta + \frac{\alpha}{\rho} + \frac{W}{m_e c^2} \right) \psi_e(\rho, \tau), \quad (13.7)$$

where the term W includes the fine and hyperfine structure. The dimensionless potential $\mathcal{V}$ is the last two terms in (13.7) and is no longer a function of (α/ρ). It is explicitly given by

$$\mathcal{V} = \frac{\alpha}{\rho} + \frac{\alpha}{2\rho^3} \frac{g_e}{2} \frac{\mathbf{L}}{\hbar} \cdot \frac{\mathbf{S}}{\hbar} - \frac{\Delta^2}{8} + \pi\alpha\delta(\rho)$$

$$+ \alpha \frac{g_p}{2} \frac{m_e}{m_p} \left\{ -\frac{1}{2\rho^3} \frac{\mathbf{L}}{\hbar} \cdot \frac{\mathbf{I}}{\hbar} + \frac{2\pi}{3} \frac{g_e}{2} \frac{\mathbf{I}}{\hbar} \cdot \frac{\mathbf{S}}{\hbar} \delta(\rho) + \frac{1}{4\rho^3} \frac{g_e}{2} \left[3 \left(\frac{\mathbf{S}}{\hbar} \cdot \mathbf{n} \right) \left(\frac{\mathbf{I}}{\hbar} \cdot \mathbf{n} \right) - \frac{\mathbf{I}}{\hbar} \cdot \frac{\mathbf{S}}{\hbar} \right] \right\}, \quad (13.8)$$

where the first line is the non-relativistic term, the second line is the the fine-structure terms (discussed further in Section 13.5.2), and the final one is the hyperfine-structure terms. Indeed, it involves only dimensionless quantities: $\alpha, \mu := m_e/m_p$, g_e and g_p. So solving this equation will result in dimensionless energies – i.e., energies in units of $m_e c^2$: $\mathcal{E} = E/m_e c^2$. The optical lines are given by

$$\mathcal{E}_n = \frac{\alpha^2}{2n^2} (1 + \mu)^{-1} \quad (13.9)$$

and the fine structure by

$$\mathcal{E}_{nlJ} = 1 + \frac{\alpha^2}{2n^2} - \frac{\alpha^4}{2n^4} \left(\frac{n}{J + 1/2} - \frac{3}{4} \right) + \cdots \quad (13.10)$$

For a full solution, of course, we also need to include the nucleus – or nuclei, in the case of heavier atoms – as in Eq. (13.5).

13.2.3 Metre Dependence on Fundamental Constants

In the international system of units, the second and the metre are defined by a hyperfine transition in ^{133}Cs. This implies that when the constants are varied, the physical systems to which the system of study is compared are also modified.

In general, the hyperfine frequency in a given electronic state of an alkali-like atom – such as ^{133}Cs, ^{87}Rb, ^{199}Hg$^+$ – is given by

$$\nu_{hfs} \simeq R_\infty c \times A_{hfs} \times g_i \times \alpha^2 \times \bar{\mu} \times F_{hfs}(\alpha), \qquad (13.11)$$

where $g_i = \mu_i/\mu_N$ is the nuclear g factor. A_{hfs} is a numerical factor depending on each particular atom, and we have set $F_{rel} = F_{hfs}(\alpha)$. The Rydberg constant R_∞ is given by $R_\infty = \alpha/4\pi a_0$ – i.e., $\alpha^2/4\pi$ in atomic units. Similarly, the frequency of an electronic transition is well approximated by

$$\nu_{elec} \simeq R_\infty c \times A_{elec} \times F_{elec}(Z, \alpha), \qquad (13.12)$$

where – as before – A_{elec} is a numerical factor depending on each particular atom and F_{elec} is the function accounting for relativistic effects, spin-orbit couplings, and many-body effects. Even though an electronic transition should also include a contribution from the hyperfine interaction, it is generally only a small fraction of the transition energy and, thus, should not carry any significant sensitivity to a variation of the fundamental constants.

It follows that the metre, defined as $L_{1m} \equiv c/\nu_{Cs}$, scales as

$$L_{1m}^{-1} = R_\infty \times A_{hfs} \times g_i \times \alpha^2 \times \bar{\mu} \times F_{Cs}(\alpha). \qquad (13.13)$$

The relativistic corrections can be characterised by introducing the sensitivity of the relativistic factors to a variation of α,

$$\kappa_\alpha \equiv \frac{\partial \ln F}{\partial \ln \alpha}. \qquad (13.14)$$

Table 13.2 summarises the values of some of them. Indeed, a reliable knowledge of these coefficients at the 1%–10% level is required to deduce limits to a possible variation of the constants. The interpretation of the spectra in this context relies, from a theoretical point of view, only on quantum electrodynamics (QED) so that we can safely obtain constraints on (α, μ, g_i), still keeping in mind that the computation of the sensitivity factors requires numerical N-body simulations.

Table 13.2 *Sensitivity of various transitions on a variation of the fine-structure constant.*

Atom	Transition	Sensitivity κ_α
^{1}H	$1s - 2s$	0.00
^{87}Rb	hf	0.34
^{133}Cs	$^2S_{1/2}(F=2) - (F-3)$	0.83
^{171}Yb$^+$	$^2S_{1/2} - {}^2D_{3/2}$	0.9
^{199}Hg$^+$	$^2S_{1/2} - {}^2D_{5/2}$	−3.2
^{87}Sr	$^1S_0 - {}^3P_0$	0.06
^{27}Al$^+$	$^1S_0 - {}^3P_0$	0.008

13.3 Influence on the Structure of Molecules

In this section, we set up a general procedure for finding the effect of a variation of fundamental constants on chemical structure.

13.3.1 Evaluating the Effect of a Change in the Constants

The starting point of the analysis is the Schrödinger equation (13.1). In most problems, it can only be solved numerically or by means of perturbation theory. The main idea is then that the full Hamiltonian H can be split as $H = H_0 + W$, where H_0 has eigenstates that can be computed, and W is a perturbation.

The existing constraints on the time variation of the fundamental constants teach us that we are looking for small effects so that perturbation theory is a well-suited approach. The Hamiltonian depends on a set of constants, c_a, the value of which is c_{a0} in the laboratory today. It can be decomposed as

$$H(c_i) = H_0 + W \tag{13.15}$$

with

$$H_0 \equiv H(c_a = c_{a0}) \tag{13.16}$$

and

$$W \equiv \sum_a \left(\frac{\delta H}{\delta \ln c_a} \bigg|_{c_a=c_{a0}} \right) \delta \ln c_a \tag{13.17}$$

since we assume that $\delta c_a / c_a \ll 1$. Given the general structure of the Hamiltonian (see Eqs. (13.5), and (13.6)), one concludes that the dependences on the fine-structure constant and gyromagnetic factor appear only in the potential while the masses appear in both the potential and kinetic terms. Hence, rescaling the perturbation potential as $\mathcal{W} = W/m_e c^2$, we get

$$\mathcal{W} = \sum_a \left(\frac{\delta \mathcal{H}}{\delta \ln c_a} \bigg|_{c_a=c_{a0}} \right) \delta \ln c_a \equiv \sum_a f_a \delta \ln c_a, \tag{13.18}$$

which defines f_a.

We shall now perform a perturbation on the value of the fundamental constants. We work with the rescaled adimensional quantities. Indeed, solving the Schrödinger equation with the Hamiltonian $\mathcal{H}_0$ may require us to use another perturbation expansion in terms of the parameter of the interaction potential. Let us assume this leads to a set of eigenmodes $\left| \varphi_p^i \right\rangle$ associated to energies $\mathcal{E}_p^0$ so that i labels degenerate states of same energy:

$$\mathcal{H}_0 \left| \varphi_p^i \right\rangle = \mathcal{E}_p^0 \left| \varphi_p^i \right\rangle. \tag{13.19}$$

As usual, the eigenstates form an orthonormal basis of the space of states,

$$\left\langle \varphi_p^i | \varphi_q^j \right\rangle = \delta_{pq} \delta^{ij}, \qquad \sum_{p,i} \left| \varphi_p^i \right\rangle \left\langle \varphi_p^i \right| = 1. \tag{13.20}$$

Perturbation theory allows one to determine the perturbed energy levels and associated eigenstates. The derivation can be found in many textbooks. In a case of a non-degenerate state, one gets, at first order,

$$\mathcal{E}_n = \mathcal{E}_n^0 + \left\langle \varphi_p | \mathcal{W} | \varphi_p \right\rangle \tag{13.21}$$

for the energy and

$$\left| \psi_p \right\rangle = \left| \varphi_p \right\rangle + \sum_{n \neq p} \sum_i \frac{\left\langle \varphi_n^i | \mathcal{W} | \varphi_p \right\rangle}{\mathcal{E}_p^0 - \mathcal{E}_n^0} \left| \varphi_n^i \right\rangle \tag{13.22}$$

for the eigenstate.

13.3.2 Effect on Bond Length

A molecule with N atoms and P electrons has a state that is specified by $\vec{R}_I$ for the position of the atoms and $\vec{r}_i$ for the position of the electrons so that we are looking for a wave function $\psi(\vec{R}_I, \vec{r}_i; c_a)$. The bond length is then given by

$$\ell_{IJ} \equiv \left\langle \psi | R_{IJ} | \psi \right\rangle \tag{13.23}$$

for a given bond between nuclei I and J with $\vec{R}_{IJ} \equiv \vec{R}_J - \vec{R}_I$. However, this has units. We can define two dimensionless quantities, either the bound length compared to the metre (the value of which depends on the value of the constants; see Section 13.2.3, or in units of the Bohr radius a_0,

$$\tilde{\ell}_{IJ} \equiv \frac{\ell_{IJ}}{L_{1m}}, \qquad \lambda_{IJ} \equiv \frac{\ell_{IJ}}{\ell_a}. \tag{13.24}$$

Both are dimensionless quantities, the variation of which characterises a change in the geometry of the molecule. Another quantity that satisfies this criteria is the ratio between different bond lengths,

$$\alpha_{IJ;KL} \equiv \frac{\ell_{IJ}}{\ell_{KL}}. \tag{13.25}$$

Given the preceding description, we get that

$$\delta \lambda_{IJ}^{(p)} = \sum_a \left(\sum_{n \neq p} \sum_i \frac{\left\langle \varphi_n^i | f_a | \varphi_p \right\rangle}{\mathcal{E}_p^0 - \mathcal{E}_n^0} \left\langle \varphi_p | R_{IJ} | \varphi_n^i \right\rangle + c.c. \right) \delta \ln c_a \tag{13.26}$$

for the state p. What is usually called the bound length refers to the fundamental state. It follows that the sensitivity of the bond length to the variation of the constant c_a is

$$\frac{\delta \lambda_{IJ}}{\delta \ln c_a} = \sum_{n \neq p} \sum_i \frac{\langle \varphi_n^i | f_a | \varphi_p \rangle}{\mathcal{E}_p^0 - \mathcal{E}_n^0} \left\langle \varphi_p | R_{IJ} | \varphi_n^i \right\rangle + c.c., \tag{13.27}$$

where p is the fundamental state. This expression depends only on the energies and eigenstates of the standard molecule so that it can be computed from existing codes.

13.3.3 Effect on the Geometry

Angles are dimensionless so that a good indicator of the change of the geometry is

$$\beta_{\widehat{IJK}} \equiv \left\langle \psi \left| \frac{\vec{R}_{IJ} . \vec{R}_{IK}}{R_{IJ} R_{IK}} \right| \psi \right\rangle. \tag{13.28}$$

As before, we deduce the sensitivity of the angle to the constant c_a as

$$\frac{\delta \beta_{\widehat{IJK}}}{\delta \ln c_a} = \sum_{n \neq p} \sum_i \frac{\langle \varphi_n^i | f_a | \varphi_p \rangle}{\mathcal{E}_p^0 - \mathcal{E}_n^0} \left\langle \varphi_p \left| \frac{\vec{R}_{IJ} . \vec{R}_{IK}}{R_{IJ} R_{IK}} \right| \varphi_n^i \right\rangle + c.c. \tag{13.29}$$

13.4 Applications

The simplest applications to test the effect of changing the fine-structure constant on molecular structure are the hydrogen atom (Section 13.4.1) and hydrogen molecule (Section 13.4.2). We thank Peter Atkins for discussions and help with this section. Atkins's book *Molecular Quantum Mechanics* [1] (MQM) is a great resource as regards concepts and calculational methods. The next simplest application is the water molecule (Section 13.4.3), which is of considerable biological importance.

13.4.1 Hydrogen Atom

The constants needed in what follows, chosen in order to simplify the equations, are

$$\alpha = e^2/4\pi\epsilon_0\hbar, \ \beta = \hbar/(2m_e c), \ \gamma = c\hbar, \tag{13.30}$$

which imply

$$e^2/4\pi\epsilon_0 = \alpha c\hbar = \alpha\gamma, \ \hbar^2/(2m_e) = \beta\hbar c = \beta\gamma. \tag{13.31}$$

The energies (see [1], p. 89; $Z = 1$) are

$$E_n = -\frac{m_e e^4}{32\pi^2\epsilon_0^2\hbar^2} \cdot \frac{1}{n^2} = -\frac{m_e}{2\hbar^2} \left(\frac{e^2}{4\pi\epsilon_0}\right)^2 \cdot \frac{1}{n^2} = -\frac{\gamma\alpha^2}{4\beta} \cdot \frac{1}{n^2}. \tag{13.32}$$

The orbital of lowest energy ($1s$; $n = 1, l = 0, m_l = 0$) is ([1], p. 88 and Y from p. 78)

$$\Psi_{1s}(r,\theta,\phi) = \frac{1}{\pi^{1/2}}\left(\frac{1}{a_0}\right)^{3/2} e^{-r/a_0} = \frac{1}{\pi^{1/2}}\left(\frac{\alpha}{2\beta}\right)^{3/2} e^{-2\beta r/\alpha}. \tag{13.33}$$

The most probable distance of the electron from the nucleus is the Bohr radius a_0:

$$a_0 = (4\pi\epsilon_0\hbar^2)/(m_e e^2) = \frac{2\beta}{\alpha} \Rightarrow \frac{\partial a_0}{\partial\alpha} = -\frac{2\beta}{\alpha^2} = -\frac{a_0}{\alpha}. \tag{13.34}$$

That is,

$$\partial \ln a_0/\partial\alpha = -1/\alpha = -1/137 = -0.0007. \tag{13.35}$$

As expected, an increase in α (an increase in charge) decreases the size of the atom.

13.4.2 Hydrogen Molecule

The Hamiltonian for the hydrogen molecule has terms for the two protons and the two electrons, all interacting though a Coulomb potential,

$$\mathcal{H} = -\frac{1}{2}\sum_{i=1,2} P_i^2 - \frac{1}{2\mu_p}\sum_{I=1,2} P_I^2 + \frac{\alpha}{R_{12}} + \frac{\alpha}{r_{12}} - \sum_{I,i} \frac{\alpha}{r_{Ii}}, \tag{13.36}$$

from which we extract the dependence on α and μ_p separately:

$$f_\alpha = \frac{\alpha}{R_{12}} + \frac{\alpha}{r_{12}} - \sum_{I,i} \frac{\alpha}{r_{Ii}}, \qquad f_{\mu_p} = +\frac{1}{2\mu_p}\sum_{I=1,2} P_I^2. \tag{13.37}$$

To determine specific values, we follow Peter Atkins thus: consider the hydrogen molecule ion (H_2^+). From [1], p. 265 in the LCAO approximation, for an internuclear separation R, with $s = R/a_0 = \alpha R/2\beta$, one has

$$E_+ - E_{1s} = \frac{\alpha\gamma}{R} - \frac{j' + k'}{1 + S} \tag{13.38}$$

$$j' = \frac{j_0}{R}\{1 - (1+s)e^{-2s}\} = \frac{\alpha\gamma}{R}\left[1 - \left(1 + \frac{\alpha R}{2\beta}\right)e^{-\alpha R/\beta}\right] \tag{13.39}$$

$$k' = \frac{j_0}{a_0}1 + se^{-s} = \frac{2\alpha^2}{\beta}\left[1 + \frac{\alpha R}{2\beta}\right]e^{-\alpha R/2\beta} \tag{13.40}$$

$$S = \left(1 + s + \frac{1}{3}s^2\right)e^{-s} = \left[1 + \frac{\alpha R}{2\beta} + \frac{1}{3}\left(\frac{\alpha R}{2\beta}\right)^2\right]e^{-\alpha R/2\beta}. \tag{13.41}$$

Exploring numerically how the energy varies with r for different values of α results in Figures 13.1–13.3. The key outcome is the result,

Variation of the fine-structure constant by about 6% gives a change in the bond length of the hydrogen ion molecule by about 6%.

This agrees with the above result (Eq. (13.35)) for the hydrogen atom.

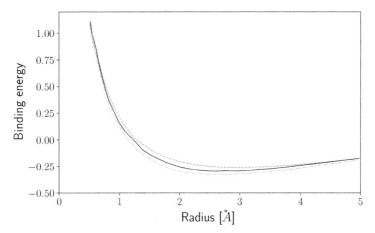

Figure 13.1 Graph of energies of the system as a function of radius. Units: Angstroms (but note that the metre depends on α). Binding will correspond to a minimum. Various values of α are shown. The middle one is the physical value of α. The upper and lower are $\pm6\%$ change in the binding radius.

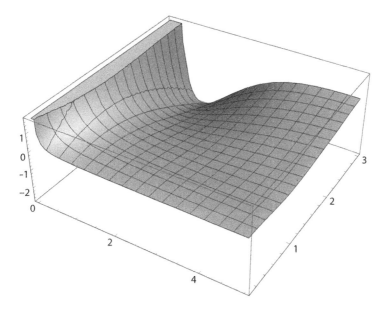

Figure 13.2 Variation of energy with radius (on the x-axis marked 0–4) against change in the fine-structure constant α on the y-axis. For each value of α, the minimum energy gives the binding radius.

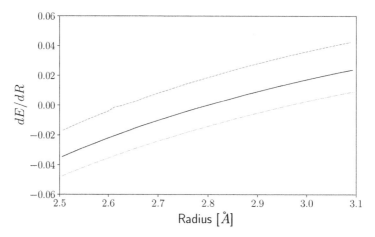

Figure 13.3 Finding the minimum. Graph of dE/dr as a function of radius. Various values of α are shown. The middle line is the physical value of α, the lowest curve is α reduced by 5.6%, and the highest curve is α increased by 6.4%. Both give ±6% change in the binding radius, which may be the anthropic bound (Section 13.6).

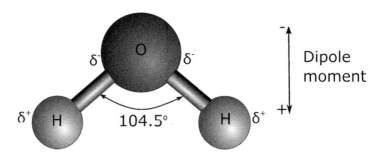

Figure 13.4 The water molecule has a crucial dipole moment because of its H-O-H angle of 104.5°. From [18], Figure 2.5.

13.4.3 Water Molecule

The water molecule H_2O [3] is a significant biological molecule because it is an excellent solvent that is liquid at room temperature, but freezes from the top because ice floats (frozen water is less dense than liquid water).

Water forms a liquid at room temperature because of its H-O-H angle of 104°, resulting in a dipole moment (the positive charges on the hydrogen atoms and negative charge on the oxygen atom are not aligned; Figure 13.4). Electrical attraction between water molecules due to this dipole reduces intermolecular distances, increasing the interaction energy and so raising its boiling point. Water is also a good solvent, because this polarity enables it to be strongly attracted to other polar molecules such as NaCl and, indeed, to disrupt them into their component atoms.

Hamiltonian

The Hamiltonian for the water molecule is given by the general formula for any molecule. Let i label the electrons in a molecule and I the nuclei, with masses m_i and M_I; then the non-relativistic equation is

$$\mathcal{H}\Psi = E\Psi, \tag{13.42}$$

where

$$\mathcal{H} = -\frac{\hbar^2}{2m_e}\sum_i \nabla_i^2 - \frac{\hbar^2}{2m_e}\sum_i \nabla_I^2 + \frac{e^2}{4\pi\epsilon_0}\sum_{(I,J)}\frac{Z_I Z_J}{r_{IJ}}$$

$$-\frac{e^2}{4\pi\epsilon_0}\sum_{(i,I)}\frac{Z_I}{r_{iI}} + \frac{1}{2}\frac{e^2}{4\pi\epsilon_0}\sum_{(i,j)\neq)}\frac{1}{r_{ij}}, \tag{13.43}$$

where all positions are determined with respect to the oxygen nucleus. r_{IJ} are the inter-nuclear distances, r_{iI} the electron to nucleus distances, and R_{ij} the inter-electron distances.

In the case of the water molecule with the most abundant isotope of oxygen, there are two hydrogen nuclei (protons of atomic mass 1), one oxygen nucleus (atomic mass 16), and 18 electrons.

Standard Lore

The general geometric form of molecules of the form $AX_m E_n$ – where A is the central atom, X is the other atom, and E is the free electron pairs – is depicted in Figure 13.5. The sum $m + n$ determines the geometry. Water, and all its isotopes, have $m + n = 4$ and are of form f in Figure 13.5. Because of the attraction by the nuclei of the X atom, a bonding pair is always further away from the central atom than a free electron pair (i.e., a non-bonding pair). The electronic interaction, always repulsive, is then always stronger between two non-bonding pairs than between two bonding pairs. Hence, the interactions can be ordered as *non-bonding/non-bonding* > *non-bonding/bonding* > *bonding-bonding*. Hence, in practice, the existence of a non-bonding doublet results in the deformation of the angle of the molecule because of the competition between the different replsive actions of the doublets. This induces a dipolar moment, which is important for the interaction between the molecules. It follows that

A free doublet leads to the opening of the E-A-X angles and the closure of the angle X-A-X.

Water enjoys a tetrahedral symmetry but with only two summits occupied by a hydrogen atom. It follows that its opening angle is 104.45° instead of 109.47°, which it would be if the symmetry were exact (and as it actually is for methane CH_4). To compare, NH_3 is similar but has an opening angle of 107° (see Figure 13.6). This shows that while the angle

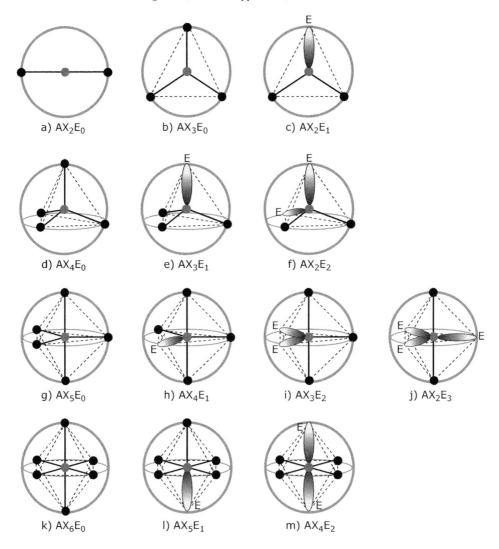

Figure 13.5 Configuration of molecules of the type $AX_m E_n$.

is determined in the first approximation by pure geometry, it also depends on the electric forces and the competition between bonding and non-bonding pairs. The more non-bonding the pairs, the smaller the angle.

13.4.4 Effect of the Variation of the Constants

On the numerical side, King *et al.* [15] have already investigated the effect of variation of the fundamental constants on water in detail via massive computer calculations. They

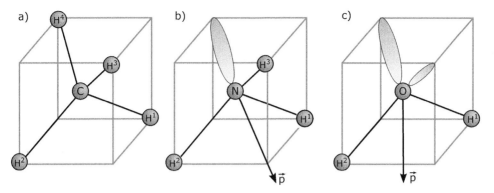

Figure 13.6 Tetragonal molecules. The opening angle changes when an atom is replaced by an electronic pair. It is perfect for methane (a), $\theta = 109.47^{\circ}$, and descreases for amonia (b), $\theta = 107^{\circ}$, and water (c), $\theta = 104.47^{\circ}$.

state, 'The dramatic advances of recent decades in electronic structure methods, numerical algorithms, and raw computing power permit the determination of solutions very close to the ab initio limit for molecular systems of reasonable size'.

Their method is as follows. For non-relativistic chemistry and the Schrödinger equation, the relevant equation is (13.43). To solve it, one can use the Born-Oppenheimer approximation, whereby the electronic part of the Schröedinger equatin is first solved with clamped nuclei [15]. Then,

$$\hat{\mathcal{H}}_0\Psi_e = \left[-\frac{\hbar^2}{2m_e} \sum_i \nabla_i^2 + \frac{e^2}{4\pi\epsilon_0} \sum_{(I,J)} \frac{Z_I Z_J}{r_{IJ}} - \frac{e^2}{4\pi\epsilon_0} \sum_{(i,I)} \frac{Z_I}{r_{iI}} + \frac{1}{2}\frac{e^2}{4\pi\epsilon_0} \sum_{(i,j\neq)} \frac{1}{r_{ij}} \right] \Psi_e$$

$$\tag{13.44}$$

$$= \epsilon_e(\rho_{nuc})\Psi_e, \tag{13.45}$$

yielding a potential energy surface (ρ_{nuc}) for the motion of the nuclei as a function of the scaled nuclei positions ρ_{nuc}. The resulting nuclear wave equation is then solved for the final rovibronic energy levels:

$$\left[-\frac{1}{2}\sum_I \frac{\beta}{\mu_I} \nabla_I^2 + \epsilon_e(\rho_{muc}) \right] \Psi_{nuc} = \epsilon\Psi_{nuc}. \tag{13.46}$$

The topography of the surfaces $\epsilon_e(\rho_{nuc})$ provides the basis for ascribing geometrical structure to molecules [15]. The local minima of this multi-dimensional wave surface correspond to the three-dimensional molecular structures of chemistry, allowing bond lengths and angles to be estimated. The results depend on basis sets chosen.

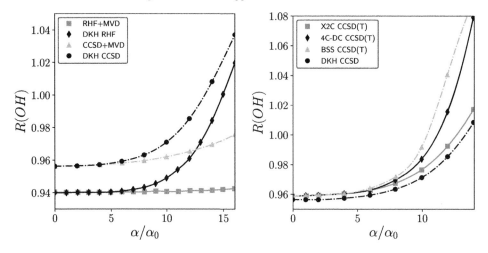

Figure 13.7 The water molecule bond length (Angstroms) as a function of the fine-structure constant α, computed for various basis sets.

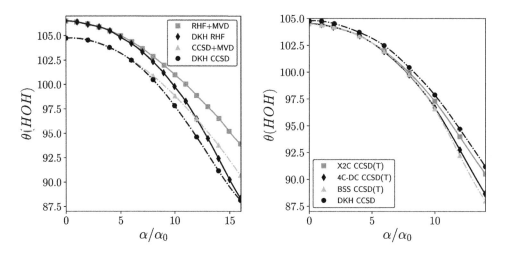

Figure 13.8 The water molecule angle (degrees) as a function of the fine-structure constant α, calculated using various basis sets.

Figure 13.7 shows the water bond length as a function of α computed for various basis sets (for details, see [15]). Figure 13.8 shows the water molecule angle as a function of α, and Figure 13.9 shows the resulting variation of the dipole moment as a function of α. Finally, Figure 13.10 shows the bond length and angle as a function of the electron-proton mass ration μ.

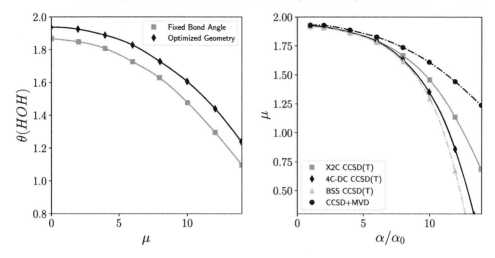

Figure 13.9 The water molecule dipole moment in Debyes as a function of the fine-structure constant α, calculated using various basis sets.

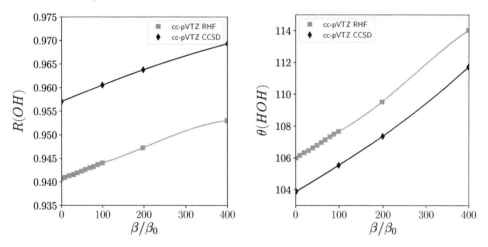

Figure 13.10 The water molecule bond length in Angstroms (left) and angle in degrees (right) as a function of the electron-proton mass ratio μ.

These interesting figures show that, indeed, there is an effect on key molecules if the variation in α and μ is large enough. Note that unlike the H_2 molecule, the radius of separation for water appears to increase as the coupling, through α, increases. This

13.5 The Issue of Scaling

However a key issue now arises: what if what we have been working out is just a similarity transformation, that leaves biological function unchanged? (Section 13.5.1). That leads to a consideration of relativistic effects in biology (Section 13.5.2).

13.5.1 What about Conformal Invariance?

Maybe a variation of α is just a change in size or, equivalently, of units (cf. Section 13.2.3). To explore this, consider a simple quantum mechanical system which has only a single force law acting, with a homogeneous potential term $V(\lambda r) = \lambda^n V(r)$, which has a linear dependence on some coupling constant C – e.g., $V(r) = C|r_1 - r_2|^n$.

Use this in the Schrödinger equation for a system of particles:

$$\frac{-\hbar^2}{2m}\nabla^2 \psi(r) = (V(r) + E)\psi(r), \tag{13.47}$$

If we change C, we can find a second solution to our equations by defining a second wave function $\xi(r) = \psi(\lambda r)$, which will solve the Schrödinger equation with a different eigenvalue (energy level) but the same shape; altering C does not affect the angular modes that come into our solution.

To see this, suppose we look at a Coulomb potential, $V(r) \propto C/r$, and take a solution to the Schrödinger equation $\psi(r)$. If change our coordinate system to $R = 2r$, and let $\xi(R) = \psi(2r)$, then in this system, $\nabla_R^2 = \nabla_r^2/4$, and $\hat{V}\xi = \hat{V}\psi/2$. Thus, if we reduce C by a factor of two and call the new potential V, we find that

$$\frac{-\hbar^2}{2m}\nabla^2\xi(R) - V'(R)\xi(R) + = \frac{1}{4}\left(\frac{-\hbar^2}{2m}\nabla^2\psi(r) - V(r)\psi(r)\right) = \frac{E}{4}\psi(r) = \frac{E}{4}\xi(R).$$

$$\tag{13.48}$$

So in reducing the coupling by a factor of two, we find that there is a solution of exactly the same shape as before, but with all distances doubled and one quarter of the energy. This leads us to consider conformal changes.

Conformal change: By rescaling the potential by C, it just changes all lengths together: it's just a larger molecule. Perhaps its biological function is unchanged?

But firstly, energies matter also. It is not clear whether one would obtain a rescaled system with the same conformational properties and energies rescaled to allow the same biological function but with different size systems or not. Specifically, some key effects related, for example, to chlorophyll and rhodopsin depend on photon energies $E = h\nu$ that are not related to the Schrödinger equation and will not scale with α. Secondly, there are spin effects in biology that are not conformally invariant. This has to account for the changes in angles shown in the results of (13.8) of King et al. [15], which clearly demonstrate that the molecular changes with variation of α do not just result in a change of scale. QED and relativistic effects matter (Figure 13.11).

13.5.2 Relativistic Effects

According to Likhtenshtein [17], 'The pivotal role of electron spin interactions in Nature cannot be overestimated'. It can be shown, for example, to have significant effects on brain function [25].

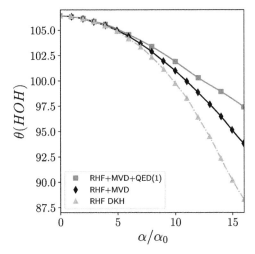

Figure 13.11 QED and relativistic effects on the bond angle (degrees) of the water molecule as a function of the fine-structure constant [15].

The relativistic terms for the hydrogen atom including fine and hyperfine structure were given in (13.8); we now discuss the fine-structure terms given there. This structure can be determined using perturbation theory; that is, by setting

$$H = H_0 + W.$$ (13.49)

The spin-orbit interaction is described in dimensionless form by

$$\frac{W_{S,O}}{m_e c^2} = \frac{\alpha}{2\rho^3} \frac{g_e}{a} \left(\frac{\mathbf{L}}{\hbar} \cdot \frac{\mathbf{S}}{\hbar} \right),$$ (13.50)

where g_e is the electron gyromagnetic factor. The second correction arises from the $(v/c)^2$ relativistic terms and is of the form

$$W_{rel} = -\frac{\mathbf{P}^4}{8m_e^3 c^2}.$$ (13.51)

The third and last correction is known as the Darwin term and arises from the fact that in the Dirac equation, the interaction between the electron and the Coulomb field is local. But the non-relativistic approximation leads to a non-local equation for the electron spinor that is sensitive to the field in a zone of order the Compton wavelength centred on $\mathbf{r}$. It follows that

$$W_D = \frac{\pi \hbar^2 q^2}{m_e^2 c^2} \delta(\mathbf{r}).$$ (13.52)

The average in an atomic state is of order

$$\langle W_D \rangle = \pi \hbar^2 q^2 / (2m_e^2 c^2) |\psi(\mathbf{0})|^2 \sim m_e c^2 \alpha^4 \sim \alpha^2 H_0.$$ (13.53)

Its dimensionless version is (not forgetting that the Dirac distribution has a dimension)

$$\frac{W_D}{m_e c^2} = \pi \alpha \delta(\rho). \tag{13.54}$$

These interactions will no longer be scale invariant under change of α.

In the calculations of King *et al.* [15], they state the following: the dimensionless ratios that have consequences for chemistry are the fine-structure constant $\alpha = e^2/(\hbar c)$ and the electron-proton mass ratio $\mu = m_e/m_p$. In conventional, non-relativistic quantum chemistry within the Born-Oppenheimer approximation, it is assumed that both ratios are negligibly small. The most important relativistic effects in chemistry can be investigated by means of the Cowan-Griffin Hamiltonian in which $\hat{H}_0$ is augmented with one-electron mass-velocity and Darwin terms:

$$\hat{H}_1 = \alpha^2 \left[-\frac{1}{8} \sum_i \nabla_i^4 + \frac{\pi}{2} \sum_{I,j} \delta(\rho_{Ii}) \right]. \tag{13.55}$$

The consequences of finite m_e/m_p ratios on chemical systems can be probed by means of the diagonal Born-Oppenheimer correction (DBOC)

$$E_{DBOC} = -\frac{\mu}{2} \sum_I \frac{1}{\mu_I} \left\langle \Psi_e | \nabla_I^2 | \Psi_e \right\rangle. \tag{13.56}$$

This is what is represented directly in Figure 13.11 and indirectly through the (Figures 13.8 and 13.9), where the angle and dipole moment depend on the fundamental constants.

13.6 Extrapolations and Conclusions

Can we extend these results to key organic molecules?

13.6.1 Scaling to Biology: Organic Molecules

The structure of DNA has very tight constraints in order for it to function. In order for the DNA helix structure to exist, one needs the equality between the lengths of $A + T$ and $G + C$ to be equal. According to Reference [4], 'The distance between adjacent sugars or phosphates in the DNA chain is 6 Angstroms. It must be between 5.5 and 6.5 Angstroms for it to work'. This means that we require

$$\delta \left(\frac{\ell_{A+T}}{\ell_{G+C}} \right) < 10\%. \tag{13.57}$$

How this relates to a constraint on the fine-structure constant is unclear without detailed study using the methods of King *et al.* [15]. This may not be as hopeless as it sounds because of the very regular structure of the molecule in its double-helix form, so there will be symmetries that can be exploited. However, one might hope that the results will be indicated by those for the water molecule, given that the underlying equation is the same. This constraint on a dimensionless parameter that specifies the geometry and thus the possibility to build a helix should be of the order of between 6% and 10% variation in α. However, this rough estimate could be completely wrong.

13.6.2 *Proteins: Voltage Gated Ion Channels*

As mentioned at the beginning of this chapter, what one really wants is to study the viability of molecules such as voltage gated ion channels, which play a key role in biology [12], under variation of the fundamental constants. This is much more difficult than DNA because their tertiary and quaternary structure is so much more complex [5, 20]. However, it may not be completely hopeless because they are made of subunits such as α and β helices that have a more regular form. This may be hopeless, but it is worth stating as a long-term goal.

13.6.3 *A General Theorem?*

Peter Atkins has provided the following argument for the general case. Let i label the electrons in a molecule and I, the nuclei. Using constants as in (13.30), the Hamiltonian can be written as

$$H = -\beta\gamma \sum_i \nabla_i^2 - \alpha\gamma \sum_{(i,I)} \frac{Z_I}{r_{iI}} + \frac{1}{2}\alpha\gamma \sum_{(i,j?)} \frac{1}{r_{ij}} + \frac{1}{2}\alpha\gamma \sum_{I,J\neq I} \frac{Z_I Z_J}{r_{IJ}}. \tag{13.58}$$

Taking partial derivatives,

$$\frac{\partial H}{\partial \alpha} = -\gamma \sum_{(i,I)} \frac{Z_I}{r_{iI}} + \frac{1}{2}\gamma \sum_{(i,j)} \frac{1}{r_{ij}} + \frac{1}{2}\gamma \sum_{I,J\neq I} \frac{Z_I Z_J}{r_{IJ}} = \frac{H}{\alpha} + \frac{\beta\gamma}{\alpha}\sum_i \nabla_i^2. \tag{13.59}$$

$$\frac{\partial H}{\partial \beta} = -\gamma \sum_i \nabla_i^2. \tag{13.60}$$

From the first of these expressions,

$$\left\langle \frac{\partial H}{\partial \alpha}\right\rangle = \frac{1}{\alpha}\langle H\rangle + \frac{\beta\gamma}{\alpha}\left\langle \sum_i \nabla_i^2\right\rangle. \tag{13.61}$$

And then from the second,

$$\left\langle \frac{\partial H}{\partial \alpha}\right\rangle = \frac{1}{\alpha}\langle H\rangle - \frac{\beta}{\alpha}\left\langle \frac{\partial H}{\partial \beta}\right\rangle. \tag{13.62}$$

From the Hellmann-Feynman theorem [13, 14], which states that for a parameter P and exact, normalised wave function,

$$\left\langle \frac{\partial H}{\partial P}\right\rangle = \frac{d}{dP}, \tag{13.63}$$

this expression, with $\langle H \rangle = E$, becomes

$$\frac{dE}{d\alpha} = \frac{E}{\alpha} - \frac{\beta}{\alpha}\frac{dE}{d\beta} \qquad (13.64)$$

or

$$\frac{\alpha}{E}\frac{dE}{d\alpha} + \frac{E}{\beta}\frac{dE}{d\beta} = 1 \;\Leftrightarrow\; \frac{d\ln E}{d\ln\alpha} + \frac{d\ln E}{d\ln\beta} = 1, \qquad (13.65)$$

This equation with $f = \ln E$, $x = \ln\alpha$, and $y = \ln\beta$ – has the form $df/dx + df/dy = 1$ and, therefore, has the solution (with a constant of integration $\ln k$) $f = ax + by + \ln k$, with $a + b = 1$. That is,

$$\ln k E = a\ln\alpha + b\ln\beta + \ln k = \ln k\alpha^a\beta^b. \qquad (13.66)$$

It then follows that, regardless of the identity of the molecule,

$$E = k\alpha^a\beta^b \text{ with } a + b = 1. \qquad (13.67)$$

This is consistent with the H atom, where $E_n \propto \alpha^2/\beta$, for which $a = 2, b = -1$.

Dimensional analysis helps to take this further. Write $k = \gamma\kappa$, with $\gamma = c\hbar$ (as in (13.30)). Then,

$$E = \kappa\gamma\alpha^a\beta^b. \qquad (13.68)$$

The dimensions are as follows:

$$[E] = [\kappa][\gamma][\alpha^a][\beta^b] : \; [ML^2T^{-2}] = [\kappa][ML^3T^{-2}][1]^a[L]^b = [\kappa][ML^{3+b}T^{-2}], \qquad (13.69)$$

which implies that

$$[\kappa] = L^{-(b+1)}. \qquad (13.70)$$

One can then try the following. The only combination of fundamental constants in the Hamiltonian that has dimensions of length form the combination β. Therefore, κ must be proportional to $\beta^{-(b+1)}$. However, κ is independent of β (it is proportional to the constant k). Therefore, the only value of b that is allowed is $b = -1$, with the implication that $[\kappa] = [1]$ and $a = 2$. Therefore,

$$E = \kappa\gamma\alpha^2\beta^{-1}. \qquad (13.71)$$

for some constant κ, which is

$$E = 2m_ec^2\kappa\,\alpha^2, \;\Leftrightarrow\; \mathcal{E} = \frac{E}{m_ec^2} = \kappa\,\alpha^2, \qquad (13.72)$$

which is what we get for the hydrogen atom. And if the argument is valid, this is the case for all atoms and molecules, regardless of their identity (solids too, which are just big molecules). A puzzle is the result for H_2^+, where this is not the case, but this may be because the LCAO treatment of H^+ there is an approximation, whereas this result (depending as it does on the Hellmann-Feynman theorem) is for exact energies and wave functions.

If an argument like this works, it will suggest how energies scale with α (and other constants), which is key to one of the major themes of biology: metabolism [32]. As mentioned before, there cannot just be a scaling of metabolic effects with α because some, such as photosynthesis, are independent of α.

13.6.4 Other Scales

Finally, we return to considering the relation of this work to other scales.

- Nuclear scale: stable nuclei possible.
- Stellar scale: existence of stable nuclei due to stellar processes. Requires the Hoyle state to exist.
- Galactic to stellar scale: existence of galaxies, stars, and planets due to cosmological and astrophysical processes. Requires restricted cosmological parameters.

The fine-structure constant α and mass ratio μ are not the only relevant constants for these effects to be favourable; thus, there, in principle, can be many physical situations where life cannot occur (e.g., because carbon does not come into being), even though the organic molecules would functon fine, were they to come into existence.

13.6.5 Fine-Tuning Relations

The effect of fine-tuning on biomolecules has importance in the understanding of the fine-tuning of the Uxniverse. We can ask a series of questions:

1. Are constraints from physics (stability of nuclei) stronger than those of biology?
2. Are constraints propagating?
3. Can we argue that if the constraints for the formation of carbon-12, etc., are satisfied, then conditions for life (at least on those parameters) are satisfied? Or do they set further non-trivial constraints?

We can ask, *which sets the tighter limits: physics/astrophysics or biology?*

It appears from the preceding that it is likely that the physics constraints are indeed tighter than the constraints from water and biomolecules. As regards α:

- Stellar nucleosynthesis limits, holding everything else fixed, give limits $\simeq 10^{-3}$.
- This seems much tighter than the likely limits from water.
- This may or may not be tighter than limits from DNA and proteins, but an initial estimate based on the preceding argument is that it will indeed be tighter.

As regards μ: there are only combinations of limits. Limts on μ per se are unclear.

    If physics uniquely leads to molecules compatible with life in this way, then in some sense, physics foresees the existence of life.

Why? This is a conundrum: *physics seems fine-tuned to expect life.*

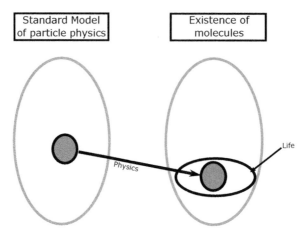

Figure 13.12 Fundamental physics and biology: the Standard Model of particles physics might have been implemented with a variety of fundamental constants (oval on left). The set that actually occurred (small circle on left) is in the small subset of these possibilities that leads to possible existence of molecules compatible with life (open circle on right).

13.7 Conclusion

It is a legitimate and important question to quantify the dependence of higher levels of complexity on the parameters of a fundamental theory. In particular, it allows one to discuss the level at which the fine-tuning on the fundamental constant is the more stringent. This can be compared to Darwinian selection, which takes place simultaneously but with different timescales, on the scale of the gene, the cell, the organ, the animal, and the population.

Two final issues remain. First, what about other forms of life? Is it too restrictive to only allow the consideration of water and organic molecules? Our viewpoint is that it is only organic molecules, and specifically proteins [22], that have the capacity to produce the complexity of living systems that respond to the environment in an adaptive way and allow biology to emerge from physics [12]. We do not believe life could come into existence without being based in carbon and its huge variety of organic molecules that allow extraordinary complexity of function, as characterised by the possibility spaces of molecular biology [32].

Second, we have not considered the relation to a more fundamental theory that might underlie the Standard Model of particle physics. We do not believe enough is known to make reliable estimates in this regard.

References

[1] Atkins, P., and Friedman, R. 1997. *Molecular Quantum Mechanics*. Oxford University Press.
[2] Barrow, J. D., and Tipler, F. J. 1986. *The Anthropic Cosmological Principle*.
[3] Cabane, B., and Vuilleumier, R. 2005. 'The Physics of Liquid Water'. *Comptes Rendus Geoscience*, **337**(Feb.), 159–171.

[4] Calladine, C. R., and Drew, H. R. 1992. *Understanding DNA: The Molecule & How It Works*. Academic Press.

[5] Catterall, W. A. 1993. 'Structure and Function of Voltage-Gated Ion Channels'. *Trends in Neurosciences*, **16**(12), 500–506.

[6] Watson, J. D. 2004. *Molecular Biology of the Gene*. Pearson Education.

[7] Ekström, S., *et al.* 2010. 'Effects of the Variation of Fundamental Constants on Population III Stellar Evolution'. *Astronomy and Astrophysics*, **514**(May), A62.

[8] Ellis, G. F. R. 2005a. 'Physics and the Real World'. *Physics Today*, **58**(7), 49–54.

[9] Ellis, G. F. R. 2005b. 'Physics, Complexity and Causality'. *Nature*, **435**(June), 743.

[10] Ellis, G. F. R., and Uzan, J. P. 2005. '*c Is the Speed of Light, Isn't It?*.' *American Journal of Physics*, **73**(Mar.), 240–247.

[11] Ellis, G. 2016. *How Can Physics Underlie the Mind?: Top-Down Causation in the Human Context*. The Frontiers Collection. Springer.

[12] Ellis, G. F. R., and Kopel, J. 2018. *Wandering towards a Goal: The Key Role of Biomolecules*. Springer International Publishing. Pages 227–243.

[13] Feynman, R. P. 1939. 'Forces in Molecules'. *Physical Review*, **56**(Aug.), 340–343.

[14] Hellmann, H. 1937. *Einführung in Die Quantenchemie: Texte Imprimé*. F. Deuticke.

[15] King, R. A., *et al.* 2010. 'Chemistry as a Function of the Fine-Structure Constant and the Electron-Proton Mass Ratio'. *Physical Review A: General Physics*, **81**(4), 042523.

[16] Lewis, G. F., and Barnes, L. A. 2016. *A Fortunate Universe*. Cambridge University Press.

[17] Likhtenshtein, G. 2014. *Electron Spin Interactions in Chemistry and Biology: Fundamentals, Methods, Reactions Mechanisms, Magnetic Phenomena, Structure Investigation*. Biological and Medical Physics, Biomedical Engineering. World Scientific Publishing Company Incorporated.

[18] Lodish, H. 2008. *Molecular Cell Biology*. 6th edn. W. H. Freeman and Company.

[19] Luo, F., Olive, K. A., and Uzan, J. P. 2011. 'Gyromagnetic Factors and Atomic Clock Constraints on the Variation of Fundamental Constants'. *Physical Review D*, **84**(9), 096004.

[20] Magleby, K. L. 2017. 'Structural Biology: Ion-Channel Mechanisms Revealed'. *Nature*, **541**(7635), 33.

[21] Nakamura, K., and Particle Data Group. 2010. 'Review of Particle Physics'. *Journal of Physics G Nuclear Physics*, **37**(7), 075021.

[22] Petsko, G. A., and Ringe, D. 2004. *Protein Structure and Function*. Vol. Primers in Biology. New Science.

[23] Rees, M. 2000. *Just Six Numbers: The Deep Forces That Shape the Universe*. Basic Books.

[24] Rees, M. 2003. *Our Cosmic Habitat*. Princeton University Press.

[25] Turin, L., Skoulakis, E. M. C., and Horsfield, A. P. 2014. 'Electron Spin Changes during General Anesthesia in Drosophila'. *Proceedings of the National Academy of Science*, **111**(Aug.), E3524–E3533.

[26] Uzan, J. P. 2003. 'The Fundamental Constants and Their Variation: Observational and Theoretical Status'. *Reviews of Modern Physics*, **75**(Apr.), 403–455.

[27] Uzan, J. P. 2011. 'Varying Constants, Gravitation and Cosmology'. *Living Reviews in Relativity*, **14**(Mar.), 2.

[28] Uzan, J. P. 2013. 'Models of the Cosmos and the Emergence of Complexity. In Ellis, G. F. R., Heller, M., and Pabjan, T., eds., *The Causal Universe*. Copernicus Center Press.

[29] Uzan, J. P. 2017. *Emergent Structures of Effective Field Theories*. Cambridge University Press.

[30] Uzan, J., and Lehoucq, R. 2005. *Les constantes fondamentales*. Belin Sup. Belin.

[31] Villani, C., Uzan, J., and Moncorgé, V. 2017. *The House of Mathematics*. Le Cherche Midi.

[32] Wagner, A. 2014. *Arrival of the Fittest: Solving Evolution's Greatest Puzzle*. Penguin.

[33] Wilczek, F. 2007. 'Fundamental Constants'. *arXiv e-prints*, Aug.

Index

A, anthropic domain of, 14, 50
ab initio models, 300
AC measure. *See* Anderson-Castano measure
Active Galactic Nuclei, 181, 230
adiabatic expansion, 177
adiabatic perturbations, 360
advanced civilisations, 459
Affleck-Dine mechanism, 189
ALMA molecular mapping, 396
alpha clusters, 284
alpha decay, 304
amorphous carbon (AC), 466
Anderson-Castano measure (AC measure),
    321–322
Andromeda galaxy, 486
Andromeda nebulae, 349
Andromeda VIII, 486
angles
    Cabibbo, 314
    CP, fine-tuning of, 309
    electroweak mixing, 249
    quark CKM matrix, 241, 514t
    strong CP, 309
    Weinberg, 241
anisotropy, 143–144
    CMB, 112, 129, 131, 164
    temperature, 123, 345
annihilation
    cross section, 70, 183, 185, 354
    dark matter cross section for, 354, 366, 367
    of DM, 357
    gamma rays and anti-matter, 180
    nucleon-anti-nucleon, 180–181
    p-wave, 356, 366
    s-wave, 355, 366
    of WIMPs, 367
anthropic principle (AP), 25, 60, 243. *See also* strong
    anthropic principle; weak anthropic principle,
    339–340, 488

constraints and, 20
cosmological constant fine-tuning and,
    340
curvature and, 100–101
distribution of Λ and, 99–100
ensemble reasoning and, 102
fine-tuning and, 21, 309
fine-tuning types and, 28
interpretations of, 56–58
many universes and, 98
multiverse and, 100–102, 340
reasoning of, 13–16
string landscape and, 340
trivial, 77–78
vacuum and, 97–102
vacuum density and, 484
anti-baryons, 80, 187
anti-matter
    cosmic rays and evidence of, 181
    CP violation and, 8
    detecting, 180
    ISM lifetime of, 180
anti-nuclei, cosmic rays and, 181
anti-protons
    cosmic rays and, 180
    discovery of, 175
    WIMP detection and, 369
antiquarks, 182
anti-stars, 180
AP. *See* anthropic principle
Archean era, 491
asteroid belts, 412
    mass of, 425
    orbits of objects in, 428
    resonance locations and, 431
    water snow line and, 422–423, 425
asteroid impacts
    effects on inner Solar System, 428–436
    effects on outer Solar System, 436–437

asteroid impacts (cont.)
  numerical simulations of, 426–428
  super-Earths and, 434, 436
  on terrestrial planets, 425–437
  water and, 421
ASTrO mission, 475
atomic
  cooling, 403
  isotopes, formation sites of, 299
  physics, 515–516
  scale, 29
  theory, 22
axial symmetry, 334
axions, 52, 93, 308, 309, 332, 338, 339, 351

background radiation density, galaxy
        formation and, 49
BAO. *See* Baryon Acoustic Oscillations
Barbieri-Giudice criterion (BG), 320, 321
Bardeen potential, 132
baryogenesis, 188, 189
Baryon Acoustic Oscillations (BAO), 6, 11, 119
baryon asymmetry, 8, 80, 174–176, 307, 352
  alternate universes with different, 195–199
  BBN and, 195–196
  cold thermal relics and, 352
  CP violation and, 333
  fine-tuning of, 174, 178–179
  Hubble constant and, 193
  large-scale structure and, 196–199
  linear regime and changes in, 196–198
  in matter-antimatter symmetric universe, 177
  natural value of, 174
  nonlinear regime and changes in, 198–199
  nucleosynthesis and, 189–192
  observable cosmological parameters,
        and, 192–194
  particle physics models for generating, 187–189
baryon density, 69, 87, 365
baryon fraction, 99, 102, 488
baryon mass, 3
baryon mass density parameter, 193
baryon number, 80
  Big Bang and, 8
  conservation of, 334
  inflation and, 187
  non-conservation, 175, 177, 179, 200
  non-conservation of, 179
  Sakharov conditions and, 187, 188
baryon-dark matter entropy perturbation,
        216 217
baryon-only Universe, 361–364
baryons, 6, 11, 176, 187
  cosmic, 106
  decoupling and, 362
  equations of state, 117
  Jeans mass and, 361

  means-free paths of, 362
  Silk damping and, 216
baryon-to-dark matter ratio, 215
baryon-to-photon ratio. *See* photon-to-baryon ratio
Bayes' theorem, 72
Bayesian approaches, 71–74, 136
beryllium, 256, 302
  binding energies, 300
  Hoyle state and, 273
  lifetime of, 304
  lithium production and, 276
  photodissociation of, 277
  production of, 42, 271–272
beyond standard model (BSM), 309, 328
BG. *See* Barbieri-Giudice criterion
Bianchi I model, 143
Bianchi identities, 114
BICEP/Keck data, 136
Biermann batteries, 391
Big Bang, 4, 5, 8, 24, 458
  fine-tunings and evolution of Universe after, 54
  fossil radiation from, 384
  hot, 48, 111, 117, 120
  large-number coincidence and, 79
  multiple, 13
  Pyramid of Complexity and, 55
  standard, 49
Big Bang Nucleosynthesis. *See* nucleosynthesis
binding energies, 254–256, 292, 300
biochemical reactions, 446
biologically important molecules, 512
  energy and, 530
  fine-tuning and, 535
  relativistic effects and, 530–532
  scaling and, 529–532
biomass, 13
biophilic universes, 13
biosignatures, 458–460, 496
Birch-Murnagham EOS, 472
Birkoff's theorem, 222
black holes, 4, 55, 393, 394
  feeding, 405
  mass of, 31
  Q and, 50
  scales of structure and, 29–31
  supermassive, 400–407
    astrophysics of, 404–405
    feedback, 405–407
    growth of, 402–403
    radio jets and, 401–402
    seed massive black holes and, 403–404
black-body radiation, 5, 98, 205
blue giants, 39
Bohr radius, 516, 520
Boltzmann
  brain, 75
  distributions, 366

equation, 183, 354, 355, 364
   factor, 354
Bondi accretion rate, 401, 402
Bonnor-Ebert mass, 389
Born-Oppenheimer approximation, 527, 532
Bose-Einstein condensates, 351
   chemical potential and, 352
   kinetic equilibrium and, 352
Bose-Einstein gases, 262
bosons
   gauge, 244, 323, 327
   NGB, 330, 334
   pNGB, 330, 338
   SUSY and, 328
   vector, 245, 326
   Z-boson, 371
Brownian motion, 468
BSM. *See* beyond standard model
bubble universes, 75

C violation, 187
Cabibbo angle, 314
Cabibbo-Kobayashi-Maskawa (CKM), 241, 316,
      336, 514t
carbides, 464
carbohydrates, 512
carbon, 256
   abundance, 276, 470
   binding energies, 300
   in circumstellar discs, 472
   computation for modeling production of, 283–284
   disc composition and, 467
   graphitisation of, 472
   life existence and, 35
   nuclear constants and production of, 289–291
   nuclear parameter sensitivity of production of,
      286–288
   population III star production of, 281–291
   production of, 42, 271, 272
   stellar evolution and production of, 288–289, 385
   supernova explosion models and, 466
   3α reaction and production of, 284–286
carbon planets, 466–474, 476
carbon-based life, 77, 81
carbon-enhanced metal-poor stars (CEMP stars),
      463–476
Carter's argument, 444–448
Casimir factors, 326
CBR. *See* cosmic background radiation
CEMP stars. *See* carbon-enhanced metal-poor stars
CEMP-no stars, 465–469
central molecular zone (CMZ), 395
C-grains, 466
Chandrasekhar mass, 32, 78
charge, 245
   vacuum fluctuations and, 311
charged particles, WIMP detection and, 369

charge-parity problem (CP problem), 307
charginos, 372
charm quark, 315–316
CHeB. *See* core He burning
chemical elements, mass-energy density and, 346
chemical equilibrium, cold thermal relics and,
      352–353
chemical modeling, for water formation, 479
chemical reactions, 258–260
chemistry, SAP and constraints from, 42–45
chiral
   fermions, 328
   limit, 251
   symmetry, 317, 318
   transformations, 337
chirality, 245
   CP violation and, 333
   fermions as eigenstates of, 244
chlorophyll, 530
circumgalactic medium, 401
circumstellar discs, 469
   carbon in, 472
CKM. *See* Cabibbo-Kobayashi-Maskawa
cluster model
   for binding energies, 255
   MN force and, 300
CMB. *See* cosmic microwave background
CMZ. *See* central molecular zone
CNO
   cycle, 271, 286, 288
   nuclides, 191–192, 195, 268, 271–277, 281
CO snow line, 438–442
COBE. *See* COsmic Background Explorer
coincidence problem, 496
   vacuum, 90
cold dark matter, 205, 364
cold thermal relics, 350–359
   freeze-out, 354, 355
   number density of, 354
   WIMPs, 356
collisional damping, 362
collisionless Boltzmann equation, 364
colour, 244, 334
Coma cluster, 349
comets, life and, 438
compactified extra spatial dimensions, 60
complex molecules, 231
complexity
   fluctuations and, 9–10
   prerequisites for, 7–11
   Pyramid of, 54
   strong anthropic tunings and, 29
complexity principle, 25
   anthropic principle as, 58
   cosmology and, 54–56
Compton
   cooling, 462

Compton (cont.)
  emission, 180
  scattering, 120, 215
  wavelength, 4, 31, 531
conformal invariance, 530
consciousness
  quantum mechanical wave-function and, 57
  Universe existence and, 57
conservation of energy, 92, 114, 162, 224–225, 407
constants, 3. *See also* physical constants; *specific topics*, 511
  fine-tuning of, 4
  hydrogen atoms and, 521–522
  hydrogen molecules and, 522
  mass determination and, 29
  meter dependence on, 517–518
  primordial nucleosynthesis parameters and, 268
  in standard models, 26, 513
  water and variations of, 526–529
continuity equation, 209
convective stars, 40
Copenhagen interpretation, 82
core He burning (CHeB), 288–291
COROT mission, 486
CoRot mission, 475
COsmic Background Explorer (COBE), 112
  normalisation for data from, 141
cosmic background radiation (CBR)
  perturbation amplitude and, 364
  temperature anisotropy of, 345
cosmic baryons. *See* baryons
cosmic coincidence, 90–91
cosmic expansion rate, 11, 24
cosmic horizons, 12, 17
cosmic inflation. *See* inflation
cosmic microwave background (CMB), 5, 111
  anisotropy of, 129, 131, 164
  baryon asymmetry parameter and
      temperature of, 192
  density perturbations and, 208
  discovery of, 175
  Doppler peaks in, 165
  entropy and, 48
  fluctuations in, 9, 127
  habitability and, 463
  as heating source, 461
  Hubble constant inference from fluctuations
      in, 346
  isotropy and homogeneity of, 205
  isotropy of, 47, 123
  life and, 178
  observed isotropy of, 47
  observer selection and, 74–75
  observing dark energy properties and, 87–89
  perturbations and, 213
  power spectrum, 190
  reheating and, 131, 140

structure formation and, 205
  temperature of, 48, 111
  thermal coupling to, 462
cosmic neutrino background, 263
cosmic plasma, 182, 263
cosmic rays
  anti-matter evidence from, 181
  anti-nuclei and, 181
  anti-protons and, 180
  hydrogen ionisation by, 476, 478
  interstellar gas collisions with, 181
  as ionisation source, 441
Cosmic Uroborus, 22–26, 60
cosmic web, 203
cosmological background evolution, 360
cosmological constant, 82, 488
  anthropic arguments and, 339
  anthropic principle in fine-tuning, 340
  expansion rate and, 212
  fine-tuning and, 69, 70
  fine-tuning of, 340
  modified gravity and, 96
  probability distribution for life and, 494
  SAP and, 50
  testing for anthropic origin of, 484–489
cosmological effects of vacuum, 82–86
cosmological expansion rate, 46
  acceleration of, 50
cosmological horizon, 207
cosmological nucleosynthesis, 46–47, 55
cosmological parameters, 12
  baryon asymmetry parameter and, 192–194
  CMB fluctuations and, 6
  observer selection and, 74–75
  SAP and, 38, 46
cosmological perturbations, inflationary, 131–135
Cosmological Principle, 359
Coulomb
  barrier, 270, 302
  field, 310, 531
  forces, 473
  functions, 302
  interaction, 186
  potential, 522, 530
coupling constants, 3, 28, 53, 57, 241, 245, 248,
      514t, 514
Cowan-Griffin Hamiltonian, 532
CP angle, fine-tuning of, 309
CP conservation, 335
CP problem. *See* charge-parity problem
CP violation, 8, 187
  baryon asymmetry and, 333
  baryon density and, 69
  chirality and, 333
  QCD and, 333
CPT invariance, 188
critical density of universe, 193, 345

critical energy density, 118
curvature
    anthropic vacuum and, 101
    perturbations in, 132
    spatial, 3, 16, 206
        critical density and, 345
        perturbations and, 361
curvature fluctuations, 9–10

dark energy, 6, 50, 232
    approaches to, 68
    cosmological effects of vacuum and, 82–86
    density of, 3
    energy density and, 217
    equations of state, 217
    large-scale structure and, 87, 111
    mass-energy density and, 347
    microstructure of space and, 15
    models for dynamical, 89–94
    modified gravity vs, 95–97
    observing properties of, 87–89
    relativistic theory of gravity and, 95
dark matter (DM), 6, 176, 196, 232, 345
    annihilation cross sections of, 354, 366, 367
    annihilation of, 357
    axions and, 308
    categories of, 205, 350–351
    cold, 205, 364
    collisionless, 364
    decoupling and, 364
    discovery of, 348–350
    energy density, 119
    fine-tuning and, 377–378
    fine-tuning for structure formation, 366
    freeze-out and, 357
    growth of perturbations, 364–366
    Higgs mass and, 328
    hot, 205, 364
    large-scale structure and, 7, 348, 366, 377
    non-WIMP, 376
    percentage of universe, 348
    production cross sections of, 354
    SAP and, 51–52
    scalar fields as, 93–94
    structure formation and, 359–366
    subclumps of, 370
    symmetric universe and, 186
    taxonomy of particle, 351
    transfer function for, 220
    warm, 205
dark matter halos, 104, 198, 226, 230, 373
    density profiles, 368
    galaxy formation and, 226
    structure formation and, 370
    virialised, 203
dark-matter halos, 368
Darwin term, 531

Darwinian evolution, 446
darwinism, 76–77
DBOC. *See* diagonal Born-Oppenheimer correction
de Sitter
    eternal inflation and, 154
    expansion, 156
    invariance, 162
    quantum scalar field in, 153
    space, 102
dead zones, 440–442
debris discs, 438
decoupling, 51, 263
    baryons and, 362
    dark matter and, 364
    field, 249
    neutrino, 178
    photon-matter, 205
    scale, 167, 238, 513
    Silk scale and, 363
    structure formation and, 363
density contrast, 220
density fluctuations, 461
    generation of, 207–209
    gravity and, 7
    large-scale structure and, 24
    normalisation of, 102
    primordial, 50, 197, 396
    quantum origin of, 25
    SAP and, 50
density parameter, SAP and, 47
density perturbations, 198, 208
deuterium, 190, 191, 258
    binding energy of, 254
    fragility of, 266
    role of burning, 388–390
deuterons, 45, 253, 258, 265
DFSZ model. *See* Dine-Fischler-Srednicki-Zhitnitsky
        model
diagonal Born-Oppenheimer correction (DBOC), 532
diffusion damping. *See* Silk
dimensional regularisation, 327
Dine-Fischler-Srednicki-Zhitnitsky model
        (DFSZ model), 339
dipole moment, of water, 524
diprotons, 45, 258
Dirac
    coincidence, 395
    equation, 531
    fermionic representations, 249
    large-number hypothesis, 78
    quantisation, 163
discrete symmetries, 176
dissipation region, 399
dissociative recombination, 476
distance-ladder technique, 346
DM. *See* dark matter

DNA, 512
  structure of, 532–533
dust shielding, 478
dust-settling
  timescale, 469
  velocity, 467
dwarf galaxies, 230, 399, 488
  habitable planets and, 485
  observations of, at high-redshifts, 487
dwarf spheroidal galaxies, 370
dwarf stars, 460
dynamical dark energy, models for, 89–94

EAGLE galaxy formation simulation, 105
Earth, 447
  age of, 77
  gravitational settling and, 421
Earth-mass planets, 460
Eddington
  limit, 49, 392
  luminosity, 36, 406, 461
  mass outflow rate, 406
  ratio, 402
EDM. *See* electric dipole moment
EELT. *See* European Extremely Large Telescope
effective theories, 238, 513
effective viscosity parameter, 441
eigenfrequencies, of Saturn, 435
Einasto profile, 368
Einstein equations, 115, 116
  anisotropic initial conditions and, 145
  perturbed, 131
Einstein field equations, 359
Einstein gravity, 96
Einstein-de Sitter growth rates, 212, 214
Einstein-de Sitter model, 80, 86, 105
  background density and, 224
Einstein-Hilbert action, 113
  Starobinsky model and, 141
electric dipole moment (EDM), 336
electric fine structure constant, 26
electromagnetic
  binding energy, 250, 270
  coupling constant, 246
  field, 246
electromagnetic force, 282
  non-trivial chemistry and, 11
electromagnetic interaction, 252
  CP conservation and, 335
electromagnetism, 22, 25, 323
electron degeneracy energy, 33
electron field, energy of, 310–312
electron mass, 310, 317, 318
  fine-tuning of, 310
  quantum mechanics and, 310
electronic transition, 518

electron-photon interaction, 311–312
electron-positron pair emission, 305
electrons, 176, 244–246
  self-energy of, 69
  WIMP detection and, 369
electron-scattering opacity, 35, 39
electrostatic Coulomb forces, 473
electrostatic energy, 43
electroweak
  baryogenesis, 189
  fine-tuning in, 323–333
  interactions, 243–245
  mixing angle, 249
  scale, 307, 308, 331
  unification, 24
electroweak symmetry breaking (EWSB), 309, 323,
      326, 339
  technicolour models and, 329
elementary Higgs boson, 329
elementary particles, dark matter as, 351
energy budget, 117
  scalar fields and, 126
energy conservation, 114
  virial theorem and, 224
energy density
  critical, 118
  dark energy and, 217
  dark matter, 119
  matter and, 211
  neutrino, 119
  photon, 117
  power spectrum shape changes and, 219
energy scales, 97
energy momentum conservation, 310
energy-momentum tensors, 206
ensemble reasoning, 75–77
  anthropic vacuum and, 102
  SAP and, 82
entropy
  conservation of, 184
  microwave background and, 48
  neutrinos and, 263
  per baryon, 48
  per particle, 99
  per photon, 179
EOG. *See* Extreme Outer Galaxy
EOS. *See* equations of state
Epstein regime, 467
equations of state (EOS), 117
  solid planet internal structure and, 472
eternal inflation, 12, 13, 15, 154–157, 160, 161
  de Sitter space and, 154
  multiverse and, 164
Euler equation, 209, 210
European Extremely Large Telescope (EELT), 487
EUV. *See* extreme ultraviolet

event horizon
  de Sitter space and, 103
  of Universe, 12
evolutionary biology, 77
EWSB. *See* electroweak symmetry breaking
exoplanet statistics, 448
exoplanets, 75, 412
  atmospheres of, 459
  carbon, 476
  close-in planets or debris and, 416
  compositions of, 471
  detection of, 474, 475
  elemental abundances and, 464
  giant planets and, 426
  mass distribution of, 413
  super-Earths and, 416, 426
  warm dust and, 421
extragalactic high luminosity sources, 181
extragalactic planet searches, 485–487
extra-Solar intelligent civilisations, 444–448
extrasolar planets, 412
Extreme Outer Galaxy (EOG), 467
extreme ultraviolet (EUV), 467
Extremely Large Telescope, 448

Faber-Jackson relation, 399
far-ultraviolet (FUV), 477
  photodissociation by, 481, 482
  water and, 481
Fermi
  constant, 28, 314
  model, 314
  Paradox, 412, 444
Fermi-Dirac gases, 262
  chemical potential and, 352
  kinetic equilibrium and, 352
fermions, 244, 257, 314, 327
  chiral, 328
  cold thermal relics and, 351–353
  EWSB and, 329
  Majorana, 375
  mass of, 317
  SUSY and, 328
55 Cancri e, 464
filaments, 203
fine-structure constant, 69, 270, 519, 522
  DNA and, 533
fine-tuning, 3
  anthropic, 21
  anthropic arguments and, 309
  of baryon asymmetry parameter, 174, 178–179
  biological molecules and, 511
  biologically important molecules and, 535
  cosmological constant and, 69, 340
  of CP angle, 309
  of dark matter, for structure formation, 366
  dark matter amounts, 348

dark matter and, 377–378
defining, 68–71
electron field energy, 310–312
of electron mass, 310
electro-weak sector, 323–333
galaxy formation and, 383
inflation and, 139–142
life and, 178
measures for particle physics, 319–323
naturalness in particle physics and, 68–69
nuclear physics and, 257–260
particle physics problems of, 309
quantifying, for particle physics, 317–323
Standard Model puzzles, 120
in strong sector, 333–339
of structure formation, 231–233
types of, 28
Universe evolution and, 54
first metal enrichment epoch, water formation during, 476–484
first planets, 461–463
Fischer-Tropsch-type reactions (FTTs), 466, 474
flat Universe, 462
flatness problem, 101, 124–125
  inflation and, 126
flavour symmetry, 318
FLRW form. *See* Friedmann-Lemaître-Robertson
fluctuations, 9–10
  in CBR, 346
  CMB, 127
  density, 7, 24, 50, 207–209, 461
  power spectrum of, 150
  quantum, 75
  stochastic, 157
Fokker-Planck equation, 162–163
Fourier modes, 211
free-fall time, 397, 399
free-streaming distance, 364
freeze-out
  cold thermal relics and, 354, 355
  DM and, 357
  temperature, 46
  time, 48
  of WIMPs, 357
freeze-out time, 48
Friedmann equation, 37, 83, 85, 91, 116, 128, 140, 214
  baryon asymmetry parameter and, 192–193
  cosmological dynamics and, 260
  critical energy density and, 118
  Hubble flow parameter and, 129
  linear power spectrum and, 219
  matter perturbation growth and, 217
Friedmann-Lemaître-Robertson-Walker form (FLRW form), 114, 115, 131, 145, 206, 218, 260, 360
FTTs. *See* Fischer-Tropsch-type reactions

fundamental constants
   nuclear physics and, 240–242
FUV. *See* far-ultraviolet

galactic centre, 368, 370, 400, 401, 404, 477
galactic habitable zone, 447
galactic rotation curves, 6, 349, 400
galactic scales, 396–400, 535
galaxies
   black holes at centers of, 400
   characteristic mass of, 396–398
   distribution of, 203
   gravitational potential well, 6
   gravity and formation of, 7–8
   luminosity function of, 230
   mass of clusters of, 205, 230, 349
   mergers of, 489
   motions of, 6
   number of, 12
   rotation curves of, 6, 205
   semi-anthropic formation of, 102–106
   Seyfert, 181
   x-ray emitting clusters of, 181
galaxy formation, 33, 49, 232
   numerical simulations and, 383–384
galaxy scale, 33
gamma rays, 180
gamma-ray bursts (GRBs), 447
Gamow energy, 272, 274, 305, 389
gas accretion events, 405
gas giants, 473, 475
   locations of, 423–425
   mass of, 414
gas-phase H_2 formation, 480, 481
gauge
   coupling constants, 248, 257
   couplings, 283, 329
   electroweak bosons, 323
   hierarchy problem, 328–333
   invariance, 323
gauge bosons, 244, 327
GCM. *See* Generator Coordinate Method
Gell-Mann matrices, 244
Gell-Mann–Nishijima relation, 245
gene expression, 512
General Relativity, 360
   credibility of, 13
   gravity and, 113
   Planck scale and, 4
Generator Coordinate Method (GCM), 300
Geneva stellar code, 288
geophysical processes, 447
geothermal energy, 463
g-factors, 251–252
Giant Magellan Telescope (GMT), 486, 487
giant molecular clouds (GMCs), 395

giant planets
   locations of, 423–425
   orbital properties of, 436
   snow line and, 421
GIM mechanism, 316
glaciation episodes, 447
global PQ symmetry, 338
globular clusters, 487
gluinos, 372
GMCs. *See* giant molecular clouds
GMT. *See* Giant Magellan Telescope
grand unified theory (GUT), 24, 176, 249
   baryogenesis and, 189
   Higgs mass and, 325
   inflation and scale of, 125, 127, 140
   scale, 24, 189
   unification energy, 7
graphite, 464
gravitation, 240, 463, 513
gravitational collapse, 131
gravitational constant, 26, 53, 193, 206, 242, 345,
      356, 387, 408, 511
gravitational instability
   inhomogeneities and, 360
   structure formation and, 207
gravitational lensing, 6, 386
   galaxy cluster mass estimation from, 205, 349
gravitational magnification, 487
gravitational microlensing, 485
gravitational potential, 83, 96
   linear growth phase and, 209–210
   Poisson equation for, 209
   scaling relations and, 407
gravitational potential well, 6
gravitational waves, 131, 393
   discovery of, 351
   inflation and, 133
gravitino, 351
gravity, 8, 22, 458
   constraints on, 7–8
   dark energy vs modified, 95–97
   density fluctuations and, 7
   Einstein, 96
   galaxy formation and, 7
   General Relativity and, 113
   Lagrangian for, 127
   mass and, 4
   Newtonian, 209
   quantum, 27, 116
   quantum vacuum fluctuations and, 131
   relativistic theory of, 95
   stellar evolution and, 279
GRBs. *See* gamma-ray bursts
ground states, anthropic arguments and, 340
GUT. *See* grand unified theory
gyromagnetic factors, 251–252, 519, 531

habitability, 458, 460, 463
habitable epoch, 458, 461–463
habitable planets, 485, 489–492, 494
habitable zone, 75, 412, 489, 494, 496
hadron colliders, 371–372
halo cooling, 195
halo mass function, 226–228, 230
haloes, star-forming, 462
Hamiltonians, 515, 517, 519
   Cowan-Griffin, 532
   for water, 525
Harrison-Zeldovitch power spectrum (HZ power
     spectrum), 165, 219, 365
Hawking temperature, 31
heat death of Universe, 54
heavy elements
   asteroids and, 421
   binding energies and, 255
   disc composition and, 467
   life and, 458
   supernovae and, 464
heavy nuclei, 81
Heisenberg uncertainty principle, 310
helium
   abundance of, 479
   after BBN, 267
   burning, 285
   cosmological nucleosynthesis and, 46
   mass-energy density and, 347
   primordial, 196, 292
   primordial nucleosynthesis and levels of, 203
   in stars, 385
helium mass fraction, 191, 195
helium-burning phase, 42
Hellmann-Feynman theorem, 533, 534
Hertzsprung-Russel diagram (HR diagram), 283, 288
hierarchy problem, 69, 326, 328–333
Higgs
   boson, 34, 189, 371
      electroweak fine-tuning and, 325
      elementary, 329
      EWSB and, 329
      pNGB and, 330
   coefficients, 241, 514t
   couplings, 325
   doublets, 258, 338
   field, 141, 166, 215, 247, 324
   hierarchy problem, 329
   mass, 28, 34, 309, 320
      higher-order corrections to, 325–328
      inflation and, 332
      quantum corrections to, 325
      SUSY and, 329
   mechanism, 243
   scalar doublet, 247
   vev, 250, 255, 268, 283, 324, 327, 340
high-redshift galaxies, 487–488

Hill radius, 427
homogeneity, 206, 360
   Cosmological Principle and, 360
homogeneous initial conditions, 142–143
horizon, 120–123, 125. *See also* event horizon
   CMB observation and, 87
   cosmic, 12, 17
   cosmological, 207
   particle, 207
   quantum fluctuations and, 205
hot Big Bang model, 48, 111, 117, 120
hot dark matter, 205, 364
Hoyle level, carbon production and, 282
Hoyle state, 253, 272–274, 302, 515
HR diagram. *See* Hertzsprung-Russel diagram
Hubble drag, 93, 211, 212
Hubble expansion rate, 211, 222, 268
Hubble flow, 103, 209, 222
Hubble parameter, 11, 147, 153, 183, 193, 346,
     447, 462
Hubble radius, 78, 111, 207, 214
   dark matter and, 361
   inhomogeneities smaller than, 149
   perturbations and, 361
Hubble Space Telescope, 486
Hubble volume, 12, 462
hydrogen, 258, 517
   cloud cooling and, 403, 462
   constant and atoms of, 521–522
   constant and molecules of, 522
   cosmic ray and x-ray ionisation of, 476, 478
   ionisation potential of, 80
   mass-energy density and, 347
   radius of, 27
   recombination and, 120
   relativistic effects and, 531
   in stars, 385
   thermal photon ionisation of, 70
   water formation and, 476, 478–479
hydrostatic equilibrium, 278, 472, 473
hypercharge, 245–247
hyperfine frequency, 518
HZ power spectrum. *See* Harrison-Zeldovitch
     power spectrum

ice, asteroids and, 421
IMBH. *See* intermediate-mass black hole
IMF. *See* initial mass function
industrial pollution, 459
inflation, 111
   advantages, 165–166
   avoiding self replication and, 157–160
   constraints on, 135–139
   doubts and criticisms about, 112
   end of, 130
   eternal, 13, 15, 154–157, 160, 161
   field theory and, 126

inflation (cont.)
  fine-tuning and, 139–142
  flatness problem and, 125
  gravitational waves and, 133
  GUT scale and, 125, 127, 140
  Higgs mass and, 332
  horizon problem and, 125
  large field, 142
  large-scale structure and, 131
  multiverse and, 160–165
  perturbations and, 131–135, 360
  quantum fluctuations and, 98, 205
  quantum fluctuations of field, 75
  realising phase of, 126–131
  scalar fields and, 98
  stochastic, 152–153, 160
  trans-Planckian problem of, 150
  UV sensitivity of, 166
inflationary cosmological perturbations, 131–135
inflationary initial conditions, 142–149
inflationary phase, 12, 25
inflationary potential, free parameters of, 139–142
inflaton
  field, 111–112, 126–128, 130, 131, 140–141,
    207–208
    Higgs field and, 166
    Hubble parameter and, 153
    inhomogeneous distribution of, 148
    Klein-Gordon equation and, 152, 153
    quantum behavior of, 153
  self-coupling, 148
initial conditions
  anisotropic, 143–144
  homogeneous, 142–143
  inflationary, 142–149
  inhomogeneous, 145–149
  for perturbations, 149–152
initial mass function (IMF), 390, 491–492
instantons, 334, 335
Integrated Sachs-Wolfe effect, 6
intelligent life, 79
intergalactic clouds, 401
intermediate-mass black hole (IMBH), 407
interstellar medium (ISM), 180, 406, 464,
    476, 478
intra-cluster gas, 181
ionisation
  cosmic rays and x-rays as source of, 441
  fraction, 440
  hydrogen potential for, 80
  normalisation of rate of, 482
  re-ionisation, 131
  temperature, 39
  thermal photons and, 70
ion-molecule sequences, 476, 478, 480, 481
iron-poor stars, 464
isentropic perturbations, 360

ISM. *See* interstellar medium
isospin, 334
isotropy, 206
  Cosmological Principle and, 359
  of microwave background, 47, 123

James Webb Space Telescope, 448, 487
Jeans
  criterion, 361
  length, 212, 214–216
  mass, 49, 361–363, 390, 488
Jupiter, 424–426
  asteroid impacts and, 434–435
  asteroid precession rate and, 435
  Saturn orbit and, 434, 435, 437
JWST mission. *See* James Webb Space Telescope

Kaluza-Klein
  excitations, 331
  particle, 351
Kelvin-Helmholtz time-scale, 394
Kepler Space Observatory, 7, 17, 448, 460, 475,
    486, 494
Keplerian velocity, 468
k-essence, 91–93
Kim-Shifman-Vainstein-Zakharov model (KSVZ
    model), 339
kinetic equilibrium, cold thermal relics and, 352–353
Kirkwood gaps, 426
KK excitations. *See* Kaluza-Klein
Klein-Gordon equation, 128, 129, 145, 152, 153
KSVZ model. *See* Kim-Shifman-Vainstein-Zakharov
    model
Kuiper belt, 412, 438
Kullback-Leibler divergence, 138

Lagrangians
  dark energy and, 97
  for electron-photon interaction, 311–312
  for gravity, 127
  for *k*-essence, 92
  for non-gravitational interactions, 245, 247
  of QCD, 244, 334, 337
  scalar field, 89
  zero levels, 97
Lambda Cold Dark Matter (ΛCDM), 25, 35, 104, 105,
    120, 195, 346, 448, 458
landscape hypothesis. *See also* string landscape, 99
Langevin equation, 153, 154, 156, 158
Large Electron Positron collider (LEP), 320, 329
Large Hadron Collider (LHC), 5, 320, 328, 341
  SUSY particles and, 329
  WIMP testing with, 357, 371–372
Large Magellanic Cloud (LMC), 486
large-number hypothesis, 78
large-scale structure, 345
  baryon asymmetry parameter and, 196–199

coincidences of time and, 70
  dark energy and, 87, 111
  dark matter and, 6, 348, 366, 377
  density fluctuations and, 24
  inflation and, 131
  linear regime and, 196–198
  nonlinear regime and, 198–199
  quantum fluctuations and, 131
Laser Interferometer Gravitational Observatory
  (LIGO), 52, 351, 393
last scattering surface, 111
laws of nature, 3
  multiverse and, 12
LEP. *See* Large Electron Positron collider
leptogenesis, 189
leptons, 244, 246, 307
  generations of, 308
Levi-Civita tensor, 245
LHC. *See* Large Hadron Collider
life
  artificial spread of, 459
  asteroid belts and, 412–444
  carbon existence and, 35
  comets and, 438
  emergence of, 3, 458
  fine-tuning and, 178
  heavy elements and, 458
  radiation and, 446
  relative likelihood for, 489–497
  search for, 460
  thermal gradients and, 463
light, speed of, 24
light elements, synthesis of, 265, 292
LIGO. *See* Laser Interferometer Gravitational
  Observatory
Lilly-Madau diagram, 103
linear growth phase, 209–211
linear perturbation theory, 218, 220, 224
linear power spectrum, 218–220
linear regime
  baryon asymmetry parameter changes in,
    196–198
  structure formation and, 209–220
lithium
  abundance of, 267, 275
  primordial, 189–190
  reactions influencing, 276
little Higgs models, 330
LMC. *See* Large Magellanic Cloud
Local Group, 103
local thermodynamic equilibrium (LTE), 350, 351
logical operations, 512
low-metallicity discs, 467
LTE. *See* local thermodynamic equilibrium
Lyman Werner radiation, 478
Lyman-alpha-cooling primordial clouds, 402
Lyα emitters, 487

MACHOs. *See* MAssive Compact Halo Objects
Magellanic Clouds, 486
magnetically-driven feedback, 390–391
magneto-rotational instabilities (MRI), 391, 439, 440
Magorrian relation, 406
main-sequence phase, 32, 288, 444, 446
Majorana fermions, 375
Mars, 420, 427
mass, 250–251, 291
  of asteroid belt, 425
  of black holes, 31
  of celestial objects, 4
  of clusters of galaxies, 205, 230, 349
  constants determining, 31
  eigenstates, 314
  electron, 310, 317, 318
  fermion, 317
  of galaxies, 396–398
  gravity and, 5
  habitable planets and stellar, 490–492
  of neutrinos, 248, 349
  of pions, 254
  proton, 5
  quarks and, 251, 255, 257, 334
  of scalars, 328
  of seed massive black holes, 403–404
  solid planet internal structure and
    conservation of, 471
  of stars, 4, 5–494
mass-energy density
  chemical elements and, 346
  dark energy and, 347
  dark matter and, 347
  hydrogen and helium, 347
  neutrinos and, 346
  radiation and, 346
MAssive Compact Halo Objects (MACHOs), 351
mass-radius relationships, for carbon planets, 471–474
mass-to-light ratio, 348
matter. *See also* anti-matter; dark matter; *specific types*
  budget, 69
  density, 70, 80, 193
    matter types and, 205
    WAP and, 80
  distribution, 203
  energy density and, 211
  fields, 513
  perturbation growth, 216
  radiation decoupling from, 361
matter-antimatter asymmetry. *See* baryon asymmetry
matter-antimatter symmetry, 177
matter-radiation equality, 69, 80, 122
  cosmic coincidence and, 90
maverick WIMPs, 372
$M_{BH-\sigma}$ scaling relation, 405–407
M-dwarf stars, 494
measure problem, 67

Mendeleev table, 238, 292
Mercury, 420, 443
MERCURY orbital dynamics package, 426
Mészáros effect, 214, 219
metacosmology, 61
metallicity, 443, 464, 477, 479, 482, 487
    CEMP stars and, 474
    disc lifetime and, 467, 469
    early stars and, 394–395
    of gas environments, 477
    primordial nucleosynthesis and, 267
    stellar evolution and, 280, 282, 284
metal-poor stars, 464
meter, dependence on fundamental constants, 518–519
methane, 525
methanol, 438
metric tensor, 113, 131, 206
microphysics
    limits of testing of, 13
    macrophysics connections to, 24
microstructure of space, dark energy and, 15
microwave background. *See* cosmic microwave
        background
Milky Way, 7, 12, 384, 395, 405, 412, 444, 478, 491
minimum mass solar nebular (MMSN), 417–418
minimum-mass extrasolar nebular (MMEN), 417, 419
minisuperspace approximation, 164
Minkowskian momentum, 313
MMEM. *See* minimum-mass extrasolar nebular
MMSN. *See* minimum mass solar nebular
modified gravity, dark energy vs, 95–96
MOdified Newtonian Dynamics (MOND), 350, 351
molecular biosignatures, 496
molecular structure
    constant and, 519–521
    hydrogen, 522
    water, 524–525
Møller flux, 354
momentum conservation, 406
MOND. *See* MOdified Newtonian Dynamics
Moon, 180
moons
    life emergence and, 496
    weather extremes and, 421
MRI. *See* magneto-rotational instabilities
M-theory, 21, 25
    string landscape variant of, 50
Mukhanov's potential, 157, 164
multiverse, 12–15, 21, 57–58, 60, 166
    anthropic arguments and, 82, 340
    anthropic vacuum and, 99
    avoiding self replication and, 157–160
    ensemble reasoning and, 75–77
    eternal inflation and, 154–157, 164
    inflation and, 160–165
    stochastic fluctuations and, 157
    stochastic inflation and, 152–153

string landscape and, 164
    vacuum density and, 484
muons, 180, 244, 245

NACRE rate, 286, 301, 304
Nambu-Goldstone boson (NGB), 330, 334
natural selection, 76
naturalness, 317–319
    BG and, 320, 321
    low-energy states and, 319
    measures of, 321–322
    in particle physics, 68–69
    quantitative formalisation of, 320
    supersymmetry models and, 320
Navarro-Frenk-White profile (NFW), 368
n-body simulations, 426–428
NEC. *See* Null Energy Condition
net baryon asymmetry, 187
neutral-neutral reactions, 477, 480, 483
neutrinos, 244, 245, 247, 351
    cosmic plasma equilibrium with, 263
    decoupling temperature of, 178
    energy density of, 118
    entropy and, 263
    equations of state, 118
    flavours of, 177, 178
    mass of, 247, 349
    mass-energy density and, 347
    muon, 315
    non-zero masses of, 328
    nucleon-anti-nucleon annihilation and, 181
    primordial nucleosynthesis and, 263–264
    scatting of, 373
    sterile, 351
    supernovae and, 40–41
    types of, 346
    WIMP detection and, 369
neutron stars, 393
neutrons, 43
    EDM for, 336
    Pyramid of Complexity and, 54
    quantum degeneracy pressure of, 393
neutron-to-proton ratio, 46, 264
Newton constant, 113, 241, 268, 514
    effects of changes in, 395
Newtonian
    gravity, 209
    mechanics, 360
Newton's gravitational constant, *See* gravitational
        constant
NFW. *See* Navarro-Frenk-White profile
NGB. *See* Nambu-Goldstone boson
Noether
    current, 334
    theorem, 89
non-anthropic tunings, 28, 34
nonlinear perturbation growth, 198

non-linear growth, 220–231
non-trivial chemistry, 10–11
non-WIMP dark matter, 376
nuclear
    binding energy, 250
    constants, 289–290
    fusion, 10–11
    ignition, 32
    interaction, 252
    parameters, 268
        carbon production and, 286–288
        effects of variation of, 271–277
        element diversity and, 292
    physics, 237, 240, 291–292, 515
        fine-tuning and, 256–260
        gyromagnetic factors and, 251–252
        stellar evolution and, 279
    reactions
        binding energies and, 254–255
        resonance energies of, 256, 305–306
    scale, 535
nucleic acids, 446, 512
nucleons, 55, 176, 182, 252
    density, 186
    interactions of, 252, 253
        carbon production and, 282
        parameterisation of, 274
nucleosynthesis, 5, 10, 24, 45, 120, 177, 238, 240,
        258, 260, 515
    baryon asymmetry parameter and, 189–192,
        195–196
    carbon production and, 271, 272
    CNO production in, 271
    cosmological, 46–47, 55
    deuterium destruction after, 266
    larger baryon asymmetry range and, 191–192
    non thermal, 267
    p-process, 239
    parameters for, 268–271
    primordial, 190–192, 260–277
        cosmological dynamics and, 260
        early phases of Universe and, 260–264
        helium levels and, 203
        light elements from, 292
        mechanism of, 264–265
        neutrinos and, 263–264
        observations on, 266–268
        parameters for, 268–271
        radiation era at thermodynamical
            equilibrium, 262
        thermal history and, 261
    resonance energies and, 256
    r-process, 238
    s-process, 238
    standard, 190–191
    stellar, 278–291, 515

steps of, 264
    in supernovae, 238
Null Energy Condition (NEC), 161
number density
    of cold thermal relics, 354, 355
    of water, 477
    of WIMPs, 356, 367
numerical integration of cross sections, 301–302
numerical simulations
    for asteroid impacts on terrestrial planets, 426–428
    of star and galaxy formation, 383

observable Universe
    during inflation, 205
    limits of, 12
    number of protons in, 37
    size of, 35
observables, 516–518
observer selection, 74–75
OGLE group, 486
Oort cloud, 438
opacity, 120
open universe, 101
    perturbations in, 213
orbital eccentricity, 14, 412
orbital radii, of potential carbon planets, 466–471
organic molecules, 512, 532
oxygen
    atmospheric concentration of, 446
    disc composition and, 466
    production of, 446
    stellar production of, 282, 288

panspermia, 459
parity, 176
particle physics
    anthropic arguments and, 339–341
    constants for standard model of, 248
    extensions to standard model, 248–250
    fine-tuning measures for, 319–322
    fine-tuning problems in, 309
    hierarchy problem and, 69
    historical examples of fine-tuning problems in,
        309–317
    matter-anti-matter asymmetry generation with
        models from, 187–189
    naturalness in, 68–69
    quantifying fine-tuning for, 317–323
    standard model of, 26, 176, 240, 243–248, 257,
        484, 513
    technical naturalness and, 317–319
    unification and, 248
Pauli exclusion principle, 262, 392
PBHs. *See* primordial black holes
Peccei-Quinn
    mechanism, 337–339
    symmetry, 52, 309

Peccei-Quinn-Weinberg-Wilzcek axion, 338
periastron, 423
Periodic Table. *See also* Mendeleev table
perturbation theory, 520
   relativistic effects and, 530
photodissociation, 477, 478, 482
photoevaporation, 467
photon-matter decoupling, 205
photons
   cross-section of, 120
   energy density of, 119
   equations of state, 117
   hydrogen ionisation by thermal, 70
   means-free paths of, 362, 363
   nucleon-anti-nucleon annihilation and, 180
   WIMP detection and, 369
photon-to-baryon ratio, 48–50, 70, 80, 174, 176, 199
photospheric solar abundances, 479
physical constants, 22, 26–28, 511
pions
   mass difference in charged and neutral, 312–314
   mass of, 254
   neutral, 312–314
   nucleon-anti-nucleon annihilation and, 180
Planck
   constant, 26, 241, 511, 514
   density, 50
   length, 4, 24, 27, 78
   map, 123
   mass, 27, 70, 114, 320, 330, 331
   pressure, 389
   satellite, 6, 11, 112, 136, 140, 190, 269
   scale, 29, 35, 307
      anthropic vacuum and, 97
      electron mass and, 317, 318
      EW scale and, 308
      general relativity and, 4
      reduced, 330
      vacuum energy density at, 485
planetisimals, 421, 422
planets
   Earth-mass, 460
   first, 461–463
   maximum and minimum size of, 33
   production efficiency, 488
   rocky, 461, 462, 489
   searching for extragalactic, 485
   timescale for formation of, 467
plateau models, 164
PLATO mission, 475
pNGB. *See* pseudo Nambu-Goldstone boson
pocket universes, 13, 101, 156, 157, 164–165
Poincaré group, 240
Poisson's equation, 83, 209
Population II stars, 464
Population III stars, 394
   carbon production in, 281–291

SNe ejecta from, 465
   supernovae of, 465
positrons, 175, 369
p-process nucleosynthesis, 239
PQ mechanism. *See* Peccei-Quinn
Press-Schechter
   apparatus, 99
   mass function, 390
   theory, 226–231
primordial abundance, 190–192, 265–269, 274–277
primordial black holes (PBHs), 52
primordial density fluctuations, 50, 198, 205,
      218, 396
primordial nucleosynthesis. *See* nucleosynthesis
probabilistic frameworks
   Bayesian approaches for, 71–74
   cosmology examination with, 71–77
   ensemble reasoning, 75–77
   observer selection and, 74
proteins, 512, 533
protons
   decay, 179
   mass, 5
   number of, in observable Universe, 37
   Pyramid of Complexity and, 55
   scale, 29
   stability of, 260
proton-to-electron mass ratio, 29
protoplanetary discs, 417, 421, 426
   CO snow line and, 438–440
   with dead zone, 440–442
   graphitisation in, 472
   low-metallicity, 467
   turbulent, 438–440
   water snow line evolution in, 422
protoplanets, 475
protostars, 389
protostellar fragment. *See* protostars
Proxima Centauri, 460, 496
pseudo Nambu-Goldstone boson (pNGB), 330, 338
p-wave, 356, 366
Pyramid of Complexity, 54

Q, 9
   anthropic domain of, 14, 50
QCD. *See* quantum chromodynamics
QED. *See* quantum electrodynamics
QSOs, 181
quantum chromodynamics (QCD), 239, 240, 259
   axion, 309
   binding energies and, 255
   chiral limit of, 251
   CP violation and, 333
   instantons, 334
   Lagrangian of, 244, 334, 337
   nuclear physics and, 291–292
   scale, 27

strong interactions and, 334
technical naturalness and, 318
vacuum phase, 241, 514
quantum corrections, 69
quantum degeneracy pressure, 393
quantum electrodynamics (QED), 311, 334, 518
quantum field theory, 97, 248, 323, 485
quantum fluctuations, 75, 98, 131, 205
quantum foam, perturbation emergence from, 150
quantum gravity, 24, 116
anthropic vacuum and, 97
Planck scales and, 29
quantum numbers, cold thermal relics and, 351
quantum physics, 23
electron mass and, 310
wave-function, consciousness and, 57
quark/gluon jets, 372
quark-hadron transition, 182
quarks, 13, 244, 307
charm, 315–316
flavours of, 249
generations of, 308
light, 317
mass and, 251, 255, 257, 334
massless, 337
Pyramid of Complexity and, 54
top, 318
up, 315
quasars, 60, 267, 401, 402
quintessence, 89–91

radiation
era at thermodynamical equilibrium, 262–263
fossil, 384
life and, 446
mass-energy density and, 346
matter decoupling from, 361
pressure, 386
radiation-dominated Universe, 213–214
radiative stars, 40
radiative transport, 39
radio jets, 401–402
radioactive energy, 463
Randall-Sundrum models (RS models), 331
recombination, 80, 120, 121, 197, 215–217, 462, 476
red dwarfs, 39
red giants, 42, 282
reheating, 130, 138, 140
re-ionisation, 131
relativistic cosmology, 113–117, 360
relativistic theory of gravity, 95
relics
abundance of, 182–183, 186, 352
density of, 70
nonthermal, 351
thermal, 351
renormalisation, 248, 249, 319, 322

Resonating Group Method (RGM), 300
RGM. *See* Resonating Group Method
Ricatti equation, 183
Ricci
curvature tensor, 114, 330, 359
scalar curvature, 359
Riemann
curvature tensor, 359
tensor, 114
zeta function, 353
RNA, 512
rocky planets, 461, 463, 489
rotation curves, 6, 349, 400
R-parity, 371
r-process, 239
RS models, *See* Randall-Sundrum models
Runge-Kutta scheme, 473
Rydberg
constant, 518
energy, 389

Sagittarius A, 404
Saha equation, 120, 285
Sakharov conditions, 8, 175, 187–188
SAP. *See* strong anthropic principle
Saturn, 426
asteroid impacts and, 434–437
eigenfrequency of, 435
orbital location, 434, 436, 437
precession rates and, 434–435
SBBN. *See* Standard Big Bang Nucleosynthesis
scalar fields, 89, 91, 93–94, 98, 126, 131, 145, 153, 154. *See also* inflaton
scalar perturbations
amplitude of, 132
power spectrum of, 132
scalar spectral index, 136
scale decoupling, 167, 238, 513
scale factors, 122, 148, 359
scales
cosmic, 4
Cosmic Uroborus and, 22–24
forces at different, 22
galaxy, 33
hierarchy of structure and, 512
of panspermia, 459
Planck, 5
range of, 20
subatomic, 4
of supersymmetry breaking, 97
Schrödinger equation, 252, 253, 515–516, 519
conformal invariance and, 530
water and, 526
Schwarzschild radius, 4, 31
scientific theory
effective, 238, 513
hierarchy of, 512–513

SCUBA, 465
SDSS. *See* Sloan Digital Sky Survey
seconds, defining, 517
secular resonances, 426, 430
SEGUE survey, 464
self-gravity, 4
self-lensing events, 486
semi-anthropic galaxy formation, 102–106
SFI. *See* Small Field Inflation
sheets, 203
sigma model, 254
silicates, 464, 473
Silk
    damping, 198, 215, 216, 362
    mass scale, 363
Sloan Digital Sky Survey (SDSS), 464
slow-roll regime, 151
SM. *See* Standard Model
Small Field Inflation (SFI), 142
Small Magellanic Cloud (SMC), 486
small-scale inhomogeneities, 360
SMBHs. *See* black holes
SMC. *See* Small Magellanic Cloud
snow line, 421
    asteroid belt location and, 425
    CO, 438
    water, 422–423
snowball Earth, 447
Solar System, 24, 412, 442
    asteroid orbits in, 427
    disc with dead zone model and, 441
    effects of inner system architecture, 428–436
    effects of outer system architecture, 436–437
    lack of close-in planets or debris, 416
    super-Earths and, 416, 418–420
    water in, 421
solar wind, 180
solitons, 351
Sommerfeld enhancement, 186
space telescopes, 486
spacecraft, 459
spatial dimensions, 60
Special Relativity, 83
speed of light, 24, 26, 113, 121, 206, 241, 511, 514
speed of sound, baryon density and, 87
spherical Bessel function, 153
spherical collapse model, 224
spherical top-hat model, 220–226
spin-orbit interaction, 531
spin-spin interaction, 252
Spite plateau, 267
squarks, 372
s-process, 238
Standard Big Bang Nucleosynthesis (SBBN), 190–191
standard evolution equation (SEE), 183
Standard Model (SM), 3, 189, 325
    constants in, 26, 513

of cosmology, 113–120, 490
CP violation and, 8
fine-tuning puzzles, 120–125
flatness problem and, 124–125
of particle physics, 26, 176, 240, 243–248, 257,
    484, 513
    constants for, 248
    extensions to, 248–250
QCD interactions in, 329
quark and lepton generations in, 308
SAP and, 45
strong sector of, 336
WIMPs and, 357
Yukawa couplings and, 249
standard-candle method, 346
star-forming halos, 462
Starobinsky
    model, 138, 141–143, 149, 157, 164, 167
    potential, 128, 140
stars. *See also* carbon-enhanced metal-poor stars;
        white dwarfs
    carbon production in, 35, 281–291
    characteristic mass of, 388
    deuterium burning role in, 388–390
    dwarf, 460
    evolution, 7, 384
        carbon production and, 288
        first stars, 394–395
        of most massive stars, 392–394
        Newton constant and, 395
        stages of, 280
        star formation physics, 395–396
        stellar physics basics, 278–279
        system of equations for, 279–280
        white dwarfs, 392–395
    evolution stages of, 279
    first, 394–395
    formation of, 32, 385, 386, 395–396, 405
    gravity and, 8
    habitable planets and mass of, 490
    iron-poor, 464
    lifetime of, 35, 45, 79, 492
    luminosity of, 35–36, 79
    magnetically-driven feedback in, 390
    main-sequence phase of, 32, 36
    mass of, 4, 31, 79, 496
    mass scales, 386–396, 535
        characteristic mass of stars, 388–391
        minimum mass of stars, 391
        minimum protostellar fragment mass, 387–388
    M-dwarf, 494
    metal-poor, 464
    minimum mass of, 391
    minimum protostellar fragment mass, 387–388
    most massive, 392–394
    numerical simulation of formation of, 383–384
    oxygen production in, 282

physics of, 278–279
Population II, 464
Population III, 281–291, 394
self-regulation, 390
surface temperatures, 39
thermally pulsing asymptotic giant branch
    stars, 282
steady-state theory, 79
Stefan-Boltzman constant, 459
stellar nucleosynthesis, 278–281, 515
stochastic inflation, 152–153, 160
stochastic scalar field, 154
stress-energy tensor, 260, 359
string landscape, 21, 50, 99
    anthropic arguments and, 340
    multiverse and, 164
    vacua, 484
strong anthropic principle (SAP), 20, 29, 59
    astrophysical coincidences, 38–45
    chemistry constraints and, 42–45
    convective and radiative stars and, 39–40
    cosmological constant and, 50
    cosmological nucleosynthesis and, 46–47
    dark matter and, 51–52
    density fluctuations and, 50
    density parameter and, 47
    ensemble of universes and, 82
    number of dimensions and, 53
    photon-to-baryon ratio and, 48–50
    supernovae and, 40–42
    triple-alpha and, 42
    tunings, 22, 28
strong CP
    angle, 309
    massless quarks and, 337
    Peccei-Quinn mechanism and, 337–339
    problem, 307, 308
    $U(1)_A$ and, 333
strong interactions, 3, 22, 28, 43, 243, 244,
    252–253
    coupling constant, 28
    CP conservation and, 336
    fine-tuning in, 333–339
    Higgs field and, 247
    non-trivial chemistry and, 11
    nucleon formation and, 182
    QCD and, 333
    standard model kinetic terms for, 245
structure formation, 101, 102, 111, 203
    baryon asymmetry parameter and,
      196–198
    CMB and, 208
    dark matter and, 359–361, 365
    decoupling and, 363
    fine-tuning of, 231–233
    gravitational instability and, 207
    linear power spectrum and, 218–220

linear regime and, 209–220
non-linear growth and, 220–231
perturbations and, 208, 213
recombination and, 215–217
spherical top-hat model, 220–226
$SU(2)$, 243, 245, 247, 513
$SU(3)$, 243, 244, 249, 513
Sun, 385, 418, 443
Sunyaev-Zeldovich effect, 106
super-Earths, 443
    asteroid impacts and, 434, 437
    formation of, 417–421
    lack of, 416, 428
supernovae, 40, 393, 464, 466
    neutrinos and, 40–41
    nucleosynthesis in, 237
    Population III, 465
    SAP and, 40–42
    SNe ejecta, 465
    Type Ia, 6, 220
superstring theories, 25
supersymmetry (SUSY), 328, 331
    baryogenesis and, 188
    naturalness and, 320
    scale of breaking, 97
    SUSY-WIMP hypothesis, 375
    unbroken, 97
    WIMP and, 93, 329
    WIMPs and, 351, 371–372
SUSY. *See* supersymmetry
s-wave annihilation, 355, 366
Seyfert galaxies, 181
symmetric universe, 180–187
symmetry-breaking, 243
    anthropic principle and, 57
    axions and, 332
    electroweak, 309, 323, 327, 329, 338
    energy scale of, 52
    particle physics and, 248
    pNGBs and, 330
    PQ and, 339

tardigrades, 459
technicolour models, 329, 330
telescopes, 357, 368, 370, 448, 486
temperature
    anisotropy, 123, 345
    freeze-out, 46
    Hawking, 31
    ionization, 39
    of neutrinos decoupling, 178
    of reheating, 140
    of star surface, 39
    of Universe, 262
temperature fluctuations, 9
terrestrial planets, 421, 425–437
TESS mission, 448, 475, 486

TeVeS, 350
TFD. *See* Thomas-Fermi-Dirac EOS
theory of everything (TOE), 3, 25
    multiverse and, 57
thermal coupling, 462
thermal equilibrium
    in carbon production, 288
    radiation era at, 262
    Sakharov conditions and, 187–188
thermal gradients, 463
thermal history, 261
Thirty-Meter Telescope (TMT), 486, 487
Thomas-Fermi-Dirac EOS (TFD), 472
Thomson scattering, 215, 392
t'Hooft criterion, 308, 317, 319, 332,
        336, 341
TMT. *See* Thirty-Meter Telescope
TOE. *See* theory of everything
Toomre parameter, 441
top quark, 318
topological defects, 207
translucent clouds
    water abundance in, 481
    water formation in, 476
trans-Planckian
    physics, 166
    problem of inflation, 150, 164
TRAPPIST-1, 496
triple-alpha, 42, 284
trivial anthropic principle, 77–78
Tully-Fisher relation, 399
Type Ia supernova, 6, 484

$U(1)_A$, 336
ultraviolet (UV radiation), 446
    cutoff Λ, 313, 319
    EUV, 467
    FUV, 477, 481, 482
    inflation sensitivity to, 166
    photodissociation from, 481, 482
unification, 248, 258
Universe
    biophilic, 13
    complexification of, 278
    composition of, 118, 345–348
    critical density of, 193
    energy budget, 117
    ensemble of, 82
    habitability of, 458, 460
    heat death of, 54
    pocket, 12, 101, 156, 164–165
    structure emergence in, 203, 205–208
    temperature of, 262
    wave-function of, 163
up quarks, 315
UV radiation. *See* ultraviolet

vacuum, 82–86
    anthropic, 97–102
    density, 70, 83, 90, 97, 100, 118, 484–485
    expectation value, 324, 327, 333, 338
    fluctuations, 131, 310
    perturbations in, 92
vector bosons, 245, 326
vector symmetry, 334
Venus, 180, 413, 414
Virgo cluster, 484
virial theorem, 32, 33, 224, 349
virialisation, 203, 224, 462
virtual particles, 248
voids, 203
voltage gated ion channels, 533

WAP. *See* weak anthropic principle
warm dark matter, 205
Warm-Hot Intergalactic Medium (WHIM), 106–107
water, 512
    asteroid belts and, 422
    asteroid impacts and, 421
    dipole moment of, 524
    first metal enrichment epoch formation of, 476–484
    gas phase abundance of, 482, 483
    Hamiltonian for, 525
    in ISM, 476
    ion-molecule sequences forming, 476
    molecular structure, 524–525
    phase diagram of, 458
    snow lines, 422–423, 425
    variations of constants and, 526–529
wave function, 520
    of Universe, 163
weak anthropic principle (WAP), 20, 22, 29, 35–38,
        49, 58, 78
    Dicke's argument and, 35–37
    intelligent life and, 79
    large-number hypothesis and, 78
    sufficiency of, 37–38
weak interactions, 3, 22, 28
    cosmological nucleosynthesis and, 45–46
    cross-section for, 41
    electron mass and, 68
    neutrino decoupling temperature and, 178
    supernovae and, 40
weak isospin, 245
Weakly-Interacting Massive Particles (WIMPs),
        51, 356
    annihilation cross-section, 367
    cold thermal relics and, 356
    collider production and detection of, 371–372
    direct detection of, 372–375
    freeze-out of, 357
    indirect detection of, 368–371
    LHC experiments for, 358
    nonrelativistic annihilation of, 367

nucleon coupling, 373
nucleon scattering rates, 367
nucleus couplings, 373
nucleus scattering, 367, 373
number density of, 356, 368
Standard Model and, 357
supersymmetric relic, 93
WIMPzillas, 351
Weinberg angle, 241, 514
Wentzel-Kramers-Brillouin approximation
    (WKB), 163
WFIRST mission, 448
Wheeler-De Witt equation, 163
WHIM. *See* Warm-Hot Intergalactic Medium
white dwarfs, 32, 392

Wick rotation, 313
Wilkinson Microwave Anisotropy Probe (WMAP),
    6, 7, 118, 158
WIMPs. *See* Weakly-Interacting Massive Particles
WKB. *See* Wentzel-Kramers-Brillouin
WMAP. *See* Wilkinson Microwave Anisotropy Probe

x-rays, 181, 441, 467, 476, 478

Yukawa couplings, 241, 250, 255, 283, 325–327, 514

Z-boson, 371
zero-age main sequence (ZAMS), 288
zero-point energy, 97
Zircons, 491